ADVANCES IN COMPOSITE MATERIALS FOR MEDICINE AND NANOTECHNOLOGY

Edited by **Brahim Attaf**

Advances in Composite Materials for Medicine and Nanotechnology
Edited by Brahim Attaf

Published by InTech
Janeza Trdine 9, 51000 Rijeka, Croatia

Publishing Process Manager Ivana Lorkovic

Technical Editor Teodora Smiljanic

Cover Designer InTech Design Team

Additional hard copies can be obtained from orders@intechopen.com

Advances in Composite Materials for Medicine and Nanotechnology,
Edited by Brahim Attaf
p. cm.
ISBN 978-953-307-235-7

For EU product safety concerns:
IN TECH d.o.o., Prolaz Marije Krucifikse Kozulić 3, 51000 Rijeka, Croatia,
info@intechopen.com or visit our website at intechopen.com.

Meet the editor

Dr. Brahim Attaf is an expert/researcher in composite materials & structures. He obtained his engineer's degree from the National Polytechnic School of Algiers (ENPA) in 1985, and then his PhD degree from the University of Surrey (UK) in 1990. He has previously worked as a lecturer/researcher (Maître de Conférences) within the Departments of Aeronautics, Mechanics and Civil Engineering at the University of Blida in Algeria (1990-2000) and as a responsible of design office and quality assurance with composite firms in France (2000-2003). Since 2003 he has been independently involved in sustainability research actions leading to ecodesign of composite materials & structures through the integration of environmental and health aspects into product lifecycle. His research and teaching interests focus on vibration and stability analyses of composite structures subjected to hygrothermomechanical loading using finite-element approach and experimental investigations. He has published many scientific papers and had successfully trained/supervised many engineers and postgraduate students (Masters and PhD).

Contents

Preface

Initially, the field of composite materials concerns essentially the aerospace industry and other engineering applications. Due to their good mechanical characteristics in terms of stiffness and strength coupled with mass-saving advantage and other attractive physico-chemical properties, their use is seen to be growing rapidly and now extends in a wide range of various fields among them: medicine and nanotechnology. Some imaginable techniques towards the application of nanotechnology in medicine have become realities, where significant progress has been made in recent years. Other research studies including "innovative techniques" are still in their various stages of development.

For instance, in orthopedic and stomatological applications, implants of repair and reconstruction of bones made of composite materials have been produced with similar properties to the real bones. Some of these composite bone grafts are now implanted in the skeletal system of human body. In addition, dental amalgams for dental filling used before to repair tooth structure using a mixture of mercury, silver and other metals are now replaced by composite materials (ceramic and synthetic resin): a new challenge in the field of aesthetic dentistry. Furthermore, important scientific advances in the field of liposomes as drug carriers and medical stents using composite materials were carried out successfully in various research laboratories. Their beneficial impact on the human being will be proved in the coming years.

Much of the book focuses on advanced activities and results conducted within the field of nanotechnology and related materials. These results are obtained using theoretical and/or experimental methods. In addition, the confidence in the developed theories is established by a simple comparison of results. The resulting discrepancy provides information on tolerance data before undergoing rigorous qualification tests on the material, process and product. On the other hand, the assessment and optimization of material characteristics will lead to select the right and safe nano-material for use in specific applications such as electronics, mechanics, optics, biology, chemistry, and so on. Some examples, aiming to design and produce new devices, are presented in this book. However, results are achieved when multidisciplinary approaches are involved.

But taking into account ethical and ecological considerations, researchers and scientists involved with use of composite materials in medicine applications and/or nanotechnology are strongly encouraged to integrate health and environmental aspects in the analysis and design of a new composite product development, providing then a better assessment of health and environmental performances. In addition, adopting the strategy of ecodesign with the object to green products can also contribute to the enhancement of human well-being and living conditions for present and future generations.

Given the importance of the two key disciplines (i.e., medicine and nano-technology), the material published in this book has been grouped into two main parts, for which the 24 chapters composing the book have been divided into the following parts:

- *Medicine, dental and pharmaceutical applications*: research was focused on the use of bio-composites for bone cancer treatments, bone grafts, artificial prostheses, to replace or repair tissues in the human body (i.e., medical stents), dental fillers and prosthetic dentistry. Recent studies with liposomes (nano-spherical vesicles) as drug carriers are presented.

- *Nanocomposites for energy efficiency*: investigations were undertaken on renewable energy techniques to achieve clean and high energy conversion efficiency using photovotaic cells, solar cells, nanoparticles, nanowires, nanotubes, nanofibers and nano-porous materials for hydrogen storage.

- *Characterization and fabrication*: studies were conducted on mechanical characterization of plasma and coating to sustain stress and corrosion, to lock radiation and on the production of nanocarbon materials.

With the collaboration of all authors in their best research work, I am very pleased to have this opportunity to edit this important book, opening a new challenge in the world of composite materials and the achieved advancements of specialists in their respective areas of research in medicine and nanotechnology.

I would like to thank the authors of the chapters of this book for their remarkable contributions in their areas of expertise and I genuinely appreciate the resulting synergy between theory and practice coming from all around the world. Without this rich variety of contributions, the existence of the Book titled *"Advances in Composite Materials for Medicine and Nanotechnology"* would not have been possible.

I also wish to acknowledge the help given by InTech Open Access Publisher staff, in particular Ivana Lorković for her assistance and support.

January 2011

Dr Brahim Attaf
Marseille,
France

Part 1

Medicine, Dental and Pharmaceutical Applications

1

Advances in Collagen/Hydroxyapatite Composite Materials

Anton Ficai, Ecaterina Andronescu, Georgeta Voicu and Denisa Ficai
Politehnica University of Bucharest, Faculty of Applied Chemistry and Materials Science
Romania

1. Introduction

The annually necessary human bone grafts are in continuous grown due to the increasing of fractals, congenital and non-congenital diseases. Based on statistical reports (Murugan and Ramakrishna 2005), only in USA about 6.3 million fractures occur every year and about 550.000 of these require bone grafting. The most frequently fractures occur at the level of hip, ankle, tibia and fibula. Due to the higher physical effort, the men are more exposed to fracture than women (2.8% in the case of men comparing to 2.0% in the case of women). The number of fractures increases year by year and consequently many researchers from different research fields co-operate in order to develop new bone graft materials. Also, it is important to mention that, in the present, the bone diseases are overlapped only by hearth diseases.

The history of bone grafting is starting in 1913 when Dr. D.E. Robertson assays a piece of cat's bone and a piece of human bone for bone grafting into dogs (Gallie and Toronto 1914). The microscopic analysis of implanted graft after 20 days shows that the space between graft and living bone is filled with new cancellous bone. These early works made the premises for the development of the bone grafts.

Due to the increasing of the necessarily bone grafts, autografts and allografts can not cover the overall need of bone grafts. For compensate this gap, artificial (synthetic) grafts are necessary and, consequently used. The use of synthetic grafts has some advantages versus allografts, autografts and xenografts: the possibility to obtain unlimited number/quantity of synthetic grafts, more safety use of artificial bone grafts without disease transmission risk, pain limitation by elimination of some secondary surgical intervention.

The need of bone grafts materials lead to the synthesis of many kind of materials with different properties. Function of the nature of these materials and the relation between these grafts and the host tissue, these materials can be divided into 4 generations (Fig. 1).

The components of the first generation of bone grafting biomaterials have remarkable mechanical properties but they are neither bioresorbable nor bioactive. More than, the use of these kind of bone grafts have limited lifetime (usually less than 10-15 years) and need to be extracted and replaced surgically. Some of the most representative biomaterials from the first generation of bone grafting biomaterials are: the iron, cobalt, chromium, titan or their alloys: steel (especially 316 L), cobalt or titan based alloys (Corces 2002; Corces and Garcia 2007) etc.

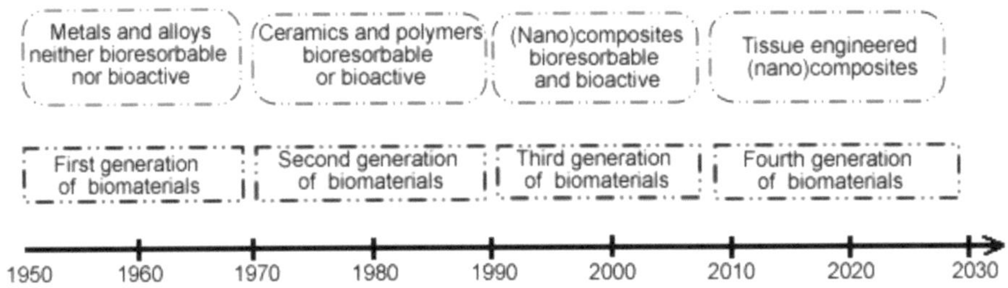

Fig. 1. Biomaterials evolution in the field of bone grafting

The components of the second generation of bone grafting biomaterials are at least bioresorbable or bioactive and they do not require to be replaced in time. The most representative biomaterials from the second generation of bone grafting biomaterials are: calcium phosphates (especially hydroxyapatite and tricalcium phosphate), the bioglasses (Hench et al. 2004), alumina (Sedel et al. 1994); zirconia (Clarke et al. 2003) and the following polymers: poly ε-caprolactone (Oláh et al. 2007), polyurethanes (Bonzani et al. 2007 ; Guelcher et al. 2004) etc.

The components of the third generation of bone grafting biomaterials are both bioresorbable and bioactive and have superior properties. It has to mention that these biomaterials present higher specific properties than the first two generations of bone graft materials and short time after implantation these materials are resorbed and in time, in the place of the bone graft the new bone is formed. The properties of these (nano)composite materials is strongly influenced by the nature of components, the composition and the morphology. That is why many researchers tried to obtain not only compositional similitude with natural bones from mineralogical and morphological point of view. The most representative biomaterials of the third generation of bone grafting biomaterials are: (nano)hydroxyapatite/collagen (Wahl and Czernuszka 2006), (nano)hydroxyapatite/collagen/hyaluronic acid (Bakos et al. 1999), hydroxyapatite/poly-L-lactic acid (Kesenci et al. 2000), hydroxyapatite/chitosan (Wang and Li 2007).

The fourth generation of bone grafting biomaterials is similar with the third generation materials but improved by the presence of bonny cells, growth factors, bone morphogenetic proteins etc.

One of the most important characteristic of bone grafts materials is the osteointegration. The osteointegration (and also osteoconductivity) of the grafts is related to the degree of porosity and pore size (applicable especially to the last three generation of bone grafting biomaterials) (Develioglu et al. 2005). Based on the literature data and also based on the size of osteoblasts (which vary up to 20-25 μm) (Chang et al. 2000; Develioglu et al. 2005; Gauthier et al. 1998) the optimum pore size was found to be 50-550μm.

The natural bones contain mainly collagen and hydroxyapatite. That is the reason because many researchers try to understand and obtain (nano)hydroxyapatite/collagen composite for hard tissue repairing. The composition of bones varies function of many factors such as: specie, sex, age, bone type, location etc.. The relative composition of bone is presented in the Table 1, while the most important biomedical properties of the bones are presented in Table2. The *in vivo* bone biosynthesis is controlled by many factors such as: BMP (bone morphogenetic proteins) (Abe et al. 2000), transforming growth factors (Tashjian et al. 1985), cytokines (de Vernejoul et al. 1993), hormones (Bollerslev et al. 1991 ; De Vernejoul et al.

1990; Hock and Gera 1992), transcription factors (Cui et al. 2003; Ogawa et al. 2000), adhesion molecules (Miyake et al. 1991) and so on.

Components	wt %
Mineral phase	
Hydroxyapatite	60–66
Carbonate (mostly as carbonated hydroxyapatite)	~ 4
Citrate	~ 0,9
Na^{+}	~ 0.7
Mg^{2+}	~ 0.5
Others	Traces
Organic phase	
Collagen	20–25
Non-collagenous proteins: (osteocalcin, osteonectin, osteopontin, sialoprotein, BMP)	2 - 3
Others: (polysaccharides, lipids, cytokines)	Traces
Water	**8 - 9**

Table 1. The composition of healthy bone

Property	**Cortical Bone**	**Cancellous Bone**
Young's (Tensile) Modulus (GPa)	7–30	0.05–0.5
Compresive strength (MPa)	100–230	2–12
Flexural Strength (MPa)	50–150	10–20
Fracture toughness (MPa $m^{1/2}$)	2–12	0.1
Strain to failure	1–3	5–7
Apparent density (g/cm^3)	1.8–2.0	0.1–1.0
Surface area/volume ratio (mm^2/mm^3)	2.5	20

Table 2. Biomechanical properties of healthy bone

Bone formation and remodelation is controlled by many factors: physical, chemical, hormonal, growth factors, and anti-mineralization agents (Karsenty 2000; Karsenty et al. 2009; Wallach et al. 1989). The bone tissue piezoelectricity is the key element which assists the bone formation, remodelation and growth, especially in the early stages of the bone formation and. The piezoelectricity of the bone is due, among the rests, due to the bone tissue anisotropy. The main elements which induce the anisotropy of the collagen based structures are: the anisotropy of the collagen molecules themselves and the anisotropy of collagen fibrils and fibres as well as the oriented morphology of the fibrils and fibres disposed in the isotropic extrafibrillar space (Hellmich et al. 2004).

Trying to obtain such composite with same composition and structure with the natural bone, the researchers elaborated many synthesis methods (Wahl and Czernuszka 2006) such as: *in vitro* collagen mineralization (Lawson and Czernuszka 1998), thermally - triggered assembly of hydroxyapatite/collagen gels (Pederson et al. 2003), vacuum infiltration of collagen into a ceramic matrix (Werner et al. 2002), enzymatic mineralization of collagen sheets (Yamauchi et al. 2004), freeze drying and supercritical point drying (Pompe et al. 2001), biomimetic synthesis (Li and Chang 2008).

One of the most promising bone graft material seems to be the collagen/nanohydroxyapatite composite (Li et al. 2006) due to its very good compositional and structural similarity with natural bone (Yunoki et al. 2007; Yunoki et al. 2006). The role of each component is not very well known. It is generally accepted that mineral phase, mainly containing hydroxyapatite, provide toughness and rigidity while the organic matrix provide tensile strength and flexibility of bone. In composite materials, collagen and hydroxyapatite play the same roles as they play in natural bone.

The mineral phase is deposited on the organic phase through electrostatic interactions between the carboxyl groups from collagen and Ca^{2+} from hydroxyapatite, but there is no unison about the mineralization sites. Many researchers assert that hydroxyapatite is deposited only onto the non-collagenous proteins and citrate anions (Rhee and Tanaka 1998) while others assert that mineralization occurs also on the pure collagen (Lin et al. 2004; Zhang et al. 2003). It is well known that materials properties are influenced not only by composition but also by morphology. This is the reason because even for similar composition of many bones their properties differ very much. Starting from these hypothesis, scientists tried to improve or induce new properties of the COLL/HA composite materials by addition of third component (or even more components) or by inducing different morphology. The influence of the morphology can be easily marked out by comparing the mechanical properties of compact and cancellous bone (Table 2) (Hench and Wilson 1993).

Collagen mineralization occurs due to the interactions which appear between collagenous structures and hydroxyapatite nanocrystals. In fact, these interactions occur between carboxylate groups and Ca^{2+} cations and can be illustrated as presented in Fig. 2. This hypothesis is supported by FTIR data and was explained based on the spectral shifts of C-O and C=O bands in pure collagen and COLL/HA composites (Ficai et al. 2009a).

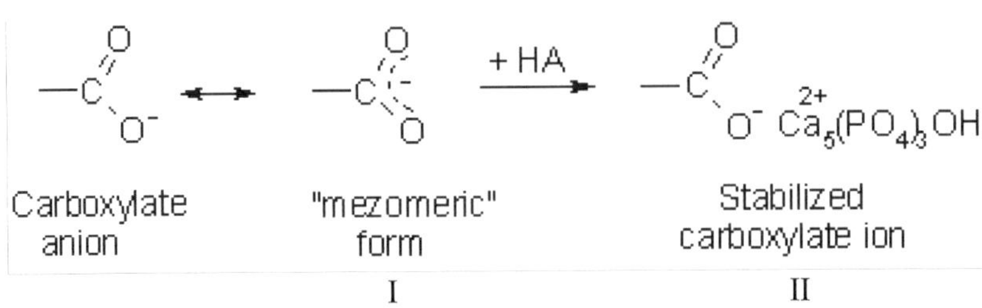

Fig. 2. Mezomeric form stabilization of carboxylate group due to mineralization

As a result of these interactions, hydroxyapatite particles are preferentially deposited onto the collagenous support forming so-called nucleation centers. Once formed, the nucleation centers increase and can lead even to the formation of a thin film. If the amount of deposited HA is low enough, the mineralization centers can be visualized by scanning electron microscopy (Fig. 3).

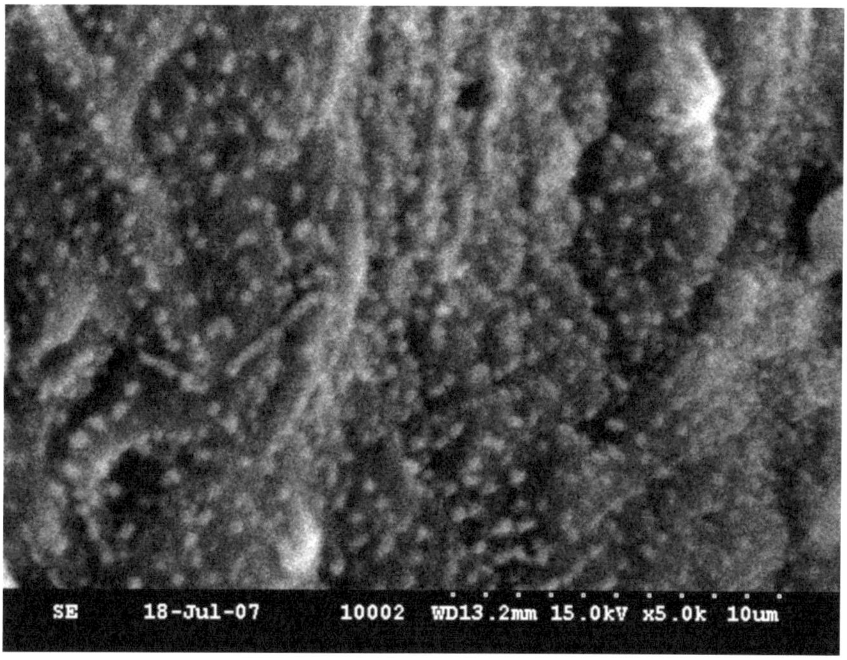

Fig. 3. The SEM micrograph of mineralized collagen fibres

In order to obtain good bone substitutes, researchers have to understand the biosynthesis of natural bone and to control some parameters such as composition, hydroxyapatite shape and size, collagen fibrils and hydroxyapatite blade orientation.

The synthesis of COLL/HA composite materials with different porosity/density is very important in order to obtain materials with tailored properties. COLL/HA composite materials with tailored ceramic properties can be easily obtained by combining the controlled air drying with freeze drying. The controlled air drying followed by freeze drying can be easily used for the synthesis of COLL/HA composite materials with any composition. In this scope, the wet composite materials are let, under controlled atmosphere for certain time (controlled temperature and humidity) followed by a final drying, realized by freeze drying. The use of mixed drying method (controlled air drying combined with freeze drying) is technically easy to control and economically sustainable. The ceramic properties of the COLL/HA composite materials can be easily controlled by the air drying step conditions (especially drying time); the longer controlled air drying time lead to composite materials with lower porosity and increasing density.

Scanning electron microscopy can be used for qualitative analysis of these materials obtained by controlled air drying followed by freeze drying (Fig. 4). Analysing the SEM images of the materials obtained by air drying (increasing drying time) followed by freeze drying it is stand out a mile that the increasing air drying lead to denser materials.

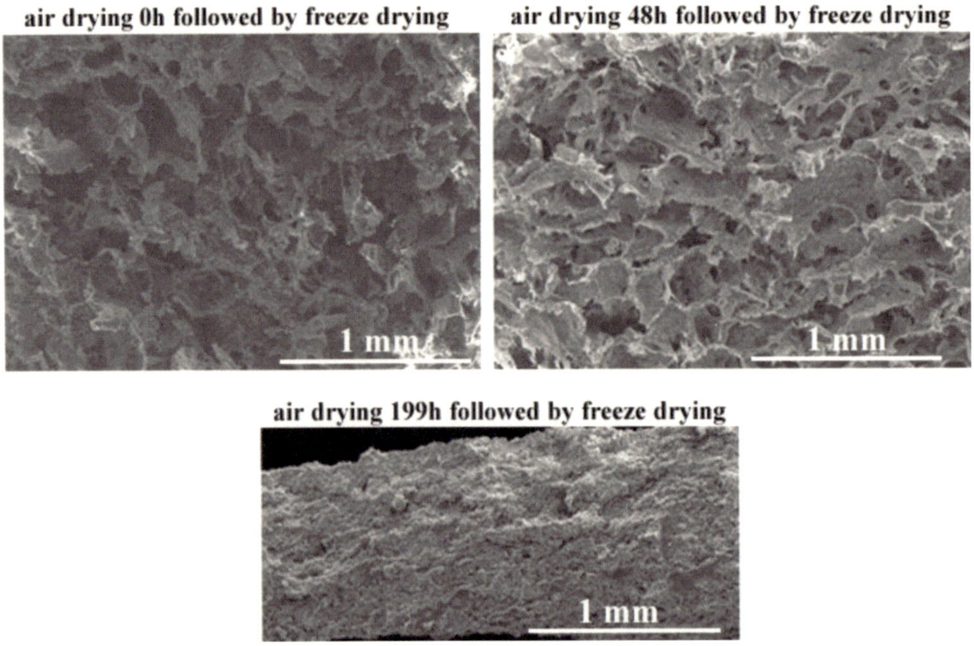

Fig. 4. SEM images of COLL/HA composite materials obtained by controlled air drying (0, 48 and 199h) followed by freeze drying

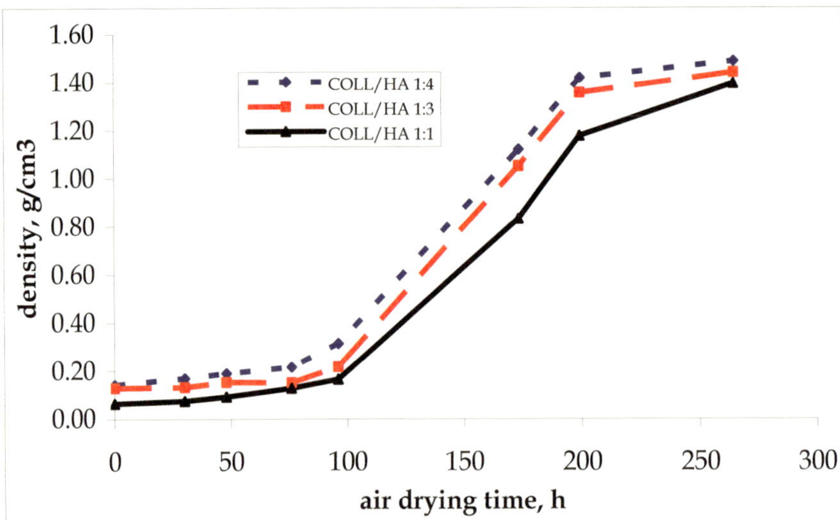

Fig. 5. Ceramic properties of the three series of COLL/HA composite materials versus air drying time

It has to mention that during the air drying process, simultaneous with the water evaporation the morpho-structural restructuring of the composite happened which leads to more compact materials, the density of the COLL/HA composite materials being 0.15-1.6g/cm^3 while the open porosity 25-95% (Fig. 5). As a general rule, the density of the

composite materials increases with the increase of the air drying time. This method is very useful because allow to tailor the ceramic properties of the materials without modify the composition.
The orientation of collagen/hydroxyapatite composite is induced by the mutual interaction between collagen and hydroxyapatite in aqueous solutions. For better orientation, it is possible to use electric and / or magnetic field. Cunyou Wu and co-workers (Wu et al. 2007) report unidirectional oriented hydroxyapatite/ collagen composite using 10 T magnetic field starting from calcium containing collagen solution and phosphate solution at 37 ^{0}C. If magnetic field is imposed, perpendicularly to the rotation axis, unidirectional oriented collagen fibrils and hydroxyapatite crystals have obtained due to their magnetic susceptibility anisotropy (Hellmich et al. 2004).
Due to the importance of these composite materials thousands of papers and many comprehensive reviews are published yearly. Some of the most comprehensive reviews about bones and bone grafts materials based on collagen and/or hydroxyapatite were published in the last years by Cui at al (Cui et al. 2007), Dorozhkin (Dorozhkin 2009), Murugan and Ramakrishna (Murugan and Ramakrishna 2005) and by Wahl and Czernuszka (Wahl and Czernuszka 2006).
Some of the most recent advances in the field of COLL/HA composite materials are systematically presented below.

2. The influence of organic and inorganic species on the COLL/HA composite materials

Recent studies made by Ficai et al. (Ficai et al. 2010b; Ficai et al. 2010e) reveal the influences of the PVA, denaturated collagen (hydrolysate collagen and ionic species on the collagen/hydroxyapatite composite materials. These works are essential because neither *in vitro* nor *in vivo* mineralization is readily understood, and the synthesis of materials that mimic the characteristics of bone has not been previously accomplished.
The properties of COLL/HA composite materials can be tailored by addition of different organic or inorganic components; well studied in the literature being that ions which naturally occur in natural bone: citrate, fluoride, chloride, carbonate, magnesium or some ions which accidentally occur in natural bones and, usual is responsible for different diseases (Pb^{2+}, Sr^{2+} etc.).
The synthesis of complex COLL/HA composite materials which also contain PHA, collagen hydrolysate or other ionic species than calcium and phosphate were synthesized by co-precipitation of hydroxyapatite on collagenous matrices (Ficai et al. 2009b). The mineralization process consisted of two successive stages. In the first stage, a $Ca(OH)_2$ suspension was added drop-wise and allowed to interact for 24 h. In the second stage, a stoichiometric quantity of NaH_2PO_4 solution was added drop-wise. The quantities of collagen matrix and HA precursors were used at a ratio of 1:4. After the hydroxyapatite precursors were added, the pH was adjusted to 9 using a solution of NaOH, in order to assure brushite-free HA precipitation. During the synthesis and drying of the composite material, the temperature was maintained at approximately 37 ^{0}C.
To study the effect of the citrate ions on the mineralization process, the collagen matrices were dipped into a dilute citrate solution (5 ‰) for 30 minutes and then mineralized as presented above. Upon completion of this process, the resulting materials were analyzed.

In this chapter, two collagenous matrices that contained 5% and 70% of the hydrolysate were mineralized following the aforementioned procedure. The COLL/PVA and COLL-PVA/HA hybrid materials were obtained starting from collagen gel and PVA solution, the mixing ration being COLL:PVA = 1:2. the as obtained gel was than mineralized as presented above.

In order to obtain fluoride-substituted hydroxyapatite, a modified mineralization method was used. For this purpose, in the second stage of mineralization, a corresponding amount of NaF was added to the phosphate solution to assure the transformation of 5%hydroxyapatite into fluoroapatite.

In order to design COLL/HA composite materials and to obtain a bone-like morphology, it is important to understand the influence of each component on the morphology of the material.

The influence of the collagen hydrolysate, PVA and ionic species were analyzed by XRD, FTIR and SEM (-EDS) from the point of view of composition and morphology. The synthetic conditions and characteristics of the resulting hydroxyapatite were strongly influenced by the organic support.

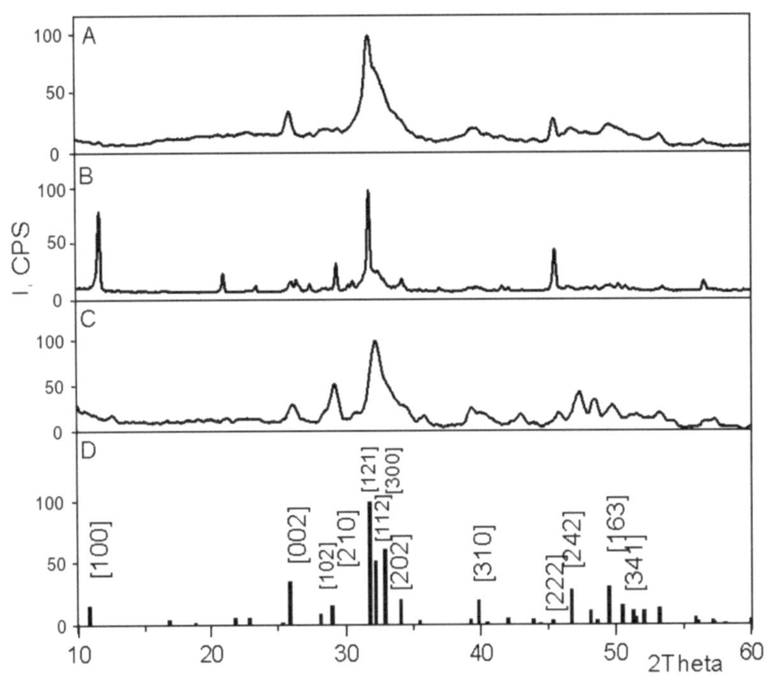

Fig. 6. XRD patterns of different collagen/hydroxyapatite-based composite materials: A. mineralized cross-linked matrix (glutaraldehyde, 1%); B. mineralized cross-linked matrix (glutaraldehyde, 1%), the mineralization occurs in the presence of F^- (5 % (molar) reported to HA); C. mineralized hydrolysate-enriched collagen matrix; and D. crystalline HA (ASTM 74-5966)

X-ray diffraction patterns (Fig. 6) were recorded in order to confirm the synthesis of hydroxyapatite and for the qualitative estimation of preferred crystallization directions. In the XRD diffraction pattern of citrate-enriched COLL/HA composite materials, no difference was observed when compared to COLL/HA or COLL-PVA/HA (Ficai et al. 2010e). The diffraction patterns of citrate-enriched COLL/HA and COLL-PVA/HA are not presented herein, as the results were similar to the diffractogram of pure COLL/HA composite materials presented in Fig. 6A.

The results of the four diffractograms lead to the conclusion that the organic phase composition and the presence of ionic species greatly influences the mineralization process. From the comparison of the diffractograms displayed in Fig. 6A and C, it can be observed that hydrolysate induced a preferred direction for the growth of HA [210] and that the other crystalline phases exhibited approximately the same intensity.

The presence of F^- had a strong influence on apatite crystallization (Fig. 6B). Among all of the crystallization directions, only a few were characteristic of partially substituted apatite ($Ca_5(PO_4)_3OH_{1-x}F_x$) and were present at a relevant intensity: [121], [100], [210] and [222]. Additionally, no other fluoride salts could be identified from the diffractograms. The main crystallization directions presented in crystalline HA (ASTM 74-5966) are presented in Fig. 6D.

Scanning electron microscopy was employed in order to study the morphology of the composite materials.

SEM images were recorded at different magnifications as a function of the morphology of each composite and the size of the deposited HA particle (Fig. 7). The morphology of the hydroxyapatite varied and was dependent on the collagenous support and on the presence of different ionic species.

In order to understand the influence of different collagen-based supports and ionic species on the morphology of COLL/HA composite materials, all data were compared to the morphology of pure COLL/HA composite materials (Fig. 7D).

After the mineralization of a pure collagen matrix, a homogenous HA deposition was observed. In mineralized hydrolysate-enriched matrices, the morphology was dependent on the hydrolysate content (Fig. 7A and B). An increase in the amount of hydrolysate induced a higher content of fine acicular HA, which is unusual in pure collagen matrix mineralization.

The presence of F^- during the deposition of HA onto the pure collagen matrix induced a thin lamellar deposition (Fig. 7C). The lamellar HA had a mean size of 50x20x1 μm.

The presence of citrate ions induced a special kind of HA mineralization in the collagen matrices (Fig. 7E and F). The deposited HA displayed a similar morphology to those obtained in the presence of hydrolysate but at a different scale. Citrate ions appeared to induce a greater amount of organization of HA into the acicular form, as compared to hydrolysate. This finding is important because natural bones contain a small amount of citrate. In this case, the rods of HA had a mean length of ~10 μm and a mean diameter of less than 0.4 μm.

When PVA is present, the microstructure is strongly influenced for both COLL-PVA and COLL-PVA/HA materials.

The scanning electron microscopy has a decisive significance for the complete characterization of such materials. Therefore, the SEM images were recorded at different magnifications on all COLL-PVA/HA materials (with the component ratio of 1:2:0-blank, non-mineralised sample and 1:2:3-mineralised sample obtained by freeze drying and controlled drying).

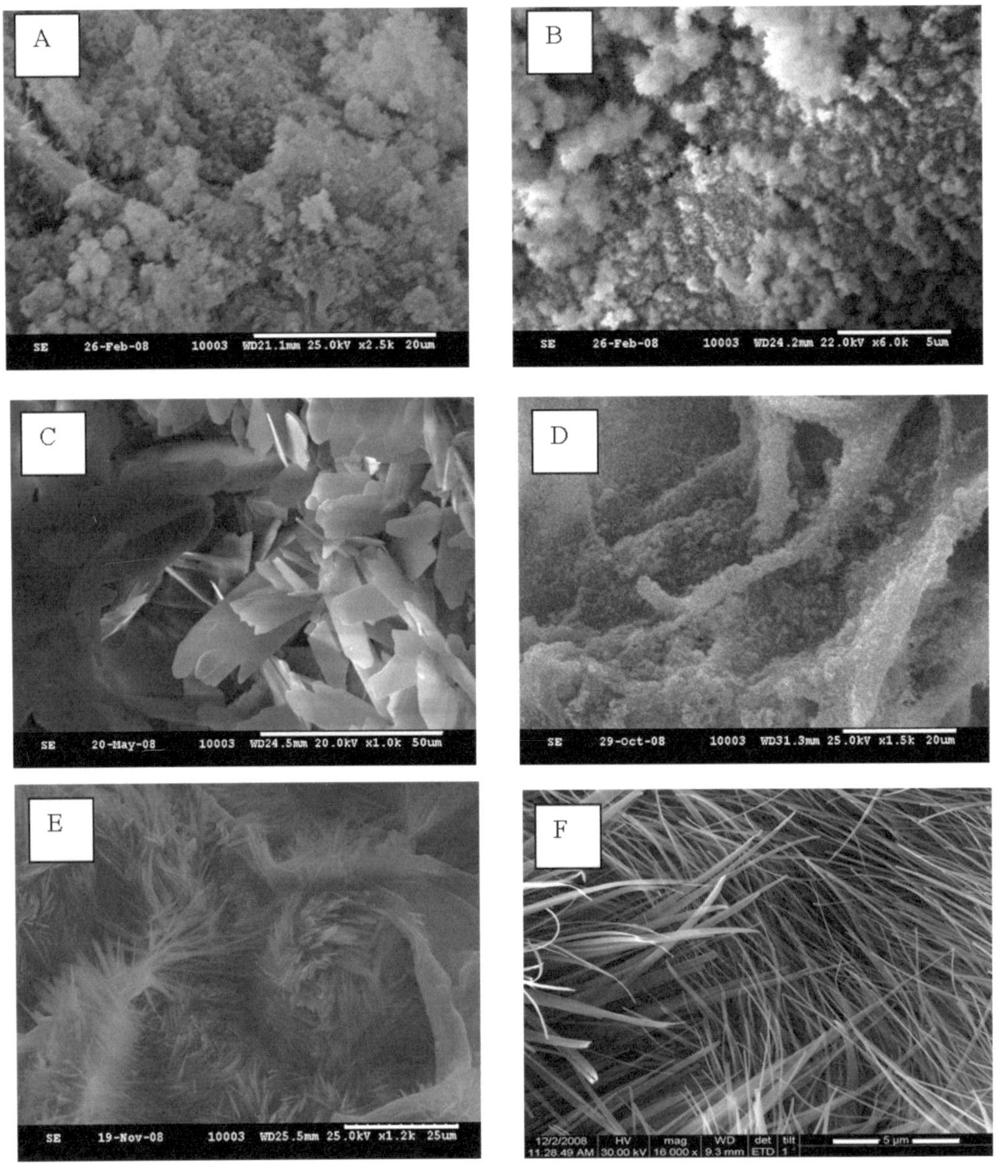

Fig. 7. SEM images of different collagen and hydroxyapatite composite materials: A. composite obtained by mineralization of a collagen-hydrolysate matrix (5% of hydrolysate); B. composite obtained by mineralization of a collagen-hydrolysate matrix (70% of hydrolysate); C. composite obtained by mineralization of a pure collagen matrix (the mineralization was achieved in the presence of F^- (5%-molar reported to HA); D. mineralized (pure) collagen matrix; E, F. composite obtained by mineralization of a collagen-citrate matrix (citrate less than 2%)

The SEM images have given some morphological information about the four materials. Fig. 8a, b and Fig. 9a, b show the morphology of the two materials obtained by controlled drying in air at 30°C; they are stratified but compact materials. Fig. 8c, f and Fig. 9c, f show the morphology of the freeze dried composite materials; such materials exhibiting stratified but porous morphologies. Based on the SEM images, the freeze dried COLL-PVA/HA 1:2:0 hybrid material has the mean distance of 60-120μm between the sheets while for the freeze dried COLL-PVA/HA 1:2:3 composite, the hybrid material has the mean distance of 40-70μm between the sheets.

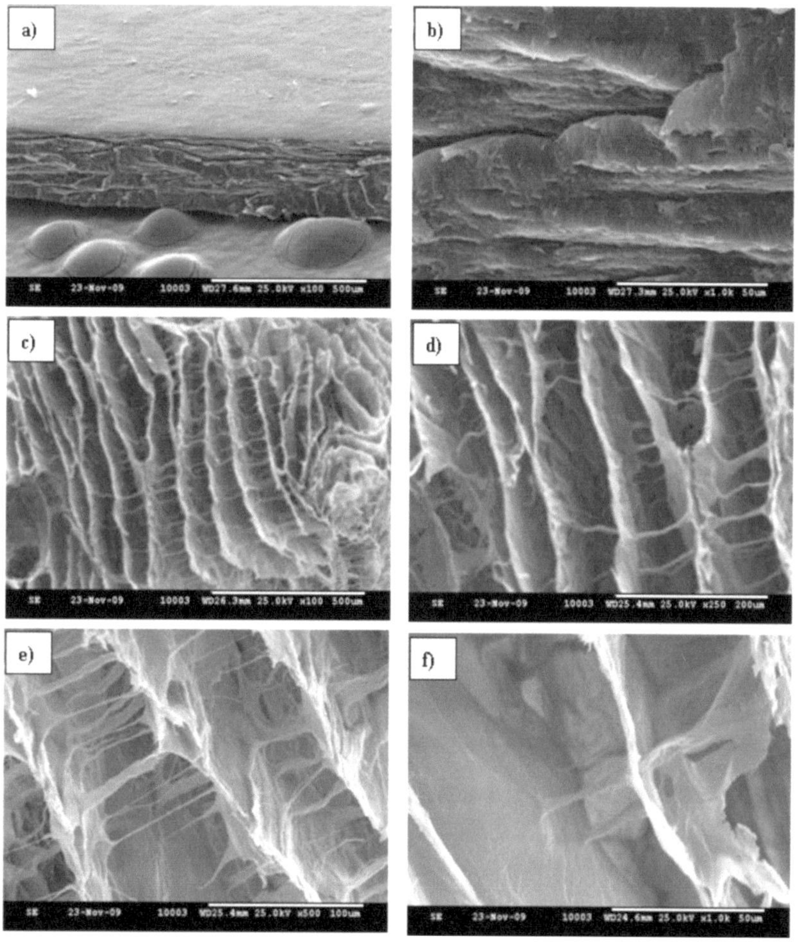

Fig. 8. SEM images of the COLL–PVA hybrid materials (weight ratio of 1:2) obtained by (a and b) controlled drying and (c–f) freeze drying

In fact, the two morphologies are similar, the compact materials resulted from the continuous remodelling and restructuring until it becomes compact; if the freeze drying is used, once the materials are frozen the restructuring is blocked and the materials retain their initial high porosity.

For both compositions, the controlled air drying at 30°C has led to parallel layers bounded each other by fibres of different diameters (from less than 1μm up to 10μm).

A higher magnification (Fig. 9f, for instance) has revealed the very good homogeneity and compatibility of the three components practically, no free HA agglomerates being visible.

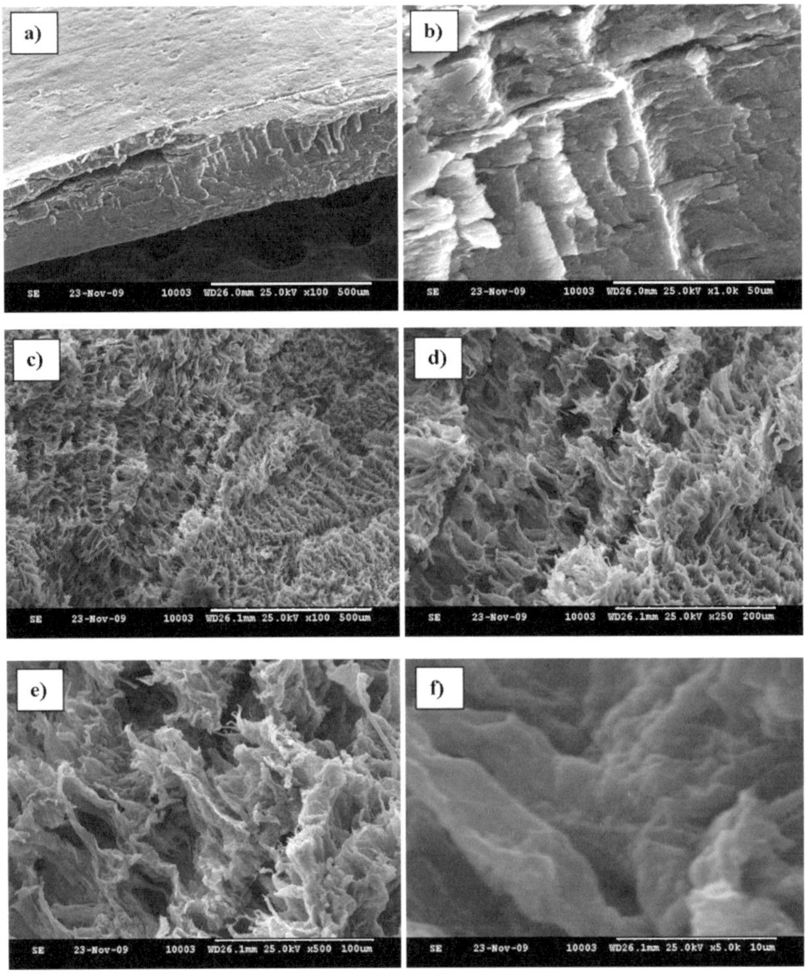

Fig. 9. SEM images of the COLL–PVA/HA hybrid composite materials (weight ratio of 1:2:3) obtained by (a and b) controlled drying and (c–f) freeze drying

The osteointegration of the bone graft materials is induced, among the rests, by the porosity. The density, porosity and absorption (Table 1) were measured by the Arthur method. In all cases three replicates were done, the experimental error being less than 1-2%. The results are in very good agreement with the SEM observation. The highest porosities correspond to the freeze dried materials and have reached about 80 and 90%. For the materials obtained by controlled drying at 30°C in air, the porosity was less than 20%. The xylene absorption is proportional with the porosity while the density is in inverse proportion. The presence of HA has induced the density increase and a decrease in the porosity and xylene absorption.

Hybrid materials with intermediary properties were obtained by a mixed drying method which involves controlled drying followed by freeze drying (not presented in this paper).

Sample	Density, g/cm³	Porosity, %	Absorption, %
COLL-PVA 1:2, freeze drying	0.10	92.7	90.0
COLL-PVA 1:2, controlled drying	0.75	19.3	19.0
COLL-PVA/HA 1:2:3, freeze drying	0.44	79.2	78.9
COLL-PVA/HA 1:2:3, controlled drying	0.88	13.7	13.0

Table 3. Ceramic properties of the COLL-PVA 1:2 (wt) hybrid materials and COLL-PVA/HA 1:2:3 (wt) composite materials

In order to understand the influence of citrate ions on the morphology of acicular and non-acicular zones, the EDS spectra of these two zones were recorded (Fig. 10). The results indicated that the ratio of Ca to P differed between these two zones, where the Ca:P ratio was 1.6 (similar to stoichiometric HA) in the non-acicular zone (Fig. 10A) and 1.1 in the acicular zone (Fig. 10B). The acicular zone occurs due to a higher content of Na^+ and citrate.

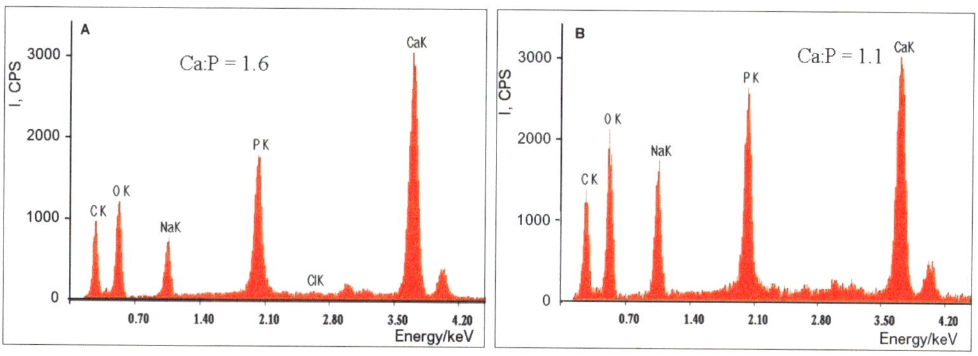

Fig. 10. EDS spectra of a) a non-acicular and b) an acicular reach zone

Infrared spectroscopy (Fig. 11) is an efficient tool for the investigation of organic-inorganic interactions. The differences in morphology may be explained by these strong interactions and are evidenced in the shift of peaks or in peaks duplications. In order to understand organic-inorganic interactions, the IR spectrum of a pure collagen matrix was recorded (Fig. 11A). The most important peaks in the pure matrix were the amide peaks (Chang and Tanaka 2002; Ficai et al. 2009b). After collagen matrix mineralization with HA precursors, the main amide peak of pure collagen (1630 cm^{-1}) was shifted to a higher wave number for the COLL/HA composite (1650 cm^{-1}), while the phosphate peaks appeared at 1030, 609 and 564 cm^{-1} (Fig. 11B).

The infrared spectrum of the mineralized hydrolysate-enriched collagen matrix (Fig. 11C) was characterized by the duplication of each main collagen peak and led to the differentiation of the phosphate peak and the appearance of a shoulder at about 1110 cm^{-1}. These results can be explained by the degree of condensation between collagen and collagen hydrolysate, which induces interactions of varying strength between mineral and organic components. As reported in the literature (Silva et al. 2001), the degradation of collagen leads to a decrease in the absorption band at 1240 cm^{-1}. In composites obtained with hydrolysate, this band being less intense than that in pure collagen.

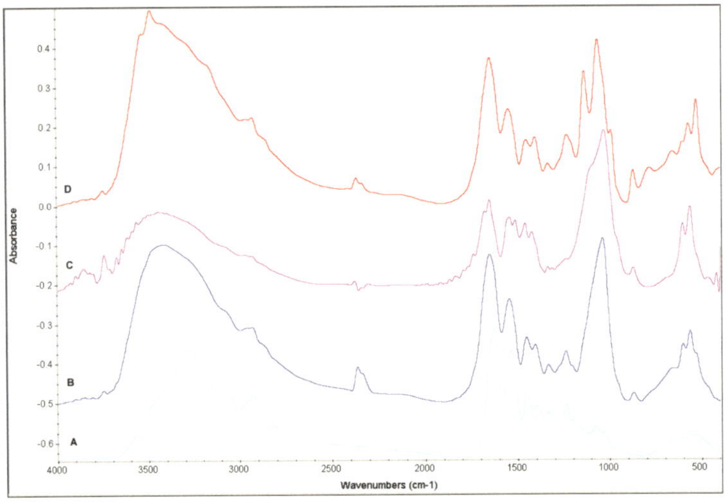

Fig. 11. FTIR spectra of: A. a collagen matrix (cross-linked with glutaraldehyde, 1%); B. a mineralized cross-linked collagen matrix (glutaraldehyde, 1%); C. a mineralized hydrolysate-enriched collagen matrix; and D. a mineralized cross-linked collagen matrix achieved in the presence of F^- (5 % (molar) reported to HA)

When the mineralization of the collagen matrix occurs in the presence of fluoride, the collagen peaks were not shifted (compared to the COLL\HA composite), but the phosphate peaks of HA were split due to the partial substitution of hydroxyl groups with fluoride (Fig. 11D).

3. The synthesis of COLL/HA composite materials with oriented structure

In this field of orientation some works were published in the last 5 years. The synthesis of COLL/HA composite materials with oriented structure can be induced by self-assembly, electric field orientation or by using a high magnetic field (Wu et al. 2007), best results being reported by self-assembly (Ficai et al. 2010c) and pulsed electric field orientation (Ficai et al. 2010a). The magnetic field orientation is possible due to the magnetic anisotropy of the hydroxyapatite (Wu et al. 2007).

3.1 Self-assembling of COLL/HA composite materials

The self-assembly is one of the most easy to realize orientation which not imply any external influences but require slow drying at pH=6.9-9 which can easily induce denaturation of the collagen.

The synthesis of COLL/HA composite materials by self-assembling consists by two successive stages but have some particularities. In the first stage, the collagen gel is treated with the desired amount of $Ca(OH)_2$ suspension, drop-wise and magnetically stirred for 24 h and let to interact. In the second stage, the stoichiometric quantity of H_3PO_4 solution was added also dropwise. $Ca(OH)_2$ was used in order to assure the necessary basic pH. During the first stage of mineralization the pH was maintained at pH $\simeq$ 9 by addition of HCl.

The ratio of collagen, $Ca(OH)_2$ and H_3PO_4 was so chosen that the final ratio of COLL:HA to be 20:80 (wt) and at the end of synthesis the concentration of collagen to became 1.66%. After the H_3PO_4 solution is added, the pH was adjusted at about 9 using NaOH solution, in order to assure pure HA precipitation. During the synthesis and drying of the composite material, the temperature was kept at ~37 ^{o}C.

pH 9 was choose based on two characteristics of the collagen molecules:

a. at pH= 6.9-8 the collagen molecules are in an extended conformation (length of collagen molecule 180-200 nm) and,
b. the fibrillogenesis is increasing in the range of 6.6-9.2.

The main processes which occur during the synthesis are represented in Fig. 12. The collagen molecules have different conformation, function of the pH can be linear (at pH≥9) or crimpy (for pH< 7); at intermediate pH the collagen molecules being crimpy but with a more pronounced linear aspect. In order to be sure that collagen molecules are in elongated form (linear) the working pH was set at 9.

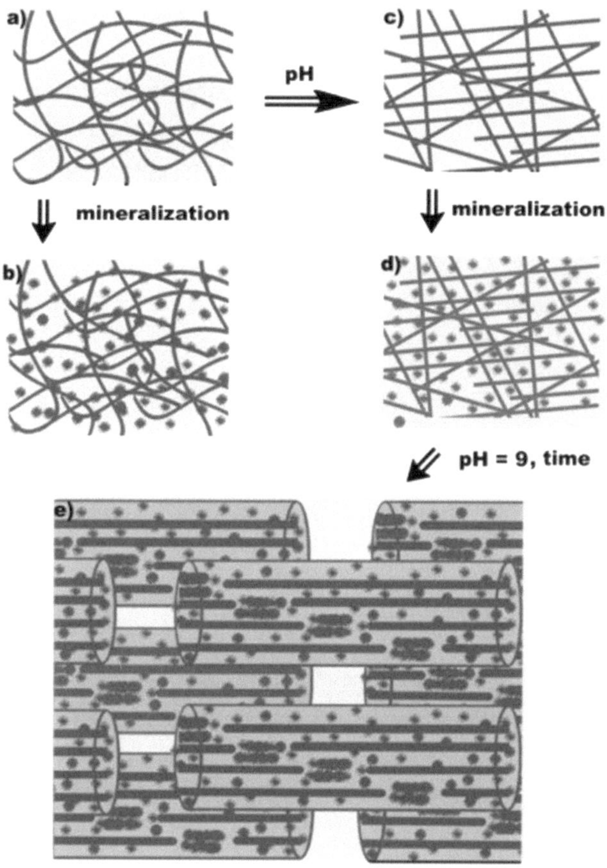

Fig. 12. Specific processes occurred in solution during the self-assembly of COLL/HA composite materials

Function of the mineralization pH, we can obtain different morphologies (Fig. 12b and d). If the pH increases, the crimpy structure of collagen molecules from gel and also from

composite can be modified into the elongated one. If the drying time is slowly enough, these linear (elongated) collagen molecules can self-assembly and form cylindrical fibrils and fibres (Fig. 12e), otherwise they will form fibrils and fibres, growth tri-dimensional, with crimpy collagen molecules (for pH< 7) or with linear collagen molecules (for pH ≥ 9). The structure presented in Fig. 12b can be transformed into the structure presented in Fig. 12d by increasing of pH at an adequate value. In this case, the pH must be higher than in case of transformation of the structure illustrated in Fig. 12a into the structure presented in Fig. 12c, probably due to the interaction between HA and collagen.
The self-assembling process was characterized especially by XRD and SEM.
XRD was used to point out the mineralization process. The XRD spectrum, in Fig. 13, shows the formation of HA.

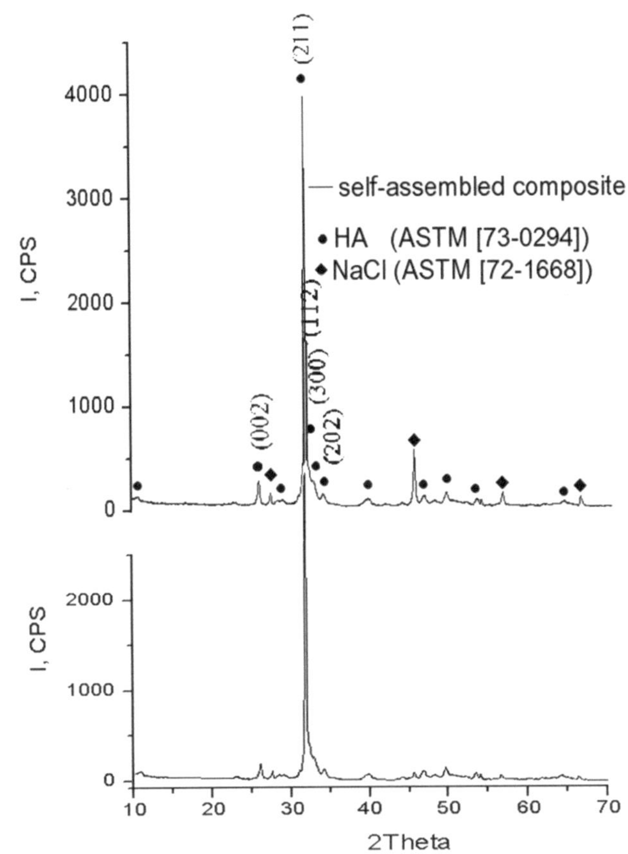

Fig. 13. XRD pattern of COLL/HA composite material (a) before sodium chloride removal and (b) after sodium chloride removal

In the case of COLL/HA composite obtained by self-assembly comparing with the HA obtained by precipitation, the X-ray diffraction pattern exhibit a much higher intensity for the 2 1 1 peak reported to the intensity of the other peaks characteristic to HA. This result can be attributed to a preferential growth of the HA crystals in the 2 1 1 direction due to the collagen influence. As it can see, the composite material also contains NaCl (Fig. 13a). The removal of NaCl can be easily made by washing the composite materials with distilled water, following the next procedure: after drying, the COLL/HA composite material with uniaxial orientation of the constitutive fibres is crosslinked with glutaraldehyde, washed with plenty of water and dried. Following this procedure, the chloride was completely removed (Fig. 13b) without altering the mineral phase, especially from the point of view of crystallinity and preferential crystallization direction.

The self-assembling structure of collagen molecules and hydroxyapatite particles can be proved by SEM (Fig. 14). The samples were analyzed in perpendicular and parallel section reported to mineralized collagen fibres, in order to study the formation and the orientation of collagen fibrils and fibres. The sodium chloride removal not alters the composite morphology; the SEM images recorded before and after sodium chloride removal being similar. Whatever the analysis section, the recorded SEM images show the formation of collagen fibrils and fibres, which are mineralized with HA.

Fig. 14. The SEM images of collagen/hydroxyapatite composite materials, recorded at different magnification; (a) parallel view with the fibres, (b–c) perpendicular view with the fibres, (d) high resolution SEM at 80,000× magnification

The SEM image presented in Fig. 14a is recorded in a fibres perpendicular section. This image shows a homogenous arrangement of the fibres. Fig. 14b and c are recorded in a fibres parallel profile and shows, at different magnification the stratified structure of the composite materials, the fibrils and fibres being organized in layers. It can also observe highly oriented fibres and fibres bundle. The homogenous arrangement can be evidentiated also in fibres parallel profile. Analyzing the rupture profile, it can be observed a shift between the rupture point, comparing different neighbor adjacent layers. In Fig. 14c, recorded at higher magnification, it can better observe the mineral deposition. It can conclude, that at the end of collagen fibres, the HA density is higher. At broken, the rupture seems to follow the highly mineralized zone of interfibrilar gap which exist at the end of two successive fibres. Finally, at a much higher magnification (Fig. 14d) the SEM images of fibres' end allow the estimation of the dimensions of HA, these particles being in the nanometric range. At high magnification from SEM image, it can be seen that the dimensions of the HA crystals are in the nanometric range (4–40 nm), with an elongated morphology. These results are in good agreement with the observation reporting to natural HA from the natural bone.

The SEM images obtained at different magnification, in the rupture, on parallel and perpendicular direction of collagen fibres, are showing a homogenous microstructure, with fibres organized in layers, with high orientation. Taking also in consideration the way in which the rupture took place, and also the fact that we have a high concentration of HA grains in the rupture points we might conclude that the COLL/HA composite material synthesized has a very similar structure with that of the natural long bone, that have been obtained through auto-assembling process.

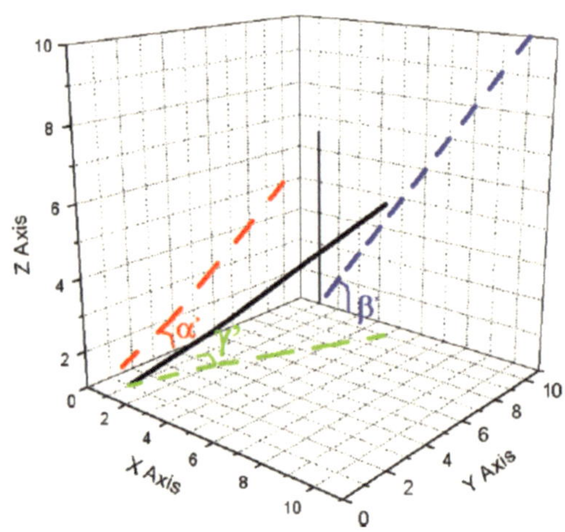

Fig. 15. Straight line projection onto the three dimensions and visualization of the three angles formed between the projections and the three dimensions

The orientation degree can be quantified function of the deviation angle of each collagen fibers from the direction of the applied electric field. The orientation degree of mineralized collagen fibers is highly influenced by magnification; with the increasing magnification the orientation degree increase. At a magnification of 1000× the average of deviation degree is less than 5% while the orientation degree is more than 95%. The average fiber deviation and also the orientation degree of collagen fibers can be quantified based on the following equations:

$$\bar{D}(\%) = \frac{100}{N} \cdot \sum_{i=1}^{N} \frac{DA_i}{90} \text{ and the } OD(\%) = 100 - \bar{D}$$

where, N is the number of fibers; DA_i – correspond to deviation angle of each fiber; $\overline{D}$ (%) correspond to the average deviation of the fibers toward the applied electric field and OD(%) correspond to the orientation degree.

Obviously the orientation is difficult to quantify and only a few methods permit this kind of measurements (and usually these are indirect methods); the orientation will be quantify based on the SEM images.

Hydroxyapatite was obtained by co-precipitation in the presence of collagen gel. In fact, starting from collagen gel and hydroxyapatite precursors, in certain conditions, due to the interactions between collagen, hydroxyapatite and water it was obtained self-assembled, highly oriented composite materials. It can note that, many authors published a lot of papers dealing with COLL/HA composite materials which start from collagen gel and calcium hydroxide and ortophosphoric acid as precursors, but due to the inadequate processing conditions they do not obtained the uniaxial orientation of the constitutive fibres.

In aqueous solution the collagen molecules and hydroxyapatite precursors have the capacity to induce synthetic bone formation by self-assembling. The obtained composite material is more similar with compact bones morphology; the recorded analysis being very similar with these bones.

Very important is that we also propose a new way to estimate the average deviation of fibres and we determined it, assuming a 2D model. The average deviation is 2.54±0.2% which means that the degree of orientation is 97.46±0.2%.

3.2 Electric field orientation of COLL/HA composite materials

Also, orientation can occur due to different external factors (magnetic or electric fields), most efficient and easy to realize being the electric field orientation. The electric field orientation occur really fast (less than 1h) at low electric field (<1V/cm), the best electric field being the pulsatory electric filed. The magnetic orientation can be realized but require very high magnetic field (10T). First time in the literature, the degree of orientation was mathematically quantified using a very simple equation, based on the SEM images.

The synthesis of COLL/HA composite materials via electric field orientation is realized in similar conditions with the mention that, before drying different kinds of electric fields are applied. By short, the collagen mineralization was conducted in two stages, in order to mimic the in vivo osteosynthesis. First stage is the calcium deposition onto the collagen that takes 24 h. The second stage consists in phosphate addition leading to hydroxyapatite precipitation (Ficai et al. 2009a). After 24 h, at 37^{0}C and pH=9–10, the mineralization can be considered completed and the orientation process can be initiated.

For orientation purpose of the mineralized collagen gel, there were considered two types of electric field: pulsatory field (0.93 V/cm) and superposed field (direct field -0.67 V/cm and pulsatory field -0.93 V/cm). In order to be able to compare the obtained results, all the samples were obtained in similar hollow mold with the dimensions of W×L×H=2×3×2 cm^3; if electric field is applied, the electrodes are fixed in the hollow mold at 3 cm distance.

The mineralized collagen gel is maintained for 1 h in the desired electric field. After that, the samples were introduced in the freeze drier and frozen at -35^{0}C. The freezing of the material is compulsory, in order to preserve the obtained structure. After freezing, the electrodes can be removed and the freeze drying process is started.

When no electric field is applied, the collagen molecules dipoles are randomly disposed, without any orientation. If an electric field is imposed, the collagen dipoles are becoming oriented, due to the interaction between collagen and the electric field (Fig. 16).

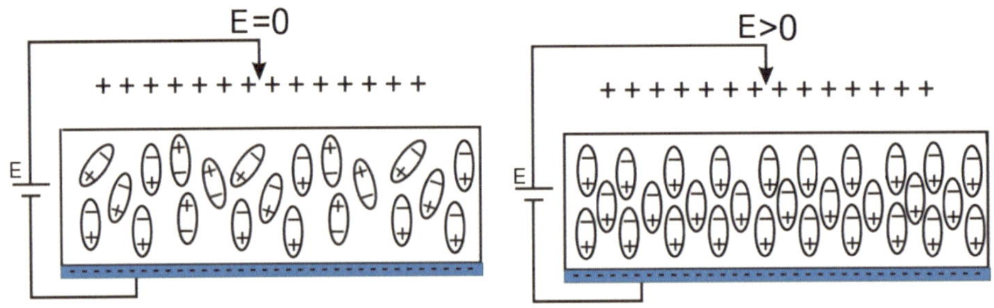

Fig. 16. Electric field orientation of collagen dipoles

The electric field is obtained with a "home-made" device, consisting in two different electric generators, each with its own role. One is responsible for the direct current, while the other for the pulsatory current generation. These two generators can work separately, at a maximum potential of 30 V.

The SEM micrographs (Fig. 17) show relevant difference between the samples obtained in the presence and respectively in the absence of electric field. If different types of electric field are applied, different degrees of orientation are achieved. Fig. 17a, d show the microstructure of collagen/hydroxyapatite composite material obtained without electric field, Fig. 17b, e show the microstructure of collagen/ hydroxyapatite composite material, obtained by orientation in a pulsatory electric field and Fig. 17c, f show the microstructure of the composite material obtained in a combined (superposed) electric field.

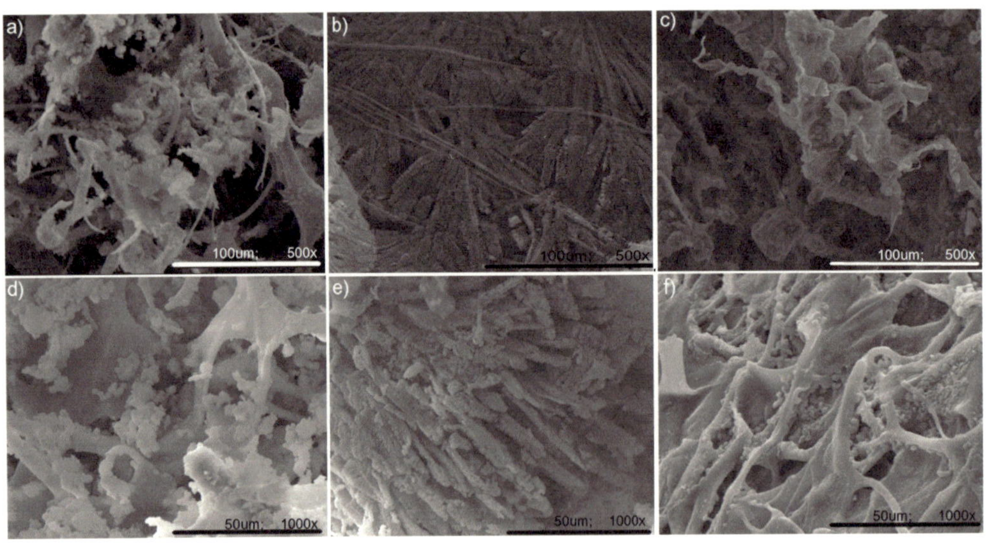

Fig. 17. SEM images of coll/HA composite: a, d) control – no electric field is applied; b, e) pulsatory electric field is applied, 0,93 V/cm, frequency = 1 Hz; c, f) superposed electric field: 0,67 V/cm direct electric field and 0,93 V/cm and 1Hz pulsatory electric field

The mineralization process of collagen gels leads to composite materials with a structure similar to that of spongy bone (Fig. 17a, d), without any orientation of the collagen fibers. If a pulsatory electric field is applied the collagen molecules and fibril are becoming oriented and the structure became more compact (Fig. 17b, e), like compact bone. The superposed electric field induces an intermediary degree of orientation of the composite material (Fig. 17c, f). The presence of the direct electric field induces not only orientation but also migration processes of the collagen molecules and due to the collision with the collagen molecules or hydroxyapatite particle the orientation degree will be worse comparing with pure pulsatory electric field.

If a combined electric field is applied, the influence of the electric field on the microstructure of samples is not that evident, the structure exhibiting a very low degree of orientation (Fig. 17c, f).

It is well known that the natural osteogenesis is assisted by the collagen piezoelectricity. Briefly, the piezoelectricity can be defined as the translation of the mechanical stress into electric field (Ferreira et al. 2008; Noris-Suarez et al. 2007). The osteogenesis is a very complex process which consists of continuous collagen deposition and mineralization. Piezoelectricity influence not only the mineralization process but, due to the electrical surface charge induced by the mechanical stimuli, also the collagen deposition is assisted. The electric field produced by piezoelectricity is similar with the pulsatory electric field (the frequency of the pulsatory electric field being of less than 5-10 Hz) and due to this similarity the results are the same: in the cortical bone the microstructure is orientated (these bones are exposed to mechanical stress) as well as in the case of in vitro, pulsatory electric field assisted COLL/HA composite materials synthesis.

4. The synthesis of COLL/HA composite materials by "layer by layer" method

The LbL deposition of hydroxyapatite (HA) onto a collagen matrix involves HA synthesis on the collagen matrix starting from electrically charged support. (Ficai et al. 2009c).

The mineralization process was performed directly on the collagen matrix starting from $Ca(OH)_2$ suspension and NaH_2PO_4 solution. In all cases, the collagen matrices weighed between 0.06 and 0.07 g in order to obtain reproducible depositions. The precursors were prepared in order to obtain a 0.026 M $Ca(OH)_2$ suspension and a 0.015 M NaH_2PO_4 solution. Nine pieces of collagen matrix of similar mass were used for the mineralization process.

The layer by layer method is schematically represented in Fig. 18.

The mineralization process was performed as follows. In the first stage, the collagen matrices were bound on a stainless steel net and immersed into the mechanically stirred $Ca(OH)_2$ suspension for 20 minutes. In the second step, the matrices were immersed in the phosphate solution for 10 minutes. When the matrices were immersed in the phosphate solution, hydroxyapatite precipitation occurred due to the retained Ca^{2+} ions. Between the two immersions, the matrices were allowed to stay exposed to air for 5 minutes to allow for water draining.

The mineralization process can be analyzed by SEM, FTIR and ATD-TG.

Scanning electron microscopy images were recorded in order to study the mineralization, the morphology and the porosity of the obtained deposition products.

The recorded SEM (Fig. 19) images revealed that better mineralization occurred when a larger number of hydroxyapatite layers were deposited. The increased number of HA layers led to an increased amount of deposited HA. When studying different composites with an increasing number of deposited HA layers, we can conclude that the composite materials have become denser, due to the filling of the matrix holes with HA.

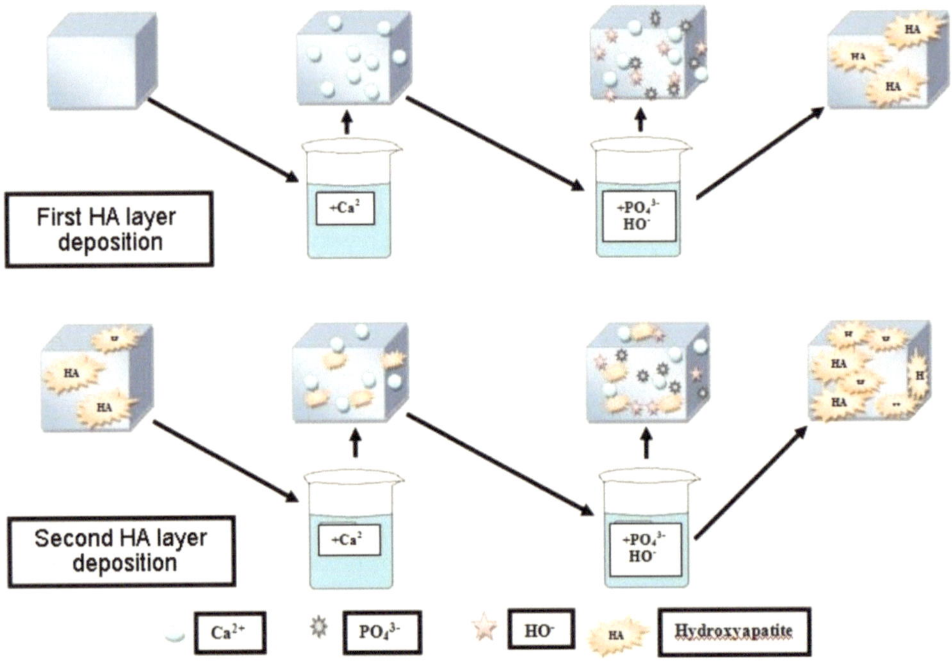

Fig. 18. Schematically representation of layer by layer deposition method

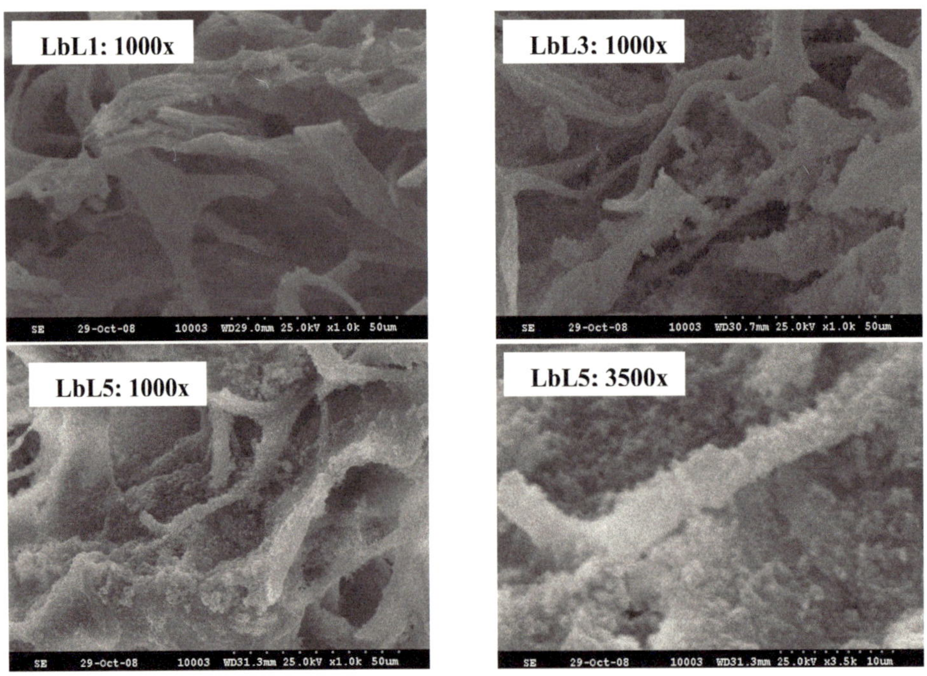

Fig. 19. SEM images of layer by layer mineralized collagen matrix: 1, 3, 5 layers of HA

The amount of deposited HA is dependent on the mineralization process, and particularly on the citrate ions present within the collagenous materials. The amount of HA is seen to be directly proportional to the number of deposited HA layers.

At higher SEM resolutions, the size of HA agglomerates was determined. The typical size of the HA agglomerates increases with the number of layers from less than 100 nm for the composite obtained by LbL1 up to several micrometers in the case of composite obtained by LbL5.

Infrared spectroscopy (Fig. 20) confirmed the composite structure of the obtained materials. IR analysis also confirmed the increasing amount of deposited HA by showing increasing PO_4^{3-} to carbonyl group peak intensity ratios with increasing numbers of HA layers. For visual quantification of the mineralization process, the IR spectra of collagen matrix and mineralised matrices were worked up to obtain the same intensity of the carbonyl peaks (1637 cm^{-1}) without modifying the peak ratios. In this case, it can be assumed that the ratio between the intensity of HA peaks for each multilayered composites is proportional to the ratio of the deposited HA amounts.

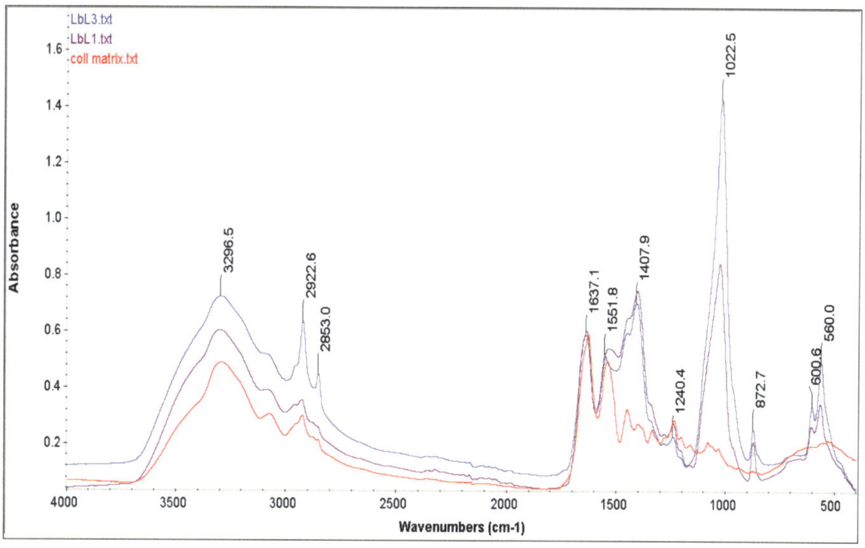

Fig. 20. FT-IR spectra of pure and mineralized collagen matrix with 1 and 3 layers of HA

When the collagenous matrix is immersed for 24 h into $Ca(OH)_2$ suspension and than into NaH_2PO_4 solution, about 32% of HA is deposited onto the matrix (Fig. 21, dotted line). The same results were obtained when alternately immersing the matrix three or four times into a $Ca(OH)_2$ suspension (for 20 min) and a NaH_2PO_4 solution (for 10 min).

The amount of deposited HA was determined by gravimetric analysis by measuring the initial mass of collagen matrices and the mass of the dried mineralised matrices. The amount of deposited HA is quite linear with respect to the number of deposited layers. This relationship was quantified and the root-mean-square deviation was found to be 0.9857.

The layer by layer deposition method was successfully applied in order to increase the amount of mineral phase that is deposited on the collagen matrix. The root-mean-square-deviation is very close to 1, meaning highly linear layer growth was observed for even six HA layers.

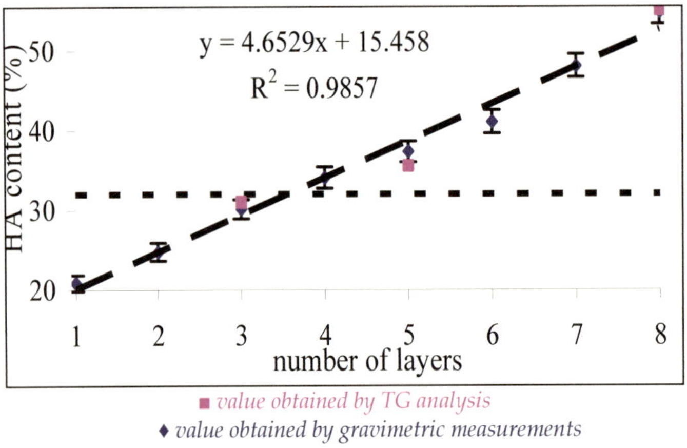

Fig. 21. Gravimetric variation of deposited HA with the number of deposited layers

The classical mineralization method conducted by dipping a collagen matrix in $Ca(OH)_2$ suspension for 24 h followed by dipping the collagen matrix in NaH_2PO_4 solution for a further 24 h can be improved by using the LbL deposition method. Deposition of six HA layers obtained by alternately soaking in a $Ca(OH)_2$ suspension for 20 min and then in a NaH2PO4 solution for 10 min resulted in a 12% increase in HA deposition. Extrapolating from the obtained results, it can be estimated that for a collagen matrix, such as the ones used in these experiments, one would need to deposit about 14 layers of HA to obtain a composite material similar to whale bone, while antler bone composition can be obtained if seven HA layers are deposited.

The obtained composite materials can be assumed to be hydroxyapatite matrices reinforced with mineralized collagen fibers as they have a structure very similar to the bone structure suggested by Hellmich et al. (Hellmich et al. 2004).

The good linearity of the quantity of deposited HA with respect to the number of layers is due to interactions occurring not only between collagen and HA, but also between deposited HA and Ca^{2+} ions.

From the parameters that influence the deposition of HA, we can distinguish three main categories: support dependent, solution/suspension dependent and processing parameters. While the first two categories have been extensively studied and can be quantified, the third category is very difficult to be quantified.

Under similar mineralization conditions, the amount of deposited HA is greater in the presence of citrate ions, but the LbL method applied to a pure collagen matrix can increase the amount of deposited HA even more than in the case of one layer deposition (24+24 h) onto citrate enriched collagen matrices.

5. The synthesis of complex COLL/HA+Fe_3O_4 composite materials

Magnetite is a mineral with multiple roles in both medical (Ito et al. 2005; Mornet et al. 2006; Zhang and Misra 2007) and non-medical applications (Ficai et al. 2010d; Ju and Bian 2006). The addition of magnetic nanoparticles (especially magnetite nanoparticles) induces new properties to the COLL/HA composite materials (Andronescu et al. 2010). The bone regenerative effects are due to the presence of COLL/HA while the anti-tumoral effects are due to the presence of magnetite which can produce hyperthermia when an electromagnetic field is applied. These systems, even at low magnetite concentration (5%) can be used for curative purpose because can generate the necessary hyperthermia and consequently induce tumoral cell apoptosis. It is important to mention that the presence of magnetite even at low concentration (1-2%) may induce hyperthermia but, in order to be useful for medical applications (hyperthermia - cancer treatment) at least 5% of magnetite is required. One of the most important advantage of the use of magnetite based composite materials is that hyperthermia can be activated only when is necessary and consequently the side effects is limited comparing with chemotherapy, for instance. As a matter of course these materials will be improved by the addition of other antitumoral agents such as silver or gold nanoparticles, cytostatics or other drugs for pain managements.

6. Conclusions

Collagen/hydroxyapatite composite materials are the most similar synthetic grafts with bone from many points of view, bone being composed from collagen and hydroxyapatite as main components and few percent of other components.

The morphology and subsequent the properties of the composite materials is strongly influenced by the presence of different components, even when they are present in small proportions. The reason that perfect bone graft materials have not been successfully synthesized is due to the limited number of components used in the synthesis of bone graft materials, which typically include only collagen and hydroxyapatite or carbonated apatite.

It is worth to mention that all commercially available collagen forms can be converted into COLL/HA composite materials with dense or spongious microstructure. If collagen gel can be easily converted in dense or spongious materials by a proper choice of the drying method, collagen

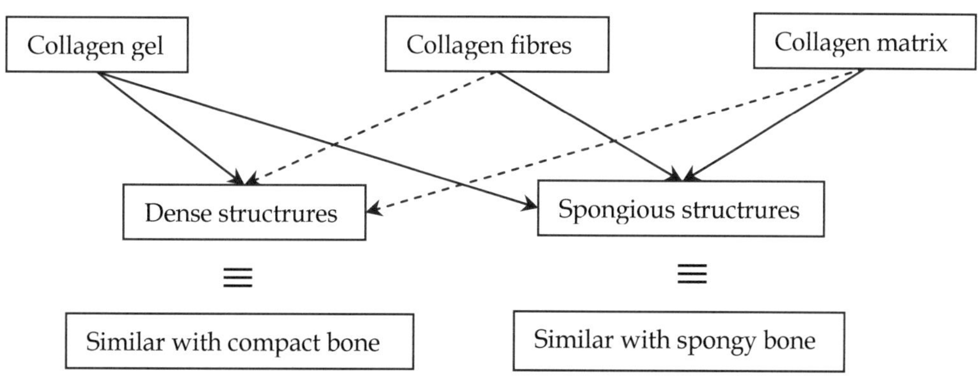

Fig. 22. The influence of collagen form on the composite materials microstructure

Based on the presented results, it can be concluded that the presence of additional components (which are usually found in natural bone in small concentrations) is of significant importance. For instance, the presence of fluoride induces a higher crystallinity in the deposited mineral phase. The morphology of the apatite phase was found to be lamellar by SEM, where visible pores were not observed, even at relatively high magnification.

The shape of the mineral phase of the composite material obtained by *in vitro* co-precipitation in the presence of fluoride is biomimetic and similar to the mineral phase of natural bone. The main difference between the mineral phases of bone and composite materials obtained in the presence of F^- is the size of the crystal. In order to obtain natural-sized crystals in the mineral phase, crystallization inhibitors may be used. Under these conditions, we expect to reduce the size of the crystals.

Not only the presence of different components can induce morpho-structural modifications but also the synthesis route. For instance, the proper, applied electric or magnetic field or the drying method corroborated with ionic strength and pH lead to the formation of highly oriented COLL/HA composite materials. The synthesis of COLL/HA composite materials with oriented morphology of the mineralized collagen fibrils and fibres is an essential step to obtain bone grafts of the long bones. The orientation degree, based on SEM images, was of great importance and allows the quantification of the orientation. Based on the existing data, best orientation can achieve with self-assembling, the mean orientation degree being of ~ 97%.

7. Aknowledgements

Authors recognize financial support from the European Social Fund through POSDRU/89/1.5/S/54785 project: "Postdoctoral Program for Advanced Research in the field of nanomaterials and from Romanian Authority for Scientific Research through the project 72-198. We also thank to Elsevier for the amability to reuse some parts of text or figures.

8. References

Abe, E., Yamamoto, M., Taguchi, Y., Lecka- Czernik, B., O'Brien, C. A., Economides, A. N., Stahl, N., Jilka, R. L. and Manolagas, S. C. 2000. Essential requirement of BMPs-2/4 for both osteoblast and osteoclast formation in murine bone marrow cultures from adult mice: antagonism by noggin. *J Bone Miner Res* 15:663-673.

Andronescu, E., Ficai, M., Voicu, G., Ficai, D., Maganu, M. and Ficai, A. 2010. Synthesis and characterization of collagen/hydroxyapatite:magnetite composite material for bone cancer treatment. *Journal of Materials Science - Materials in Medicine* 21(7):2237-2242.

Bakos, D., Soldan, M. and Hernandez-Fuentes, I. 1999. Hydroxyapatite-collagen-hyaluronic acid composite. *Biomaterials* 20(2):191-195.

Bollerslev, J., Gram, J., Nielsen, H., Brixen, K., Storm, T., Larsen, H. and Mosekilde, L. 1991 Effect of a short course of 1,25-dihydroxyvitamin D3 on biochemical markers of bone remodeling in adult male volunteers. *Bone* 12 339-343.

Bonzani, I. C., Adhikari, R., Houshyar, S., Mayadunne, R., Gunatillake, P. and Stevens, M. M. 2007 Synthesis of two-component injectable polyurethanes for bone tissue engineering. *Biomaterials* 28(3):423-433.

Chang, M. C. and Tanaka, J. 2002. FT-IR study for hydroxyapatite/collagen nanocomposite cross-linked by glutaraldehyde. *Biomaterials* 23:4811–4818.

Clarke, I. C., Manaka, M., Green, D. D., Williams, P., Pezzotti, G., Kim, Y.-H., Ries, M., Sugano, N., Sedel, L., Delauney, C. and others. 2003. Current Status of Zirconia Used in Total Hip Implants. *Journal of Bone and Joint Surgery* 85(Supplement 4):73-84.

Corces, A. 2002. Metal Alloys. *Proc. III Miami Symposium for Total Joint Replacement*, South Miami Hospital, Miami, Florida.

Corces, A. and Garcia, M. 2007. Metallic Alloys. [Online] Available: http://www.emedicine.com/orthoped/TOPIC610.HTM [2008].

Cui, C. B., Cooper, L. F., Yang, X., Karsenty, G. and Aukhil, I. 2003. Transcriptional Coactivation of Bone-Specific Transcription Factor Cbfa1 by TAZ. *Mol Cell Biol* 23(3):1004-1013.

Cui, F.-Z., Li, Y. and Ge, J. 2007. Self-assembly of mineralized collagen composites. *Materials Science and Engineering* R 57:1-27.

de Vernejoul, M., Cohen-Solal, M. and Orcel, P. 1993 Bone cytokines. *Curr Opin Rheumatol* 5:332-338.

De Vernejoul, M., Pointillart, A., Bourdeau, A., Morieux, C., Modrowski, D., Miravet, L. and Caulin, F. 1990. Effect of calcitonin administration on young pig trabecular bone remodeling. *Bone* 11:29-33.

Develioglu, H., Koptagel, E., Gedik, R. and Dupoirieux, L. 2005. The effect of a biphasic ceramic on calvarial bone regeneration in rats. *Journal of Oral Implantology* 31(6):309-312.

Dorozhkin, S. V. 2009. Calcium orthophosphate-based biocomposites and hybrid biomaterials. *J Mater Sci* 44:2343-2387.

Ferreira, A. M., Noris-Suarez, K., Bello, A., Marquez, A. H., Feijoo, J. L. and Lira-Olivares, J. 2008. Effect of type I collagen piezoelectricity on cellular adhesion. Iv Latin American Congress on Biomedical Engineering 2007, *Bioengineering Solutions for Latin America Health*, Vols 1 and 2 18(1,2):659-662.

Ficai, A., Andronescu, E., Ghitulica, C., Voicu, G., Trandafir, V., Manzu, D., Ficai, M. and Pall, S. 2009a. Colagen/Hydroxyapatite Interactions in Composite Biomaterials. *Materiale Plastice* 46(1):11-15.

Ficai, A., Andronescu, E., Trandafir, V., Ghitulica, C. and Voicu, G. 2010a. Collagen/hydroxyapatite composite obtained by electric field orientation. *Materials Letters* 64(4):541-544.

Ficai, A., Andronescu, E., Voicu, G., Ghitulica, C. and Ficai, D. 2010b. The influence of collagen support and ionic species on the morphology of collagen/hydroxyapatite composite materials. *Materials Characterization* 61(4):402-407.

Ficai, A., Andronescu, E., Voicu, G., Ghitulica, C., Vasile, B. S., Ficai, D. and Trandafir, V. 2010c. Self assembled collagen/ hydroxyapatite composite materials. *Chemical Engineering Journal* 160(2):794-800.

Ficai, A., Andronescu, E., Voicu, G., Manzu, D. and Ficai, M. 2009b. Layer by layer deposition of hydroxyapatite onto the collagen matrix. *Materials Science and Engineering:* C 29(7):2217-2220.

Ficai, A., Andronescu, E., Voicu, G., Manzu, D. and Ficai, M. 2009c. Layer by layer deposition of hydroxyapatite onto the collagen matrix. *Materials Science & Engineering C-Materials for Biological Applications* 29(7):2217-2220.

Ficai, D., Ficai, A., Voicu, G., Vasile, B. S., Guran, C. and Andronescu, E. 2010d. Polysulfone based Membranes with Desired Pores Characteristics. *Materiale Plastice* 47(1):24-27.

Maria Ficai, Ecaterina Andronescu, Denisa Ficai, Georgeta Voicu, Anton Ficai; Synthesis and characterization of COLL-PVA/HA hybrid materials with stratified morphology; *Colloids and Surfaces* B: Biointerfaces; 2010:81(2): 614-619

Gallie, W. E. and Toronto, M. B. 1914. The histrory of a bone graft. *The Journal of Bone and Joint Surgery* s2-12:201-212.

Guelcher, S., Patel, V., Gallagher, K., Connolly, S., Didier, J., Doctor, J. and Hollinger, J. 2004. Synthesis of Polyurethane Foam Scaffolds for Bone Tissue Engineering. *Proc. AIChE Anual meeting Austin*, Texas.

Hellmich, C., Barthelemy, J. F. and Dormieux, L. 2004. Mineral-collagen interactions in elasticity of bone ultrastructure - a continuum micromechanics approach. *European Journal of Mechanics A/Solids* 23(5):783-810.

Hench, L. L., Hench, J. W. and Greenspan, D. C. 2004. BIOGLASS®: A SHORT HISTORY AND BIBLIOGRAPHY. *Journal of Australian Ceramic Society* 40(1):1-42.

Hench, L. L. and Wilson, J., (eds.) 1993. *An Introduction to Bioceramics*. World Scientific Publishing Co., Singapore.

Hock, J. and Gera, I. 1992. Effects of continuous and intermittent administration and and inhibition of resorption on the anabolic response of bone to PTH. *J Bone Miner Res* 7:65-72.

Ito, A., Shinkai, M., Honda, H. and Kobayashi, T. 2005. Medical application of functionalized magnetic nanoparticles. *Journal of Bioscience and Bioengineering* 100(1):1-11.

Ju, D. Y. and Bian, P. 2006. Development of ferrite magnetic materials with high strength by a low-temperature sintering method. *Science of Engineering Ceramics Iii* 317-318:893-898.

Karsenty, G. 2000. How many factors are required to remodel bone? *Nature Medicine* 6:970 - 971.

Karsenty, G., Kronenberg, H. M. and Settembre, C. 2009. Genetic Control of Bone Formation. *Annual Review of Cell and Developmental Biology* 25:629-648.

Kesenci, K., Fambri, L., Migliaresi, C. and Pişkin, E. 2000. Preparation and properties of poly(L-lactide)/hydroxyapatite composites. *Journal of Biomaterials Science-Polymer Edition* 11(6):617-632.

Lawson, A. and Czernuszka, J. 1998. Collagen-calcium phosphate composites. *Proc Inst Mech Eng* [H] 212:413-425.

Li, X. K. and Chang, J. 2008. Preparation of bone-like apatite-collagen nanocomposites by a biomimetic process with phosphorylated collagen. *Journal of Biomedical Materials Research Part A* 85A(2):293-300.

Li, X. M., Feng, Q. L., Liu, X. H., Dong, W. and Cui, F. H. 2006. Collagen-based implants reinforced by chitin fibres in a goat shank bone defect model. *Biomaterials* 27(9):1917-1923.

Lin, X. Y., Li, X. D., Fan, H. S., Wen, X. T., Lu, J. and Zhang, X. D. 2004. In situ synthesis of bone-like apatite/collagen nano-composite at low temperature. *Materials Letters* 58(27-28):3569-3572.

Miyake, K., Medina, K., Ishihara, K., Kimoto, M., Auerbach, R. and Kincade, P. W. 1991. A VCAM-like Adhesion Molecule on Murine Bone Marrow Stromal Cells Mediates Binding of Lymphocyte Precursors in Culture. *The Journal of Cell Biology* 114:557-565.

Mornet, S., Vasseur, S., Grasset, F., Veverka, P., Goglio, G., Demourgues, A., Portier, J., Pollert, E. and Duguet, E. 2006. Magnetic nanoparticle design for medical applications. *Progress in Solid State Chemistry* 34(2-4):237-247.

Murugan, R. and Ramakrishna, S. 2005. Development of nanocomposites for bone grafting. *Composites Science and Technology* 65:2385–2406.

Noris-Suarez, K., Lira-Olivares, J., Ferreira, A. M., Feijoo, J. L., Suarez, N., Hernandez, M. C. and Barrios, E. 2007. In Vitro Deposition of Hydroxyapatite on Cortical Bone Collagen Stimulated by Deformation-Induced Piezoelectricity. *Biomacromolecules* 8:941-948.

Ogawa, S., Harada, H., Fujiwara, M., Tagashira, S., Katsumata, T. and Takada, H. 2000 Cbfa1, an essential transcription factor for bone formation, is expressed in testis from the same promoter used in bone. *DNA Research* 7(3):181-185.

Oláh, L., Filipczak, K., Czvikovszky, T., Czigánya, T. and Borbás, L. 2007. Changes of porous poly(ε-caprolactone) bone grafts resulted from e-beam sterilization process. *Radiation Physics and Chemistry* 76(8-9):1430-1434.

Pederson, A. W., Ruberti, J. W. and Messersmith, P. B. 2003. Thermal assembly of a biomimetic mineral/collagen composite. *Biomaterials* 24(26):4881-4890.

Pompe, W., Gelinsky, M., Hofinger, I. and Knepper-Nicolai, B. 2001. Functionally graded collagen-hydroxyapatite materials for bone replacement. *Functionally Graded Materials* 2000 114:65-72.

Rhee, S. H. and Tanaka, J. 1998. Hydroxyapatite coating on a collagen membrane by a biomimetic method. *Journal of the American Ceramic Society* 81(11):3029-3031.

Sedel, L., Nizard, R. S., Kerboull, L. and Witvoet, J. 1994. Alumina-Alumina Hip Replacement in Patients Younger Than 50 Years Old. *Clinical Orthopaedics & Related Research* 298:175-183.

Silva, C. C., Thomazini, D., Pinheiro, A. G., Aranha, N., Figueiro, S. D., Goes, J. C. and Sombra, A. S. B. 2001. Collagen-hydroxyapatite films: piezoelectric properties. *Materials Science and Engineering B-Solid State Materials for Advanced Technology* 86(3):210-218.

Tashjian, A. H., Voelkel, E. F., Lazzaro, M., Singers, F. R., Roberts, A. B., Derynck, R., Winklerii, M. E. and Levine, L. 1985. alpha and beta human transforming growth factors stimulate prostaglandin production and bone resorption in cultured mouse calvaria. *Proc Natl Acad Sci* 82:4535-4538.

Wahl, D. A. and Czernuszka, J. T. 2006. Collagen-hydroxyapatite composites for hard tissue repair. *European Cells & Materials* 11:43-56.

Wallach, S., Avioli, L. V. and Carstens, J. H. J. 1989. Factors in bone formation. *Calcified Tissue International* 45(1):4-6.

Wang, L. and Li, C. Z. 2007. Preparation and physicochemical properties of a novel hydroxyapatite/chitosan-silk fibroin composite. *Carbohydrate Polymers* 68(4):740-745.

Werner, J., Linner-Krcmar, B., Friess, W. and Greil, P. 2002. Mechanical properties and in vitro cell compatibility of hydroxyapatite ceramics with graded pore structure. *Biomaterials* 23(21):4285-4294.

Wu, C. Y., Sassa, K., Iwai, K. and Asai, S. 2007. Unidirectionally oriented hydroxyapatite/collagen composite fabricated by using a high magnetic field. *Materials Letters* 61(7):1567-1571.

Yamauchi, K., Goda, T., Takeuchi, N., Einaga, H. and Tanabe, T. 2004. Preparation of collagen/calcium phosphate multilayer sheet using enzymatic mineralization. *Biomaterials* 25(24):5481-5489.

Yunoki, S., Ikoma, T., Monkawa, A., Ohta, K., Kikuchi, M., Marukawa, E., Sotome, S., Shinomiya, K. and Tanaka, J. 2007. Fabrication of three-dimensional porous hydroxyapatite/collagen composite with rubber-like elasticity. *Journal of Biomaterial Science*, Polymer Edition 18(4):393-409.

Yunoki, S., Ikoma, T., Monkawa, A., Ohta, K., Kikuchi, M., Sotome, S., Shinomiya, K. and Tanaka, J. 2006. Control of pore structure and mechanical property in hydroxyapatite/collagen composite using unidirectional ice growth. *Materials Letters* 60(8):999-1002.

Zhang, J. and Misra, R. D. K. 2007. Magnetic drug-targeting carrier encapsulated with thermosensitive smart polymer: Core-shell nanoparticle carrier and drug release response. *Acta Biomaterialia* 3(6):838-850.

Zhang, W., Liao, S. S. and Cui, F. Z. 2003. Hierarchical self-assembly of nano-fibrils in mineralized collagen. *Chemistry of Materials* 15(16):3221-3226.

2

Composite Scaffolds for Orthopaedic Regenerative Medicine

John P. Gleeson[1,2] and Fergal J. O'Brien[1,2]
[1]*Royal College of Surgeons in Ireland,* [2]*Trinity Centre for Bioengineering*
Ireland

1. Introduction

The goal of tissue engineering is to synthesise substitutes that mimic the natural environment to help guide the growth of new functional biological tissue in vitro or in vivo. Tissue engineering relies heavily on the use of porous 3D scaffolds to provide a supportive environment for the regeneration of tissues and organs, acting primarily as a template for de novo tissue formation. However, new advances in fabrication technologies and composite materials are facilitating the rapid development of many novel composites that are beginning to play a more active role in directing the regenerative process. It has been long recognised that the combination of two or more characteristically-distinct materials can often yield composite materials that possess many of the constituent materials mechanical and biological advantages with few of their disadvantages. When applied to regenerative medicine, these new composite materials are beginning to show real potential as bioactive, biodegradable substitute materials, capable of facilitating rapid orthopaedic tissue regeneration, while degrading in parallel with the advancing tissue repair process. These idealised tissue regenerative aids could finally offer clinicians the potential to completely regenerate damaged orthopaedic tissue, leaving no evidence that the tissue was ever damaged in the first instance.

At a simplistic level, biological tissues consist of cells, signalling mechanisms and extracellular matrix (ECM). Tissue engineering technologies are based on this biological triad and consist of (i) the scaffold that holds the cells together to create the tissue's physical form, (ii) the cells that create the tissue, and (iii) the biological signalling mechanisms (such as growth factors or bioreactors) that direct the cells to express the desired tissue phenotype (Figure 1). In native tissues, cells are held within an ECM which guides development and directs regeneration of the tissue. The ECM serves to organise cells in space and provides them with environmental signals to direct cellular behaviour. Consequently, the ECM is responsible for two of the three components in this tissue engineering triad, highlighting the critical role that this extracellular environment plays on tissue formation. What is becoming increasingly evident is that the combination of biomaterials into novel composite scaffolds can result in an engineered biomimicry of this extracellular environment that provide all the environmental cues to promote rapid development of de novo tissue. Consequently, these composites can be designed to act not only as carriers and supporting structures for the associated cells but to play a more active role in the initiation and development of the repair tissue.

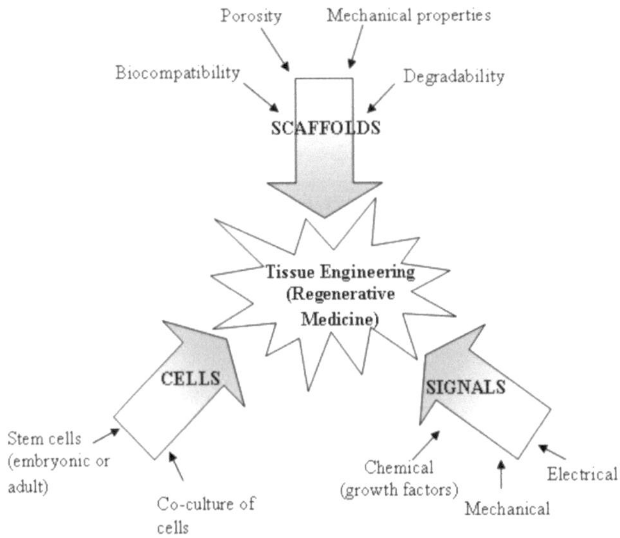

Fig. 1. The tissue engineering triad; factors that need to be considered when designing a suitable structure for tissue engineering applications

2. Scaffold requirements

Traditionally, scaffolds designed for tissue engineering attempted to meet a small number of common requirements that would allow them to be used safely for the in vitro production of engineered tissue or alternatively for in vivo implantation as regenerative aids. These included (i) providing an environment conducive to the facilitation of desirable cell-matrix interactions (e.g. cellular infiltration, attachment, proliferation and differentiation), (ii) support nutrient and waste product transport and encourage the movement of biochemical signals throughout the structure/matrix, facilitating long-term cellular survival and proliferation, (iii) biodegrade at an appropriate rate in parallel with the body's own natural healing process (i.e. supporting the healing phase while in tandem ensuring that the scaffold does not act as a barrier to the regenerative process, and (iv) to be bioinert and provoke a minimal inflammatory or immunological response as a result of its implantation or population with cells.

When designing or evaluating a scaffold for applications in tissue regeneration, there are a number of scaffold architectural, compositional and physical characteristics that have a deterministic influence on these four overlying tissue engineering scaffold requirements;

1. Biocompatibility

   The word *biocompatibility* was mentioned for the first time in peer-review journals and meetings in 1970 by RJ Hegyeli and CA Homsy (Homsy et al., 1970). The definition of biocompatibility is "the ability of a material to perform with an appropriate host response in a specific application" and is a critical criterion for scaffold design. Not only must the scaffold material itself be biocompatible, but so too must its degradation products in vivo. It must not elicit toxic or injurious responses within biological systems once implanted or at any point throughout the degradation of the material in vivo. This characteristic alone rules out many synthetic or man-made materials as suitable components of tissue engineering matrices.

2. Biodegradability
   Biodegradability refers to the ability of the physiological environment to breakdown or degrade an implanted material. Biodegradation is a critical characteristic of materials that are designed to regenerate tissue, as opposed to acting simply as inert substitutes for the native tissue. The overall goal of cutting edge regenerative tissue engineering therapies is to act as "smart" biomaterials by supporting the initial healing processes while also beginning to degrade in parallel with the advancing production of newly formed tissue matrix, ideally with no evidence of implantation once sufficient tissue regeneration has occurred, negating the need for subsequent clinical interventions to remove the implant from the body.
3. Bulk Mechanical Properties
   The scaffold should provide an environment that is capable of surviving the implantation process and surgical manipulation required as part of the clinical procedure. It is widely believed that these materials must possess adequate mechanical integrity to survive the normal physiological loading environment at the site of implantation but this is currently an area of contention within the field. Traditional tissue engineering scaffolds prioritise a mechanically-competent scaffold, capable of supporting load bearing immediately upon implantation. Unfortunately, the characteristics of host tissue-like mechanical strength and the levels of porosity, permeability and pore interconnectivity, necessary for long-term scaffold in vitro and in vivo viability, are incompatible from a biomaterials perspective. Consequently current scaffolds are utilising cutting-edge advancements in composite biomaterials technology in an attempt to balance provision of bulk mechanical properties suitable for implantation and cellular support while retaining a material porosity high enough to encourage cell infiltration via diffusion throughout the scaffold.
4. Substrate Stiffness
   Substrate mechanical properties of tissue engineering scaffolds plays a critical role in controlling and regulating a number of factors involved in directing cellular activity (Engler et al., 2006). Recent unpublished work from our laboratory has demonstrated that collagen-based scaffolds with a bulk stiffness of approximately 4 kPa exhibit increased cell attachment, proliferation and migration compared to less stiff scaffolds. Interestingly, recent studies have investigated the bulk and localised mechanical properties of highly porous scaffolds (Harley et al., 2007) and shown that the nature of high porosity structures means that their bulk mechanical properties are dramatically different to the mechanical properties of the individual struts within the open foam network. As a result, the substrate stiffness that a cell 'feels' while attached to one or multiple struts within the porous scaffold can be significantly higher than that predicted by bulk assessment of the material. Based on their study, it was estimated that the substrate stiffness experienced by a cell attached to a pore within a highly porous scaffold exhibiting a bulk stiffness of approximately 4 kPa would be of the order of approximately 50 to 100 MPa. Therefore, the effect of local substrate stiffness in a three-dimensional environment such as a porous tissue engineering scaffold is still an area that requires significant future investigation.
5. Pore Size and Pore Size Distribution
   Pore size (Figure 2) is cell type specific (Murphy et al., 2010) and is arguably the most critical factor in the design of a tissue engineering scaffold optimised for a repair or regeneration of a specific tissue type. Pore size has a dramatic effect on cell seeding

efficiency within the scaffold (O'Brien et al., 2005) which can result in improved in vitro performance. If scaffold pore size is too small, cells are unable to rapidly infiltrate into the scaffold centre or homogenously populate the matrix. Densification of cells around the matrix periphery occurs, acting as a barrier to further cell infiltration and leading to avascular necrosis within the scaffold centre (Phelps 2010, Ko 2007). Some investigators argue that having larger pores in the centre of an implant can help to support vascularisation of the implant when grafted onto a patient. It is believed that increased pore size with increasing depth is desirable (Mc Kegney et al., 2001). On the other hand, pores that are too big result in a significantly reduced specific surface area within the construct (Byrne et al., 2008).

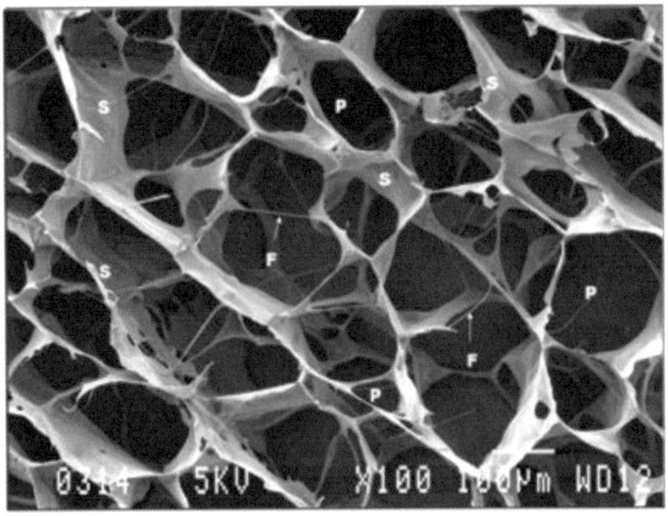

Fig. 2. SEM micrograph of a scaffold defining its pore structure. Wessels et al., 2008; S. Afr. j. sci. vol.104 no.11-12 Nov/Dec 2008

Cells interact with the pore substrate via ligands, chemical binding sites naturally associated with extracellular materials such as collagen. The availability of these ligands for promoting cell binding is directly related to the specific surface area, which is related to mean pore size. Consequently, pore size must be sufficiently large to provide the ideal ligand density to allow binding of a critical proportion of cells to occur. While the mean pore size is critical, the distribution of pore size range around this mean also plays an important role. Given that pore size is cell type specific, tissue engineering scaffolds with heterogeneous pore size distribution provide an environment with heterogeneous optimisation of the structure. Local ligand densities will differ substantially throughout the matrix, as will the ability of cells to infiltrate and attach throughout the construct. Therefore, development of a homogenous structure with predictable cell adhesion, proliferation and differentiation characteristics requires a high degree of pore size homogeneity throughout and this is a characteristic offered by only a limited number of construct manufacturing processes. Figure 2 is an example of a porous structure of a collagen-based scaffold. An ideal scaffold is one which is open and has an interconnected pore network and a high degree of porosity (>90%), as it is easy for the scaffold to interact and integrate with the host (Freyman et al., 2001).

Clearly the optimisation of pore size is a critical design characteristic to allow optimisation, and consequently in vitro and in vivo performance, and the integration of cells within the developing extracellular matrix.

6. Porosity, Pore Interconnectivity and Permeability
Porosity is defined as the percentage of void space in a solid and its importance in tissue engineering scaffolds cannot be overstated. The degree of porosity in a scaffold with a given mean pore size will have a direct effect on the interconnectivity of the porous architecture and consequently the permeability/fluid mobility within the scaffold. These characteristics play an important role not only on the amount and rate of cell and fluid infiltration into the constructs but also facilitate the transport of nutrient and waste products throughout the cell-seeded construct, as well as construct vascularisation (Kuboki et al., 1998), for the duration of de novo tissue formation. These properties not only encourage complete tissue formation within the construct but also support the integration and mechanical interlocking of the implant. A scaffold which possesses an open and interconnected pore network, coupled with a high degree of porosity (>90 %) is ideal for cellular interaction and de novo tissue integration with the existing host tissue (Freyman et al., 2001).

3. Scaffold biomaterials for orthopaedic tissue regeneration

The first generation of biomaterials specifically designed for implantation into the human body appeared around the 1960s and 1970s. This first generation was characterised by attempting to "achieve a suitable combination of physical properties to match those of the replaced tissue with a minimal toxic response in the host" (Hench, 1980). These materials were designed primarily around the principal of bio-inertness i.e. the idea of causing as little disruption to the physiological environment while facilitating a primarily structural role. Fuelled by the initial success of many of these devices, the field rapidly began to focus on improving on the concept of bio-inertness and began to aspire to creating biomaterials that exhibited a degree of bioactivity i.e. to elicit a positive and controlled response within the implanted physiological macro-environment that would aid the healing or regenerative process. It was at this point that a large move towards the use of ceramic based materials occurred.

3.1 Ceramics

Ceramics (inorganic, non metallic materials) include the calcium phosphates, bioglasses and glass-ceramics (Hench, 1998). Bioceramics can be further classified as being osteoconductive (supporting bone growth) or osteoinductive (stimulating bone growth). While osteoconductivity is common to all types of bioceramics, relatively few are osteoinductive, a property that although extremely coveted, is not fully understood or easily replicated in synthetic materials (Barrere et al., 2008). The calcium phosphate based bioceramics, bioglasses and glass-ceramics are commonly used as scaffolds for bone tissue engineering as they have a compositional similarity to the mineral phase of bone (Hing et al., 2005). Hydroxyapatite (HA) and tri-calcium phosphate (TCP) are two of the most commonly used calcium phosphate bioceramics in tissue engineering applications. TCP is commonly used as the basis of biodegradable scaffolds due to its relatively rapid degradation rate (Ducheyne et al., 1993) and its osteoconductivity. HA, the mineral that occurs naturally in bone tissue, has also been used extensively as a tissue engineering scaffold material due to its

osteoinductivity. It is typically used for coating biomedical implants to induce bone regeneration, allowing the implant to integrate with the surrounding tissue. While HA was originally popular for use as a scaffold for tissue engineering, it is non-resorbable in bulk form which has limited its popularity as a bone graft substitute material (Figure 3). However, recent work on micro- and nano-sized particles has led to a paradigm shift regarding the degradability of this material and has re-ignited interest in this ceramic as a critical component of tissue engineering composite scaffolds optimised for bone tissue regeneration.
Ceramic materials offer a facilitative environment to bone forming cells, offering mechanical support that promotes mineralisation in the hope of achieving stability equal to the normal anatomical tissue. Synthetic calcium phosphates have been popular for a number of different applications, ranging from a simple coating layer on prosthetic devices (Klein et al., 1993) to being the implantable device itself, as porous bone graft substitutes in the repair or augmentation of bony defects. Many of these devices have been used clinically with some degree of success and are still widely used as bony void fillers. However, their clinical applications have been limited because of their brittleness and difficulty of shaping for implantation (Wang et al., 2003), low porosity and long term mechanical integrity issues (Bohner, 2010). Difficulties also exist in controlling the degradation rate of ceramics so as to ensure optimal resorption (Tancred et al., 1998). Although HA is a primary constituent of bone and might seem ideal as a bone graft substitute, problems include a slow degradation rate (Marcacci et al., 2007), poor mechanical properties and new bone formed in a porous HA network cannot sustain the mechanical loading needed for remodelling (Wang et al. 2003).

3.2 Synthetic polymers

Due to the numerous drawbacks of ceramic-based biomaterials, significant advances were made towards the development and use of bioresorbable second generation materials. These materials are polymeric-based and include many different biocompatible and bioresorbable materials, such as PLA and PGA scaffolds for use as implantable devices (Athanasiou et al., 1998). Polymeric-based biomaterials have a number of advantages such as high mechanical strength and biodegradability. Their mechanical, physical and biological properties can be tailored to give a wide range of properties that are desirable for bone tissue regeneration. In addition, their degradation rates can be controlled, as can their degradation by-products (Hennick and Van Nostrum, 2002). Among the many biodegradable synthetic polymers used for tissue engineering applications, there are numerous reports on the use of polylactic acid (PLA), polyglycolic acid (PGA) and their copolymers poly (DLlactic- co-glycolic acid) (PLGA), which are approved by the US Food and Drug Administration (FDA). These polymers degrade by hydrolytic mechanisms and are commonly used because their degradation products can be removed from the body as carbon dioxide and water. Unfortunately, they are also associated with a number of problems, ranging from issues involving device rejection and resisting adequate resorption to promote vascularisation and ingrowth of new bone. Localised lowering of pH within the region of degradation can result in inflammatory responses. While these materials showed some promise, it is clearly evident from the lack of synthetic polymer-based products currently in use clinically as bone graft substitutes that synthetic polymers alone simply do not provide a sufficient degree of long-term biocompatibility or performance required for the clinically-successful regeneration of bone tissue.

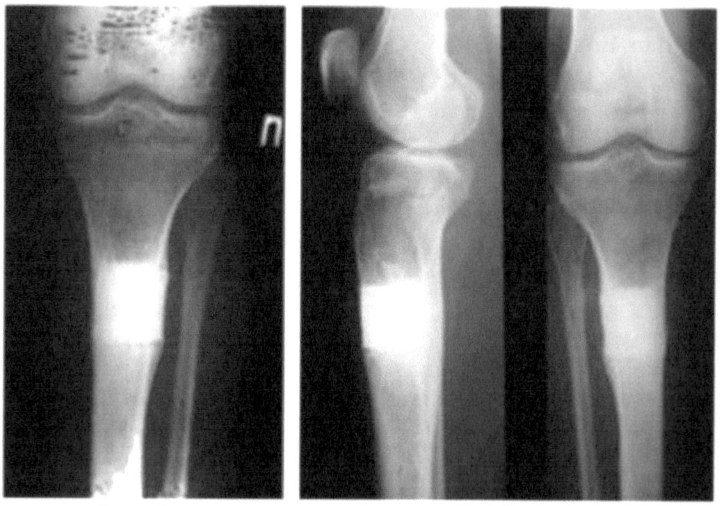

Fig. 3. Porous HA implant used to fill bony void but no resorption after 2.5 years (Cancedda et al., 2007)

Polymeric materials however do offer a number of significant advantages over ceramics from a composite scaffold point of view. Biodegradable polymers can be used to encapsulate biologics and/or growth factors and can be designed to control the release kinetics of these substances via tailored polymer degradation properties for specific tissue regenerative applications. This ability presents a range of potentially controllable processes including cell growth, tissue regeneration and host response that can be influenced by developing suitable biopolymers (Hutmacher, 2000) for incorporation into novel composite materials. Appropriate selection and development of certain synthetic polymers for applications involving controlled biologic or growth factor release kinetics can be further advanced by the incorporation of 'surface eroding polymers', which biodegrade only at their surfaces. Poly(anhydrides), poly(orthoesthers) and polyphosphazene exhibit this property and offer numerous advantages over bulk degradation polymers (Rezwan et al., 2006).

3.3 Natural polymers

Natural polymers offer a number of significant advantages over synthetically-derived polymers due to their biocompatible, biodegradable and bioactive nature. By using materials that form the basic building blocks of organic systems, host response due to immunocompatibility issues can be drastically, and sometimes completely, reduced. This approach also ensures that these materials are easily biodegradable via the body's own metabolic processes and that the resulting degradation by-products are non-toxic and can be easily assimilated or expelled from the tissue. Alternatively, natural polymers can usually be crosslinked using a number of means including physical or chemical crosslinking techniques. Physical crosslinking methods include UV radiation and dehydrothermal treatments, whilst cross-linking agents such as glutaraldehyde and carbodiimides (EDAC) can be used to produce chemically cross-linked natural polymers, allowing a customised degradation rate for different regenerative applications. This makes them ideal biomaterials for developing scaffolds with tailored biodegradation rates that can match the formation of de novo tissue. Natural polymers used in bone tissue engineering include alginate, chitosan, fibrin, silk, glycosaminoglycans and collagen. Most natural polymers are biocompatible, degradable, and readily solubilised in physiological fluid (with exception of chitosan which is soluble under mild acidic conditions) (Seeherman and Wozney, 2005). This simple biomimetic approach means they can more closely mimic the natural extracellular matrix of tissues and can play a bioactive role in substrate cell interactions, such as promoting cell adhesion and proliferation.

Alginate and chitosan are two natural polysaccharides that are not found within the human body but due to their structural similarity to the glycosaminoglycans (GAGs), they have been the focus of a significant body of scientific work evaluating their applications in regenerative medicine (Terada et al., 2005; Chenite et al., 2000). While these materials are attractive due to their low toxicity and biocompatibility, there are significant issues with a lack of load bearing ability (Suh and Matthew, 2000) which has limited their clinical applications as single phase materials but chitosan still plays a major role as a composite component, due to its flocculating and chelating properties (Zhang et al., 2010) and its role in composite mechanical stabilisation (Jiang et al., 2006). Structural proteins such as fibrin have also been used in tissue engineering applications. Fibrin can be used as a natural wound healing material, and is most commonly used clinically as a sealant or adhesive. However, fibrin has very poor mechanical properties, degrades rapidly in vivo and cannot withstand physiological loading long-term when implanted into orthopaedic defects (Barrere et al., 2008). In spite of this, fibrin hydrogels have been used in muscle (Cummings et al., 2004) and cartilage tissue engineering (Hunter et al., 2004) in vitro. Silk has also been utilized as a tissue engineering scaffold for stem cell osteogenic differentiation of MSCs (Meinel et al., 2006). Silk as a biomaterial combines slow biodegradability, excellent mechanical properties, and biocompatibility Native silk fibers are some of the strongest natural fibers known, rival synthetic materials such as Kevlar in terms of tensile strength (Cuniff et al., 1994) and have been used as suture materials for over 20 years (Vepari and Kaplan, 2007). Silk scaffolds have previously been shown to support tissue engineering of bone in vitro (Altman et al., 2003). Glycosaminoglycans are found in the natural extracellular matrix of tissues i.e. skin, bone, and blood vessels and are a form of proteoglycan; organic polymers that are found in cells and that are a major component of the natural ECM. Given the importance of GAGs in stimulating normal tissue growth, the use of GAGs as components of a scaffold for tissue engineering appears to be a logical approach for scaffold development. Glycosaminoglycans (GAGs) are long, unbranched polysaccharides that do not elicit an immune response and have been used extensively for tissue engineering applications and can be copolymerized with collagen to increase the stiffness and toughness and decrease its degradation rate.

Hyaluronic acid is one of the largest GAG components found in the natural extracellular matrix of all soft tissues and synovial fluid of joints (Drury and Mooney, 2003) and has been used in tissue engineering applications due to its role in structural organisation of the ECM components. However, its use has been limited due to rapid dissolution in water and fast biodegradation in biological environments. However, it can be chemically modified to produce a more hydrophobic molecule, thus reducing its solubility in water. Hyaluronic acid is a major component of cartilage matrix and is a viscoelastic material that forms random coils which entangle with each other at low concentrations and at high concentrations contains a very high viscosity dependant on shear stress making it pseudoplastic. Chondrocytes in cartilage have a high affinity for Hyaluronic acid via CD44 receptors and RHAMM (Lisignoli, 2001) and consequently it plays a more important role in cartilage tissue engineering constructs.

Scaffolds fabricated from type I collagen and a glycosaminoglycan have been used to study cell migration and contraction in vitro (Sethi et al., 2002) as well as to induce regeneration of the skin, conjunctiva, and peripheral nerves in vivo (Harley et al., 2004; Yannas, 2001). Collagen as a scaffold material is ideal as it possesses all the biological prerequisite for successful implantation such as biocompatibility, immunogenicity, cell adhesion and proliferation (Murphy et al., 2010; Byrne et al., 2008; Berry et al., 2004; O'Brien et al., 2005). Collagen is the most abundant ECM protein in the human body and is readily isolated and purified from various animal species by enzyme treatment. Since collagen type I is the main organic component that in human bone tissue and it is the substrate on which bone mineralisation occurs during osteogenesis, collagen has been extensively used as the material of choice in nearly all commercial orthobiologic bone tissue engineering applications currently used clinically in the repair and regeneration of bone tissue. Two of the biggest advantages of using collagen-based scaffolds for tissue engineering applications are that (i) they provide an extremely attractive substrate for cell adhesion and proliferation and (ii) collagen scaffolds do not alter the phenotype of seeded cells. Another significant attribute is the recent FDA approval and clinical success of collagen-based scaffolds used for skin and nerve regeneration (Yannas, 2001). However, despite the excellent biocompatibility, collagen, like all other natural polymers, is insufficient mechanically for orthopaedic applications and this has limited its use in load-bearing bone tissue defects.

4. Composite scaffolds for bone tissue regeneration

4.1 Synthetic polymer and ceramic composites

As a result of the problems associated with the use of single phase synthetic scaffolds, advances in the development and fabrication of composite scaffolds offered new materials using a combination of synthetic polymeric and ceramic phases. The development of these synthetically-based composites was believed to facilitate the development of biomaterials with all of the advantages of these single phase materials with none of their disadvantages. In spite of these efforts, these second generation bone graft substitutes have enjoyed limited clinical success (Ratcliffe, 2008) and simply do not possess all of the prerequisite characteristics of an ideal bone graft substitute (biocompatible, bioresorbable, osteoconductive, osteoinductive, structurally similar to bone, easy to use and cost-effective). The inability of synthetic materials to respond or adapt to changing physiological conditions means that they will always represent a compromise when used in the repair or regeneration of human tissue. As a result, this second generation of biomaterials have not been able to act as a viable clinical alternative to the gold standard, autogenous bone. Numerous innovative synthetic polymer-ceramic composites have been developed with porosities high enough to ensure cell infiltration and an environment potentially conducive to osteogenesis. While some have shown evidence of osteoinductivity when implanted into ectopic bone formation models (Barbieri et al., 2010), in in vitro conditions (Li et al 2010); and evidence of potent bioactivity (Deplaine et al., 2010), only a handful of these materials have been evaluated in large pre-clinical bone tissue defects.

4.2 Natural polymer and ceramic composites

The combination of natural polymers with a reinforcing and bioactive ceramic phase has shown significant promise over the last decade with a number of these composites progressing to widespread clinical use (Lee and Goodman, 2009; Carter et al., 2009; Kitchel, 2006; Muschler et al., 2005; Scabbia and Trombelli, 2004). While these materials fulfil many of the requirements of synthetic bone void fillers, none have shown conclusive clinical evidence of being a superior alternative to the clinical gold standard of autogenous bone. Such an ideal bone graft substitute must be capable of promoting rapid osteogenesis in vivo, encourage de novo bone formation and remodelling of the defect to restore anatomical normality and mechanical integrity, and biodegrade at the same rate as the progressing tissue regeneration process. Recent advances in composite biomaterials have led to a paradigm shift in this area towards the development and use of biomimetic composite scaffolds for orthopaedic regenerative medicine. Biomimicry, both in terms of composite composition and fabrication process may provide a compromise between the competing mechanical and the biological prerequisites needed to rapidly promote healing of bone tissue defects (Gleeson et al., 2010). The use of materials found naturally to occur within the human body allows the implantation of materials that are easily integrated, processed and degraded by the body. These materials form part of the normal "building blocks" of the human system and offer extremely favourable biological interactions. Advances in the processing and composite engineering of new composite materials comprised of these naturally-occurring materials offer exciting possibilities for not only meeting a compromise between the mechanical and biological prerequisites for implantation and bioactivity, but also for directing and controlling the chemical, biological and mechanical events that occur during the regenerative process.

There are a large number of natural materials currently used as the components of composite scaffolds for orthopaedic regenerative medicine. While there are many different materials used to promote bone tissue repair, bone's native composition of predominantly type I collagen and hydroxyapatite makes these materials an obvious choice as the basis for a composite biomaterial capable of supporting and promoting the bone regenerative process (Dawson et al. 2008). Recent studies have shown that the interaction between osteoblasts and PLLA scaffolds can be improved by the application of a collagen-HA coating (Jiashen et al., 2010) clearly demonstrating the potential of a composite material composed of only collagen and hydroxyapatite for use as a bioactive bone graft. These composite scaffolds were traditionally fabricated using a number of different techniques involving some form of pre-processing step, with the scaffold architecture subsequently formed via a number of distinct processes, all of which possess numerous advantages and disadvantages (particulate leaching methods, phase separation, lyophilisation, foaming, emulsion templating, and solid free form (SFF) fabrication).

Current advances in composite fabrication have been driven by the desire to replicate the hierarchical scale of the naturally-occurring tissues. Bone is composed of mineralised tropocollagen molecules, arranged in a distinct quarter stagger arrangement, with nanometre-sized hydroxyapatite crystal intimately associated with these fibrils. The nature of these nanohydroxyapatite particles endows them with a number of distinct regenerative advantages relative to the micron-sized particles and this has been a driving factor in the move towards the hypothesis that biomimetic scaffolds fabricated using biomimetic processes may allow the development of composite scaffolds optimised to promote rapid osteogenesis in vivo. As a result, many recent studies have utilised biocompatible or bioactive dispersants, such as chitosan (Zhang et al., 2010b) or biomimetic fabrication methods for the *in situ* mineralisation of collagen-HA scaffolds during the fabrication process (Zhang et al., 2010a; Xu et al., 2010; Yoshida et al.,2010; Kikuchi et al., 2004).

One of the earliest attempts to combine biologically-occurring composite constituents with biomimetic fabrication processes to form a bone graft substitute material was carried out by Kikuchi (Kikuchi et al., 2004). This approach involved a self-organisation mechanism designed to synthesis a bone-like collagen/hydroxyapatite nanocomposite material for use as a bone tissue engineering scaffold. There are many motivating factors that support the use of nano HA, most critically increased in vivo resorption rate, increased osteoblast adhesion and proliferation and increased bone formation in vivo. By using a self-organisation mechanism that involves the in situ mineralisation of a collagen solution, Kikuchi and colleagues were able to develop a collagen/HA composite that exhibited bone-like orientation of nano-sized hydroxyapatite crystals, aligned along the length of the collagen fibrils. However, although in vitro and preliminary in vivo data were promising, their application to orthopaedic regenerative medicine remains to be elucidated. In addition, control and regulation of this process and the resulting nature of the fabricated HA can be difficult with implications on the purity and crystallinity of the resulting mineral phase. Given that HA crystallinity and purity plays a significant role in promoting bone tissue formation *in vivo* (Zhang et al., 2010a; ter Brugge et al., 2002), the ability to produce pure collagen-HA scaffolds of high purity and crystallinity is desirable from a tissue engineering perspective.

However, a significant flaw that exists in the biomimetic fabrication approach is the assumption that an ideal optimised bone graft substitute material must mimic the composition and structure of the final bone tissue. This postulation does not consider that a truly biomimetic composite scaffold for orthopaedic regenerative therapies must match the idealised environment that supports the earliest stages of osteogenesis. This presents the biggest current challenge within the field of composite scaffolds for orthopaedic regenerative medicine, namely to develop a composite material that meets the multitude of prerequisite characteristics necessary to induce, promote and support the process of osteogenesis. Such a composite scaffold must (i) be highly porous to facilitate rapid and complete cellular infiltration once implanted (ii) provide surfaces and an environment that encourages cellular adhesion and proliferation (iii) be highly permeable to facilitate the exchange of nutrients and waste products throughout the scaffolds and preventing avascular necrosis (iv) be sufficiently strong to withstand surgical manipulation and the implantation procedure (v) exhibit a high degree of pore interconnectivity (vi) must be osteoconductive and ideally osteoinductive (vii) biodegradable, biocompatible and bioactive and (viii) possess an optimised pore size and pore size distribution to facilitate homogenous mineralisation of the composite scaffold in vitro and in vivo.

Many of the tradition methods for fabricating scaffolds for tissue engineering are not ideal for the development of highly biocompatible scaffolds with the prerequisite pore network, porosity and pore interconnectivity characteristics that are increasingly being recognised as determining factors in the long-term in vivo viability of tissue engineering construct. Problems associated with these techniques include, but are not limited to, poor control over internal architecture and a limited range of pore sizes, residual solvent and residual porogens, porosity limits of about 70% and highly heterogeneous nature of the pore structure, limitations the development of photopolymerisable and biocompatible, biodegradable liquid polymer materials, high processing temperatures, use of toxic organic solvents and lack of mechanical strength. In addition, one of the major barriers to the successful development of a collagen-HA scaffold using these techniques is the difficulty in achieving a homogenous distribution of the HA throughout polymer-based matrices (Supova, 2009), an issue that can have a significant effect on a collagen-HA biomaterial's *in vivo* vascularisation and production of newly formed bone tissue (Lyons et al., (2010); Zhang et al. (2010a)). Recent attempts to overcome this issue have employed additional naturally-derived dispersants in an attempt to achieve a homogenous dispersion of the osteoinductive HA particles throughout the fabricated matrix (Zhang et al., 2010b) but radiological results do not provide evidence of the ability of these scaffolds to heal critically-sized bone defects.

Freeze-drying is a process which can potentially solve many of these composite scaffold fabrication issues and is ideally suited to organic biomaterials such as collagen. This technology has a number of distinct advantages with regard to the production of high porosity, highly interconnected, homogenous biological constructs. Currently, our Tissue Engineering Laboratory produces a range of highly porous collagen-based scaffolds, using a constant cooling lyophilisation process, with specific applications including bone and cartilage tissue repair (Murphy et al., 2010, Farrell et al., 2009). These scaffolds have been optimised for bone tissue healing and have recently been shown to provide ideal substrates for supporting the process of osteo- and chondro-genesis *in vitro* (Farrell et al., 2006). Our laboratory's approach has involved the optimisation of these collagen-based scaffolds in terms of composition (Tierney et al., 2009), cross linking density (Haugh et al., 2009) and pore architecture (Murphy et al., 2010) for use in bone tissue engineering applications.

Case study: Highly porous collagen hydroxyapatite scaffolds for bone tissue repair

We have recently developed the ability to combine our highly porous, optimised collagen scaffolds with an osteoinductive HA phase in an effort to improve their osteogenic potential and provide the prerequisite mechanical integrity to promote rapid in situ bone tissue regeneration (Gleeson et al., 2010) (Figure 4). By combining the two primary constituents of human bone tissue, namely type 1 collagen and high purity, highly crystalline hydroxyapatite particles using a novel mixing process (WO200896334A2), a highly porous composite tissue engineering scaffold with a high degree of pore interconnectivity, improved mechanical strength, permeability and cellular bioactivity was developed, overcoming traditional HA dispersion problems and exhibiting a homogenous distribution of the osteoinductive HA phase. The combination of the extremely biocompatible and biodegradable collagen scaffold with an osteoinductive mineral component (Barrere et al., 2003) provides an ideal mechanical and biological environment to facilitate cell recruitment and maintain pore structure in order to promote healing.

By optimising these compliant and highly porous collagen-hydroxyapatite scaffolds to promote mineralisation upon implantation (Hutmacher et al., 2000), these materials have the potential to rapidly produce de novo bone tissue with a load bearing capacity within the newly mineralised bone tissue graft. These scaffolds possess all the ideal prerequisite characteristics of a biodegradable scaffold optimised for bone tissue regeneration, are comprised only of bone's natural constituent materials, and have demonstrate their ability to promote osteogenesis in vitro and in vivo repair of critical-sized bone defects (Figure 4).

These highly porous collagen hydroxyapatite scaffolds have also been used pre-clinically in the repair of critical-sized segmental bone defects in long bones and show a rapid return to anatomical normality (as evident by formation of a continuous marrow cavity through the regenerated segmental defect), as well as evidence of de novo tissue remodelling and rapid restoration of load bearing ability after only 6 weeks implantation (unpublished data). These materials appear to demonstrate real potential as a bone graft substitute materials, capable of facilitating and promoting osteogenesis in vivo.

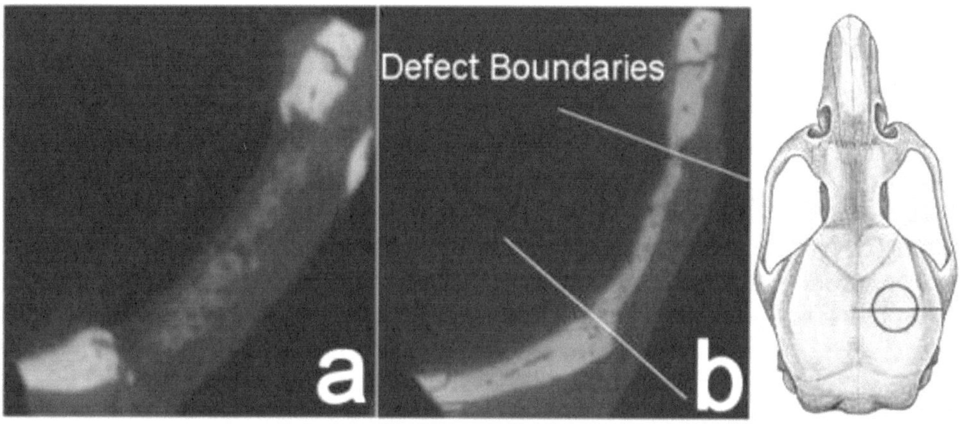

Fig. 4. MicroCT slice of representative level of mineralisation within rat calvarial defect centre showing defect boundary edges in **(a)** empty defect group, **(b)** collagen HA scaffold group after 28 days implantation. Almost complete defect bridging was observed in the collagen HA group, with mineralisation level comparable to surrounding native calvarial bone tissue (Gleeson et al., 2009)

5. Scaffold biomaterials for osteochondral tissue regeneration

Articular cartilage is a highly specialised tissue found covering the surfaces of the bony ends of all synovial joints in the human body. Its function is to lubricate joint movement and absorb small shock impacts within a joint. Articular cartilage is primarily composed of 70-80% water, 15% collagens (80% of which is type II collagen) and 5% cells. In synovial joints, the overlying lubricating articular cartilage is attached to the bone via a specialised tissue unit, known as osteochondral tissue. The structure and composition of osteochondral tissue is made up of a number of distinct but seamlessly integrated layers which vary in composition and structure according to their function and this serves to transfer mechanical forces at articulating surface down to the stiffer underlying subchondral bone via an intermediate tidemark layer, known as articular calcified cartilage. Cartilage is significantly different to bone from a regenerative point of view as it has a low regenerative capacity and damage to this tissue is irreparable and almost inevitably leads to the development of osteoarthritis (OA) within a joint and the eventual replacement (Arthroplasty) of the affected joint. Traditionally, the inherent lack of a natural regenerative capacity meant that autografting, allografting and joint replacement therapies offered the best potential outcomes for patients but recent advances in composite scaffolds are offering new regenerative possibilities.

Tissue engineering via in situ tissue regeneration may provide the prospect of regenerating cartilage and osteochondral tissue using a combination of an optimised tissue engineering scaffold and a source of potent progenitor cells used because of their ability to differentiate into a specific cell type and provide a source of extracellular matrix. This approach to cartilage and osteochondral repair is based on creating access to the most abundant source of nearby progenitor cells which reside in the underlying bone marrow. It is believed that this multi-tissue approach has the potential to offer improved restoration of the entire osteochondral unit and may potentially be more successful than simply trying to regenerate the avascular and largely non-regenerative mature articular cartilage tissue alone.

There are a number of critical factors that must be taken into consideration when designing tissue engineering scaffolds optimised for osteochondral repair via in situ tissue regeneration (Frenkel and Caesare, 2003). An ideal osteochondral graft substitute must (i) be biocompatible, (ii) provide adequate mechanical properties to withstand the implantation procedure and the subsequent mechanical and hydrodynamic loading within the joint, (iii) possess sufficiently high levels of porosity to allow ingrowth of host tissue and/or seeded cells, (iv) be retained at the site of implantation (Beris, 2005), (v) support and direct regeneration of the two predominant tissue types, namely bone and cartilage (Hutmacher, 2000) to ensure overlying repair cartilage is adequately supported, (vi) be optimised for cell attachment to favour colonisation by native cells (Coutts et al., 2001), (vii) promote integration with the existing tissues (Beris, 2005).

Composition is elementary in creating biologically active materials that can induce synthesis of new tissue in vitro and in vivo. As with bone graft substitutes, there exists a wide array of materials that have been used in an attempt to produce an optimised osteochondral graft substitute. Synthetic polymers have been widely used for tissue engineering. Polymers such as polylactide (PLA), polyglycolide (PGA) and their copolymers (poly(D,L-lactide-co-glycolide) are commonly used due to their degradability and US Food and Drug Administration (FDA) approval for clinical use. They have been investigated as scaffolds for cartilage tissue engineering since the early 1990s (Vacanti et al., 1991) and have formed cartilage-like tissue with good mechanical properties (Ma and Langer, 1999). The major advantages of the synthetic polymers include a wide range of tailored structural properties and the lack of disease transmission. PLA scaffolds have shown some promise in vivo but de novo tissue was found to be biochemically inferior to native cartilage tissue (Dounchis et al., 2000). PLA and PLA co-polymer composites have also been investigated for use in cartilage tissue repair (Niederauer et al., 2000) and shown evidence of hyaline cartilage and good bony restoration. However, there are many disadvantages with these materials, predominantly associated with the biological interactions that take place between the host and these synthetic materials over time and ultimately the translation of these synthetic polymer-based constructs into human clinical applications has not been seen due to concerns about their ability to illicit foreign body host responses and concerns of localized and systemic effects due to their toxic degradation by-products in vivo (Stoop, 2008)

Scaffolds developed using components of the ECM are generally more favourable than artificial polymers due to their ability to regulate cell function through specific cell-matrix interactions through ligand-integrin associations. Fibrin has been used extensively in the development of tissue engineered constructs for cartilage tissue engineering but the biochemical and morphologic features were not consistent with those of normal articular cartilage and exogenous fibrin may trigger an immune response (Kawabe and Yoshinao,

1991). Currently the use of fibrin is primarily limited clinically to securing perichondrial scaffold grafts. Agarose is another material that has been used for cartilage repair but does not resorb well and has been shown to elicit a foreign body giant cell response in vivo (Rahfoth et al., 1998). Alginate as a matrix for supporting chondrogenesis show improved biological performance seen histologically but resulting repair has been shown to be biochemically distinct form native tissue (Dausse et al., 2003).

Collagens and polysaccharides are the most commonly used components of the ECM in Tissue Engineering (TE). The ability of these materials to be fabricated into highly porous scaffolds allows enhanced diffusion of culture medium as well as an even distribution of ligands present for cell association and cellular migration (Murphy et al., 2010). The use of natural materials in TE reduces the risk of having products of wear and degradation that may elicit host tissue immune response. Collagen is commonly used in TE due to its abundance and ubiquitous nature which allows cellular biocompatibility. From a biomimetic point of view, osteochondral graft substitutes based on collagen and proteoglycans may facilitate the development of natural substrates capable of directing and regulating the chondrogenic process. Collagen based matrices have been shown in a number of studies to be capable of maintaining a differentiated phenotype of chondrocytes and promoting appropriate proteoglycan synthesis in vitro (Wakitani et al., 1994). In combination with seeded chondrocytes and a standard microfracture technique, collagen-based scaffolds have been shown to be capable of regenerating hyaline-like cartilage tissue in an ovine model (Dorotka et al., 2004). This has also been seen in rabbit model where chondrocytes in collagen fibers induced a hyaline-like repair that was biochemically and mechanically similar to native tissue after 6 months (Frenkel et al., 1997) but hyaline-like tissue formation has long been accepted as only a temporary solution (Buckwalter and Mankin, 1998). In addition, integration with the existing host tissue in studies using single phase collagen type I scaffolds has been notably disappointing (Wakitani et al., 1998; Frenkel et al., 1997) Alternatively, collagen type II, a primary component in native articular cartilage, has been shown to promote a more chondrocytic phenotype when used as an in vitro substrate for chondrogenesis results in increase DNA and GAG content, compared to collagen type I-based scaffolds when seeded with chondrocytes while type I matrices show a more fibroblastic phenotype (Nehrer et al., 1997). Type II collagen has also been shown to provide chondroinductive signalling resulting in chondrogenic differentiation of adipose tissue-derived stem cells (Lu et al., 2010) and when combine with type I collagen and GAG, these scaffolds have outperformed ACI repair in pre-clinical trials (Breinan et al., 2001)

Hyaluronic acid (HyA), abundant in the synovial fluid and ECM, plays an important role in structural organisation of the ECM components, maintenance of ECM space, transport of ions and nutrients, and maintenance of tissue hydrodynamics. Studies have shown that HyA addition in collagen scaffolds has resulted in a change of the matrix stiffness, fibrillogenesis and matrix viscoelasticity (Tang et al., 2007). Hence, the addition of HyA in collagen scaffolds initiates biophysical cues that influence cell response. The properties and roles of hyaluronic acid are however dependant on its molecular weight. Low molecular weight hyaluronic acid has been shown to induce angiogenesis whereas high molecular weight hyaluronic acid is more chondroinductive in vivo (Loken et al., 2008). A number of commercial scaffolds are currently available and are based predominantly on Hyaluronic acid and have shown very positive short term results. Hyalograft C (Fidia Advanced Biomaterials) is a 3-dimensional engineered scaffold made of Hyaff 11, the benzyl ester of hyaluronic acid and has been used in clinical studies and has resulted in over 95% of patients returning with normal or nearly normal cartilage after arthroscopic examination after 4 years with another study showing no statistical significant between patients receiving ACI and Hyalograft C (Grigolo et al., 2005). Hyaluronic acid has been used in combination with Chondroitin sulphate and gelatin, forming tri-copolymer scaffolds and used to investigate in vitro chondrogenesis using porcine chondrocytes. These scaffold show excellent results, with even distribution of chondrocytes, evidence of newly secreted ECM and collagen type II, and good phenotype retention for up to 5 weeks (Chang et al., 2003).

While many of these materials possess the basic requirements of biocompatibility and can be manipulated so as to provide adequate mechanical support, porosity, cell attachment and retention at the site of implantation, the regeneration of osteochondral tissue requires a multiplicity of biological and biochemical functions that are difficult to provide when using only a single phase material. Indeed, the possibility of a single phase scaffold promoting both chondro- and osteo-genesis and replicating the natural anatomical structure of native osteochondral tissue seems highly problematic (Niederauer et al., 2000). Native osteochondral tissue exhibits a multi-phase anatomical structure, comprised of multiple layers seamlessly interwoven and integrated on a molecular level. Given this more multiphase anatomical structure, it would seem clear that a more advanced biomimetic strategy may be required to comprehensively regenerate the tissue and ensure long-term repair of a defect. This has been recognised within the field and more and more we are seeing the emergence of multi-layered scaffolds, attempting to regenerate multiple tissue types in vivo. The use of bi-layered scaffolds allows the development of optimised, tissue-specific biological environments within each respective layer via variations in mechanical, structural, and chemical properties (O'Shea et al., 2008). These scaffolds can be designed to better mimic the native ECM for each tissue type independently rather than trying to fabricate a construct that attempts to compensate for the functional requirements of both cartilage and bone in a single structure but fail to address the mechanical and biological need for an intermediate calcified cartilage phase to comprehensively integrate these distinct tissue types in vivo.

A number of recent studies provide convincing evidence supporting the need for a scaffold comprised of a number of distinct constituents, contained within a gradient scaffold architecture. Gradient scaffold pore architectures have been used to investigate the effect of a gradient structure on in vitro chondrogenesis (Woodfield et al., 2005). In vitro culture on these scaffolds can lead to zonal distributions of glycosaminoglycans (GAGs) and collagen type II while changes in the permeability or fluid mobility of scaffolds (that would occur within scaffolds with varying pore size gradients) have also been shown to preferentially favour chondrogenic differentiation of BMSCs and cartilaginous ECM production of chondrocytes (Kemppainen and Hollister, 2010). Computational models of osteochondral repair (Kelly et al., 2006) using a mechano-regulation algorithm point to the importance of a depth-dependent mechanical properties and permeability as optimum for osteochondral repair. The use of bi-layered scaffolds was first pioneered by Shaefer and colleagues (Schaefer et al., 2002) and provided promising results as the first attempt at using multi-phase scaffolds to regenerate osteochondral defects in rabbit knees. The results showed good integration with the underlying bone but not with the peripheral cartilage, although the constructs did show evidence of engineered cartilage that remodelled into osteochondral tissue. Subsequent studies applying this biomimetic approach have resulted in distinct tissue healing within respective layers of a bi-layered construct (Tampieri et al., 2008) with evidence of a mineralised interface (Schek et al., 2004).

Although multi-layered scaffolds can be fabricated by the combination of individually fabricated layers, standard techniques for combining these layers (such as suturing and gluing) are problematic from both a mechanical and a biological point of view. Currently, bi-layered scaffolds with a seamlessly integrated structure are only available composed of synthetic materials (Ghosh et al., 2008) with one exception (Chondromimetic, Tigenix) and although these multi-layered constructs have shown promise in vitro, a significant amount of in vivo data needs to be gathered regarding their clinical efficacy (O'Shea et al., 2008). Consequently, the ability to fabricate multi-layer natural scaffolds exhibiting layer-specific composition, porosity, pore size, mechanical properties, degradation rate and permeability as part of a seamlessly integrated construct is of significant interest. Our laboratory has recently developed a novel multi-layered natural scaffold for osteochondral tissue repair, developed using our existing collagen-based technologies previously used as novel bone graft substitutes. These novel multi-layered scaffolds for use as osteochondral graft substitutes are designed to promote regeneration that replicates the structure and composition of healthy anatomical osteochondral tissue. These scaffolds exhibit a seamless integration between the distinct layers, ensuring rapid cellular infiltration, creation of optimised, tissue-specific biological environments in each respective layer via variations in mechanical, structural, and chemical properties and are designed to better mimic the native ECM of each tissue type independently. These scaffolds may provide a characteristic interfacial region that may help to inhibit the phenomenon of growth factor–induced angiogenesis and the consequential up growth of osseous tissue into the cartilage region. In addition, the use of collagen as a basic component in all three layers provides the opportunity to load distinct bioactive molecules within each distinct scaffold layer.

Case study: Multilayer composite scaffolds for osteochondral tissue regeneration

Our laboratory's approach has been the development of a number of novel lyophilisation fabrication techniques that facilitate a large degree of control over the manufacture of each layer, while still resulting in a completely integrated construct, free from interfacial barriers to cell infiltration and migration. This multi-layered polyphasic scaffold exhibits a structure and composition designed to replicate the environments of the three major layers of anatomical osteochondral tissue, namely subchondral bone, calcified cartilage and cartilage. This material is currently being optimised to provide a substrate for high quality hyaline cartilage repair tissue, seamlessly integrated with a supporting calcified cartilage and bone layers. Each of the three layers possesses a unique composition. The top or cartilage layer contains type I collagen, type II collagen and glycosaminoglycan. Glycosaminoglycans can easily be cross-linked to collagen and the level of cross-linking can alter the mechanical properties of these scaffolds as well as improving their degradation rate (Lee et al., 2001). Chondroitin sulphate (CS) is the most abundant type of GAG in the natural cartilage tissue and is fundamental in the maintaining the structure of cartilage tissue as well as generating electrostatic repulsion via highly charged sulphate groups in the structure. CS manipulates chondrocyte morphology, proliferation and proteoglycan production due to the GAG molecules acting as ligands for the regulation of metabolism and gene expression in chondrocytes. This biochemical is widely used both in vitro and in vivo due to its anti-inflammatory activity as well as the decrease in catabolic activity of proteolytic enzymes such as Nitric Oxide which degenerate cartilage matrix (van Susante et al., 2001). Ko and colleagues (Ko et al., 2009) investigated the effect of collagen-chondroitin sulphate-hyaluronan (CCH) composites on chondrogenesis and their results demonstrated that CCH scaffolds showed upregulated cartilage specific gene expression of collagen-2 and aggrecan compared to scaffolds with no GAGs. Furthermore their results revealed that the CCH scaffolds provided the best microenvironment for the preservation of chondrocyte phenotype.

The intermediate or calcified cartilage layer contains type I collagen, type II collagen and hydroxyapatite while the bottom or bone layer is identical in composition to our optimised bone graft substitute material (Gleeson et al., 2010), specifically type I collagen and hydroxyapatite. We have shown previously that the inclusion of the osteoinductive hydroxyapatite phase in discrete amounts within the collagen-based scaffolds allows a degree of control on the osteogenic process in vitro. The composition of the intermediate layer is designed to act as an interface between the cartilage and bone layers while at the same time facilitating a degree of integration between the tissues engineered within each layer. The fabrication process produces a seamlessly integrated construct. Mechanical testing of the interfacial strength has shown excellent interfacial bonding, with no delamination occurring at scaffold yield and evidence of collagen fibre pullout at the yield point. Interestingly, failure of these construct in tension does not occur at the interface but within the mechanically weakest layer, strong evidence of the seamless integration at the layer interfaces.

We have shown that the addition of the GAGs to the top cartilage layer appears to increase the levels of cartilage ECM specific sulphated GAG production when these constructs are cultured for up to 28 days in vitro. In addition, the type of GAG added to the top layer of the multi-layered construct has a dramatic effect on cellular infiltration, specifically the addition of hyaluronic acid significantly improves cell distribution throughout the constructs when seeded with MSCs and cultured in chondrogenic media (Matsiko et al., 2010). We have also shown that the addition of type II collagen results in increased production of sulphated GAG assessed using Safrinin-O staining within the constructs after culture with rat MSCs in chondrogenic media for up to 28 days (Levingstone et al., 2010). These constructs have also been investigated using a preliminary in vivo pre-clinical rabbit osteochondral defect model in a small cohort of animals (n=2). This work was carried out with ethical approval from the RCSI research ethics committee. A critical-size osteochondral defect (3 mm diameter, 4 mm deep) was created in the femoral condyle of 2 New Zealand White Rabbits. The multi-layered scaffold was cut to size and placed into the osteochondral defect with no fixatives and following the procedure the rabbits were allowed to weight bear with appropriate analgesia. 12 week healing compared to empty defects can be seen in the MicroCT images in Figure 5.

The femoral condyles are currently being assessed histologically to investigate the level of healing and composition of the regenerated tissue. Although MicroCT does not allow a detailed analysis of the overlying chondral repair tissue in these samples, it is clear that a significant amount of subchondral bone healing has taken place and that the defect has been completely sealed at the condylar surface. Additional images taken of the defect surface after 12 weeks (Figure 5) show evidence of integration between the in situ regenerated tissue within the defect and the peripheral hyaline cartilage and the macroscopic appearance of this tissue closely matches the surrounding cartilage in terms of colour and glistening appearance.

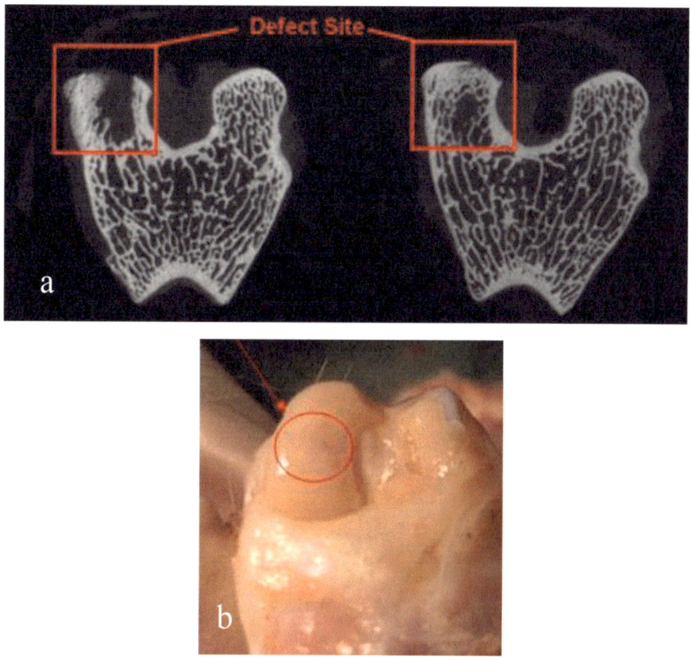

Fig. 5. (a) MicroCT slice of empty control (Left) and scaffold-treated group (Right) after 12 weeks implantation. Note the extensive subchondral remodeling and appearance of cartilage-like repair tissue at joint surface (b) Appearance of treated condyle after 12 weeks implantation. Good repair tissue integration with surrounding cartilage and hyaline-like appearance of de novo tissue

While there remains much work left to do on these constructs and indeed within the field of multi-layered constructs for osteochondral tissue repair, preliminary in vitro and in vivo studies have been extremely promising. The natural composition of the structure, in combination with the layer specific control of composition, structure and mechanical properties endow these materials with great potential for facilitating the in situ repair of osteochondral tissue and may finally offer a clinical therapy that may finally offer patients a real regenerative possibility for damaged or degenerated osteochondral tissue.

6. Conclusions and future directions

The regenerative capabilities of mature bone and osteochondral tissue differ significantly and this has resulted in the two distinct approaches, summarised in this chapter, in an attempt to repair or regenerate these tissues either in the lab or within the patient themselves. Interestingly, despite their final differences, the genesis of these tissues begins in a common cartilaginous anlage present during limb development. The future of bone and osteochondral tissue repair may therefore lie in our understanding of the process of endochondral ossification, and an ability to control and direct this process using improved advances in composite materials and technology. Angiogenesis plays a critical role in endochondral ossification, during which the avascular cartilaginous tissue precursor is gradually transformed into vascular osseous tissue by the migration of blood vessels via cartilage channels. This ingrowth and advancement of vasculature provides the first set of cells capable of disintegrating the cartilage ECM, preventing avascular necrosis and supporting osteogenesis and subsequent bone development and growth. The ability to direct or control tissue vasculogenesis using cutting-edge composite materials and scaffolds may lead to new ways to develop both de novo bone and osteochondral tissue, possibly from a common composite scaffold.

7. References

Altman, G.H.; Diaz, F.; Jakuba, C.; Calabro ,T.; Horan, R.L.; Chen, J.; Lu, H.; Richmond J. & Kaplan D.L. (2003). Silk-based biomaterials. *Biomaterials*, 24, 3, 401–416

Athanasiou, K.A.; Agrawal, C.M.; Barber, F.A.; & Burkhart, S.S. (1998). Orthopaedic applications for PLA-PGA biodegradable polymers, *Arthroscopy*, 14, 7, 726-737

Barbieri, D.; Renard, A.J.S.; de Bruijn, J.D. & Yuan H. (2010). Heterotopic bone formation by nano-apatite containing poly(d,llactide) composites, *Eur Cells Mater*, 19, 252-262

Barrere, F.; Mahmood, T.A.; de Groot, K. & van Blitterswijk, C.A. (2008). Advanced biomaterials for skeletal tissue regeneration: Instructive and smart functions, *Mat Sci Eng*, 59, 38-71

Barrere, F.; van der Valk, C.M.; Dalmeijer, R.A.; Meijer, G.; van Blitterswijk, C.A.; de Groot, K. & Larolle, P. (2003). Osteogenecity of octacalcium phosphate coatings applied on porous metal implants. *J Biomed Mater Res A*, 66: 779-788

Beris, A.E.; Lykissas, M.G.; Papageorgiou, C.D. & Georgoulis, A.D. (2005). Advances in articular cartilage repair, *Injury*, 36, S14–S23

Berry, C.C.; Campbell, G.; Spadiccino, A.; Robertson, M. & Curtis, A.S. (2004). The influence of microscale topography on fibroblast attachment and motility, *Biomaterials* 25, 5781-5788

Bohner, M. (2010). Design of ceramic-based cements and putties for bone graft substitution, *Eur Cells Mater* 20, 1-12

Breinan, H.A.; Minas, T.; Hsu, H.P.; Nehrer, S.; Shortkroff, S. Spector, M. (2001). Autologous chondrocyte implantation in a canine model: change in composition of reparative tissue with time, *J Orthop Res*, 19, 3, 482-492

Byrne, E.M.; Farrell, E.; McMahon, L.A.; Haugh, M.G.; O'Brien, F.J.; Campbell, V.A.; Prendergast, P.J. & O'Connell, B.C. (2008). Gene expression by marrow stromal cells in a porous collagen-glycosaminoglycan scaffold is affected by pore size and mechanical stimulation. *J Mat Sci Mat Med*, 19, 3455-3463

Carter, J.D.; Swearingen, A.B.; Chaput, C.D. & Rahm, M.D. (2009). Clinical and radiographic assessment of transforaminal lumbar interbody fusion using HEALOS collagen hydroxyapatite sponge with autologous bone marrow aspirate, *Spine J*, 9, 434–438

Chang, C.; Liu, H.; Lin, C.; Chou, C & Lin, F. (2003). Gelatin–chondroitin–hyaluronan tri-copolymer scaffold for cartilage tissue engineering, *Biomaterials*, 24, 4853–4858

Chenite, A.; Chaput, C.; Wang, D.; Combes, C.; Buschmann, M.D.; Hoemann, C.D.; Leroux, J.C.; Atkinson, B.L.; Binette, F. & Selmani, A. (2000). Novel injectable neutral solutions of chitosan form biodegradable gels in situ, *Biomaterials*, 21, 2155–2161

Coutts, R.D.; Healey, R.M.; Ostrander, R.; Sah, R.L.; Goomer, R.; Amiel, D. (2001). Articular cartilage repair: tissue engineering, cells, scaffolds, and growth factors: matrices for cartilage repair, *Clin Orthop Relat R*, 391, S271-S279

Cummings, C.L.; Gawlitta, D.; Nerem R.M. & Stegemann J.P. (2004). Properties of engineered vascular constructs made from collagen, fibrin, and collagen-fibrin mixtures, *Biomaterials* 25, 3699-3706

Dausse, Y.; Grossin, L.; Miralles, G.; Pelletier, S.; Mainard, D.; Hubert, P.; Baptiste, D.; Gillet, P.; Dellacherie, E.; Netter, P. & Payan, E. (2003). Cartilage repair using new polysaccharidic biomaterials: macroscopic, histological and biochemical approaches in a rat model of cartilage defect, *Osteoarth Cartilage*, 11, 1, 16-28

Dawson, J.I.; Wahl, D.A.; Lanham, S.A.; Kanczler, J.M.; Czernuszka, J.T. & Oreffo, R.O.C. (2008). Development of specific collagen scaffolds to support the osteogenic and chondrogenic differentiation of human bone marrow stromal cells. *Biomaterials*, 29, 3105-3116

Deplaine, H.; Gómez Ribelles, J.L. & Ferrer, G.G. (2010). Effect of the content of hydroxyapatite nanoparticles on the properties and bioactivity of poly(L-lactide) - Hybrid membranes, *Compos Sci Technol*, 70, 1805–1812

Drury J.L. & Mooney, D.J. (2003). Hydrogels for tissue engineering: Scaffold design variables and applications, *Biomaterials*, 24, 4337-4351

Dounchis, J.S.; Bae, W.C.; Chen, A.C.; Sah, R.L.; Coutts, R.D. & Amiel, D. (2000). Cartilage repair with autogenic perichondrium cell and polylactic acid grafts, *Clin Orthop Relat R*, 377, 248-264

Ducheyne, P.; Radin, S. & King, L. (1993). The effect of calcium phosphate ceramic composition and structure on in vitro behavior. I. Dissolution, *J Biomed Mater Res*, 27, 25–34

Engler, A.J.; Sen, S.; Sweeney, H.L. & Discher, D.E. (2006). Matrix elasticity directs stem cell lineage specification, *Cell* 126, 677-689

European Musculoskeletal Review (2010). Volume 3, Issue 2, Touch Group PLC. www.touchbriefings.com

Farrell, E.; van der Jagt, O.P.; Koevoet, W.; Kops, N.; van Manen, C.J.; Hellingman, C.A.; Jahr, H.; O'Brien, F.J.; Verhaar, J.A.; Weinans, H. & van Osch, G.J. (2009). Chondrogenic Priming of Human Bone Marrow Stromal Cells: A Better Route to Bone Repair?, *Tissue Eng Part C Methods*, 15, 2, 285-95

Frenkel, S.R. & Di Cesare, P.E. (2004). Scaffolds for Articular Cartilage Repair, *Ann Biomed Eng*, 32, 1, 26–34

Frenkel, S.R.; Toolan, B.; Menche, D.; Pitman, M.I. & Pachence, J.M. (1997). Chondrocyte transplantation using a collagen bilayer matrix for cartilage repair, *J Bone Joint Surg Br*, 79, 5, 831-836

Freyman, T.M.; Yannas, I.V.; Yokoo, R.L. & Gibson J. (2001). Fibroblast contraction of a collagen-GAG matrix, *Biomaterials* 22, 2883–2891

Ghosh, S.; Viana, J.C.; Reis, R.L.; & Mano, J.F. (2008). Bi-layered constructs based on poly(l-lactic acid) and starch for tissue engineering of osteochondral defects. *Mater Sci Eng C* 28, 80-86

Gleeson, J.P.; Plunkett, N.A. & O'Brien, F.J. (2010). Addition of hydroxyapatite improves stiffness, interconnectivity and osteogenic potential of a highly porous collagen-based scaffold for bone tissue regeneration, *Eur Cells Mater*, 20, 218-230

Grigolo, B.; Roseti, L.; De Franceschi, L.; Piacentini, A.; Cattini, L.; Manfredini, M.; Faccini, R. & Facchini, A. (2005). Molecular and immunohistological characterization of human cartilage two years following autologous cell transplantation. *J Bone J Surg Am*, 87, 46 – 57

Harley, B.A.; Leung, J.H.;, Silva, E.C.C.M. & Gibson, L.J. (2007). Mechanical characterization of collagen glycosaminoglycan scaffolds. *Acta Biomater* 3, 463-474

Harley, B.A.; Spilker, M.H.; Wu, J.W.; Asano, K.; Hsu, H.P.; Spector, M. & Yannas, I.V. (2004). Optimal degradation rate for collagen chambers used for regeneration of peripheral nerves over long gaps, *Cells Tissues Organs*, 176, 153–165

Haugh, M.G.; Jaasma, M.J. & O'Brien FJ. (2009). The effect of dehydrothermal treatment on the mechanical and structural properties of collagen-GAG scaffolds. *J Biomed Mater Res A*, 89, 2, 363-369

Hench, L.L. (1998). Bioceramics, *J Am Ceram Soc*, 81, 1705-28

Hench, L.L. (1980). Biomaterials, *Science*, 208, 826-831

Hennick, W.E. & Van Nostrum, C.F. (2002). Novel crosslinking methods to design hydrogels, *Adv Drug Deliver Rev*, 54, 13-26

Hing, K.A. (2005). Bioceramic bone graft substitutes: Influence of porosity and chemistry, *Int J Appl Ceram Technol*, 2, 184-199

Homsy, C.A. (1970). Bio-compatibility in selection of materials for implantation, *J Biomed Mater Res.*, 4, 341-356

Hunter, C.J.; Mouw J.K. & Levenston, M.E. (2004). Dynamic compression of chondrocyte-seeded fibrin gels: Effects on matrix accumulation and mechanical stiffness, *Osteoarthr Cartilage* 12, 117-130

Hunziker, E.B. (2002). Articular cartilage repair: basic science and clinical progress. A review of the current status and prospects, *Osteoarthr Cartil.* 10, 432–463

Hutmacher, D.W. (2000). Scaffolds in tissue engineering bone and cartilage, *Biomaterials*, 21, 2529-2543

Jiang, T; Abdel-Fattah, W.I. & Laurencin, C.T. (2006). In vitro evaluation of chitosan/poly(lactic acid-glycolic acid) sintered microsphere scaffolds for bone tissue engineering, *Biomaterials*, 27, 4894–4903

Kawabe, N. & Yoshinao, M. (1991). The repair of full-thickness articular cartilage defects. Immune responses to reparative tissue formed by allogeneic growth plate chondrocyte implants, *Clin Orthop Relat Res*, 268, 279-293

Kelly, D.J. & Prendergast, P.J. (2006). Prediction of the Optimal Mechanical Properties for a Scaffold Used in Osteochondral Defect Repair, *Tissue Eng*, 12, 9, 2509-2519

Kemppaine, J.M. & Hollister S.J. (2010). Differential effects of designed scaffold permeability on chondrogenesis by chondrocytes and bone marrow stromal cells, *Biomaterials*, 31, 279–287

Kikuchi, M.; Ikoma, T.; Itoh, S.; Matsumoto, H.N.; Koyama, Y.; Takakuda, K.; Shinomiya, K. & Tanaka, J. (2004). Biomimetic synthesis of bone-like nanocomposites using the self organization mechanism of hydroxyapatite and collagen. *Compos Sci Technol*, 64, 819-825

Kitchel, S.H. (2006). A preliminary comparative study of radiographic results using mineralized collagen and bone marrow aspirate versus autologous bone in the same patients undergoing posterior lumbar interbody fusion with instrumented posterolateral lumbar fusion, *Spine J*, 6, 405–412

Klein, C.P.A.T.; Wolke, J.G.C. & deGroot, K. (1993). Stability of calcium phosphate ceramics and plasma sprayed coating, In: *An Introduction to Bioceramics*, Hench, L. L. & Wilson, J. (Eds.) , 11, 199-223, World Scientific, 9810214006, London

Ko, C.S.; Huang, J.P.; Huang, C.W. & Chu, I.M. (2009). Type II collagen-chondroitin sulfate-hyaluronan scaffold cross-linked by genipin for cartilage tissue engineering, *J Biosci Bioeng*, 107, 2, 177-82

Ko, H.C.H.; Milthorpe, B.K. & McFarland, C.D. (2007). Engineering thick tissues - the vascularisation problem. *Eur Cells Mater*, 14, 1-19

Kuboki, Y.; Takita, H.; Kobayashi, D., Tsuruga, E., Inoue, M., Murata, M., Nagai, N.; Dohi, Y. & Ohgushi, H. (1998). BMP-induced osteogenesis on the surface of hydroxyapatite with geometrically feasible and nonfeasible structures: topology of osteogenesis. *J Biomed Mater Res*, 39, 2, 190–1999

Laurencin, C.T.; Khan, Y.; Kofron, M.; El-Amin, S.; Botchwey, E.; Yu, X. & Cooper Jr., J.A. (2006). The ABJS Nicolas Andry Award: tissue engineering of bone and ligament: a 15-year perspective, *Clin Orthop Relat Res*, 447, 221–236

Lisignoli, G.; Grassi, F.; Piacentini, A.; Cocchini, B.; Remiddi, G.; Bevilacqua, C. & Facchini, A. (2001). Hyaluronan does not affect cytokine and chemokine expression in osteoarthritic chondrocytes and synoviocytes, *Osteoarthr Cartilage*, 9, 2, 161-168

Lee, C.H.; Singla, A. & Lee Y. (2001). Biomedical applications of collagen, *Int J Pharm*, 19, 221, 1-22

Lee, K. & Goodman, S.B. (2009). Cell Therapy for Secondary Osteonecrosis of the

Femoral Condyles Using the Cellect DBM System: A Preliminary Report, *J Arthroplasty*, 24, 1, 43-48

Levingstone, T.J.; Dickson, G.; O'Brien, F.J. & Gleeson, J.P. (2010). A multi-layer collagen-based scaffold for osteochondral defect repair, *Eur Cells Mater*, 20, S47

Li, J.; Chen, Y.; Mak, A.F.T.; Tuan, R.S.; Li, L. & Li, Y. (2010). A one-step method to fabricate PLLA scaffolds with deposition of bioactive hydroxyapatite and collagen using ice-based microporogens, *Acta Biomaterialia*, 6, 2013–2019

Løken, S.; Jakobsen, R.B.; Arøen, A.; Heir, S.; Shahdadfar, A.; Brinchmann, J.E.; Engebretsen, L. & Reinholt, F.P. (2008). Bone marrow mesenchymal stem cells in a hyaluronan scaffold for treatment of an osteochondral defect in a rabbit model, *Knee Surg Sports Traumatol Arthrosc*, 16, 10, 896-903

Lu, F.; Doulabi, B.Z.; Huang, C.; Bank, R.A. & Helder, M.N. (2010). Collagen type II enhances chondrogenesis in adipose tissue–derived stem cells by affecting cell shape, *Tissue Eng Pt A*, 16, 1, 81-90

Lyons, F.; Al-Munajjed, A.; Kieran, S.; Toner, M. & O'Brien, F.J. (2010). The healing of bony defects by cell-free collagen based scaffolds compared to stem cell-seeded tissue engineered constructs, *Biomaterials*, 31, 35, 9232-9243

Ma, P. & Langer, R. (1999). Morphology and mechanical function of long-term in vitro engineered cartilage, *J Biomed Mater Res*, 44, 217-221

Marcacci, M.; Kon, E.; Moukhachev, V.; Lavroukov, A.; Kutepov, S.;Quarto, R.; Mastrogiacomo, M. & Cancedda, R. (2007). Stem cells associated with macroporous bioceramics for long bone repair: 6- to 7-year outcome of a pilot clinical study. *Tissue Eng*, 13, 947–955

Matsiko, A.; Levingstone, T.J.; O'Brien, F.J. & Gleeson, J.P. (2010). Addition of hyaluronic acid improves cellular infiltration and cartilage-specific extracellular matrix synthesis in a porous collagen scaffold, *Eur Cells Mater*, 20, S49

McNickle, A.G.; Provencher, M.T. & Cole B.J. (2008). Overview of existing cartilage repair technology, *Sports Med Arthrosc*, 16, 196-201

Meinel, L.; Hofmann, S.; Betz, O.; Fajardo, R.; Merkle, H.P.; Langer, R.; Evans C.H.; Vunjak-Novakovic, G. & Kaplan, D.L. (2006). Osteogenesis by human mesenchymal stem cells cultured on silk biomaterials: Comparison of adenovirus mediated gene transfer and protein delivery of BMP-2, *Biomaterials*, 27, 4993–5002

Murphy, C.M.; Haugh, M.G. & O'Brien, F.J. (2010). The effect of mean pore size on cell attachment, proliferation and migration in collagen glycosaminoglycan scaffolds for tissue engineering, *Biomaterials*, 31, 461-466

Muschler, G.F.; Matsukura, Y.; Nitto, H.; Boehm, C.A., Valdevit, A.D.; Kambic, H.E.; Davros, W.J.; Easley, K.A. & Powell, K.A. (2005). Selective retention of bone marrow-derived cells to enhance spinal fusion, *Clin Orthop Relat Res*, 432, 242–251

Nehrer, S.; Breinan, H.A.; Ramappa, A.; Shortkroff, S.; Young, G.; Minas, T.; Sledge, C.B.; Yannas, I.V. & Spector, M. (1997). Canine chondrocytes seeded in type I and type II collagen implants investigated in vitro, *J Biomed Mater Res*, 38, 2, 95-104

Niederauer, G.G.; Slivka, M.A.; Leatherbury, N.C.; Korvick, D.L.; Harroff, H.H.; Ehler, W.C.; Dunn, C.J. & Kieswetter, K. (2000). Evaluation of multiphase implants for repair of focal osteochondral defects in goats, *Biomaterials*, 21, 24, 2561-2574

O'Brien, F.J.; Harley, B.A.; Yannas, I.V. & Gibson, L.J. (2005). The effect of pore size on cell adhesion in collagen-GAG scaffolds. *Biomaterials*, 26, 4, 433-441

O'Shea, T.M. & Miao X. (2008). Bilayered Scaffolds for Osteochondral Tissue Engineering, *Tissue Eng Pt B-Rev*, 14, 4, 447-464

Phelps, E.A.; Landázuri, N.; Thulé, P.M.; Taylor, W.R. & García, A.J. (2010). Bioartificial matrices for therapeutic vascularization, *Proc Natl Acad Sci U S A*, 107, 8, 3323–3328

A.G. and Atala, A. (Eds.), 463-472, Artech House Publishers, 1596931116, Boston

Rahfoth, B.; Weisser, J.; Sternkopf, F.; Aigner, T.; von der Mark, K. & Bräuer, R. (1998). Transplantation of allograft chondrocytes embedded in agarose gel into cartilage defects of rabbits, *Osteoarthr Cartilage*, 6, 1, 50-65

Ratcliffe, A. (2008). Industrial approaches in tissue engineering, In: *Translational approaches in Tissue Engineering and Regenerative Medicine*, Mao, J.J.; Vunjak-Novakovic, G.; Mikos

Rezwan, K.; Chen, Q.Z.; Blaker, J.J.; Boccaccini, A.R. (2006). Biodegradable and bioactive porous polymer/inorganic composite scaffolds for bone tissue engineering, *Biomaterials*, 27, 3413–3431

Scabbia, A. & Trombelli, L. (2004). A comparative study on the use of a HA/collagen/chondroitin sulphate biomaterial (Biostites) and a bovine derived HA xenograft (Bio-Osss) in the treatment of deep intra-osseous defects, *J Clin Periodontol*, 31, 348–355

Schaefer, D.; Martin, I.; Jundt, G.; Seidel, J.; Heberer, M.; Grodzinsky, A.; Bergin, I.; Vunjak-Novakovic, G. & Freed L.E. (2002). Tissue-engineered composites for the repair of large osteochondral defects, *Arthritis Rheum*, 46, 9, 2524-2534

Seeherman, H. & Wozney, J.M. (2005).Delivery of bone morphogenetic proteins for orthopedic tissue regeneration, *Cytokine Growth F R*, 16, 329–345

Sethi, K.K.; Yannas, I.V.; Mudera, V.; Eastwood, M.; McFarland, C. & Brown, R.A. (2002). Evidence for sequential utilization of fibronectin, vitronectin, and collagen during fibroblast-mediated collagen contraction, *Wound Repair Regen*, 10, 397–408

Stoop, R. (2008). Smart biomaterials for tissue engineering of cartilage, *Injury*, 39S1, S77–S87

Suh, J. & Matthew, H. (2000). Application of chitosan-based polysaccharide biomaterials in cartilage tissue engineering: a review, *Biomaterials* 21, 2589–2598

Supova, M. (2009). Problem of hydroxyapatite dispersion in polymer matrices: a review, *J Mater Sci Mater Med*, 20, 1201–1213

Tampieri, A.; Sandri, M.; Landi, E.; Pressato, D.; Francioli, S.; Quarto, R. & Martin, I. (2008). Design of graded biomimetic osteochondral composite scaffolds, *Biomaterials*, 29, 26, 3539-3546

Tancred, D.C.; McCormack, B.A. & Carr, A.J. (1998). A synthetic bone implant macroscopically identical to cancellous bone, *Biomaterials*, 19, 2303-2311

Tang, S.; Vickers, S.M.; Hsu, H & Spector, M. (2007). Fabrication and characterization of porous hyaluronic acid–collagen composite scaffolds, *J Biomed Mater Res*, 82, 2, 323-335

Terada, S.; Yoshimoto, H.; Fuchs, J.R.; Sato, M.; Pomerantseva, I.; Selig, M.K.; Hannouche, D. & Vacanti, J.P. (2005). Hydrogel optimization for cultured elastic chondrocytes seeded onto a polyglycolic acid scaffold. *J Biomed Mater Res Pt A*, 75, 907–916

ter Brugge, P.J.; Wolke, J.G.C.; Jansen, J.A. (2003). Effect of calcium phosphate coating composition and crystallinity on the response of osteogenic cells in vitro, *Clin Oral Impl Res*, 14, 472–480

Tierney, C.M.; Haugh, M.G.; Liedl, J.; Mulcahy, F.; Hayes, B. & O'Brien, F.J. (2009). The effects of collagen concentration and crosslink density on the biological, structural and mechanical properties of collagen-GAG scaffolds for bone tissue engineering, *J Mech Behav Biomed Mater*, 2, 202-209

Vacanti, C.A.; Langer, R.; Schloo, B. & Vacanti, J.P. (1991). Synthetic polymers seeded with chondrocytes provide a template for new cartilage formation. *Plast Reconstr Surg*, 88, 753–9

van Susante, J.L.C.; Pieper, J.; Buma, P.; van Kuppevelt, T.H.; van Beuningen, H.; van Der Kraan, P.M.; Veerkamp, J.H.; van den Berg, W.B. & Veth, R.P.H. (2001). Linkage of chondroitin-sulfate to type I collagen scaffolds stimulates the bioactivity of seeded chondrocytes in vitro, *Biomaterials*, 22, 17, 2359-2369

Vepari, C. & Kaplan, D.L. (2007). Silk as a biomaterial. *Prog Poly Sci*, 32, 991-1007

Wakitani, S.; Goto, T.; Pineda, S.J.; Young, R.G.; Mansour, J.M.; Caplan, A.I. & Goldberg, V.M. (1994). Mesenchymal cell-based repair of large, full-thickness defects of articular cartilage, *J Bone Joint Surg Am*, 76, 579–592

Wakitani, S.; Goto, T.; Young, R.G.; Mansour, J.M.; Goldberg, V.M. & Caplan, A.A. (1998). Repair of large full-thickness articular cartilage defects with allograft articular chondrocytes embedded in a collagen gel, *Tissue Eng*, 4, 4, 429-444

Wang, M. (2003). Developing bioactive composite materials for tissue replacement, *Biomaterials*, 24, 2133-2151

Woodfield, T.B.; Van Blitterswijk, C.A.; De Wijn, J.; Sims, T.J.; Hollander, A.P. & Riesle, J. (2005). Polymer scaffolds fabricated with pore-size gradients as a model for studying the zonal organization within tissue-engineered cartilage constructs, *Tissue Eng*, 11, 1297-1311

Xu, Z.; Neoh, K.G. & Kishen A. (2010). A biomimetic strategy to form calcium phosphate crystals on type I collagen substrate. *Mat Sci Eng C*, 30, 822–826

Yannas, I.V. (2001). *Tissue and Organ Regeneration in Adults*, Springer, 0387952144, New York

Yoshida, T.; Kikuchi, M.; Koyama, Y. & Takakuda, K. (2010). Osteogenic activity of MG63 cells on bone-like hydroxyapatite/collagen nanocomposite sponges, *J Mater Sci Mater Med*, 21, 1263–1272

Zhang, L.; Tang, P.; Xu, M.; Zhang, W.; Chai, W. & Wang, Y. (2010a). Effects of crystalline phase on the biological properties of collagen–hydroxyapatite composites, *Acta Biomater*, 6, 2189-2199

Zhang, L.; Tang, P.; Zhang, W.; Xu, M. & Wang, Y. (2010b). Effect of chitosan as a dispersant on collagen hydroxyapatite composite matrices. *Tissue Eng Pt C-Meth*, 16, 71-79

3

Composite Material Stent Comprising Metallic and Non-metallic Materials

Yuzo Shomura, MD.
Senri Chuo Hospital, Department of Surgery
Japan

1. Introduction

Stents that are used for the purpose of canceling stenotic lesions in luminal organs are classified into three types according to the materials used. The first type is ones comprising silicone, and the second type metal. A composite of silicone and metal is employed in the third type. Recent years have seen an increase in the number and range of products of the last type. I am one of those who hold that this is because a stent consisting of a composite material complements those consisting of single materials. The advantage of the silicone stent is the easy removal from the body. The disadvantages include its weak expansile force, requirement for general anesthesia when implanting it in the body, and unstableness. In contrast, the metallic stent is easily inserted in the body and offers strong expansile force while it is difficult to remove once it is inserted. For another disadvantage, it may cause damage to peripheral tissues by superdistention. In order to complement these disadvantages of stents made of single materials, stents comprising a composite of these materials are currently being developed.

I as well as other researchers list the following conditions for an ideal stent.

First, as for the mechanical properties as a structure for canceling stenotic lesions, a stent must:

1. have sufficient expansile force to cancel stenotic lesions,
2. minimize the stress on normal tissues by adjusting its shape according to organs, and
3. not damage the physiologic functions of organs such as secretion and peristalsis.

Second, as a measure against temporal and/or spacial changes of a lesion, a stent must:

1. be removed and reinserted easily when necessary, and
2. retain its properties for an extended period of time.

Third, as for additional properties as a treatment device, a stent must:

1. be able to contain and release therapeutic medicine.

There is a long way to go before a stent is developed that fulfills all these conditions. For the time being, it is more realistic to select one which better meets the specific requirement of each case. We devised a method of knitting metallic and non-metallic materials with a view to modifying the structural properties of the knitted metallic stent. This method enables modification of stent properties in accordance with a specific lesion by changing the non-metallic material to be used with the metal wire. For instance, for improved stent removability, adopting a stent structure with a mesh of bio-absorbing material knit together with metallic wire makes a stent easy to remove. After the bio-absorbing material dissolves, the metallic component remains as a simple spiral, which can be straightened and removed easily by pulling it at one end. Also, compared with metallic materials coated with therapeutic medicine, non-metallic materials such as bio-absorbing textile can contain medicine more easily so it can be designed to optimize the release of the medicine. In addition, as mentioned above, the metallic part can be removed easily after the medicine is released.

The mechanical properties of a composite material stent of metallic and non-metallic materials can be adjusted as well. Metallic stents are composed of metals such as nickel titanium (Ni-Ti) and stainless steel, and the physical properties are defined by the design of the mesh and by the material properties of the metals used. Stents with strong expansile force generally show high elasticity against bending. For this reason, if an indwelling stent with strong expansile force is inserted in a curved luminal organ, the restoring force caused by bending yields excess surface pressure to tissues, which may result in ulceration of and perforation in the tissues. (Fransen et al. 2003) Also, when a stent is bent to an acute angle, adjacent struts come into contact and transform themselves in such a way that they protrude inward, which may result in kinks. (Mori et al. 2004) In a composite material stent of metallic and non-metallic materials, mechanical properties such as expansile force and restoring force are determined almost exclusively by the material and structural properties of the metallic component, as the materials have significantly different degrees of bending stiffness. That is, the structural properties of a hybrid stent are similar to those of metallic spirals. Therefore, a composite material knitted stent realizes both strong expansile force and applicability to curved luminalis.

For this experiment, two types of composite material knitted stents were developed: one using stainless wire and PLA fiber, and the other using Ni-Ti wire and poly-phenylene-benzobisoxazole (PBO) fiber. The former was developed to evaluate the removability after the dissolution of the PLA fiber. The latter was used to evaluate mechanical properties such as the radial expansile force, restoring force against bending, and resistance against kinks.

2. Composite material stent comprising metallic wire and bioabsorbable fiber: mechanical characteristics and retrievability

2.1 Background

Metallic knitted stents are characterized by strong radial force, thin wall structure, and simple stenting procedure under local anesthesia. The major disadvantage of these stents, however, is the difficulty in removal from the body (Noppen et al., 2005). Once placed in a luminal organ, it is often difficult to fold and withdraw the expanded stent. Song et al. reported a metallic stent retrieval procedure using a hook, which allows easy stent removal by pulling the hook after attachment to the stent body (Song et al., 1999). The retrieval hook is useful for certain types of metallic stent structure, but may not be applied to all types of stent textile pattern as some cannot be folded up sufficiently by pulling the end of wire outwards.

One of the reasons why conventional knitted metallic stents are difficult to remove from the body is that they are made with a wire knitted in a cross-linked textile pattern. When stent removal is necessary, it is often difficult to fold up the complex cross-linked wire structure. The only way to remove the stent is to remove it as a whole structure.

We examined the textile patterns of present metallic stents to develop a simple unlocking procedure, and found that cross-linking of the metallic wire can be avoided if the textile is knitted from two different wires, one of which comprising a non-metallic material. This concept can be applied to a composite material stent comprising a metallic wire and a bioabsorbable fiber.

In this study, we made a composite material stent comprising a metallic wire and a Poly Lactic Acid (PLA) fiber (hybrid stent). With such a stent, metallic wire is cross-linked only with bioabsorbable fiber, not with itself, and it can easily be deformed and removed after bioabsorbable fiber has degraded.

The purpose of this study is to evaluate mechanical characteristics of the composite material stent and to examine its retrievability after bioabsorbable fiber degradation.

2.2 Material and methods

2.2.1 Stent structure

The composite material stent was produced from stainless steel wire and poly lactic acid (PLA) fiber. They were cross knitted in the same textile pattern of an commercial Ultra Flex stent (Fig. 1). American National Standard of Industry (ANSI) 304 grade stainless steel wire of 0.2 mm in diameter was knitted with a 0.23mm PLA monofilament fiber (520T; Chukoh Chemical Industries, Ltd. Fukuoka, Japan: S-P stent). Stent size is 20 mm in diameter and 60 mm in length.

For comparison of mechanical strength, identical test stent of the PLA monofilament fiber only was made (P-P stent).

Commercial metallic stents, Ultra Flex (Boston Scientific Co., Natick, USA) of the same size were also prepared for comparison.

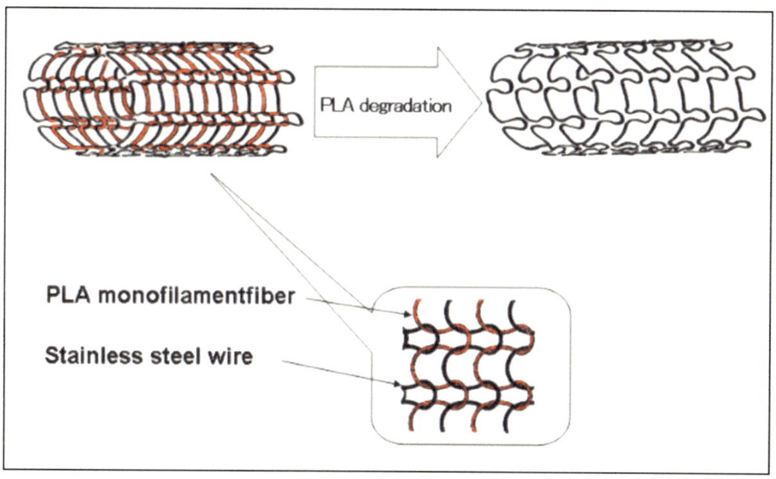

Fig. 1. Schematic structure of composite material stent; A stainless steel wire and a PLA monofilament fiber were knitted in the same textile structure as Ultra Flex stent. Non-cross-linked metallic wire can be straightened after PLA fiber degradation, and easily removed from the body

2.2.2 Evaluation of mechanical properties

1. Radial expansile force

The radial expansile forces of the test stents were measured as the resilient force response against circumferential shrinkage stress load, as described previously (Miura et al., 2003). In summary, a test stent covered with a non-stretchable film was placed on a board fixed to one end of the covering film. The other end of the film was connected to a push-pull gauge (Imada Co.,Toyohashi, Japan), and shrinkage load was added to the stent by pulling the film. The load added to the stent was recorded for stent diameter change (Fig. 2 (a)).

2. Lateral stent rigidity

Mechanical strength against lateral compression, defined as stent rigidity in this study, was measured as the resilient force response to lateral compression load. The test stent was set in a load cell, and lateral compression stress was added to the point when stent diameter reduced to 50% of the original dimension. The resilient force at this point was recorded (Fig. 2 (b)).

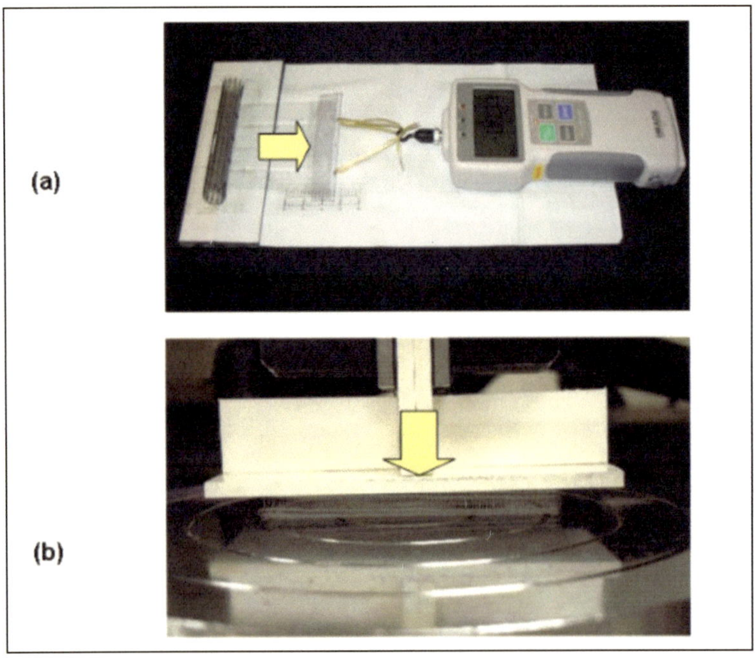

Fig. 2. Evaluation of mechanical properties; (a) Radial expansile force was evaluated as the resilient force response against circumferential shrinkage stress load applied to the test stent. (b) Resilient force against lateral compression load was measured at 50% reduction of the original diameter

3. Change in radial expansile force after PLA fiber degradation

To estimate change of mechanical strength after PLA fiber degradation, an artificial PLA degradation process was added to S-P stent by immersing it into a water bath at 80°C for 48 hours. The change in radial expansile force was compared to the original values P-P stent was also tested in the same way for comparison.

2.3 Retrievability of metallic core component

S-P stent underwent the heating process mentioned above was set in a silicone tube. One end of the metallic core component was grasped using forceps and pulled outwards to unlock spiral structure. Torsion of the metallic wire, stent displacement, resistance in the unlocking process and deformities in the core were visually evaluated.

2.4 Results

2.4.1 Mechanical properties

1. Radial expansile force

S-P stent exhibited linear radial force response up to the point when the diameter was 85% of the original dimension. This linearity was comparable to that of the same size UltraFlex stent. The lumen of P-P stent was obstructed at 15 percent diameter shortening, and did not exhibit radial force increase over this point (Fig. 3).

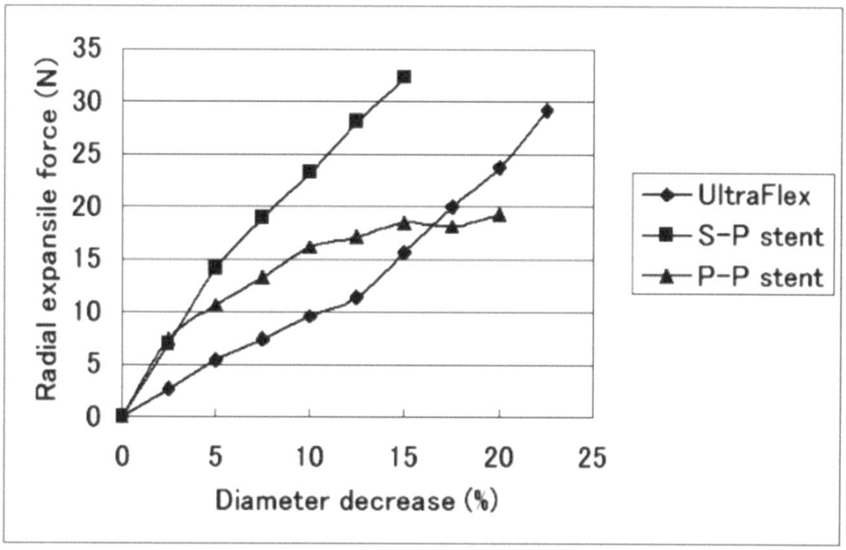

Fig. 3. Radial expansile force response against circumferential shrinkage stress load; The S-P stent presented linear radial force increase within the range of 15 percent shortening of the original diameter. In P-P stent, linear response was lost when the diameter decreased by 15%

2. Stent rigidity

Resilient force of a S-P stent against lateral compression was 86.8 kN and 21.1kN for S-P stent at the point when stent diameter reduced to 50% of the original dimension. It was approximately four times than that of the same structural S-S stent. (Table 1).

Stent	Resilient force (kN)
S-P stent	86.8
P-P stent	21.1

Table 1. Resilient force of test stents against lateral compression at the point when stent diameter reduced to 50% of the original dimension

3. Change in radial expansile force after PLA fiber degradation

S-P stent undergoing the heat process showed a slight decrease in radial expansile force, but it was not more than 5% within the range of a 15 % reduction in original stent diameter. In contrast, P-P stent did not exhibit an effective radial force, and was destroyed at 12% diameter reduction (Fig. 4). S-P stent maintained its original shape even after PLA degradation. In contrast, P-P stent was easily crushed by slight compression loads (Fig. 5).

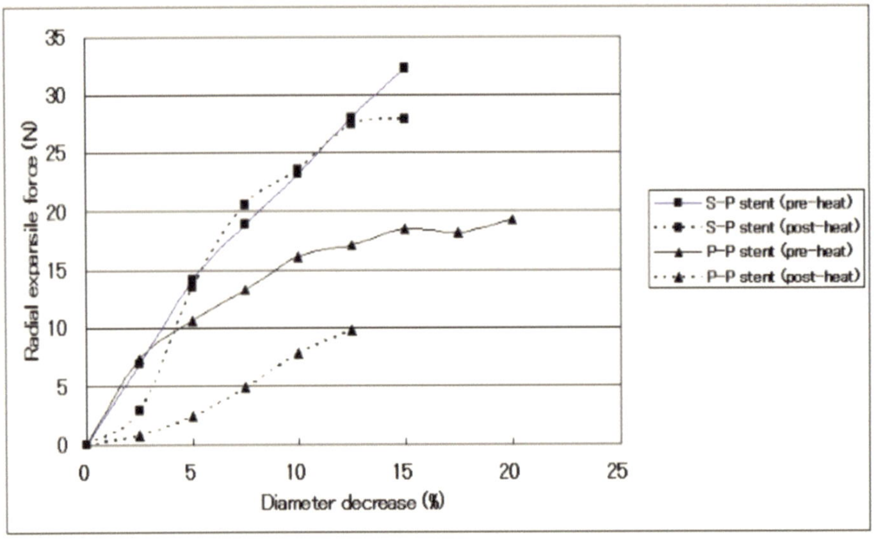

Fig. 4. Radial expansile force responseof the heat processed stents; The radial force decrease of the S-P stent was within 5 percent of the original force. PLA stent did not exhibit effective radial force after the process

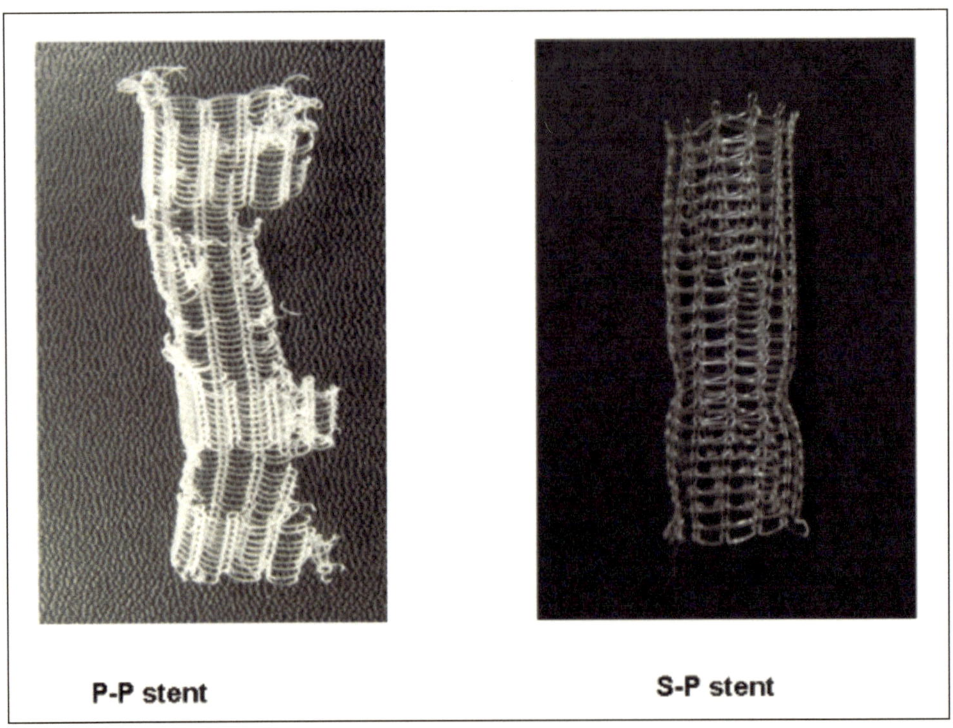

Fig. 5. Gross appearance of the heat processed test stents; The S-P stent maintained its original shape, even after PLA degradation, but the P-P stent was destroyed with slight compression

2.4.2 Retrievability of metallic core component

After the heat process, the non-cross-linked metal core of the S-P stent was easily straightened by pulling the end of the stent wire. There was slight displacement of the stent in the silicone tube while pulling the wire, but no unusual resistance, deformity or torsion of the core was observed. As a result, the residual metal core component was removed from the silicone tube without difficulty (Fig. 6).

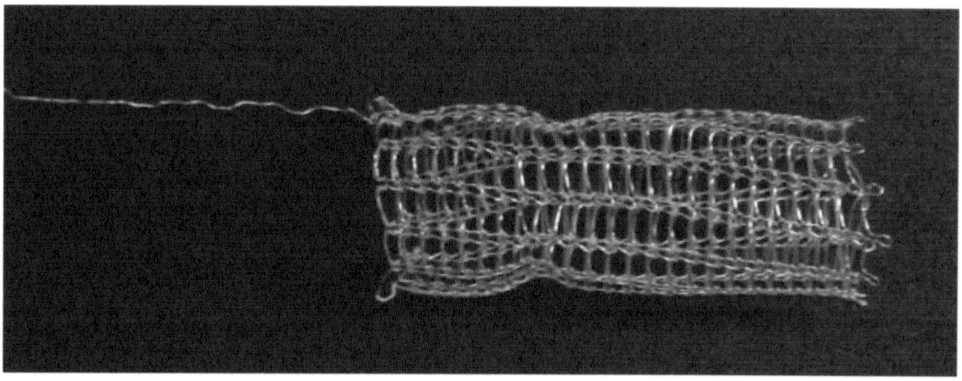

Fig. 6. Unlocking performance of metal component of the heat processed S-P stent; The non-cross-linked metal core was easily straightened by pulling the end of the stent wire, and was removed from a silicone tube without difficulty

2.5 Discussion

Silicone are characterized by high elasticity and retrievability, and are often used for temporary stenting, such as strictures of benign etiology (Schildge et al., 2001), or regions where flexibility is required due to various organ movement, such as the trachea (Puma et al. 2000). However, they also have disadvantages related to the complex stenting procedure and structural problems. Rigid fiberscopy under general anesthesia is required for stenting; the stent wall is too thick to create effective radial force, which often results in stent migration or displacement. Therefore, the clinical applications of silicone stents are limited when compared to metallic stents.

So-called bioabsorbable stents have been developed for temporary stenting purposes. They are designed to degrade in the body at certain periods of time after deployment (Korpala et al., 1999). A common bioabsorbable material for this purpose is poly-l-lactic acid (PLLA), which has been widely used as a surgical suture material, and its biocompatibility and degradability in the human body has been established (Isotalo et al., 2002; Zilberman et al. 2005). However, the physical properties of PLLA fibers, such as tensile strength and elasticity, are inferior to metallic wires (Grabow et al., 2005); therefore, the potential uses of this material in stents are limited.

Under these circumstances, the purpose of our study was to improve the retrievability of metallic knitted stents, and for this purpose, we produced a composite material stent comprising bioabsorbable fiber and metallic wire. Combination of dissimilar materials with different physical properties in stent design has been in practice in the dynamic stent (Freitag et al., 1994), for which a silicone plate and elastic rubber were successfully combined to achieve excellent conformability with tracheal wall movement. Our composite material stent is another attempt for metallic knitted stent to use different materials in order to improve its retrievability.

Mechanical strength of the composite material stent is dependent on the metallic spiral component, and the bioabsorbable fiber component acted as a spacer by which the metallic core can be properly introduced, positioned and fixed at the target site. Therefore, apparent reduction of radial expansile force was not observed between before and after artificial PLA degradation process. Fuerthermore, the fact that mechanical strength of the composite material stent, as a whole, is influenced by that of the metallic spiral core component means that stent mechanical characteristics can be designed simply by dimensional factors of a metallic wire, such as diameter, circumferential loop density, and length of a loop leg as well as physical properties of the material used. Without taking the effect of cross–linkage of stiff metallic wire with itself into account, stent structural parameters can be optimized.

The changes made to the composite material stent structure may have additional benefit to its characteristics; excellent kink resistance and low axial resilient force. Upon our observation of why stent kinking occurs in conventional metallic knitted stents, it is brought about by the collision of adjacent metallic loops during bending of the stent, resulting in compressed loops protruding inward. In contrast, because the metallic loop is surrounded by soft loops of non-metal fibers in the hybrid stent, collision between metallic loops is not likely to occur, and therefore kinking may be avoided. Furthermore non-cross-linked metallic loop structures produce only resilient torsial force at stent bending, and thus the straightening force of the stent is minimized and excellent conformability may be achieved.

Our study was limited by the fact that PLA degradation process was not evaluated in vivo. The bioabsorbable fiber component is expected to act as a spacer by which the metallic core can be properly introduced, positioned and fixed at the target site, and to be degraded after metallic core has been fixed. However, once placed in the body, it is not possible to control reaction speed of PLA hydrolyzation reaction. Loco-regional settings of the target site, such as existence of bacteria, influence the reaction, which may result in insufficient fixation of the metallic core due to unexpectedly early PLA degradation, or in disturbance in stent removal due to persistence of unhydrolyzated PLA fiber. Thus, further experimental studies in vivo are warranted.

In conclusion, the composite material stent comprising metallic wire and bioabsorbable fiber can preserve effective physical properties comparable with the identical metallic stent. And it also has high retrievability after bioabsorbable fiber degradation.

3. Composite material stent comprising Ni-Ti wire and PBO fiber, and its mechanical characteristics

3.1 Background

An ideal stent has not only expansile force to enlarge a stenotic lumen but also elasticity to follow flexibly various forms of external force such as bending and compression caused by movement of the body (Fransen et al., 2003).

A stent's characteristics such as the mechanical strength and transformation due to loading are determined by the physical properties of the materials and by the design of the mesh pattern (Flueckiger et al., 1994). As for the metallic stent, it shows strong elasticity against external force such as bending and compression with its strong expansile force.

Caution is required when implanting a metallic stent in a curved luminal organ because it may cause ulceration or perforation in tissues by giving excess surface pressure as it unwinds to the linear form.

With a view to developing a stent with sufficient expansile force against stenosis while following flexibly bends in luminal organs, stents knit in tiers were created using wires of different degrees of bending stiffness. The results showed that expansile force and flexibility can be gained simultaneously by an appropriate combination of materials and textile design. The purpose of this experiment was to evaluate the physical properties of a composite material stent comprising metallic wire and polymeric fiber.

3.2 Materials and methods

3.2.1 Stent structure

For the metallic wire, nickel titanium alloy (Ni-Ti) wire, 0.12mm in diameter, was chosen. For the non-metallic wire, multifilament fiber (ZYRON®-AS; TOYOBO, Osaka, Japan) consisting of poly-phenylene-benzobisoxazole (PBO), 0.18mm in diameter, was selected. These materials were knit at the ratio of 1 to 1 into a cylinder having the same textile pattern as the Ultra Flex Stent (Fig. 7). The size of the stent was 15mm or 8mm in diameter and 80mm in length (N-Z stent). For control, a metallic stent was created using Ni-Ti only but having the same textile pattern (N-N stent). After the stents were knit, they were heated for thirty minutes at 400 degrees Celsius, for a shape memory treatment of the Ni-Ti alloy. The loop pitches of the Ni-Ti wires after the shape memory treatment measured 1.84mm in the N-Z stent and 1.92mm in the N-N stent. Also, for comparison, three stents of 8mm in diameter that are already in clinical use were prepared: a Wall stent (Boston Scientific Co., Natick, USA), a Spiral Z stent (Medico's Hirata, Tokyo, Japan), and a SENDAI stent (Piolax Medical Device Co., Yokohama, Japan).

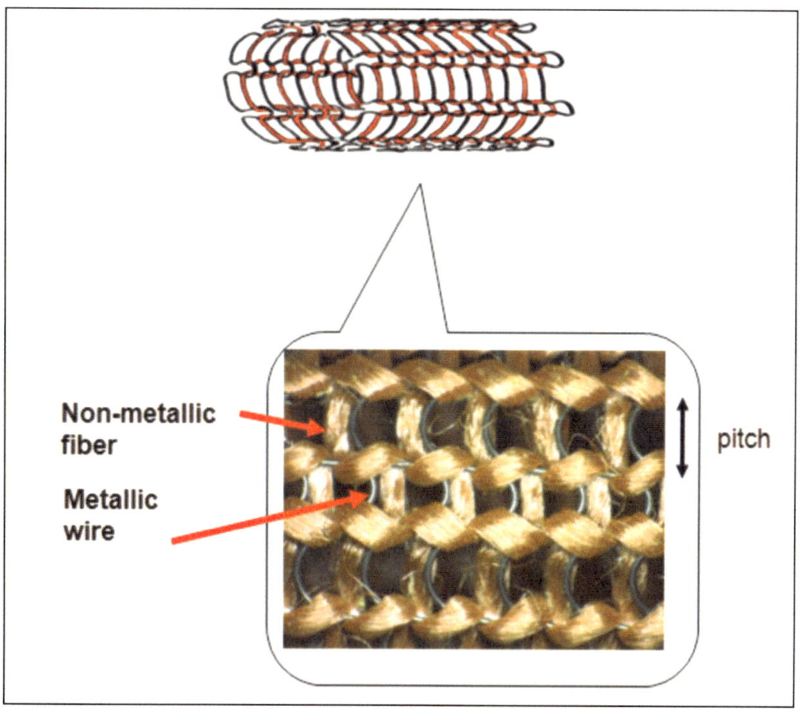

Fig. 7. A composite material stent comprising Ni-Ti wire and PBO multifilament fiber. The photo was taken prior to the shape memory heat treatment of the Ni-Ti

3.2.2 Evaluation of mechanical properties

1. Radial expansile force

An N-Z stent and an N-N stent of 15mm in diameter were chosen as the objects of study. One metallic plate was attached to the top of a universal testing machine (RTC-1350A, ORIENTEC Co., Ltd., Tokyo, Japan) and another to the bottom. The stents were set between the plates. The initial point of measurement was determined when the upper metallic plate was lowered to a point where the diameter of the stent was 6mm (40% of the original). The pressure was released by raising the upper metallic plate at 5mm/min. The autograph connected to the upper plate measured the restoring force of the stent sequentially to the ratio of the amount of transformation of the stent. When the value of the force reached zero, the upper plate was lowered at 5mm/min to the initial point. This process was also measured. A cycle was defined by this up-and-down, and the radial expansile force was calculated on the basis the average score after two cycles (Fig. 8).

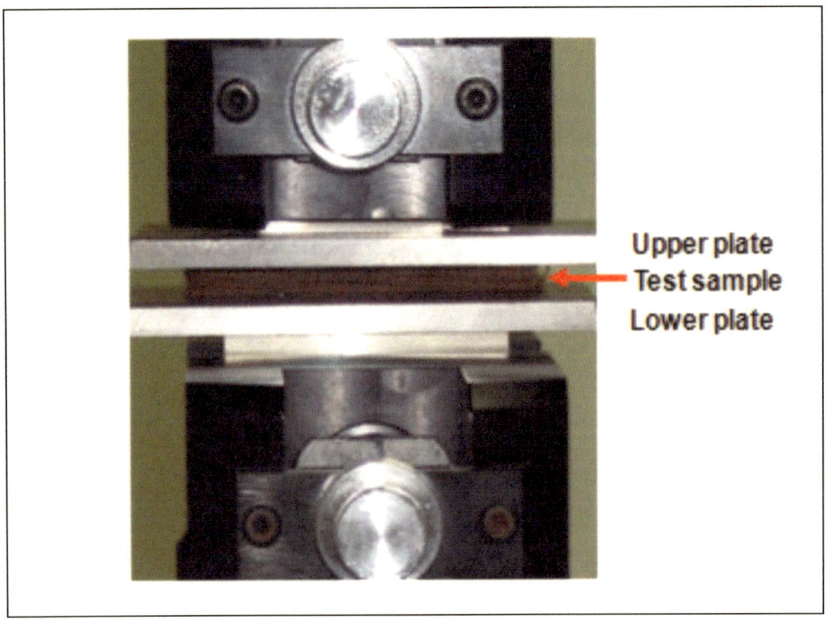

Fig. 8. A method of measuring radial expansile force. The restoring force of the stent on the upper plate when the stent was given a compressive weight was measured as the stent's radial expansile force.

2. Bending force

An N-Z stent and an N-N stent of 8mm in diameter, a Spiral Z stent, a SENDAI stent, and a Wall stent were chosen as the objects. A mandrel was inserted 4cm deep at one end of each stent so the stents were fixed on one side. The free ends of the stents were contacted by a push-pull gauge (IMADA Co., Aichi, Japan) in the direction of the diameter. External force was given to the stents through the gauge until they were bent to an angle of 90 degrees. With the stents bent at an angle of 90 degrees, the weight of the load was measured on the push-pull gauge. The bending force was calculated on the basis of the average values after three measurements (Fig. 9).

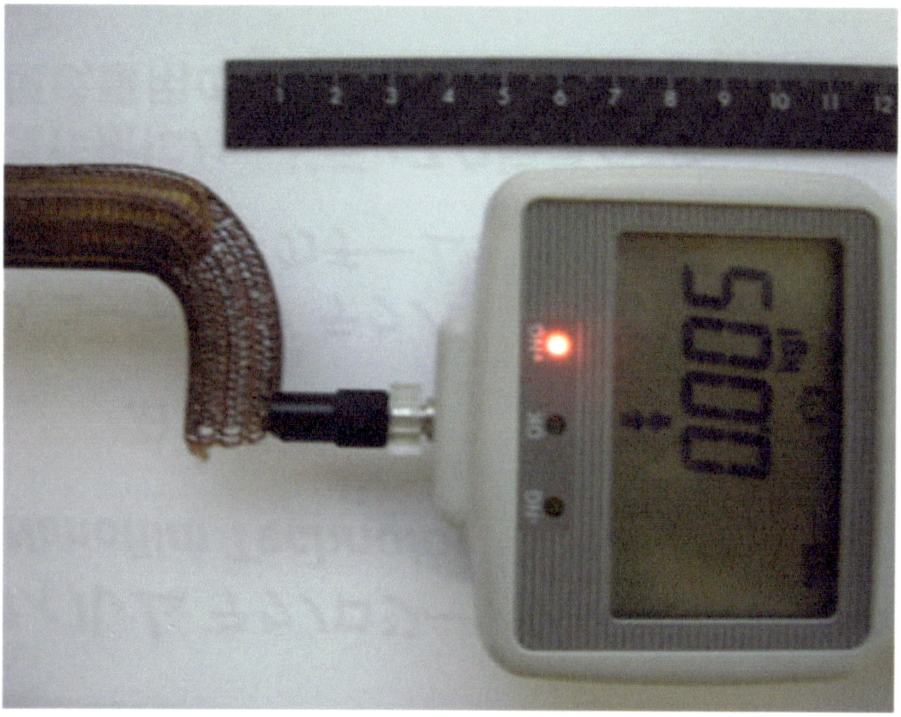

Fig. 9. With a mandrel inserted at one end of each stent, the other end was contacted by a push-pull gauge (IMADA Co., Aichi, Japan) on the surface of the stent, and was given external force until the stents were bent to an angle of 90 degrees. The value of the weight on the gauge was defined and measured as the bending force of the stent

3. Kink resistance

The objects were an N-Z stent, an N-N stent, a Wall stent, and a Spiral Z stent. A mandrel was inserted 20mm deep at either end of each stent. The stents were bent manually via the mandrels on a mat measuring angles. The bending angles were measured on the basis of the angles the two mandrels made, and the outer diameters of the most stenosed part of the kinks were measured. The bending angles of the stents were measured at an angle of 0 through 180 degrees with an interval of 10 degrees.

3.3 Results

1. Radial compressive force

For the N-Z and the N-N stents, the radial expansile forces increased in a virtually linear manner when they were loaded and compressed to 40% of the original diameter (9mm compression). At this point, the radial expansile force measured 6.1N in the N-Z stent and

5.0N in the N-N stent (Fig. 10). In other words, compared with the metallic stent of the equivalent mesh density, the hybrid stent showed greater expansile force.

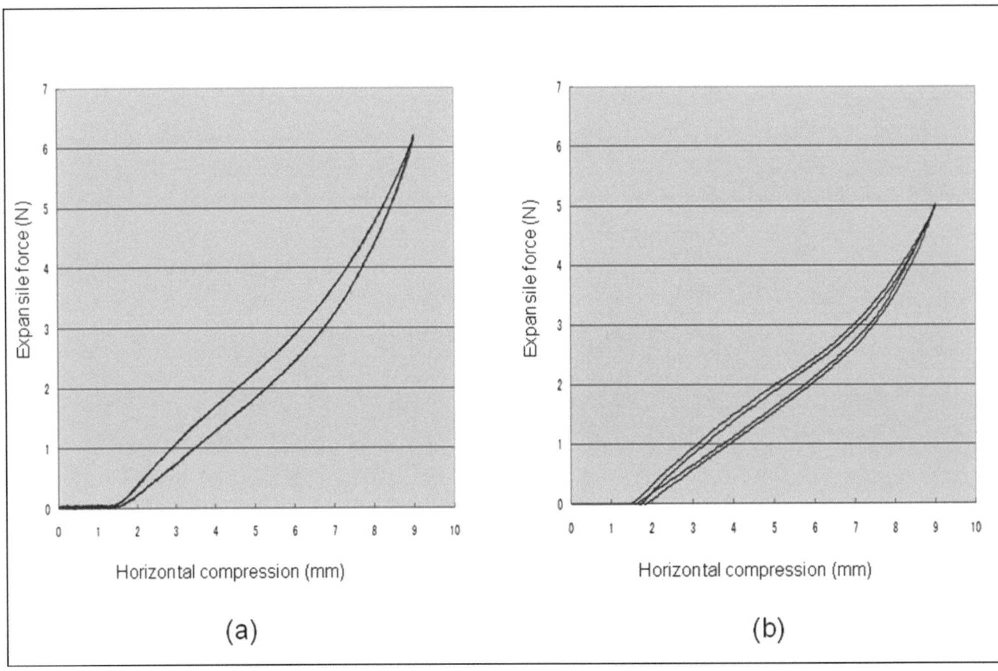

Fig. 10. The relation of the amount of horizontal compression to expansile force in the N-Z stent (a) and N-N stent (b). In the range of 9mm compression (40% of the original diameter), the expansile force increased in proportion to the amount of compression for both stents

2. Bending force

When the stents were bent at the angle of 90 degrees, the bending force measured 0.003N in the N-Z stent, 0.034N in the N-N stent, 0.09N in the Spiral Z stent, 0.11N in the SENDAI stent and 0.03N in the Wall stent.

3. Kink resistance

Kinks emerged at an angle of about 60 degrees in both the N-N and the SENDAI stents. As the bending angle increased the kinks showed further development until the lumen was stenosed almost completely at an angle of 150 degrees or more. A slight kink was formed at a bending angle of 150 degrees or more in the N-Z stent, while the kink remained small even when the stent was bent further. In the Spiral Z stent, the largest kink was observed at an angle of 120 degrees. The Wall stent didn't have any kinks on the central part at a bending angle of 100 degrees or more. Rather it showed a slight increase in the diameter (Fig. 11). When they were bent to an angle of 180 degrees, the lumens of the N-N, the SENDAI and the Spiral Z stents were stenosed almost completely, while the kink on the N-Z stent was not significant (Fig. 12).

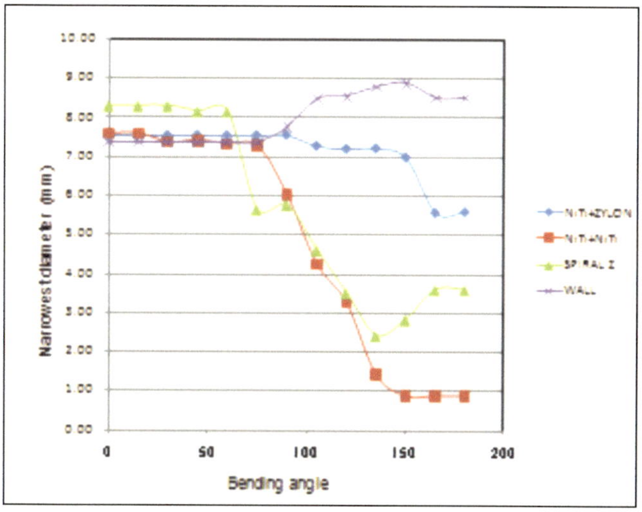

Fig. 11. The relation of the bending angle of the stent to the narrowest diameter of kinks. For spiral Z stent and N-N stent, kinks emerged at an angle of about 60 degrees, abd grew larger as the bending angle increased. On the N-Z stent, however, no apparent kinks emerged until the stent was bent to an angle of 150 degrees. No kinks emerged that would cause stenosis of the lumen as the stent was bent further

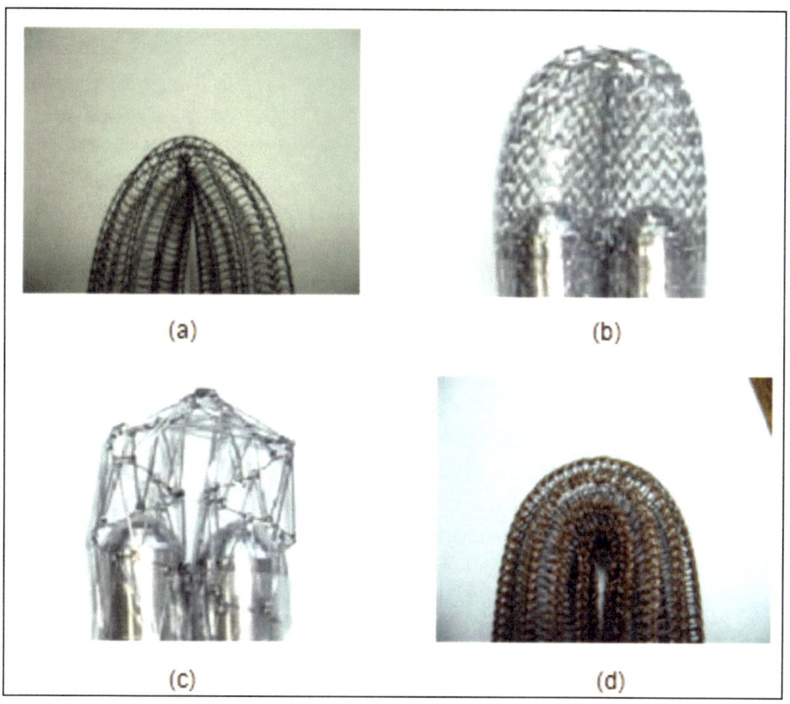

Fig. 12. Occurrence of kinks on bending to an angle of 180 degrees. In the N-N (a), the SENDAI (b), and the Spiral Z stents (c), the lumen was completely stenosed, while the kink was minor in the N-Z stent (d)

3.4 Discussion

Metallic stents can be classified according to their structure into the following types: ones made with a metallic wire knit into a cylinder (e.g., the UltraFlex stent), ones using two or more metallic wires forming a braid (e.g., the Wall stent), hollow tubes molded with a laser beam (e.g., the Cyphor stent) and others. In knitted stents, the loops of the mesh form the hollow cylindrical structure. In the case of stents molded with laser, a regular array of closed and semi-closed polygonal structural units called cells play the role of the loops. The adjacent loops of knitted stents are fixed by confounding, whereas the cells in laser molded stents are connected with linking components. It is one of the necessary conditions for loops or cells to maintain a rigid array if a stent is to yield effective expansile force. At the same time, however, it will also be a cause of restoring force against bending transformation. It is difficult to design a stent of this type that has low restoring force while retaining strong expansile force.

In contrast, if stents receive a strong bending transformation, adjacent loops or cells may come into contact at unexpected parts. Or they may yield an irregular external force through the joints. These may result in transformation and deviation of loops and cells, with part of them protruding inside the lumen of the stents.

In their discussion on the mechanism of emergence of kinks in stents using the Finite Element Method, Mori et al. state that stent meshes or struts contacting or colliding with each other will cause decrease in their flexibility and lead to formation of kinks (Mori et al., 2004).

As a solution to interference of adjacent loops in knitted stents in bending transformation described above, we devised knitted stents whose two wires differed significantly in bending stiffness. Each of the wire intersected regularly at the small loops, forming a structure of two spirals of different materials knit together.

This structure reacts differently to external force from the same structure consisting of one single metal. In bending transformation, the fiber loops, coupled with the metal loops, transform themselves, keeping the transformation of the metal spiral loops to a minimum. This yields little elasticity caused by the restoring force in bending transformation of the metallic spiral component, making the restoring force related to the twisting stiffness of the metallic material the only cause of the restoring force. Also, kinks are less likely to occur as the metallic loops do not come into contact with each other. As for the load in the direction of the diameter, materials with low stiffness function as a spacer to maintain the spiral structure of the material of high stiffness enabling to produce effective radial expansile force. Our approach also prevents the metallic spiral structure from significant transformation or divergence by weight.

It is in this point that our approach differs from loose spiral structures. This is the reason why the composite material stents yield effective expansile force. The expansile force can be adjusted by the bending stiffness of the material of high bending stiffness and the pitch and height of the loops. The stents developed for this experiment had a knitted structure generally in use such as in the Strecker and the Ultraflex stents. However, our hybrid materials approach has made it possible to adjust expansile force, anti-kink characteristics, and axial restoring force independently, which was not possible in metallic or laser molded stents.

In conclusion, a composite material knitted stent, compared with a metallic stent of the same pattern, allows strong radial compressive force while keeping bending force low. This hybrid approach is also superior in the anti-kink characteristics to one using a single metallic material.

4. Future research

In this experiment we demonstrated that the properties of a composite material knitted stent using metallic and non-metallic materials can be manipulated easily. As we have shown, a composite material stent comprising metallic and a bio-absorbing textile can offer advanced removability from the body without losing the structural properties of the stent. It can also be used as a drug eluting stent by filling the bio-absorbing textile with therapeutic medicine such as an anticancer drug.

Most of the drug eluting stents currently in use release therapeutic medicine from polymers covering the surface of a metallic stent. Our hybrid materials stents can contain multiple agents by selecting an appropriate textile.

The combination of a metal and a fiber can also change the structural properties of a stent. Realizing strong expansile force and flexibility at a high level was hard in metallic stents. In the combination of PBO fiber and Ni-Ti wire, our stents realized a stronger expansile force and a greater conformity against bending than metallic stents of the same textile pattern. They also achieved a superior anti-kink characteristics.

The concept of employing multiple materials in a stent is not limited to the knitted stent structure we used in this experiment, but it can be applied to other stent structures as well. Existing stents such as the Spiral Z and the Wall stents can be expected to show improvement in removability and be designed to contain therapeutic medicine by employing multiple materials (Fig. 13).

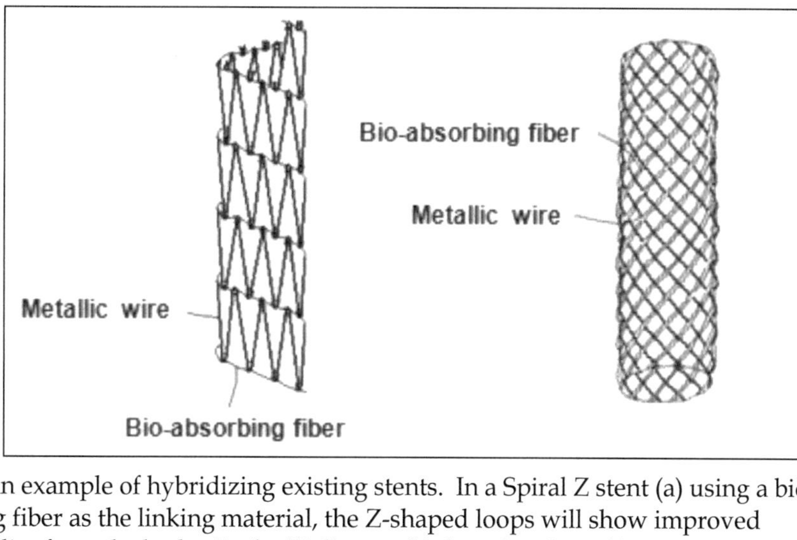

Fig. 13. An example of hybridizing existing stents. In a Spiral Z stent (a) using a bio-absorbing fiber as the linking material, the Z-shaped loops will show improved removability from the body. In the Wall stent (b), bio-absorbing fiber can be woven between the wires. By filling the textile with therapeutic medicine, the medicine can be released as the textile dissolves

5. References

Flueckiger F, Sternthal H, Klein GE, Aschauer M, Szolar D, Kleinhappl G. (1994). Strength, elasticity, and plasticity of expandable metal stents: in vitro studies with three types of stress. *J Vasc Interv Radiol*, 5, 745-750.

Fransen GA, Desgranges P, Laheij RJ, Harris PL, Becquemin JP. (2003). Frequency, predictive factors, and consequences of stent-graft kink following endovascular AAA repair. *J Endovasc Ther,* 10, 913-918.

Freitag L, Eicker R, Reuter S. Theoretical and experimental basis for the development of a dynamic airway stent. (1994). *Eur Respir J,* 7, 2038-2045

Grabow N, Schlun M, Sternberg K. (2005). Mechanical properties of laser cut poly(L-lactide) micro-specimens: implications for stent design, manufacture, and sterilization. *J Biomed Eng,* Feb 127(1), 25

Isotalo T, Talja M, Valimaa T. (2002). A bioabsorbable self-expandable, self-reinforced poly-L-lactic acid urethral stent for recurrent urethral strictures: long-term results. *J Endourol,* Dec, 759-62

Korpala A, Aarnio P, Sariola H. (1999). Bioabsorbable self-reinforced poly-L-lactide, metallic, and silicone stents in the management of experimental tracheal stenosis. *Chest,* 115(2), 490-495

Miura S, Yoshioka T, Furuichi K, Tanaka T, Kichikawa K, Ohishi H. (2003). Mechanical Properties of Biliary Metllic Stents: An Experimental Comparison. *NIPPON ACTA RADIOLOGICA,* 63, 201-209

Mori K, Saito T. (2004). Effects of stent structure on stent flexibility measurements. *Ann Biomed Eng,* 63, 733-742.

Noppen M, Stratakos G, D'Haese J, Meysman M, Vinken W. (2005). Removal of Covered Self-Expandable Metallic Airway Stents in Benign Disorders: Indications, Technique, and Outcomes. *Chest,* 127:482-487

Puma F, Farabi R, Urbani M. (2000). Long-term safety and tolerance of silicone and self-expandable airway stents: an experimental study. *Ann Thorac Surg* , 69, 1030-1034

Schildge J, Hauk H. Successful treatment of benign tracheal stenosis with a silicone stent (Polyflex-Stent) over 5years. (2001). *Pneumologie,* 55, 152-158

Song HY, Shim TS, Kang SG. Tracheobronchial strictures: treatment with a polyurethane-covered retrievable expandable nitinol stent-initial experience. (1999). *Radiology,* 213, 905-912

Zilberman M, Nelson KD, Eberhart RC. Mechanical properties and in vitro degradation of bioabsorbable fibers and expandable fiber-based stents. (2005). *J Biomed Mater Res B Appl Biomater,* 74(2), 792-9

4

Composite Materials for Some Biotribological Systems

Jan R. Dabrowski, Piotr Deptula and Joanna Mystkowska
Bialystok Technical University
Poland

1. Introduction

Complex functions connected mainly with directed motion, power transmission, and material forming, are performed in biotribological systems. This concerns both technical biotribological systems (machines and devices) as well as human biosystems - especially joints, and teeth and parodontium. The basic function of joints is to transmit complex dynamic loadings. Within the area of teeth and parodontium - an element of a stomatognathic system - taking nutrients and preparing them for digestive processes take place. Durability of such biotribological systems is directly reflected in the functioning of the whole human organism. Lesions and mechanical injuries of joints lead to unfavourable changes which often impede normal locomotion in humans. Similarly, in tooth systems, deterioration of hard tooth tissues, most often because of attrition and caries processes, makes normal functioning of stomatognathic systems much difficult. One of more and more popular methods applied in repairing human joints is endoprosthetics, i.e. a surgical treatment consisting in replacing a diseased joint with an artificial apparatus - an endoprosthesis. In the case of defects in hard tooth tissues, the defects are either filled or the damaged tooth is replaced with an implant. Biomaterials used for these purposes, beside having a number of biofunctional features, should have very good tribological characteristics - low friction coefficients and antiwear performance.

Progress in reconstructive surgery of human joints and in conservative dentistry, in particular selecting appropriate biomaterials and constructions, depends to a large extent on getting to know these biotribological systems, which facilitates producing more reliable functional and structural models of the systems, and thus results in developing 0more correct methods of assessing their biotribological characteristics.

Fig. 1. shows a graphical representation of a general functional model of a biotribological system, in which, beside the classical external environment, the environment of an autonomous character is marked together with the feedback of parameters characteristic of the input-output function. It is a proposal of a system functional model of the "black box" type.

It has to be emphasised, however, that transferring achievements of the methodology of biotribological research on technical systems to biological systems has to be preceded by analysing the differences between these two types of systems. Beside selecting basic parameters for making such an analysis, mainly parameters concerning kinematics of motion and loading dynamics, phenomena and their effects occurring in the autonomous

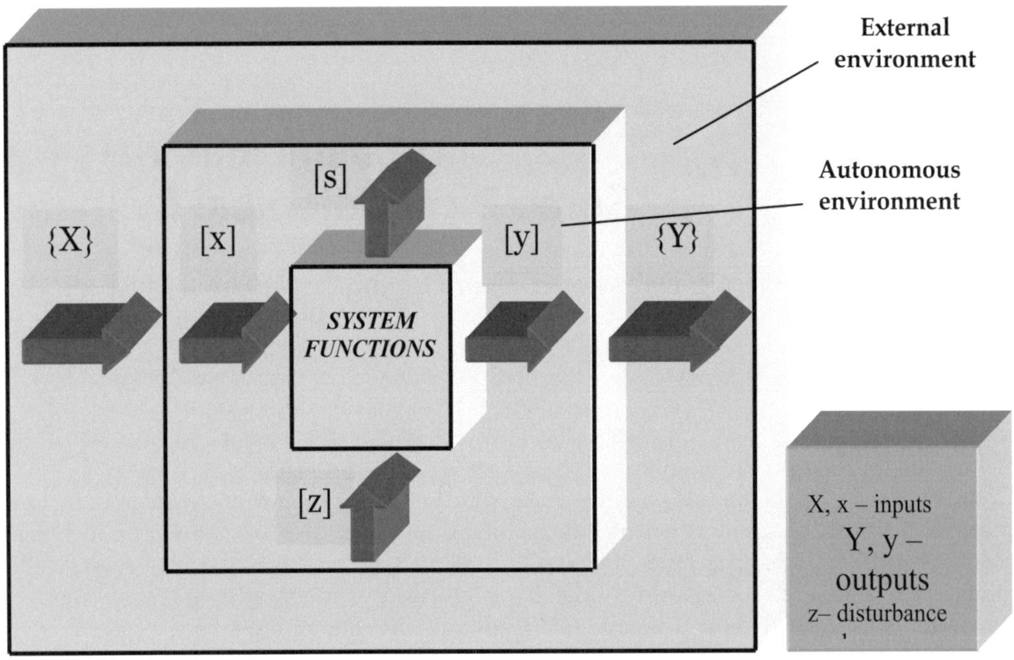

Fig. 1. General functional model of a biotribological system [5]

environment of a particular biotribological system need to be taken into consideration. An autonomous environment is characteristic of the phenomenon of homeostasis (feedback), open to incoming information and closed to energy and matter. For example, chondrocytes play a significant role as centres which regulate exchanging matter and energy within articular cartilage. The reaction of joints in degenerative arthritis, consisting in increasing the joint area and thus decreasing the pressure per unit area, is widely known. Morphological reconstruction of joint areas might also be a reaction of a healthy joint to too much mechanical loading. Reactions of morphological elements of teeth to the influence of external stimuli, characteristic of the process of homeostasis, are quite well-known: processes of reparation, or tertiary dentin forming in case of excessive tooth attrition, can be mentioned as an example.

The complexity and dissimilarity of biotribological systems in comparison to technical ones do not facilitate research experiment planning. Biotribological research and inference cannot therefore be based exclusively on theories and methodologies developed for technical biotribological systems. For that reason, special devices are constructed for biotribological research experiments, i.e. wear simulators which reflect motion kinematics and dynamics of loadings in the analysed system as closely as it is possible. Descriptions of known wear simulators of joints can be found in , and these of teeth in 2 Materials for friction pairs of artificial joints

2. Materials for hip endoprotheses

2.1 Introduction

A hip joint endoprosthesis is composed of a metal stem, fixed in the femoral bone, an acetabular cup (most often a polyethylene or ceramic one, placed in the pelvis), and a head

(metal or ceramic) frictionally cooperating with the acetabular cup. A diagram of an endoprosthesis is presented in Fig. 2. Selecting appropriate materials, based on the concept of Charnley's 'low friction arthoplasty', was a breakthrough in endoprosthetics of human joints. Applying a metallic head (austenitic steel, implant alloys of cobalt) and an acetabular cup made of high-molecular polyethylene (UHMWPE) made it possible to construct commercial endoprostheses which have been most widely used so far. However, neither patients nor doctors find the average time of using endoprostheses, which is 10-15 years, fully satisfactory. Moreover, unfavourable influence of wear products (toxic, immunogenic, mutagenic, etc.) of endoprostheses, osteolysis around implants, and finally, surgical difficulties of re-alloplastics (repeated implantation) are very important reasons for further research in the field of construction material solutions to be applied in human joint endoprostheses. Much hope is placed in this aspect in the development of composite materials for endoprosthetics. A particularly significant research task is connected with applying new titanium alloys - without toxic additions of vanadium or aluminium, and with better tribological characteristics. It has to be underlined that titanium alloys are not used in friction nodes of endoprostheses due to the high susceptibility of these alloys to tribological wear and their frictional resistance.
The subject of the authors' own research was the development of new composite materials based on implant alloys of cobalt and titanium, with special emphasis on shaping the biotribological characteristics of these materials.

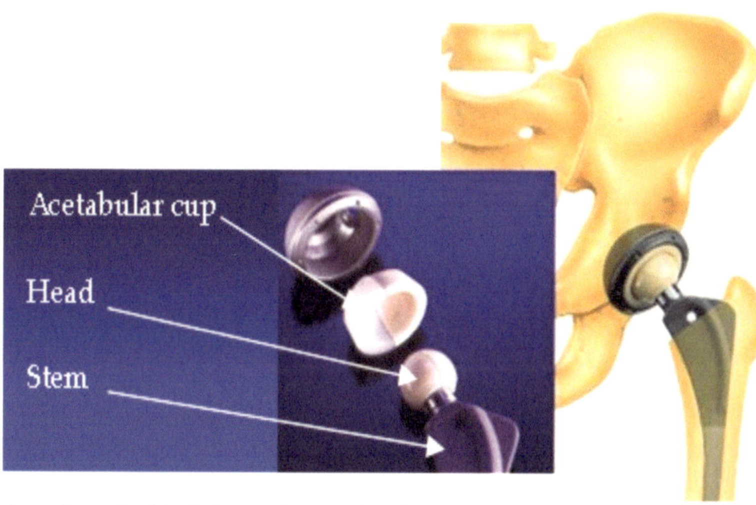

Fig. 2. Construction of a hip joint endoprosthesis

2.2 Investigated materials

The specimens were fabricated by means of the powder metallurgy method. Composite materials based on Co-Cr-Mo powders as well as titanium powders were used for the tribological tests. Co-Cr-Mo powders with 10% weight fraction of $Ca_2P_2O_7$, BN, Si_4N_4 and titanium powder with 10% and 20% volume fraction of graphite (Ti + C) as well as 10% and 20% volume fraction of titanium carbide (Ti + TiC) and 2% volume fraction of yttrium were investigated (Table 1.). The basic powders with friction modifiers were dry-mixed in a Pulverisette 6 planetary ball mill for 15 minutes. Some of the initially mixed powder was milled in a SPEX 8000D high-energy shaker mill in a protective argon atmosphere (mechanical alloying). All the powder mixtures were cold-pressed in a matrix under the pressure of 600 MPa. The mouldings were sintered in a tube furnace in a protective atmosphere (a vacuum of 10^{-5} mbar) for 3 hours at the pipe temperature of 1230°C. The mouldings based on Cr-Cr-Mo powders were sintered for 1 hour. The specimens were cooled down naturally to the ambient temperature inside the furnace.

Material			
Co-Cr-Mo sinters		Ti sinters	
Symbol	*Composition*	*Symbol*	*Composition*
S0	Co-Cr-Mo alloy	**T0**	Pure titanium
S1	sinter	**T1/10**	Ti+10%C
S2	sinter+$Ca_2P_2O_7$	**T1/20**	Ti+20%C
S3	sinter+BN	**T2/10**	Ti+10%TiC
S4	sinter+ Si_3N_4	**T2/20**	Ti+20%TiC
-	-	**T3**	Ti+2%Y_2O_3

Table 1. Chemical composition of the investigated composites based on Co-Cr-Mo and titanium

2.3 Tribological tests

The microstructures of the composite materials were observed under a Hitachi S-3000N scanning electron microscope with an X-ray NSS microanalyser. Maps of the location of the elements, a line analysis of the elements appearance, and a chemical composition analysis were also made by means of scanning microscopy. An X - ray diffraction analysis for the powder samples after different times of mixing, heating in a calorimeter and consolidation were made with a Rigaku MiniFlex II, applying the CuKα (λ = 1.54178 Å) radiation. To assess the average size crystallites and the average amount of net defects, the Williamson-Hall method was used. Hardness was measured with the Brinnel method, while microhardness was measured with the Vickers method with a Neophot-21 ($\mu HV_{0,1}$) optical microscope.

The tribological tests were performed with a special tribometer using a reciprocating ring-on-disc system. Schematic diagrams of the tribometer are presented in Fig. 3. The research was conducted during the periodically variable motion with low velocity and variable values of pressure. The tribological tests were performed with tribometer using reciprocating ring-on-disc system with an angular displacement f = 0.24 and a frequency of 1 Hz. The research was conducted during the periodically variable motion with low velocity and pressure variable values. The ring was loaded along its axis, and the loading was changing within the range of 0 to 350N (the maximum contact pressure p_{max} = 2MPa). The maximum value of sliding speed was v_{pmax} = 0.018m/s.

The friction coefficients were calculated from the maximum values of the friction force to describe extreme resistances to motion. The geometry and dimensions of the tribological pair are shown in Fig. 3. The investigated sintered composites (with 10 and 20% addition of graphite, 10 and 20% addition of titanium carbide, and composites with yttrium) were used as discs, while the counterspecimen in the form of a ring was made of a bulk titanium alloy.

In the case of Co-Cr-Mo sinters, the couterspecimen was a bulk Co-Cr-Mo alloy. The tribological tests for the composites based on Co-Cr-Mo were carried out in a distillation water environment and a CMC environment. The quantitative research was performed by two methods - on the grounds of the roughness measurements and by volume measurements for the sinters based on Co-Cr-Mo powders and by weight measurements for the composites based on titanium. In this paper, a T8000-R60-400 Hommeltester profilograph with the possibility of friction simulation on investigated surfaces was used.

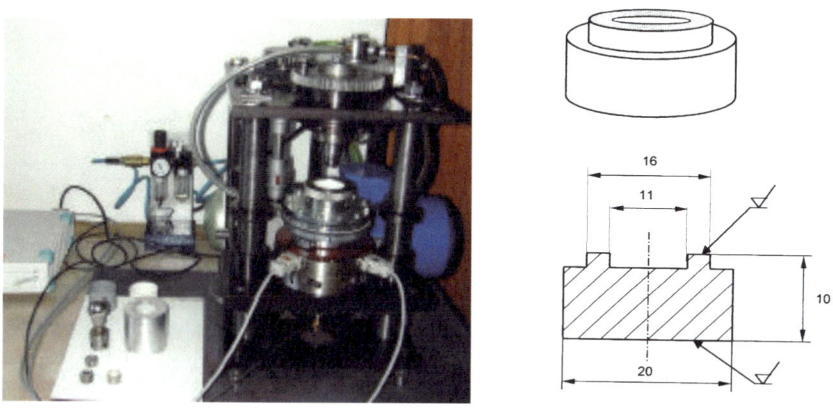

Fig. 3. The tribometer and the geometry of the tribological pair

The device may be used for testing head/acetabular cup or ring/disc kinematic systems. A kinematic pair is placed in a specially constructed container providing for continuous lubrication (forced circulation of the lubricating fluid). The loading was applied as a sinusoidal cycle, with the head (ring) moving in a reversible way, and the specified initial interspace between the friction faces. The possibility to pre-set the size of the initial interspace between the friction faces corresponds to physiological conditions influencing the functioning of the human hip joint. It is also possible to select the point of application of loadings normal to motion coordinates and to regulate the period and amplitude of loadings. The presence of the interspace between the friction faces and its size are some of the most important parameters which influence creating appropriate conditions of lubricating artificial joints.

The prepared samples of porous implantation materials were compressed with an INSTRON 8502 strength testing machine and further the compression curves were analysed.

2.4 Results and discussion

For composites based on Co-Cr-Mo powder, the *Ra* initial surface roughness of the composite samples based on Co-Cr-Mo powder was determined to be of the order of ~ 0.06...0.11 μm. The loading was applied in the first part of the friction cycle (the first half of the sinusoidal loading period), reaching the maximum pressure of 10 MPa. The size of the initial interspace was determined to be of the order of 0.2...0.3mm. Changes in the rubbing speed and in the values of loadings as well as these in the friction force were recorded in a continuous way. After the test was finished, the lubricating fluid was collected, and the contents of chromium and cobalt in the fluid were determined using Atomic Absorption Spectrometry (AAS, Z5000 Hitachi). The friction faces were observed with the use of a scanning microscope (SEM, Hitachi 3000N).

The observed changes in the roughness of these surfaces were assessed using a Talysurf 10 (Taylor-Hobson) profilometer. The volume wear was assessed with a planimeter from the profilograms for the friction surfaces of the plates in the radial direction. R = 7mm was adopted as the (r) mean friction radius. Biotribological characteristics of the following groups of materials obtained on the basis of powder III were determined.

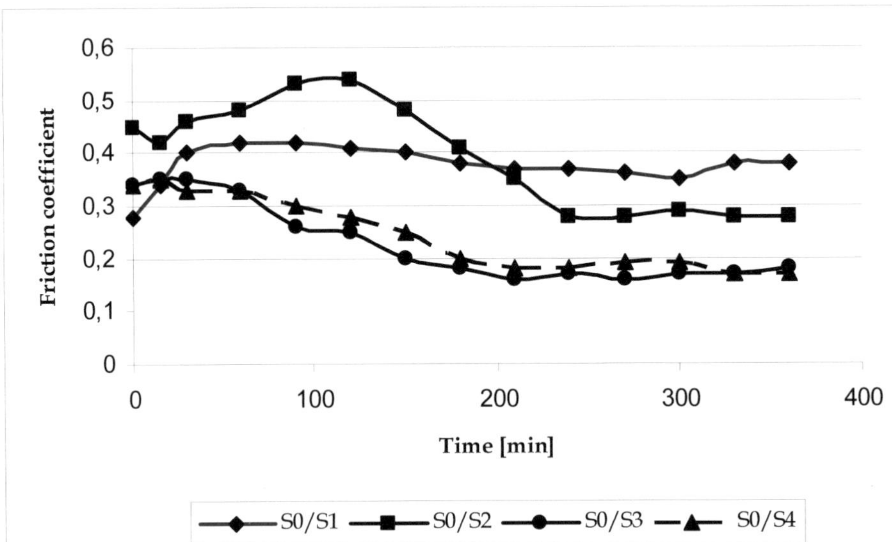

Fig. 4. Changes in friction coefficients of the tested materials in the time function

As it can be seen from Fig. 4., after the running-in period, stabilisation of the friction coefficient takes place in the tested materials. The presented data indicate that the friction coefficients for the composites with the addition of silicon nitride and boron nitride were twice as small as those for the moulding/pure sinter pair. However, in terms of tribological efficiency, the results for calcium pyrophosphate were distinctly worse in comparison to these of the tested friction modifiers.

The tribological wear of the materials under research was assessed on the basis of the cobalt and chromium contents in the lubricating fluid as well as on the basis of profilograms for the friction surfaces of the moulding plates (the volume of wear products). The results of these tests are listed in Table 2.

Friction pairs	Content of elements, [ppm]		Volume wear, x 10^{-3} mm^3
	cobalt	*chromium*	
S0 / S1 (pure sinter)	7.16	2.72	1.29
S0 / S2 ($Ca_2P_2O_7$)	9.51	10.20	2.05
S0 / S3 (BN)	3.68	2.72	1.07
S0 / S4 (Si_3N_4)	1.95	1.99	0.65
lubricant	6.45 x 10^{-3}	5.62 x 10^{-3}	-

Table 2. Results of wear measurements

The data presented in Table 2. are indicative of a favourable influence of boron nitride and silicon nitride on decreasing wear in comparison to the data for pure sinter. The composites containing silicon nitride have the best anti-friction and anti-wear characteristics, whereas the addition of calcium pyrophosphate - contrary to the expectations - unfavourably influenced both the motion resistance and the wear of the tested materials.

Tribological characteristics of the obtained composite materials depend to a large extent on the anti-friction and anti-wear efficiency of the introduced additions, the material cohesion, and the surface morphology. The performed microscopic observations show that composites with the addition of silicon nitride have the most favourable structure of top layers (Fig. 5a). An analysis of photographs of the surface of this composite indicates that the top layer material is characterised by good cohesion. The surface is smoothed, with small irregular pores and clearly visible light areas containing silicon nitride. Photographs of friction surfaces of composites with calcium pyrophosphate and boron nitride are quite dissimilar. Observations of the surfaces of sinters with the addition of calcium pyrophosphate (Fig. 5b) show that this addition unfavourably influences the processes of sintering cobalt powder particles, and thus the cohesion of the obtained composites. In photographs of surfaces of such composites, the presence of additions surrounded by non-sintered particles of cobalt alloy powder is clearly visible. This decreases the material cohesion and thus increases the possibility of metallic particles to be crumbled out from the surface during friction. A loss of calcium pyrophosphate from surface pores in the friction process is also observed.

It is most probably caused by a low internal cohesion of this phase and its elution by the lubricating fluid. Similar surface structures were observed in boron nitride composites, with characteristic numerous discontinuities occurring at the phase border between the metal particles and the additive, which is indicative of a low interphase cohesion of these materials.

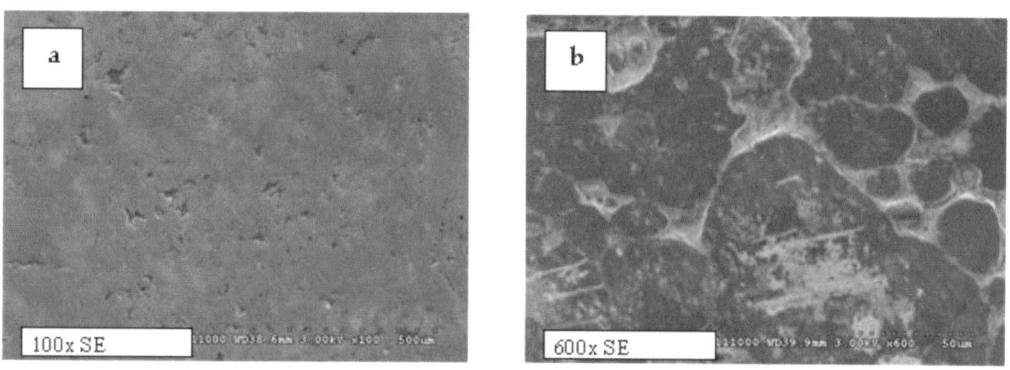

Fig. 5. Microstructure of friction trails: a) S4 (Si_3N_4), b) S2 ($Ca_2P_2O_7$)

The results of the performed tests reveal that the composites containing silicon nitride have the most favourable tribological characteristics and therefore silicon nitride seems to be an effective anti-friction and anti-wear additive to porous materials based on a Co-Cr-Mo alloy. The idea of using such composites to construct friction nodes of endoprostheses might be supported by a wide application of materials of this type in tribotechnics and by promising literature data concerning applying silicon nitride composites or even its sinters in friction nodes of endoprostheses.

For the titanium composites, the first experimental part included structural tests on the alloys manufactured with traditional powder metallurgy methods without alloying powder mixtures. The green specimens of composites with graphite, titanium carbide, and the addition of Y_2O_3 before sintering were characterized by good compactibility. The relative density of the received Ti+C molder was about 83%, which could be explained by a good lubricating effect of graphite during the compaction process. The material with the addition of graphite of 93% showed the largest compactibility after sintering. This indicates the possibility of a thermally activated reaction taking place between the components during sintering. The relative density of other composites was minor in each case (Fig. 7).

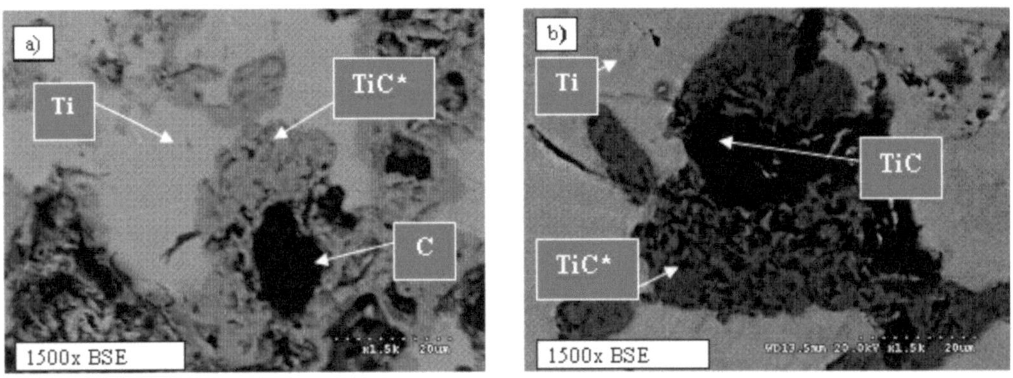

Fig. 6. Microstructures of composites: Ti + graphite sintered in 1230°C, b) Ti + titanium carbide sintered in 1230°C

The microstructures of the composite materials with titanium modifiers confirmed this fact. As a result of diffusion, a new phase titanium carbide appeared on the border between the titanium grains and graphite. The obtained materials were composed of a soft titanium warp, graphite or titanium carbide residuals, and a secondary hard phase of carbides - TiC* (Fig. 6). It may be supposed that such a material would have very good tribological properties.

Fig. 6. shows that interactions between the metallic basis and modifier (graphite and titanium carbide) during sintering were observed.

At the second stage, samples of sinters after mechanical alloying were investigated. The titanium and graphite powder mixtures underwent the mechanical alloying processes. The mechanical alloying favourably influenced the solubility of carbon in titanium and the creation of a large amount of secondary carbides. The obtained powders underwent consolidation through the processes of pressing and sintering. In the SEM image of the samples of the obtained sinters, sintered titanium powder grains with dissolved carbon as well as precipitations of titanium carbides on the grain borders are clearly visible. These results were confirmed by rentgenostructural tests. Composites after mechanical alloying were characterised by a lesser compaction. This results from the fact that friction between nanoparticles was much bigger during compaction, which causes less compaction in the process of pressing. As a result of the mechanical alloying, a significant fragmentation of the powder structure occurred. The X - ray crystallography and microscopy (TEM) studies show that the powder grain size was 20 - 120nm.

A good correlation was found between the structural analysis and the hardness of the investigated composites. The results of macrohardness measured with the Brinnel method, as well as the average microhardness of the material, showed a significant influence of the sintering temperature on the composite properties.

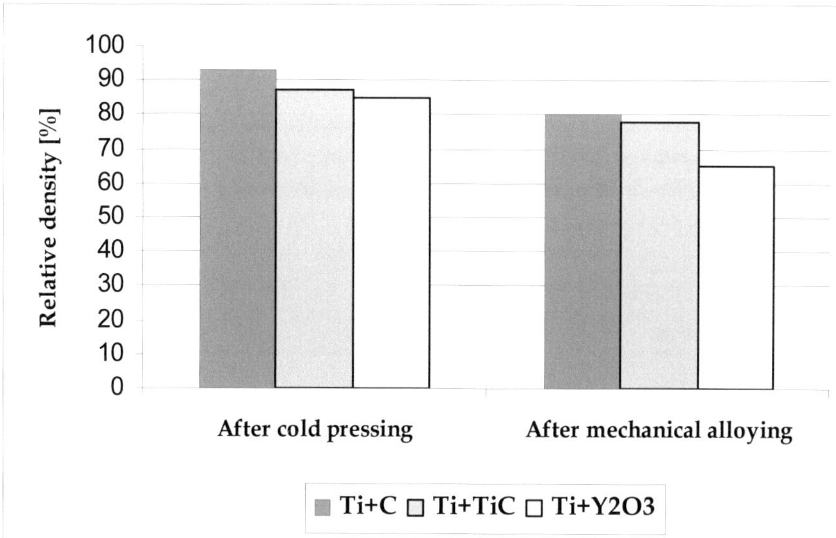

Fig. 7. Results of relative density measurements for investigated composites

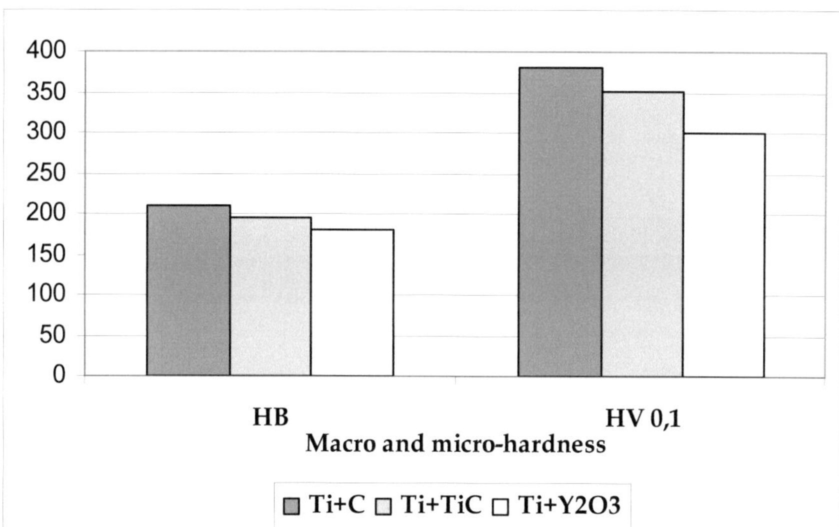

Fig. 8. Results of macro- and microhardness measurements of specimens

The macrohardness of the titanium-graphite composites sintered under the temperature of 950°C was approximately 90HB, whereas the hardness of pure bulk titanium amounted to 83,4HB. The macrohardness of the titanium-titanium carbide composites sintered under the temperature of 950°C was approximatly 85HB. The hardness value of the composites with graphite sintered under 1230°C was similar to the value for high-carbon normalized steel (Fig. 8).

The macro-hardness micro-hardness test in the neighbourhood of graphite confirms that it indeed is a hard phase. The obtained material was composed of a soft titanium warp, graphite residuals and a secondary hard phase of carbides (Fig. 6). It may be supposed that such a material would have very good tribological properties.
The results of the microhardness tests on the obtained sinters, listed in Fig. 9., and Table 3., indicate that mechanical alloying favourably influenced the increase in the microhardness of the produced materials. The composite materials with the addition of graphite were characterised by the greatest hardness. It has to be emphasised that the microhardness of such composites, after the mechanical alloying process, increased to very high values of the range of 900 HV0.1.

Microhardness HV 0.1	**Ti+C**	**Ti+TiC**	**Ti+Y_2O_3**
Powder metallurgy	380	352	300
Mechanical alloying	820	-	450

Table 3. Results of microhardness measurements of specimens after P.M. and mechanical alloying

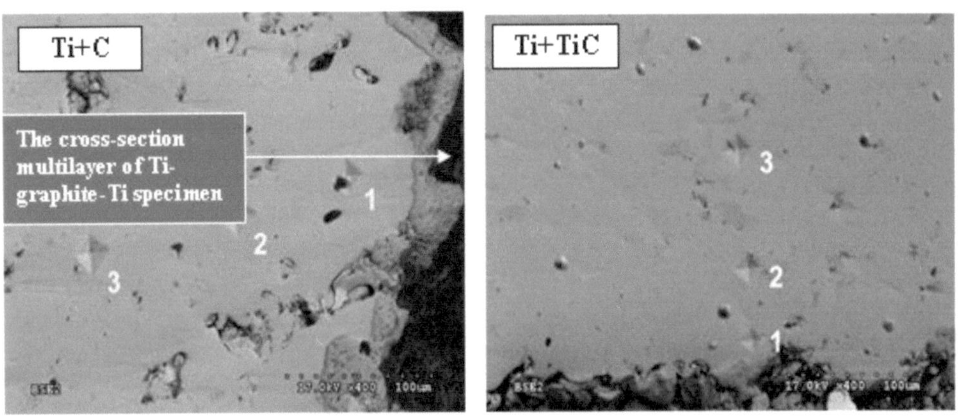

Sample	**Temperature of sintering [°C]**	**Point №**		
		1	**2**	**3**
Ti+C	1230	376	299	270
Ti+TiC		835	665	530

Fig. 9. Influence of carbon diffusion on microhardness of titanium samples

A comparison of the friction coefficient values for the investigated materials is shown in Fig. 10. The analysis of the obtained results confirmed a good influence of graphite on tribological properties of the sinters.
The friction coefficients for the composites with graphite reduced over twice compared with friction of pure titanium samples. The values of the motion resistance are also lower for the composites with titanium carbide, although to a lesser degree than in the case of the composites with graphite. In two cases, the influence of modifiers is clearly visible – an

increased volume of the fillers caused a decrease of the friction coefficients. The quantitative research was performed on the basis of the roughness measurements and by weight measurements.

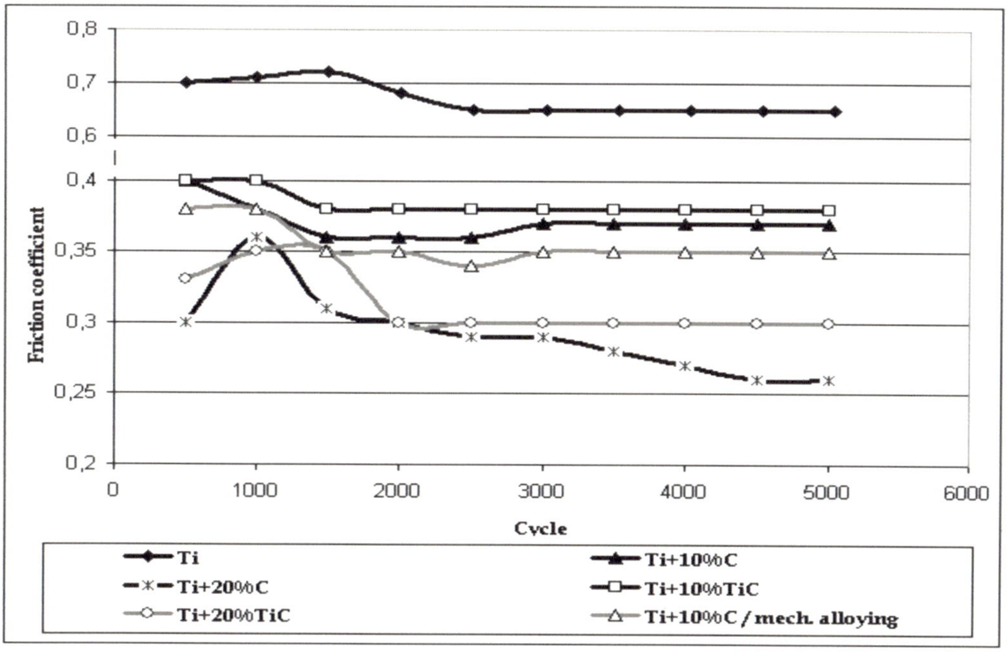

Fig. 10. Influence of graphite and titanium carbide on composite tribological properties of investigated composites

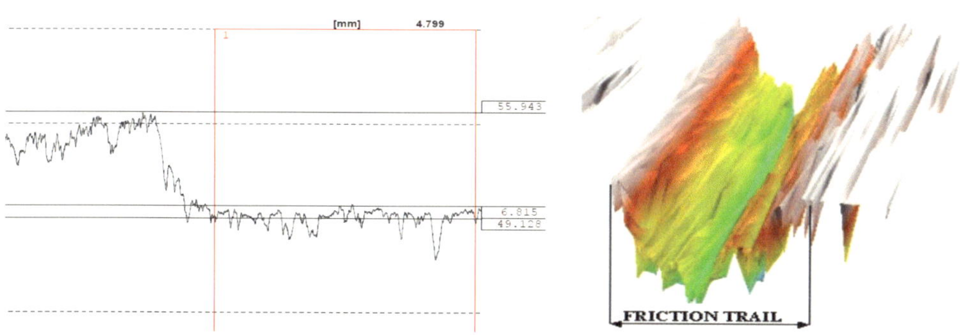

a) b)

Fig. 11. Friction trail of Ti+20%TiC composite: a) roughness measurement, b) wear simulation

Examples of profilograms and wear simulation of the friction surface of Ti+20%TiC are presented in Fig. 11., whereas Table 4. shows the wear of the investigated materials. Significant decreases in the wear for the samples with graphite, and increases in the wear for the samples with titanium graphite in comparison to pure titanium sinters can be observed. For the composite with 20% volume fraction of graphite, the friction coefficient after 1h of tribological testing was at the level of 0.27, which confirmed good lubrication conditions during friction. The wear for this composite, measured after the tribological test was of the range of about 9.8 g $\cdot 10^{-3}$ (Table 4). This material had the best tribological properties.

Sample	Wear loss	
	Linear wear [µm]	*Weight wear *10^{-3} [g]*
Pure titanium	30.4	21.5
Ti + 10% C	16.6	11.5
Ti + 20% C	9.8	7.2
Ti + 10% TiC	33.9	23.7
Ti + 20% TiC	49.1	35.8

Table 4. Results of wear tests

A significant influence of the volume of the fillers on the wear volume is visible. In the case of the Ti+C composites, the wear decreases, while in the case of the Ti+TiC composites it increases. An advantageous influence of the graphite addition on decreasing the motion resistance and wear of the obtained composites should be underlined, which will have a significant meaning in technical aspects.

The less advantageous tribological characteristics of Ti+TiC composites results from lower cohesion of this type of materials and demolition of hard TiC particles, which is connected with this. This demolition might cause intensifying of secondary wear processes, which was confirmed by the observation of friction surfaces of the samples after friction.

The abrasive and fatigue wear is dominating with the visible deformation areas and spalling of the surface layer in the friction surfaces. In the case of the Ti+TiC sample, small particles of filler loosely connected with the surface, can be seen. This might cause intensification of wear processes.

The obtained results are indicative of a better compatibility and more advantageous mechanical properties of the Ti+C composites. The presence of reactive phases between the titanium and graphite particles has an advantageous influence on the cohesion and properties of this type of composites.

The performed tribological studies also indicate a favourable effect of mechanical alloying on the tribological properties of the alloy. The sinters made of the powders after mechanical alloying were characterized by a lower coefficient of friction (Fig. 10) and higher tribological wear resistance. It is an important feature of this kind of implant materials as regards capabilities of applying them in friction nodes of biomedical devices.

3. Materials for dental fillings

3.1 Introduction

Amalgams, glass-ionomers, compomers and composites are materials for dental fillings. Composites stand out against the group and are often applied on teeth mainly due to their good mechanical and aesthetic properties. Dental composites are built of both organic (i.e. a system of monomers, polymerization initiators, stabilizers, pigments, accelerators etc.) as well as inorganic components such as powder and fiber fillers. A silane coupling agent connects these two phases and plays a key role in enhancing the adhesion of the interface between the inorganic powder and organic polymer. On the other hand, silane, due to its low reactivity, brings about a decrease in material polymerization shrinkage. The constitution of the organic matrix and fillers also plays an important part in defining final properties of materials. It is generally accepted that the properties of composites are mainly dependent on the type, size, spacing and volume fraction of the utility filler. The latest literature available on the subject points to a significant influence of fluoride sources, nanoparticles and friction additives on the structure and utilization properties of composites. It has been found that reducing the particle size to a nanoscale level has already reached a significant degree of efficiency. Among dental composites, nanosilica appears to have become the most commonly used nanopowder.

The wear of dental materials is one of the major problems concerning almost every kind of dental filling. As far as dental materials are concerned, there are a few main types of wear processes: abrasive, adhesive and fatigue. Generally, the wear of a filling material should be comparable with the wear of human enamel. The ultimate goal of advanced dental material studies is to produce a material that can be used in all circumstances as an amalgam replacement material which revealed an optimal wear resistance. The best way of developing such a material is to combine it with composite materials for dental fillings. In order to reduce the wear and coefficient frictions of the analyzed materials, the composition of the applied fillers is modified. The addition of hard corundum particles, aluminium and glass fibres have a positive effect on mechanical properties only. An evident improvement of the tribological properties of a material can be obtained by using a suitable filler, called friction modifier. In the case of inorganic fillers, the wear of the material is probably connected with the sum of damage caused to the silane coupling agent. The composites with the highest degree of coupling show the highest wear rate. Recently, ceramic whiskers were used to reinforce dental resin composites. Xu and others demonstrate that silica-fused whisker reinforcement produced dental resin composites that exhibited high resistance to wear with smooth wear surfaces. Generally, inorganic particles are well known to enhance the mechanical properties of polymers, which have been widely investigated in the past decades. Different organic fillers are also applied. For example, the PTFE is applied when a reduction of the adhesion of the investigated material to the counterpart material is needed. Different methods are available for the quantification of wear, using different loads, times of investigations and sizes of samples. The type of the used wear method has a distinct effect on receiving results. The pin-on-disc is the most frequently applied tester system. Various lubricants for in vitro wear testing are also used.

Among many desirable requirements of materials for dental fillings, fluoride release is also of a significant quality. Fluoride ions play a key role in dental and enamel remineralisation. Many cases of studies show that fluorine can also be an inhibitor of a cariogenic effect of microorganisms. It decreases the risk of caries that forms in tissues directly adjacent to the filling material. Numerous investigations are made with the aim to obtain a material which would perform those tasks and would have high mechanical, physicochemical, and tribological characteristics at the same time. The best possibility for developing such a material is combining it with composite materials for dental fillings. The quickness of fluoride release depends on many factors, e.g. the kind and volume of the used fluoride source and the environment into which it is released. Fluoride ions are introduced into dental composite materials in the form of fluoridated glass or pure compounds, such as CaF2, NaF, YbF3, YF3, SrF2. However, due to an innovative approach to this problem, there is little literature data concerning the influence of dynamical loading on fluoride release kinetics. The fluoride release may change the material structure. Therefore, the tribological, mechanical, and physicochemical properties as well as analyses of material surfaces are described in literature.

The investigations of mechanical, tribological, and physicochemical properties of some originally made composite materials for dental fillings were performed at the Department of Materials and Biomedical Engineering (Faculty of Mechanical Engineering, Bialystok University of Technology, Poland. The aim of these studies was to compare the properties of composite materials for dental fillings consisting of organic fillers of polyethylene (PE) and polytetrafluoroethylene (PTFE) with those obtained with inorganic friction modifiers of silicon nitride ($Si3N4$) and boron nitride (BN). In the next stage, the influence of the used loading sample on fluoride release, the surface roughness and the material structure was examined.

3.2 Tribological characteristics

At this stage of the tests, five ceramic-polymer microfilled composites containing friction modifiers were tribologically tested. The sample without a friction modifier was a standard one. Each composite consisted of a polymer matrix (40 vol. %) in which the mixture of the following organic resins was used: Bis-GMA resin {2,2-bis[p-(2'- hydroxy-3'-methacryloxypropoxy)-phenyl] propane}, TEGDMA {tri(ethylene glycol) dimethacrylate}, DEA-EMA {2-(diethylamino) ethyl methacrylate}, BHT, CQ {D,L-Camphorquinone} and a photoinitatior. All the components were provided by the Sigma-Aldrich Company. The remaining part of the composite (60 vol. %) was a mixture of inorganic powders: fluorine source and friction modifiers. The fluoridated glass (J-20 symbol) consists of: SiO2- P2O5-Al2O3-BaO-SrO-Na2O-F^-. The glass was worked out and prepared at the Institute of Glass, Ceramics, Refractory and Construction Materials (Warsaw, POLAND).

Fig. 12. SEM pictures of particle structure of used fillers: (a) PE, (b) Si3N4

The friction modifiers including Si3N4, BN, PE, and PTFE were also delivered from the Sigma-Aldrich Company. Each of the listed powder fillers is distinguished by various forms and sizes of the fillers. PE, PTFE and BN have irregular grains of different particle dimensions. Silicon nitride whiskers were used with mean diameters of 0.5 µm and mean lengths of 7 µm. SEM pictures of some selected modifiers are presented in Fig. 12.

The particle sizes of all the used powder fillers are presented in Table 5.

The surface treatment of all inorganic fillers was treated with functional silane (3-(trimethoxysilyl)propyl methacrylate (the Sigma-Aldrich Company). The procedure involved dissolving this liquid in toluene. The aim of the silanization process is to absorb the active silane groups on the powder surface in a vacuum evaporator. After the silanization process, reactive silane is combined with an inorganic filler and can co-polymerize with the polymer network.

Filler	**J-20 glass**	**PE**	**PTFE**	**BN**	**Si_3N_4**
Particle size [mm]	1-3	5	39	5	*Dimension: 0.5* *Length: 7*

Table 5. Particle dimensions of fillers

Composites were prepared as shown in Table 6. All the organic constituents were weighed to a porcelain mortar and precisely mixed until uniform polymer paste was obtained. The particles of J-20 glass (57 vol. %) and friction modifiers (3 vol. %) were added to an organic matrix, which had been prepared in advance. Composite A contained an organic matrix and the J-20 filler particles only. The other materials contained also organic or inorganic friction modifiers (3 vol. %) beside containing fluoridated fillers. These organic-inorganic mixtures were then homogenized in a porcelain mortar for 10 minutes. The paste was later placed in cylindrical aluminium moulds (diameter: 3mm; length: 2mm), and it was cured for 40 seconds for each 2-millimetre layer using a Vivadent light-curing system at a wavelength of 420 nm. In order to uncover the analyzed material, the 2-millimetre aluminium envelope was removed. The restored surface was finished by wet grinding with an 800-grit silicon carbide paper. Six specimens were prepared for each composite material.

Material symbol	**Component content, [vol. %]**		
	Organic matrix	*J-20 glass*	*Friction modifier (3%)*
A	40	60	-
B	40	57	PE
C	40	57	PTFE
D	40	57	Si_3N_4
E	40	57	BN

Table 6. Composition of the materials

The influence of the load and the filler contents on the friction coefficients and wear was estimated. The wear tests were conducted using a pneumatic control pin-on-disk tribotester, which is described in detail elsewhere. The wear machine used in this study was equipped with a spring-sine-cam system allowing cyclic loading of specimens. The samples were the composites presented in Table 6. The samples had average surface areas of 3mm^2 with the Ra roughness of about 0.1μm. A hardening stainless-steel plate with the hardness of 64 HRC was used as the counterface material due to its relatively high hardness as compared with the hardness of the analyzed materials. The average Ra surface roughness of the counterpart amounted to about 0.1μm. The normal loads were: 1, 5, and 10 MPa. That wear cycle was estimated at the frequency of 1.5 Hz. The wear time was 3 hours, and a sliding wear track of 2.5mm was applied. Prior to the wear tests, the samples and the counterface had been cleaned in distilled water. The wear tests were conducted by means of a special tribotester in the presence of a phosphate buffer (pH of natural saliva - 6.8) as a lubricant. The wear was estimated using the gravimetric method. All the measurements were made after the samples

had been dried in an exsiccator for 24 hours. The mass wear was determined with digital scales to the accuracy of 0.01mg.
The measurements of the friction coefficients using 1 MPa [presented in Fig. 13a] and 10 MPa [presented in Fig. 13b] show that they depend on the kind of filler and loading on the sample.

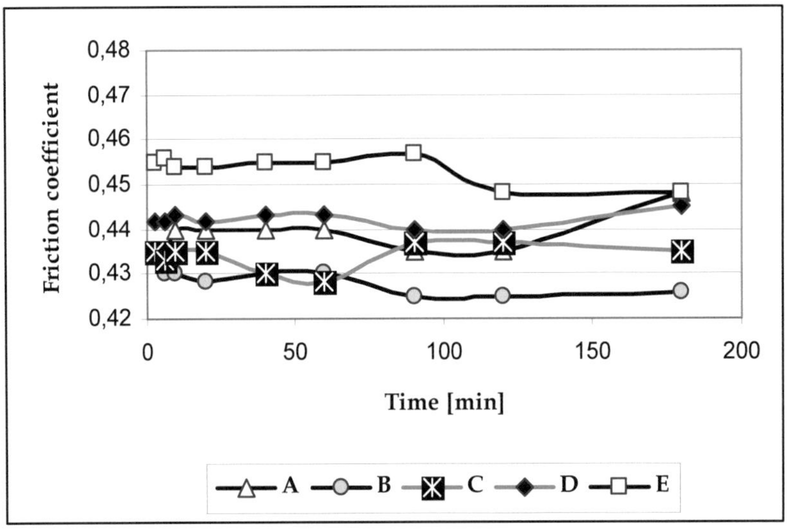

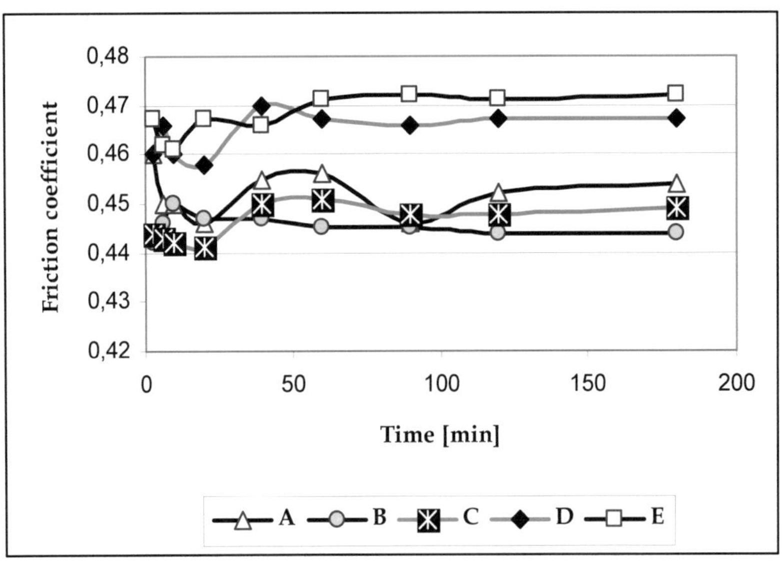

Fig. 13. Influence of time on friction coefficients of investigated materials: a) p=1MPa, b) p=10MPa

The figure illustrates the wear results of the five analyzed composite materials. The first stage of the work consisted in determining the influence of the fluoridated glass filler on the tribological properties. The friction coefficient of this composite material was about 0.45. In the next stage, the composites with the fluoridated glass filler (J-20) and friction modifiers listed earlier were investigated. At the beginning of the friction processes, the friction coefficient magnitudes ranged from 0.42 to 0.46 for the composites to which the loading of 1 MPa was applied. In the case of the 10 MPa loading, the coefficients ranged from 0.44 to 0.47. This reveals that the friction coefficient increases with the increase of the load. The results show that the friction coefficients are strongly dependent on the kind of the used powder filler. The lowest friction coefficients were observed for the composites with organic friction modifiers (PE and PTFE) , while the biggest friction coefficients were observed for inorganic fillers. The addition of 3 vol.% BN or 3 vol.% Si3N4 increases the composite friction coefficients by about 0.2, while the addition of 3 vol.% PE or PTFE decreases the analyzed parameter. This advantageous effect can be explained by a better adhesion of organic fillers to the organic matrix. The structure of these materials is more enhanced. It is probably a result of these filler grain forms and bonding forces between the fillers and the organic matrix. The improvement of the adhesion of the inorganic glass to the organic resin is achieved through the silanization process. The preparation process applying active silane groups on the powder surface was a process in which active silane groups on the powder surface were drifted. According to some literature sources, such a process does not always improve the mechanical properties of investigated composites. Luo, who analyzed the nanoporous filler structure at inorganic-organic composites, says that application of a silane coupling agent results in a negative effect on the composite wear resistance.

The weight wear of the materials for dental fillings presented in Fig. 14 is the lowest in the case of the composites with organic friction modifiers. The results are correlated with earlier inferences that organic fillers are more firmly connected with the both fluoridated glass filler and the organic polymer matrix. It can be a result of stronger bonding between the polymer basis and organic fillers, due to their original chemical structure. Thus, these particles have some difficulty to spall from the composite material. In addition, the adhesion to the organic matrix was greater. Beside that, the improvement of the wear behaviour of polymeric materials is achieved through applying organic fillers, especially polytetrafluoroethylene, which reduces the adhesion of the material to the counterpart. Inorganic friction modifiers after the silanization process are also obliged to connect with the polymer basis in a chemical way. However, as the results show, such bonding does not suffice and filler particles were chipped during the friction tests. Apparently, there were certain areas in the composites where the bonding of both phases was heterogeneous. These uninterpreted areas are of a brittle nature of unreinforced gel and allow for a greater in-depth crack propagation. The form of the filler particles also influences that process. Therefore, the grinding organic matrix uncovered the filler grains and, as a result of natural abrasive wear, the particles were pulled out from the composite. Additionally, the filler particles in the formulas are significantly harder than the matrix, leading to preferential loss during friction. As it was earlier described, the Si3N4 particles have elongated grain structures, so the materials with the addition of it were the most destructible. Insufficient adhesion of inorganic fillers to organic polymer was probably the other reason for material destruction. The results are different from the inferences of Xu et al., where relatively high wear resistance of the whisker composites were obtained. The results of the wear weight of all the composites suggest that the two-body wear behaviour of these materials depends on the load applied during the test. The wear weight increased by one order of magnitude when the loading increased from 1 MPa to 10 MPa. The wear of the investigated materials increased with the applied loading.

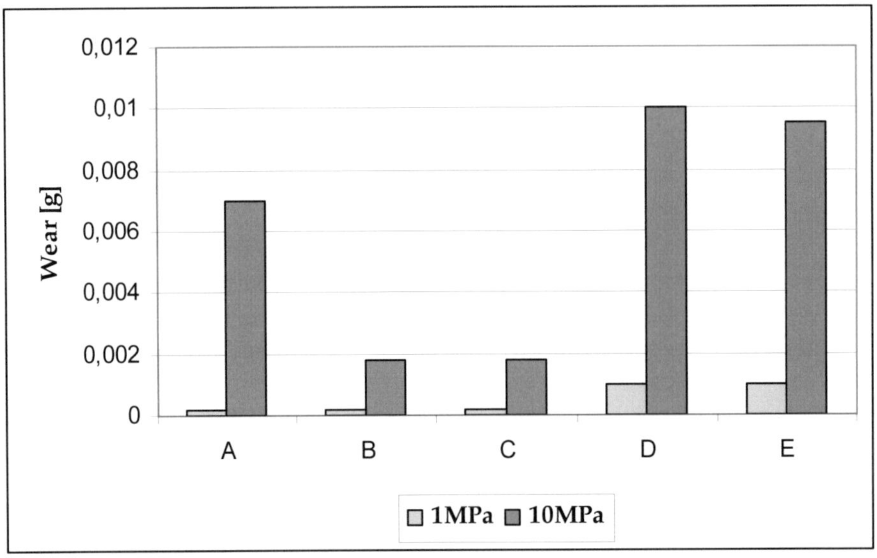

Fig. 14. Wear of investigated materials at loading of 1 and 10MPa

The tribological characteristics of Bis-GMA resin-based composite materials with the addition of fluoridated glass and friction modifiers (PE, PTFE, BN, Si3N4) were investigated. On the basis of the performed examinations, the following conclusions were formulated:

- the wear and friction coefficients of the analyzed materials depend on the filler type and the value of the loading applied to the sample,
- the composites basing on organic fillers as modifiers have the lowest friction coefficients,
- an increase of the loading induces intensification in the wear and friction coefficients.

3.3 Fluoride ions release

At this stage of work, four ceramic-polymer composites containing fluorine sources and nanosilica were tested. All the materials used in this study consisted of 40 vol. % of light-cured organic matrix whose composition was described earlier. The other part of the composite (60 vol. %) was a mixture of inorganic powders: fluorine sources (fluoridated glass J-20, ytterbium fluoride YbF_3) and nanosilica (n-SiO_2). The compound of YbF3 has radiopaque properties additionally. The powders of YbF_3 and nanosilica were delivered from the Sigma-Aldrich Company. Ytterbium fluoride has irregular particles whose mean dimension values were 1µm. The average particle size of nanosilica was 10nm.

The composition of the materials marked with A-D symbols is presented in Table 7. Three specimens were prepared from each composite material for each kind of study. Composite A contains an organic matrix and the J-20 filler powder only. The other materials, beside containing the fluoridated glass, contain also YbF3 or/and nanosilica in the amount of 10 vol. %. All the organic constituents were weighed to a porcelain mortar and precisely mixed until uniform polymer paste was obtained. The J-20 glass particles and other fillers were added to an organic matrix prepared in advance. These organic-inorganic mixtures were then homogenized in the porcelain mortar for 10 minutes. The paste was then placed on PTFE plates (diameter: 11mm; length: 1mm), and it was cured for 40 seconds using a Vivadent light-curing system at a wavelength of 420 nm. The restored surface was finished by wet grinding with an 800-grit silicon carbide paper.

Materials' symbol	**Component content, [vol. %]**			
	Organic matrix	*J-20 glass*	*YbF3*	*n-SiO*$_2$
A	40	60	-	-
B	40	50	10	-
C	40	40	10	10
D	40	-	10	-

Table 7. Composition of the materials

For the purpose of determining the content of fluorine released from the examined materials, the method of direct potentiometry with combined fluoride ions selective electrode was used. The prepared materials were placed in 10 ml of a buffer solution of pH of 6.8 in polyethylene containers and the temperature was maintained at 37^0 C. The tests were performed after 1, 4, 7, 14, 30, and 60 days of keeping the samples in the agent solution. TISAB buffer was added to the analyzed solutions before the examination in order to stabilize the pH and eliminate the influence of foreign ions during the examination. Three specimens were prepared of every composite material for each measurement.

The influence of the load and filler content on the fluoride ions release process was estimated. The wear tests were conducted using a pneumatic control pin-on-disk tribotester, which is described in detail elsewhere. The wear machine used in this study was equipped with a spring-sine-cam system allowing cyclic loading of specimens. The samples were the composites presented in Table 7. The samples had average surface areas of 3 mm^2 with the Ra roughness of about 0.1μm. A hardening stainless-steel cylinder with the hardness of 64 HRC was used as the counterface material due to its relatively high hardness as compared with the hardness of the analyzed materials. The average Ra surface roughness of the counterpart amounted to about 0.1μm. The normal loads were: 10, 15, and 20 MPa. That wear cycle was estimated at the frequency of 1.5 Hz. The wear time was 5 hours, and the sliding wear track of 2.5mm was used. Prior to the wear tests, the samples and the counterfaces had been cleaned in distilled water. The friction process was conducted by means of a special tribotester in a phosphate buffer environment with pH=6.8 corresponding to the pH of saliva. Examinations of fluorine release were performed for A, B, C, D materials. The results of the surface roughness research of the specimens were directly measured with a Talysurf 10 profilometry device manufactured by Taylor Hobson. In order to obtain the structure differences before and after fluoride release, studies of the surfaces were performed using a Hitachi S 3000N scanning electron microscope with an attachment for X-ray microanalysis.

The first stage of the work consisted in determining the influence of the various fluoride sources on the amount of fluoride release to the buffer solution. The results of fluoride release from the investigated materials are presented in Fig.15. The results were presented as the fluoride amount released from 1 mm^2 of the composite material as a function of time. All the tested materials showed accumulative increase in the amount of fluoride in the solutions. It showed, that the highest amounts of the fluoride ions emission were observed for materials A and B.

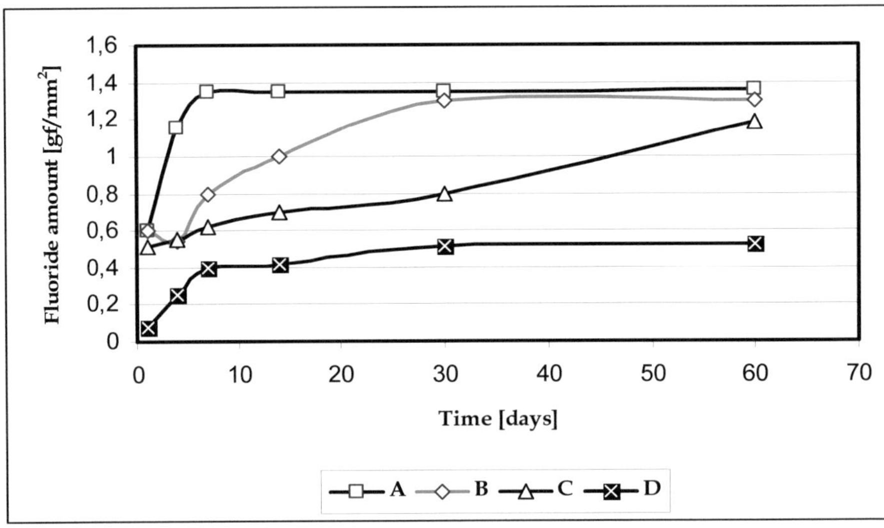

Fig. 15. Amount of fluoride release from tested composites

The maximum peak of fluorine for composite A (about $1.3x10^{-6}g/mm^2$) was obtained on the seventh day of the examinations, with a simultaneous stabilization of its level in the solution. It should be noted that the amount of the fluoride release was higher for the composites A and B due to the higher amount of fluoride sources in the materials. At the same time, the fluorine release from the composite with nanosilica (material C) was still growing. For materials C and D, a lower level of fluoride release was observed.

Therefore, the tendency of the emission of fluoride ions from these both composites begins on the 1st day and lasts until the 60th day. In order to estimate the fluoride release after 60 days, a further increase in the exposition time of the composite in the solution is necessary.

The roughness tests of the composite samples after fluoride release reveal some significant differences between the materials. The results presented in Fig. 16 show that the composites with nanoparticles have lower surface roughness (about 0.15 µm) in relation to the composites without that filler.

This can result from the fact of achieving, for this content of nanosilica, a uniform distribution of particles in the composite structure and forming of a more homogenous structure. Consequently, the smaller the filler particles, the lower the surface roughness is. Similar results were obtained in the work of Tagtekin et al. [46]. The estimated parameter is sustained at the limit of 0.11 - 0.19µm for composites A and B, and 0.30µm for composite D. It was also determined that the surface roughness is strongly dependent on the time of the analysis. It could be a result of higher destruction of the material in contact with the buffer solution.

The second stage of the work concerned the estimate of fluoride release from the samples which had earlier undergone the friction process. The main goal of that part of the research was to determine the influence of cyclic loading on fluoride release from the composites. The measurements of fluoride ion contents in the solutions after the friction tests are presented in Fig. 17.

The results show that the cyclic loadings increase the amount of fluoride ions released. The composites without any loading on them evidently release much lower fluorine levels in contrast to the materials which were tested at different loads. Moreover, the obtained results also show that cyclic loadings intensify fluorine release processes at a later time, when they are immersed in a solution prior to the friction test.

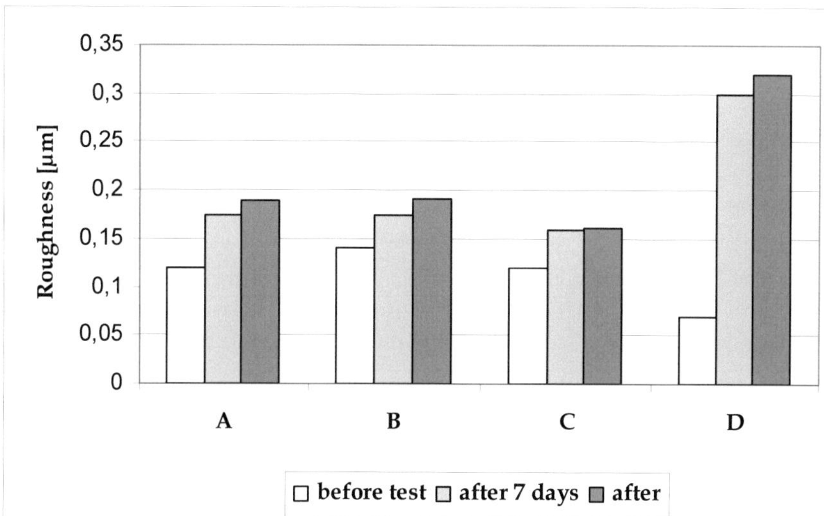

Fig. 16. Surface roughness of tested materials

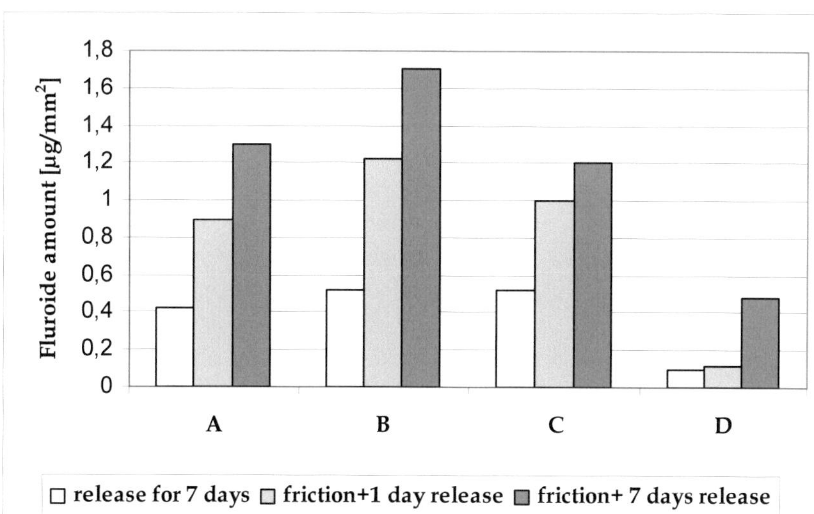

Fig. 17. Influence of cyclic loading and time (p=15MPa) on fluorine release from analyzed composites

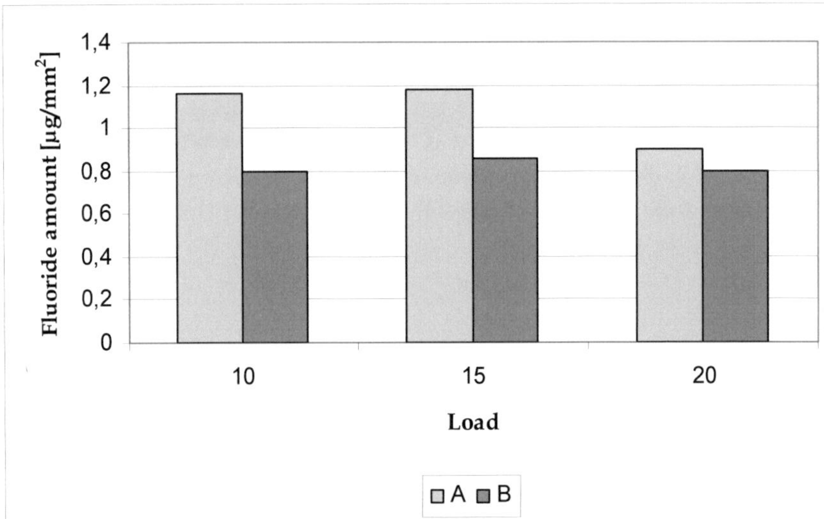

Fig. 18. Influence of loadings (p=10, 15, 20MPa) on fluorine release from analyzed composites

During the 7 days of fluorine emission following the friction process, more concentration was observed in comparison to the 1st day after the friction. It is possible that the wear test changed the internal structure of the materials, and led to the formation of some microcracks. Those discontinuities of the composites can be ways of diffusion for fluoride ions. In consideration of the fact that composites B and C released the highest amount of fluoride ions, they were investigated in successive examinations. The main goal of the planned research was to estimate the influence of the load value on fluoride release. In order to simulate natural forces that occur in the human oral cavity (4-18MPa), three various loads were used: 10, 15, and 20 MPa. The results are presented in Fig. 18.

As the research showed, there is no evident influence of the load value on the amount of fluoride release during the friction process. The amounts of the emitted fluoride ions are comparable in all three value loads. The composite content is more significant. Thus, the presence of nanosilica in a material decreases the fluoride release. It is probably an effect of a low summary content of fluoride sources in composite C in comparison to composite B. Additionally, nanosilica, due to its nanoparticles, could block the diffusion paths for fluoride from the material to the environment solution. Figure 19a shows the surface of composite C with a uniform distribution of the filler particles in this material. The fluoridated glass (symbol J-20) and YbF_3 particles before the fluorine release are visible. Figure 19 b shows the structure of composite C after the fluoride ions emission. It proves that numerous microcracks are formed after the friction process.

The observed microcracks may be associated with the interfaces between the filler particles and the resin matrix, which was not enhanced very much. Thus, it can be concluded that during the cyclic loadings, the surface is plastically deformed, and that leads to generating microcracks within the material. These discontinuities, as it was earlier mentioned, are potential paths for diffusing the fluoride ions.

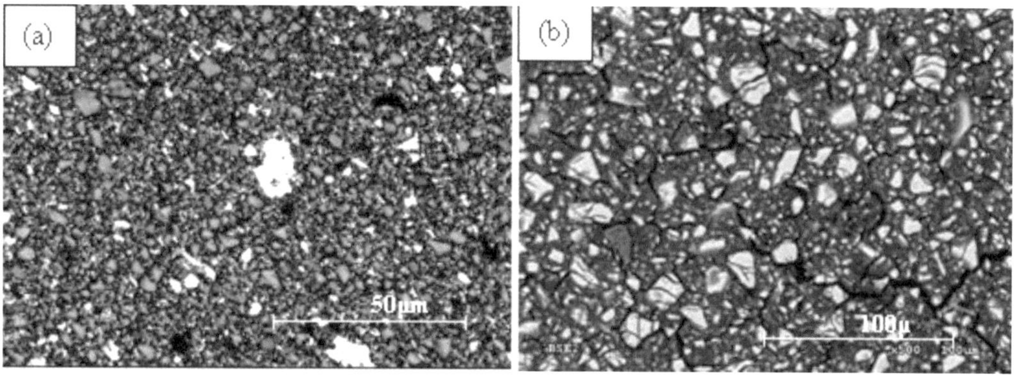

Fig. 19. SEM micrographs of the structure of composite C before (a) and after (b) fluoride release

The influence of the selected fillers and the usage of various loads on releasing the fluoride ions from the analyzed materials were estimated. Taking into account the results of the research, the following conclusions may be formulated:

- the fluoride ions release was observed for all the composite materials,
- the highest level of the fluoride emission occurred in the case of materials based on fluoridated glass,
- nanosilica addition reduces the composite surface roughness during fluoride release as compared to the roughness of the materials without it,
- the fluoride release increases strongly after the cyclic loading process for the analyzed materials, there is no evident relationship between the value of loads applied to the sample and the fluoride release level,

4. References

Dumbleton J.H.: Tribology of natural and artificial Joints. Tribology series 3. Els.Sci.Publ.Co., Amsterdam-Oxford-New York 1983

Dąbrowski J.R.: Tribological aspects of human joints. Tribologia, 1, 1994, 54-63

Grippo J.O., Simring M., Schreiner S.: Attrition, abrasion, corrosion and abfraction revisited. A new perspective on tooth surface lesions. J. Am. Dent. Assoc., 135, 2004, 1109-1118

Cordey J.: Biofunctionality and biomechanics of implant. In: Biomaterials - hard tissue repair and replacement. Els.SciPubl., 1992, 235-245

Dąbrowski J.R., Sajewicz E.: Tribological characteristics of biomaterials. Symp. 'Biomaterials for Medicine and Veterinary Medicine', Kraków (Poland) 1999, pp.245-251

Ungethuem M.: Technologische und biomechanische Aspekte der Hueft-und Kniealloartroplastik. Verlag Haus Huber, Bern-Stuttgart-Wien 1978

Popko J., Dąbrowski J., Dudarew A., Sajewicz A., Iwaszkiewicz B.: Badania tribologiczne stawu biodrowego po założeniu endoprotezy połowiczej w obserwacjach eksperymentalnych. Chir. Narz. Ruchu Ortop. Pol., LXI Supl.3A, 1996, 127-131

Heintze S.D., Zappini G., Rousson V.: Wear of ten dental restorative materials in five wear simulators-results of round robin test. Dent. Mater., 21, 2005, 304-317

Plitz W.: Technologie des kunstlichen Gelenkenersatzes. Expert Verlag, Renninges-Malmsheim 1994

Kramer K.-H.: Implants for surgery - a survey on metallic materials. In: Materials for medical engineering, Band 2, Euromat'99, Viley-VCH Verlag, Munchen 1999, 9-29

Bondarenko W.P.: Tribotiechniczeskije kompozyty z wysokomodulnymi napołnitielami. Izd. Naukowa Dumka, Kijów 1987

Jin Z.M., Dowson D., Fisher J.: Lubrication mechanisms In metal-on-metal hip joint replacements. Proc. 5th World Biomaterials Congress, Toronto 1996, 787

Vancoille E., Celis J.P., Roos J.R.: Tribological and structural characterization of a physical vapour deposited TiC/Ti(C,N)/TiN multilayer. Tribology Int., 26, 1993, 115-119

Mc Kellop H., Park S.-H., Chiesa R., Lu B., Normand P., Doorn P., Amstutz H.: Twenty-year wear analysis of retrieved metal-metal hip prostheses. Proc. 5th World Biomaterials Congress, Toronto 1996, 854

Dorn P.F., Mirra J.M., Campbell P.A., Dorr L.D., van Sambeck K.J.M., Amstutz H.: Pathology of metal on metal total hip prostheses. Proc. 5th World Biomaterials Congress, Toronto 1996,180

Grądzka-Dahlke M., Deptuła P., Dąbrowski J.R.: Properties of composite material based on titanium with graphite addition. 8th World Biomaterials congress, Amsterdam, May 28-June 1, 2008-CD-ROM

S. Klapdohr, N. Moszner: *New Inorganic Components for Dental Filling Composites.* Monatshefte fur Chemie, 136, 2005, pp. 21-45.

J. Luo, R. Seghi, J. Lanutti: *Effect of silane coupling agents on the wear resistance of polymer nanoporous silica gel dental composites.* Mat Sci and Eng, C, 5, 1997, pp. 15-22.

K.S. Wilson, K. Zhang, J.M. Antonucci: *Systematic variation of interfacial phase reactivity in dental nanocomposites.* Biomaterials, 26, 2005, pp. 5095-5103.

X. Xu, J. Burgess: *Compressive strength, fluoride release and recharge of fluoride-releasing materials.* Biomaterials, 24, 2003, pp. 2451-2461.

M. Atai, M. Nekoomanesh, S.A. Hashemi, S. Amani: *Physical and mechanical properties of an experimental dental composite based on a new monomer.* Dental Materials, 20 2004, pp. 663–668.

X.S. Xing, R.K.Y. Li: *Wear behavior of epoxy matrix composites filled with uniform sized submicron silica particles.* Wear, 256, 2004, pp. 21-26.

A.A. Zandinejad, M. Atai, A. Pahlevan: *The effect of ceramic and porous fillers on the mechanical properties of experimental dental composites.* Dental Materials, 22, 2006, pp. 382–387.

K. Friedrich, Z. Zhang, A. Schlarb: *Effects of various fillers on the sliding wear of polymer composites.* Comp Sci and Techn, 65, 2005, pp. 2329-2343.

S. Beun, T. Glorieux, J. Devaux, J. Vreven, G. Leloup: *Characterization of nanofilled compared to universal and microfilled composites.* Dental Materials, 23, 2007, pp. 51–59.

J. Siejka-Kulczyk, J. Mystkowska, M.Lewandowska, J.R.Dabrowski, K.J. Kurzydłowski: *The influence of nano-silica on the wear-resistance of ceramic-polymer composites intended for dental fillings,* Solid State Phenomena, 151, 2009, pp.135-138.

V. Nagarajan, S. Jahanmir, V. Thompson: *In vitro contact wear of dental composites*, Dental Materials, 20, 2004, pp. 63-71.

M. Sulong, R. Aziz: *Wear of materials used in dentistry: A review of the literature*, Journal of Prosthetic Dentistry, 63, 1990, pp. 342-349.

P. Vale Antunes, A. Ramalho: *Study of abrasive resistance of composites for dental restoration by ball-cratering*, Wear, 255, 2003, pp. 990-998.

A.C. Shortall, Q. Hy Xiao, P.M. Marquis: *Potential countersample materials for in vitro simulation wear testing*, Dental Materials, 18, 2002, pp. 246-254.

H.H.K. Xu, J.B. Quinn, A.A. Giuseppetti, F.C. Eichmiller, E.E. Parry, G.E. Schumacher:*Three-body wear of dental resin composites reinforced with silica-fused whiskers*, DentalMaterials, 20, 2004, pp. 220-227.

S.D. Heintze, A. Cavalleri, M. Forjanic, G. Zellweger, V. Rousson: *A comparison of three different methods for the quantification of the in vitro wear of dental materials*, Dental Materials, 22, 2006, pp. 1051-1062.

X. Hu, E. Harrington, P.M Marquis, A.C. Shortall: *The influence of cyclic loading on the wear of a dental composite*, Biomaterials 20, 1999, pp. 907-912.

J. Mystkowska, A. Niewczas, P. Kordos, J.R. Dąbrowski: *Fatigue stress resistance of some composite materials for dental fillings*, Journal of Vibroengineering, 11, 2009, pp. 717-724.

S.D. Heintze, G. Zappini, V. Rousson: *Wear of dental restorative materials in five wear simulators-Results of a round robin test*, Dental Materials, 21, 2005, 304-317.

C.P. Turssi, J.J. Faraoni, M. de Menezes. M.C. Serra: *Analysis of potential lubricants for in vitro wear testing*, Dental Materials, 22, 2006, pp. 77-83.

E.A. Glasspoole, R.L. Erickson, C.L. Davidson: *A fluoride-releasing composite for dental applications*, Dental Materials, 17, 2001, pp. 127-133.

G. Vermeersch, G. Leloup, J. Vreven: *Fluoride release from glass-ionomer cements, compomers and resin composites*, Journal of Oral Rehabilitation, 28, 2001, pp. 26-32.

G. Furtos, V. Cosma, C. Prejmerean, M. Moldovan, M. Brie, A. Colceriu, L. Vensenyi, L. Silaghi-Dumitrescu, C. Sirbu: *Fluoride release from dental resin composites* Mat Sci Eng, 25, 2005, pp. 231-236.

T. Itota, O. Al-Naimi, T.E. Carrick, M. Yoshiyama, J.F. McCabe: *Fluoride release from aged resin composites containing fluoridated glass filler*, Dental Materials, 21, 2005, pp. 1033-1038.

J. Mystkowska, G. Marczuk-Kolada, K. Leszczyńska, J.R. Dąbrowski, J. Karaś: *Fluoride release and antibacterial activity of self-made composite materials for dental fillings*, Solid State Phenomena, 147-149, 2009, pp. 801-806.

J. Mystkowska, G. Rokicki, J. Sidun, J.R. Dąbrowski: *Mechanical and physicochemical properties of some originally made composite materials for dental fillings*, Solid State Phenomena, 165, 2010, pp.142-146.

J. Mystkowska, J.R. Dąbrowski: *The influence of selected powder fillers on tribological properties of composite materials for dental fillings*, Solid State Phenomena, 144, 2009, pp. 33-38.

E. Sajewicz, Z. Kulesza: *A new tribometer for friction and wear studies of dental materials and hard tooth tissues*, Tribology International, 40, 2007, pp. 885-895.

J. Mystkowska: *Fluoride release from composite material for dental fillings*, Solid State Phenomena, 144, 2009, pp. 27-32.

D.A. Tagtekin, F.C. Yarikoglu, F.O. Bozkurt, B. Kologlu, H. Sur: *Selected characteristics of an Ormocer and a conventional hybrid resin composite,* Dental Materials, 20, 2004, pp.487-497.

5

Liposil, a Promising Phospholipid/Silica Composite Material

Begu Sylvie, Aubert-Pouëssel Anne,
Lerner Dan A and Devoisselle Jean-Marie
Institute Charles Gerhardt Montpellier, Advanced Materials for Catalysis and Health
Montpellier
France

1. Introduction

Since their discovery by Bangham (Bangham, 1965), liposomes have become paradigms for biomembranes, models of self-assembling colloids as well as vehicle for pharmaceutical, diagnostic and cosmetic agents. The word 'liposome' was coined by Weissmann (Sessa, 1968) according to Bangham's expression (Bangham, 1995) and is derived from the Greek: 'lipo' referring to their fatty constitution and 'soma' referring to their structure. Liposomes are constituted by phospholid bilayer (s) surrounding an inner aqueous volume. Due to this constitution, they are used as drug-carrier for hydrophilic, lipophilic or amphiphilic molecules (Gregoriadis, 1976). Depending on the method of preparation (Vemuri, 1995), liposomes can vary widely in term of size (0.02–10 mm) and number of lamellae (uni or multi lamellae). Usually, liposomes are classified into three categories on the basis of their size and lamellarity (number of bilayers): small unilamellar vesicles (SUVs) or oligolamellar (OLVs), large unilamellar vesicles (LUVs) and multilamellar vesicles (MLVs). Interest in liposomes results from this vesicular structure limited by one or more outer bilayer lipid and an inner water pool. The application of liposomes was first an artificial model for the biological membrane (Chapman, 1984). Forming lipid bilayers through hydrophobic interaction, liposomes are then considered as excellent platforms for the delivery of hydrophobic and hydrophilic drugs. These phospholipid dispersions in water solutions were able to trap and release solutes, to which they were selectively permeable. Then, active compounds have been loaded either in the aqueous volume, if they were water-soluble, or in the lipid membrane, if they were lipid soluble. Liposomes have been administered in a variety of ways but intravenous injection is the most practical route because liposomes present considerable persistence in the blood. This feature facilitates efficient drug delivery to target tissues. Lipids are characterized by different fatty acid chain lengths, different head groups, and different melting temperatures. Consequently, temperature (Jeong, 2009) or pH sensitive (Obata, 2010) liposomes can be assembled by manipulating the formulation. The effectiveness of 1-methylxanthine (1MTX) as a radiosensitizer and the in vivo efficiency of the temperature sensitive liposomal 1-methylxanthine (tslMTX) were evaluated when combined with regional hyperthermia and ionizing radiation (Jeong, 2009). Intraperitoneal injection of the tslMTX inhibited tumor growth in the mouse xenograft tumor model.

Moreover, the combination of tslMTX with regional hyperthermia and ionizing radiation clearly inhibited tumor growth. Most recently, to target leukemic cells, pH sensitive immunoliposomes including a terminally alkylated N-isopropylacrylamide (NIPAM) in the bilayer, were coupled with the anti-CD33 monoclonal antibody (Simard et al., 2009). The pH sensitive immunoliposomes exhibited high cytotoxicity against HL60 cells, suggesting that these immunoliposomes could be active in acute myeloid leukemia therapy. Commercial liposomes have already gained approval from US Food and Drug Administration (FDA). The typical examples are doxorubicin encapsulating liposomes (Doxil® ALZA Corporation), which has shown a strong antitumor activity against a wide range of cancers and amphotericin B encapsulating liposomes (AmBisome®, Nexstar Pharmaceuticals, Inc) for the treatment of fungal infections refractory to amphotericin B.

Despite all the work done, liposomes still have not attained their full potential as drug and gene delivery vehicles (Guo et al., 2003) and their use was mostly limited to pH close to neutrality and above, due to their destabilization in acidic media.

Recently, the self-assembly of organized nanoscopic structures has been the subject of a strong interest in both colloidal and materials science. Vesicle templating presents an unique opportunity to obtain hollow submicrometer particles (Hubert et al., 2000; Hentze et al., 2003). Templating agents used in the synthesis of organized materials range from lyotropic lamellar phases of non-ionic and ionic surfactants, micro-emulsions, phospholipid tubules to polymer particles. Among recent examples, one may cite novel vesicular particles described by Katagiri (Katagiri et al., 2003). The walls of these hybrid particles consist of a bilayer built from dialkyl amphiphiles modified by alkoxysilanes. Cha et al. used polyelectrolyte homopolymers to spontaneously obtain micron size vesicles using citrate stabilized quantum dots as a template (Cha et al., 2003). A careful balance between charge control and hydrogen bond leads to the self-assembly.

Other spherical particles made from various materials have been obtained, some being multilamellar (El Rassy et al., 2005). Some of them were used to encapsulate fluorescent dyes, enzymes, polymer particles (Schmidt & Ostafin, 2002). In a different approach, the polymerization of trialkoxysilane bearing a double alkyl chain lead directly to hybrid analogs of liposomes, named "cerasomes" (Katagiri et al., 2003). Liposomes were also trapped into sol–gel silica materials and loaded with a fluorescent probe to build a pH sensor (Besanger et al., 2002), with proteins (Li & Tip, 2005) or with a transmembrane peptide ion channel (gramicidin A) to control ion diffusion in sensors or in drug screening devices (Besanger & Brennan, 2003). The stability of supramolecular structures such as vesicles and microemulsions depends on pH, ionic strength and on the presence of organic solvents in the reaction mixture, and as such, they were often affected by the reaction conditions.

In our work, we have shown that nanoscale liposomes could be used as templates for the deposition of silica to create hollow silica nano-shell systems called "liposils" (Bégu et al., 2003) in which the trapped unilamellar liposomes maintained their fundamental properties. The conditions for the template directed assembly were fulfilled due to electrostatic charge density matching or hydrogen bonding between template and inorganic precursor. The silica walls being non-porous, the dried silica spheres retained the inner aqueous pool. The use of zwitterionic phospholipids (non toxic) as templates required a modified approach since the liposome structure was sensitive to low pH and high ionic strength and organic solvents were not allowed for pharmaceutical applications.

The non-porous and amorphous silica shell of liposils protected the trapped liposomes from pH and temperature variations and maintained the specific properties of the phospholipid bilayer. Furthermore, the synthesis of these hybrid nanoparticles allowed loading the liposomes with hydrophilic or lipophilic drug (Bégu et al., 2004).
The stability of drug-loaded liposils was tested at two different pH (1.2 and 7.4), in a flow cell according to the USP 28 norm. Characterizations of liposils were done at various steps of the kinetics. The stability observed for liposils make them good starting material for drug storage and release schemes. For instance, functionalization of their external surface should improve their capture by cells whereby drug release could then be induced by a pH variation after endocytosis. In these conditions, the stability observed made them potential candidates for various drug storage and release schemes. Another release process was developed using triggered release of the content by either low-frequency ultrasound or by microwave treatments (Steinberg et al., 2007). The bursting of a capsule by using an external signal has considerable potential applications, wherever remote activation of content-release is needed. This concept has been also demonstrated for polymeric capsules (Radziuk et al., 2007) and for soft liposomes (Schroeder et al., 2007) and micelles, (Husseini et al., 2000) leaving the idea of external triggered release from rigid ceramic capsules relatively unexplored.
The objective of the present chapter aimed at the study (i) of the morphological characterization of the hybrid material (ii) of the behaviour of the liposome phospholipids and (iii) of the ability of liposils to act as drug delivery system. The morphological characterization, which was expected to reveal the presence of the silica coating, was provided by different methods such as electron microscopy (TEM and SEM), FTIR measurements and nitrogen adsorption. The presence of the entrapped materials (phospholipids) was studied by FTIR, elemental analysis and fluorescence anisotropy measurements.
Keywords: Liposomes; Liposils; Silica nanospheres; Drug delivery

2. Methods and tools

1. Liposils synthesis

Liposils synthesis involved two steps. Large unilamellar liposomes were prepared first and then the silica spheres enclosing these liposomes were formed (Rapuano & Carmona-Ribero, 2000). Two liposome samples were prepared according to the method described by Bangham (Bangham et al., 1995). For the first one, a suitable amount of L-alpha-dipalmitoyl phosphatidylcholine, DPPC (phase transition temperature = 41°C), was dissolved in chloroform. Small amounts of cholesterol were added to DPPC for the second sample, leading to a liposome composition with a DPPC:cholesterol molar ratio of 7:3. After complete removal of the chloroform (at 40°C under reduced pressure), a 5,6-carboxyfluoresceine (CF) solution (100mM in PBS, 50 mM, pH 7.4) was added to both dry phospholipid films in order to obtain a 10 mg/ml lipid suspension in the form of large multilamellar vesicles loaded with the water soluble label in their aqueous core.
These suspensions of multilamellar vesicles were then extruded using an extruder (Lipex Biomembranes Inc.) equipped with polycarbonate membranes (mean size diameter: 400, 200 and finally 100 nm, Nucleopore) in order to reduce and homogenize the liposomes size distribution. The extrusion of the lipids was done above their respective transition temperature, 41°C for DPPC and 45°C for the DPPC:chol mixture (Hamilton, 1980).

After extrusion, the external aqueous phase still contained the fluorescent label. So the liposomes were dialyzed in cylindrical membranes (12,000 Da, SpectraPor) for 1 day with PBS (150 mM, pH 7.4) in order to eliminate any 5,6-CF present in the external aqueous phase. After the dialysis step, 5,6-CF was only present in the internal aqueous core of the liposomes.

To determine the phase transition temperature of both types of liposomes, the hydrophobic fluorescent probe diphenylhexatriene (DPH) was loaded in the phospholipid bilayer instead of 5,6-CF to prepare two new sets of liposomes (DPH added in tetrahydrofuran to obtain a 1:200 DPH:lipid ratio) (Bégu et al., 2004). The protocol was then the same as described above. For these experiments, and for each of the two samples, a small amount of DPH loaded liposomes suspension was diluted in PBS (150mM) and the resulting suspension was placed in a thermostated spectrofluorimeter cell. The cell temperature was slowly increased from 20°C to 60°C (1°C/min) while monitoring the fluorescence intensity (λex = 360 nm and λem = 430 nm).

The last step of the synthesis consisted in the silica shell formation. To that aim, the inorganic precursor tetraethoxysilane (TEOS, Fluka) was first hydrolyzed in the PBS buffer (150mM, pH 7.4) for 2 days at 40°C under stirring. The silicate solution was added drop wise to the liposome suspensions (DPPC or DPPC:chol, with a TEOS:DPPC molar ratio 8:1) at room temperature (20°C) under stirring. The final mixture was gently stirred for 1 day. Sodium fluoride (NaF, Sigma Aldrich) at a 4% molar ratio with respect to TEOS was then added to both samples, and the reaction medium was stirred at room temperature (48 h), protected from light.

The resulting samples were then controlled by dynamic light scattering (DLS), filtered (pore size 200 nm) and the residual solid was dried at 40°C for 24 h to give free flowing silica coated particles .

Two silica hybrid samples were prepared: the first referred to as sample 1 contains DPPC liposomes and the second, sample 2, contained DPPC:chol 7:3 liposomes. Both solid samples, loaded with 5,6-CF in the liposomes aqueous core, were washed twice with water and then with an isopropanol/acetic acid (0.1 M) solution. This washing was a precautionary step done in case some 5,6-CF and phospholipid were freed after disruption of some of the liposomes during the silica condensation step and could adsorb onto the external surface of the silica shell. The same experimental conditions, including the washing step, were followed to prepare the two sets of DPH loaded liposils used to determine the phase transition temperature of the trapped bilayer.

2. Characterization of the hybrid material

   a. Dynamic light scattering (DLS)

Liposomes and liposils size measurements were performed by DLS at a 90° angle, using a spectrogoniometer equipped with a He/Ne laser and a photon correlator (Sematech, SM633/RTG). In DLS, nanometric particles diffuse in the small volume of a sample hit by a laser beam and scatter the laser light. The movement of the particles across that volume is due to thermal activation (Brownian diffusion) which is such that the probability of a translation step of amplitude ΔR at time t ($\Delta R = R(0) - R(t)$) is described by a gaussian distribution. It results that the square of the average displacement ΔR^2 is given by the following (Eq. (1)):

$$\Delta R^2(t) = 6D_0 t \quad (1)$$

D_0 is the diffusion coefficient that is used in turn to define a relaxation time for the diffusion. In short, DLS theory shows that the latter was quantified by an experimental autocorrelation function. So an analysis of the correlogram of scattered light will give access to relaxation times, hence to a diffusion coefficient, from which the hydrodynamic diameter R_h is deduced using the Stokes-Einstein relationship (Eq. (2)) :

$$D_0 = \kappa T / 6\pi\eta R_h \quad (2)$$

b. TEM and SEM

The morphological characterization was done on the samples by conventional transmission (JEOL 1200 EXII) and scanning electron microscopy (Hitachi S 4500) following standard preparation. The samples were first washed with an isopropanol acetic acid mixture and then dried.

c. TGA DSC

The experiments were done on as-synthesized and on isopropanol/acetic acid washed liposil (Perkin Elmer, DSC 111). The temperature was increased from 50 to 550°C at 5°C/min for washed and unwashed material.

d. DPPC determination (enzymatic).

After crushing the sample, the concentration of DPPC in the suspension was quantified by an enzymatic determination of phospholipids (PAP 150, Biomerieux, France). Phospholipids are hydrolized by phospholipase D and the liberated choline is measured by the Trinder reaction (Takayama et al. 1977).

e. Elemental analysis.

For the determination of the liposils composition, liposomes were prepared according to the method described above using NaCl 9% solution instead of the phosphate buffer solution and the C, P, N, Si and O analysis have been done.

f. Nitrogen adsorption

For liposil sample 1 and sample 2, the characterization of the material was carried out on the as-synthesized material immediately after synthesis, then after a drying step at 40°C, and finally after the stability test at the two different pH values after a drying step. The specific surface area of the samples was assessed by nitrogen adsorption measurements according to the Brunauer-Emmett-Teller standard method (Brunauer et al., 1938). Measurements were carried out at 35°C during 12 h under 10^{-2} Torr using an ASAP 2010 adsorption analyzer from Micromeritics.

g. Fluorescence anisotropy measurements

A fully equipped spectrofluorimeter (Shimadzu RF 5310), was used for the steady state fluorescence depolarization experiments. A sample of liposil loaded with DPH was specially prepared to that effect. The excitation and emission wavelength were set to 360 nm and 430 nm respectively and the relative intensities for the four combinations of vertically and horizontally polarized excitation and emission beams were recorded in the ratio mode in order to eliminate source intensity fluctuations. The steady state emission anisotropy was calculated as:

$$r = 2p/(3\text{-}p) \quad (1)$$

with

$$r = (I_{//}\text{-}I_{\perp}.G)/(I_{//}+2I_{\perp}.G) \quad (2)$$

The transmission factor G was taken into account by the method of Azumi and McGlynn, as usual. This factor took care of all polarization effects due to the optical components in the excitation path, including those due to the grating (Azumi & McGlynn, 1962). The experiment was done after the treatment in the acidic medium. No contribution due to light scattering was present since a dilution of the vesicles labelled with DPH had no effect on the measured fluorescence anisotropy.

h. Dissolution kinetics

The tests were done in a flow-through dissolution cell (USP apparatus 4, Sotax; the norm chosen was USP 28-NF-23 (2005)) which was a dissolution testing unit for solid or powder dosage form, using the USP/EP flow-through method.

The cell was used in the experiments with a thermostated bath, and may be described as having 3 parts: an inverted cone at the bottom, the cylindrical portion of the cell in the middle and the filter holding head, on top (Fig.1).

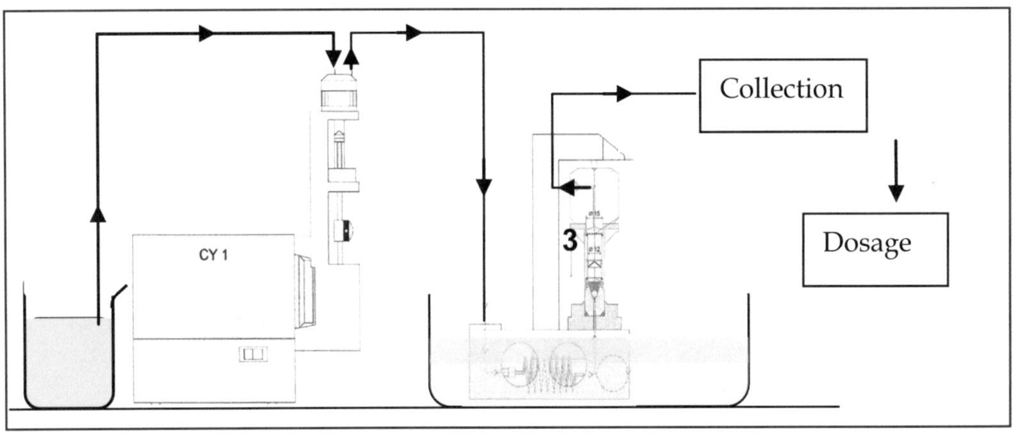

Fig. 1. Scheme of the flow cell apparatus used for kinetic dissolution tests

The dissolution medium entered the cone from underneath through a capillary bore and flowed upwards through the cell. The cone was separated from the cylindrical section by a 40-mesh screen holding a glass microfiber filter. The filter head on the top part of the cylinder also held a glass microfiber filter. At the bottom of the conical part, a single 6 mm diameter glass bead regulated the distribution of the solution jet entering the cell and the rest of the cone volume was filled with 1 mm diameter round glass beads. The role of the glass microfiber filter placed on the bottom part and on the top part (Whatman GF/B and GF/D) of the cylindrical section was to prevent some material particles from being drawn outside the cell. This detailed description was necessary since recent publications mentioned that the release was affected by the position of the glass beads (Bhattachar et al., 2002).

The sample to test was deposited in the flow-through cell, between the filters sitting at the bottom and top of the cylindrical part (Fig. 2). For each experiment, an amount of about 50 mg of the sample (the drug was featured here by the test molecule 5,6-CF) was needed. The flow rate of the medium, kept at 37 °C, was set at a value close to 4 ml/min. In a first experiment, the behaviour of liposil was tested in an acidic medium (pH 1.2) during 30 min. In a second independent experiment, a fresh sample was tested in a pH 7.4 medium for 1 h and a half. These experiments, done in triplicate, were carried out using a closed loop setup connected in line to a UV-visible spectrophotometer in order to monitor the OD variation at λ=430 nm during the kinetics at pH 7.4.

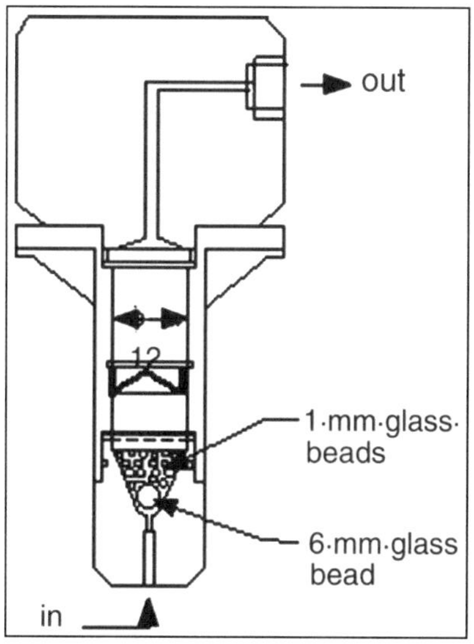

Fig. 2. Scheme of the flow-cell used in this study for the stability tests (the specific localization of the various glass beads are indicated by arrows)

3. Results

The liposomes used for the synthesis of the silica material were constituted by zwitterionic phospholipids: dipalmitoylphosphatidylcholine (Fig. 3a) presenting a 16 C long fatty acyl chain and a choline head group. The synthesized liposomes have a diameter of 80 nm (polydispersity: 0.181) for DPPC and 84 nm (polydispersity: 0.192) for DPPC:chol. They correspond to small unilamellar liposomes (Fig. 3b) with a narrow size distribution and a DPPC concentration of 9.5 mg/mL. The TEOS:DPPC molar ratio (8:1) was chosen to keep the amount of released ethanol under 4% (v/v) during the TEOS hydrolysis. The strong electrostatic interaction between the positively charged lipid head groups (quaternary ammonium) and the negatively charged silica monomers and oligomers was the driving force in the templating. The lateral mobility of the lipids in the outer leaflet of the bilayer (Fig. 3b) can explain efficient charge compensation. The resulting silica cladding is dense and non-porous.

The hybrid material was originally mostly obtained as clusters of monodispersed but aggregated spheres (Bégu et al., 2003). In the last synthesis process (Bégu et al., 2007), an improvement of the former, a pre-hydrolyzed silica precursor was added to the liposomes suspension. Improvements have also been done using PEG phospholipid (Avnir et al., 2007). From DLS measurements on freshly prepared samples of liposils in suspension (i.e. before the drying step), the size distribution of the liposil spheres was found to be centred at 90 nm with a low polydispersity index (0.20). DLS measurements done at each individual step of the synthesis showed that as long as the liposils remained in suspension in the buffer solution, there was no aggregation of the liposil particles. Interestingly, the liposils size polydispersity reflected the size distribution of the liposome template.

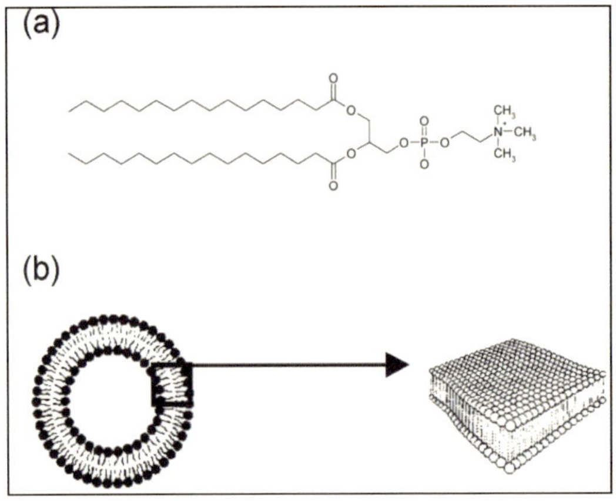

Fig. 3. a: Dipalmitoylphosphatidylcholine structure; b: Schematic representation of an unilamellar liposome and the organization of the phospholipid bilayer

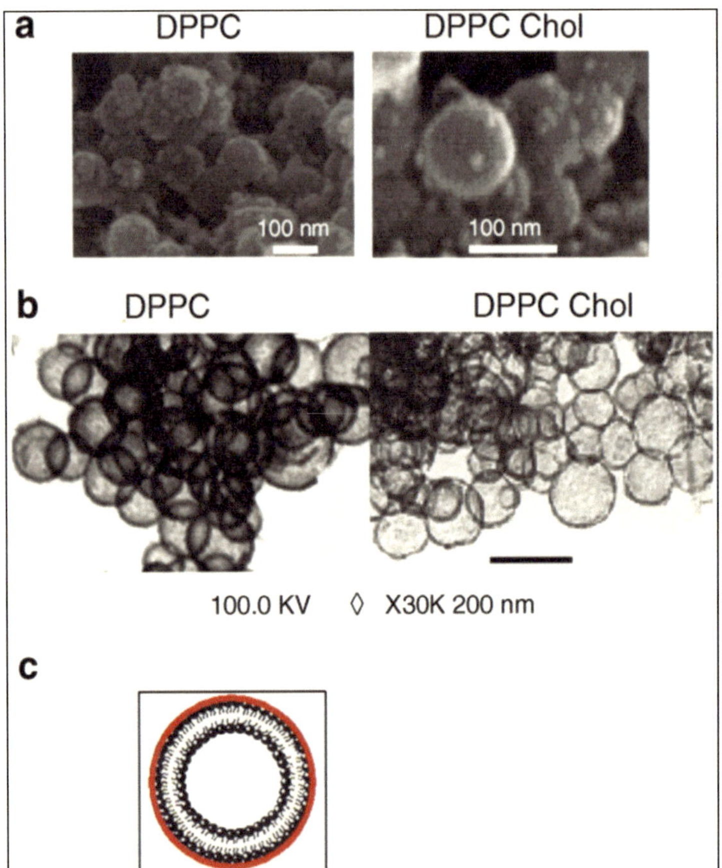

Fig. 4. SEM micrographs (a) and TEM micrographs (b) of dried DPPC and DPPC:chol liposils; schematic representation of a liposil (c) in which the silica shell is figured by the outermost thick line

After the drying step, these liposils still displayed the presence of spheres mostly grouped in clusters, with individual liposils retaining their perfectly spherical shape and wall integrity (Fig. 4a). SEM controls were done after the drying step. The reason was that before the latter, the presence of amorphous silica in excess of the amount needed for the coating process made the observation impossible. Aggregation occurred during the drying step.

TEM data clearly showed silica walls of even thickness of about 4 nm (Fig. 4b). The increase in particle size going from liposomes to liposils occurred during the condensation phase after addition of hydrolyzed TEOS, as a consequence of the formation of the silica shell. Aggregates formation resulted from interactions occurring between the OH groups of the silanols when the individual particles were brought into contact during the drying step (Park et al., 2002).

Here, aggregation generated microparticles which consisted of aggregated but well defined nanocapsules, each its own well delimited spherical shell. The improvement was seen first in the reduced size polydispersity, the thinner silica shell and a better homogeneity of the latter.

Furthermore, after the drying step, the particles from aggregates did not share a whole portion of their surface with their neighbours, but appeared to have only a small contact area, maintaining so the individuality of the silica shells. Many techniques, such as spray drying (Cheow et al., 2010) or freeze-drying, were tested in our laboratory to limit the aggregating effect of this drying step but they ended up with the breaking of a large number of the liposils, even if some of them appeared as free particles. SEM micrographs (Fig. 4a) showed particles with a size of one to several microns consisting in aggregated spherical nanoparticles. TEM results (Fig. 4b) were in agreement with the SEM observations and confirmed the presence of hollow nanocapsules with a mean diameter close to that of the unilamellar liposomes (80 nm).

The silica wall of these nanocapsules was thin enough (close to 4 nm) to be partially transparent to the electron beam. The material morphology was the same whatever the liposome composition used (1 or 2).

TGA-DSC studies show that between 25 and 150°C a very small amount of adsorbed water is desorbed from the surface of unwashed particles compared to the amount seen for washed particles. This means that the external surface of the former particles is almost totally coated by a single layer of lipid (as 70 percent of the total lipids are found inside the liposils), and that the washing step efficiently removes the latter. Between 160 and about 260°C, under an air flux, the DPPC coating the outer surface of the micro-spheres is burnt whereas the mass loss in this region is almost zero for the washed nanospheres (Fig. 5). For those, the inner bilayer is shielded from oxygen by the compact silica shell. Above 260°C, the particles explode under the pressure from vaporized water trapped in their core. Afterwards, the combustion of the now exposed lipid from the templating liposomes can be followed up to about 500°C.

The nanocapsules size was consistent with the liposome diameter augmented by twice the silica shell thickness. The synthesis conditions and specially the silicon alkoxide/phospholipid ratio (8:1) ensured a homogeneous deposition of the prehydrolyzed silica on the vesicle surfaces that were fully covered (refer to the SEM micrograph). During the latter process, the phospholipid bilayer integrity as well as the shape of the liposomes were maintained. This stability resulted from the choice of a phospholipid with a high phase transition temperature (Bégu et al., 2004; Bégu et al., 2007).

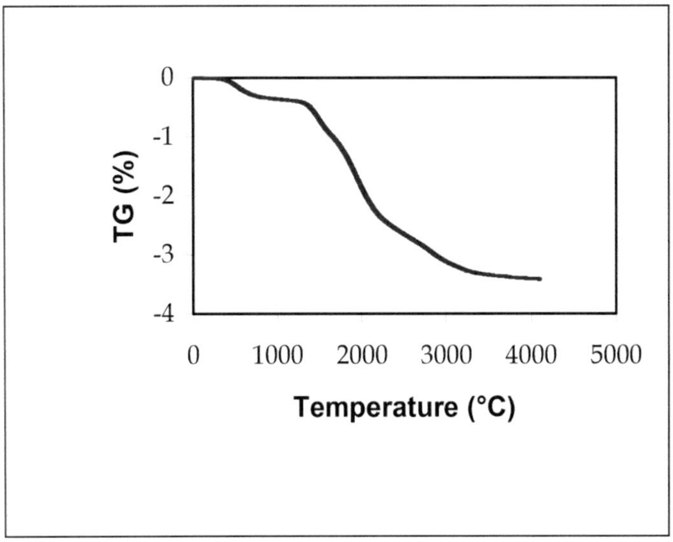

Fig. 5. TGA/DSC scans of liposils

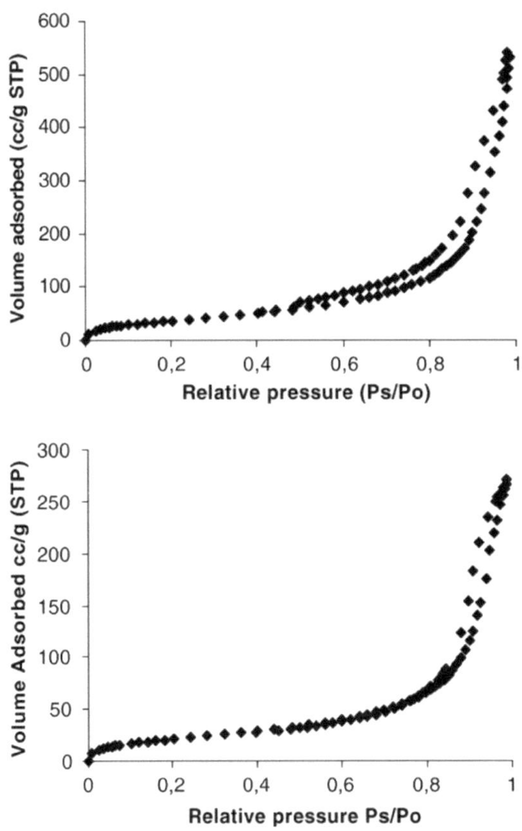

Fig. 6. Nitrogen adsorption isotherms for dried nanoparticles: a) top graph, as synthesized material; b) bottom graph, after the kinetics in acidic fluid

Elemental analysis carried out on unwashed particles at 200 and 350°C reveals a content of carbon of 11.2% and 1.6% respectively. This is in perfect agreement with the previous data concerning the early removal of the external lipid layer and the combustion of the inner bilayer after rupture of the vesicles.

Nitrogen adsorption isotherms (Fig. 6, top) showed that the as-synthesized materials had neither micro nor mesoporosity (type III isotherm). Further, measurements of the surface area of these hybrid solids lead to a small computed BET surface of 100 m^2/g (the porosity observed was mostly interparticular). These data confirmed the non-porous nature of the silica coating of the liposomes.

It is interesting to compare experimental data and results from a purely geometrical analysis. In a typical preparation, 15.53 mg DPPC are encapsulated in the silica liposil material. The total calcination of the latter leaves 22 mg of pure SiO_2. With an outer diameter of about 110 nm and an estimated shell thickness of 10 nm, the volume of the non-porous silica shell amounts to 4 10^{-16} cm^3 and its mass to 8.8 10^{-16} g based on a value of 2.2 $g.cm^{-3}$ for the density of amorphous silica. The total surface for the two layers of the included liposome amounts to 6.3 10^{-14} m^2.

Considering an accepted area of 0.60 nm^2 for the polar head group of the lipid in the film (Pidgeon & Hunt, 1981) the number of DPPC molecules and their expected mass (for a single particle) should be close to 1.1 10^5 and 1.3 10^{16} g respectively. Using the above values, 22 mg of silica and 15.5 mg of DPPC would correspond to 2.5 10^{13} and 12 10^{13} particles respectively. The difference between these computed values reflects a number of uncertainties, the largest being certainly the shell thickness that results from a visual estimation of TEM data. Taking the average of the two values as a reasonable guess, and the computed surface of a particle as 3.8 $10^{-16}m^2$, one ends up with a computed surface area of about 73 m^2/g of the core shell liposil. This is close to the 55 m^2 per gram found experimentally. The difference cannot be totally attributed to the aggregation of the shells, since the shell thickness and amount of aggregation is not precisely known. However these results point to the coherence of all the data.

Slight modifications of the material were observed after the kinetics in acidic fluid. The surface area increased from 100 to 140 m^2/g as a result of the partial destruction of the aggregates and breaking of a small fraction of the liposils induced by the dynamics of the flow in the cell (Fig. 3, bottom). This result was in agreement with the SEM and TEM micrographs which confirmed the integrity of most of the 1 and 2 liposil nanoparticles after the kinetics in acidic medium.

Liposils may be schematically represented as seen in Figure 4c. However, no particle was left intact after the complete kinetics at pH 7.4, and amorphous silica chunks only were seen, originating from the broken and partly hydrolyzed silica shells.

The behaviour of the liposil particles at two different pH values (1.2 and 7.4) was also tested using the hydrophilic probe 5,6-CF (Weinstein et al., 1983) expected to diffuse away from the aqueous core across the phospholipid bilayer. The integrity of the bilayer of encapsulated liposomes was also tested as described in the original synthesis (Bégu et al., 2004), using the emission polarization of the linear fluorescent probe DPH. This probe was only soluble in the inner region of the bilayer and helped monitor the rigidity of the lipids alkyl chains. When the temperature of the samples was varied across a region encompassing the transition temperature from the solid gel state to the liquid gel state, a typical decrease in the polarization of the emission was observed.

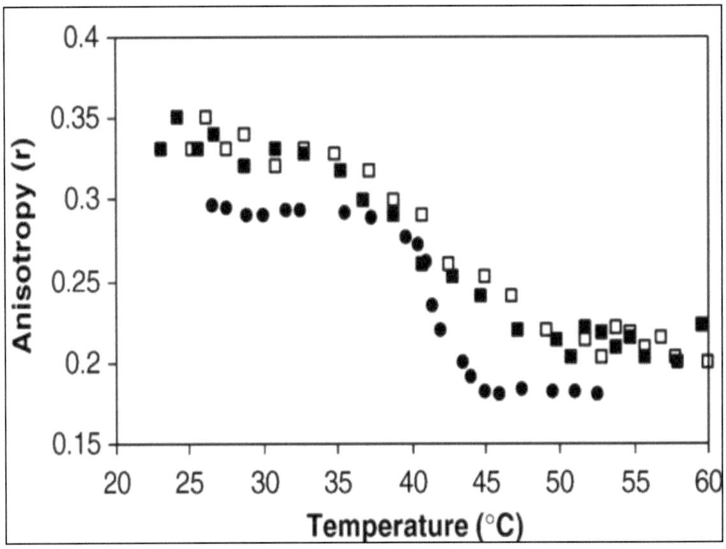

Fig. 7. Liposomes transition temperature from DPH fluorescence polarization: a) DPPC liposomes (•); b) as-synthesized DPPC liposils (▪); c) DPPC liposils after the kinetics in acidic fluid (□)

This effect is clearly seen for free DPPC liposomes for which the transition temperature was 41°C (Fig. 7a) [14]. In as-synthesized liposils, that transition temperature was shifted towards a higher value as the external leaflet of the bilayer interacted with the silica wall in a way that increased its cohesion (Fig. 7b). Measurement of the fluorescence anisotropy was also carried out on DPH loaded liposil after the 30 min stability step at pH 1.2 and it could be seen that the temperature of the phase transition of the template phospholipid bilayer was not affected by that treatment (Fig. 7c).

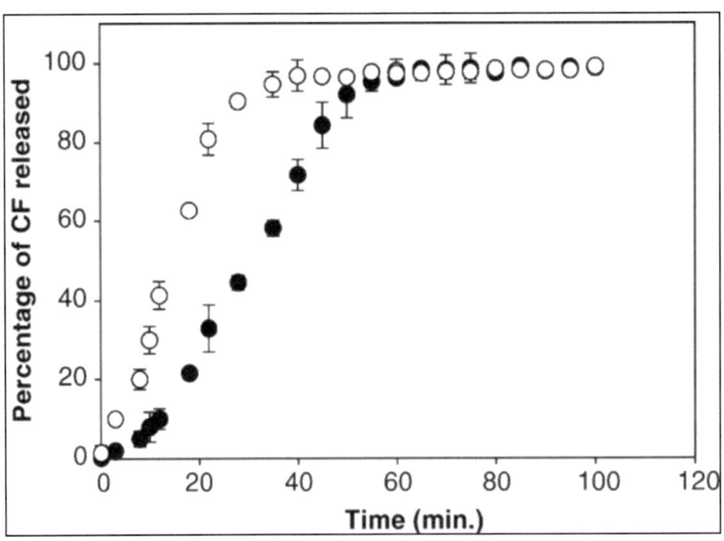

Fig. 8. Kinetics profiles for the release of 5,6-CF from the composite silica nanoparticles in suspension at pH 7.4 for 90 min: a) DPPC liposils (–o–); b) DPPC:chol liposils (–•–)

Samples were left for 30 min in contact with the acidic dissolution medium (pH 1.2), in order to test the stability of the silica shell in these conditions. At the end of this first step, the amount of released 5,6-CF in the medium was measured. That determination was done after raising the pH to 7.4 since 5,6-CF, in the form of a lactone at pH 1.2, would have been insoluble at that pH. It revealed that at most 5% of the total amount of the fluorescent probe associated to the hybrid solid was solubilized. The origin of that 5,6-CF could be due to residual adsorbed molecules that resisted the washing step. But it was most probably due to some probe molecules being trapped in the few closed inter-grain spaces formed as a result of the aggregation of the liposils, that were released in a breakage induced by the shearing stress. This was confirmed by the fact that further immersion of the treated particles in solution did not release 5,6-CF any more, and by the almost unchanged overall fluorescence intensity of the particles. These observations were identical for particles 1 and 2, and confirm the non-porous nature of the silica shell. This conclusion was also in agreement with the TEM and SEM micrographs results, which revealed at most minor changes in the silica shell of the treated liposils nanoparticles.

In the second part of the test, a kinetics run was done at pH 7.4 (Fig. 8) with another aliquot of the same liposil samples. In these conditions, a release of the probe occurs, due to leakage from the liposil core towards the pH 7.4 outer medium. To improve the signal value, the experiment was done with a closed loop setting; this was possible since any freed 5,6-CF was shown not to be readsorbed onto the silica surface. In the 37 °C bath, if the 5,6-CF probe could exit the liposomes at pH 7.4, that is, if a disruption of the liposomes had occurred concomitantly with the rupture of the silica shell, a burst should have been seen. This phenomenon was not observed whether or not the liposomes bilayer included cholesterol. A consequence of this observation was that the dissolution of the silica shell should initially free intact liposomes. This interesting hypothesis was also supported by the different rates of release of 5,6-CF observed for the two types of liposils, as the kinetics corresponding to the DPPC:chol templated material is the slowest. This reflects the known relative stability of the two different types of liposomes used as templates, the ones with cholesterol being more stable. Of course, due to the hydrodynamic shear forces in the flow cell, the liposomes were finally destroyed and released their content. The latter event defined the observed release kinetics. In physiological conditions, it was shown that liposomes were stable at pH 7.4 over a time stretch much longer than the 2h used for the kinetics of the release during which only a very limited release of 5,6-CF occurs (Agarwall et al., 1986; Semple et al., 1996). In any case, these tests cannot be assimilated to test for the standard release from liposomes. The shell was viewed as a protection towards acidic media, temperature and the presence of enzymes, all factors susceptible to destabilize bare liposomes. The test of the release of 5,6-CF in an acidic medium was only a control of the integrity of that silica shell. Later these liposil particles will be used for drug release in slightly improved form (functionalized external surface of the shell, etc.).

4. Conclusion

The stability of liposils, silica-based composites obtained via liposome templating was tested at pH 1.2 and pH 7.4 at the constant temperature of 37 °C. Two samples, one with DPPC and the second with DPPC and cholesterol, were carefully studied. The non-porous amorphous silica cladding of liposils protects the trapped liposomes which retain the fundamental properties of their bilayer. Carboxyfluorescein, 5,6-CF, simulating a typical hydrophilic drug was loaded in the liposomes aqueous phase before the templating process. The stability tests were done using a flow cell, according to the USP 28 norm. As expected, at the acidic pH 1.2, the non-porous silica coating was stable and prevented the rapid degradation generally observed for free liposomes in these conditions. At a pH value of 7.4, the silica shell was hydrolyzed, and at first the intact liposomes were freed. This interesting feature was revealed by the release kinetics and was observed independently of the fact that the bilayer contained cholesterol or not. In turn, at that pH, the bilayers are destabilized in these conditions. The stability observed for liposils makes them good candidates for drug storage and release schemes. Their size and shape should facilitate their capture by cells and the presence of the water pool in the trapped liposome should allow for an externally triggered release using ultrasonic waves or microwaves. This would provide a fast release, complementing the slow release associated to erosion of the liposil particle wall. Further work is in progress to prevent the individual liposil nanospheres from aggregating when going to the dry state. Optimizing the latter factors will of course lead to a better control of drug release.

5. References

Agarwall, K.; Bali, A. & Gupta, C.M. (1986) Influence of the phospholipid structure on the stability of liposomes in serum, *Biochim. Biophys. Acta.* 856, 36–40.

Azumi, T. & McGlynn, S.P. (1962) Polarization of the luminescence of phenanthrene, *J. Chem. Phys.* 37, 2413–2420.

Bangham, A.D. (1965) Diffusion of univalent ions across unilamellar of swollen phospholipids. *J. Mol. Biol.*, 13, 238–252.

Bangham, A.D. (1995) Surrogate cells or Trojan horses. The discovery of liposomes. *Bioassays.* 17, 1081-1088.

Bhattachar, S.N.; Wesley, J.A.; Fioritto, A.; Martin, P.J. & Babu, S.R. (2002) Dissolution testing of a poorly soluble compound using the flow-through cell dissolution apparatus, *Int. J. Pharm.* 236, 135–143.

Bégu, S.; Durand, R.; Lerner, D.A.; Charnay, C.; Tourne-Peteilh, C. & Devoisselle, J.-M. (2003) Preparation and characterization of siliceous material using liposomes as template. *Chem. Commun.* 640–641.

Bégu, S.; Charnay, C.; Tourné-Peteilh, C.; Lerner, D.A. & Devoisselle, J.M. (2004) Characterisation of a phospholipid bilayer entrapped into non-porous silica nanospheres, *J. Mater. Chem.* 14, 1316–1320.

Bégu, S.; Aubert-Pouëssel, A.; Lerner, D.A. ; Tourne-Péteilh, C. & Devoisselle, J.M. (2007) Liposil, a promising composite material for drug storage and release. *Journal of Controlled Release*, 118, 1-6.

Besanger, T.R.; Zang, Y. & Brennan, J.D. (2002) Characterization of fluorescent phospholipid liposomes entrapped in sol-gel derived silica. *J. Phys. Chem.* 106, 10535–10542.

Besanger, T.R. & Brennan, J.D. (2003) Ion sensing and inhibition studies using the transmembrane ion channel peptide gramicidin A entrapped in sol-gel derived silica. *Anal. Chem.* 75, 1094–1101.

Brunauer, S.; Emmett, P.H. & Teller, E. (1938) Adsorption of gases in multimolecular layers, *J. Am. Chem. Soc.* 60, 309–319.

Cha, J.; Bartl, M.H.; Wong, M.S.; Popitsch, A.; Deming, T.J. & Stucky G.D. (2003) Microcavity lasing from block peptide hierarchically assembled quantum dot spherical resonators *Nano Letters*, 3, 7, 907–911.

Cheow, W. S.; Li S. & Hadinoto K. (2010) Spray drying formulation of hollow spherical aggregates of silica nanoparticles by experimental design. *Chemical Engineering Research and Design,* 88, 673–685.

Chapman D. (1984) Physicochemical properties oh phospholipids and lipid-water systems. In : Gregoriadis G, editor. *Liposome technology*. Boca Raton, FL : CRC press. 1-18.

El Rassy, H.; Belami, E.; Livage, J. & Coradin, T. (2005) Onion phases as biomimetic confined media for silica nanoparticle growth. *Langmuir* 21, 19, 8584–8587.

Guo, X.; Francis, C. & Szoka, J.R. (2003) Chemical approaches to triggerable lipid vesicles for drug and gene delivery. *Acc. Chem. Res.* 36, 5, 335–341.

Gregoriadis, G. (1976) The carrier potential of liposomes in biology and medicine. *New England J Med.,* 295, 765-770.

Hamilton, R.L. (1980) Unilamellar liposomes made with the French pressure cell: a simple preparative method, *J. Lipid Res.* 21, 981–992.

Hentze, H.P.; Raghavan, S.R.; McKelvey, C.A. &Kaler, E.W. (2003) Silica hollow spheres by templating of catanionic vesicles, *Langmuir, 19,* 4, 1069–1074.

Hubert, D.H.W.; Jung, M. & German, A.L. (2000) Vesicle templating, *Adv.Mater.* 12, 17, 1291–1294.

Husseini, G. A.; Myrup, G. D.; Pitt, W. G.; Christensen, D. A. & Rapoport, N. Y. (2000) Combined cancer therapy by micellar-encapsulated drug and ultrasound. *J. Control. Release,* 69, 43-52.

Jeong, S.Y.; Yi, S.L. & Lim, S.K. (2009) Enhancement of radiotherapeutic effectiveness by temperature sensitive liposomal 1m ethylxanthine *Int J Pharm,* 372, 1-2, 132-139.

Katagiri, K.; Hamasaki, R.; Ariga, K. & Kikuchi, J.-I. (2003) Preparation and surface modification of novel vesicular nano-particle "Cerasome" with liposomal bilayer and silicate surface, *J. Sol-Gel Sci. Technol.* 26, 393–396.

Li, Y. & Yip, W.T. (2005) Liposomes as protective capsules for active silica sol–gel biocomposite synthesis, *J. Am. Chem. Soc.,* 127, 37, 12756–12757.

Obata, Y.; Tajima, S. & Takeoka, S. (2010) Evaluation of pH responsive liposomes containing amino acid based zwitterionic lipids for improving intracellular drug delivery in vitro and in vivo. *J Control. Release,* 142 2, 267-276.

Park, K.; Kim, K.D. & Kim, H.T. (2002) Preparation of silica nanoparticles: determination of the optimal synthesis conditions for small and uniform particles, *Colloids Surf., A Physicochem. Eng. Asp.* 197, 7–17.

Pidgeon, C. & Hunt, C.A. (1981) Calculating number and surface area of liposomes in any suspension, *J. Pharm. Sci.,* 70, 2, 173–176.

Radziuk, D.; Shchukin, D. G.; Skirtach, A.; Mohwald, H. & Sukhorukov, G. (2007) Synthesis of silver nanoparticles for remote opening of polyelectrolyte microcapsules. *Langmuir, 23,* 4612-4617.

Rapuano, R. & Carmona-Ribeiro, A.M. (2000) Supported bilayers on silica. *J. Colloid Interface Sci.,* 226, 299-307.

Schmidt, H.T. & Ostafin, A.E. (2002) Liposomes directed growth of calcium phosphate nanoshells. *Adv. Mater.,* 14, 7, 532–535.

Schroeder, A.; Avnir, Y.; Weisman, S.; Tzemach, D.; Najajreh, Y.; Gabizon, A.; Talmon, Y.; Kost, J. & Barenholz, Y. (2007) Controlling liposomal drug release with low frequency ultrasound: mechanism and feasibility. *Langmuir, 23,* 4019-4025.

Semple, S.C.; Chon, A. & Cullis, P.R.; (1996) Influence of cholesterol on the association of plasma proteins with liposomes, *Biochim. Biophys. Acta,* 35, 2521–2525.

Sessa,G. & Weissmann, G. (1968) Phospholipid spherules (liposomes) as a model for biological membranes. *J. Lipid. Res.,* 9, 310-318.

Simard, P. & Leroux, J.C. (2009) pH sensitive immunoliposomes specific to the CD33 cell surface antigen of leukemic cells. *Int. J. Pharm.,* 381, 2, 86-96.

Steinberg, Y.; Schroeder, A.; Talmon, Y.; Schmidt, J.; Khalfin, R.L.; Cohen, Y.; Devoisselle, J.M.; Begu, S. & Avnir, D. (2007) Triggered release of aqueous content from liposome-derived sol-gel nanocapsules, *Langmuir,* 23, 12024-12031.

Takayama, M.; Itoh, S.; Nagasaki, T.; & Tanimuzi, I. (1977) A new enzymatic method for determination of serum choline-containing phospholipid. *Clin. Chim. Acta,* 79, 93-98.

Vemuri, S. & Rhodes, C.T. (1995) Preparation and characterization of liposomes : a review. *Current Sci.* 68, 715-724.

Weinstein, J.N.; Ralston, E.; Leserman, L.D.; Klausner, R.D.; Dragsten, P.; Henkart, P.; & Blumenthal, R. (1983) Self quenching of carboxyfluorescein fluorescence: uses in studying liposome stability and liposome cell interaction, *Liposome Technol.* 13, 183–204.

Part 2

Nanocomposites for Energy Efficiency

6

Advances in Thermoelectric Energy Conversion Nanocomposites

Lusheng Su and Yong X. Gan
Department of Mechanical, Industrial and Manufacturing Engineering
College of Engineering, University of Toledo, Toledo, OH 43606, USA

1. Introduction

The painstaking exploring for alternative energy resources is at the research leading edge because energy plays an irreplacable role in the whole world. As a result, how to achieve high-efficient energy conversion rate has become more and more crucial under the condition of limited resources. Thermoelectric (TE) nanocomposite materials, which can generate electricity from heat, could be an alternative solution for global sustainable energy. Whether the power generation by TE nanocomposite materials could be employed as a reliable substitution for getting limited energy resources or not is contingent upon the fact that synthesized TE nanomaterials possess higher thermoelectric conversion efficiency than traditional bulk materials. The inherent characteristics of the promising TE materials like potential high efficiencies and environmentally clean from the use of geothermal and solar heat made the substitution a reality (Weidenkaff et al., 2008). Moreover, as of the rapid development of the modern fabrication and characterization techniques, especially the emerging of nanoscale composite materials, a brand-new period of manufacturing nanocomposites is approaching. The enormous needs for sustainable energy combined with recent advances in thermoelectrics inspire an increasing excitement as always.

Through several decades' endeavors of the researchers, TE nanocomposites have been applied widely to an advanced level. Practical applications have been found such as renewable energies (Robert et al., 2007), electrical power generation, such as thermoelectric generators providing power in remote terrestrial and extraterrestrial like deep space exploration (Snyder & Toberer, 2008), air-warming systems (Cosnier et al., 2008), cooling systems (Lineykin & Sam, 2007) like solid-state Peltier coolers with precise thermoelectric effects for optoelectronics and passengers' seat in aotomobiles, and like thermoelectric micro-coolers with high cooling power density, short response time and device scalability which are very suitable for high-power and compacted microelectronic devices (Liao & She, 2007) and self-powering sensors (biomedicine, environmental monitoring, gas sensing, radio frequency field detector, infrared rays detector such as $SrSi_2$ (Hashimoto et al., 2007) and mobile phones in the future (Dragoman & Dragoman. 2007)).

High efficient utilization of waste heat energy from the surrouding environment is attributed to TE technology application. Considerable quantity of power is generated by heat energy with typical efficiencies in the range from 30% to 40% efficiency. At such rates, it is estimated that 15 terawatts of the power energy generated by heat energy is lost to the environment. To efficiently convert lost heat into electricity, thermoelectric nanocomposite materials can be used to accomplish this mission. Automobile exhuast and industrial manufacturing processes all generate an enormous amount of unused waste heat that could be high-efficiently converted to electricity by using silent, scalable and reliable TE devices which can scavenge waste heat from the environment. This technology can be employed in the vehicle field for the new generation vehicles with electrical power from waste engine heat from exhaust systems and radiators. One critical merit of thermoelectric nanocomposite is their scalability. For example, waste heat (Gelbstein et al., 2008) and co-generation sources could be as small as a home water heater, or as large as inductrial sources. This technology of energy conversion from waste heat into electrical energy is very important to develop alternative energy technologies in the reduction of our dependence on fossil fuels. To better understand the thermoelectric nanocomposites, knowledge in the fields such as solid-state chemistry, solid-state physics, thermal transport measurement and high-temperature electronics is required.

1.1 Development of thermoelectric materials

Thermoelectrics have long been considered inefficient for its most applications. TE nanocomposites have not resurged until last century when theoretical studies predicted that thermoelectric efficiency could be greatly improved through quantum confinement of the electron charge carriers. The electron energy band in a quantum confined structure is narrow and thus producing effective masses and therefore high Seebeck coefficient. In addition, heterostructured nanocomposites may decouple the Seebeck coefficient and electrical conductivity because of electron filtering which could progressively gain enhanced value of ZT. From the mid-1950 to the present, most of the bulk thermoelectric materials (like clathrates, skutterudites and Zintl phases) were found with high-efficiency. The pioneering work was performed and put into practice by Professor Abram (Nolasa et al., 1998). Professor Abram Ioffe and his collaborators did precursory work (Dresselhaus et al., 1999) on introducing semiconductors as promising thermoelectric materials. Then people gradually selected heavy elements from lower right part of the periodic table to synthesize semiconducting compounds. As a result, the lattice thermal conductivity can be reduced through manufacturing mixed or doped crystals. The more efficient thermoelectric materials at that time existed in the $Bi_{2(1-x)}Sb_{2x}Te_{3(1-y)}Se_{3y}$ family (Koga et al., 1997) right because of their lower thermal conductivity. Professor Abram's contributions to more efficient semiconducting materials paved a way for a quite active time when people are expressing great interests in thermoelectric materials. The representative ones are Bi_2Te_3, Bi_2Se_3, and Sb_2Te_3. The high potential of Bi_2Te_3 as a thermoelectric material made it still the basis in the thermoelectric field up to present.

Over the past 30 years, $Bi_{2(1-x)}Sb_{2x}Te_{3(1-y)}Se_{3y}$ and $Si_{1-y}Ge_y$ alloys have been extensively studied and optimized as TE materials for their applications like refrigeration and power generation. These conventional TE materials have experienced extensive investigation and there appears little room for future improvement in common bulk materials. As time goes by, a new class of TE compounds emerged and was investigated due to its higher performances. The comparison of ZT values between the traditional TE materials ($CsBi_4Te_6$, Bi_2Te_3, PbTe and $Si_{1-x}Ge_x$) and the new TE materials ($AgPb_mSbTe_{2+m}$, Zn_4Sb_3, $Yb_{0.19}Co_4Sb_{12}$ and $CeFe_{4-x}Co_xSb_{12}$) can be seen from Fig. 1.

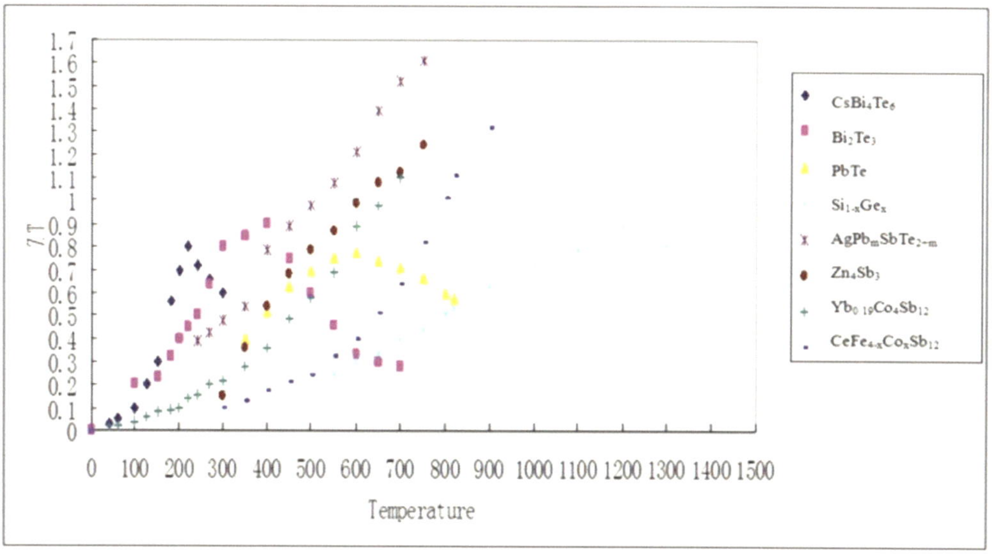

Fig. 1. Relations between temperatures (K) and ZT values. Data source from (Tritt et al., 2006)

During later 90s, researchers discovered that low dimension structures, as occurs in quantum wells (2D) or quantum wires (1D) can enhance the thermoelectric figure of merit rather than their bulk counterparts like two-dimensional PbTe quantum-well and 1D or 2D bismuth which are poor thermo-electrics in 3D. Although the low-dimensional TE materials like superlattices and nanowires could have a chance to improve ZT values due to the decoupling of the three parameters, it is still hard to employ superlattices and nanowires in large-scale energy conversion applications dut to their limitations in cost and heat transfer issures. Recently, many researchers used nanocomposite materials (polymers and inorganic materials) as the novel TE materials because composite materials have the properties from both the inorganic materials and polymers which could have higher values of figure of merit.

1.2 Thermoelectric effect

The thermoelectric effect is the direct conversion of thermal energy into electric energy through a themoelectric installation and vice versa. The Seebeck effect is the direct conversion of thermal energy into electrical energy. This effect was revealed by an Estonian-German physicist Thomas Johann Seebeck. The explanation of the thermoelectric mode (TE) is that the applied temperature gradient facilitates charged carriers, say electrons and holes in semiconductors, in the materials to diffuse from one end to another: hot charge carriers in the materials diffuse from the hot end to the cold end because of a lower density of hot carriers at the cold end and cold charge carriers diffuse from the cold end to the hot end. Consequently, the current is induced from the teperature diffecence between two ends of the materials. The reasons for the phenomenon of thermoelectric effect is that there are too many free charge carriers which will carry both charge and heat in metals and semiconductors. A net charge will be generated due to the build-up of charge carriers diffusing from the hot end to the cold end when a temperature gradient applied to a material. A net charge will produce an electrostatic potential. This phenomenon, known as Seebeck effect, is the basis of thermoelectric energy generation. (Snyder & Toberer, 2008)

The converse energy conversion process (ET) is well known as Peltier effect, which converts electrical energy into thermal energy.This phenomenon was discovered in 1834 by a French physicist called Jean-Charles Peltier who observed that when an electrical current is passed through the junction of two dissimilar materials, heat is either rejected or absorbed at the junction depending on the direction of the current. The Peltier effect is due largely to the difference in Fermi energies of the two materials. So the thermoelectric effect may also be called the Peltier-Seebeck effect. The thermoelectric effects have wide and practical applications across the whole world. The devices employing Seebeck effect called thermogenerators can convert various heat sources into electrical energy: turning waste heat from power plants and automobiles into electrical power, conversing radioisotope energy to support space probes and enhancing energy conversion efficiencies of solar cells by utilizing low frequency heat from the sun to a large extent. The devices using Peltier effect can create a compact refrigerator etc.

1.3 Measurements of thermoelectric nanocomposites

This section will first introduce the general measurements of the thermoelectric effect. Then this section emphasizes what sources can generate the thermal conduction and thus states the ways to reduce thermal conductivity. What's more, the internal conflicting factors that limit the maximum themoelectric performance is explained with several principles. Following that the relation of the carrier concentration and thermoelectric performance will be stated. Then, other equivalent measurements of thermoelectric performance will be introduced. At last, this section will talk about the thermoelectric devices' efficiency.

The main measurement of the performance of the thermoelectric nanocomposites is the dimensionless quantity ZT (see Eqn. 1). As a measurement standard, ZT is taken as the general expression of thermoelectric materials' effectiveness. Let T be the temperature (in Kelvin) and Z be the thermoelectric figure of merit. Then the dimensionless figure of merit has the following form of expression:

$$ZT=\alpha^2\sigma T/\kappa \tag{1}$$

In this equation, α stands for Seebeck coefficient measured in volts per kelvin (V/K), σ denotes the electrical conductivity measured in siemens per meter (S/m) and κ denotes the thermal conductivity measured in watts per kelvin per meter (W/K·m). α is the measurement of the induced voltages in response to temperature differences across the materials. It is called the thermoelectric power, or thermopower, or Seebeck coefficient. Typically the unit of the Seebeck coefficient is volts per kelvin, while in practice, it is more prevalent to use microvolts per kelvin (μV/K). The Seebeck coefficient can be represented by α which has the expression of delta V over delta T. Meanwhile, the Seebeck coefficient can also be defined as the following equation (Eqn. 2):

$$\alpha = \frac{8\pi^2 k_b^2}{3eh^2} m^* T \left(\frac{\pi}{3n}\right)^{\frac{2}{3}} \tag{2}$$

Where n is the carrier concentration; m* is the effective mass of the carrier; T is the temperature; e is the elementary charge; k_b is Boltzmann's constant; h is Planck's constant. Moreover, the Seebeck coefficient can also be defined as the following equation (Eqn. 3):

$$\alpha = \frac{\pi^2 k_b T}{3e}\left(\frac{\partial \ln \sigma}{\partial \varepsilon}\right)_{\varepsilon = E_F} \tag{3}$$

Where k_b is Boltzmann's constant; T is the temperature; e is the elementary charge; σ is the electrical conductivity and E_F is the Fermi energy.
Thermal conductivity stems from two aspects: κ_e (electrons and holes transporting heat) and κ_l (phonons travelling through the lattice). This means $\kappa=\kappa_e+\kappa_l$. Most of the κ_e is directly related to the electrical conductivity based on Wiedemann-Franz Law: the higher the electrical conductivity, the higher κ_l becomes. The Wiedemann–Franz law (see Eqn. 4), states that the electrical thermal conductivity (κ_e) is proportional to the electrical conductivity and the temperature (T).

$$\kappa_e = \sigma LT = ne\mu LT \text{ (For semiconductor)} \tag{4}$$

In above equation, L is the Lorenz factor with the value of 2.4×10^{-8} $J^2K^{-2}C^{-2}$ for free electrons; σ is the electrical conductivity; T is the temperature; n is the carrier concentration; e is elementary charge; μ is carrier mobility.
Based on the above stated Weidemann-Franz law, it is necessary to minimize κ_l instead of κ_e. In semiconductors, electron thermal conductivity is less than lattice thermal conductivity. Therefore, it is easier to decrease lattice thermal conductivity while maintaining the electrical conductivity. A phonon-glass and electron-crystal approach is proposed by G. A. Slack stating that phonons which are responsble for thermal conductivity must experience the material as they would in a glass meaning a high degree of phonon scattering, the lower the thermal conductivity; while electrons must experience the material as a crystal meaning little scattering maintaining the electrical conductivity. As a result of these two methosds, figure of merit will be optimized. This phonon-glass and electron crystal could also be explained as materials possessing electrical properties of a crystalline structure and thermal properties of an amorphous or glass-like material.
The higher value ZT of a material, the more preferable it is as a thermoelectric material for greater thermodynamic efficiency. It is very difficult to simultaneously achieve increased $\alpha^2\sigma$ and reduced κ due to interdependence of the three parameters. In other words, the three parameters are all functions of carrier concentration and are interrelated with each other. As a general rule, a rise in thermopower implies a decrease in electrical conductivity due to carrier density considerations which could be proved by the Seebeck coefficient equation, and an increase in electrical conductivity results in an increase in thermal conductivity based on Wiedemann Franz law. To assure a large Seebeck coefficient, there should be only one single type of carrier rather than mixed n-type and p-type conduction resulting in cancelling out the induced Seebeck voltages. In general, semiconductors have larger Seebeck coefficient than metals because doped semiconductors bear excessive electrons or holes, the contributions to Seebeck coefficient of which can not be cancelled out like metals. That is the main reason that semiconductors have a promising future as the thermoelectric materials. This could be demonstrated by the Fig. 2.
Figure 2 indicated the striking balance between large Seebeck coefficient and high electrical conductivity in TE nanocomposites which should be complied with maximizing the figure of merit ZT. The peak value is shown approximately at carrier concentration between 10^{19} and 10^{21} carriers per cm^3 with very high mobility. This is just right falls in the field between metals and semiconductors (Snyder & Toberer, 2008) which is the primary reason that

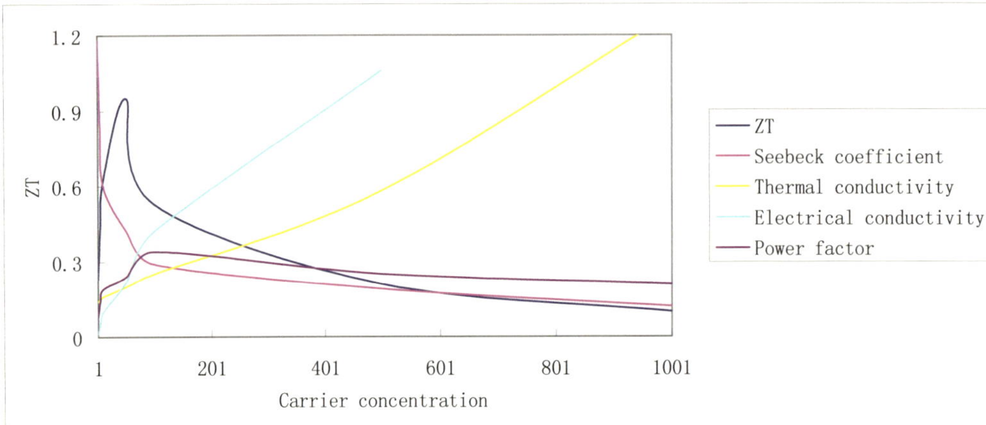

Fig. 2. Relations between Carrier concentration ($10^{18}/cm^3$) and related ZT factors. Data source from (Snyder & Toberer, 2008)

doped semiconductors were found with promising applications in TE nanocomposites. Generally speaking, maximizing the thermoelectric figure of merit refers to a compromise of thermal conductivity, Seebeck coefficient and electrical conductivity. From Fig. 2, thermoelectric power factor will maximize at higher carrier concentration than ZT.

To reduce the lattice thermal conductivity, the approaches fall into three ways. The first method is to separate the electron-crystal from the phonon-glass to guarantee not destroying the crystallinity of the electron-transport part; the second method is to form superlattices by scattering phonons at interfaces of nanostructured materials; the last method is to scatter phonons within the unit cell by generating rattling structures, vacancies or points defects.

There are some other equivalent measurements for TE nanocomposites. The first equivalent measurement of ZT (Eqn. 5) is to replace the denominator in Eqn. 1.

$$ZT=T\sigma\alpha^2/(\kappa_l, \kappa_e) \tag{5}$$

In Eqn. 5, σ,T and α are the same terms, κ_l and κ_e are the lattice contributions to the thermal conductivity and electronic contributions to the thermal conductivity respectively.

The second equivalent measurement of ZT is determined by Eqn. 6.

$$ZT=m^{1.5}\mu T/\kappa_l \tag{6}$$

In Eqn. 6, m is the effective mass, μ is the carrier mobility, κ_l is the lattice thermal conductivity and T is the absolute temperature. A large m results in a high Seebeck coefficient, large μ is good for high electrical conductivity, and the thermal conductivity κ_l relies heavily on the crystal structure.

For materials with high electrical conductivity or very low lattice thermal conductivity, the ZT can be defined by Eqn. 7:

$$ZT=\frac{\alpha^2/L}{1+\frac{\kappa_l}{\kappa_e}} \tag{7}$$

In Eqn. 7, L is Lorenz factor and κ_l/κ_e<<1.

Moreover, the power factor (see Eqn. 8), is used more and more frequently to measure the energy conversion efficiency.The definition of power factor is the square of Seebeck coefficient divided by electrical resistivity (ρ).

$$PF=\alpha^2/\rho \tag{8}$$

Effective thernmoelectric nanocomposites should be crystalline semiconductors that can scatter phonons without significantly interrupting the electrical conductivity because crystalline semiconductors have lower carrier mobility due to lower effective masses and increased electron scattering. And glasses will usually exhibit lower lattice thermal conductivity. However, glasses are not used as TE materials because they lack the needed electron-crystal properties. As a result, thermoelectric nanocomposites should bear the properties of phonon-glass and electron-crystal. The electron-crystal can compromise the conflicting Seebeck coefficient and electrical conductivity for crystalline semiconductors. The phonon-glass requires the lattice thermal conductivity as low as possible. This new field of vision to a large extent solves the traditional TE materials' problem. Traditional thermoelectric materials employed alloying with isoelectronic elements to conserve the crystalline structure at the price of disrupting the phonon path. Many experiments successfully demonstrated specific methods to synthesize electron-crystal and phonon-glass nanocomposite materials (Snyder & Toberer, 2008).

Thermoelectric devices consisted of many couples containing n-type and p-type thermoelectric elements assembled in parallel. For instance, a TE generator employs heat flow to power an electric load through the external circuit when applied a temperature gradient. Then a voltage is generated due to the Seebeck coefficient, or the heat flow derives the electrical current. So, it is important to know the measurement of the TE devices. A thermoelectric device's conversion efficiency η (see Eqn. 9), and modified figure of merit Zτ (see Eqn. 10), can be defined as:

$$\eta = \frac{T_h - T_c}{T_h} * \frac{\sqrt{1+Z\tau}-1}{\sqrt{1+Z\tau}+\frac{T_c}{T_h}} \tag{9}$$

$$Z\tau = \frac{(S_p - S_n)^2 \tau^2}{[(\rho_n \kappa_n)^{0.5} + (\rho_p \kappa_p)^{0.5}]^2} \tag{10}$$

In Eqn. 9 and Eqn. 10, T_h is the temperature at the hot junction, T_c is the temperature at the surface being cooled, Zτ is the modified figure of merit, ρ is the electrical resistivity, τ is the average temperature between the hot and cold surface and subscripts n and p stand for n-type and p-type semiconducting materials.

In conclusion, semiconductor nanomaterials, which can get higher Seebeck coefficient and elevated electrical conductivity and lower thermal conductivity simultaneously or at least amelioration of some of the above three goals, give contributions to higher values of ZT, that means higher energy conversion efficiencies. To date, the best reported ZT value is between 2 and 3. Higher values of ZT for thermoelectrics are more suitable for future competition especially through nanocomposite materials.

2. Fabrication of thermoelectric materials

Before the discussions of thermoelectric materials' advancement, it is necessary to introduce the themoelectric materials fabrication methods to obtain a comprehensive and clear panorama. Then the thermoelectric, photoelectric, electromagnetic, electrochemical and mechanoelectric characteristics of any material can be well investigated. The 'fabrication' here has two different meanings: chemical fabrication and physical fabrication. It is obvious that chemical fabrication means that a new substance has been produced from the starting materials, while physical fabrication stands for a transformation among different physical states like gas, solid or liquid. Consequently, the following parts will explain the commonly used approaches of fabricating thermoelectric materials.

2.1 Fabrication from gas to solid phase

Numerous techniques refer to the concentration of gas compounds on solid-state materials. The most straightforward fabrication way is to use the phase or state transformation process from gas state to solid state to deposit a thin film or multilayers on an appropriate substrate. Among various methods based on phase transformations, the typical methods are Physical Vapor Depostion and Chemical Vapor Deposion

Physical vapor depostion

Physical vapor deposition depicts a purely physical process such as high temperature vacuum evaporation or plasma sputter bombardment, and this process can deposit thin films on a wafer or substrate through condensation of vaporized materials without any chemical reaction. PVD generally consists of evaporative deposition, electron beam PVD, sputter depostion, cathodic arc deposition, pulse laser deposition and molecular-beam epitaxy. And the most prevalent methods are the sputter deposition and molecular-beam epitaxy (MBE). MBE it the fabrication technique that can permit precise control over deposition composition. It can totally solve the basic problem involved in the general PVD. That is, it is hard to control the deposition composition especially when the starting reaction material is a complicated compound. Molecular-beam epitaxy happened in high vacuum or ultra high vacuum and most apprarent aspect of MBE is the low thin film growth rate(typically bewteen 300 nm and 1000 nm/hour) which in turn promted the films to grow epitaxially. As a rule in solid source MBE, ultra-pure elements such as gallium and arsenic are heated in different quasi-knudsen effusion cells until they begin to gradually sublimate, then the gaseous elements of such materials will condense on the substrate. Sputtering is another commonly used PVD method. It is a process in which atoms are expelled from a solid target by bombardment with energetic ions from an electrically excieted, low-pressure plasma. Owing to collisions, ejected ions or clusters of atoms will be deposited on the specific film substrate.

It is reported that ErAs nanoparticels were epitaxially doped into $[InGaAS]_{1-x}[InAlAs]_x$ using molecular beam epitaxy (MBE) (Zeng et al., 2009). One common example of sputtering is the preparation of bismuth telluride thin films. Many methods have been attempted to prepare the bismuth telluride(n-type and-228 μV/K, p-type 81 μV/K) thin films, including pulsed laser flash evaporation, ion-beam sputtering, metal organic chemical vapor deposition, molecular beam epitaxy, and electrodeposition techniques, while sputtering is the most easy and economic way. Both p-type and n-type bismuth telluride thin films were prepared by sequential sputter deposition of elemental Bi and Te layers and post-annealing treatment. Periodic Bi/Te multilayer structures were sputter-deposited at room temperature and transformed into bismuth telluride via Bi/Te interfacial reaction during post thermal annealing (Liao & She, 2007).

Chemical vapor deposition

Compared with Physical Vapor Deposition (PVD), Chemical Vapor Deposition (CVD) gained much attention with its high growth rate and low cost of reaction equipment and it is the most helpful microfabrication technique which is used to produce high-purity and high-value thin films of semiconducting materials. In a typical CVD microfabrication process, gaseous precursors will react or decompose on the surface of a substrate or wafer to acquire the desired deposition thin film. Then the redundant gas or volatile by-products will be removed by gas flow through the reaction chamber. The deposition forms of CVD method include monocrystalline, polycrystalline, amorphous and epitaxial such as nanofibers, filaments, nanotubes, and various semiconductor compounds. The deposition rate is temperature-dependent in a typical CVD microfabrication process. When the temperature is high, mass-transfer will decide the whole deposition rate; on the other hand, if the temperature is low, the deposition rate will be determined based on chemical reaction. As a result, it is quite important to maintain a uniform temperature distribution to guarantee homogeneous deposition. Moreover, there is one special deposition technique called vapor-liquid-solid(VLS) deposition which involves multiphase catalyst-assisted heterogeneous chemical reactions. In this process, the reactants come from gas phase, while the liquid catalyst expedites the the deposition on the substrate. Similar to VLS, laser-assisted catalytic growth is another promising technique used for fabrication of nanostructure deposition. This method employs catalyst-assisted, multiphase chemical reaction which is the same with VLS. The differences between these two methods are laser-assisted catalytic growth not only hailing from gases but also from bombardment of laser beams. Catalyst such as Fe, Ni, and Au can be prepared on the substrate. Consequently, nanostructures and core/shell coaxial nanotubes can be synthesized through this technique.

Here we only introduce some common classicifications. According to operation pressure, CVD will be divided into atmospheric pressure CVD (APCVD), low-pressure CVD (LPCVD) and ultrahigh vaccum CVD (UHVCD). In view of plasma methods, the CVD includes microwave plasma-assisted CVD (MPCVD), plasma-enhanced CVD (PECVD), remote plasma-enhanced CVD (RPECVD), atomic layer CVD (ALCVD) combustion chemical vapor deposition(CCVD) rapid themal CVD (RTCVD) and vapor phase epitaxy (VPE).

The following examples are given to illustrate the applications of CVD. Zinc-doped cubic boron nitride thin films were prepared by films deposited in pure Ar (Nose & Yoshida, 2007). The crystals of $FeSb_2$ were prepared by the self-flux method or by the chemical vapor transport method (Sun et al., 2009). A key application of LPCVD is employed for fabricating polycrystalline Si by heating silane in a low pressure reactor through pyrolysis, in this process, the by-product H_2 will escape from the growing film of solid silicon. This reaction can be expressed as $SiH_4= Si\downarrow+2H_2\uparrow$.

Another prominenet use of LPCVD is to fabricate SiO_2 nanoscale particals. In a CVD chamber, silane and oxygen can react with each other through the reaction $SiH_4+2O_2=SiO_2\downarrow+2H_2O$. Moreover, using LPCVD can synthesize the inert silicon nitride like Si_3N_4 through equation $3SiCl_2H_2+4NH_3=Si_3N_4\downarrow+6H_2\uparrow+6HCl\uparrow$. PECVD can enhance CVD through electrical plasma;

this process can also be called glow-discharge decomposition. This method is generally used to produce thin films of amorphous or polycrystalline semiconductors. Microcrystalline Si can be manufactured through the dilution of SiH_4 by H_2 at higher wafer temperature. This technique is simultaneously used to fabricate crystalline diamond thin films from a hydrocarbon-hydrogen gas mixture. Also SiO_2 can be obtained through the disassembling of tetraethoxysilane or through the chemical reaction of dichlorosilane and nitrous oxide following the equation of $SiCl_2H_2+2N_2O=SiO_2\downarrow+2N_2\uparrow+2HCl\uparrow$.

As for the VLS, due to the influence of catalyst, one-dimensional nanostructure such as nanofibers and nanowires generated from the interface between the liquid and solid. Au-catalyzed VLS growth of phosphorus-doped silicon nanowires, indium nitride nanowires in a dissociated ammonia environment and germanium nanowires with great crystallographic orientations at 350 °C were well fabricated with fabulous shape and orientation. Perhaps the most obvious advantage of laser-assisted catalytic technique is the fabrication of ordered longitudinal herostructure such as herojunctions and superlattices of Si/SiGe.

2.2 Fabrication from liquid to solid phase

Crystalline and non-crystalline materials can be fabricated starting from the liquid phase when the desired materials are in big volumn, which means materials manufacturing takes place in the liquid phase in which it is easy to manipulate and achieve. This section will introduce the methods of liquid phase deposition and Sol-gel deposition.

Liquid phase deposition

Using liquid phase deposition onto substrate or wafer is a common way of fabricating nanoparticles, nanorods and nanotubes. Yttrium aluminium garnet nanopowder was synthesized with an average grain size of 80 nm. Single crystalline ZnO nanorods with smallest diameter of about 5 nm were fabricated at ambient temperature in liquid phase.Also, CdS/TiO_2 hybrid coaxial nanocables using liquid phase deposition were synthesized by soaking porous anodic aluminum oxide template into mixed solutions. Prior literature has suggested that the fabrication of PbS, CuS and CdS nanorods can occur either by electrodeposition or injection of reactants within the channel pores of either anodic aluminum oxide or mesoporous silica templates. Parallel nanowires of CdS and ZnS (Zhang et al., 2009) with the diameter as small as 3 nm templated was another example synthesized by hexagonal liquid crystals (Li et al., 1999).

Sol-gel deposition

Sol-gel deposition which is well known as chemical solution depostion is widely used in the field of materials fabrication like metal oxide. This method's starting material is a sol, which is a colloidal dispersion of small paticles suspended in liquid. Then under appropriate conditions, the sol can form a continous network of connected particles called a gel, which is discrete particle. The sol-gel process can fabricate gel materials in a wide range of forms and structures. The precusor sol can be deposited on a wafer to be cast into specific desired shape in a container. This approach that allows the great control over the product's composition is a cheap and low-temperature technique.This method can typically be used in the synthesis of refractory oxide at relatively low temperature like fabrication of SiO_2 glass at operating temperature of 1200 degree centigrade. Nanocrystals like TiO_2, ZnO and SnO_2 oxides were fabricated recently employing sol-gel method.

2.3 Fabrication from solid to solid phase

This section will introduce the fabrication ways from solid state to solid state: hot pressing, spark plasma sintering, electrodeposition and doping technique will be recounted.

Hot pressing

Hot pressing is used for fabricating powder or powder compact with high pressure and high temperature as a result of which sintering process was begetted. Hot pressing is primarily employed to synthesize hard and brittle materials. One of prevalent examples is the reinforcement of cutting tools and technical ceramics. The hot pressing technique can be found in the form of induction heating, indirect resistance heating and direct hot pressing. In the process of inductive heating, a high frequency electromagnetic field by electronic generator and induction coil is exerted on the mould so that the the heat will be generated wihtin the mould. The merit of it is the pressure and inductive power is not dependent. The disadvantages are the the high expenditure of high-frequency generator and the dependence on satisfactory thermal conductivity and inductive coupling. The indirect resistance heating technique placed the mould in a graphite heating chamber which will cause the heat transferred into the mould. The benefit of this method is indenpendent from mould's conductivity, heat and pressure. Also a high temperature is easy to attain. But the demerit is it will take longer time to heat up the mould and heat transfer from furnace to the mould surface. The direct hot pressing is under developing. New methods called Spark Plasma Sintering (SPS) or Field Assisted Sintering Technique (FAST) have been established. These methods will be introduced in the following paragraghs.

The examples of fot pressing here are the partially filled ytterbium skutterudite materials and n-type $CoSb_3$ nanocomposites. In an argon atmosphere at 26 ksi and 650 °C for 2 h, densification of the powdered specimens into dense polycrystalline pellets was accomplished by hot pressing using graphite dies to acquire partially filled ytterbium skutterudite materials (Nolasa et al., 2000). Under the usage of hot pressing the mixture of nanoscale and micronized $CoSb_3$ powders synthesized by solvothermal method using $CoCl_2$ and $SbCl_3$ as the precursors and $NaBH_4$ as the reductant in an ethanol solution and then melting, then n-type $CoSb_3$ nanocomposites were prepared by this method (Mi et al., 2007).

Spark plasma sintering

In the process of Spark Plasma Sintering (SPS), the mould is directly connected to electrical power to shorten cycle time and save energy, and SPS has higher heating conversion rate. This helps increase the speed of sintering of metal power and reduce the threshold sintering teperature and pressure compared to traditional sintering process. In addition to these, in contrast to conventional hot pressing where heat is provided by external heating, SPS 's heat is generated internally. Therefore, the sintering process is fast. The TE compound In_4Se_{3-x} was fabricated by SPS through a solid state reaction at 550 °C under the uniaxial pressure of 70 MPa.

There are a lot of examples of SPS for TE nanocomposites. Through SPS and the subsequent thermal treatment, the polycrystalline copper aluminates $Cu_{1-x-y}Ag_xNi_yAlO_2$ and $Cu_{1-x-y}Ag_xZn_yAlO_2$ were synthesized. These nanocomposites could be used as high temperature thermoelectric materials by partial substitution of Ag, Ni and Zn for Cu site in $CuAlO_2$ (Yanagiya et al., 2010). Bi-Te based TE p-n modulus were fabricated through SPS solid bonding for measuring their TE performances (Kima et al., 2006). The polycrystalline samples such as $BaSi_2$, $SrSi_2$, and LaSi were prepared by SPS. Among these SPS generated materials, $SrSi_2$ exhibited a higher power factor than BaSi2 and LaSi. Also, after heating the pressed pellets of pure mixture of Sb (99.9999%), Co (99.99%) and fullerene (C_{60}) 99.5%, and then heating the mixed nanocomposites to 943 K under an Ar atmosphere for about 150 h, fine TE ground powders were obtained through SPS at 848 K for 15 min (Shi et al., 2004). It is reported that $Ba_yCo_4Sb_{12}$ nanocomposites with dispersed fullerene or barium fulleride have been successfully fabricated by the SPS and solid state reaction (Shi et al., 2007). Also, $SiO_2/\beta\text{-}Zn_4Sb_3$ core-shell nanocomposite particles with the low thermal conductivity were synthesized by SPS method (Ruan & Xiao, 2007). Moreover, $CoSb_{3-x}Te_x$ skutterudite polycrystals were synthesized by SPS along with mechanical alloying (Liu et al., 2007). Furthermore, Zn-incorporated $Ba_8Ga_{16}Zn_xGe_{30-x}$ clathrates were fabricated with the methods of SPS and solid-state reaction (Deng et al., 2007). To enhance the bulk material s' densities, highly aligned $Ca_3Co_4O_9$ and similar formula ceramics can be treated by SPS to achieve this purpose (Zhou et al., 2003).

Electrodeposition

Electrodeposition is the process that employs electrical current to reduce cations of a specific material within the solution and deposits nanoparticles on the surface of cathodes to synthesize a thin layer of the material with desired properties. This process bestowed many merits like depositing on complicated surfaces and synthesizing functionally-gradient materials.

Fig. 3 shows how the electrodeposition worked. The set-up for electrodeposition includes the anode, the cathode and a specific solution. They are connected to a power supply such as battery. When the external power supply is switched on, the metal at the anode will be oxidized with a positive charge. These cations will meet anions in the solution, and thus they are reduced at the cathode to deposit in the specific metal. Taking an acid solution as an example, copper is oxidized at the anode becoming Cu^{2+} by losing two electrons. Then

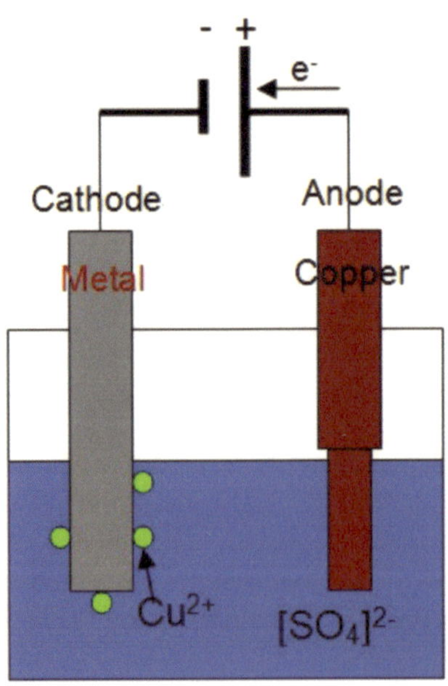

Fig. 3. Electroplating of a metal with copper in a copper sulfate bath.

the Cu^{2+} will meet up the anion SO_4^{2-} in the solution to form copper sulfate. At the cathode, the Cu^{2+} is reduced to metallic copper by gaining two electrons. The result is the effective transfer of copper from the anode source to a specific metallic cathode.The reaction mechanism of electrodeposition has been extensively studied recently to gain expected and improving properties of various materials.

The application of electrodeposition has an extensive range.The most cheap and easy way to synthesize polyaniline is utilizing the electrodeposition method and using a stainless steel as the template. It is reported that electrodeposition of nanometer-sized alumina particles in a copper matrix was prepared and well studied. It is found that PbTe can be deposited on copper and nanoporous nickel through electroplating (Madhavaram et al., 2009). Also, a multitude of nanoparticles were inserted into metallic or ceramic coatings: submicron polymeric particles were electrodeposited into copper; SiC particles were incorporated into Ni matrix. Simultanesouly, the morphology and mechanical properties of electrodeposited Al_2O_3 nanoparticles that solidated copper matrix were studied. Additionally, nanostructured devices such as microgears and pulp systems were possible when using electrodeposition method. Here is the most popular thermoelectric material Bi_2Te_3 that has been studies for years. Extensive studies on Bi_2Te_3 doping have been carried out by many researchers for the optical, thermoelectric and electrical properties. And this material's growth technique was achieved by vacuum deposition technique, sol–gel, spray pyrolysis, screen printing etc. While the most economical method of fabricating bismuth telluride thin film is by cathode electrochemical deposition technique. Thin films of bismuth telluride can be grown by cathode electrochemical deposition technique on conducting glass and Mo sheet substrates. They can easily be used as a low band gap dopant in the materials for large-scale photovoltaic applications. The quality and composition of the deposited films can be easily controlled by controlling electrochemical parameters. The as-deposited bismuth telluride films are highly crystalline and compositionally and morphologically uniform throughout. This accomplishemt suggests that electrochemical deposition is a cost effective, efficient, non-hazardous process for the preparation of bismuth telluride films and large area depositions. The feasibility of preparation of bismuth telluride films was demonstrated in the aqueous electrochemical deposition process (Golia et al., 2003).

Doping technique

Doping is generally referred to the process of intentionally adding impurities to extremely pure semiconductors. Hence, the doping technique is used in the semiconductors to change the material' electrical properties or other characteristics. The doped-materials have a promising application for the thermoelectics. It is known that perovskite-type metal oxides are potential materials to enhance the efficiency of thermoelectric devices and cobaltite with p-type conductivity and n-type manganates are considered for the development of a ceramic thermoelectric converter. Sintered pellets and thin PLD films with the composition $La_{1-x}Ca_xMO_3$ (x= 0, 0.3, 0.4) (M=Co, Mn) were synthesized and characterized concerning their thermoelectric properties in a broad temperature range recently. It was found that similar to polycrystalline samples the electrical conductivity of $LaCoO_3$ increases significantly with 40% Ca-substitution due to the formation of Co^{4+} ions while the thermopower decreases. The thermopower values of the $La_{0.8}Ca_{0.2}MnO_{3-d}$ films have a negative sign, but become large and positive at temperatures of 1000 K. Doping technique is suitable for the production of homogeneous metal or metal alloy powders. Silicon– germanium (Si–Ge) has been known as a high temperature thermoelectric element material, which has the function

of thermoelectric energy conversion by making use of a temperature difference between both ends of the element (Otakea et al., 2004).
Furthermore, a route, alternative to the conventional chemical doping, to modify a wide spectrum of physical properties of nanoscale sized films, by dynamically varying the concentration of the carriers by an applied electric field (Electric field effect doping-EFED), has been proposed recently. This theory of an electric-field-effect (EFE) induced increasing of the thermoelectric figure of merit of a thin film thermoelectric TE plate in a capacitive structure. EFED is applicable for thin films of narrow-gap semiconductors (e.g. PbTe, SnTe, PbSe, PbS, etc.) and semimetals (such as Bi) (Sandomirsky et al., 2003).

3. Major methods to enhance conversion efficiencies

A long pursued goal is to achieve larger ZT. Expectations appeared a decade ago, when it was theoretically proposed that superlattices and nanowires can have a much higher figure of merit than traditional bulk materials. Since then, important theoretical and experimental steps have been made, bringing us closer to the practical realization of the theoretically envisioned breakthrough. Here are three primary ways employed to improve the thermoelectric figure of merit based on (Eqn. 1): to enhance the Seebeck coefficient (thermopower) or power factor, to increase the electrical conductivity and to lower the thermal conductivity. The power factor can be enhanced by means of quantum confinement effects. The Seebeck coefficient can be increased via energy filtering and electron scattering. Typical good Seebeck coefficient's value should be larger than 200 $\mu V/K$. The low lattice thermal conductivity could be achieved from the increasing of phonon scattering by a mixed or complex structure, rattling scattering, grain-boundary scattering and huge surfaces and interfaces in nanocomposite materials. The electrical conductivity can be improved by doping or using metal-like conducting semiconductors, or in other words high-mobility carriers and very small gap will result in high electrical conductivity. Typical good electrical conductivity values should be larger than $10^3\ \Omega^{-1}\ cm^{-1}$. And the thermal conductivity can be lowered using doping technique of nanopartices and superlattice heterostructures.Typical good thermal conductivity value should be less than 2.0 $W/m{\cdot}K$. Slack predicted that the low thermal conductivity can be achieved is between 0.25 $W/m{\cdot}K$ to 0.5 $W/m{\cdot}K$. The following part will elaborate all the three methods, but not limit to any single method. For example, some thermoelectric nanocomposite mainly possesses low thermal conductivity, but at the same time, it may be provided with higher conductivity or higher themopower.

3.1 To increase electrical conductivity

The first method to increase ZT value is through the enhancement of electrical conductivity. Electrical conductivity is also called specific conductance, which is used to measure a specific material's capacity to conduct an electric current. The definition of electrical conductivity σ is the ratio of current density J to the electric field strength E with the expression $J= \sigma E$. So the improvement of electrical conductivity is to enhance the ability to conduct electric current. This part will narrate from three aspects: doping nanocomposite materials, Co-based oxide ceramics and metal-like nanocomposite all leading to high electrical conductivity.
Improvement of electrical conductivity by doping was quite effective to increase figure of merit value. As the concentration of zinc dopant increased, the electrical conductivity of films deposited in pure Ar can be increased. For example, The electrical conductivity of zinc-

doped cubic boron nitride (cBN) thin films, prepared by sputter deposition in pure Ar, increased as the concentration of zinc dopant increased. It was found that the conductivity increment in such films was accompanied by a linear increase in the B/(B+N) ratio. It was also found that both Al_2O_3, TiO_2 or ZrO_2 doped ZnO and TiO_2 doped Fe_2O_3 showed a large power factor in the wide temperature range. Fe_2O_3's ZT value was smaller compared with ZnO doped nanocomposites. Additionally, Ag and Zn doped $CuAlO_2$ was found to enhance the electrical conductivity (Tsuchida et al., 1996). α-Al_2O_3 with doped Mg nanocomposite was investigated for its electrical conductivity under the direct and alternating current with different concentrations of Mg in the temperature range 250-800 K. Al_2O_3 crystal is a high-temperature and good mechanical semiconductor rather than MgO doped with lithium which is brittle as a semiconductor of the p-typed. Although there are many impurities existed in the Al_2O_3 crystals, the electrical conductivity of Al_2O_3 with dopant Mg increases linearly with the concentration of Mg in both DC and AC fields (Tardı́o et al., 2001).
Oxide ceramics possess good TE properties mostly because of their thermal stability. It has already reported that (Ca, Bi, and Sr)-Co-O single crystal whiskers have the ZT values higher than 1.2 and they are suitable for practical applications. But the size of the materials limit their applications as a TE device. While the $Ca_3Co_4O_9$ (See Fig. 4) and $NaCo_2O_4$ are reported to bear great TE performances. Thus they are more favorable than traditional semiconducting materials. Moreover, the transport performance of $Ca_3Co_4O_9$ is increased significantly because of its layered structure as followes.

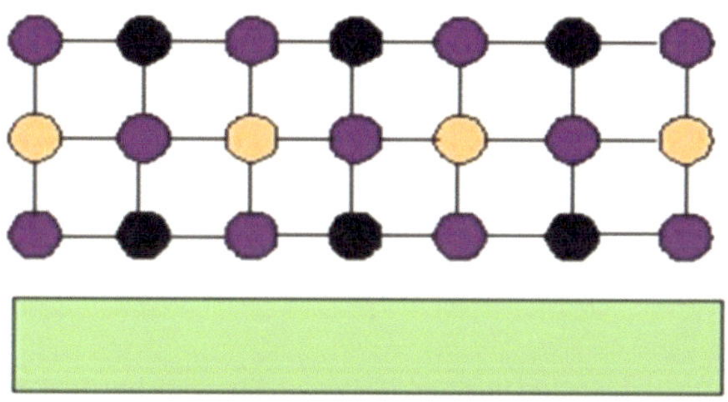

Fig. 4. Layered view of $Ca_3Co_4O_9$ after (Zhou et al., 2003).

The in-plane electrical conductivity is much larger than that of the out-of-plane electrical conductivity, showing that grain-aligned ceramics could have higher TE properties than randomly oriented ceramics with the same compositon. As a result, the fabrications of the highly grain-aligned densified TE ceramics are quite crucial for the application of polycrystalline layered structure. A ZT value of 0.8 is obtained for oxide $NaCo_2O_4$ at 1000 K. Furthermore, cation substitution can improve the ZT value. For example, an improved ZT value of $NaCo_2O_4$ can be proved from Cu-substitution for Co site; an enhanced ZT value of $Ca_3Co_4O_9$ can be proved from Bi and rare earth elements for Ca site due to increased Seebeck coefficient. $Ca_3Co_4O_9$ showed higher electrical conductivity with grain-aligned structure and had no apparent effect on the Seebeck coefficient which could reach 180 μV/K at 700 °C (Zhou et al., 2003). Polycrystalline bulks with aligned grains of the misfit-layered cobalt oxides like $[(Bi_{1-x}Pb_x)_2Ba_2O_4]_{0.5}CoO_2$ was reported to show higher electrical conductivity. The electrical conductivity of this material is four times higher than that of traditional sintered bulks. And its Seebeck coefficient is 120 μV/K at ambient temperature and its thermal conductivity is 1 W/m·K (Motohashi et al., 2008).

Some nanocomposites like green covellite copper sulfide possess metal-like electrical conductivity and chemical sensing capabilities. For example, CuS films maintain transmittance in the infrared, low reflectance in the visible and relatively high reflectance in the near-infrared region, which are ideal characteristics for solar energy adsorption. Moreover, CuS transforms into a superconductor at 1600K and has recently been used as cathode material in lithium rechargeable batteries (Zhang & Wong, 2009).

3.2 To enhance thermopower or Seebeck coefficient

The second way of improving the TE properties is through the increase of thermopower, or the power factor This part will first elucidate the nanocomposites with higher thermopower in carbon nanotubes, $X_3Au_3Sb_4$, $Bi_2Sr_2Co_2O_y$ whiskers, $LaCoO_3$ system, lead chalcogenide and $Ag_nPb_{18+x}SbTe_{20}$ bulk material. Then introducing some promising nanocomposites with higher power factor which is equal to the square of thermopower divided by electrical resistivity, will be stated afterwards like LaPdSb, GdPdSb, $YbAl_3$ and In_4Se_{3-x}.

The nanotube can be viewed as a graphene sheet rolled into a seamless cylinder with a length over diameter ratio greater than 1000. Generally, carbon nanotubes can be divided into two types. One is called single walled nanotubes (SWNTs). The other is called multiwalled nanotubes (MWNTs). Nanotubes normally can be synthesized from three methods: chemical vapor deposition, pulsed laser vaporization and electric arc discharge.

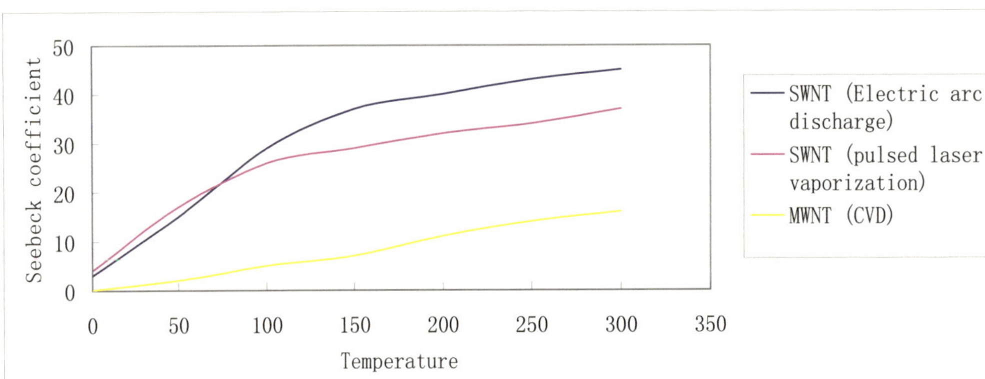

Fig. 5. Relations between temperature (T) and Seebeck coefficient (μV/K) for different carbon nanotubes. Data source from (Apparao et al., 2006)

Fig. 5 shows the general comparison of Seebeck coefficient carton nanotubes with three different fabrication methods: SWNT synthesized by the electric arc discharge and pulsed laser vaporization and MWNT produced by CVD. Clearly, the SWNT (45 μV/K) has a much higher Seebeck coefficient than MWNT (17 μV/K). The positive of the Seebeck coefficient proves both SWNTs and MWNTs are p-type materials (Apparao et al., 2006).

Taking carbon nanotube as an example, ZT in nanocomposite materials could be larger than 2.5 due to electron confinement resulting from a high density of states near the Fermi level E_F thus leading to high power factor, while κ decreases due to the suppression of phonon boundary scattering. On the contrary, ZT could near 0.8 as a maximum for bulk thermoelectric materials. Such nanocomposite materials are easily assembled into different shapes and can be scaled up for commercial applications. Wireless sensors in the 1-10 m range require very low-power consumption, which does not exceed few hundreds of microwatts. For longer distances the wireless sensor networks are designed to work in a multihop mode, while large distances are covered with multiple miniaturized sensor nodes characterized by low-power consumption and low cost. The modern architecture based on a large number of wireless nodes is possible only if each node is self-powered when working inside the network. In the advanced wireless sensor and portable device architectures carbon nanotubes (CNTs) could play a double role as sensors and harvesters, generating DC voltages necessary for the operation of small wireless sensors in the multihop mode. CNTs are very effective sensors, with many applications in environmental monitoring, biomedicine, or automotive industry. Smart biological devices could be implemented using this harvesting principle, but the generated power is of few picowatts, which is not enough for wireless applications. However, thermoelectric power devices based on CNTs could be the key elements of self-powering devices of wireless sensors as well as of any low-consumption portable device (Dragoman & Dragoman. 2007).

The relatively earlier nanocomposites with notably performance of thermoelectricity are ternary antimonides and stannides with half-Heusler structure. The recently reported TE material like $X_3Au_3Sb_4$ has a more complex crystal structure than those earlier nanocomposites. The specific formula is $Ln_3Au_3Sb_4$ for Ln= Nd, Sm, Gd and Ho with lower thermal conductivity in the range of 1.4-1.8 W/m·K at ambient temperature because of heavy masses of the component atoms and complex structure. The highesst Seebeck coefficient ever reported is 100-200 μV/K, showing that the dominant carriers are holes enhancing with increasing temperature. The resisitivities and Seebeck coefficient can be greatly affected by the ratio of Au:Sb. Moreover, there is no influence of the electrical resisitivity resulted from the improved Seebeck coefficient for nanocomposites with mixtures of lanthanides on the large atom position (Young et al., 1999).

There are other alloys with higher Seebeck coefficient. $Bi_2Sr_2Co_2O_y$ whiskers possess high performance of thermoelectricity. The Seebeck coefficient of this material can reach 300 μV/K at 973 K. The thermal conductivity is measured about 2 W/m·K because of the phonon-phonon interactions. The ZT value is more than 1.1 which stands for about one tenth energy conversion at 973 K (Funahashi & Shikano. 2002). Due to the high Seebeck coefficient of 600 μV/K at room temperature, the $LaCoO_3$ system is a promising thermoelectric nanocomposite despite of its high electrical resistivity which lower the whole conversion efficiency. But as the temperature goes up to 750K, the electrical resistivity will drop down which again enhances the conversion efficiency. Furthermore, different substitution of a specific site will cause promotion of ZT. Substitution in the $LaCoO_3$ system like $La_{0.8}Ca_{0.2}MnO_3$ and $La_{0.6}Ca_4CoO_3$, represents a semiconducting-like behavior, which can improve the conversion efficiency. Another example of the $LaCoO_3$ system is La-Sr-Co-O compound. The polycrystalline $La_{0.8}Sr_{0.2}Co_{1-x}Mn_xO_3$ was synthesized by citrate complex method. And the TE propertied can be tuned by adjusting Co site substitution by Mn atoms. The experimental results showed that both the Seebeck coefficient and electrical resistivity will increase with the increasing Mn content. The Seebeck coefficient is positive in the temperature range between 90 K and 290 K, and the electrical resistivity indicated a metallic semiconducting temperature behavior. Also a power factor of 18 $\mu W/K^2$ is achieved with x=0.08 (Moreno et al., 2008).

Prevalent examples of wide applications with higher Seebeck coefficient existed in the lead chalcogenide which has been studied for years due to their special physical characteristic and easy fabrication within ambient pressure. The synthesized lead chalcogenide possesses the advantages of a short growth time, high yield and cost effective. Lead chalcogenide exhibits extensive quantum-size effects in nanocrystalline form because of its smaller bandgap and larger Bohr excitation radius. Up to now, a variety of forms such as nanorods, nanowires, nanotubes, nanocubes and nanoflowers have been successfully fabricated under controllable conditions. Parkin and co-worders used a method to fabricate lead chalcogenides in liquid ammonia at room-temperature. Later, it is reported that the synthesized size of PbS, PbSe and PbTe nanotubes is 200-300 nm, 50-120 nm and 30-60 nm respectively fabricated in hydrazine hydrate saturated alkalline solution under the general experimental conditions. And the Seebeck coefficient value of the three composite nanotubes are 154.4 μV/K, 199.8 μV/K and 451.1 μV/K respectively (Wan et al., 2010). Through the TEM images and electron diffraction patterns, lead chalcogenide is single crystal nanotube and PbTe nanotube has the smallest size with a round-angle. It is concluded that PbTe nanotube bears the higher thernoelectric property. Additionally, PbTe nanorod's Seebeck coefficient could be up to 679.8 μV/K at room temperature. The thermopower of PbTe coating on copper and nickel was found much higher than either pure copper or copper-nickel alloy. PbTe on nanoporous nickel shows higher Seebeck coefficient than on the pure copper. It is about 50% higher on nanoporous support. The possible explanation is that the electron confinement effect of nanoporous materials causes higher thermoelectric potential (Madhavaram et al., 2009). Because of the high potential usage of PbTe as TE material, various PbTe nanocomposite materials have been successfully synthesized by solvothermal/hydrothermal fabrication, sonoelectrochemical synthesis, chemical bath method and high temperature solution-phase synthesis (Wang et al., 2009). Addtionally, n-type TE material $Ag_nPb_mSbTe_{m+2n}$ system (ZT>2) and p-type TE material Ag_9TlTe_5 system are reported with easier fabrication methods like mechanical alloying and SPS as powder metallurgy processes. Specifically, $Ag_nPb_{18+x}SbTe_{20}$ bulk materials with micrometer sized grains were synthesized by the powder metallurgy process. A maximum power factor of 1.766 mW/ mK^2 was obtained at 673 K for the $Ag_{0.8}Pb_{22}SbTe_{20}$ corresponding to a high ZT value of 1.37. The experimental results also indicated that the TE properties are highly related to the composition especially the content of Pb atom (Wang et al., 2006).

There are also some TE materials with high power factor attributing to high Seebeck coefficient. LaPdSb showed a large power factor of 50 μW/cm K^2 at 327 K. And GdPdSb also indicated relatively large power factor with high Seebeck coefficient and low electrical resistivity resulted from high Hall mobility (Sekimoto et al., 2006). Moreover, both n-type $YbAl_3$ and p-type $CePd_3$ showed high Seebeck coefficient values. $YbAl_3$ exihibits a high power factor as high as 5.4 W/m·K at 300 K which is nearly 4 to 5 times larger than optimized Bi_2Te_3 based TE materials. However, the high thermal conductivity in $YbAl_3$ limits its total TE performance. Therefore, further investigation should be done on how to reduce the thermal conductivity to achieve high ZT value. The Se-deficiency polycrystalline compound In_4Se_{3-x} has the property of increasing the poewr factor and decreasing the semiconducting band gap. The experimental results showed that small Se deficiency will lower the thermal conductivity resulting from the random disorder phonon scattering. On the other hand, the thermal conductivity will increase with the increasing Se deficiency. Moreover, the power factor will increase with increaing Se deficiency between 300 K and 750 K. The power factor of In_4Se_{3-x} measured may be lower than that of double-doped skutterudites because of relativiely high electrical resistivity of $In_4Se_{3-x.}$ In the same temperature range, the ZT value of In_4Se_{3-x} due to high Seebeck coefficient and low thermal conductivity is comparable to the n-type PbTe and $CoSb_3$ (Rhyee et al., 2009).

3.3 To lower thermal conductivity

The third way to improve the TE performance or the ZT value is through reducing the total thermal conductivity. Lowering the thermal conductivity is the most focused way to obtain the high performance of TE materials. This part will elucidate the acquired accomplishments from Zintl compounds, skutterudite family, Zn_4Sb_3, clathrate, telluride alloys and InSb. Among these TE materials, rattlers atoms could be inserted into the voids to further lower thermal conductivity in cage-like materials skutterudites and clathrates.

Zintl compound

Nanocomposite Zintl have recently arised as a new class of thermoelectric materials for it has complex crystal structures. Zintl phase is the product of a reaction between alkali metals, alkaline earthes and post transition metals, metalloids. Zintal phases were named for the German chemist Eduard Zintl who investigated them in the last 30s. Zintl phases can provide the great properties for TE materials: they will form small band gap semiconductors and possess complex structures. As of the their valence compound traits, zintl phases have the semiconducting nature (Zhang et al., 2008). Zintl phases have the structural properties used for electron-crystal and phonon-glass characteristics which result in higher ZT value. They combine different regions of covalent bonding for electron-crystal properties with ionically bonded cations that can be subtstituted for tuning electronic traits and bringing disorder which generate low lattice thermal conductivity due to phonon-glass properties. Improved thermoelectric performance will be expected due to enhanced structural modifications that allow the increase in the fundamental transport parameters such as carrier mobility, carrier concentration and effective mass. The low thermal conductivity value is gained in Zintl thermoelectric materials on accounts of low velocity of optical phonon modes and point defect scattering.

A Zintl compound consists of both covalently and ionically bonded atoms. Then, the general ionic cations will give their electrons to the covalently anionic species. And Zintl phases comprise of electropositive cations (Group 1 or Group 2) which will give their electrons to electronegative anions; which in turn will form bonds to suffice for valence. Assuming the presence of both ionic and covalent contributions to the bonding type is the reason of the structural requirements of Zintl phase. The purely ionic materials possess less mibility of the charge-carrier species than covalent bonding. The conplex nanostructures with multiple strucural units in the same structure are resulting from the combination of the bonding types. It is reported that due to the reason that zintl phases possess highly electropositive alkali and alkaline-earth elements, they are air and moisture sensitive. To overcome this drawback, the air-stable antimonide analogs of zintl phases are synthesized: $X_{14}MPn_{11}$, $Yb_{14}MnSb_{11}$, $YbZn_2Sb_2$, $YbZn_{2-x}Mn_xSb_2$, $BaZn_2Sb_2$, $EuZn_2Sb_2$, $YbCd_2Sb_2$, $YbCd_{2-x}Zn_xSb_2$, $Ca_xYb_{1-x}Zn_2Sb_2$, $X_{11}Sb_{10}$ (X=Ca, Yb, and Eu), $Ca_{11}AlSb_9$, $CaZn_2Sb_2$, $BaGa_2Sb_2$, Mo_3Sb_7, filled skutterudite and Zn_4Sb_3 etc.

$X_{14}MPn_{11}$ expression conforms to zintl formulism. In this expression X means a heavy or alkaline earth metal; M is a transition or main group metal and Pn is group 15 elements such as P, As, Sb or Bi. Doping on different sites will improve the TE properties through electrical and thermal conductivities. For example, incorporating on M site will adjust the electronic parameter and incorporating on X sites will tune the disorder scattering of phonons and carrier concentration. $Yb_{14}MnSb_{11}$ is the appropriate candidate for polar intermetallic compared with the transition metal zintl compound $Ca_{14}AlSb_{11}$. And thus $Yb_{14}MnSb_{11}$ made ideal candidates for thermoelectric materials because it possesses the properties of electron-crystal and phonon-glass. The $Yb_{14}MnSb_{11}$ has extremely low lattice thermal conductivity and high Seebeck coefficient. The experimental results showed that Seebeck coefficient is positive and will increase with increasing temperature from 300 K to 1300 K; the resistivity increases with increasing temperature and gets a maximum value at 1200 K; a notable low thermal conductivity is between 0.7-0.9 W/m·K for temperatures from 300 K to 1275 K; substitution of La^{3+} for Yb^{2+} on X site and substitution of Mn^{2+} with Al^{3+} will result to a high ZT value at lower temperature (Kauzlarich et al., 2007). $Yb_{14}MnSb_{11}$ is comprised of distinct anionic units like $Yb_{11}Sb_{10}$ rather than structrue like $CaZn_2Sb_2$ shown later. This type of zintl material consists of polyatomic $[Sb_3]^{7-}$ anions, $[MnSb_4]^{9-}$ tetrahedra in addition to isolated Sb^{3-} cations and Yb^{2+} cations. This compound is isostructural to zintl phase $Ca_{14}AlSb_{11}$. A high temperature melting point is anticipated due to anionic units ionically bonded together via cations. Thus, $Yb_{14}MnSb_{11}$ has a much complex structure than $CaZn_2Sb_2$. The Fig. 6 is the general $X_{14}MPn_{11}$ layered view. In this figure X means 2+ cation with, M means metal and Pn means P, As, Sb or Bi. The green, blue and red spheres denote X, Pn and M atoms respectively and yellow spheres denote X atoms which are not in the same plane.

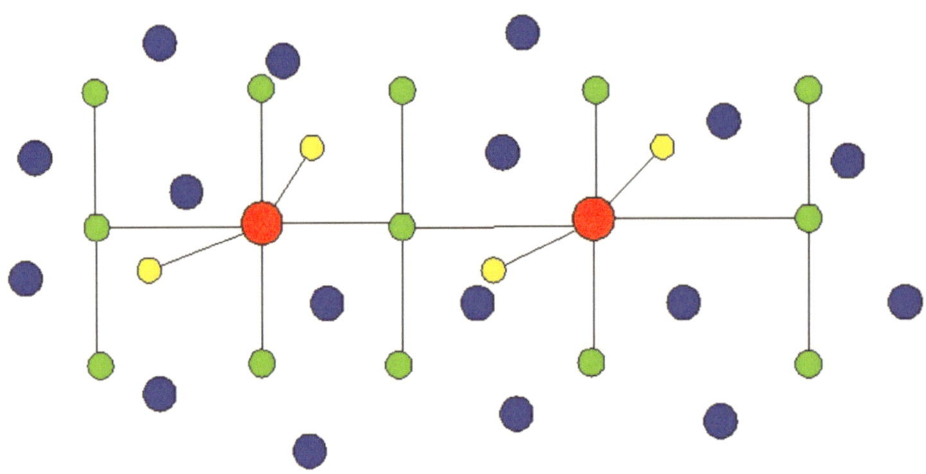

Fig. 6. Layered view of $X_{14}MPn_{11}$ structure after (Kauzlarich et al., 2007)

It is reported that $Yb_{14}MnSb_{11}$ has the ZT value of 1.0 at 900 °C along with large Seebeck coefficient and high electrical conductivity. The $Yb_{14}MnSb_{11}$ possesses a ZT value more than 1.0 up to 1275 K. In comparison, the ZT value of $Yb_{14}MnSb_{11}$ is proven that two times of the SiGe nanocomposite which is used prevalently before. The Fig. 7 showed ZT values of several p-type TE materials (Zn_4Sb_3, TAGS means $(GeTe)_{0.85}(AgSbTe_2)_{0.15}$, $CeFe_4Sb_{12}$, $Yb_{14}MnSb_{11}$, $CuMo_6Se_8$ and SiGe) at different temperature range. It indicated that $Yb_{14}MnSb_{11}$ possesses highest ZT value at higher temperatures. While the some other materials (Zn_4Sb_3, TAGS and $CeFe_4Sb_{12}$) indicated high ZT values at low temperatures.

What's more, it is reported that Zintl compound $YbZn_2Sb_2$ with isoelectronic substitution of Zn by Mn in anionic $(Zn_2Sb_2)^{2-}$ system possesses lowered thermal conductivity. The attained value of figure of merit value is between 0.61 and 0.65 at 726 K compared with the

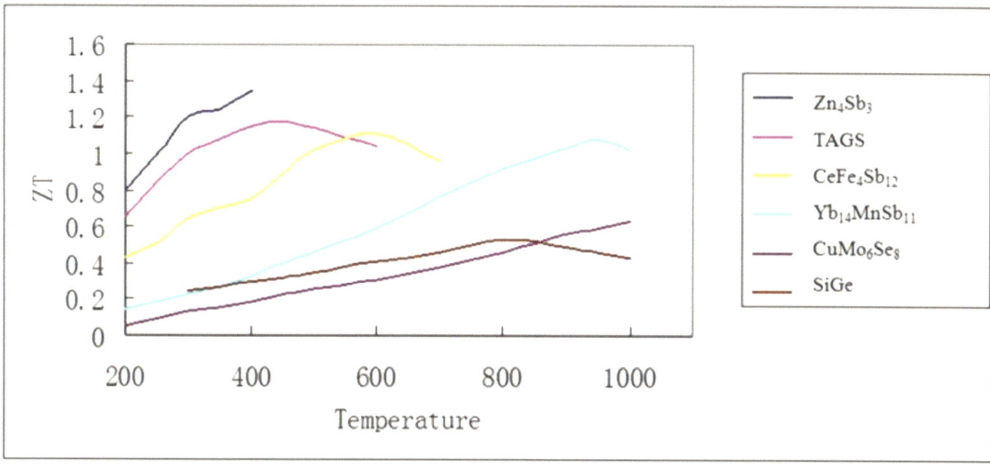

Fig. 7. Figure of merit as a function of temperature (°C) for a number of p-type materials. Data source from (Kauzlarich et al., 2007)

unsubstituted $YbZn_2Sb_2$'s ZT value of 0.48 (Zhu et al., 2008). Yb site substitution can well tune the TE properties in $YbZn_2Sb_2$. Here, through a solid-state reaction followed by certain cooling, annealing, grounding and hot-pressing densification process, the p-type $YbZn_{2-x}Mn_xSb_2$ were synthesized. The reason of the lowered thermal conductivity and significantly increased Seebeck coefficient is because the Mn substitution. As a result, ZT value of 0.65 is obtained at 726 K for x=0.05-0.15. In $YbZn_{2-x}Mn_xSb_2$ structure, the cell refinement revealed the lattice constant a increased linearly with increasing Mn content, while the lattice constant c showed a relatively small change. That means the unit cell expands in the x-y plane but little changes in the z-axis with increasing Mn contents. It can be illustrated from the Fig. 8 (Yu et al., 2008).

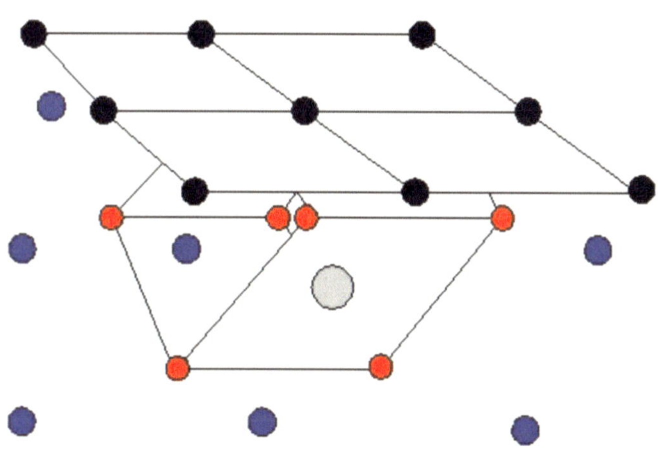

Fig. 8. $YbZn_{2-x}Mn_xSb_2$ structure view after (Yu et al., 2008). Black spheres denote Yb atoms, red spheres denote Sb atoms, blue spheres denote Zn atoms and gray sphere denotes Mn atom.

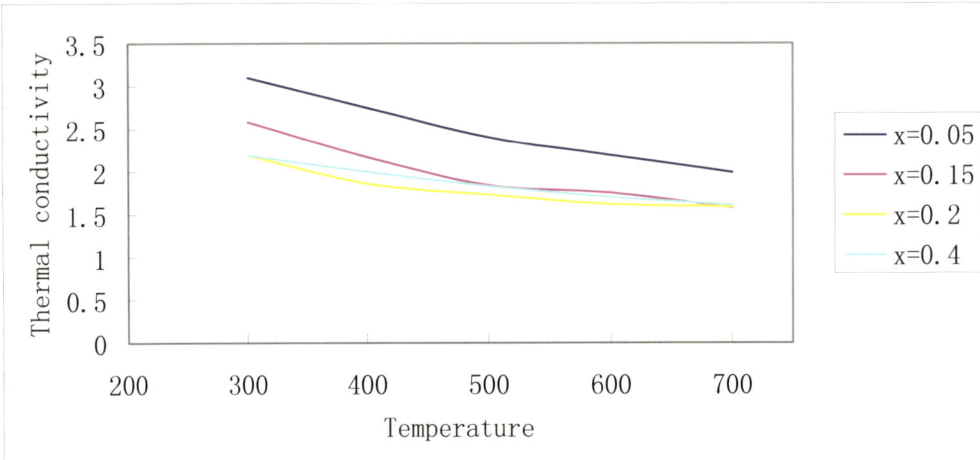

Fig. 9. Relations between temperature and thermal conductivity for $YbZn_{2-x}Mn_xSb_2$. Data source from (Yu et al., 2008)

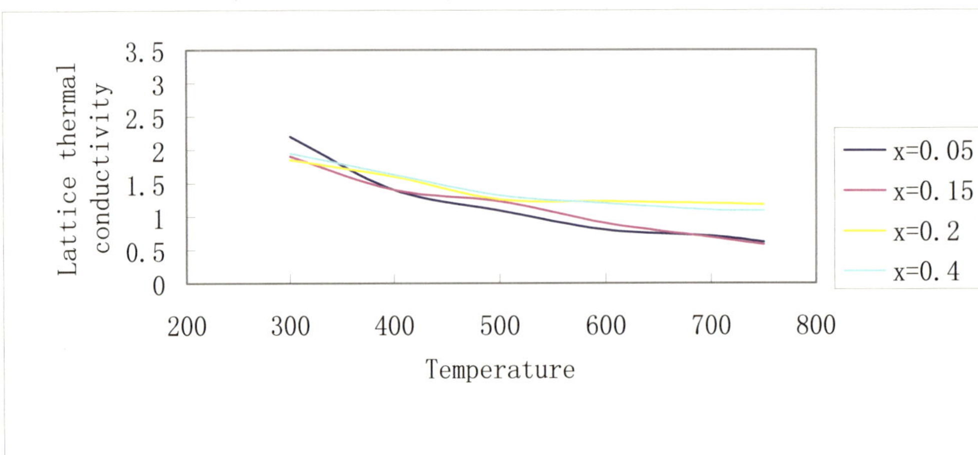

Fig. 10. Relations between temperature and lattice thermal conductivity for $YbZn_{2-x}Mn_xSb_2$. Data source from (Yu et al., 2008)

The thermal conductivity and lattice thermal conductivity at different temperatures for $YbZn_{2-x}Mn_xSb_2$ nanocomposites are shown above (See Figs. 9 and 10). It can be seen that thermal conductivity will decrease with increasing Mn substitution at ambient temperature and indicate a trend that they will converge at higher temperatures.

Meanwhile, polycrystalline sample of the title compound ($BaZn_2Sb_2$,p-type) was prepared and its thermoelectric properties from 2 to 675 K were studied. This Zintl compound shows rather low thermal conductivity, 1.6 W/m·K. The value of its thermoelectric figure of merit ZT reaches 0.31 at 675 K. A few new simple ternary Zintl phase compounds contain alkali-earth and rare-earth elements such as $Ba_4In_8Sb_{16}$, $Yb_5In_2Sb_6$, $Eu_5In_2Sb_6$, and $BaCu_2Te_2$ were synthesized and considered as potential TE materials. For the Zintl compound with low thermal conductivity, we care more about their electronic structures which could provide some useful information such as the energy gap, effective mass, electronic components, and so on (Wang et al., 2007). Additionally, a higher figure of merit ZT at 700 K is achieved as high as more than 1 from Cd-rich nanocomposite $YbCd_2Sb_2$ and Zn substitute of Cd nanocomposite $YbCd_{2-x}Zn_xSb_2$ which have low thermal conductivity and high ZT value (Wang et al., 2009). The polycrystalline TE material $EuZn_2Sb_2$ is reported with its high Seebeck coefficient as high as 181 μV/K, low thermal conductivity as low as 0.4 W/ m K and high electrical conductivity ranging from 524-1137 S/cm. The acquired TE figure of merit is 0.92 (Zhang et al., 2008). $Ca_xYb_{1-x}Zn_2Sb_2$ expressed good thermoelectric properties as a zintl phase. The structure of $CaZn_2Sb_2$ indicates the bonding within the Zn-Sb layers between planes of Ca^{2+}. The covalent bonding in the $[Zn_2Sb_2]^{2-}$ layers give the electron-crystal traits like high mobility, on the other hand, the substitution like Yb can be made at the Ca position for tuning of the electron concentration and generating disorder for phonon-glass properties.

These materials comprise of electron precise Zn-Sb anionic sheets. Because both Zn and Sb have the similar electronegativity, the bonding within should have mostly covalent character. The $[Zn_2Sb_2]^{2-}$ system possesses good hole mobility because of its low polarity. The bonding of the cation is ionic and the $CaZn_2Sb_2$ is a semiconductor with a band gap of one quarter eV. Doped $YbZn_2Sb_2$ is an extremely p-type semiconductor with metal-like transport properties when less electropositive Yb cation makes it a little electron deficient. As a result, $Ca_xYb_{1-x}Zn_2Sb_2$ forms a doped polar intermetallic. The electronic band structure of the conducting Zn_2Sb_2 will not change notably when different components in $Ca_xYb_{1-x}Zn_2Sb_2$ except for carrier concentration. Thus, a gradually tuning of the hole concentration resulting in improvement of ZT value will emerge by alloying isoelectronic species. More importantly, the lattice thermal conduvtivity will apparently be reduced due to disorder scattering by alloying Ca atom and Yb atom on the cation position. It is also reported that Ru doped Mo_3Sb_7 indicated higher ZT value than pure Mo_3Sb_7 over 300 to 700 K. The fabricated $Mo_{3-x}Ru_xSb_7$ exhibited a complex crystalline structure related to improved thermoelectric performance, while the weak TE performance of undoped material is because its metallic nature (Candolfi et al., 2009).

A phonon-glass property can be acquired from binary zintl nanocomposite of complex structure. The Fig. 11 indicates $X_{11}Sb_{10}$ as a good example of such complex structure. This kind of structure contains 22 X^{2+} cations with five $[Sb_2]^{4-}$ dumbbells and 10 Sb^{3-} anions in each unit cell. It is proven that this complicated structure has notably low lattice thermal conductivity for a high melting point around 1200 K. This structure is like a semiconductor catering for electron-crystal needs. $Yb_{11}Sb_{10}$ is similar to a heavily doped semiconductor and the $Ca_{11}Sb_{10}$ is like a lightly doped semiconductor. So $X_{11}Sb_{10}$ nanocomposites indicated properties of bipolar conduction resulting in low thermopower and therefore low ZT value.

The zintl anions provide the electron-crystal electronic structure through the covalently bonded network of the complicated anions or metalloids. The following three figures indicated the covalently bonding regions separate from the ionic regions of antimony containing zintl phases. The covalent bonded regions supply electron-crystal traits like high high mobility, while the ionic region can be doped for tuning of the electrion concentration and creating disorder to give phonon-glass properties.

Complex anions ($Ca_{11}AlSb_9$), tunnel structure ($BaGa_2Sb_2$) and a layered net ($CaZn_2Sb_2$) are all semiconductors.. $CaZn_2Sb_2$ has already been explained before. $Ca_{11}AlSb_9$ comprised of 11 Ca^{2+} cations which give their electrons to $[CaSb_4]^{9-}$ tetrahedron, 3 isolated anions Sb^{3-} and

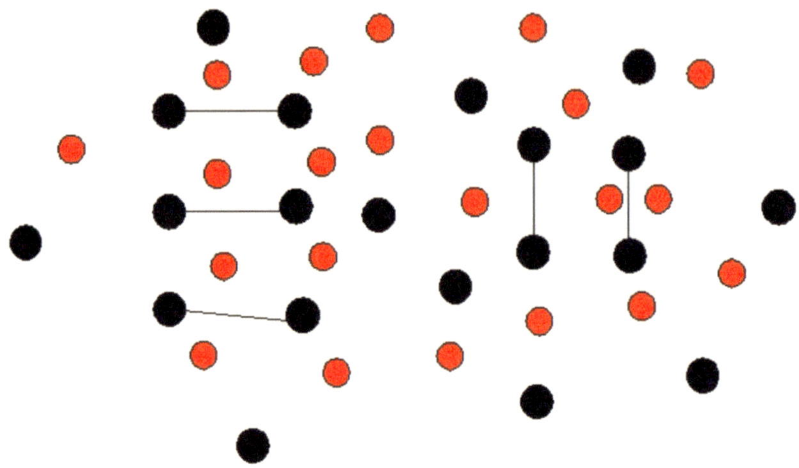

Fig. 11. Layered view of three distinct types of Sb atoms in $X_{11}Sb_{10}$ structure after (Kauzlarich et al., 2007). X is denoted by red sphere and Sb is denoted by black sphere.

$[Sb_2]^{4-}$ dumbbell. The Sb anions are six coordinated by alkaline earth cation. This nanocomposite complex structure possesses low lattice thermal conductivity. $BaGa_2Sb_2$ consists of Ba^{2+} and $[Ga_2Sb_2]^{2-}$ network which is comprised of $[Ga_2Sb_{6/3}]^{2-}$ units considered as a ethane-like Ga-Ga. These kinds of $[Ga_2Sb_{6/3}]^{2-}$ units are connected in a way that leaves holes in the structures as shown above. Such as electron compound will form bonding and antibonding bands that will cause necessary semiconducting band-gap. Because the band gap emerges on accounts of the separation of bonding and antibonding states in the anions, the existence of nonbonding or partially bonding states can fill the gap region with states and make the compound metallic. On the other hand, more electronegative elements such as fulfur will reduce mobility. Thus, zintl anions coming from heavier, softer atoms will of course gain lower lattice thermal conductivity. The cations in Zintl phases provide regions that can be doped precisely to control electron concentration and interrupt phonon transport through alloying scattering. In this way, optimized characteristics of zintl phases can be achieved. (Kauzlarich et al., 2007)

Mo_3Sb_7 has been indicated a promising TE material because of low thermal conductivity and high Seebeck coefficient. Mo_3Sb_7 is a complex 3-D structure (See Fig. 12) of atoms consisted of antimony dimers, square antiprisms and empty Sb_8 cubes. The Sb_8 cubes give more room for incorporating dopants. Moreover, substitution of Te for Sb site will improve the TE property. For example, $Mo_3Sb_{7-x}Te_x$ showed a decreased thermal conductivity and inceased Seebeck coefficient leading to a ZT value of 0.8 at 1050 K. The Fig. 12 is the layered view of Mo_3Sb_7 structure.

Skutterudite family

Skutterudite is the cobalt arsenide mineral that has variable amounts of nickel and iron with the general formula like (Co, Ni, Fe)As_3. The thermoelectrical performances highly rely on the compositions of skutterudites. The abundant compositional types permit various physical properties for its potential applications. The general structure of skutterudite is a cubic form with space group comprising of eight corner-shared octahedra (Co,Rh or Ir)(P, As or Sb)$_6$.

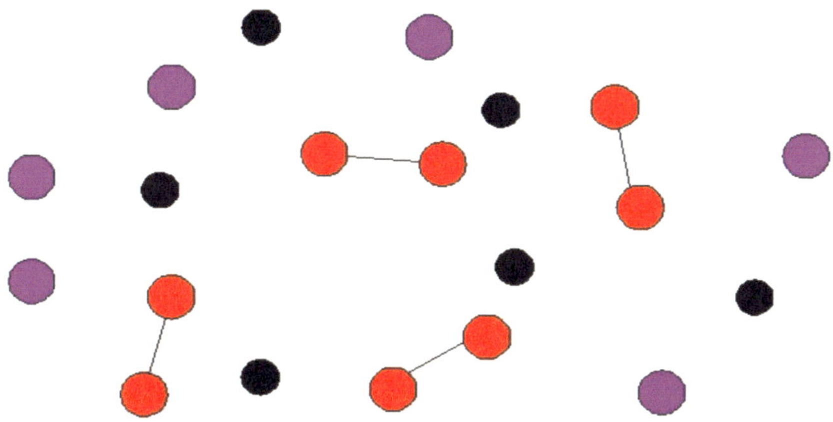

Fig. 12. A layered view of Mo_3Sb_7 after (Kauzlarich et al., 2007). Red spheres denote Sb-Sb bonds, purple spheres denote isolated Sb atoms and black spheres denote Mo atoms.

General skutterudites compounds showed high Seebeck coefficient and high electrical conductivity and high thermal conductivity. So how to lower the thermal conductivity had become the key point.The nanocomposite morphologies are effective in reducing thermal conductivity more than electrical conductivity, and Such nanocomposite materials are easily assembled into a variety of desired shapes and can be scaled up for commercial applications. The reason why skutterudite family of compounds could be regarded as thermoelectric nanocomposites is because the low thermal conductivity can be obtained when filling the voids with small diameter, large-mass interstitials. The filled skutterudites have reduced thermal conductivity than pure skutterudites because of rattling disorder of the void-filling ions that substantially affect the phonon scattering through the lattice. As of the more loosely bound rattlers producing local vibrational modes of lower frequency, and thus they are more effective in scattering the lower-frequency, heat-carrying phonons. The heavier and smaller the ion is in the voids, the larger the disorder that is produced and, therefore, the larger the reduction in the lattice thermal conductivity. The skutterudite antimonides are provided with large voids and therefore of ideal candidate for TE material. This concept, first introduced by Slack, is corroborated by the large atomic displacement parameters that have been observed in alkaline-earth and lanthanide-filled skutterudites.

At present, the reported studies showed that the filled skutterudites have the following fillers: rare earth elements, alkaline-earth elements (Ca, Sr, and Ba), foreign atoms (e.g. La, Ce, Nd, Sm, Eu etc) and others (Y, Tl, Sn and Ge). The continuing effort in improving the thermoelectric properties of skutterudite compounds has resulted in attempts to fill the voids in the crystal structure with ever-differing atoms.Synthesis approaches are at present underway in order to form ever more varied compounds in this diverse materials system.This section will elucidate $CoSb_3$ and the filled skutterudite materials especially the $CoSb_3$ based skutterudite materials .

$CoSb_3$ is one of the skutterudite family materials due to their similar formula. $CoSb_3$ is narrow-band gap semiconductor with a parabolic bottom of the conduction band. Due to its specific band structure, n-type $CoSb_3$ should have a high Seebeck coefficient while p-type

$CoSb_3$ should have high carrier mobility. However, the thermal conductivity of $CoSb_3$ is almost ten times higher than other state-of-the-art TE materials. Hence, the prepared $CoSb_3$ by hot pressing and SPS showed reduced thermal conductivity compared with that of the specimen fabricated by the traditional way like melting annealing/hot-pressing method which have some interial drawbacks.The slvothermal route without high temperature synthesized nanosized materials ranging from 150 nm to 250 nm and this method is economical and high efficient. An experimental ZT value of 0.61 is obtained for undoped $CoSb_3$ fabricated by SPS (Mi et al., 2007). N-type $CoSb_3$ fabricated by solvothermal method and melting is reported. Nanostructured and microstructured $CoSb_3$ powders indicated reduced thermal conductvity more than electrical conductivity. A ZT value of 0.71 is achieved for nanocomposite material with about 40% nanopowder inclusions (Mi et al., 2007).

The disorder within the unit cell can help to reduce thermal conductivity in structures containing void spaces. The skutterudite family material $CoSb_3$, containing corner-sharing $CoSb_6$ octahedra (See Fig. 13), which can be considered as a distorted variant of the ReO_3 structure. From the Fig. 13, it is seen that void spaces which can be filled with rattling atoms energes due to tilted octahedra.

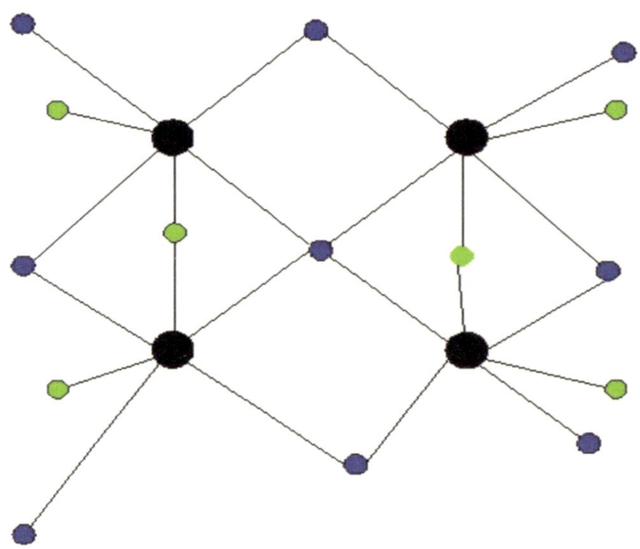

Fig. 13. The $CoSb_3$ structure as a typical skutterudite family material after (Snyder & Toberer, 2008). Black sphere denotes Co atom, blue sphere denotes Sb atom which stays in the same plane as Co atom and green sphere denotes Sb atom which is perpendicular to the plane.

In nanocomposites such as $CoSb_3$ and $IrSb_3$, which contain a high degree of covalent bonding, great electron-crystal properties can be expected because of high carrier mobilities. But the defect of this kind of strong bonding results in high lattice thermal conductivity. As a result, doping $CoSb_3$ by alloying either transition metal or the antimony site could lower the lattice thermal conductivity for this kind of skutterudite. The Fig. 14 can better show high value of thermal conductivity $CoSb_3$ can be reduced when alloying the $Co(Ru_{0.5}Pd_{0.5}Sb_3)$ or $Sb(FeSb_2Te)$ sites or filling the void spaces with $CeFe_3CoSb_{12}$ ranging from 100 °C to 500 °C.

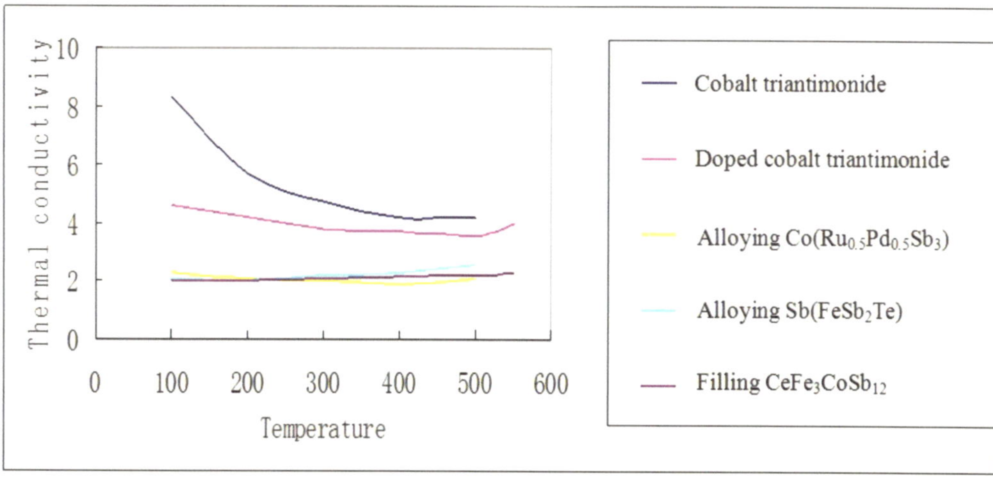

Fig. 14. Reduced thermal conductivity (W·K^{-1}·m^{-1}) with different doping and filling materials at different temperatures (°C). Data source from (Snyder & Toberer, 2008).

The following contents will elaborate the filled $CoSb_3$-based skutterudite materials including $YCoFe_3P_{12}$, $LaRh_4Sb_{12}$, $CoSb_{2.875}Te_{0.125}$, $CoSb_{2.8}Te_{0.2}$, $CoSb_{2.85}Te_{0.15}$, $M_yFe_{4-x}Co_xSb_{12}$, $Ba_{0.3}Co_{3.95}Ni_{0.05}Sb_{12}$, $Ba_xYb_yCo_4Sb_{12}$, $Sr_yCo_4Sb_{12}$, $Tl_xCo_{4-y}Fe_ySb_{12}$, $Tl_xCo_4Sb_{12-y}Sn_y$, $Yb_{0.19}Co_4Sb_{12}$, $Yb_yCo_4Sb_{12}$/ Yb_2O_3, $Yb_{0.2}Co_4Sb_{12+y}$, $Yb_{0.3}Co_4Sb_{12.3}$ and Fullerene-filled $CoSb_3$.

The general filled skutterudites materials reported include $YCoFe_3P_{12}$ and $LaRh_4Sb_{12}$. For example, the Y-filled compounds have larger rattling amplitudes than the La-filled, but are less stable; only the completely filled $YCoFe_3P_{12}$ is estimated to be stable at 0 K. Many other Y-filled compounds are close to being stable, however, and could possibly be stabilized by temperature and entropy effects. It is expected that P-based skutterudites like (La, Y)$CoFe_3P_{12}$ and (La, Y)$_{0.67}Co_2Fe_2P_{12}$ are potential TE materials in the future (Mangersnes et al., 2008). Additionally, filled $RhSb_3$ is also a promising TE material like $LaRh_4Sb_{12}$. It can be anticipated that doping La in the $RhSb_3$ will yield lower lattice thermal conductivity by means of La's vibration yielding considerable amplitudes (Fornari & Singh. 1999).

Many reports have also writen the filled $CoSb_3$-based skutterudite materials with improve TE properties. Doping partially by Co-site or Sb-site incorporation, filling voids and multisite substitution is an effective way to enhance TE performance. Te-doped $CoSb_3$ has a dramatically improvement on TE properties: a ZT value of 0.72 is achieved at 700 °C from fabricated $CoSb_{2.875}Te_{0.125}$ by melting annealing method as well as SPS; a ZT value of 0.83 is acquired at 427 °C from fabricated $CoSb_{2.8}Te_{0.2}$ by encapsulated induction melting method; Te-doped polycrystalline $CoSb_3$ like $CoSb_{2.85}Te_{0.15}$ with fine grains and an average grain size of 160 nm was synthesized by Mechanical alloying and SPS showing improved thermoelectric properties. The $CoSb_{2.85}Te_{0.15}$ TE material possessed the lower thermal conductivity and higher power factor leading to an enhanced ZT value of 0.93 at 547 °C (Liu et al., 2007). The mischmetal skutterudites material like $M_yFe_{4-x}Co_xSb_{12}$ was studied ranging from 300 K to 800 K. Mischmetal is an alloy of rare earth elements in various naturally-occuring propertions. A typical composition will consist of 50% cerium, 25 lanthanum and small amounts of neodymium and Praseodymium. Enhanced TE properties were found from decreasing size from micro scale to nano-scale (Zhang et al., 2010).

It is reported that Barium-filled skutterudites are among the most promising n-type materials with lower lattice thermal conductivity compared to pure $CoSb_3$, and it possess great electrical transport properties. A ZT value of 1.2 is achieved for the compound $Ba_{0.3}Co_{3.95}Ni_{0.05}Sb_{12}$ which is one of the highest experimentally determined values reported for n-type skutterudite (Zhao et al., 2006). And a ZT value of 1.3 is obtained through $Ba_yCo_4Sb_{12}$-based composites with dispersed fullerene or barium fulleride fabricated by SPS and solid state reaction. And the experimental results indicated that both the electrical and thermal conductivity will simultaneously decrease with the increasing impurity contents attributing to enhanced grain-boundary scattering of charge carriers and phonons. Moreover, a ZT value of 1.3 at 800 K is obtained from the double filled skutterudites $Ba_xYb_yCo_4Sb_{12}$. An improved lattice thermal conductivity is observed due to the reason of the combination of Ba and Yb fillers inside the voids of the skutterudite structure leading to a wide range of resonant phonon scattering (Shi et al., 2008).

Addtionally, strontium-filled skutterudites $Sr_yCo_4Sb_{12}$ have been fabricated by a melting method. The lattice parameters can increase linearly with the increase of the Sr content. The experimental resulted indicated that the thermal conductivity of $Sr_yCo_4Sb_{12}$has been decreased significantly compared with undoped $CoSb_3$. The ZT value of 0.9 of these TE materials is obtained at 850 K. Moreover, filled skutterudites such as $Tl_xCo_{4-y}Fe_ySb_{12}$ and $Tl_xCo_4Sb_{12-y}Sn_y$ have been studied for their electrical and thermal transport properties. Thallium atom is an unusual element because it is chemically silimar to other heavy metals like lead atom in some manner. It may have less effect on electrical transport than the rare earths because electronegativity of Tl is near to that of Sb. The substitution of Tl in the TE material means substantial rattling of the Tl about its equilibrium position. The resonant scattering of acoustic phonons by the Tl rattlers is considered the main reason of the quick decrease in the lattice thermal conductivity when incorporating small amout of Tl into the voids of the skutterudite structure. The Tl compounds have higher electron mobilities and lowered thermal conductivity. The maximum of ZT value of 0.8 is acquired at ambient temperature for Tl-doped Co_4Sb_{12} like $Tl_{0.22}Co_4Sb_{12}$ compound (Sales et al., 2000).

A relatively high value of figure of merit in a polycrystalline skutterudite partially filled with ytterbium ions was reported. The small-diameter yet heavy-mass Yb atoms partially filling the voids of the host $CoSb_3$ system exhibit low values of thermal conductivity while they still possess the quite favorable electronic properties which are not substantially perturbed by the addition of Yb. This combination is ideal for thermoelectric applications exemplifying the "phonon-glass electron-crystal" concept of a thermoelectric material, resulting in ZT=0.3 at room temperature and ZT=1 at 600 K for $Yb_{0.19}Co_4Sb_{12}$ (Nolasa et al., 2000). Additionally, Yb-filled skutterudite materials incorporated with Yb_2O_3 is prepared by in situ reactio method. In this $Yb_yCo_4Sb_{12}$/ Yb_2O_3 structure, some Yb_2O_3 particles reside at the grain boundaries and others are dispersed within $Yb_{0.2}Co_4Sb_{12}$ grains as nanoscale inclusions. A remarkable reduction in thermal conductivity is acquired on accounts of the combination of the rattling of Yb ions inside the voids of $CoSb_3$ and the phonon scattering of the oxide defects. A maxiumu ZT value of 1.2 is obtained for $Yb_{0.21}Co_4Sb_{12}$/ Yb_2O_3 compound and a maximum ZT value of 1.3 is gained for $Yb_{0.25}Co_4Sb_{12}$ compound at 850 K (Zhao et al., 2006). Nanostructured $Yb_{0.2}Co_4Sb_{12+y}$ was fabricated by melt spinning techniqu with SPS. An average grain size of 150 nm was acquired, and with the increase of value y, the grain size will increase accordingly. Just because of the nanostructure of $Yb_{0.2}Co_4Sb_{12+y}$, a lowered thermal conductivity is obtained. Also, because of the moderately excessive Sb atoms, electrical transport properties are enhanced. A ZT value of 1.26 is achieved at 800 K for $Yb_{0.2}Co_4Sb_{12+y}$ (Li et al., 2008). A method called synthetic route that can remarkably reduce the fabrication time of $Yb_{0.3}Co_4Sb_{12.3}$ is proposed leading to improved TE properties. As a result, a ZT value of 1.3 is obtained at 800 K for this single-phase, fully densified filled skutterudites. The notably reduced manufacturing time along with the improved TE performance can be employed in the practical application of skutterudite-based TE materials (Li et al., 2008).

Fullerene-filled $CoSb_3$ was studied of its structural, chemical, and transport properties Fullerene is a 60-atom carbon molecule that forms microsize clusters between the grain boundaries of $CoSb_3$. An experimental approach has proved that partial void filling is the optimizing method to improve thermoelectric properties. The advantage of partial fillingis that a random alloy mixture of filling atoms and vacancies can be established which accelerates effective point-effect scattering. Furthermore, soft phonon modes and rattling moedes which can efficiently reduce the value of lattice thermal conductivity can be created from large space for the filling atom in skutterudites and clathrates. Filling the voids with ions generates an additional source of lattice disorder like Fe^{2+} frequently employed to substitute Co^{3+} in pure $CoSb_3$. The result showed that dominant scattering mechanism in the electrical transport changes from impurity scattering to grain-boundary scattering near a C_{60} content of mass percentage from 5% to 6%, and that thermal conductivity decreases with increasing C_{60} content. The result also implies a transition from ionized impurity scattering to grain-barrier scattering occurs when the C_{60} concentration increases above mass percentage of 5%. A significant increase in the thermoelectric figure of merit is achieved for doped C_{60} compared to the pure $CoSb_3$ (Shi et al., 2004). At last, La doped $IrSb_3$ indicated enhanced TE efficiency by the decreasing lattice thermal conductivity because of filled La atoms rattling in the structural vacancies of the skutterudite crystal structure (Kim et al., 2004).

Zn_4Sb_3

The extremely high value of figure of merit acquired from Zn_4Sb_3 is coming from the remarkably glass-like thermal conductivity. The low thermal conductivity of Zn_4Sb_3 is coming from high levels of interstitials and corresponding local structural distortions and from domains of interstitial ordeirng, leading to disorder at multiple length scales.

In Zn_4Sb_3 structure, about one fifth of the Zn atoms are on three crystallographically distinct interstitial sites in ambient temperature. These interstitials are along with local lattice distortions with Zn diffusion rates as high as that of superionic conductors (Snyder & Toberer, 2008). Zn_4Sb_3 structure (See Fig. 15) contains following types of antimony atoms: Sb-Sb bonded dumbbels, $[Sb_2]^{4-}$ and isolated Sb^{3-} without bonding to other antimonyatoms. The general zintl components can not precisely reflect the true strucuture which can be depicted by a complicated and delocalized multicenter bonding. Zn_4Sb_3 phase keep this delocalized multicenter bonding. However, Zn atoms will be disordered at longer length scales. Consequently, in ambient temperature with symmetry, one fifth of the Zn atoms will be in interstitail positions. Just due to the complex Zn interstitial structure, Zn_4Sb_3 possesses a higher ZT value to a large extent on accounts of lower lattice thermal conductivity which is less than 1 Wm·K.

It is reported that Zn_4Sb_3 can be fabricated using solid state reaction (Cadavid & Rodríguez, 2008), conventional quench method and zone-melting technique etc (Pedersen & Iversen, 2008). Only the Zn_4Sb_3 fabricated by the last method is more thermal stable for more promising applications. The synthesized Zn_4Sb_3 by other two methods will partly

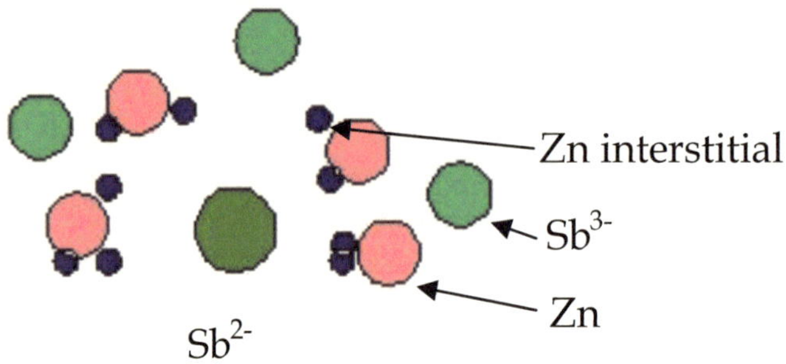

Fig. 15. A layered view of the structure of Zn_4Sb_3 containing electron density map indicating the existence of interstitial Zn atoms after (Kauzlarich et al., 2007)

decompose to ZnSb, ZnO and Zn below expected stability range which resulted in a notable degradation of the ZT value. Therefore, the Zn_4Sb_3 prepared by the zone-melting technique is a high-performance TE material. Also, compacted Zn_4Sb_3 synthesized by SPS is investigated with grain size less than 45 μm. The experimental and measurement result showed that there is an obvious correlation between the density and thermoelectric properties. The results indicated that changes in Zn_4Sb_3 compaction conditions will have a larger influence of ZT than doping (Pedersen et al., 2006). It is not sure to say that doping can improve the electron crystal traits of Zn_4Sb_3 But doping with small amount of metal dopants changes the subtle balance between the Zn deficiency and Zn disorder. As a resutl, it is estimated that doping with metals can enhance its carrier concentrations (Litvinchuk et al., 2008).

It is known that Zn_4Sb_3 possesses at least four crystalline phases, among which, only β phase Zn_4Sb_3 is an efficient TE material from 263 K to 765 K. β phase Zn_4Sb_3 possesses hexagonal rhombohedric crystal structure and its highest ZT value achieved is 1.4 at 400 °C . Single phased polycrystalline specimen was fabricated by melting and well homogenizing Zn and Sb in closed quartz ampoules. The acquired ingots were ground into powders and then hot pressed to generate crack-free specimen (Nolas et al., 2006). A method of disordering is emerging as a efficient way to lower the thermal conductivity as well as loosely bonded atoms and point defects. β phase Zn_4Sb_3 is a structurally disordered intermetallic compound at moderate temperature range 450 K- 650 K. So β phase Zn_4Sb_3 is a promisng nanocomposite material which possesses potential low thermal conductivity. The great TE properties of p-type β–Zn_4Sb_3 fills the gas in the ZT values between the low temperature state-of-the-art materials Bi_2Te_3-based alloys and the intermediate teperature materials PbTe-based alloys and Te-Ge-Sb-Ag alloys. β–Zn_4Sb_3's stability is remaining stable up to 670 K under static vacuum and argon. Much effort is devoted to producing n-type materials and other isostructural compounds such as Cd_4Sb_3. Meanwhile, a series of SiO_2/β–Zn_4Sb_3 core-shell nanocomposites with different thickness were fabricated by coating β–Zn_4Sb_3 pariticles with SiO_2 nanoparticles prepared by hydrolyzing the tetraethoxysilane in alcohol-alkali-water solution. The experimental results and analysis indicated that SiO_2 nanoparticles are coated with uniformly β–Zn_4Sb_3. During the coating process, both the thermal conductivity and electrical conductivity gradually decrease, and at the same time, the thermopower increases compared to β–Zn_4Sb_3 bulk material. The results also demonstrate that the themopower will increase dramatically in high temperature range. It showed that thermal conductivity of SiO_2/β–Zn_4Sb_3 nanocomposite material with thickness of 12 nm isn only

0.56 W/m-1K-1 at 460 K. Consequently, the ZT value of $SiO_2/\beta-Zn_4Sb_3$ nanocomposite is 0.87 at 700 K (Ruan & Xiao, 2007).

Clathrate

Clathrates are a series of materials which possess teetrahedrally bonded atoms form of a framework of cages that can encircle relatively large metal atoms. Clathrate compounds consisting of polyhedral cages and guest atoms encapsulated in the cages are expected to exhibit relatively high electrical conductivity and Seebeck coefficient, in addition to very low thermal conductivity from their crystal structures comprising of polyhedral cages doped with a guest atom.There is a potential way to further reduce thermal conductivity for clathrates which consist of rattling atoms and large cages, that is via disorder within the unit cell containing void spaces (Snyder & Toberer, 2008). The polyhedral cages consist of group-IV and/or -III elements while the guest atoms are typically alkali metals or alkali-earth metals. The "rattling" of guest atoms in oversized cages is considered to scatter heat-carrying phonons efficiently, resulting in low thermal conductivity, while electrical conductivity remains relatively high because electronic conduction mainly takes place through the cage framework. Therefore, clathrate compounds have been investigated in recent years as promising thermoelectric materials, Clathrate compounds form in a variety of structure types, depending on the combination of different constituting cages. The rest part of this section will emphasize on the type-I clathrate and type III clathrate.

One of prominent properties of clathrates is its low glass-like thermal conductivity in type-I clathrates. A large number of clathrate types belong to the type-I with formula $X_8Y_{16}Z_{30}$ (Bentien et al., 2004). Polycrystalline $Sr_8Ga_{16}Ge_{30}$, in which Sr atom is encapsulated by the $[Ga_{16}Ge_{30}]$ framework, of single crystal of the type-I clathrate compounds are reported in the temperature range from 5 K to 300 K with stability. The compounds exhibit n-type semiconducting behavior with relatively high Seebeck coefficients and electrical conductivity, and bear carrier concentrations in the range of 10^{17}–10^{18} cm^{-3} at room temperature. An enhanced ZT value of more than 1 is achieved at more than 700 K. thus exceeding that of most known materials. At the same time, the $Sr_8Ga_{16}Ge_{30}$ compound's thermal conductivity is drastically reduced, to nearly that of the theoretical minimum, while good electronic conductivity in not affected (Nolasa et al., 1998). The $Ba_8Ga_{16}Ge_{30}$ has nearly equal coefficient of thermal expansion with $Sr_8Ga_{16}Ge_{30}$ is also of the type-I clathrate and they both showed higher performance at higher temperatures. In comparison, $Sr_8Ga_{16}Ge_{30}$ has lower thermal conductivity than $Ba_8Ga_{16}Ge_{30}$ due to more notably rattling of Sr atoms in the cages (Okamoto et al., 2008). A ZT value of 1 is reported for poly crystalline $Sr_8Ga_{16}Ge_{30}$ and a ZT value of 1.7 is reported for polycrystalline $Ba_8Ga_{16}Ge_{30}$ with optimized compositions previously. And improved thermoelectric properties are expected from polycrystalline crystal $Ba_8Ga_{16}Ge_{30}$. The experimental results indicated that polycrystalline specimen of $Ba_8Ga_{16}Ge_{30}$ with optimization showed higher TE properties than reported single crystal $Ba_8Ga_{16}Ge_{30}$ (Martin et al., 2008). To date, the relatively high ZT value of 1.35 is obtained through Czochralski method at 900 K for $Ba_8Ga_{16}Ge_{30}$. Additionally, n-type $Ba_8Ga_{16}Zn_xGe_{30-x}$ was fabricated by solid-state reaction and SPS. The electrical conductivity and carrier mobility will increase with the decreasing Zn content, while the carrier concentration will increase with the increasing Zn content. Specifically, Seebeck coefficient of $Ba_8Ga_{16}Zn_{3.2}Ge_{26.8}$ will reach 250 μV/K at 300 K or 307 μV/K at around 700 K. The thermal conductivity will decrease with increasing Zn content. And it is expected that electrical conductivity and carrier concentration will increase significantly by slightly adjusting Ga content (Deng et al., 2007).

It is also reported that Ba_8Ge_{43} is considered as type-I clathrate compound. The crystal structure of Ba_8Ge_{43} is quite different from other type-I clathrate compound. But type-I clathrate compound is a supperlattice structure according to ususl type-I clathrate structure. The experimental indicated that this kind of material is not a good TE material because of its high electrical resistivity attrbuting from the Ge vacancies (Okamoto et al., 2006). In comparison, tellurium doped type-I clathrate compound $Ge_{30}P_{16}Te_8$ possesses lower thermal conductivity as low as of $Sr_8Ga_{16}Ge_{30}$ and $Ge_{38}Sb_8I_8$. Because the general clathrates materials are n-type, so the fabrication of this p-type clathrate compound is very crucial. $Ge_{30}P_{16}Te_8$ has a thermal conductivity of 0.9 W/m·K at ambient temperature (Kishimoto et al., 2006).
Clathrate compounds with the formula like $X_{24}(Y/Z)_{100}$ or $X_{24}Y_{32}Z_{68}$ is defined as type III clathrate compounds in which, Y and Z are tetrahedrally and trigonally bonded cage atoms, respectively. $Ba_{24}Ge_{100}$ has been reported to exhibit low lattice thermal conductivity of about 1 W/m·K but also to show a high value of electronic thermal conductivity and a small value of Seebeck coefficient. The high value of electronic thermal conductivity and the small value of Seebeck coefficient are considered to originate from the high electron concentration of $Ba_{24}Ge_{100}$. There exist 16 extra electrons per $Ba_{24}Ge_{100}$ unit formula corresponding to a higher electron concentration than usual one. As a result, the reduction of the extra electrons can enhance Te performance. The addition of Al and Ga doped into clathrate compound can decrease the thermal conductivity and increase the Seebeck coefficient. Furthermore, thermal conductivity in clathrate compounds like Ba–In–Ge ($Ba_{24}In_XGe_{100-X}$) system generally decreases with the increase in the In content because of the decrease in the number of excess electrons. And the system Ba-Al-Ge and Ba-Ga-Ge systems have the similar traits (Kim et al., 2007). It is reported that n-type $Ba_8Ga_xGe_{100-x}$ of type- III clathrate compound synshesize by arc melting, achieved a ZT value of 1.25 at 903 K. (Deng et al., 2007).

Tellurides alloys

This section introduces binary telluride alloys, ternary telluride alloys and many entities telluride alloys. This section will firstly state the binary telluride alloys PbTe, In_2Te_3 and Ga_2Te_3, then narrate the doped binary telluride alloys, after that recount the ternary telluride alloys and quaternary alloys, and at last will compare some TE nanocomposites with lower thermal conductivity and state the state-of-the-art telluride superlattice with highest ZT value achieved in tellurides alloys.
Synthesized lead telluride based alloys are considered as n-type and p-type thermoelectric materials with lower thermal conductivity. The achieved great thermoelectric characteristics are greatly relying on the preparation technology and the carrier concentration. PbTe has a high melting point, good chemical stability. Good chemical strength, low vapor pressure and high figure of merit. Lead telluride can be operated at 900 K and is an intermediate thermoelectric power generator which is widely employed in the army, space crafts and batteries. The outstanding property of this TE material is its low lattice thermal conduvtivity. The TE material PbTe can be acquired from the decomposition of metastable $Pb_2Sb_6Te_{11}$ into PbTe and Sb_2Te_3. A layered microstructure is obtained of PbTe and Sb_2Te_3. The inter-layered space can be adjusted by the temperature and time of the decomposition process. The ajacent PbTe and Sb_2Te_3 layers are crystallographically oriented, resulting in high-quality epitaxy like interfaces. Average layer spacings are 180 nm compared to a PbTe layer thickness of 40 nm. These nanoscae multilayers stands for the thin-film superlattice thermoelectric materials which have indicated notably high thermoelectric efficiency (Ikeda et al., 2007). Addtionally, the single crystalline PbTe nanowires indicated decreased thermal

conductivity with the decreasing nanowire diameter. The lower thermal condutivity is 1.29 W/m K for a 182 nm nanowire at 300 K which is half value of bulk PbTe. This kind of single crystalline PbTe nanowires can be synthesized by chemical vapor transport method with prepared length ranging from 182-436 nm in diameter. The Figs. 16 and 17 show the relations between temperature and thermal conductivity of different nanowire diameters. Based on these figures, it can be seen that the thermal conductivity will decrease as the diameter shrinks (Roh et al., 2010).

Ga_2Te_3 is another example telluride alloy with low thermal conductivity due to effective scattering phonons in the 2-D vacancy plane. In_2Te_3 was compared with Ga_2Te_3 for their TE performances. They have the same crystal structure and their thermal conductivities were compared in the Fig. 18. It can be seen that In_2Te_3's thermal conductivity decreases with

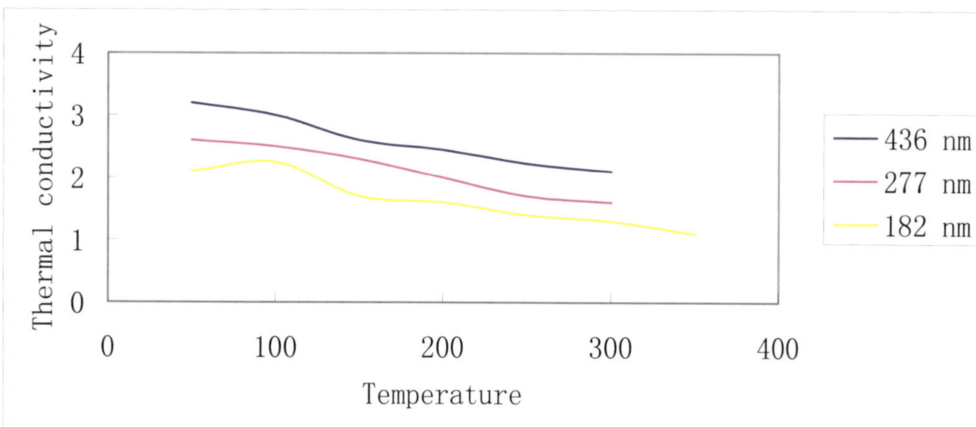

Fig. 16. The measured thermal conductivity of individual single-crystal PbTe nanowires with different diameters at different temperatures. Data source from (Roh et al., 2010)

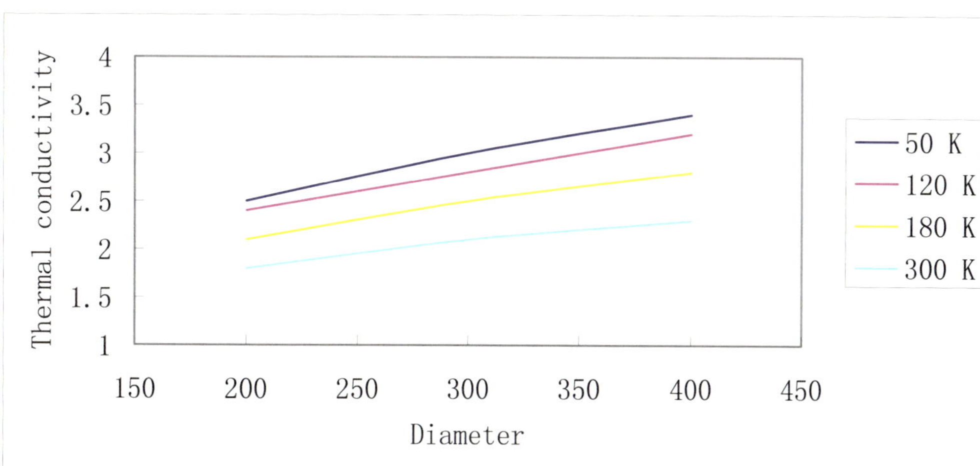

Fig. 17. The measured thermal conductivity of individual single-crystal PbTe nanowires at different temperatures with different diameters. Data source from (Roh et al., 2010)

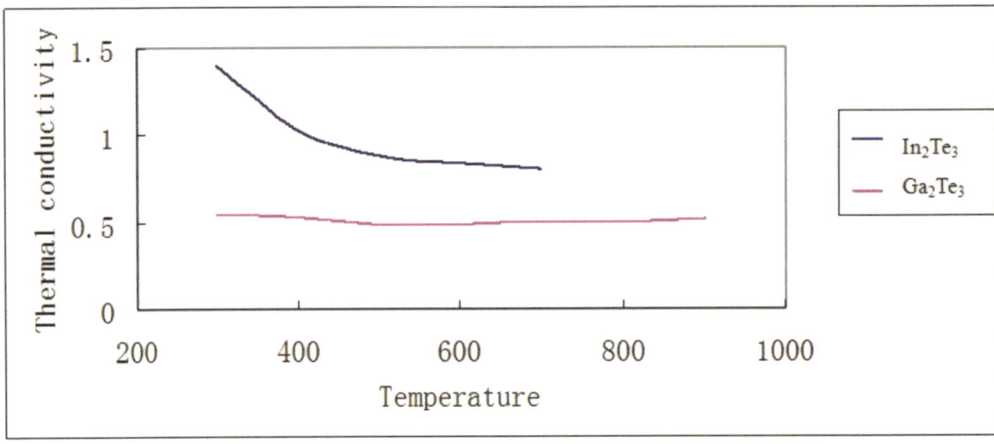

Fig. 18. Measured relations between temperature (K) and thermal conductivity (W/m·K) for Ga_2Te_3 and In_2Te_3. Data source from (Kurosaki et al., 2008)

increasing temperature, while Ga_2Te_3 showed a nearly flat temperature dependence. It is obvious to see that Ga_2Te_3 exhibits a lower thermal conductivity than In_2Te_3, and it has a rather low thermal conducitivity of as low as 0.5 W/m·K (Kurosaki et al., 2008).

Alloying binary tellurides (Bi_2Te_3, Sb_2Te_3, PbTe, GeTe and Ga_2Te_3) with other isoelectronic cations and anions can reduce the thermal conductivity rather than lower the electrical conductivity. The traditional TE materials without doping generally have a maximum value of ZT up to 1 at ambient temperature relying on the band gap of the materials. For example, the nanocomposites like Bi_2Te_3, PbTe and SiGe alloys (Yamashita, 2004) have ZT value of 0.9 at 400 K, 0.8 at 650 K and 0.9 at 1200 K, respectively. Because of the inherent flaws involved in these materials like their cost and scale-up, many other new preparation methods of nanocomposite have emerged like sinodal decomposition, matrix encapsulation and eutectic system. A eutectic system is a mixture of chemical compounds or elements that has a single chemical composition that freezes at a lower temperature than any other composition. This composition is knows as the eutectic composition and the temperature is knows as the eutectic temperature. The reported entectics falls in the range of semiconductor and metal compounds. Here, we will introduce the new-explored eutectic material PbTe-GeSi. The incorporation of Ge into the PbTe lowers the brittleness notably. The composition PbTe-Ge (4:1) was studied indicating that the lattice thermal conductivity was not significantly improved than PbTe itself. While when the composition of Ge is lower than 20 %, the incorporated PbTe showed an reduced thermal conductivity. Furthermore, when incorporating Ge and Si simultaneously, a remarkably reduction was observed. An enhanced ZT value up to 1.3 was achieved via PbTe-$Ge_{0.8}Si_{0.2}$ (19:1) (Sootsman et al., 2009). A maximum figure of merit as high as 3.2×10^{-3}/K is achieved from p-type alloys with the composition $Bi_{24}Sb_{68}Te_{158}Se_6$. The lattice thermal conductivity is found to decrease when the telluride content is increasing (Bergvall & & Beckman, 2003). An improvement in thermoelectric property is attained of $CsBi_4Te_6$ which is a complex variant of Bi_2Te_3. It has a lamellar structure with slabs of $(Bi_4Te_6)^{1-}$ alternating with layers of Cs^+ ions. Low doping degrees will greatly influence the charge transport properties. Due to few Bi-Bi bonds and complexity of the Cs layers in $CsBi_4Te_6$, $CsBi_4Te_6$ has a lower lattice thermal conductivity than Bi_2Te_3. Meanwhile, an enhanced Seebeck Coefficient with slightly damage to the mobility will be resulted in because of the anisotropic effective mass in layered $CsBi_4Te_6$. A ZT value of 0.8 is obtained at 225 K which is 40% greater than that of Bi-Sb-Te-Se alloys. Transition metal pentatellurides of the orthorhombic polytype structure $ZrTe_5$ and $HfTe_5$ exihibits large resistivity anomalies. These compoumds have the similar structure with Bi_2Te_3, with van der Waals gaps between the individual layers. Doped pentatellurides showed high power factors at low temperatures, surpassing the optimized Bi_2Te_3 value, but the high thermal conductivity is relatively high (4W/m·K-8W/m·K). Eventually, pentatellurides still need to be tuned to make them high performance TE materials. Doped crystals of the pentatellurides such as $Hf_xZr_{1-x}Te_5$ showed reduced thermal conductivity on accounts of phonon scattering (Zawilski et al., 2000).

Alloyed bismuth telluride also possesses such properties. Solid state alloying improves the thermoelectric figure of merit by decreasing lattice thermal conductivity without affecting electrical properties. Alloying engenders part disorder in the film network that may increase the scattering of phonons with minor effect on the charge carriers which have long wavelength as compared to phonons. Bismuth telluride could be easily synthesized by cathodic electrochemical deposition. The deposition process can be controlled by bridling electrochemical parameters. The as-deposited bismuth telluride films are crystalline and morphologcally smooth in the whole area.

Bismuth telluride could be employed as a thermoelectric power generator with other doped materials. It has reported that thermoelectric power generator modules contain of 0.8 mm Bi_2Te_3 and 50 μm thick ErAs (60% of volume): $[InGaAs]_{1-x}[InAlAs]_x$. A output power of 6.3 W was measured when the heat source temperature was at 610 K. When ErAs nanoparticles are doped into $[InGaAs]_{1-x}[InAlAs]_x$, a bending potential barrier is created at the interface between the particle and semiconductor. The Seebeck coefficient can be increased by meas of the electron filtering effects of these potential barriers. The performance of a thermoelectric generator module can be effectively enhanced by employing segmented element structures with materials whose thermoelectric properties are optimized in successive temperature ranges. A notable improvement in TE properties has been seen in the Bi_2Te_3/Sb_2Te_3 superlattices deposited on GaAs. Supperlattices provide the chances to enhance the ZT value by increasing Seebeck coefficient and decreasing the thermal conductivity compared to bulk materials (Touzelbaev et al., 2001). The experimental results indicated that via incorporating ErAs nanoparticles in the compound InGaAlAs, the thermal conductivity was lowered than that of $[InGaAs]_{0.8}[InAlAs]_{0.2}$. The thermal conductivity of this nanocomposite is lowered by one quarter when increasing the ErAs concentration from 0.3% to 3% (Zeng et al., 2009). An improved ZT value up to 1.04 was obtained after incorporating the n-type $Bi_2[Te_{0.94}Se_{0.06}]_3$ with I and Te by Bridgman method and anealed for 2 hours at 473 K. This speciman possesses a remarkably high power factor of 6.57×10^{-3} W/mK^2 at 298 K which is the highest in bismuth tellurides. In comparison, $Bi_2[Te_{0.94}Se_{0.06}]_3$ doped with I, Te and CuBr fabricated by the same method indicated a lower ZT value of 0.91 at 298 K. The reason may be attributed to the addition of CuBr (Yamashita & Tomiyoshi, 2004). The Te doped p-type $(Bi_{0.25}Sb_{0.75})_2Te_3$ was synthesized by the Bridgman method at a rate of 6 cm/h. After the annealing and vacuum processes, the maximu ZT value is 1.41 at 308 K (Yamashita et al., 2003).

Neither Si nor Ge is a good TE material because Si's lattice thermal conductivity value is 150 W/m·K and Ge's lattice thermal conductivity value is 63 W/m·K which is quite large for TE materials. However, through alloying, the lattice thermal conductivity could be exceedingly reduced, like $Si_{0.7}Ge_{0.3}$'s thermal conductivity is 10 W/m·K. Using nanostructure method, a

ZT value of 1.3 is achieved for n-type silicon germanium (SiGe) bulk alloy at 900 K. The enhancement of TE performance is due to notably reduction in thermal conductivity resulted from large phonon scattering (Wang et al., 2008). Moreover, it is reported that the power factor ZT value of n-bype $Si_{1-x}Ge_x$ nanowires are larger than those p-type counterparts with the same Ge content and incorporating concentration. A maximum ZT value of 1 is achieved for n-type $Si_{0.5}Ge_{0.5}$ nanowires (Shi et al., 2010).
The charge density wave material like $CeTe_2$ indicated relatively low thermal conductivity1.25 W/m·K at 300 K and high Seebeck coefficient as high as 477 μV/K at 300 K. And also polycrystalline $Ce_{1-x}Cu_xSe_2$ series nanocomposites have been investigated. The experimental results showed a large Seebeck coefficient value of 344 μV/K and low thermal conductivity value of 0.71 W/m·K. With the increment of Cu doping concentrations, the Seebeck coefficient will decrease, referring to increase in the charge carrier concentration from Cu doping. $Ce_{1-x}Cu_xSe_2$ indicated a lower power factor than Bi_2Te_3 and its alloys due to the higher electrical resistivity. From the band structure calculation, the high Seebeck coefficient for the $Ce_{0.9}Cu_{0.1}Se_2$ compound is due to the localized Ce *f*-band near the Fermi level. The Figs. 19 and 20 showed the crystal structure of the stoichiometric $CeSe_2$ and $Ce_{1-x}Cu_xSe_2$ with Cu as a dopant. From the energy formation calculation, the dopant Cu at the Ce-Se block is rather advantageous than the interstitial or the Se position doping. The doped Cu atom modified the atomic coordination as shown by the red arrow (Rhyee et al., 2010).
It is reported that a high thermoelectric performance can also be obtained from the ternary alloys. Thallium compounds indicated low thermal conductivity thus exhibiting high TE performance. Particularly, bulk polycrystalline material Ag_9TlTe_5 showed a ZT value of 1.23 at around 700 K. What's more, polycrystalline sintered specimen of Tl_2GeTe_3, Tl_4SnTe_3 and Tl_4PbTe_3 showed thermoelectric figure of merits as follows: 0.29 at 473 K, 0.74 at 673 K and 0.71 at 673 K largely because of their low thermal conductivities. Although Tl_2GeTe_3, Tl_4SnTe_3 and Tl_4PbTe_3 did not indicate an optimized electrical conductivity, their low

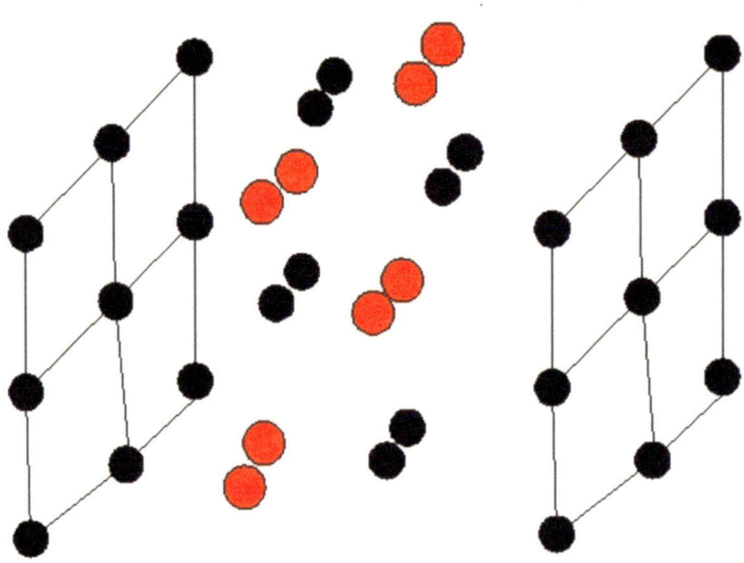

Fig. 19. Crystal structure of $CeSe_2$ after(Rhyee et al., 2010). Black spheres denote Se atoms and red spheres denote Ce atoms.

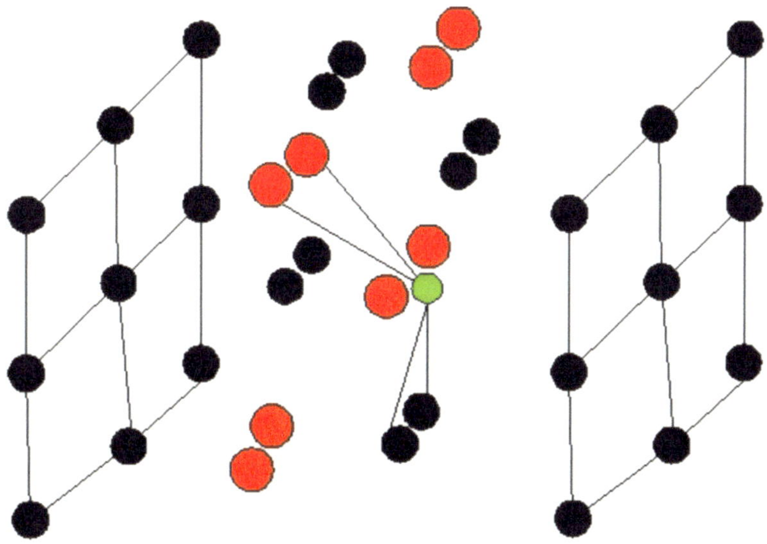

Fig. 20. Crystal structure of $Ce_{1-x}Cu_xSe_2$ after(Rhyee et al., 2010). Black spheres denote Se atoms, red spheres denote Ce atoms and green sphere denotes Cu atom.

thermal conductivity resulted from weak bonding of atoms and complicated crystal structure stands for their potential utilizations (Kosuga et al., 2006). Furthermore, $TlGaTe_2$ and $TlInTe_2$ showed semiconductorlike properties with low thermal conductivities and relatively high ZT values. And TlTe indicated an exceptionally high thermal conductivity due to its large electronic contributions and high lattice thermal conductivity (Matsumoto et al., 2008). This maybe explained from their structure (See Figs.21 and 22) : $TlGaTe_2$ and $TlInTe_2$ have the same structure, while TlTe has a more complex structure than $TlGaTe_2$ and $TlInTe_2$. Both $TlGaTe_2$ and $TlInTe_2$ have 16 atoms per unit cell, and TlTe has 32 atoms per unit cell.

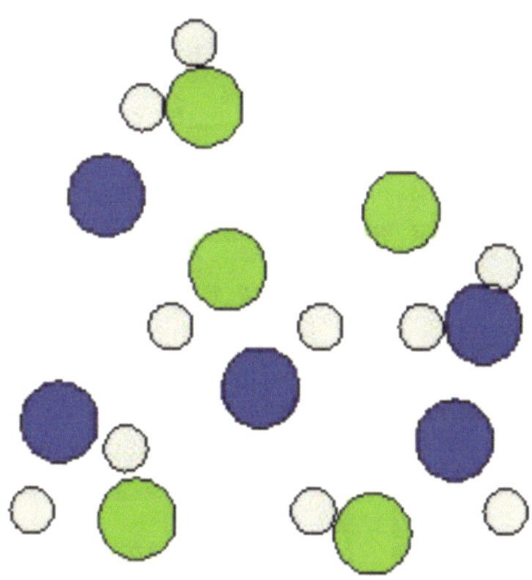

Fig. 21. Plane view of $TlGaTe_2$ and $TlInTe_2$ after (Matsumoto et al., 2008). The green, blue and gray balls represent Ga or In, Tl and Te atom respectively.

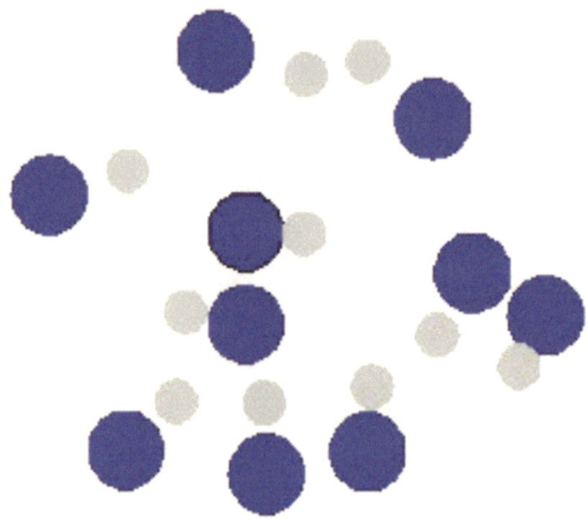

Fig. 22. Plane view of TlTe after (Matsumoto et al., 2008). The green, blue and gray balls represent Ga or In, Tl and Te atom respectively.

Although the conventional low dimension thermoelectric nanocomposites such as superlattive Bi_2Te_3/Sb_2Te_3 and quantum dot superlatticve $PbSe_{0.98}Te_{0.02}/PbTe$ indicated improved ZT to a large extent due to the lattice thermal conductivity, it is still hard to scale up these supperlattices for large energy conversion applications because of their limitations (Ma et al., 2008). So how to apply the TE materials into practical applications have become important. A high ZT value is achieved from silver antimony lead telluride fabricated from zone melting and crystal growth technique. A ZT value of 1.56 is obtained at 300 K from the p-type $Bi_{0.52}Sb_{1.48}Te_3$ bulk material with microstructures. This bulk material is fabricated by melt spinning technique and SPS. This microstructure contains nanocrystalline domains implanted in amorphous matrix and 5-15 nm nanocrystals with coherent grain boundary. The achieved higher ZT value is attributed to the lower thermal conductivity, and this value is more than a half enhancement of Bi_2Te_3 ingot material (Xie et al., 2009). It is also reported that at the optimized doping concentrations assuming that scattering rate is the same as a function of doping by using a constant relaxation time and an averaged thermal conductivity, a ZT value of 1.8 can be achieved for the p-type doping in $(Sb_{0.75}Bi_{0.25})_2Te_3$ compound (Lv et al., 2010).

A ZT value of 1.4 is acquired from ternary alloyed bismuth antimony telluride (BiSbTe) at 100 °C through hot pressing nanopowders that were ball-milled from crystalline ingots under intert conditions than traditional achieved ZT value of 1 for several decades. The high performance of thin nanocomposite is attibuted to the sligh increase of electrical conductivity and large decrease of the thermal conductivity. The lower value of ZT is acquired at higher lattice disorder for $Bi_{0.5}Sb_{0.5}Te$. And experimental results and measurements indicated that a low thermal conductivity is gained due to increased phonon scattering by grain boundaries and defects. This kind of nanocomposite can be used as cost-saving and high-performance TE material for cooling and power generation (Poudel et al., 2008). Moreover, from the TEM picture of $Bi_{0.5}Sb_{0.5}Te$, it can be seen that both nanosized and microsized grains are engendered. And between the nanograins, the interface regions have a 4 nm thickness in addition to nanoprecipitates. A ZT value of 1.4 is acquired for this material due to its low thermal conductivity and slightly improved electrical conductivity (Lan et al., 2009).

Except for ternary alloys Bi-Sb-Te, some other ternary nanostructured homologous serious like $Ga_mSb_nTe_{1.5(m+n)}$ is studied for its valuable performance in TE properties. $Ga_mSb_nTe_{1.5(m+n)}$ was studied over the temperature range of 318-412 K, and a measured ZT value of 0.98 is achieved with m:n=1:10 at 482 K which is higher than that of undoped Sb_2Te_3 at the same temperature. This improvement is on accouts of phonon scattering caused by amorphous structure, amorphous structure and lattice distortion leading to remarkable reduction in thermal conductivity. With increase of m:n ratio (See figure 23), both the thermal conductivity and electrical conductivity will decrease and the Seebeck coeffcient will increase. The ZT values with different ratio for $Ga_mSb_nTe_{1.5(m+n)}$ are shown in the Fig. 23: ZT value will increase with m:n ratio until m:n=1:10, after that, will decrease. In comparison, a ZT value of 0.65 is obtained fro Ga doped telluride alloy like $Ga_xBi_{0.5}Sb_{1.5-x}Te_3$ at 318 K (Cui et al., 2009).

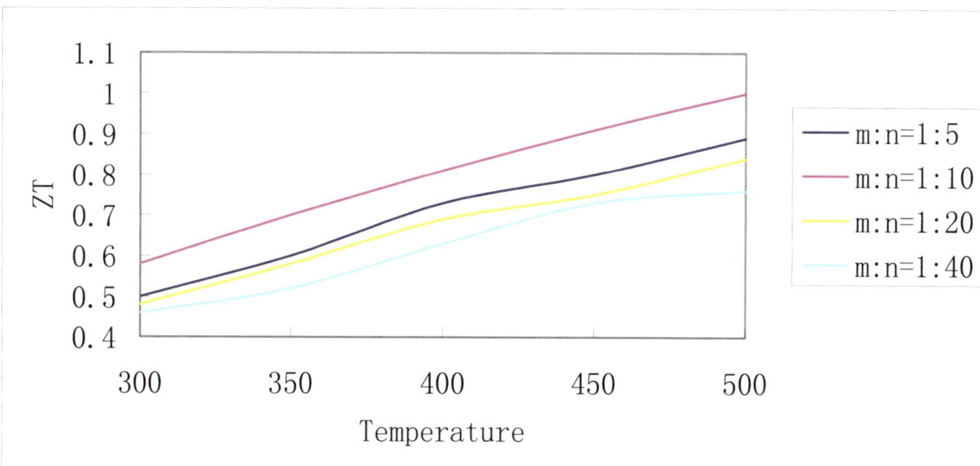

Fig. 23. The relations between temperature (K) and ZT value for $Ga_mSb_nTe_{1.5(m+n)}$ alloys. Data source from (Cui et al., 2009).

The quaternry telluride based alloys with formula $Ag_nPb_mSb_nTe_{m+2n}$ and $Ag_nPb_mSi_nTe_{m+2n}$ were reported with high ZT value. For example, $AgPb_{10}SbTe_{12}$ indicated an exceptionally high ZT value more than 2.0 at enhanced temperature arising from very low total thermal conductivity.

Recently, unit cell disorder is employed as a effective method to reduce lattice thermal conductivity through interstitial sites, rattling atoms or partial occupancies as well as the disorder inherent in the alloying materials. The following figure shows the lower thermal conductivities recently found in nanocomposites compared with previous state-of-the-art thermoelectric alloys such as Bi_2Te_3, PbTe and SiGe.

From the results in Fig. 24, the novel nanocomposites have lower thermal conductivity than traditional thermoelectric materials. The quaternary alloy $Hf_{0.75}Zr_{0.25}NiSb$ in the above figure with high thermal conductivity has a simple cubic structure rather than large and complicated unit cells related to low thermal conductivity. For $La_{3-x}Te_4$, because of large number of random vacancies in rare-earth chalcogenides $La_{3-x}Te_4$ with similar structure like Th_3P_4, rare-earth chalcogenides have a lower lattice thermal conductivity. It can be seen that Ag_9TlTe_5 has lower lattice thermal conductivity. Low lattice thermal conductivity can also be attained from thallium-base thermoelectric materials like Tl_9BiTe_6. The possible explanation of the low value of lattice thermal conductivity is because extremely soft thallium bonding can be observed in low elastic modulus.

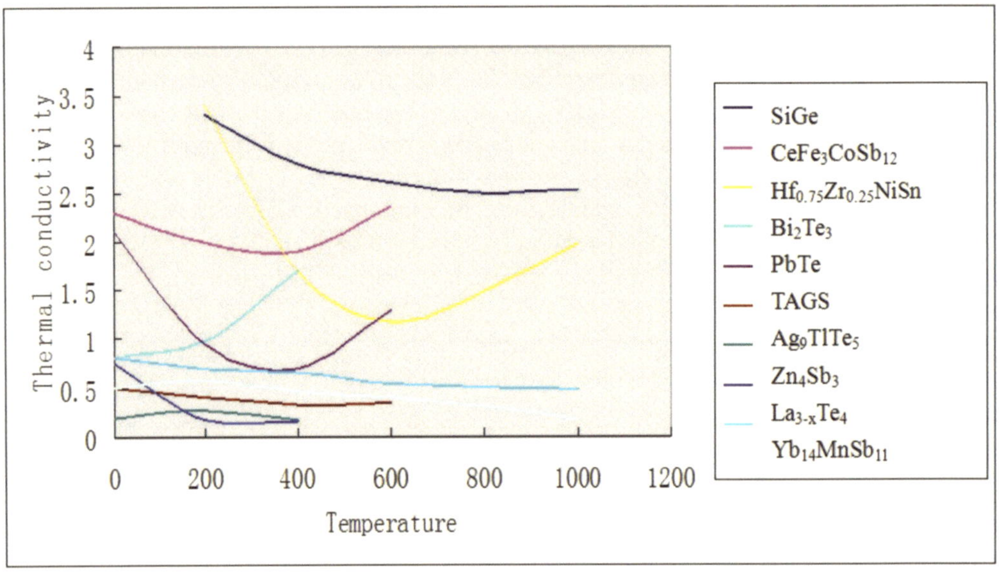

Fig. 24. Relations between temperature (K) and lattice thermal conductivity (W/m·K). Data source from (Snyder & Toberer, 2008).

A ZT value of 2.4 is obtained at 330 K at Research Triangle Institute through the manufacturing superlattice Bi_2Te_3/Sb_2Te_3. The significant improvement lies in the created 'nanoengineered' material which is effective in thermal insulation while maintaining the electrical conductivity. The thermal insulation comes from a intricate localized property for phonons, while the electron transmission is facilitated by optimal choice of band offsets in these semiconductor heterostructures. So the improvement in superlattice materials seems to be more promising for a reduction in lattice thermal conductivity than an increase in power factor. (Tritt et al., 2006)

InSb

Originally, by using a relaxation time approximation to the electron scattering, an unlimited increase of ZT was predicted as the thickness of the nanowire decreased. However, the calculation is via exact solution of the Boltzmann transport equation, which is devised by Ludwig Boltzmann, describing the statistical distribution of one particle in a fluid, including all the fundamental electron scattering mechanisims: polar optic, acoustic deformation potential, piezoelectric and ionized impurity scattering. More recently, the exact solution of the Boltzmann transport equation showed that, if all frequency dependent electron scattering mechanisms are properly taken into account, there is an effective limit to the maximum enhancement of the power factor, that can be obtained with nanowires of a given material. This limit occurs because the acoustic phonon–electron scattering increases dramatically as the thickness of the nanowire is reduced. As a result, it was predicted that a more power factor enhancement of ten times or less would be possible for GaAs quantum well superlattices, and only a two-fold enhancement might be achieved for PbTe. These findings led to speculation that the main contribution to ZT enhancement might come from the lattice thermal conductivity reduction.

There exist large differences between different III–V materials regarding their suitability for nanowire based thermoelectric applications. Only InSb seems to be a promising candidate for which nanowires around 10 nm thick might suffice to obtain a reasonably high figure of merit. GaAs and InP nanowires appear to be of little practical use at room temperature, down to 1 nm thickness. InP nanowires yield the lowest power factor and ZT within the materials studied, and InAs lies in between the InSb and GaAs cases. Enhancements of the power factor of only 10–20 times are predicted in the narrow wire limit, while much of the ZT increase in these III–V nanowires occurs due to κ_l reduction. Two separate sets of calculaltlions are carried out: one for the electronic transport properties, and another one for the lattice thermal transport, for nanowires made of GaAs, InP, InSb and InAs.The result showed that InSb nanowires around 10 nm thick is standing out as a great nanocomposite to exhibit higher power factor and lower lattice thermal conductivity, when compared to GaAs, InP and InAs, as a great nanocomposite to exhibit higher power factor and lower lattice thermal conductivity (Mingoa, 2004).

Recently, In doped $CoSb_3$ based skutterudite nanocomposites have drawn much attention due to their potential TE performance. A ZT value of 1.2 at 575 K is obtained from $In_{0.25}Co_4Sb_{12}$ compound. The single bulk crystal Te doped InSb has been studied for its TE properties. The maximum ZT value is 0.6 at 673 K and the maximu output power is 216 μV at 195 K (Matsumoto et al., 2005). The combined doping influence of In and Ge leads to high properties in n-type skutterudite $In_xCe_yCo_4Sb_{12}$ which was synthesized by a melt-quench-anneal-spark plasma sintering method. The doped $In_xCe_yCo_4Sb_{12}$ showed a nanostructured InSb phase with the grain size from 10 nm to 80 nm which is averagely dispersed on the boundaries of the skutterudite matrix. The evenly dispersed InSb nanophase at the boundaries is believed to notably lower the lattice thermal conductivity and enhance the thermopower. The nanostructured InSb phase also possesses a significant effect on phonon scattering resulting in remarkable reduction in lattice thermal conductivity. Additionally, the lattice thermal conductivity is further reduced due to doped Ce atoms. A reported ZT value of 1.43 was proven at 800 K in the $In_xCe_yCo_4Sb_{12}$ compound (Li et al., 2009).

Other TE materials with low thermal conductivity

Because of the good thermoelectric properties and available component elements, Mg_2Si and Mg_2Sn compounds have indicated potential utilization for TE applications. The high thermoelectric performance $Mg_2Si_{0.4-x}Sn_{0.6}Sb_x$ alloy synthesized by directly melting stoichiometric amount of atom Si, Sn and Sb under argon atmosphere, showed a ZT value of 1.1 with x=0.075 with relatively low thermal conductivity (Zhang et al., 2008). Additionally, oxygen-deficient lanthanum doped strontium titanate $Sr_{1-x}La_xTiO_3$ films showed significant reduction in thermal conductivity due to randomly dispersed and clustered vacancies which are effective to scatter phonons. Specifically, the thermal conductivity could be reduced to 0.5 W/ m K at 300 K for $S_{r0.98}La_{0.02}TiO_3$ via doping other phonon suppression methods like implanting various size nanostructures (Yu et al., 2008).Tedrahedrally bonded stannite (Cu_2FeSnS_4) compound $Cu_2ZnSuSe_4$ showed a narrow band gap of 1.44 eV as a result of which indicating a high TE performance at intermediate temperature. A maximum ZT value of 0.95 at 850 K for $Cu_2ZnSn_{0.9}In_{0.1}Se_4$ is achieved by means of relatively low thermal conductivity due to distorted structure and tunable electrical traits through doping (Shi et al., 2009). In the end, chalcopyritelike ($CuFeS_2$) quaternary chalcogenides Cu_2ZnSnS_4 and

$Cu_2ZnSnSe_4$ were studied for their TE properties with a wide band gap. It has shown that doping can generate more charge carriers thus improving electrical conductivity and distorted structure and quaternary components can lead to low lattice thermal conductivity. Specifically, a ZT value of 0.9 is obtained for $Cu_{2.1}Zn_{0.9}SnS_4$ at 860 K (Liu et al., 2009).

4. Applications of thermoelectric materials

This part will introduce the TE materials' applications at high temperatures and at low temperatures, nanocomposites with organic and inorganic materials and potential TE material graphane.

4.1 Applications at high temperatures

High temperature thermoelectric nanocomposites could be widely used in the field of air-warming system, Aeronautics and astronautics etc. It is an active realm driven by its applications. As a result, to acquire thermoelectric nanocomposite which possesses high energy conversion rate is fundamental for its applications. This process needs improved figure of merit from high electrical conductivity, high Seebeck coefficient, low thermal conductivity or any combinations of the above. This section will introduce perovskite-type oxides, double perovskites, half-Heusler alloy, boron icosahedra, zintl compound, copper aluminate and other oxides with high performance in high temperature. Besides, high temperature materials such as skutterudites and clathrates were narrated earlier in low thermal conductivity section, and chalcogenides were recounted in both low thermal conductivity section and high Seebeck coefficient section.

Perovskite-type oxides

Classic thermoelectric materials suffer from high toxicity, low stability and low efficiency while ceramics have been recently recognized as good thermoelectric with high stability even at elevated temperatures and low production costs. The perovskite is calcium titanium oxide mineral composed of calcium titanate, with the formula $CaTiO_3$. Perovskite-type oxides like Ln(Co, Ni)O_3 (Ln= La, Pr, Nd, Sm, Gd and Dy) compounds can exhibit higher ZT with low thermal conductivity in more than1200 K. The $LaCoO_3$ system is a promising thermoelectric material due to its high Seebeck coefficient of 600 mV/K at room temperature. The thermopower of $LaCoO_3$ is positive due to the partial disproportionation 2 $Co^{3+} = Co^{2+} + Co^{4+}$. Nevertheless the electrical resistivity is rather high which lowers the conversion efficiency. The amount of charge carriers and thus the electrical conductivity and thermoelectric properties in this system can be tuned by suitable Co-site and La-site substitution. Similarly, the compositions $La_{0.8}$ $Ca_{0.2}MnO_3$ and $La_{0.6}$ Ca_4CoO_3 were successfully grown by Weidenkaff et al. (Weidenkaff et al., 2008). These kinds of compounds could be used in regaining heat from the wastes. Still among perovskite-type oxides, some ceramic oxide materials such as $Na_xCo_2O_4$, site-substituted $SrMnO_3$ bear good electric conductivity and other perovskite-type oxides possess large thermopower (Robert et al., 2007). Besides, their chemical stability can prevent them from being oxidized. So, they are promising thermoelectrical materials due to simple manufacturing processes and low production costs in addition to above advantages.

Double perovskites

Double perovskites thermoelectric materials were reported showing high Seebeck coefficient at both high temperature and ambient temperauture. Sr_2RuYO_6 and Sr_2ErRuO_6 are such materials. Fig. 25 indicated changes in Seebeck coefficient with temperature. From the results of Fig. 25, it can be seen that the values are negative in the whole temperature range due to majority of the charge carriers are electrons. The thermoelectric figure of merit values are large when the Seebeck coefficient of Sr_2RuYO_6 is -475 μV/K at 500 K and Seebeck coefficient of Sr_2ErRuO_6 is -400 μV/K at 500 K. From the figure 28, for Sr_2ErRuO_6 compound, the Seebeck coefficient is almost constant between the temperature range of 620 K and 950 K.

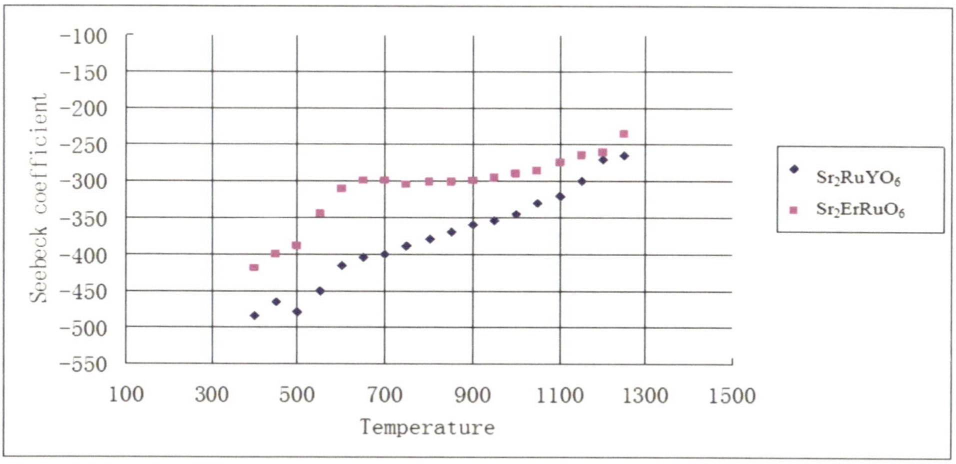

Fig. 25. The relations bewteen temperature (K) and Seebeck coefficient (μV/K) for Sr_2RuYO_6 & Sr_2ErRuO_6. Data source from (Aguirre et al., 2005).

Fig. 26 shows the relations between the total thermal conductivity and temperature for the these two compounds. The data points in Fig. 26 indicated that the trend of the thermal conductivity decreases with the increasing temperature. Obviously, Sr_2RuErO_6 has a lower thermal conductivity than Sr_2RuYO_6 attributing to that Er atom is larger than Y atom leading to a more efficient scattering center for phonons.

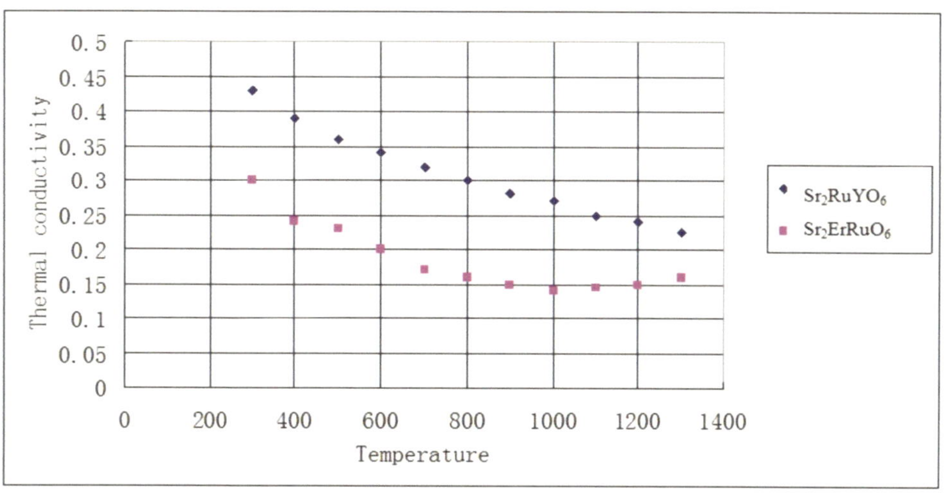

Fig. 26. The relations bewteen temperature (K) and total thermal conductivity ($W \cdot K^{-1} m^{-1}$) for Sr_2RuYO_6 & Sr_2ErRuO_6. Data source from (Aguirre et al., 2005).

Half-Heusler (HH) alloy

A Heusler alloy is a ferromagnetic metal alloy based on a Heusler phase which is intermetallic with particular composition and face-centered cubic crystal structure. This term is named after a German mining engineer and Chemist Friedrich Heusler, who studies such an alloy in 1903. Heusler alloy has the formulism Cu_2MnSn. While the half-Heusler alloys are considered as high-performance thermoelectric materials for high temperature. Specifically, (Ti, Zr, Hf)NiSn system consisting of three interpenetrating fcc sublattices indicated higher Seebeck (100-500 μV/K at 300 K) coefficient and relatively low resistivity. To achieve higher power factor, Sb could be incorporated in Sn site which is the main provider of carriers for charge transport for half-Heusler TiNiSn at high temperature. As a result, The (Ti, Zr, Hf)NiSn system is becoming a semimetallic material. The extrinsic charge carrier due to doping is fundamental for higher TE efficiency. Therefore, it is believed that properly alloying for these half-Heusler materials could enable these compounds to show higher TE performance at high temperature by tuning crystal structure and carrier concentrtion (Kim et al., 2004).
For example, for TiNiSn compound, incorporating Hf into Ti site and incorporating Sb into Sn site will reduce thermal conductivity. The highest ZT value of 0.78 is obtained at 770 K for $Ti_{0.95}Hf_{0.05}NiSn_{0.99}Sb_{0.01}$ compound; incorporating both Hf and Zr into Ti site and incorporating Sb into Sn site will cause an improved ZT value of 1.4 at 700 K for $(Zr_{0.5}Hf_{0.5})_{0.5}Ti_{0.5}NiSn_{1-y}Sb_y$ compound. Moreover, for n-type ZrNiSn-based alloys, incorporating Hf into Zr site and incorporating Sb into Sn site will cause an improved ZT value of 0.7 at 800 K for $Zr_{0.5}Hf_{0.5}Ni_{1-x}Pd_xSn_{0.99}Sb_{0.01}$ compound, and an enhanced ZT of 0.78 at 1070 K for $Hf_{0.75}Zr_{0.25}NiSn_{0.975}Sb_{0.025}$ compound (Culp et al., 2006). What's more, The Bi containing HH alloys has lower thermal conductivity compared to general HH alloys. A Ni doped ZrCoBi prepared by solid state reaction showing improved TE performance was reported (Ponnambalam et al., 2007). It is also reported that incorporating at the Ni site is vital for the optimization of the TE efficiency. The concentration of Ti, Zr and Hf as well as Pd or Pt substitution at Ni site can reduce the lattice thermal conductivity through mass fluctuations and strain field effects. Moreover, incorporating Pt on the Ni site of the disordered ZrNiSn-based compounds will improve both mass fluctuation and strain field fluctuation on accounts of Pt has a larger atomic mass and metallic radius than Pd. HH alloy with the formula $Zr_{0.5}Hf_{0.5}Ni_{0.5}Pd_{0.5}Sn_{0.99}Sb_{0.01}$ was such kind material with reduced thermal conductivity, thus indicating enhanced TE properties (Yang et al., 2004). A study deliberates a series of half-Heusler compounds XNiPn in which X stands for atoms such as Y, La and Bu, Pn means a pnicogen As, Sb and Bi. The results indicated that all these compounds are narrow-gap semiconductors and could be utilized as potential candidates for high properties TE materials (Larson et al., 1999).
(Zr,Hf)Co(Sb,Sn) alloys can be fabricated by mixing elemental pieces in proper quantities, and then melting several times in an arc furnace under flowing. An lowered thermal conductivity is found between 4.1 $W/m^{-1} K^{-1}$ and 3.6 $W/m^{-1} K^{-1}$ ranging from 300 K to 1000 K for $Zr_{0.5}Hf_{0.5}CoSb_{0.8}Sn_{0.2}$. An obvious improved ZT value is observed due to decreasing in thermal conductivity as well as increasing power factor beyond 1100 K. A ZT value of 0.51 is found for $Zr_{0.5}Hf_{0.5}CoSb_{0.8}Sn_{0.2}$ which is higher than SiGe at the same temperature 1000 K (Culp et al., 2008). Nearly single-phase NbCoSn alloys was synthesized by directional solidification employing optical floating zone melting method. This material is generally hard to fabricate because the melting point of Sn is lower than other elements in the alloy.

The nearly single-phase NbCoSn indicated a Seebeck coefficient of -250 μV/K at around 900 K and showed a relatively high carrier concentration of 4.82×10^{26} m^{-3}. Also NbCoSn exhibited an excellent power factor of 2.5 mW/MK at around 650 K without adjusting electrical properties (Kimura et al., 2008). HH alloys $Zr_{1-x}Ti_xCoSn_ySb_{1-y}$ is also investigated for their possible use as high temperature p-type TE material. In this material, the Zr site is mainly substituted with Ti to lower thermal conductivity, which the content of Sn is varied to optimize the TE properties. Fig. 27 indicates the relation betweens temperature and thermopower. It showed that, thermopower values are high at higher temperatures due to high effective mass (Ponnambalam et al., 2008).

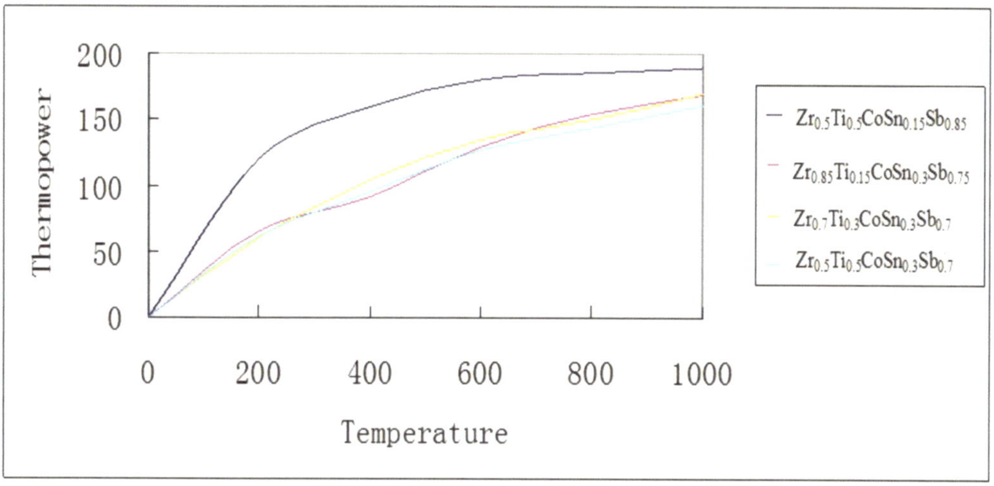

Fig. 27. Relations between thermopower (μV/K) and temperature (K) for Ti content varied $Zr_{1-x}Ti_xCoSn_{0.3}Sb_{0.7}$ series and Sn content reduced $Zr_{0.5}Ti_{0.5}CoSn_{0.15}Sb_{0.85}$ alloy. Data source from (Ponnambalam et al., 2008)

Boron icosahedra

Numerous novel nanocomposites were studied for the low thermal conductivity. Boron-rich cluster compounds are attractive materials because of their stability under high temperature, typically exhibiting melting points of above 2300 K, and more importantly, they also have intrinsic low thermal conductivity (Mori et al., 2007). Among the rare-earth-containing boron icosahedra compounds, only RB_{66} (R=rare earth) has been investigated over a wide temperature region. YB_{66} was found to exhibit low thermal conductivity with an amorphous like temperature dependence. A homologous series of n-type boron icosahedra compounds $RB_{17}CN$,$RB_{22}C_2N$ and RB_{66} (R=rare earth) without doping were discovered to express low thermal conductivity in the temperature range of 300–1000 K rather than traditionally synthesized boron cluster compounds like boron carbide, RB_{66} and RB_{50} which have been found as p type (Mori et al., 2007).

Zintl compound

Moreover, it is reported that C_{60}-containing $Ba_{0.44}Co_4Sb_{12}$ possessing great thermal and electrical properties showed lower themal conductivity as low as 1.7 W/mK at around 650K.Also, the maximum output of the ceramic-layer-inserted p-n-joined $Si_{0.8}Ge_{0.2}$ thermoelectric elements was 48 mW/K starting from 500K to 1300K synthesized by pulse-current sintering compared with dense bodies from gas-atomized powders. Obviously, the thermal conductivities of fine-grained Si–Ge dense bodies from ball-milled powders were greatly lowered (Otakea et al., 2004). But there is some special TE nanocomposite with lower thermal conductivity showing higher ZT value. It is reported that the compound $Ba_{0.24}Co_4Sb_{12}$ shows a ZT value as high as 1.3 at 850 K with decreased electrical conductivity and thermal conductivity due to increasing impurity contents. The reason is because of enhanced grain-boundary scattering of charge carriers and phonons, and barium fullerides formed from apart of barium in the filled skutterudite reacted with fullerene (Shi et al., 2007).

Copper aluminate

Another thermoelectric promising nanocomposite which is still stable at high temperatures up to 1400 K is copper aluminate ($CuAlO_2$). This TE nanocomposite could be used in thermoelectric devices and optoelectric applications dut to that $CuAlO_2$ is a transparent semiconductor and has a direct band-gap of 3.5eV. Many researchers use $CuAlO_2$ as a basic material to fabricate nanocomposite with higher value of ZT. The electrical conductivity and the Seebeck coefficient can be enhanced with the substitution of Ca for Al like $CuAl_{0.9}Ca_{0.1}O_2$.The nanocomposite $CuAl_{0.9}Fe_{0.1}O_2$ was obtained at 1140 K showing higher value of power factor up to $1.1(10^{-4})$ W/mK. Recently, the newly prepared doped $CuAlO_2$ showed the highest power factor of $1.26(10^{-4})$ with formation of $Cu_{1-x-y}Ag_xZn_yO_2$ at 1060 K, expressing a large enhancement compared with pure $CuAlO_2$. This could be proved by the XRD measurements demonstrating that the substitution of Ag and Zn for Cu improved the figure of merit. XRD also verified if using the Ag and Ni for the substitutions, the chemical reaction will be hindered between the staring materials (Yanagiya et al., 2010).

Other Oxides

Indium oxide indicated special TE properties when partly substituted with other cations. And indium oxide possesses great electrical conductivity and has a cubic bixbyite-like structure. Bixbyite is a manganese iron oxide meneral with a formula like $(Mn, Fe)_2O_3$. Doping SnO_2 into indium tin oxide bixbyite structure indicated an notably improved electrical conductivity and this kind of material is widely used for optoelectric application. Moreover, Ge incorporated In_2O_3 matrix with $In_2Ge_2O_7$ inclusion showed an enhanced ZT value of 0.46 at 1000 K which is the higher value obtained from polycrystalline oxides at high temperature. Additionally, $In_{1.6}Zn_{0.2}Sn_{0.2}O_3$ nanocomposite achieved a ZT value of 0.3 at 1273 K with a reduced lattice thermal conductivity resulting from mass fluctuation scattering of the phonons (Bérardan et al., 2008). What's more, Ca_3CoRhO_6, Ca_3CoIrO_6, $Sr_4Co_3O_9$ and $Ca_3Co_2O_6$ series were studied for their TE performances. Both Ca_3CoRhO_6 and Ca_3CoIrO_6 indicated a Seebeck coefficient of 150 μV/K at around 1000 K; Ca_3CoRhO_6 showed a lowerer power factor than $Sr_4Co_3O_9$; While Ca_3CoRhO_6's power factor is more than two times than $Ca_3Co_2O_6$ at 1073 K. As a result, Ca_3CoRhO_6 could be considered as high promising TE material at high temperatures (Takami & Ikuta, 2008). Moreover, $Ca_3Co_4O_9$ fabricated by, chemical substitution, SPS and hot pressing indicated improved performance of TE effect. Many researches showed that substitution for Co site with Fe and substitution for Ca site with Na, Bi, Ag and Eu is an effective way to improve TE performances. Also, using the method of partially doping heavier ions in the Co site can lower the thermal conductivity. As a result, Ga ion with trivalence could be served for improving TE properties. $Ca_3Co_{4-x}Ca_xO_9$ oxides showed an enhanced ZT value than undoped $Ca_3Co_4O_9$ specimen because Ga doped specimen possesses lower thermal conductivity than that of undoped one at around 1100 K (Nong et al., 2010).

4.2 Applications at low temperatures

There are many nanocomposite materials possess some high-efficient thermoelectric characteristics within low temperatures. This section will introduce such materials with Te, Sb or Bi based alloys, $FeSb_2$ and TiS_2, B-based compounds and cobalt oxide.

The incorportaed Te, Sb or Bi based alloys showed improved TE performance than traditional Bi-Sb-Te-Se alloys. It is reported that the $CsBi_4Te_6$ showed a high value ZT up to 0.8 at 225 K when incorporating appropriately. At cryogenic temperatures, the themoelectric properties of $CsBi_4Te_6$ can excel the traditional Bi-Sb-Te-Se alloys (Chung et al., 2000). Moreover, $Bi_{1.8}Sb_{0.2}Te_3$ was studied at low temperature for its Seebeck coefficient, ZT value and carrier concentration. $Bi_{1.8}Sb_{0.2}Te_3$ fabricated by gradient freeze method indicated a high Seebeck coefficient more than 500 μV/K and a high carrier concentration of 1.6×10^{19} cm^{-3}, thus achieved a ZT value of 1.1 at 200 K (Huonga et al., 2004).

$Bi_{1-x}Sb_x$ nanowires ($0<x<0.30$) was studied with diameters 10 nm$<d_W<$100 nm at 77 K for its electronic transport properties. The calculation results showed that the ZT value of $Bi_{1-x}Sb_x$ nanowires is higher than that of bulk alloy. An extremtly high ZT value is achieved for p-type $Bi_{1-x}Sb_x$ nanowires at 77 K with d_W=40 nm and x=0.13. It is also discovered that $Bi_{1-x}Sb_x$'s ZT value will be more than 1.2 for both p-type and n-type materials at 77K. The enhancement is attributed to the coalescence in energy of the subband edges of the ten hole pockets of this system and the resulting high density of states. These phenomena may not be limited to $Bi_{1-x}Sb_x$, and could serve as a strategy for the enhancement of ZT in other low-dimensional systems. It has proven that the same nanowires, under different doping conditions, can be used as either p-type or n-type legs in a TE device with ZT=1.2 at low temperature (Rabina et al., 2001). Theoretical work on transport properties in Bi was also studied. Bi nanowires have predicted that this system is a good candidate for an n-type leg for low temperature (77 K) performance. The ZT value of a Bi nanowire, oriented along the trigonal crystallographic direction and if it is properly doped, is expected to reach a ZT values of more than 1 at 77 K as the wire diameter (d_W) is decreased below 15 nm. Even smaller diameters are needed for the Bi nanowires to serve as good p-type legs.

Both the doped TiS_2 and undoped $FeSb_2$ possess the good TE properties. $FeSb_2$ was recently identified as a narrow-gap semiconductor with indications of strong electron-electron correlations. Strong electron-electron correlations are recognized to lead to a strong enhancement of the thermopower in a correlated semiconductor.

Several thermal properties of $FeSb_2$ were studied around 0 to 100K. The maximum absolute thermopower of $FeSb_2$ spans from 10 to 45 mV/K at around 10 K, greatly exceeding two reference compounds that of both $FeAs_2$ and $RuSb_2$. And $FeSb_2$ represents a promising candidate for thermoelectric cooling applications at cryogenic temperatures. Moreover, Isoelectronic substitution of antimony by arsenic $FeSb_{(2-x)}As_x$) reduces the high thermal conductivity (Sun et al., 2009). Magnesium incorporated compound $Mg_xTi_{1-x}S_2$ was studied for its thermoelectric and transport properties from 5 K to 310 K. After doping Mg for Ti in this compound, a phenomenon of a transition from metallike to semiconductorlike property because of reduction in electron concentration of Mg^{2+} substitution for Ti^{4+}, happened meaning that TiS_2 essentially is a semiconductor. The result indicated that lightly substituted compound $Mg_{0.04}Ti_{0.96}S_2$'s ZT value increased significantly mainly due to its thermopower which is nearly 1.6 times more than that of TiS_2 at 300 K (Qin et al., 2007).

For B-based compound, UB_4 's thermoelectric properties were studied along with UB_4's three different boron isotope in the temperature range from 80 K to 300 K. Each of the specimens indicated a decreased TE performance with increasing temperature. The electrical resistivity decreased with increased temperature, and the natural boron, boron-10 enriched and boron-11 enriched specimen exhibited no obvious differences in electrical conductivity (Nishi et al., 2001). Another B-based TE compound is YB_{66}. Among the rare-earth-containing boron icosahedra compounds, YB_{66} has been investigated over a wide low temperature region. And YB_{66} was found to exhibit low thermal conductivity with an amorphous like temperature dependence. What's more, thermal conductivity of a $YbB_{44}Si_2$ crystal in the low temperature (12-300K) was investigated, the result of which shows that $YbB_{44}Si_2$ exhibits higher thermal conductivity value than RB_{66} in lower temperature. As temperature goes up, the thermal conductivity of $YbB_{44}Si_2$ gradually goes down, but still a little higher than $YB_{66.}$ A layered cobalt oxide $(Ca_{0.85}OH)_{1.16}CoO_2$ was studied for its TE performances in the temperature range from 290 K to 573 K. The experimental results from x-ray diffraction, SEM indicated that $(Ca_{0.85}OH)_{1.16}CoO_2$ is a p-type semiconductor. During this temperature range, ZT value will increase with increasing temperature (Pei et al., 2009).

4.3 Applications of organic and inorganic materials

Conducting polymers are attracting more and more emphasis with respect to its low thermal conductivity compared to that of an inorganic material as a traditional TE material. Hence, to better understand the characteristics and fabrications of nanostructuring composite materials is becoming neccessary. Additionally, due to the high electrical conductivity property of some polymers such as polyaniline, polypyrrole, p-phenylenevinylene, polythiophene and their derivatives, conducting polymers have already captureed much attention of its scientific application. Moreover, the conjugated polymer is flexible to be synthesized compared with traditional inorganic TE materials. While amongst the family of conducting polymers, polyaniline is considered as a critical and most studied conducting materials of the past 20 years because of good processability, stability in the air, high redox reversibility, swift change in film color, low cost, abundant chemistry characteristics and its wider applications ,say, Li batteries, capacitors (Okamoto & Kotaka, 1998), battery electrode, immobilization of enzymes, electrochromic devices (Mu et al., 1997), chemical sensors, anticorrosion coatings and light-emitting diodes (Guo & Zhou, 2007). Moreover, Carbon nanotubes have been extensively used as filling materials for polymeric composites due to their remarkable properties such as its wide accessible surface areas, great electronic traits and low resistivity. Hence, the carbon nanotube-filled polymeric composites can be applied to power devices, sensors, solar cell and supercapacitors, among which supercapacitors played an crucial role in wider applications such as auxillary power source in combination with battery in hybrid electric vehicles, back-up power sources for computer memory and short-time power source for mobile electronic devices (Zhang et al., 2009). The conducting polymer can also be used as solar cells when combined with nanocomposite materials like lead sulfide (Watt et al., 2005).

Through the years, various methods have been approved to synthesize carbon nanotube-filled polymers. In-situ chemical oxidative polymerization and ultrasonically initiated in-situ emulsion polymerization with good thermal stability and high electric conductivity have been successfully done to fabricate carbon nanotube-filled polyaniline. The carbon nanotube-filled polyanniline composite films were fabricated by electrochemical polymerization of an aniline solution and it demonstrated higher specific capacitance, better cyclic stability. It is reported that both single-walled carbon nanotube and multi-walled carbon nanotube filled polyaniline exhibited improving electric conductivity and enhanced mechanical characteristics in supercapacitors than non-filled polymers.

In recent years, one-dimensional polyaniline have been synthesized by both chemical polymerization and electrochemical polymerization which are cheap means of gaining bulk nanofibers. The chemical fabrication consists of nanofiber seeding, electrospinning, template-based, interfacial polymerization and surfactant synthesis. But there is one big flaw involved in the chemical method: the remained surfactant and oxidants will in turn wreck sample's electrical properties. So the existing of ethanol and oligomer will reduce the effects from the dopants or current density to obtain the uniform deposited polyaniline films. While electrochemical polymerization is an environmentally friendly way to fabricate polyaniline nanofibers, and this method is a complementary way to chemical fabrication due to no surfactants and oxidants are needed (Guo & Zhou, 2007). During the production course, polyaniline chains grow on the electrode in the course of the electrochemical polymerization. Dedoped polyaniline is obtained by soaking phosphate acid doped polyaniline in ammonia solution for four hous, and their diameters range from 80 nm to 100 nm.

It is reported that aniline polymerization rate and spin activity of polyaniline are highly related to the concentration of aniline and hydrochloric acid. That is because the phenomenon that the Electron Paramagnetic resonance intensities increase obviously from 0.95 to 1.07 V during electrolysis process due to quick oxidization of aniline around 0.9 V and reduction from spinless bipolarons to polarons (Mu et al., 1997).

However, pure polymer and inorganic nanocomposite materials may not have the better TE properties than composite materials consisting of conducting polymer and inorganic TE material. The fabricated nanocomposite materials could inherit properties of both the inorganic materials and the polymers, or even with a synergistic effect. The organic-inorganic nanocomposite TE material provides the solid proof for low-cost, easy-fabrication, nontoxic and high conversion rate TE material in the future. In a two component nanostructuring composite including polymer and inorganic material, the combination of a conducting polymer and inorganic thermoelectrical material may yield high thermoelectric power and electrical conductivity. Because of the specific characteristics of polyaniline, some novel thermoelectric nanocomposite materials consisting of polyaniline (PANI) and other materials such as Bi_2Te_3 and Sb_2S_3 have been successfully synthesized using electrochemical reactions and deposited method. This kind of polymer-semiconductor composite own the traits of mechanical strengh and the solidness of semiconducting comnpounds (Subramanian et al., 2010).

For example, using inorganic bismuth nitrate as a dopant can help novel nanocomposite materials gaining higher electrical conductivity. This kind of rod-like nanocomposite thermoelectrical material has the diameter less than 100 nm. The synthesis of two component needed two solutions A and B: solution A was prepared by dissolving 200 mg of undoped PANI in 100 mg of N-dimethylformamide (MERCK) , and then mixed with 2 mg of bismuth nitrate pentahydrate as a dopant , the ratio of the PANI: dopant was 1:10. And this solution was kept under ultrasonication for 2 hours. Solution B comprised of equal volumes of bismuth nitrate solution and tellurium oxide solution. To fabricate this novel thermoelectric material, 5 ml of solution B and 20 ml of solution A were put in the electrochemical cell. ADC voltage of 20 V was applied to the solution through the electrodes. The remaining 15 ml of solution B was added drop by drop at uniform time intervals to acquire uniform layer, ensuring uniform dispersion of both the components (Chatterjee et al., 2009).

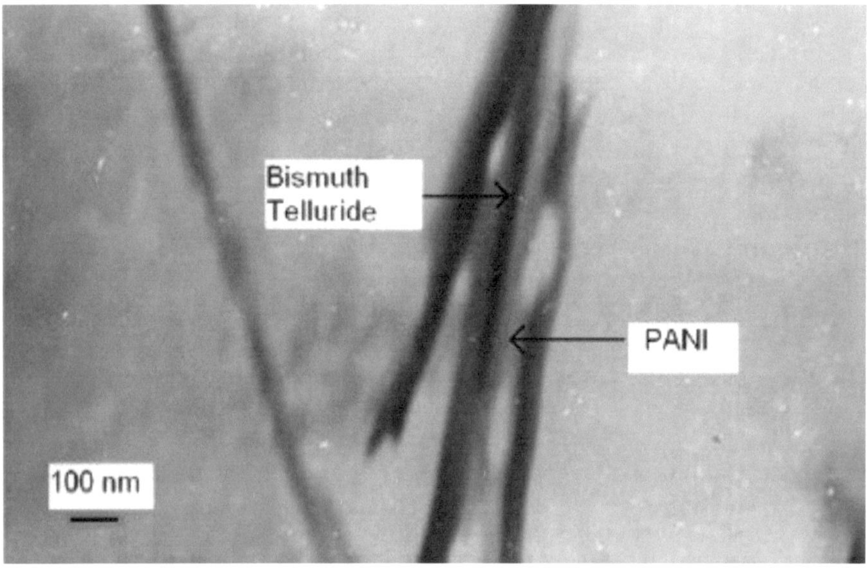

Fig. 28. TEM images of PANI/Bi_2Te_3 composite. Reproduced from (Chatterjee et al., 2009)

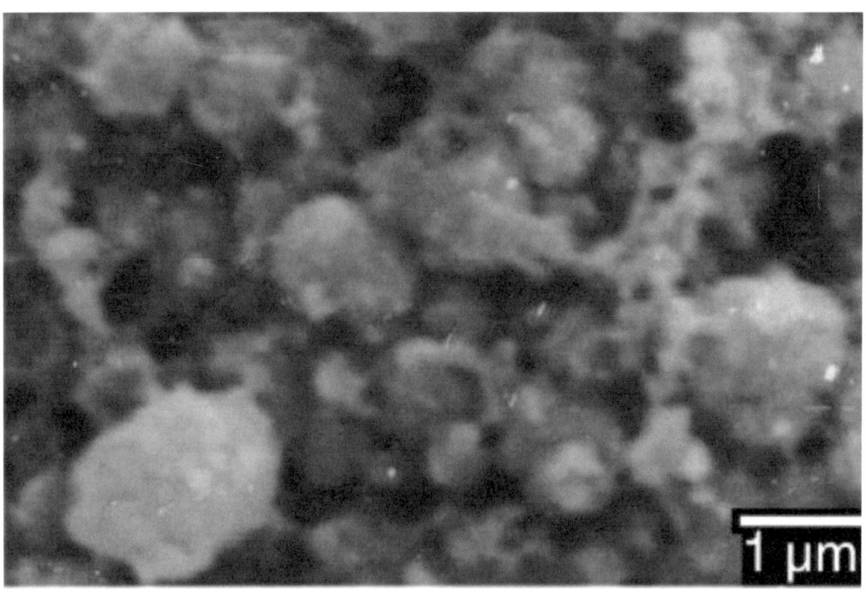

Fig. 29. SEM image of PANI/ Bi_2Te_3 (10000 magnification). Reproduced from (Chatterjee et al., 2009)

Transmission electron microscopy (TEM) (See Fig. 28) confirms the existence of polyaniline-Bi_2Te_3 and shows the rod-like nanostructure of less than 100 nm. It can be seen that the polyaniline is amorphous and transparent, while Bi_2Te_3 is dark and opaque. In comparison with TEM, Scanning electron microscopy (SEM) (See Fig. 29) demonstrates the inorganic Bi_2Te_3 and organic polyaniline have been formed and the surface of which is uniform and continuous. Through calculations, the magnitude of thermoelectric power of the composite is between its two components within 340 K - 360 K.

Another example of two component nanocomposite material is Sb_2S_3 doped in PANI. This doping has more improvement of electrical conductivity of the thin film than the Bi_2Se_3 and Bi_2S_3. Usage of Sb_2S_3 as a dopant is because this composite is applied for television cameras and possesses the weak polar semiconducting ferroelectric characteristic. Many researchers have attempted various methods to deposit this material such as electrodeposition, chemical deposition and spray pyrolysis. While the way of doping polyaniline into the electrolyte of Sb_2S_3 thin film is more promising due to its three times enhancement of conductivity than that of as-deposited Sb_2S_3 thin film and decreasement of band gap energy on PANI doping as a result of the interfacial interaction between PANI and Sb_2S_3.

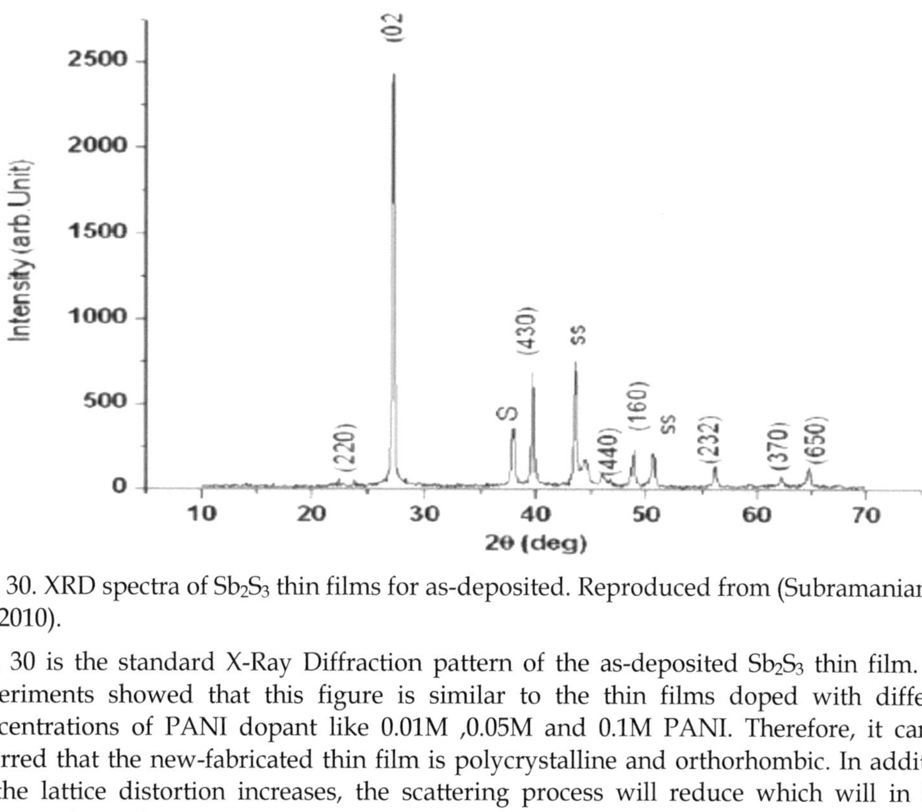

Fig. 30. XRD spectra of Sb_2S_3 thin films for as-deposited. Reproduced from (Subramanian et al., 2010).

Fig. 30 is the standard X-Ray Diffraction pattern of the as-deposited Sb_2S_3 thin film. The experiments showed that this figure is similar to the thin films doped with different concentrations of PANI dopant like 0.01M ,0.05M and 0.1M PANI. Therefore, it can be inferred that the new-fabricated thin film is polycrystalline and orthorhombic. In addition, as the lattice distortion increases, the scattering process will reduce which will in turn improve the mobility that will bring about increasement of electrical conductivity through the doping of PANI into the thin film. The figure of merit will increase on account of incorporation of PANI in the lattice of Sb_2S_3 resulting in lattice distortion.

As of the high Seebeck coefficient of the PbTe in nanocomposite material, it is very vital to well deposit inorganic TE material PbTe onto polyaniline to acquire great TE properties. Generally, it is difficult to homogenously disperse composite material into another polyer. As a result, PbTe-PANI composite could be fabricated by interfacial polymerization at room temperature. The synthesized nanocomposite material powder comprised of PbTe nanoparticles, PANI/PbTe core-shell morphology and PbTe/PANI/ PbTe three layers sphere-like morphology. As the temperature increased from room teperature to 373 K, the electrical conductivity and Seebeck coefficient will show a synergistic effect compared with that of pure PbTe and PANI.

There are still lots of fabricated organic-inorganic TE composites: Bi-PANI nanocomposite by dispersing Bi nanoparticles into PANI, Ag-poly(3-octylthiophene) nanocomposite exhibiting extremely high Seebeck coefficient 1283 μV/K, $Bi_{0.5}Sb_{1.5}Te_3$-PANI nanocomposite by mixing $Bi_{0.5}Sb_{1.5}Te_3$ powder and PANI powder and then pressing them into pellets, $NaFe_4P_{12}$-PANI whisker nanocomposite by using emulsion fabrication method and in-situ compounding and carbon nanotube-polymer nanocomposite with exceedingly electrical conductivity (Wang et al., 2009).

4.4 Potential high ZT value TE material: graphane

Graphane is composed of carbon and hydrogen with the formula unit $(CH)_n$ in which n is large. Graphane is a single-layered 2-D polymer derived from a special graphite whose witdth is only 0.0335 nm. Adding 200,000 such graphites together, it only can build up a width of a single hair. One prominent characteristic of graphane is its electrical transport properties: indicating abnormal integral quantum Hall effect. The advantages of its low cost and rapid fabrication of its nanoscale graphane derivatives made carbon-based materials a potential promising high efficiency TE nanocomposite candidate for TE materials.

Graphane is a potential TE material for its quite large predicted ZT value. It could be as high as 5.8 depending on temeperature linearly by means of randomly introducting hydrogen vacancies to the graphane nanoribbon derivatives- armchair graphane nanoribbons (Ni et al., 2009). Furthermore, intrinsic anisotropy of thermal conductance in graphane nanoribbons were studied with width ranging from 0.5 nm to 35 nm. It found that anisotropy will disappear when its width is larger than 100 nm. The intrinsic anisotropy, created from different boundary conditions at ribbon edges, can be utilized to adjust thermal conductance for higher performance (Xu et al., 2007). What's more, large Seebeck coefficient of graphane deposited on a silicon dioxide substrate is obtained reaching 30 mV/K which is reported the highest value ever. This achievement is important for TE applications such as energy harvesting and scavenging (Dragoman & Dragoman. 2007).

5. Summary

To date, the most practical ZT value achieved is less than 2.0 which has been great improved than last century. ZT value will arrive at higher value like 5.0 which is not a dream in reality through researchers' constant endeavors in near future when under ideal experimentally conditions. Moreover, for small-scale electronic and optoelectronic applications, thin films may be the most suitable materials. While for large-scale applications like refrigeration and power generation, bulk TE materials with high TE properties may be fit and will be developed.

Although researchers have achieved a great deal of advances in TE materials by employing Slack's phonon-glass and electron-crystal approach or other methods, there still numerous problems ahead of us. How to choose the potential TE materials and how to tune the TE parameters to improve TE performance play an important role in pursuing the state-of-the-art materials in commecial utilizations. Up until now, a multitude of distinct TE systems and high potential and promising TE nanocomposites are currently under investigation by a variety of global research groups. These potential nanocomposites range from semiconductors and semimetals to ceramic oxides, and from thin-film superlattice materials to large-single crystal or polycrystal materials. Consequently, there is inevitably much more work to do to obtain high-performance TE materials and devices.

Reaching higher energy efficiencies is not confined to thermoelectrical methods like creating complex structures and disorders resulting in phonon-glass properties, or leading to electron-crystal state. It is a multidisciplinary approach to quest and develop higher-efficiency TE materials and devices. As of it said, people are trying various ways to obtain higher energy conversion such as photovoltaic, mechanoelectric, electrochemical ways. And these ways are acquiring great accomplishments and are still developing. So the days when people can achieve high energy transformation rate are close by.

6. Acknowledgements

This work was supported by the research start-up fund from The University of Toledo. We also acknowledge the permissions from Elsevier B.V. Ltd through Copyright Clearance Center's Rightslink service to using Figures 28–30 in this paper.

7. References

Aguirre, M.H., Logvinovich, D., Bocher, L., Robert, R., Ebbinghaus, S.G.& Weidenkaff, A. (2008). High-temperature thermoelectric properties of Sr_2RuYO_6 and Sr_2RuErO_6 double perovskites influenced by structure and microstructure, Acta Materialis.Vol.(57): 108 - 115

Bentien, A., Christensen, M., Bryan, J.D., Sanchez, A., Paschen, S., Steglich, F., Stucky, G.D. & Iversen, B.B. (2004). Thermal conductivity of thermoelectric clathrates, Physical Review B. Vol.(69): 045107(1)- 045107(5)

Berardan, D., Guilmeau, E., Maignan, A.& Raveau, B. (2008). Enhancement of the thermoelectric performances of In_2O_3 by the coupled substitution of M^{2+}/Sn^{4+} for In^{3+}, Journal of Applied Physics. Vol.(104): 064918(1)- 064918(5)

Bergvall, P. & Beckman, O. (2003). Thermoelectric properties of non-stoichiometric bismuth-antimony-telluride alloys, Solid State Electronics. Vol.(6)(No. 12):133-136

Cadavid, D. & Rodr´guez, J.E. (2008). Thermoelectric properties of polycrystalline Zn_4Sb_3 samples preparedby solid state reaction method, Physica B. Vol.(403): 3976–3979

Candolfi, C., Lenoir, B., Leszczynski, J., Dauscher, A., & Guilmeau, E., (2009). Beneficial influence of Ru on the thermoelectric properties of Mo_3Sb_7, Journal of Applied Physics. Vol.(105): 083701(1)- 083701(4)

Chatterjee, K., Suresh, A., Ganguly, S., Kargupta, K. & Banerjeea, D. (2009). Synthesis and characterization of an electro-deposited polyaniline-bismuth telluride nanocomposite -A novel thermoelectric material, Materials Characterization. Vol.(60): 1597-1601

Chung, D.Y., Hogan, T., Brazis, P., Rocci-Lane, M., Kannewurf, C., Bastea, M., Uher, C.& Kanatzidis, M.G. (2000). $CsBi_4Te_6$: A High-Performance Thermoelectric Material for Low-Temperature Applications, Science. Vol.(287)(No. 5455): 1024 - 1027

Cosnier, M., Fraisse, G. & Luo, L.G. (2008). An experimental and numerical study of a thermoelectric air-cooling and air-heating system, International Journal of Refrigeration. Vol.(31): 1051-1062

Cui, J.L., Liu, X.L., Yang, W., Chen, D.Y., Fu, H.& Ying, P.Z. (2009). Thermoelectric properties in nanostructured homologous series alloys $Ga_mSb_nTe_{1.5(m+n)}$, Journal of Applied Physics. Vol.(105): 063703(1)- 063703(6)

Culp, S.R., Poon, S.J., Hickman, N., Tritt, T.M.& Blumm, J. (2006). Effect of substitutions on the thermoelectric figure of merit of half-Heusler phases at 800 °C, Applied Physcis Letters. Vol.(88): 042106(1)- 042106(3)

Culp, S.R., Simonson, J.W., Poon, S.J., Ponnambalam, V., Edwards, J. & Tritt, T.M. (2008). (Zr,Hf)Co(Sb,Sn) half-Heusler phases as high-temperature (>700 °C) p-type thermoelectric materials, Applied Physics Letters. Vol.(93): 022105(1)- 022105(3)

Deng, S., Tang, X.F.& Zhang, Q.J. (2007). Synthesis and thermoelectric properties of p-type $Ba_8Ga_{16}Zn_xGe_{30-x}$ type-I clathrates, Journal of Applied Physics. Vol.(102): 043702(1)-043702(5)

Dragoman, D. & Dragoman, M. (2007). Giant thermoelectric effect in graphane, Applied Physics Letter. Vol.(91):203116(1)-203116(3)

Dragoman, M., Dragoman, D. & Plana, R. (2007). Modeling of rf energy sensing and harvesting using the giant thermoelectric effect in carbon nanotubes, Applied Physics Letter. Vol.(91): 173117(1)- 173117(3)

Dresselhaus, M.S., Sun, X., Zhang, Z., Cronin, S.B. & Koga, T. (1999). Low-dimensional thermoelectric materials, Physics of the Solid State. Vol.(41)(No. 5):679-682

Fornari, M. & Singh, D.J. (1999). Prediction of room-temperature high-thermoelectric performance in n-type $La(Ru_{1-x}Rh_x)_4Sb_{12}$, Applied Physics Letters. Vol.(74)(24):3666-3668

Funahashi, R.& Shikano, M. (2002). $Bi_2Sr_2Co_2O_y$ whiskers with high thermoelectric figure of merit, Applied Physics Letter. Vol.(81)(No. 8):1459-1461

Gelbstein, Y., Dashevsky, Z. & Dariel, M. P. (2008). The search for mechanically stable PbTe based thermoelectric materials, Journal of Applied Physics. Vol.(104): 033702(1)-033702(4)

Golia, Santosh., Arora, M., Sharma, R.K.& Rastogi, A.C. (2003). Electrochemically deposited bismuth telluride thin films, Current Applied Physics. Vol(3):195-197

Guo, Y.P. & Zhou, Y. (2007). Polyaniline nanofibers fabricated by electrochemical polymerization: A mechanistic study, European Polymer Journal. Vol.(43): 2292–2297

Hashimoto, K., Kurosaki, K., Imamura, Y., Muta, H.& Yamanaka, S. (2007). Thermoelectric properties of $BaSi_2$, $SrSi_2$, and LaSi, Journal of Applied Physics. Vol.(102):063763(1)-063763(5)

Huonga, N.T., Setou , Y., Nakamoto, G., Kurisu, M., Kajihara , T., Mizukami, H.& Sano, S. (2004). High thermoelectric performance at low temperature of p-$Bi_{1.8}Sb_{0.2}Te_{3.0}$ grown by the gradient freeze method from Te-rich melt, Journal of Alloys and Compounds. Vol.(368): 44-50

Ikeda, T., Collins, L.A., Ravi, V.A., Gascoin, F.S., Haile, S.M. & Snyder, G.J. (2007). Self-Assembled Nanometer Lamellae of Thermoelectric PbTe and Sb_2Te_3 with Epitaxy-like Interfaces, Chemisty of Materials. Vol.(19)(No. 4): 763–767

Kauzlarich, S.M., Browna, S.R.& Snyder, G.J. (2007). Zintl phases for thermoelectric devices, The Royal Society of Chemistry. DOI: 10.1039/b702266b. 2099-2107

Kim, J.H., Okamoto, N. L., Kishida, K., Tanaka,K. & Inui, H.(2007). Crystal structure and thermoelectric properties of type-III clathrate compounds in the Ba–In–Ge system, Journal of Applied Physics. Vol.(102): 094506(1)-094506(9)

Kim, S.W., Kimura, Y. & Mishima, Y. (2004). Enhancement of high temperature thermoelectric properties of intermetallic compounds based on a Skutterudite $IrSb_3$ and a half-Heusler TiNiSb, Science and Technology of Advanced Materials. Vol.(5): 485–489

Kima, S.S., Yinb, F. & Kagawa, Y. (2006). Thermoelectricity for crystallographic anisotropy controlled Bi–Te based alloys and p–n modules, Journal of Alloys and Compounds. Vol.(419): 306-311

Kimura, Y., Tamura, Y. & Kita, T. (2008). Thermoelectric properties of directionally solidified half-Heusler compound NbCoSn alloys, Applied Physics Letters. Vol.(92): 012105(1)- 012105(3)

Kishimoto, K., Akai, K., Muraoka, N., Koyanagi, T. & Matsuura, M. (2006). Synthesis and thermoelectric properties of type-I clathrate $Ge_{30}P_{16}Te_8$, Applied Physics Letters. Vol.(89): 172106 (1)-172106 (3)

Koga, T., Sun, X., Cronin, S.B., Dresselhaus, M.S., Wang, K.L. & Chen, G. (1997). Models for low-dimensional thermoelectricity, Journal of Computer-Aided Materials Design, 1998 Kluwer Academic Publishers. Printed in the Netherlands. Vol.(4):175-182

Kosuga, A., Kurosaki, K., Muta, H.& Yamanaka, S. (2006). Thermoelectric properties of Tl–X–Te (X=Ge, Sn, and Pb) compounds with low lattice thermal conductivity, Journal of Applied Physics. Vol.(99): 063705(1)- 063705(4)

Kurosaki, K., Matsumoto, H., Charoenphakdee, A., Yamanaka, S., Ishimaru, M.& Hirotsu, Y. (2008). Unexpectedly low thermal conductivity in natural nanostructured bulk Ga_2Te_3, Applied Physics Letters. Vol.(93): 012101(1)- 012101(3)

Lan, Y.C., Poudel, B., Ma, Y., Wang, D., Dresselhaus, M.S., Chen, G. & Ren, Z.F. (2009). Structure Study of Bulk Nanograined Thermoelectric Bismuth Antimony Telluride, Nano Letters. Vol.(9)(No. 4): 1419–1422

Larson, P., Mahanti, S.D., Sportouch, S. & Kanatzidis, M.G. (1999). Electronic structure of rare-earth nickel pnictides: Narrow-gap thermoelectric materials, Physical Review B. Vol.(59)(No. 24):15660-15668

Li, H., Tang, X.F., Su, X.L.& Zhang, Q.J. (2008). Preparation and thermoelectric properties of high-performance Sb additional $Yb_{0.2}Co_4Sb_{12+y}$ bulk materials with nanostructure, Applied Physics Letters. Vol.(92): 202114(1)- 202114(3)

Li, H., Tang, X.F., Zhang, Q.J. & Uher, C. (2008). Rapid preparation method of bulk nanostructured $Yb_{0.3}Co_4Sb_{12+y}$ compounds and their improved thermoelectric performance, Applied Physics Letters. Vol.(93): 252109(1)- 252109(3)

Li, H., Tang, X.F., Zhang, Q.J. & Uher, C. (2009). High performance $In_xCe_yCo_4Sb_{12}$ thermoelectric materials with in situ forming nanostructured InSb phase, Applied Physics Letters. Vol.(94): 102114(1)- 102114(3)

Li, Y., Xu, D.S., Zhang, Q.M., Chen, D.P., Huang, F.Z., Xu, Y.J., Guo, G.L. & and Gu, Z.N. (1999). Preparation of cadmium sulfide nanowire arrays in anodic aluminum oxide templates, Chemistry of Materials. Vol.(11): 3433-3435

Liao, C.N. & She, T.H. (2007). Preparation of bismuth telluride thin films through interfacial reaction, Thin Solid Films. Vol.(515):8059-8064

Lineykin, S. & Sam, B.Y. (2007). User-friendly and intuitive graphical approach to the design of thermoelectric cooling systems, International Journal of Refrigeration. Vol.(30):798-804

Litvinchuk, A.P., Nylén, J., Lorenz, B., Guloy, A.M.& Häussermann, U. (2008). Optical and electronic properties of metal doped thermoelectric Zn_4Sb_3, Journal of Applied Physics. Vol.(103): 123524(1)- 123524(6)

Liu, M.L., Huang, F.Q., Chen, L.D. & Chen, I.W. (2009). A wide-band-gap p-type thermoelectric material based on quaternary chalcogenides of Cu_2ZnSnQ_4 (Q=S,Se), Applied Physics Letters. Vol.(94): 202103(1)- 202103(3)

Liu, W.S., Zhang, B.P., Li, J.F., Zhang, H.L. & Zhao, L.D. (2007). Enhanced thermoelectric properties in $CoSb_{3-x}Te_x$ alloys prepared by mechanical alloying and spark plasma sintering, Journal of Applied Physics. Vol.(102): 103717(1)- 103717(7)

Lv, H.Y., Liu, H.J., Pan, L., Wen, Y.W., Tan, X.J., Shi, J. & Tang, X.F. (2010). Enhanced thermoelectric performance of $(Sb_{0.75}Bi_{0.25})_2Te_3$ compound from first-principles calculations, Applied Physics Letters. Vol.(96): 142101 (1)- 142101 (3)

Ma, Y., Hao, Q., Poudel, B., Lan, Y.C., Yu, B., Wang, D.Z., Chen, G. & Ren, Z.F. (2008). Enhanced Thermoelectric Figure-of-Merit in p-Type Nanostructured Bismuth Antimony Tellurium Alloys Made from Elemental Chunks, Nano Letters. Vol.(8)(No. 8):2580-2584

Madhavaram, R., Sander, J., Gan Y.X. & K.Masiulaniec, C. (2009). Thermoelectric property of PbTe coating on copper and nickel, Materials Chemistry and Physics. 13419(1)- 13419(9)

Mangersnes, K., Løvvik, O.M. & Prytz, Ø. (2008). New filled P-based skutterudites-promising materials for thermoelectricity, New Journal of Physics. Vol.(10):1-16

Martin, J., Wang, H. & Nolas, G.S. (2008). Optimization of the thermoelectric properties of $Ba_8Ga_{16}Ge_{30}$, Applied Physics Letters. Vol.(92): 222110(1)- 222110(3)

Matsumoto, H., Kurosaki, K., Muta, H.& Yamanaka, S. (2008). Systematic investigation of the thermoelectric properties of $TlMTe_2$, Journal of Nuclear Materials. Vol.(104): 073705(1)- 073705(4)

Matsumoto, T., Yamazaki, J., Kaiwa, N. & Yamamoto, A. (2005). Thermoelectric properties and figure of merit of a Te-doped InSb bulk single crystal, Applied Physics Letter. Vol.(87): 201902(1)- 201902(3)

Mi, J.L., Zhu, T.J. & Zhao, X.B. (2007). Nanostructuring and thermoelectric properties of bulk skutterudite compound $CoSb_3$, The Journal of Chemical Physics. Vol.(101): 054314(1)- 054314(6)

Mi, J. L., Zhao, X. B., Zhu, T. J.& Tu, J. P. (2007). Improved thermoelectric figure of merit in *n*-type $CoSb_3$ based Nanocomposites, Applied Physics Letters. Vol.(91): 172116(1)- 172116(3)

Mingoa, N. (2004). Thermoelectric figure of merit and maximum power factor in III–V semiconductor nanowires, Applied Physics Letters. Vol.(84)(No. 14):2652-2654

Mori,T., Martin, J. & Nolas,G. (2007). Thermal conductivity of $YbB_{44}Si_2$, Journal of Applied Physics. Vol.(102): 073510(1)- 073510(4)

Moreno, L.C., Cadavid, D. & Rodríguez, J.E. (2008). Thermoelectric power factor of LSCoO compounds, Microelectronics Journal. Vol.(39): 548-550

Mori, T., Nishimura, T., Yamaura, K. & Eiji, T.M. (2007). High temperature thermoelectric properties of a homologous series of n-type boron icosahedra compounds: A possible counterpart to p-type boron carbide, Journal of Applied Physics. Vol.(101)(No. 9):093714(1)- 093714(4)

Motohashi, T., Nonaka, Y., Sakai, K., Karppinen, M. & Yamauchi, H. (2008). Fabrication and thermoelectric characteristics of $[(Bi,Pb)_2Ba_2O_{4\pm w})_{0.5}CoO_2$ bulks with highly aligned grain structure, Journal of Applied Physics. Vol.(103): 033705(1)- 033705(6)

Mu, S.L., Kan, J.Q., Lu, J.T. & Zhuang, L. (1997). Interconversion of polarons and bipolarons of polyaniline during the electrochemical polymerization of aniline, Journal of Electroanalytical Chemistry. Vol.(446): 107–112

Ni, X.X., Liang, G., Wang, J.S. & Li, B.W. (2009). Disorder enhances thermoelectric figure of merit in armchair graphane Nanoribbons, Applied Physics Letters. Vol.(95): 192114(1)- 192114(3)

Nishi, Y., Arita, Y., Terao, K., Matsui, T. & Nagasaki, T. (2001). Boron isotope effects on the thermoelectric properties of UB_4 at low temperatures, Journal of Nuclear Materials. Vol.(294):209-211

Nolasa, G.S., Cohn, J.L., Slack, G.A. & Schujman, S.B. (1998). Semiconducting Ge clathrates: Promising candidates for thermoelectric Applications, Applied Physics Letters. Vol.(73)(No. 2):178-180

Nolasa, G.S., Poon, J.& Kanatzidis, M. (2006). Recent Developments in Bulk Thermoelectric Materials, MRS BULLETIN. Vol.(31):199-205

Nolasa, G.S., Kaeser, M., Littleton IV, R.T. & Tritt, T.M. (2000). High figure of merit in partially filled ytterbium skutterudite materials, Applied Physics Letters. Vol.(77)(No. 12):1855-1857

Nong, N.V., Liu, C.J. & Ohtaki, M. (2010). Improvement on the high temperature thermoelectric performance of Ga-doped misfit-layered $Ca_3Co_{4-x}Ga_xO_{9+\delta}$, Journal of Alloys and Compounds. Vol.(491): 53 - 56

Nose, K. & Yoshida, T. (2007). Semiconducting properties of zinc-doped cubic boron nitride thin films, Journal of Applied Physics. Vol.(102):063711(1)- 063711(5)

Okamoto, H. & Kotaka, T. (1998). Structure and properties of polyaniline films prepared via electrochemical polymerization.I: Effect of pH in electrochemical polymerization media on the primary structure and acid dissociation constant of product polyaniline films, Polymer. Vol.(39)(No. 18): 4349-4358

Okamoto, N.L., Oh, M.W., Nishii, T., Tanaka, K.& Inui, H. (2006). Crystal structure and thermoelectric properties of the type-I clathrate compound Ba_8Ge_{43} with an ordered arrangement of Ge vacancies, Journal of Applied Physics. Vol.(99): 033513(1)-033513(8)

Okamoto, N.L., Nakano, T., Tanaka, K. & Inui, H. (2008). Mechanical and thermal properties of single crystals of the type-I clathrate compounds $Ba_8Ga_{16}Ge_{30}$ and $Sr_8Ga_{16}Ge_{30}$, Journal of Applied Physics. Vol.(104): 013529(1)- 013529(7)

Otakea, M., Satoa, K., Sugiyamab, O. & Kaneko, S. (2004). Pulse-current sintering and thermoelectric properties of gas-atomized silicon–germanium powders, Solid State Ionics. Vol.(172): 523-526

Pedersen, B.L., Birkedal, H., Iversen, B.B., Nygren, M. & Frederiksen, P.T. (2006). Influence of sample compaction on the thermoelectric performance of Zn_4Sb_3, Applied Physics Letters. Vol.(89): 242108 (1)- 242108 (3)

Pedersen, B.L. & Iversen, B.B. (2008). Thermally stable thermoelectric Zn_4Sb_3 by zone-melting synthesis, Applied Physics Letters. Vol.(92): 161907(1)- 161907(3)

Pei, J., Chen, G.& Zhou, N. (2009). Hydrothermal synthesis, characterization, electronic structure, and thermoelectric properties of $(Ca_{0.85}OH)_{1.16}CoO_2$, The Journal of Chemical Physics. Vol.(130): 044706(1)- 044706(5)

Ponnambalam, V., Zhang, B.,Tritt,T.M. & Poon,S.J. (2007). Thermoelectric Properties of Half-Heusler Bismuthides $ZrCo_{1-x}Ni_xBi$, Chemistry and Materials Science. Vol.(36)(No. 7): 732-735

Ponnambalam, V., Alboni, P.N., Edwards, J., Tritt, T.M., Culp, S.R. & Poon, S.J. (2008). Thermoelectric properties of p-type half-Heusler alloys $Zr_{1-x}Ti_xCoSn_ySb_{1-y}$, Journal of Applied Physics. Vol.(103): 063716(1)- 063716(5)

Poudel, B., Hao, Q., Ma, Y., Lan, Y., Minnich, A., Yu, B., Yan, X., Wang, D., Muto, A., Vashaee, D., Chen, X., Liu, J., Dresselhaus, M.S., Chen, G.& Ren, Z.F. (2008). High-Thermoelectric Performance of Nanostructured Bismuth Antimony Telluride Bulk Alloys, Science Express. Vol.(320)(No. 5876):634-638

Qin, X.Y.,Zhang, J., Li, D., Dong, H.Z. & Wang, L. (2007). The effect of Mg substitution for Ti on transport and thermoelectric properties of TiS_2, Journal of Applied Physics. Vol.(102): 073703(1)- 073703(7)

Rabina, O., Lin, Y.M.& Dresselhaus, M.S. (2001). Anomalously high thermoelectric figure of merit in $Bi_{1-x}Sb_x$ nanowires by carrier pocket alignment, Applied Physics Letters. Vol.(79)(No. 1):81-83

Rao, A.M., Ji, X.H. and Tritt, T.M. (2006). Properties of Nanostructured One-Dimensional and Composite Thermoelectric Materials, MRS BULLETIN. Vol.(31):218-223

Rhyee, J.S., Cho, E., Lee, K.H., Lee, S.M., Kim, S.I., Kim, H.S., Kwon, Y.S.& Kim, S.J. (2009). Thermoelectric properties and anisotropic electronic band structure on the In_4S_{e3-x} compounds, Applied Physics Letters. Vol.(95): 212106(1)- 212106(3)

Rhyee, J.S., Cho, E., Lee, K.H., Lee, S.M., Kim, H. & Kwon, Y.S. (2010). Thermoelectricity and localized *f*-band control by *dp*-hybridization on the $Ce_{1-x}Cu_xSe_2$ compounds, Journal of Applied Physics. Vol.(107): 053705(1)- 053705(5)

Robert, R., Aguirre, M.H., Hug, P., Reller, A.& Weidenkaff, A.(2007). High-temperature thermoelectric properties of $Ln(Co, Ni)O_3$ compounds, Acta Materialia. Vol.(55):4965-4972

Roh, J.W., Jang, S.Y., Kang, J., Lee, S., Noh, J.S., Kim, W., Park, J. & Lee, W. (2010). Size-dependent thermal conductivity of individual single-crystalline PbTe nanowires, Applied Physics Letters. Vol.(96): 103101 (1)- 0 103101 (3)

Ruan, X.F. & Xiao, W.K. (2007). Preparation and thermoelectric properties of SiO_2/β-Zn_4Sb_3 nanocomposite materials, Chemistry and Materials Science. Vol.(24)(No. 5):694-697

Sales, B.C., Chakoumakos, B.C.& Mandrus, D. (2000). Thermoelectric properties of thallium-filled skutterudites, Physical Review B. Vol.(61)(No. 4): 2475-2481

Sandomirsky, V., Schlesinger, Y., Butenko, A.V., Kolobov, I.G. & Ronen, A. (2003). Electric field effect controlled thermoelectricity in thin Bi films, Physica B. 1521–1522

Sekimoto, T., Kurosaki, K., Mut, H.& Yamanaka, S. (2006). LnPdSb (Ln=La,Gd): Promising intermetallics with large carrier mobility for high performance p-type thermoelectric materials, Applied Physics Letters. Vol.(89): 092108(1)- 092108(3)

Shi, L.H., Yao, D.L., Zhang, G.& Li, B.W. (2010). Large thermoelectric figure of merit in $Si_{1-x}Ge_x$ nanowires, Applied Physics Letters. Vol.(96): 173108(1)- 173108(1)

Shi, X., Chen, L., Yang, J.& Meisner, G. P. (2004). Enhanced thermoelectric figure of merit of $CoSb_3$ via large-defect scattering, Applied Physics Letters. Vol.(84)(No. 13):2301-2303

Shi, X., Chen, L.D.,.Bai, S.Q., Huang, X.Y,..Zhao, X.Y. & Yao, Q. (2007). Influence of fullerene dispersion on high temperature thermoelectric properties of $Ba_yCo_4Sb_{12}$-based composites, Journal of Applied Physics. Vol.(102): 103709(1)-103709(7)

Shi, X.Y., Huang, F.Q., Liu, M.L. & Chen, L.D. (2009). Thermoelectric properties of tetrahedrally bonded wide-gap stannite compounds $Cu_2ZnSn_{1-x}In_xSe_4$, Applied Physics Letters. Vol.(94): 122103(1)- 122103(3)

Shi, X., Kong, H., Li, C.P., Uher, C., Yang, J., Salvador, J.R., Wang, H., Chen, L. & Zhang, W. (2008). Low thermal conductivity and high thermoelectric figure of merit in n-type $Ba_xYb_yCo_4Sb_{12}$ double-filled skutterudites, Applied Physics Letters. Vol.(92): 82101(1)- 82101(3)

Snyder, G.J.& Toberer, E.S. (2008). Complex thermoelectric materials thermoelectric, nature materials. Vol.(7):105-114

Sootsman, J.R., He, J.Q., Dravid, V.P., Li, C.P., Uher, C. & Kanatzidis, M.G. (2009). High thermoelectric figure of merit and improved mechanical properties in melt quenched PbTe-Ge and PbTe-$Ge_{1-x}Si_x$ eutectic and hypereutectic composites, Journal of Applied Physics. Vol.(105): 083718(1)- 083718(1)

Subramanian, S., Chithralekha, P. & PathinettamPadiyan, D. (2010). Enhanced electrical response in Sb_2S_3 thin films by the inclusion of polyaniline during electrodeposition, Physica B. Vol.(405): 925-931

Sun, P.J., Oeschler, N., Johnsen, S., Iversenb, B.B.& Steglich, F. (2009). Narrow band gap and enhanced thermoelectricity in $FeSb_2$, Dalton Transactions.

Takami, T.& Ikuta, H. (2008). Magnetic and thermoelectric properties of quasi-one-dimensional oxides $A_{n+2}CoB_nO_{3n+3}$ (A=Ca,Sr, B=Co,Rh, Ir; n=1–3), Journal of Applied Physics. Vol.(103): 07B701(1)- 07B701(3)

Tardı´o, M.M., Ramı´rez, R., Gonza´ lez, R., Chen, Y.& Kokta, M. R. (2001). Enhancement of electrical conductivity in α-Al_2O_3 crystals doped with magnesium, Journal of Applied Physics. Vol.(90)(No. 8):3942-3951

Tritt, T.M., Subramanian, M.A. & Editors, G. (2006). Thermoelectric Materials, Phenomena, and Applications: A Bird's Eye View, MRS BULLETIN. Vol.(31):188-198

Touzelbaev, M.N., Zhou, P., Venkatasubramanian, R.& Goodson, K.E. (2001). Thermal characterization of Bi_2Te_3/Sb_2Te_3 superlattices, Journal of Applied Physics. Vol.(90)(No. 2):763-767

Tsuchida, K., Tanaka, Y., Ifuku, T., Nakao, Y., Matsuda, T., Nagashima, S., Maeda, H. & Kato, A. (1996). Thermoelectric properties of oxide ceramics, Journal of Chemistry. Vol(13)(No. 5): 478-481

Wan, B.Y., Hua, C.G., Xi, Y., Xu, J. & He, X.S. (2010). Room-temperature synthesis and seebeck effect of lead chalcogenide nanocubes, Solid State Science. Vol.(12): 123-127

Wang, H., Li, J.F., Nan, C.W., Zhou, M., Liu, W.S., Zhang, B.P. & Kita, T. (2006). High-performance $Ag_{0.8}Pb_{18+x}SbTe_{20}$ thermoelectric bulk materials fabricated by mechanical alloying and spark plasma sintering, Applied Physics Letter. Vol.(88): 092104(1)- 092104(3)

Wang, X.J., Tang, M.B., Zhao, J.T., Chen, H.H. & Yang, X.X.. (2007). Thermoelectric properties and electronic structure of Zintl compound $BaZn_2Sb_2$, Applied physics letters. Vol.(90):232107(1)- 232107(3)

Wang, X.J., Tang, M.B., Chen, H.H., Yang, X.X., Zhao, J.T., Burkhardt, U.G. & Yuri. (2009). Synthesis and high thermoelectric efficiency of Zintl phase $YbCd_{2-x}Zn_xSb_2$, applied Physics Letter. Vol.(94)(No. 9): 092106(1)- 092106(3)

Wang, X.W., Lee, H., Lan, Y.C., Zhu, G.H., Joshi, G., Wang, D.Z., Yang, J., Muto, A.J., Tang, M.Y., Klatsky, J., Song, S., Dresselhaus, M.S., Chen, G. & Ren, Z.F. (2008). Enhanced thermoelectric figure of merit in nanostructured n-type silicon germanium bulk alloy, Applied Physics Letters. Vol.(93): 193121(1)- 193121(3)

Wang, Y.Y., Cai1, K.F., Yin, J.L., An, B.J., Du, Y.& Yao, X. (2009). In-Situ Fabrication and Thermoelectric Properties of PbTe-Polyaniline Composite Nanostructures, Journal of Nanoparticle Research. Manuscript Number: NANO2941:1-21

Watt, A.A.R., Blake, D., HWarner, J., Thomsen, E.A., Tavenner, E.L., Halina R.D. & Meredith, P. (2005). Lead sulfide nanocrystal: conducting polymer solar cells, Journal of Physics D: Applied Physics. Vol.(38): 2006-2012

Weidenkaff, A., Robert, R., Aguirre, M., Bocher, L., Lippert, T.& Canulescu, S. (2008). Development of thermoelectric oxides for renewable energy conversion technologies, Renewable Energy. Vol.(33): 342-347

Xie, W.J., Tang, X.F., Yan, Y.G., Zhang, Q.J. & Tritt, T.M. (2009). Unique nanostructures and enhanced thermoelectric performance of melt-spun BiSbTe alloys, Applied Physics Letters. Vol.(94): 102111(1)- 102111(3)

Xu, Y., Chen, X.B., Gu, B.L. and Duan,W.H. (2009). Intrinsic anisotropy of thermal conductance in graphene nanoribbons, Applied Physics Letters. Vol.(95): 233116(1)- 233116(3)

Yamashita, O., Tomiyoshi, S. & Makita, K. (2003). Bismuth telluride compounds with high thermoelectric figures of merit, Journal of Applied Physics. Vol.(93)(No. 1):368-374

Yamashita, O. (2004). Effect of metal electrode on Seebeck coefficient of p- and n-type Si thermoelectrics, Journal of Applied Physics. Vol.(95)(No. 1):178-183

Yamashita, O.& Tomiyoshi, S. (2004). High performance n-type bismuth telluride with highly stable thermoelectric figure of merit, Journal of Applied Physics. Vol.(95)(No. 11):6277-6283

Yanagiya, S., Van Nong, N., Xu, J.X. & Pryds, N. (2010). The Effect of (Ag, Ni, Zn)-Addition on the Thermoelectric Properties of Copper Aluminate, Materials. Vol.(3): 318-328

Yang, J., Meisner, G.P. & Chen, L. (2004). Strain field fluctuation effects on lattice thermal conductivity of ZrNiSn-based thermoelectric compounds, Applied Physics Letters. Vol.(85)(No. 7):1140-1142

Young, D., Mastronardi, K., Khalifah, P., Wang, C.C.& Cava, R.J. (1999). $Ln_3Au_3Sb_4$: Thermoelectrics with low thermal conductivity, Applied Physices Letters. Vol.(74)(No. 26):3999-4001

Yu, C., Scullin, M.L., Huijben, M., Ramesh, R. & Majumdar, A. (2008). Thermal conductivity reduction in oxygen-deficient strontium titanates, Applied Physics Letters. Vol.(92): 191911(1)- 191911(3)

Yu, C., Zhu, T.J., Zhang, S.N., Zhao, X.B., He, J., Su, Z. & Tritt, T.M. (2008). Improved thermoelectric performance in the Zintl phase compounds $YbZn_{2-x}Mn_xSb_2$ via isoelectronic substitution in the anionic framework, Journal of Applied Physics. Vol.(104): 013705(1)- 013705(5)

Zawilski, B. M., Littleton IV, R. T. & Tritt, T.M. (2000). Investigation of the thermal conductivity of the mixed pentatellurides $Hf_{1-x}Zr_xTe_5$, Applied Physics Letter. Vol.(77)(No. 15):2319-2321

Zeng, G.H., Bahk, J.H., Bowers, J.E., Lu, H., Gossard, A.C., Singer, S.L., Majumdar, A., Bian, Z.X., Zebarjadi, M.& Shakouri, A. (2009). Thermoelectric power generator module of 16×16 Bi_2Te_3 and 0.6% ErAs: $(InGaAs)_{1-x}(InAlAs)_x$ segmented elements, Applied Physics Letters. Vol.(95): 083503(1)- 083503(5)

Zhao, X.Y., Shi, X., Chen, L.D. Zhang, W.Q., Bai, S.Q., Pe, Y.Z. & Li, X.Y. (2006). Synthesis of $Yb_yCo_4Sb_{12}/Yb_2O_3$ composites and their thermoelectric properties, Applied Physics Letters. Vol.(89): 092121 (1)- 092121 (3)

Zhao, X.Y., Shi, X., Chen, L.D., Zhang, W.Q., Zhang, W.B. & Pei, Y.Z. (2006). Synthesis and thermoelectric properties of Sr-filled skutterudite $Sr_yCo_4Sb_{12}$, Journal of Applied Physics. Vol.(99): 053711(1)- 053711(4)

Zhang, F. & Wong, S.S. (2009). Controlled synthesis of semiconducting metal sulfide nanowires, Chemistry of Materials. Vol.(21):4541-4554

Zhang, H., Zhao, J.T., Grin, Y., Wang, X.J., Tang, M.B., Man, Z.Y., Chen, H.H. & Yang, X.X. (2008). A new type of thermoelectric material, $EuZn_2Sb_2$, The Journal of Chemical Physics. Vol.(129): 164713(1)- 164713(5)

Zhang, J., Kong, L.B., Wang, B., Luo, Y.C. & Kang, L. (2009). In-situ electrochemical polymerization of multi-walled carbon nanotube/polyaniline composite films for electrochemical supercapacitors, Synthetic Metals. Vol.(159): 260-266

Zhang, L., Grytsiv, A., Kerber, M., Rogl, P., Bauer, E. & Zehetbauer, M. (2010). Thermoelectric performance of mischmetal skutterudites $Mm_yFe_{4-x}Co_xSb_{12}$ at elevated temperatures, Journal of Alloys and Compounds. Vol.(490): 19-25

Zhang, Q., He, J., Zhu, T.J., Zhang, S.N., Zhao, X.B.& Tritt, T.M. (2008). High figures of merit and natural nanostructures in $Mg_2Si_{0.4}Sn_{0.6}$ based thermoelectric materials, Applied Physics Letters. Vol.(93): 102109(1)- 102109(3)

Zhou, Y.Q., Matsubara, I., Horii, S., Takeuchi, T., Funahashi, R., Shikano, M., Shimoyama, J.& Kishio, K. (2003). Thermoelectric properties of highly grain-aligned and densified Co-based oxide ceramics, Journal of Applied Physics. Vol.(93)(No. 5):2653-2658

Zhu,T.J., Yu,C., He, J., Zhang, S.N., Zhao,X.B. & Tritt,T.M. (2008). Thermoelectric Properties of Zintl Compound $YbZn_2Sb_2$ with Mn Substitution in Anionic Framework , Journal of Electronic Materials. Vol.(38)(No. 7): 1068-1071

7

Metal Oxides and Their Composites for the Photoelectrode of Dye Sensitized Solar Cells

Jae-Joon Lee[1], Md. Mahbubur Rahman[2], Subrata Sarker[2], N.C. Deb Nath[2], A.J. Saleh Ahammad[2] and Jae Kwan Lee[3]

[1]*Department of Applied Chemistry, Konkuk University, Chungju 380-701,*
[2]*Department of Advanced Technology Fusion, Konkuk University, Seoul 143-701,*
[3]*Research Center for Convergence Technology, Hoseo University, Chungnam 336-795*
Korea

1. Introduction

Nowadays, solar energy is one of the most promising future energy resources in concerns to the sustenance of life on Earth and the depletion of fossil fuels. It is projected that the fossil fuel related CO_2 emissions rise from 32 to 42 billion metric tons in 2007 and 2035 (EIA., 2010). This enormous amount of CO_2 emission will leads to a severe climatic change of the world and therefore a greatest anxieties for the scientific era of 21st century. This rigorous apprehension leads the scientist for the development of solar cell that utilizes solar energy, a renewable and carbon free energy source. The solar energy strike to the earth in one hour is about $4.3x10^{20}$ J, which is higher than all the energy consumed in the planet ($4.1x10^{20}$ J). Therefore, covering 0.1% of the earth's surfaces with solar cell of 10% efficiency would satisfy the current energy demand (Grätzel., 2001).

In general, solar cells can be classified as p-n junction semiconductor solar cells and organic-based exitonic solar cells (OESCs), in which polymer solar cell (PSC), dye sensitized solar cell (DSSC) and hybrid solar cell are included. In 1991, O'Regan & Grätzel first reported the dye sensitized nanocrystalline TiO_2 solar cell (DSSC) based on the mechanism of a first regenerative photoelectrochemical processes with an efficiency of 7.1-7.9 %(under simulated solar light) (O'Regan & Grätzel., 1991). Since then, extensive researches have continued to increase the power conversion efficiency (PCE) of DSSC by incorporating n-type metal oxide semiconductors such as TiO_2, ZnO, SnO_2, Nb_2O_5, $SrTiO_3$ etc and their composites as photoelectrode materials to achieve a reasonable efficiency of DSSCs of low cost, being therefore a promising alternative to conventional p-n junction solar cell. The wide band gap (E_g > 3eV) metal oxide semiconductors having suitable band position relative to sensitizer has been employed for the fabrication of DSSCs. The high surface area of nanoporous metal oxides facilitates the improvement of light absorption with improved dye loading for improved performance of DSSC. It is evident that the metal oxides employed for the fabrication of DSSCs has solar absorption below a threshold wavelength, λ_g (where, λ_g = $1240/E_g$), i.e., they have absorption at ultraviolet region. On the other hand, dye is only responsible for the absorption of light at visible and near-infrared region. The strong absorption of light is attributed to the intramolecular charge transfer transition (ICT) from electron donating group to the anchoring acceptor group of dye. Therefore, the anchored

dye on to the metal-oxide surface helps the red shift of absorption threshold of metal oxide near infrared region (Rengaraj et al., 2005). In addition to the above physical characteristics, the inexpensiveness, natural abundancy and facile synthesis methods of metal oxides and their composites is another advantage for the application in DSSCs.
This chapter discusses the understanding the fundamental of DSSCs, different structural and technical aspect of metal oxides and their composites for the stepwise development of the performance of DSSCs. In addition, we focuses on the understanding of interfaces between metal oxides (or metal oxide composites)/dye/electrolyte, in which the electron injection and transport for improving the efficiency of DSSCs are related. Figure 1 represents the energy band diagram vs. vacuum of the metal oxides describes in this chapter.

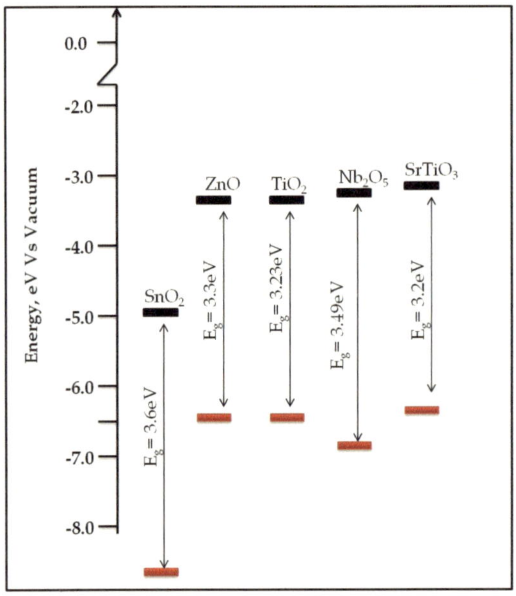

Fig. 1. Energy band diagram of the representative metal oxides applied in DSSC (vs vacuum). The lower edge of the conduction band (CB) (black) and the upper edge of the valence band (VB) (red) are presents along with the band gap in electron volts (eV). Energy levels were taken from the following references (Grätzel., 2001; Wei et al., 2008; Hod et al., 2010)

2. Dye sensitized solar cell (DSSC)

Figure 2 shows a schematic representation of the operating principle of DSSC. Based on the electron transfer mechanism from dye to TiO_2 conduction band, DSSC can be classified into two distinct classes: Type-I, and type-II. The heart of the system is mesoporous semiconducting metal oxide (mostly employed TiO_2) film that is placed in contact of a redox electrolyte (such as I^-/I_3^-, Br^-/Br_2, $SCN^-/(SCN)_2$, $SeCN^-/(SeCN)_2$, bipyridine cobalt (III/II) complexes) (Wang et al., 2005; Nusbaumer et al., 2001; Wang et al., 2004; Sapp et al., 2002; Bergeron et al., 2005) or a hole conductor (such as CuI, CuBr, CuSCN, spiro-MeOTAD) (Wu et al., 2008). A monolayer of the sensitizer (typically bipyridine ruthenium complexes) such as *cis*-diisothiocyanato-bis(2,2′-bipyridyl-4,4′-dicarboxylato)ruthenium(II) bis(tetrabutylammonium) (N719), *cis*-bis(isothiocyanato)bis(2,2'-bipyridyl-4,4'-dicarboxylato)-ruthenium(II) (N3), and *cis*-disothiocyanato-(2,2′-bipyridyl-4,4′-dicarboxylic

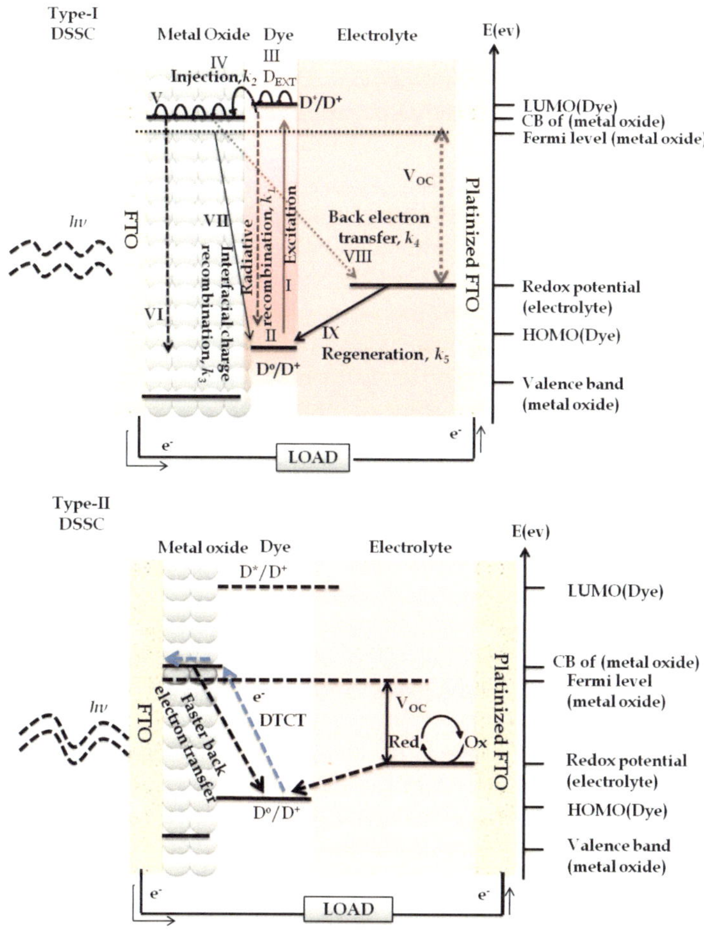

Fig. 2. Principle of operation of type-I & type-II dye sensitized solar cell (DSSC)

acid)-(2,2′-bipyridyl-4,4′-dinonyl)ruthenium(II) (Z907) etc is attached to the surface of the nanocrystalline metal oxide film in type-I DSSC (Thavasi et al., 2009). The type-I DSSC is understood as general electron transport mechanism, where the photoexcitation of dye results the injection of electron to the conduction band (CB) of the metal oxides from the excited dye, i.e. from the LUMO level. On the other hand, in type-II DSSCs, enediol group, catechol and its derivatives such as dopamine, fluorone, and anthocyanins having catechol moieties gives strong direct dye-to-metal-oxide (TiO_2) charge transfer (DTCT) bands. Therefore, electrons are injected directly from HOMO of dye to the conduction band (CB) of metal oxide rather than from the excited state (LUMO) under illumination (i.e. one-step electron injection) (Tae et al., 2005). A platinized cathode is employed to collect the electrons from the conduction band of metal oxide anode injected from the excited dye and to catalyze the redox couple *viz.* I^-/I_3^- regeneration reaction. The redox couple of the electrolyte also reduces the oxidized dye. The iodide regenerated, in turn, by the reduction of triiodide at the cathode, with the circuit being completed via electron movement through the external load. The voltage generated under illumination is the difference between the Fermi level of the electron in metal oxide and the redox potential of the electrolyte. Overall, the device generated electric power from sunlight without suffering any enduring chemical conversion (Grätzel., 2003; Bisquert et al., 2004). Whereas, in type-II DSSC faster back electron transfer limit its conversion efficiency, over and above commercial application (Huber et al., 2000; Wang et al., 2003). In this chapter, the application of metal oxide and their composites is mainly focused in type-I DSSC.

2.1 Photochemical processes in DSSCs and rate limiting steps

The photoelectrochemical process in DSSC can be expressed by the following reaction a-f. Where, electron injection (reaction b) of the photoexcited dye to the conduction band (CB) of metal-oxide (TiO_2) occurs in subpicosecond range (200 fs for coumarin dye to TiO_2) (Hara et al., 2000). Reaction (c) and (d) represents the regeneration of photoexcited dye by the oxidation of I^- to I_3^- and the reduction of I_3^- to I^- by Pt catalyst at cathode respectively. Reaction (e) and (f) correspond to the dark reaction occur during the photoconversion and do not play a significant negative impact on the photovoltaic performance of DSSC due to their sluggish reaction rate compared to the reaction (b).

Metal oxide (TiO_2) | D+hv	$\rightarrow$ Metal oxide (TiO_2) | D*	excitation	(a)
Metal oxide (TiO_2)| D*	$\rightarrow$ Metal oxide (TiO_2) | D^++$e^-_{(CB)}$	injection	(b)
Metal oxide (TiO_2)| D^++3I^-	$\rightarrow$ Metal oxide (TiO2) | D+I_3^-	regeneration	(c)
I_3- + 2$e^-_{(Pt)}$	$\rightarrow$ 3I^-	reduction	(d)
I_3- + 2e-$_{(CB)}$	$\rightarrow$ 3I^-	reception (dark reaction)	(e)
Metal oxide (TiO_2)| D^++$e^-_{(CB)}$	$\rightarrow$ Metal oxide (TiO_2)| D	recombination (dark reaction)	(f)

Photovoltaic performance is significantly influenced by the events in interfaces between the metal oxide semiconductor, the dye, and the electrolyte. The dye/metal oxide interface is such that the oxidation potential of excited dye (LUMO) is sufficiently negative to obtain the effective electron injection to the conduction band of metal oxide. Upon photoirradiation, dye molecule undergoes a $\pi \rightarrow \pi^*$ transition thereby exciting an electron to the LUMO due to the relatively weak lateral overlapping of atomic orbital's (π bonds) with loosely bound electrons (π electrons) in HOMO. These excited electrons diffuse into the dye/metal oxide interface. Hence, a built in energy gradient (ΔE) exists between the dye/metal oxide interface due to the energy difference between the LUMO state of dye and the CB of metal oxide (E_{CB}). Thus, electrons in the LUMO of photoexcited dye undergo a nonradiative transition to the conduction band of metal oxide within a few picoseconds and thereby oxidized. The oxidized dye is reduced back to its ground state by using a redox couple together with platinum as catalyst at cathode (Thavasi et al., 2000; Jose et al., 2009).

The fundamental phenomenon of photoconversion and rate limiting steps involved in DSSCs (type-I) depicted in figure 2 are listed below:

I. The rate of excitation of dye: it is determined by the absorption of the number of photon at the absorption wavelength of dye, intensity of solar radiation at that window, and the absorption cross-section of dye.

II. The relaxation/radiative recombination of the excited electron of dye from LUMO to HOMO, and the corresponding rate constant k_1: Relaxation processes occur typically in a time scale of nanoseconds. This process reduces the excited state free energy by ~ 400 meV. The typical rate of excited state decay to ground state of dye is in the range 10^7-10^{10} s^{-1} (Grätzel & Durrant., 2008).

III. Exciton diffusion length D_{EXT}.

IV. Interfacial electron transfer, i.e., the injection of excited electron from the LUMO of dye to the conduction band of metal oxide with the rate constant k_2: The kinetics of the interfacial electron transfer dynamics strongly depends on the energetic of metal oxide/dye/electrolyte interface and the density of electron in metal oxide (*i.e.* the Fermi level of metal-oxide). The interfacial charge transfer occurs in a time scale of several picoseconds. In model system studies of dye sensitized metal oxide films, electron injection rate of > 10^{12} s^{-1} have reported for a range of sensitizer (Grätzel & Durrant., 2008).

V. Injected electron transport through the mesoporous layer of metal oxide to the transparent conducting oxide (TCO). It is controlled by the diffusion coefficient of electron (D_e), and the electron lifetime (τ_e).

VI. Phonon relaxation, through which an electron loses its energy by electron-phonon recombination.

VII. Interfacial charge recombination, i.e., the capturing of electron from the conduction band (CB) of metal oxide by the oxidized dye with the rate constant k_3: Charge recombination occurred from the CB of photoelectrode to oxidized dye in the range of µs-ms. This recombination rate is strongly depends on the electron density of the photoelectrode, light intensity, and cell voltage.

VIII. Back electron transfer, i.e., the capturing of CB electrons by the oxidized mediator in the electrolyte (e.g. I_3^-) with the rate constant k_4: Back electron transfer typically very slow (in the range of ms-s) and the rate constant is strongly dependent on the concentration and viscosity of the electrolytes as well as the structure of dye (Grätzel & Durrant., 2008).

IX. Regeneration of dye, i.e., the transfer of electron to the oxidized dye from the redox mediator in the electrolyte with the rate constant k_5: The restoration of the oxidized dye takes place in the nanosecond range. It is typically 100 times quicker than any recombination reaction and about 108 times faster than the inherent lifetime of the oxidized dye (Grätzel & Durrant., 2008).

In case of N719 dye, the interfacial charge recombination and the regeneration rate constants are $k_3 = 1.4x10^3$ s^{-1} and $k_5 = 1.1x10^5$ s^{-1}, respectively. This implies the injection yield of 99%, since the regeneration step is ~100 times faster than recombination. Also, the rate of the injection of electron ($k_2 > 1.4$ X 10^{11} s^{-1}) is three order of magnitude faster than the radiative recombination (Grätzel & Durrant., 2008). Consequently, it is assumed that interfacial charge recombination and radiative recombination, the two rate limiting processes, do not limit significantly the photoconversion of DSSCs. In addition, the exciton diffusion could be neglected since the exciton is highly localized on dyes because the particle radius is significantly smaller than the Bohr radius. Thus, the rate limiting processes that contribute highly on the photoconversion efficiency are (I), (IV), (V), (VI), (VIII) and (IX). The rate limiting steps, (I) and (IX) are dependent on the dyes and electrolyte, respectively. Steps (IV) and (VIII) depend on the interface between metal oxide/dye and metal oxide/electrolyte interface. The other two steps, (V) and (VI), are completely depends on photoelectrode (Jose et al., 2009). It is clear that the metal oxides, one of the major constituent of the photoelectrode, play a significant role on the performance of DSSCs by involving with most of the operation processes in the DSSCs.

2.2 Characterization techniques of DSSCs

Photoelectric current-voltage (I-V) measurement, electrochemical impedance spectroscopic measurement (EIS), and incident photon to current conversion efficiency (IPCE) measurements are the major essential techniques for the characterization of DSSCs.

2.2.1 Photoelectric current-voltage (I-V)

The photovoltaic performance of DSSCs is estimated from the current-voltage (I-V) measurements. The photovoltaic parameters of DSSCs are open circuit potential (V_{oc}), short circuit current (I_{sc}), fill factor (FF), and efficiency (η). These parameters are measured under the standard condition of cell temperature 25 ^{0}C, incident solar radiation of 100 mW/cm^2, and the spectral power distribution of AM1.5. Figure 3A depicts the I-V characteristics of a DSSCs made with N719 sensitized TiO_2 (degusa P25) photoelectrode of active area 0.2 cm^2 (Figure 3B). It is assumed that the standard redox potential of electrolyte is constant and, therefore, the V_{oc} of a phototelectrochemical cell should also constant. However, the V_{oc} of a photoelectrochemical cell can be control by tuning the flat band (V_{FB}) of metal oxide with introducing metal oxide composites or dopants into metal oxides (Lü et al., 2010).

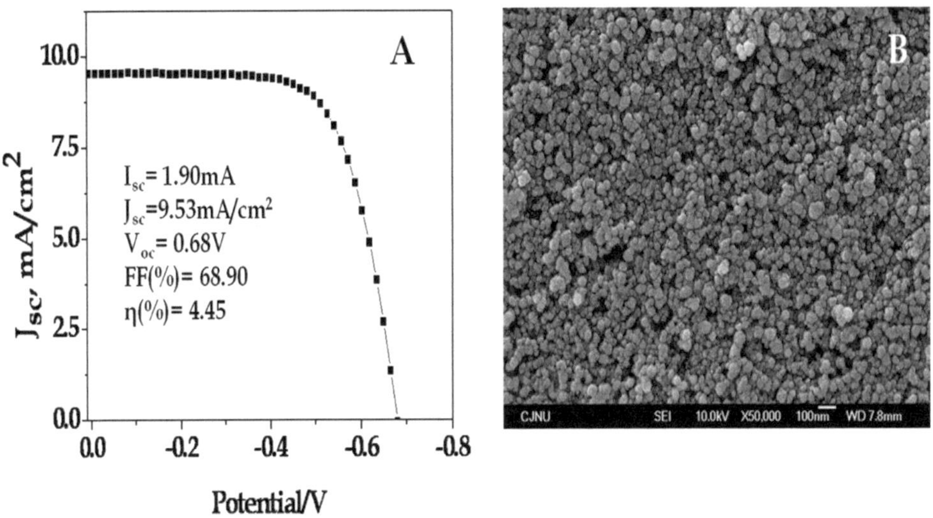

Fig. 3. (A) Photovoltaic performance of a state-of-the-art DSSC laboratory cell: *I–V* curve measured under AM 1.5 standard test conditions and (B) Scanning electron micrograph (SEM) of a sintered mesoporous TiO_2 (Degusa P25) film supported on FTO

Short circuit current density (J_{sc}) is the photocurrent per unit active area (A) of the photoelectrode when the applied potential across the DSSCs is zero, i.e. $J_{sc} = I_{sc}/A$. Under the ideal illumination condition, J_{sc} is effectively dependent on the efficiency of charge injection from the excited dye to the CB of metal oxide. The electron injection efficiency of some widely used metal oxides is found to be the following order: TiO_2 > Nb_2O_5 > SnO_2~ZnO (Thavasi et al., 2009).

Fill factor (*FF*) is the most important parameters of a photovoltaic cell. It is defined as the ratios of maximum power output ($I_{max}xV_{max}$) and the product of V_{oc} and I_{sc}. Therefore, from figure 4,

$$FF(\%) = \frac{I_{max}xV_{max}}{I_{sc}xV_{oc}} = \frac{AreaA}{AreaB} \quad (1)$$

It can be concluded that the more square like I-V curve is essential for highly desired FF. The lowering of electron loss between the FTO/electrolyte interface increases the shunt resistance, thereby increases the fill factor. On the other hand, increase of the back electron transfer and charge recombination is responsible for poor FF. Finally, comparing the ratios of total power out put (P_{out}) with the solar power input (P_{in}), efficiency (η) can be measured. i.e.,

$$\eta = \frac{P_{out}}{P_{in}} = \frac{J_{SC} V_{OC}\, FF}{P_{in}} \quad (2)$$

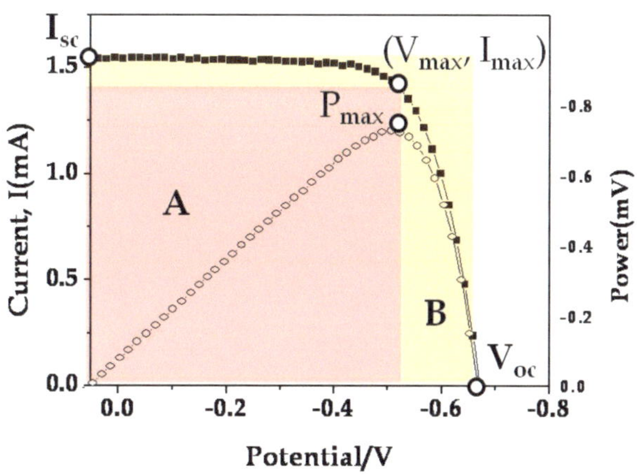

Fig. 4. I-V curve and parameters of DSSCs output current (solid line) and power (blank line) as function of voltage

2.2.2 Electrochemical Impedance Spectroscopy (EIS)

For better understanding the electron transport kinetics in DSSCs, EIS is the most powerful tools. EIS is a steady state process for evaluating the current response to the application of an ac voltage as function of frequency. EIS has been broadly employed to investigate the kinetics of electrochemical and photoelectrochemical processes including the clarification of salient ionic and electronic processes occurring in the DSSCs.

Figure 5 depicts a characteristics impedance spectrum (Nyquist plot) along with the electrical transmission line model of equivalent circuit for DSSC. The transmission line model is purely a combination of resistance (R) and capacitance (C). The series resistance, together with the sheet resistance of fluorine-doped tin oxide (FTO) glass and the contact resistance of the cell is denoted by R_S. R_{CO} and C_{CO} are the series resistance and the capacitance of the FTO/metal oxide interface, respectively. R_{TCO} and C_{TCO} are the charge transfer resistance and subsequent double layer capacitance of the FTO/electrolyte interface respectively. The transport resistance in the metal oxide film is represented by r_t. The charge transfer resistance and the charge recombination processes between electrons in the metal oxide phototelectrode and electrolyte is determined by r_{CT}. C_μ is the chemical capacitance of the metal oxide film. Z_d is the Warburg diffusion of ions in the electrolyte. Finally, R_{Pt} and C_{Pt} are the charge transfer resistance and the double layer capacitance of the platinized FTO counter electrode respectively. Therefore, the steady state process of electron transport resistance through metal oxide film, ion diffusion through electrolyte, chemical capacitance at interface of the metal oxide/electrolyte, double layer charge transfer capacitance at interface of the FTO/electrolyte, and the platinized FTO counter electrode, can be estimate by using this method (Fabregat-Santiago et al., 2007).

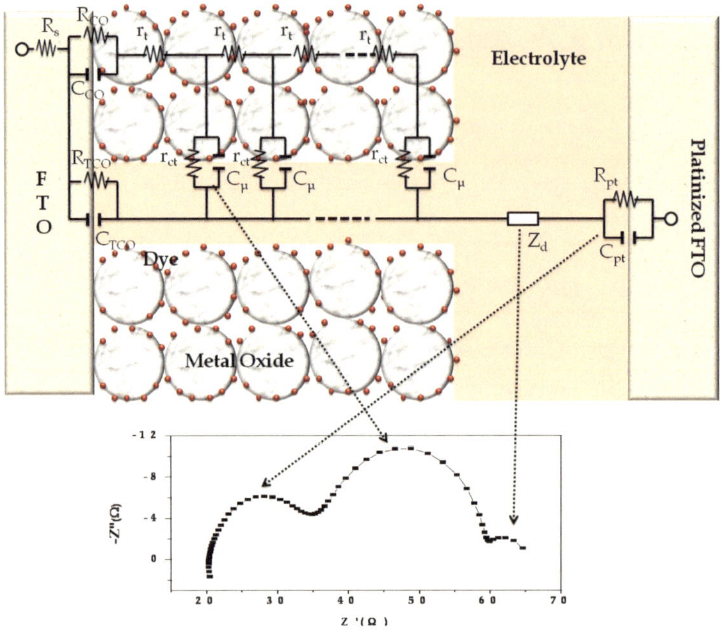

Fig. 5. Typical impedance spectra of DSSCs (Nyquist plot) and its electrical transmission line model of equivalent circuit (The transmission line model is redrawn with permission from ref. Fabregat-Santiago et al., 2007. Copyright 2010, The American Chemical Society)

2.2.3 Incident proton to current conversion efficiency (η_{IPCE})

IPCE is the measurements of spectral response of the cell that contain the partial information of quantum efficiencies of photocurrent generation processes such as the light harvesting (η_{LH}), electron injection (η_{INJ}), and charge collection efficiency (η_{COL}). Therefore,

$$\eta_{IPCE}(\lambda) = \frac{J_{SC}(\lambda)}{q\phi(\lambda)} = \eta_{LH}(\lambda)\eta_{INJ}(\lambda)\eta_{COL}(\lambda) \quad (3)$$

Where, J_{sc} is the short circuit current density of the DSSCs at incident monochromatic light with incident wavelength λ. q and φ is the elementary charge and photon flux respectively (Gentilini et al., 2010). Following each step of conversion, η_{LH} is associated with the ability of the dye in absorbing photons. η_{INJ} quantifies the charge transfer from the LUMO of dye molecule to the conduction band of metal oxide. Finally, the collection efficiency η_{COL} is the amount of electrons that effectively reach the anode, avoiding recombination. It should be noted that IPCE measurement is performed under short circuit condition, where electron lifetime is higher. It is noteworthy that the high fraction of incident photon to current conversion (IPCE) is most important for efficient DSSCs and the mesoporous metal oxide with high surface area is the prerequisite for higher IPCE value.

3. DSSC photoelectrodes with different metal oxides and their composite

As mentioned above, the different metal oxides and their composites have been investigated for the fabrication of DSSC. Herein, some of the important and highly explored metal oxides and their composites as DSSCs photoelectrode will be discussed.

3.1 Titania (TiO_2) / titania composite based photoelectrodes

Nature has abundantly the titanium dioxide (TiO_2) which is relatively inexpensive, and nontoxic. From the time of invention of its photocatalytic activity by Fujishima and Honda in 1972 (Fujishima & Honda.,1972), it has been used in a wide range of application such as paint, pigment, water splitting, hydrogen and oxygen gas sensor, polluting gas sensor, and biosensors (Rahman et al., 2010). The n-type semiconductor TiO_2 has been existed naturally as three crystalline polymorphs, explicitly anatase, rutile, and brookite as shown in figure 6. The crystalline characteristics along with the band gap of TiO_2 ploymorphs has been tabulated in table 1.

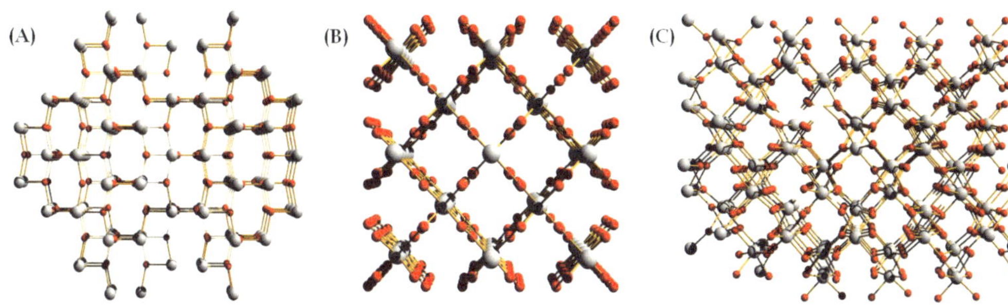

Fig. 6. Different naturally occurs crystalline polymorph of TiO_2 (A) anatase, (B) rutile, (c) brookite, where small red sphere: O^{2-}, big grey sphere: Ti^{4+}

Polymorphs	Band gap	Space group	Cell parameter (Å)	Crystal structure
Anatase	3.23 eV	$I\ 4_1/amd$	a= 3.7845, c=9.5143	Tetragonal
Rutile	3.05 eV	$P\ 4_2/mnm$	a= 4.5937, c= 2.9587	Tetragonal
Brookite	3.26 eV	$P\ bca$	a= 9.1840, b= 5.4470, c= 5.1450	Orthorhombic

Table 1. Crystal characteristics of different TiO_2 polymorphs along with their band gap

In addition, TiO_2 exists in many other metastable forms such as monoclinic, tetragonal, orthorhombic, and cubic. The electronic state of Ti in TiO_2 is Ti^{4+} ($3d^0$), in which the valence band of TiO_2 is composed of the hybridized *2p* orbital of oxygen and the *3d* orbital of Ti, while the conduction band is completely composed of pure *3d* orbital of Ti. Hence, the different parity of valence band and the conduction band of TiO_2 decreases the transition of electrons to the valence band; consequently, decrease the electron hole (e^--h^+) recombination probability. Among the polymorphs of TiO_2, it has known that rutile is the most stable and common. However, the Fermi level of anatase is 100 mV higher than that of rutile, which leads to higher open circuit potential (V_{oc}). Moreover, the greater surface area of anatase is responsible for efficient dye loading, leads the higher photocurrent, consequently, higher photovoltaic performance. Brookite is less prominent for DSSC due to its complicated and difficult synthesis process. It is noted that the oxygen vacancies and the titanium interstitials as well as the decreased recombination probability with increased surface area makes TiO_2 an attractive materials for the application in DSSCs (Li et al., 2007; Jose et al., 2009).

3.1.1 TiO_2 Photoelectrode with passivation layer and 1-D nanostructures

TiO_2 of anatase nanocrystalline form has been explored extensively in DSSCs photoelectrode due to its suitable band position, high surface area for improved dye loading, and facile synthesis procedure. However, the lack of efficient electron transfer through the FTO/TiO_2 interface increases the back electron transfer to the redox ion in electrolyte (Figure 7A) and low light harvesting with low absorption in red region limits the conversion efficiency of DSSCs. Moreover, the random electron diffusion (i.e. electron trapping) and the enhanced grain boundary density of spherical TiO_2 nanoparticles (Figure 7C) increase the loss of conversion efficiency with the increase of carrier recombination (Benkstein et al., 2003). Therefore, extensive efforts have been performed to minimize the back electron transfer/recombination. For example the carrier leakage of direct electron acceptance from the nanocrystalline TiO_2 film can be avoided through prevention of hole mediation to excited dye by addition of 4-tert-butylpyridine into redox solution (Hattori &Goto, 2007). Other impressive ways are the use of TiO_2 passivating layer/blocking layer by using rutile TiO_2 mixed with anatase TiO_2, and the fabrication of DSSCs by one dimensional (1-D) nanostructures such as nanotube, nanowire, nanorod etc(Zhu et al., 2006) .

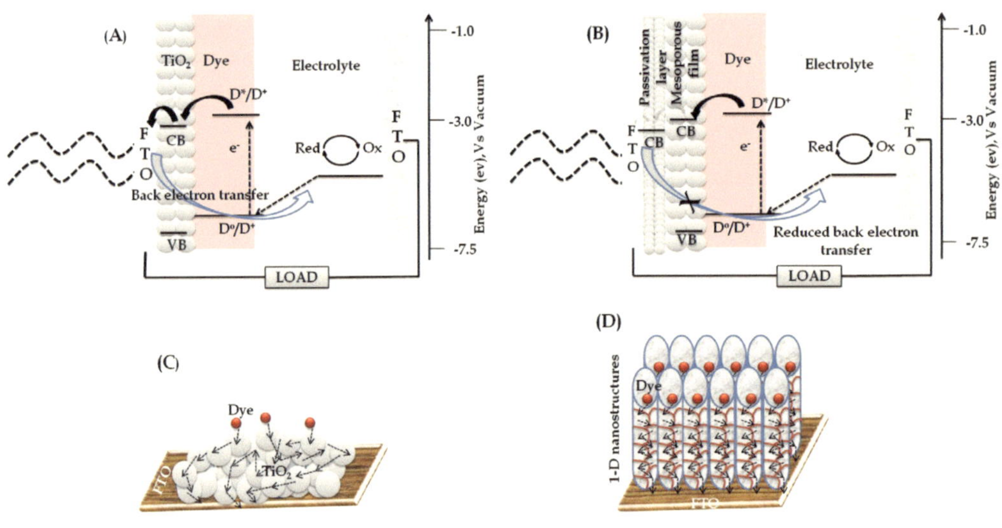

Fig. 7. Schematic representation of spherical anatase TiO_2 nanoparticles based DSSCs, where back electron transfer/recombination is the major role for loss of conversion efficiency (A), photoelectrode with lower passivation layer and upper anatase mesoporous TiO_2 to minimize the back electron transfer (B), random electron diffusion through spherical TiO_2 nanoparticles (C), and 1-D nanostructures of TiO_2 with directional electron diffusion (D)

The important routes of the back electron transfer to electrolyte are to via nanocrystalline TiO_2 and via TCO substrate. However, the back transfer from FTO to the I^-/I_3^- redox couple is negligible because the Fermi level of FTO is close to the redox Fermi level at open circuit potential. Thus, the direct route of recombination of I^-/I_3^- redox couple is to via the nanocrystalline TiO_2. Many researches are focused to prevent this back electron transfer by using a lower passivating layer (rutile/rutile mixed TiO_2) with an upper layer of mesoporous anatase TiO_2 film. The compact TiO_2 layer not only act as a shielding layer against the ionic penetration but also provoke different microstructure in the sol-gel derived TiO_2 electrodes, which in turn leads to different amount of dye absorption (Yu et al., 2009). All the DSSCs with passivating layer shows higher efficiency than that of without passivating layer. The increase of J_{SC} is generally related to the enhancement of the number of photogenerated electrons that efficiently transferred to TiO_2 electrode. The TiO_2 passivating layer has both of rutile and anatase phases (3.05 eV <Eg> 3.23 eV) and the upper mesoporuos TiO_2 film has only anatase phase (Eg = 3.23 eV). So, the injected excited electrons move in the direction of lower energy conduction band from the porous TiO_2 to the passivating layer, thus resulting in a smooth electron movement and reduce the back electron transfer (Figure 7B). In Table 2, we tabulate the comparison performance of DSSCs before and after using TiO_2 (rutile) passivating layer along with the deposition techniques of passivation layer and mesoporous layer.

Materials	Deposition techniques of passivation layer	Process for making mesoporous TiO_2 layer	Efficiency with (without) passivation layer	Ref.
A thin passivating TiO_2 (rutile) underlayer	Facing target sputtering (FTS) method	Sol-gel	2.85% (0.79%)	Hossain et al., 2008
A thin passivating TiO_2 (rutile) underlayer	Sputtering method	Sol-gel	5.25% (3.85%)	Hattori & Goto, 2007
A thin compact TiO_2 (rutile)layer	Dip-coating	Sol-gel	7.6% (5.7%)	Yu et al., 2009

Table 2. Performance of DSSCs after using passivating layer and their deposition process

Another important technical phase for the improvement of performance of DSSCs is to use 1-D TiO_2 nanostructures. Figure 7D shows the 1-D nanostructures and their directional electron diffusion. Frank and coworkers investigated the comparative study of electron transport and recombination in DSSCs between TiO_2 nanotubes (TNTs) and spherical nanoparticles (NPs) by frequency-resolved modulated photocurrent/photovoltage spectroscopy (Zhu et al., 2006). The rate of recombination in TNTs photoelctrode is much slower than that of NPs photoelectrode. This increase the charge collection efficiency in TNTs photoelectrode compare to NPs photoelectrode. The internal and external dye loading with stronger internal light-scattering effects of TNTs exhibits higher photoconversion efficiency. However, the dye loading of TNTs and charge collection efficiency is strongly dependent on the nanotube arrays. Bundling of nanotube creates additional pathway via intertube connection and decrease the internal surface area of the films accessible to dye molecules. Therefore, bundling of nanotubes increases the charge recombination and decreases the photoconversion efficiency. It was established that the high photocurrent density and photocurrent efficiency could be achieved by better-aligned nanotube arrays. Capillary stress creates the morphological disorder (e.g. bundling and microcracks) of TNTs during the solvent evaporation. Therefore, supercritical CO_2 drying process is developed for preparing crack-free and bundle-free nanotubes (Zhu et al., 2007). Many strategies have developed for the fabrication of TNTs such as direct electrochemical anodization of Ti foil, electrophoresis, hydrothermal method, and incorporation of TiO_2 into anodized alumina oxide (AAO).

1-D nanostructures	Preparation method	Efficiency of 1-D nanostructures based DSSCs	Efficiency of NPs based DSSCs	Ref.
TiO_2 nanorod	Oriented attachment of nanoparticles	6.2%	4.2%	Kang et al., 2008
TiO_2 nanorod	Solvothermal	7.9%	5.5%	Marco et al., 2010
TiO_2 nanofibre/ TiO_2 nanoparticle	Electrospinning and sol-gel techniques	10.3%	9.25%	Chuangchote et al., 2008
Macroporous TiO_2 nanowires	Hydrothermal synthesis	0.86%	-	Wang et al., 2009
TiO_2 Nanotubes	Electrochemical anodization of Ti foils	3%	-	Zhu et al., 2007

Table 3. Comparison performance of different 1-D nanostructures photoelectrode with NPs based photoelectrode for the application in DSSCs

The other 1-D nanostrucutres are nanorods, nanowires, etc. Recently, Manca and coworkers have developed a novel and cost efficient method for the preparation of anatase TiO_2 nanorod (20-30 nm × 100-300 nm) by single-step solvothermal process (Marco et al., 2010). This single step solvothermal process does not require the complete removal of organic residual that will act as a binder. This TiO_2 nanorod showed improved photoconversion performance compared with the naocrystalline counterparts due to the increased rate of electron transport arisen from the high crystallinity of the nanorod. However, many reports suggested that the poor extent of dye loading and the poor light harvesting of nanorod based DSSC is due to the low specific surface area of the nanorod. To facilitate the improved dye loading, Kang et al. synthesized TiO_2 nanorod (4 nm × 20 nm) of high specific surface area from the necking of truncated NPs by "oriented attachment approach"(Kang et al., 2008). The improved roughness factor of the nanorod (169.9 μm^{-1}) compare to nanoparticles (115.6 μm^{-1}) showed higher dye absorption ability and thereby improved conversion efficiency. Intentional integration of metal impurities into the semiconducting materials is very useful and common approach for tuning the semiconducting properties such as band gap and electrical conductivity. Feng et al. prepared successfully tantalum (Ta)-doped TiO_2 nanowire by low temperature hydrothermal method. Ta-doped TiO_2 nanowire gives a very high photovoltage (0.87 V) of comparable J_{sc} with undoped naowire in a liquid-state DSSCs (Feng et al., 2009). The high open circuit potential close to the theoretical maximum might be attributed to the negative shift of flat band potential of nanowires. In table 3 we tabulate the comparison performance of DSSCs made by 1-D TiO_2 nanostructures and TiO_2 nanoparticles.

3.1.2 TiO_2 photoelectrode with scattering layer

The efficiency of DSSCs over 11% in laboratory scale has been reported (Grätzel, 2003). The ultimate target of DSSCs technology is the up scaling of the small cell into larger modules. Therefore, many important factors have to be considered such as stability, durability, and more importantly price per watt of the module to bring the modules into market. Photosensitizing dye of high efficient DSSCs is still very expensive, so the actual cost-energy per watt of DSSCs module relies on the amount of dye usage. Many strategies such as development of organic dye, replacement of platinum by carbon materials and organic polymer have been developed for the reduction of energy cost of DSSCs. However, one remarkable strategy has been developed on the modification of TiO_2 photoelectrode is the uses of scattering (SC) TiO_2 particles (films or mixed with nanocrystalline particles). These SC layer/particles will help to develop very thin nanocrystalline TiO_2 film of high surface area that will reduce the dye usage without sacrificing the cell efficiency. Figure 8 shows the light confined effect on different TiO_2 particles. It can be seen that SC TiO_2 particles scattered the incident longer wavelength light and prolonged the optical path length. Hence, the transmitted visible light can effectively uses for photon conversion. However, light scattering property is strongly dependent on the refractive index between the active TiO_2 film, scattering layer, and on the relative size of the scattering particles (Shin et al., 2010; Arakawa et al., 2006).

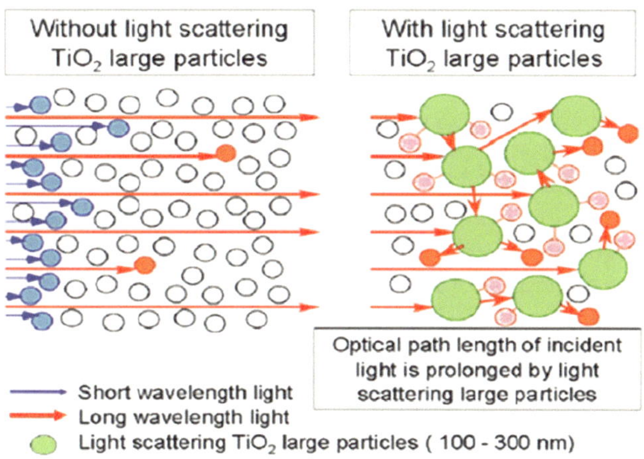

Fig. 8. Conceptual scheme of Light confined effect (reproduced with permission from ref. Arakawa et al., 2006. Copyright 2010, IEEE)

Scattering layer (SC) matrix	V_{oc} (mV) with SC (without SC) layer	J_{sc} (mAcm^{-2}) with SC (without SC) layer	FF (%) with SC (without SC) layer	η (%) with SC (without SC) layer	Ref.
TiO_2 film of 300 nm particles	735 (680)	14.5 (12.3)	No appreciable change	7.48 (5.77)	Kim et al., 2008
sub-micrometer-sized Al_2O_3	745 (730)	14.44 (12.54)	71.1 (67.7)	7.3 (6.5)	Shin et al., 2010
ZrO_2(500-1000 nm)	738 (710)	12.72 (6.42)	61.8 (70.2)	5.8 (3.2)	Hore et al., 2006
ZnO microsheets	677 (636)	8.68 (6.04)	43.98 (46.58)	2.58 (1.79)	Yi et al., 2010

Table 4. Summarization of the performance of different scattering particles based photoelectrode for the application in DSSCs

Hore et al. investigated the comparison study of the performance of different thickness (4 µm, 8 µm) of nanocrystalline TiO_2 film of high surface area with and without scattering layer (Hore et al., 2006). It was observed that the 4 µm nanocrystalline TiO_2 thin film with SC layer shows a comparable performance with 8 µm thick TiO_2 film without SC layer. Therefore, it is expected to use thin film of TiO_2 with SC layer for the industrial production of large DSSC modules, which will minimize the cost of dye by lesser consumption in the production procession. Some other metal oxide has also explored as effective SC layer on top of the active TiO_2 film such as ZrO_2, Al_2O_3, TiO_2-rutile form, ZnO etc. Table 4 summarizes the comparison of photoconversion efficiencies with different scattering particles.

3.1.3 Metal-doped titania (TiO_2) photoelectrode

Metal-ions as dopant has a significant effect on the electrical surface state modification of TiO_2. This modification leads to significant change of electrical conductivity, shifting of Fermi level potential, particle aggregation, charge transfer kinetics, and dye absorption characteristics of TiO_2. However, many reports have suggested that the performance of DSSCs at the optimized level of metal-ions doping into TiO_2 differ mainly the tuning of flat band of TiO_2. Lü et al. synthesized well-crystalline niobium (Nb) doped anatase TiO_2 (Nb-TiO_2) and investigated its effect on the performance of DSSCs (Lü et al., 2010). It was revealed that Nb doping level of 2.5 -7.5 mol% has positive effect on the short circuit current with positive shift of flat band potential. This increment of the photocurrent was attributed to the enhanced electron injection, increased charge transfer kinetics as well as the improvement of dye loading up to $5.7\pm0.2\text{x}10^{-8}$ [mol cm^{-2}] at a doping level of 7.5 mol% compared to the undoped state. The TiO_2 doped by various metals such as tungsten (W), neodymium (Nd) also exhibited the similar phenomenon (Ko et al., 2005; Yao et al., 2006), while Al doped TiO_2 showed an opposite effect (Ko et al., 2005). Lee et al. deposited a thin film of compact Nb-doped TiO_2 on FTO by pulsed laser deposition (PLD) (Lee at al., 2009). The performance of the DSSC of compact Nb-doped TiO_2 layer with an upper mesoporous TiO_2 layer is prominent compared to the compact TiO_2 layer.

Figure 9 describes the band diagram of two compact layers and the charge transfer kinetics. The upper part of the diagram is before contact with FTO and it can be seen that the Fermi level of compact TiO_2 and Nb-doped TiO_2 layers is below and above of the conduction band (E_{CB}) respectively. This is ascribed by the degeneracy of Nb-doped TiO_2 and the nondegeneracy of compact TiO_2. When compact TiO_2 and FTO are in contact (i.e. semiconductor–metal contact), a depletion region or Schottky barrier, q (φ_m-φ_s) is form by the difference of work function of FTO (φ_m) and TiO_2 (φ_s). Electrons have to overcome this barrier in order to be transferred from compact TiO_2 to FTO. Hence, the electron injection efficiency is decreased while the back electron transfer was increased. On the other hand, the contact between FTO and the compact degenerate Nb-doped TiO_2 layer form a very narrow depletion region. Electrons injected rapidly from compact Nb-doped TiO_2 to FTO by tunneling leads to ohmic contact characteristics (Lee et al., 2009). Therefore, the incorporation of compact Nb-doped TiO_2 layer remarkably reduces the interfacial charge transfer resistance and results in the enhancement of the performance of the DSSC. The performance of some of the metal-doped TiO_2 nanocompisites in DSSCs is summarized in table 5. From the experimental data of table 5, it is clearly observed that doped metal ions have a strong effect for the tuning of flat band potential, as well as the electron injection efficiency.

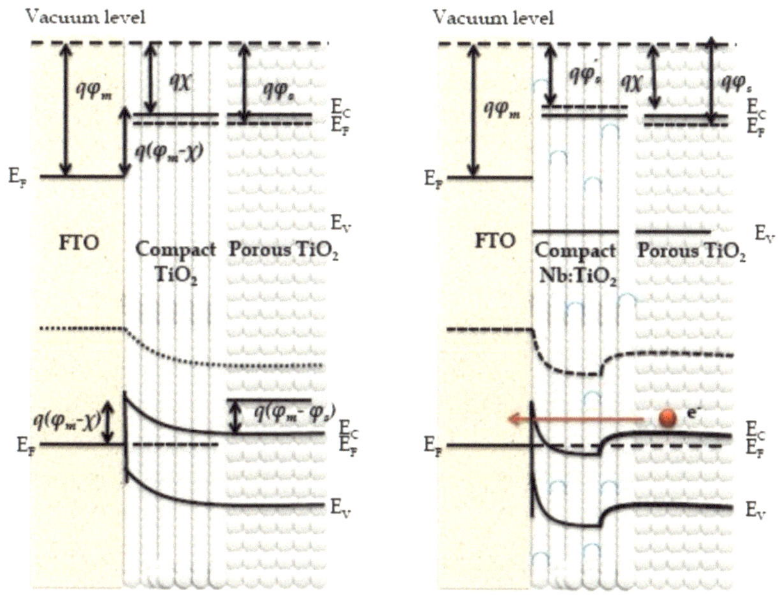

Fig. 9. Schematic illustration of band diagram of compact TiO_2 and compact Nb-TiO_2 layer before (upper) and after (under) contact with FTO (redrawn with permission from, Lee at al., 2009. Copyright 2010, The American Chemical Society)

Dopant	Synthesis techniques	V_{oc} (mV) of doped (undoped) TiO_2	J_{sc} ($mAcm^{-2}$) of doped (undoped)	FF (%) of doped (undoped)	η (%) of doped (undoped)	Ref.
5mol% Nb	Hydrothermal synthesis	700 (790)	17.67 (11.87)	63 (70)	7.8 (6.6)	Lü et al., 2010
Nd	Solvothermal reactions	701 (707)	13.1 (10.7)	40.1 (36.6)	4.4 (3.3)	Yao et al., 2006
Bi	Sol-gel	590 (570)	7.71 (5.52)	46 (38)	2.11 (1.19)	An'amt et al., 2010
Ta	Hydrothermal	870 (~750)	7.5 (8.1)	63 (68)	4.1 (4.06)	Feng et al., 2009

Table 5. Summarization of the performance of different metal-doped TiO_2 and undoped TiO_2 photoelectrodes for the application in DSSC

3.1.4 Core-shell composite of titania (TiO_2) and other metal-oxide for photoelectrode

Coating of TiO_2 with different metal oxide/hydroxide of wide band gap such as MgO, Nb_2O_5, ZnO, $SrTiO_3$, $Mg(OH)_2$, $Zn(OH)_2$ etc has received much attention for the improvement of the performance of DSSCs. Metal oxide/hydroxide coating on TiO_2 employed a core-shell structure, which is advantageous for the efficient photoconversion. Firstly, the wide band gap of shell metal oxide minimizes the electron-hole recombination by increasing the surface resistance of TiO_2 and retards the back electron transfer to the electrolyte. Secondly, the shell layer increases the dye absorption due to increased surface area leading to high photoconversion and finally tuning the conduction band of core structure (Jung et al., 2005). Figure 10 depicts the different core-shell structure of metal oxide. The use of a thin metal oxide/hydroxide layer (shell) permitted the formation of an energy barrier at the electrode/electrolyte interface. It reduced the recombination rate and thereby improved the cell performance. The core-shell structure C in Figure 10 is favorable for the diffusion of the injected electrons because it does not face any energy barriers during the electron diffusion to the FTO (Chappel et al., 2002).

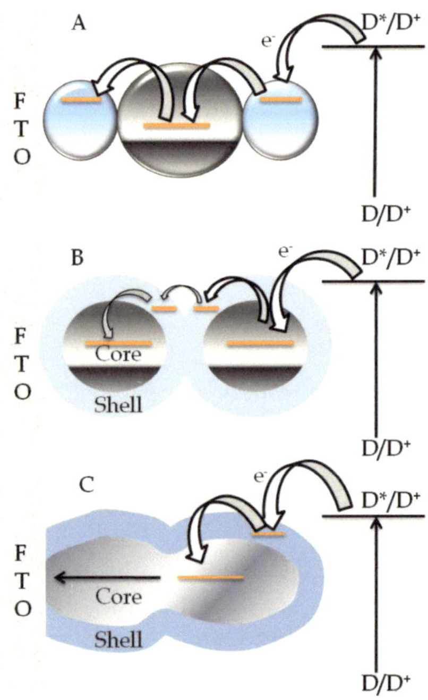

Fig. 10. Different geometries of core-shell electrodes showing that only in the core-shell matrix (c) the photoinjected electrons does not face any energy barriers during the diffusion to the TCO (redrawn with permission from, Chappel at al., 2002. Copyright 2010, The American Chemical Society)

Chen et al. prepared a new bilayer of Nb_2O_5 coated nanoporous TiO_2 photoelectrode (Chen et al., 2001). The conduction band of shell materials Nb_2O_5 is 100 mV negative than the core materials TiO_2. Thus, the Nb_2O_5 form an energy barrier to the photoelectrode and, therefore, confined the electrons to the core TiO_2 particles. This electron confinement to TiO_2 particles reduces the recombination and thereby increases the J_{sc}. The increase of the photovoltage is attributed to the shift of the Fermi level of TiO_2 towards negative direction by Nb_2O_5 (schematically describe in figure 11A). Consequently, all the cell parameter was improved and the overall cell efficiency enhanced up to 35%. Diamant et al. reported the development of $SrTiO_3$ coated TiO_2 photoelctrode. The conduction band of $SrTiO_3$ is 200 mV negative than that of TiO_2 (Diamant et al., 2003). This suitable band position of core-shell composite resulted in the increment of J_{sc} by the reduced recombination as well as the efficient electron injection to FTO. Increment of V_{oc} is attributed to the pervious phenomenon as Nb_2O_5-TiO_2 core shell photoelectrode. As a result, the overall DSSCs performance increased by 15%.

Nonmetal oxides such as $CaCO_3$ with high band energy (6.0 eV) for the formation of favorable core-shell with TiO_2 and high isoelectric point (IEP) to improve dye loading have also been explored. The increment of J_{sc} is mainly due to the high specific surface area of the $CaCO_3$ overlayer that caused the enhancement of dye loading up to 54%. The increased photovoltage is also attributed to the phenomenon described above with the more negative conduction band edge of $CaCO_3$ compare to that of TiO_2. Overall, photo conversion increased by 24.4% with this approach (Lee at al., 2007).

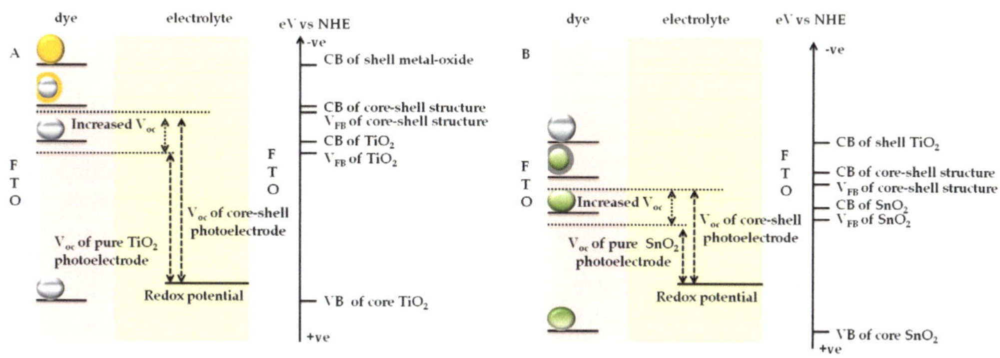

Fig. 11. Schematic illustration of energy band tuning of TiO_2 photoelectrode (A) and SnO_2 photoelectrode (B) by the introduction of core-shell structure

The lower conduction band edge of SnO_2 compare to rutile TiO_2 makes it possible to switch the core-shell structure (i.e. TiO_2 as shell matrix and SnO_2 as core matrix). Reports are available for the formation of rutile TiO_2 coated SnO_2 hollow microsphere and nanoparticles. Qian et al. fabricated rutile TiO_2 coated SnO_2 hollow microsphere (MHS) for DSSCs photoelectrode (Qian et al., 2009). The Core-shell structure was characterized by scanning electron microscope (SEM), high-resolution transmission-electron microscope (HR-TEM), and x-ray diffraction (XRD). The XRD pattern of TiO_2-MHS is well indexed with the rutile structure of SnO_2 [space group *P42/mnm* (136)] with the additional rutile TiO_2 characteristics peak observed at 2θ = 36.09, 41.22, and 54.32^0. The open circuit potential of TiO_2-SnO_2 MHS photoelectrode (V_{oc} = 0.664 V) was higher than that of the pure SnO_2 photoelectrode (V_{oc} = 0.398 V) but lower than that of pure anatase TiO_2 photoelectrode (V_{oc} = 0.697 V). This is attributed to the fact that the conduction band of SnO_2 is 300mV more positive than the rutile TiO_2. Upon TiO_2 modification by SnO_2 MHS, the electronic band shifts to a more negative value (schematically describe in figure 11B). The open-circuit potential of TiO_2-SnO_2 MHS is improved greatly. This indicates the effective suppression of interfacial electron recombination in TiO_2-SnO_2 MHS electrode. J_{sc} of the core-shell photoelectrode was observed higher than that of pure anatase TiO_2 photoelectrode. This is responsible for the formation of favorable energy barrier for efficient electron injection as depicted in figure 10C as well as the concurrent decrease of the recombination with increasing the surface resistance of SnO_2. Chappel et al. observed a similar phenomenon by using rutile TiO_2 coated SnO_2 nanoparticle (Chappel et al., 2002). The different core-shell nanostructures photoelectrode and their performance for the application in DSSCs are summarized in Table 6.

Core-shell	Synthesis techniques	V_{oc} (mV) of modified (unmodified)	J_{sc} ($mAcm^{-2}$) of modified (unmodified)	FF (%) of modified (unmodified)	η (%) of modified (unmodified)	Ref.
MgO-Coated TiO_2 NPs	Topotactic reaction	720 (640)	11.7 (10.2)	53.5 (47.3)	4.5 (3.1)	Jung et al., 2005
Nb_2O_5 Coated TiO_2	Hydrothermal	730 (661)	11.4 (10.2)	51 (56.5)	4.97 (3.62)	Wang et at., 2001)
$SrTiO_3$ coated TiO_2	Hydrothermal	708 (650)	10.2 (10.5)	58.4 (53.6)	4.39 (3.81)	Diamant et al., 2003
$CaCO_3$-Coated TiO_2	Topotactic thermal decomposition	668 (654)	21.92 (19.39)	66.1 (61.8)	9.68 (7.84)	Lee et al., 2007
$Mg(OH)_2$ coated TiO_2	Cathodic deposition	758 (696)	6.61 (6.43)	74 (73.6)	3.7 (3.3)	Yum et al., 2006
TiO_2 coated SnO_2 MHS	Sol-gel	664 (697)	14.6 (11.07)	58.3 (66.6)	5.65 (5.14)	Qian et al., 2009

Table 6. Comparison performance of different core-shell modified and unmodified TiO_2 and SnO_2 photoelectrode for the application in DSSCs

3.2 Photoelectrode based on ZnO and its composite

Zinc oxide (ZnO) is an important class of semiconductor. It is almost insoluble in water, usually appears as white powder, and exists in nature usually as zincite mineral. Many superior physical properties such as the high electron mobility (~100 cm^2/Vs), high thermal conductivity and stability, wide and direct band gap (E_g = ~3.37 eV at room temperature), large excitation binding energy (60 meV), large saturation velocity ($3.2x10^7$ cms^{-1}) and high optical gain of 300 cm^{-1} (100 cm^{-1} for GaN) at room temperature make it as an attractive materials for wide range of application such as sensor, pigments in paint, filler in rubber products, thin film transistor, photodetectors, light emitting diode(LED) and in photovoltaic device (Rahman et al.,2010). The electronic configuration of ZnO is composed by completely filled 3*d*-orbital ($3d^{10}$). The valence band of ZnO consists of only *d* orbital and the conduction band is composed by hybridized *s*–*p* orbitals. This type of electronic configuration of ZnO leads to the state of dissimilar parity with reduced (e^--h^+) recombination probability (Jose et al., 2009). Therefore, from the point of view of electronic configuration and excellencies in physical properties, ZnO is considered as a suitable material for the application in DSSC. There are three crystalline forms for ZnO: hexagonal wurtzite, cubic zincblende, and cubic rock salt. Among them, wurtzite is the most stable at ambient condition, lowest energy structure, and most common form of crystal. Cubic zincblende can also be stabilized by growing cubic ZnO on substrate. However, the cubic rock salt (NaCl) type ZnO is observed very rarely at extremely high pressure of 10 G pa (Corso et al., 1994). Figure 12 shows the structure of two crystallographic form of ZnO (wurtzite and cubic zincblende), where the zinc (Zn) and oxygen centers are tetrahedral. The conduction band energy of hexagonal wurtzite ZnO is similar to that of the conduction band of TiO_2. That's why; the hexagonal wurtzite ZnO has extensively explored for the development of efficient DSSCs and for the replacement of TiO_2 as well (Redmond et al., 1994). In this section, we will discuss on the potential application, prospects, and limitation of ZnO based photoelectode for the application in DSSCs.

Hupp and colleagues studied the electron transport properties in ZnO nanotube (64μm length) by EIS. It was observed that the electron transport is significantly faster and electron extracted more rapidly in ZnO nanotube compare to nanocrystalline TiO_2 (Martinson et at., 2009). On the other hand, Hagfeldt and coworkers suggested that the electron diffusion follows a trapping/detrapping mechanism in nanoporous ZnO. This trapping/detrapping is strongly dependent on the variation of film thickness, applied potential, and electrolyte system (Solbrand et al., 2000). It is estimated that the diffusion coefficient in ZnO photoelectrode is 10^{-6} cm^2s^{-1} at +300 mV vs Ag/AgCl in ethanol with 0.5 M $LiClO_4$ and $2x10^{-5}$ cm^2s^{-1} for nanoporous TiO_2 electrode under the same condition but in 0.7 M $LiClO_4$. Therefore, electrons are more long lived in ZnO than in TiO_2 and that the electron losses to acceptors in solution and to deep traps are less for ZnO than for TiO_2. Despite these superior physical properties of ZnO, the highest reported efficiency of ZnO based DSSCs is still very low (*ca.* 6.58%). This might be due to the formation of Zn^{2+}/dye complex that inhibits the efficient electron injection and causes the retardation of the dye regeneration. The highly efficient acidic ruthenium (II) dyes (such as N3, N719, Black dye, etc.) generally showed lower performance with ZnO based photoelectrodes. It was attributed to the reduced chemical stability of ZnO due to the dissolution of ZnO and formation of aggregates which are caused by the proton release from the acidic dye. In order to improve the performance and stability of ZnO based photoelectrode, different strategies have been introduced. It includes the design and synthesis of suitable new dyes for ZnO photoelectrode, the development of quantum dot (QD) sensitized ZnO photoelectrode, and the development of 1-D ZnO photoelectrode for efficient photoelectron collection (Quintana et al., 2007; Zhang et al., 2009).

As discussed in the figure 7D, 1-D nanostructure makes the unidirectional electron diffusion possible and therefore increases the efficiency of electron injection and collection by reducing the recombination during the interparticle hopping process. Various types of 1-D ZnO, such as hierarchical ZnO nanowire-nanosheet, nanodisk, nanowire, nanorod, nanofibre, and nanocomb have been synthesized for the application in DSSCs. Figure 13 depicts the scanning electron microscopic (SEM) images of some of the 1-D ZnO nanostructures. Umar et al. fabricated highly crystalline ZnO nanocomb directly grown on FTO by noncatalytic hydrothermal evaporation process (Umar et al., 2009). However, ZnO nanocomb photoelectrode with N719 sensitization showed very low conversion efficiency of 0.86%, due to the low dye loading and fast interfacial recombination of electron and holes. The former is attributed to the low specific surface area of 1-D nanostructures of ZnO. This concurrently results in the increases of the recombination due to the enhancement of charge recombination at the uncovered oxide surface. This effect was minimized by the development of 1-D nanostructure with high aspect ratios and the various 1-D nanostructures of ZnO used for DSSCs are summarized in Table 7. In brief, Wang et al. synthesized porous hierarchical ZnO nanodisk with higher surface area (21.8 m^2g^{-1}) by a simple low temperature hydrothermal method (Wang et al., 2010). Ramakrishna et al. developed ZnO nanofibres with even higher aspect ratio and surface area of 30 m^2g^{-1} (Zhang et al., 2009). Both showed the improvement of efficiencies up to 2.49 and 3.02% respectively mainly due to the improved dye loading.

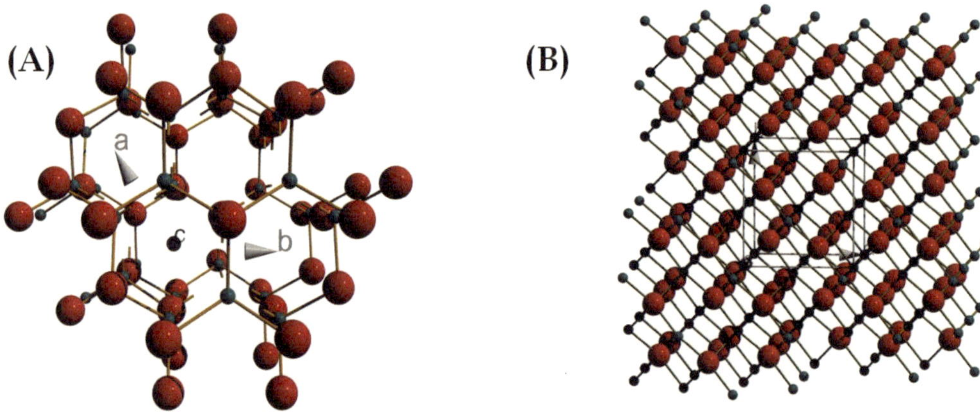

Fig. 12. Crystallographic pattern of mostly observed and common hexagonal wurtzite crystal of ZnO (space group *P 63mc*) (a) and cubic zincblende (space group *F 43m*) (b). Where big red sphere Zn^{2+}, small green sphere: O^{2-}

Fig. 13. Scanning electron microscopic (SEM) image of different 1-D nanostructures of ZnO (A) Hierarchical ZnO nanowire-nanosheet,(B) nanodisk,(C) nanowire, and (D) nanorod.(reproduced with permission for ref. Xu et al., 2010; Wang et al., 2010; Gao et al., 2007; Pang et al., 2007. Copyright 2010, The American Chemical Society)

Materials	Synthesis process	Surface area (m^2g^{-1})/ aspect ratio	Sensitizer	V_{oc} (V)	J_{sc} ($mAcm^{-2}$)	FF (%)	η (%)	Ref.
ZnO nanocomb	Noncatalytic Thermal evaporation	-	N719	0.67	3.14	34	0.68	Umar et al., 2009
ZnO nanowire	Hydrothermal	-/100-120	N719	0.54	6.79	50	1.7	Gao et al., 2007
ZnO nanorod	Hydrothermal	12.71/-	N719	0.73	3.41	74.4	1.86	Umar et al., 2009
ZnO nanodisk	Hydrothermal	21.8/-	N719	0.69	6.92	52.5	2.49	Wang et al., 2010
ZnO nanofibre	Electrospinning	30/-	N719	0.57	9.14	58	3.02	Zhang et al., 2009

Table 7. Summarization of different 1-D ZnO nanostructures, their physical properties, along with their performance in DSSCs

ZnO has been extensively explored as a composite material together with TiO_2 for the application in DSSCs. The potential application of ZnO nanostructures as a scattering layer and core-shell nanostructures composites along with TiO_2 has described in section 3.1.2 and 3.1.4. In this section, we need more focus on the hybrid nanocomposites of ZnO for the application in DSSC. In addition to the development of 1-D ZnO nanostructures with higher aspect ratio and surface area, many ZnO nanocomposites (shown in figure 14) have explored to improve the dye loading with maintaining the directional electron transport pathway as in 1-D nanostructures.

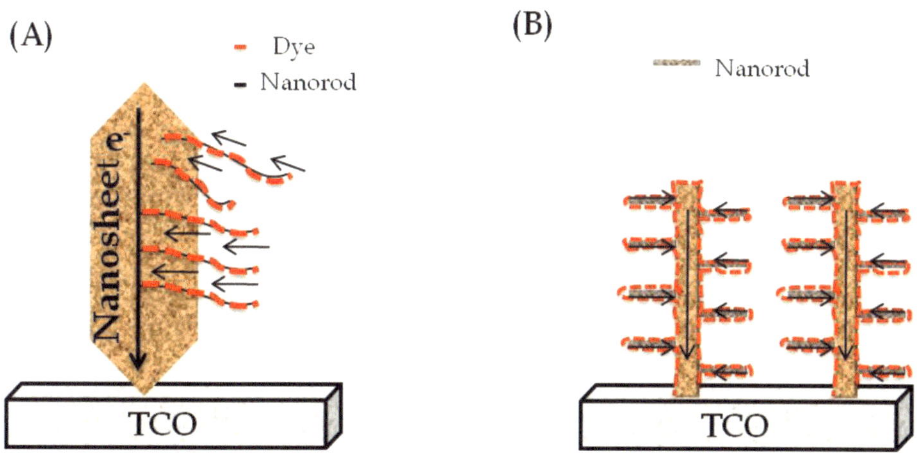

Fig. 14. Schematic illustration of different types of nanocomposites, (A) Hierarchical ZnO nanowire-nanosheet, (B) branched ZnO nanorod along with their directional electron transport. (redrawn with permission from ref. Xu et al., 2010; Cheng at al., 2008. Copyright 2010, The American Chemical Society)

Cheng et al. fabricated a branched ZnO nanowire on FTO by hydrothermal method. The fabricated branched nanowires consists perpendicular and outspread branched nanowires (Cheng et al., 2008). The phototconversion efficiency of the branched nanowires is almost twice than that of ZnO nanowire. The enhancement of photocurrent was attributed to the increase of internal surface area of the branched structure for higher dye loading without the increase of interparticle hopes. The extended nanocomposites with extended optical path length were also partly responsible for higher photoconversion efficiency due to more light scattering. Jiang et al. introduced a flower like branched ZnO nanowires, that yields higher photconversion efficiency compare to the upstanding ZnO nanowire array (Jiang et al., 2007). Subsequently, Xu et al. synthesized hierarchical ZnO nanoarchitectures, consisting of ZnO nanosheet array on indium tin oxide (ITO) glass followed by the chemical growth of ZnO nanowires on the nanosheets array (Xu et al., 2010). The photoconversion efficiency (*ca.* 4.8%) of this nanoarchitecture is more than double compare to its nanosheet counterpart photoelectrode. Table 8 shows the phoconversion parameters of some reported nanocomposites used for DSSCs.

Nanocomposites	Sensitizer	V_{oc} (mV)	J_{sc} (mAcm$^{-2)}$	FF (%)	η (%)	Ref.
ZnO nanowire array/nanoparticle	Mercurochrome	610	6.3	580	2.2	Ku & Wu, 2007
ZnO nanoflower	N719	650	5.5	53	1.9	Jiang et al., 2007
Branched ZnO Nanowires	N719	675	4.27	52.2	1.51	Cheng et al., 2008
ZnO Nanowire-Nanosheet	N719	680	10.9	65	4.8	Xu et al., 2010

Table 8. Comparison of the photoconversion performance of different ZnO nanocomposite based photoelectrode in DSSCs

3.3 Photoelectrode based on Nb_2O_5 and $SrTiO_3$

Niobium pentaoxide (Nb_2O_5) is another important type of wide band gap (E_g= ~3.49 eV) semiconductor and has been investigated for the development of photoelectrode materials for DSSCs. Nb_2O_5 is appreciably explored as composite materials with TiO_2 as a core-shell structure discussed it the section 3.1.4. The higher band position of Nb_2O_5 up to 250 mV relative to the TiO_2 makes it suitable for the formation of core-shell structure (Wei et al., 2008). Nb_2O_5 has two different crystal structures of monoclinic and orthorhombic with the space group *P2* and *Pbam,* respectively. The cell parameters of monoclinic crystal are a = 2038, b = 3.824, and c = 19.37 Å with β = 115.69^0 and for orthorhombic crystal, a = 6.168, b = 29.31, and c= 3.935 Å (Jose et al., 2009). The large unit cell dimension of Nb_2O_5 is partly responsible for the reduction of surface area. The reduced surface area of Nb_2O_5 causes the inefficient dye loading and results in the decrease of short circuit current. GuO et al. developed a single-crystal Nb_2O_5 photoelectrode for the fabrication of DSSCs (Guo et al., 1999). The low surface area of the Nb_2O_5 (25 m^{2}g^{-1}) compare to TiO_2 (104 m^{2}g^{-1}) showed relatively poor performance in DSSCs. Random movement of the injected electron into the single-crystal Nb_2O_5 might be another factor for decreased photocurrent. Therefore, the efficiency of the DSSCs (active area 0.2 cm^2) decreases from 5.0 to 2.2%. For scaling up the performance of Nb_2O_5 based DSSCs, Wei et al. synthesized 1-D Nb_2O_5 nanobelts for directional transport of electron to increase the photocurrent (Wei et al., 2008). Nevertheless, the main drawbacks of 1-D nanostructures are the drop off of surface area with decreased aspect ratios. Therefore, the decrease dye loading of the nanobelts results the poor conversion efficiency. It is noteworthy that the higher stability, higher conduction band energy of Nb_2O_5 could be makes it suitable materials for the development of high efficient DSSCs like TiO_2. Therefore, researchers should give more attention to develop the nanoporous Nb_2O_5 nanoparticles and well defined 1-D nanostructures with different nanocomposites for the fabrication of DSSCs.

$SrTiO_3$ is a binary metal-oxide with high dielectric constant (~10^4) and shares more structural similarities with anatse TiO_2. Figure 15 depicts the perovskite-type cubic structure of $SrTiO_3$ of space group *Pm3m* and lattice parameter a = 3.9. Titanium atoms are in 6-fold octahedral coordination in $SrTiO_3$, similar to the titanium arrangement in anatase TiO_2 (Lotnyk et al., 2007). The band gap of $SrTiO_3$ is ~3.2 eV and the conduction band is 0.2 eV above the conduction band of anatase TiO_2. This suitable band position of $SrTiO_3$ makes it appropriate materials for the development of DSSCs photelectrode (Burnside et al., 1999). We described the core-shell structure of TiO_2-$SrTiO_3$ in the section 3.1.4. In spite of the similar band gap of $SrTiO_3$ and anatase TiO_2, the negative positioned of the conduction band of $SrTiO_3$ leads to the negative shift of the flat band (V_{FB}) of $SrTiO_3$. Therefore, the application of pure $SrTiO_3$ as photoelectrode will increase the photovoltage of DSSC.

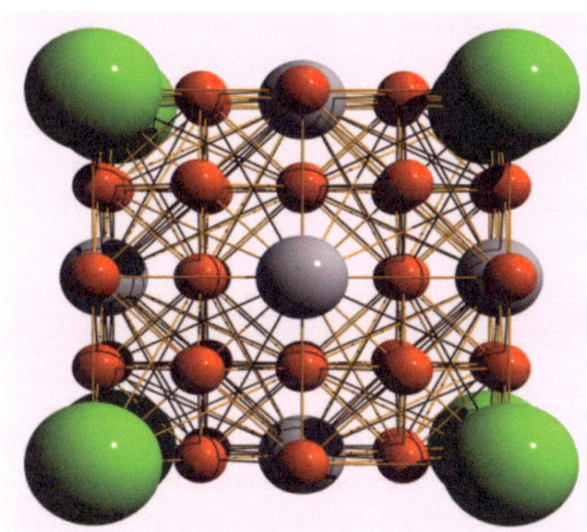

Fig. 15. Crystallographic pattern of perovskite-type cubic structure of $SrTiO_3$ (space group Pm3m), where, big green sphere: Sr^{2+}, grey sphere: Ti^{4+}, and small red sphere: O^{2-}.

Burnside et al. synthesized nanocrystalline $SrTiO_3$ by hydrothermal treatment for the application in DSSCs (Burnside et al., 1999). The prepared photoelctrode sensitized by ruthenium (2, 2'-bipyridyl-4, 4'-dicarboxylate)$_2$(NCS)$_2$) showed the expected increment of V_{oc} (789 mV) compare to TiO_2 (686 mv). However, the photoconversion efficiency is very low due to the decreased J_{sc}. This might be due to the decreased surface area of nanocrystalline $SrTiO_3$, lack of directional movement of electrons and the low dye uptake. In order to increase the performance of $SrTiO_3$ based DSSCs, research should be focused to develop new techniques for synthesizing highly porous nanoparticles of reduced size, 1-D nanostructures for the efficient injection of electrons and to retard the recombination. Development of new dyes with more negative positioned LUMO level compare to other conventional dye would be another interest of research of $SrTiO_3$ based DSSC. In this respect, Zaban and coworkers, developed a type-II DSSCs by utilizing $SrTiO_3$ coated TiO_2 for efficient electron injection by minimizing the back reaction (Hod et al., 2010). Since the conduction band of $SrTiO_3$ is more negative than TiO_2 and the LUMO of type-II sensitizer (catechol) is well above the conduction of TiO_2. Therefore, coating of TiO_2 with $SrTiO_3$ creates an energy barrier for efficient electron transfer from $SrTiO_3$ to TiO_2 and reduces the back electron transfer from oxidized dye. So an increase of J_{sc} (18%) and 70% charge collection improved the performance of DSSCs. No reports are available for the development of 1-D nanostructures of $SrTiO_3$ and its application in DSSC. Therefore, more attention should be required for the development of 1-D nanostructure of $SrTiO_3$ with high aspect ratios and exploring different dye for type-I and for type-II DSSCs.

4. Conclusion

This chapter discussed the structural aspects of various metal oxides and their composites. Interfacial electron transfer kinetics of dye/metal oxides and metal oxide/electrolyte for the conventional DSSCs were discussed. The alteration of interfacial electron transfer kinetics upon modification of the photoelectrode with metal oxide composite has also been described. In addition to the performance of different metal oxides and their composites in DSSC photoelectode, the fundamental aspects, the characterization methods, and the prospects and limitations of DSSCs were briefly explained. Various strategies for the sequential development of highly efficient DSSCs have been described and suggested nevertheless not satisfactory yet for the commercialization of DSSCs with high efficiency. The authors wish that this chapter can guide the researchers to understand the fundamental aspect and the stepwise development of different metal oxides and their composite for the photoelectrode of DSSCs.

5. Acknowledgement

This work was supported by the National Research Foundation of Korea (NRF) grant funded by the Korea government (MEST) (No. 2008-0062166) and Korea Research Foundation Grant funded by the Korean Government (MOEHRD, Basic Research Promotion Fund) (KRF-2007-211-C00029). It was also supported partly by Seoul R&BD Program (WR090671) and by the New & Renewable Energy program of the Korea Institute of Energy Technology Evaluation and Planning (KETEP) grant funded by the Korea government Ministry of Knowledge Economy.

6. References

Arakawa, H.; Yamaguchi, T.; Takeuchim, A. &Agatsuma, S. (2006). Efficiency improvement of dye-sensitized solar cell by light confined effect, *Conference Record of the 2006 IEEE 4th World Conference on Photovoltaic Energy Conversion*, pp. 36-39, 1-4244-0017-1, Waikoloa, HI,May2006, IEEE, New York.

An'amt, M.N.; Radiman, S.; Huang, N.M.; Yarmo, M.A.; Ariyanto, N.P.; Lim, H.N. & Muhamad, M.R. (2010). Sol–gel hydrothermal synthesis of bismuth–TiO_2

nanocubes for dye-sensitized solar cell. *Ceram. Int.*, 36, 7, (September2010), 2215–2220, doi:10.1016/j.ceramint.2010.05.027.

Burnside, S.; Moser, J.-E.; Brooks, K. & Grätzel, M. (1999). Nanocrystalline Mesoporous Strontium Titanate as Photoelectrode Material for Photosensitized Solar Devices: Increasing Photovoltage through Flatband Potential Engineering. *J. Phys. Chem. B*, 103, 43, (August 1999), 9328-9332, DOI: 10.1021/jp9913867.

Benkstein, K. D.; Kopidakis, N.; Van de Lagemaat, J. & Frank, A. J. (2003). Influence of the Percolation Network Geometry on Electron Transport in Dye-Sensitized Titanium Dioxide Solar Cells. *J. Phys. Chem. B*, 107, 31, (July2003) 7759–7767, DOI: 10.1021/jp022681l.

Bisquert, J.; Cahen, D.; Hodes, G.; Rühle, S. & Zaban, A. (2004). Physical Chemical Principles of Photovoltaic Conversion with Nanoparticulate, Mesoporous Dye-Sensitized Solar Cells. *J. Phys. Chem. B*, 108, 24, (May 2004), 8106-8118, DOI: 10.1021/jp0359283.

Bergeron, B. V.; Marton, A.; Oskam, G. & Meyer, G. J. (2005). Dye-Sensitized SnO_2 Electrodes with Iodide and Pseudohalide Redox Mediators. *J. Phys. Chem. B*, 109, 2, (December 2005), 937-943, DOI: 10.1021/jp0461347.

Corso, D. A.; Posternak, M.; Resta, R. & Baldereschi, A. (1994). Ab initio study of piezoelectricity and spontaneous polarization in ZnO. Phys. Rev. B, 50, 15, (June 1954), 10715-10721, DOI:10.1103/PhysRevB.50.10715.

Chen, S.G.; Chappel, S.; Diamant, Y. & Zaban, A. (2001). Preparation of Nb_2O_5 Coated TiO_2 Nanoporous Electrodes and Their Application in Dye-Sensitized Solar Cells. *Chem. Mater.*, 13, 12, (November 2001), 4629-4634, DOI: 10.1021/cm010343b.

Chappel, S.; Chen, S.-G. & Zaban, A. (2002). TiO_2-Coated Nanoporous SnO_2 Electrodes for Dye-Sensitized Solar Cells. *Langmuir*, 18, 8, (March 2002), 3336-3342, DOI: 10.1021/la015536s.

Cheng, H.-M.; Chiu, W.-H.; Lee, C.-H.; Tsai, S.-Y.& Hsieh, W.-F. (2008). Formation of Branched ZnO Nanowires from Solvothermal Method and Dye-Sensitized Solar Cells Applications. *J. Phys. Chem. C*, 112, 42, (September 2008), 16359–16364, DOI: 10.1021/jp805239k.

Chuangchote, S.; Sagawa, T. & Yoshikawa, S. (2008). Efficient dye-sensitized solar cells using electrospun TiO_2 nanofibers as a light harvesting layer. *Appl. Phys. Lett.*, 93, 3, (July2008), 033310, 0003-6951.

Diamant, Y.; Chen, S. G.; Melamed, O. & Zaban, A. (2003). Core-Shell Nanoporous Electrode for Dye Sensitized Solar Cells: the Effect of the $SrTiO_3$ Shell on the Electronic Properties of the TiO_2 Core. *J. Phys. Chem. B*, 107, 9, (February2003), 1977-1981, DOI: 10.1021/jp027827v.

EIA panel report on International energy outlook 2010 (IEO2010), July 2010, DOE/EIA-0484(2010), US department of Energy, Washington, DC 20585.

Fujishima, A. & Honda, K. (1972). Electrochemical photolysis of water at a semiconductor electrode. *Nature*, 238, (July1972), 37–38, doi: 10.1038/238037a0.

Fabregat-Santiago,F.; Bisquert,J.; Palomares,E.;Otero,L.; Kuang,D.; Zakeeruddin,S.M.; Gra1tzel,M.(2007). Correlation between Photovoltaic Performance and Impedance Spectroscopy of Dye-Sensitized Solar Cells Based on Ionic Liquids. *J. Phys. Chem.*, 111, 17, (April 2007), 6550-6560, DOI: 10.1021/jp066178a.

Feng, X.; Shankar, K.; Paulose, M. & Grimes, C. A. (2009). Tantalum-Doped Titanium Dioxide Nanowire Arrays for Dye-Sensitized Solar Cells with High Open-Circuit Voltage. *Angew. Chem. Int. Ed.*, 48, 43, 8095 –8098, DOI: 10.1002/anie.200903114.

Guo, P. & Aegerter, M. A. (1999). Ru(II) sensitized Nb_2O_5 solar cell made by the sol-gel process. *Thin Solid Films*, 351, (August 1999), 290-294, doi: 10.1016/S0040-6090(99)00215-1.

Grätzel, M. (2001). Photoelectrochemical cells. *Nature*, 414, (November2001), 338-344,_doi: 10.1038/35104607.

Grätzel, M. (2003). Dye-sensitized solar cells. *J. Photochem. Photobiol. C: Photochem. Rev.*, 4, (July 2003), 145-153, doi: 10.1016/S1389-5567(03)00026-1.

Gao, Y.; Nagai, M.; Chang, T.-C. & Shyue, J.-J. (2007). Solution-Derived ZnO Nanowire Array Film as Photoelectrode in Dye-Sensitized Solar Cells. *Crystal Growth & Design*, 7, 12, (November 2007), 2467-2471, DOI: 10.1021/cg060934k.

Grätzel, M. & Durrant, J.R. (2008). Dye-Sensitized Mesoscopic Solar Cells. In: *Nanostructured and photoelectrochemical systems for solar photon conversion*, Archer, M.D. & Nozik, A.J., 503-536, Imperial College Press Co., ISBN: 978-1-86094-255-6, London.

Gentilini, D ; D'Ercole, D.; Gagliardi, A.; Brunetti, A.; Reale, A.; Brown, T. &Di Carlo, A.(2010). Analysis and simulation of incident photon to current efficiency in dye sensitized solar cells. *Superlattices Microstruct.*, 47, 1, (January 2010), 192-196, doi:10.1016/j.spmi.2009.10.005.

Hara, K.; Sayama, K.; Ohga, Y.; Shinpo, A.; Suga ,S. & Arakawa, H. (2000). A coumarin-derivative dye sensitized nanocrystalline TiO_2 solar cell having a high solar-energy conversion efficiency up to 5.6%. *Chem. Commun.*, (March 2001), 569–570, DOI: 10.1039/b010058g.

Huber, R.; Spörlein, S.; Moser, J. E.; Grätzel, M. & Wachtveitl, J. (2000). The Role of Surface States in the Ultrafast Photoinduced Electron Transfer from Sensitizing Dye Molecules to Semiconductor Colloids. *J. Phys. Chem. B*, 104, 38, (September 2000), 8995-9003, DOI: 10.1021/jp9944381.

Hore, S.; Vetter, C.; Kern, R.; Smit, H.; Hinsch, A. (2006). Influence of scattering layers on efficiency of dye-sensitized solar cells. *Sol. Energy Mater. Sol. Cells*, 90, 9, (Mayb2006), 1176–1188, doi:10.1016/j.solmat.2005.07.002.

Hattori, R. & Goto, H. (2007). Carrier leakage blocking effect of high temperature sputtered TiO_2 film on dye-sensitized mesoporous photoelectrode. *Thin Solid Films*, 515, 20-21, (July 2007), 8045–8049, doi:10.1016/j.tsf.2007.03.079.

Hossain, M.F.; Biswas, S. & Takahashi, T. (2008). The effect of sputter-deposited TiO2 passivating layer on the performance of dye-sensitized solar cells based on sol–gel derived photoelectrode. *Thin Solid Films*, 517, 3, (December 2008), 1294–1300, doi:10.1016/j.tsf.2008.06.027.

Hod, I.; Shalom, M.; Tachan, Z., Rühle, S. & Zaban, A. (2010). $SrTiO_3$ Recombination-Inhibiting Barrier Layer for Type II Dye-Sensitized Solar Cells. *J. Phys. Chem. C*, 114, 21, (May 2010), 10015–10018, DOI: 10.1021/jp101097j.

Jung, H.S.; Lee, J.-K.; Nastasi, M.; Lee, S.-W.; Kim, J.-Y.; Park, J.-S.; Hong, K.S. & Shin, K. (2005). Preparation of Nanoporous MgO-Coated TiO_2 Nanoparticles and Their Application to the Electrode of Dye-Sensitized Solar Cells. *Langmuir*, 21, 23, (October 2005), 10332-10335, DOI: 10.1021/la051807d.

Jiang, C. Y.; Sun, X. W.; Lo, G. Q. & Kwong, D. L. (2007). Improved dye-sensitized solar cells with a ZnO-nanoflower photoanode. *Appl. Phys. Lett.*, 90, 26, (June 2007), 263501, doi:10.1063/1.2751588.

Jose, R.; Thavasi, V. & Ramakrishna, S. (2009). Metal Oxides for Dye-Sensitized Solar Cells. *J. Am. Ceram. Soc.*, 92, 2, (February 2009), 289-301, DOI: 10.1111/j.1551-2916.2008.02870.x.

Ko, K.H.; Lee, Y.C. & Jung, Y.J. (2005). Enhanced efficiency of dye-sensitized TiO_2 solar cells (DSSC) by doping of metal ions. *J. Colloid Interface Sci.*, 283, 2, (March 2005), 482–487, doi:10.1016/j.jcis.2004.09.009.

Ku, C.-H. & Wu, J.-J. (2007). Electron transport properties in ZnO nanowire array/nanoparticle composite dye-sensitized solar cells. *Appl. Phys. Lett.*, 91, 9, (August 2007), 093117, doi:10.1063/1.2778454.

Kang, S.H.; Choi, S.-H.; Kang, M.-S.; Kim, J.-Y.; Kim, H.-S.; Hyeon, T. & Sung, Y.-E. (2008). Nanorod-Based Dye-Sensitized Solar Cells with Improved Charge Collection Efficiency. *Adv. Mater.*, 20, 1, (January 2008), 54–58, DOI: 10.1002/adma.200701819.

Kim, Y.J.; Lee, Y.H.; Lee, M.H.; Kim, H.J.; Pan, J.H.; Lim, G.I.; Choi, Y.S.; Kim, K.; Park, N.-G.; Lee, C. & Lee, W.I. (2008). Formation of Efficient Dye-Sensitized Solar Cells by Introducing an Interfacial Layer of Long-Range Ordered Mesoporous TiO_2 Thin Film. *Langmuir*, 24, 22, (October 2008), 13225-13230, DOI: 10.1021/la802340g.

Li, B.; Wang, L.; Kang, B.; Wang, P. & Qiu, Y. (2006). Review of recent progress in solid-state dye-sensitized solar cells. *Sol. Energy Mater. Sol. Cells*, 90, 5, (March 2006), 549–573, doi:10.1016/j.solmat.2005.04.039.

Lotnyk, A.; Senz, S. & Hesse, D. (2007). Epitaxial growth of TiO_2 thin films on $SrTiO_3$, $LaAlO_3$ and yttria-stabilized zirconia substrates by electron beam evaporation. *Thin Solid Films*, 515, (December 2006), 3439-3447, doi:10.1016/j.tsf.2006.10.106.

Li, J.-G.; Ishigaki, T. & Sun, X. (2007). Anatase, Brookite, and Rutile Nanocrystals via Redox Reactions under Mild Hydrothermal Conditions: Phase-Selective Synthesis and Physicochemical Properties. *J. Phys. Chem. C*, 111, 13, (March 2007), 4969-4976, DOI: 10.1021/jp0673258.

Lee, S.; Kim, J.Y.; Youn, S.H.; Park, M.; Hong, K.S.; Jung, H.S.; Lee, J.-K. & Shin, H. (2007). Preparation of a Nanoporous $CaCO_3$-Coated TiO_2 Electrode and Its Application to a Dye-Sensitized Solar Cell. *Langmuir*, 23, 23, (October 2007), 11907-11910, DOI: 10.1021/la701826v.

Lee, S.; Noh, J.H.; Han, H.S.; Yim, D.K.; Kim, D.H.; Lee, J.-K.; Kim, J.Y.; Jung, H.S. & Hong, K.S. (2009). Nb-Doped TiO_2: A New Compact Layer Material for TiO_2 Dye-Sensitized Solar Cells. *J. Phys. Chem. C*, 113, 16, (March 2009), 6878-6882,_DOI: 10.1021/jp9002017.

Lü, X.; Mou, X.; Wu, J.; Zhang, D.; Zhang, L.; Huang, F.; Xu, F. & Huang, S. (2010). Improved-Performance Dye-Sensitized Solar Cells Using Nb-Doped TiO_2 Electrodes: Efficient Electron Injection and Transfer. *Adv. Funct. Mater.*, 20, 3, (February 2010), 509–515, DOI: 10.1002/adfm.200901292.

Martinson, A. B. F.; Go´es, M. S.; Fabregat-Santiago, F.; Bisquert, J.; Pellin, M. J. &Hupp, J. T. (2009). Electron Transport in Dye-Sensitized Solar Cells Based on ZnO Nanotubes: Evidence for Highly Efficient Charge Collection and Exceptionally Rapid Dynamics. *J. Phys. Chem. A*, 113, 16, (February 2009), 4015–4021, DOI: 10.1021/jp810406q

Marco, L.D.; Manca, M.; Giannuzzi, R.; Malara, F.; Melcarne, G.; Ciccarella, G.; Zama, I.; Cingolani, R. & Gigli, G. (2010). Novel Preparation Method of TiO_2-Nanorod-Based Photoelectrodes for Dye-Sensitized Solar Cells with Improved Light-Harvesting Efficiency. *J. Phys. Chem. C*, 114, 9, (February 2010), 4228–4236, DOI: 10.1021/jp910346d.

Nusbaumer, H.; Moser, J. E.; Zakeeruddin, S. M.; Nazeeruddin, K. & Grätzel, M. (2001). $Co^{II}(dbbip)_2{}^{2+}$ Complex Rivals Tri-iodide/Iodide Redox Mediator in Dye-Sensitized Photovoltaic Cells. *J. Phys. Chem. B*, 105, 43, (October 2001), 10461-10464, DOI: 10.1021/jp012075a.

O'Regan, B. & Grätzel, M. (1991). A low-cost, high efficiency solar cell based on dye-sensitized colloidal TiO_2 films. *Nature,* 353, (October 1991), 737–740, doi: 10.1038/353737a0.

Pang, S.; Xie, T.; Zhang, Y.; Wei, X.; Yang, M. & Wang, D. (2007). Research on the Effect of Different Sizes of ZnO Nanorods on the Efficiency of TiO_2-Based Dye-Sensitized Solar Cells. *J. Phys. Chem. C,* 111, 49, (November 2007), 18417-18422, DOI: 10.1021/jp0763073.

Quintana, M.; Edvinsson, T.; Hagfeldt, A. & Boschloo, G. (2007). Comparison of Dye-Sensitized ZnO and TiO_2 Solar Cells: Studies of Charge Transport and Carrier Lifetime. *J. Phys. Chem. C,* 111, 2, (December 2006), 1035-1041, DOI: 10.1021/jp065948f.

Qian, J.; Liu, P.; Xiao, Y.; Jiang, Y.; Cao, Y.; Ai, X. & Yang, H. (2009). TiO_2-Coated Multilayered SnO_2 Hollow Microspheres for Dye-Sensitized Solar Cells. *Adv. Mater.*, 21, 36, (September 2009), 3663–3667, DOI: 10.1002/adma.200900525.

Redmond, G.; Fitzmaurice, D. & Gräetzel, M. (1994). Visible Light Sensitization by cis- Bis (thioc yanato) bis (2, 2'- bipyridyl-4, 4'-dicarboxylato) ruthenium (II) of a Transparent Nanocrystalline ZnO Film Prepared by Sol-Gel Techniques. *Chem. Mater.*, 6, 5, (May 1994), 686-691, DOI: 10.1021/cm00041a020.

Rengaraj, S.; Li, X.Z.; Tanner, P.A.; Pan, Z.F.; Pang, G.K.H. (2005). Photocatalytic degradation of methylparathion—An endocrine disruptor by Bi^{3+}-doped TiO_2. *J. Mol. Catal. A: Chem.,* 247, 1-2, (November 2005), 36–43, doi:10.1016/j.molcata.2005.11.030.

Rahman, M. M.; Ahammad, A. J. S.; Jin, J.-H.; Ahn, S. J. & Lee, J.-J. (2010). A Comprehensive Review of Glucose Biosensors Based on Nanostructured Metal-Oxides. *Sensors,* 10, 5, (May 2010), 4855-4886, doi:10.3390/s100504855.

Solbrand, A.; Keis, K.; Södergren, S.; Lindström, H.; Lindquist, S. E. & Hagfeldt, A. (2000). Charge transport properties in the nanostructured ZnO thin film electrode - electrolyte system studied with time resolved photocurrents. *Sol. Energy Mater. Sol. Cells,* 60, 2, (January 2000), 181-193, doi: 10.1016/S0927-0248(99)00083-5.

Sapp, S.A.; Elliott, M.; Contado, C.; Caramori, S. & Bignozzi, C.A. (2002). Substituted Polypyridine Complexes of Cobalt(II/III) as Efficient Electron-Transfer Mediators in Dye-Sensitized Solar Cells. *J. Am. Chem. Soc.*, 124, 37, (August 2002), 11215-11222, DOI: 10.1021/ja027355y.

Shin, K.; Jun, Y.; Moon, J.H. &Park, J.H. (2010). Observation of Positive Effects of Freestanding Scattering Film in Dye-Sensitized Solar Cells. *ACS Appl. Mater. Interfaces.*, 2, 1, (December 2009), 288-291, DOI: 10.1021/am900691c.

Tae, E. L.; Lee, S. H.; Lee, J. K.; Yoo, S. S.; Kang, E. J. & Yoon, K. B. (2005). A Strategy To Increase the Efficiency of the Dye-Sensitized TiO_2 Solar Cells Operated by Photoexcitation of Dye-to-TiO_2 Charge-Transfer Bands. *J. Phys. Chem. B,* 109, 47, (November 20005), 22513-22522, DOI: 10.1021/jp0537411.

Thavasi, V.; Renugopalakrishnan, V.; Jose, R. & Ramakrishna, S. (2009). Controlled electron injection and transport at materials interfaces in dye sensitized solar cells. *Mat. Sci. Eng. R,* 63, 3, (January 2009), 81–99, doi:10.1016/j.mser.2008.09.001

Umar, A. (2009). Growth of Comb-like ZnO Nanostructures for Dye-sensitized Solar Cells Applications. *Nanoscale Res Lett.,* 4, 9, (May 2009), 1004–1008, DOI: 10.1007/s11671-009-9353-3.

Umar, A.; Al-Hajry, A.; Hahn, Y.B. & Kim, D.H. (2009). Rapid synthesis and dye-sensitized solar cell applications of hexagonal-shaped ZnO nanorods. *Electrochim. Acta,* 54, 23, (September 2009) 5358–5362, doi:10.1016/j.electacta.2009.04.015.

Wang, Z.-S.; Huang, C.-H.; Huang, Y.-Y.; Hou, Y.-J.; Xie, P.-H.; Zhang, B.-W. & Cheng, H.-M. (2001). A Highly Efficient Solar Cell Made from a Dye-Modified ZnO-Covered TiO_2 Nanoporous Electrode. *Chem. Mater.,* 13, 2, (January 2001), 678-682, DOI: 10.1021/cm000230c.

Wang, Y.; Hang, K.; Anderson, N. A.; Lian, T. (2003). Comparison of Electron Transfer Dynamics in Molecule-to-Nanoparticle and Intramolecular Charge Transfer Complexes. *J. Phys. Chem. B,* 107, 35, (July 2003), 9434-9440, DOI: 10.1021/jp034935o.

Wang, P; Zakeeruddin, S.K.; Moser, J-E.; Humphry-Baker, R. & Grätzel, M. (2004). A Solvent-Free, SeCN-/(SeCN)$^{3-}$ Based Ionic Liquid Electrolyte for High-Efficiency Dye-Sensitized Nanocrystalline Solar Cells. *J. Am. Chem. Soc.,* 126, 23, (May 2004), 7164-7165, DOI: 10.1021/ja048472r.

Wang, P.; Dai, Q.; Zakeeruddin, S.M.; Forsyth, M.; MacFarlane, D.R. & Grätzel, M. (2004). Ambient Temperature Plastic Crystal Electrolyte for Efficient, All-Solid-State Dye-Sensitized Solar Cell. *J. Am. Chem. Soc.,* 126, 42, (October 2004), 13590-13591, DOI: 10.1021/ja045013h.

Wang, H.; Li, H.; Xue, B.; Wang, Z.; Meng, Q. & Chen, L. (2005). Solid-State Composite Electrolyte LiI/3-Hydroxypropionitrile/SiO_2 for Dye-Sensitized Solar Cells. *J. Am. Chem. Soc.,* 127, 17, (April 2005), 6394-6401, DOI: 10.1021/ja043268p.

Wei, M.; Qi, Z.-M.; Ichihara, M. & Zhou, H. (2008). Synthesis of single-crystal niobium pentoxide nanobelts. *Acta Materialia,* 56, 11, (June 2008), 2488–2494, doi:10.1016/j.actamat.2008.01.049.

Wu, J.; Lan, Z.; Hao, S.; Li, P.; Lin, J.; Huang, M.; Fang, L. & Huang, Y. (2008). Progress on the electrolytes for dye-sensitized solar cells. *Pure Appl. Chem.,* 80, 11, (November 2008), 2241–2258, doi: 10.1351/pac200880112241.

Wang, H.; Liu, Y.; Li, M.; Huang, H.; Zhong, M. & Shen, H. (2009). Hydrothermal growth of large-scale macroporous TiO_2 nanowires and its application in 3D dye-sensitized solar cells. *Appl. Phys. A,* 97, 1, (August 2009), 25–29, DOI: 10.1007/s00339-009-5369-x.

Wang, J.X.; Wu, C. M. L. ; Cheung, W. S.; Luo, L. B.; He, Z. B.; Yuan, G. D.; Zhang, W. J.; Lee, C. S. & Lee, S. T. (2010). Synthesis of Hierarchical Porous ZnO Disklike Nanostructures for Improved Photovoltaic Properties of Dye-Sensitized Solar Cells. *J. Phys. Chem. C,* 114, 31, (July 2010), 13157–13161, DOI: 10.1021/jp100637c.

Xu, F.; Dai, M.; Lu, Y. & Sun, L. (2010). Hierarchical ZnO Nanowire-Nanosheet Architectures for High Power Conversion Efficiency in Dye-Sensitized Solar Cells. *J. Phys. Chem. C,* 114, 6, (January 2010), 2776–2782, DOI: 10.1021/jp910363w.

Yum, J.-H.; Nakade, S.; Kim, D.-Y. & Yanagida, S. (2006). Improved Performance in Dye-Sensitized Solar Cells Employing TiO_2 Photoelectrodes Coated with Metal Hydroxides. *J. Phys. Chem. B,* 110, 7, (February 2006), 3215-3219,_DOI: 10.1021/jp0564593.

Yao, Q.; Liu, J.; Peng, Q.; Wang, X. & Li, Y. (2006). Nd-Doped TiO_2 Nanorods: Preparation and Application in Dye-Sensitized Solar Cells. *Chem. Asian J.*, 1, 5, 737 - 741, DOI: 10.1002/asia.200600195.

Yu, H.; Zhang, S.; Zhao, H.; Will, G. & Liu, P. (2009). An efficient and low-cost TiO_2 compact layer for performance improvement of dye-sensitized solar cells. *Electrochim. Acta,* 54, 4, (January 2009), 1319–1324, doi:10.1016/j.electacta.2008.09.025.

Yi, Z.M.; Hai, W.; Yong, L.; Ming, L.; Hong, H. & Hui, S. (2010). ZnO microsheet modified TiO_2 nanoparticle composite films for dye-sensitized solar cells. *Chinese Sci. Bul.*, 55, 18, (June 2010), 1945-1948, DOI: 10.1007/s11434-009-3688-0.

Zhu, K.; Neale, N. R.; Miedaner, A. & Frank, A. J. (2006). Enhanced Charge-Collection Efficiencies and Light Scattering in Dye-Sensitized Solar Cells Using Oriented TiO_2 Nanotubes Arrays. *Nano Lett.,* 7, 1, (December 2006), 69-74, DOI: 10.1021/nl062000o.

Zhu, K.; Vinzant, T. B.; Neale, N. R. & Frank, A. J. (2007). Removing Structural Disorder from Oriented TiO_2 Nanotube Arrays: Reducing the Dimensionality of Transport and Recombination in Dye-Sensitized Solar Cells. *Nano Lett.*, 7, 12, (November 2007), 3739-3746, DOI: 10.1021/nl072145a.

Zhang, W.; Zhu, R.; Liu, X.; Liu, B. & Ramakrishna, S. (2009). Facile construction of nanofibrous ZnO photoelectrode for dye-sensitized solar cell applications, *Appl. Phys. Lett.*, 95, 4, (July 2009), 043304, doi:10.1063/1.3193661.

8

Nanocomposites for Photovoltaic Energy Conversion

Xianwu Zeng and Yong X. Gan
Department of Mechanical, Industrial and Manufacturing Engineering College of Engineering, University of Toledo, Toledo, OH 43606, USA

1. Induction

1.1 Composite materials

A composite is a multiphase solid material. It is incorporated by two or more individual materials through physical or chemical methods. The performance of different materials to complement each other will produce synergistic effects (Ivanov et al., 2008; Lu et al., 2010). The overall property of composite materials is better than that of each original material to meet different requirements. These properties include mechanical, electrical, thermal, optical, electrochemical, catalytic behaviors etc. For example (Lu et al., 2010), investigating the synergistic effect of carbon nano-fiber (CNF) and carbon nano-paper on the shape recovery of shape memory polymer (SMP) composite shows that the combination of CNF and carbon nano-paper can improve the thermal and electrical conductivities of the SMP composite, which are much better than each of the components. The individual materials in a composite are referred to as two constituent materials. The two constituent materials are matrix and reinforcement. The matrix material is a frame to support the reinforcement material. So the reinforcement material can keep its relative position. The reinforcement is a functional material which can enhance the matrix properties. Composite metal matrix (substrates) includes aluminum, copper, titanium, magnesium and its alloys and non-metallic matrix includes synthesized resin, rubber, ceramics, graphite and carbon and so on. Reinforcement mainly includes glass fiber, carbon fiber, boron fiber , aramid fiber, silicon carbide fiber , asbestos fiber, whisker, metal wire and fine grain etc. In order to make use of synergistic effect to improve composite properties, optimum combination of matrix and reinforcement should be chosen.

1.2 Nanocomposites and their properties

A nanocomposite is a special composite where one of the phases has one, two or three dimensions less than 100 nanometers (nm, 10^{-9} m). A nanocomposite also includes the material where the structures between the different phases that make up the material are in nano-scale (Beecroft & Ober, 1997). In the broadest sense, nanocomposites can also include porous media, colloids, gels and polymers, because in these materials the particles or structures are in nano scale. One nanometer is equivalent to the length to tightly line up

10~100 atoms. Nano-materials include nano-powders, nano-fibers, nano-particles and nano-thin films. Their structures are between atom (molecular) size and macro size. Materials in nano-scale have special effects such as quantum size effect, surface effect, small-size effect and macroscopic quantum tunneling effect etc.

Quantum size effect

Quantum size effect is one of the fundamental physical properties of nano-particles. When the particle size decreases to a nano-scale, the electronic energy levels of the particle atom become discrete and the energy gap becomes wider. This phenomenon is called quantum size effect. This effect does not come into play when the particles are in micro dimensions. However, quantum effects become dominant when the particles sizes decrease to nanometer, especially when the dimensions are in 100 nanometers or less. So the nanometer size dimension is also called quantum realm. The quantum size effect leads the nano-materials have very different magnetic, optical, acoustic, thermal, electrical and superconducting properties with conventional materials, such as very high non-linear optics, optical catalysis, high oxidability and reducibility etc. Due to quantum size effect, the galvanomagnetic and thermoelectric properties (electrical conductivity s, the Hall coefficient RH, charge carrier mobility m, and the Seebeck coefficient S) of nano-size PbSe thin film oscillate with the thin film thickness (3-200 nm) (Rogacheva et al., 2002). Silica glasses are doped with CdS and ZnS micro-crystals. Because of the quantum size effect, the optical absorption spectra and the photoluminescence spectra show that the optical absorption edges and the photoluminescence peaks shift to the higher energy side with decreasing the size of the micro-crystals (Zhao et al., 1994).

Surface effect

For a spherical particle, the ratio of surface area and volume is inversely proportional to the particle diameter. When particle size decreases to nano-scale, more atoms appear on surface relatively. For example, when particle diameter is 10 nm, the surface atom is about 20% of the total atoms; the diameter of 1 nm, the percentage of surface atoms increases to 99%. Almost all of the atoms are in surface. The particle surface becomes large relative the particle volume and the surface atoms are easy to combine with each other. This structure shows a high chemical activity. So, for the nano-materials, surface energy and surface binding energy are rapidly increasing compared to bulk particles. This change is called surface effect. For example, Au particles with dimension about 2 nm have no stable states; they change from octahedron, decahedron to icosahedron multicrystals. They are not liquid, not solid and can be called quasi-solid.

Small-size effect

When the particle size is close to or smaller than the light wavelength, the particle's de Broglie wavelength, semiconductor coherence length and the particle's penetration depth, the particle has new properties such as optical, thermal, acoustic, electrical, magnetic and mechanical properties etc. This phenomenon is the small size effect. For example, when the metal particle size is small than the light wavelength, the metal's color will become black. This means nano metal particles have very good light absorption and can be optothermal, photoelectrical materials. The melt temperature of bulk gold is 1064^{0}C, but 2 nm size gold particles melt temperature is only 327^{0}C. Nano-scale metal particles can be insulator.

Macroscopic quantum tunneling effect

Quantum tunneling refers to the phenomena of a particle's ability to penetrate energy barriers within electronic structures. Scanning tunneling microscope (STM) and Atomic force microscopy (AFM) are based on macroscopic tunneling effect.

1.3 Nanocomposite classification

Based on the matrix materials, the nanocomposites can be classified as ceramic-matrix nanocomposites, metal-matrix nanocomposites and polymer-matrix nanocomposites etc.

Ceramic-matrix nanocomposites

Ceramic materials such as oxides, nitrides, borides and silicides all can be the matrix of nanocomposites. Oxide ceramics include Al_2O_3, TiO_2, ZnO, MgO $\cdot Al_2O_3$, $3Al_2O_3 \cdot 2SiO_2$, PZT etc. Nitride ceramics include Si_3N_4, BN, AlN. Boride ceramics include CrB_2, ZrB_2 and TiB_2. Silicide ceramics include $MoSi_2$ etc. The reinforcement materials of ceramic-matrix nanocomposites are metals such as Fe, Al, Cu, Zn etc. Ceramic-matrix nanocomposites have very good optical, electrical and magnetic properties. Normally, preparing ceramic nanocomposite needs very high temperature. Some of the metallic components may easily react with the ceramic and thereby loses its metallic character. So, it should be careful in choosing immiscible metal and ceramic phases. There are also bi-phase ceramic/ceramic nanocomposites (Sternitzke, 1997). For example, low temperature nanocomposites Al_2O_3/SiO_2, SiO_2/MgO, Al_2O_3/TiO_2, AIN/BN; structure nanocomposites Al_2O_3/SiC, Si_3N_4/SiC, MgO/SiC etc.

Metal-matrix nanocomposites

Sometimes, or under special conditions, a metal can be the matrix of a nanocomposite. For example, Cu/Al_2O_3 metal matrix composite materials have both high strength and high conductivity. They can be used as electrodes and contact terminals. Here Cu is the matrix, and Al_2O_3 is the reinforcement. It has been identified that the Al_2O_3 particles in the consolidated composite material have the size smaller than 200 nm in diameter (Ying & Zhang, 2000).

Polymer-matrix nanocomposites

Polymers as the matrices of nanocomposites can enhance their performance, often in very dramatic degree. Nano-materials are the fillers, so, these materials are better described by the term nano-filled polymer composites. The fillers can be metals, metal oxides, polymers etc. This new class of composite materials has shown enhanced optical, electrical and dielectric properties. Polymer–inorganic nanocomposite materials are defined as inorganic nanofillers dispersed at a nanometer level in a polymer matrix. S.C. Tjong (Tjong, 2006) has reviewed polymer nanocomposites reinforced with respective layered silicates, ceramic nanoparticles and carbon nanotubes. The properties of nanocomposites depend not only on the properties of polymer matrices and nature of nanofillers but also on the way in which they are prepared. In order to achieve desired mechanical and physical characteristics, the nanofillers need disperse uniformly in the polymer matrices. Changli Lu et al (Lu et al, 2003) made ZnS–polymer nanocomposite films through chemical way. The ZnS nanoparticles were uniformly dispersed in the polymer matrix and the particles remained their original size 2.0–5.0 nm in diameter with a cubic phase structures. The resulting nanocomposite films have good thermal stability and exhibit excellent optical transparency in the visible region and higher refractive indices.

Nanocomposite materials are widely used in solar energy conversion. The application of solar cell is one of the most promising technologies in the last decade. Solar energy is clean and renewable. Converting solar energy into electricity provides a much-needed solution to the energy crisis the world is facing today. Solar cells convert sunlight into electricity based on photovoltaic effect. Even though the first generation and the second generation solar cells have become commercial products successfully, much research needs to continue to improve the energy conversion efficiency and reduce the cost of the products. Some of these efforts have produced materials that have shown very unique and important properties in contrast to traditional materials.

2. Sunlight and photovoltaic energy conversion

2.1 Sunlight and spectra

Almost all of the energy that drives the various systems (climate systems, ecosystems, hydrologic systems, etc.) found on the Earth originates from the sun. Solar energy is green power and renewable energy. Solar energy is abundant. Every day the sun radiates, or sends out, an enormous amount of energy to the earth, but only a little can be used. A sunlight or sunshine means the light radiates from the Sun and reaches to the earth. The sunlight contains energy. Based on quanta theory, the particle of light was given the name photon. A photon has an energy, E, proportional to its frequency v. This relation between the energy and frequency is called Planck relation or the Planck-Einstein equation:

$$E=hv \tag{2-1}$$

Since the frequency v, wavelength λ, and speed of light c are related by $\lambda v = c$, the Planck relation can also be expressed as

$$E=\frac{hc}{\lambda} \tag{2-2}$$

Where h is Planck's constant ($h= 6.626068\times10^{-34}\ m^2\ kg/s$), c is the speed of light ($c= 299{,}792{,}458\ m/s$). Because of the reflection of the aerosphere, about 70% energy within the light arrives at the earth and is absorbed by clouds, oceans and land masses. The sunlight is a kind of electro-magnetic wave or electromagnetic radiation. The spectrum of solar light is continuous. According to increasing order of wavelengths, the lights are ultraviolet, visible and infrared parts. The wavelengths are as follows:

Ultraviolet radiation (UV): 100-400 nm (not visible).

Visible light: 400-700 nm.

Infrared radiation (IR): 700 nm-1 mm (not visible)

On the contrary, the frequency decreases from ultraviolet, visible and infrared parts.
Figure 2.1 shows the solar radiation spectrum.
This figure shows the solar radiation spectrum for direct light at both the top of the Earth's atmosphere and at sea level. The Sun is like a blackbody and the surface temperature is about 5525 K (5250°C). The red part in the figure is the energy absorbed on sea level. Based Planck-Einstein theory, high frequency photon contains high energy. But in the sunlight the most parts are visible light and infrared radiation, only small part is ultraviolet light. Figure 2.1 shows that most of the energy is in visible and infrared spectrums. The solar energy

distribution according to wavelengths is visible light about 40%, infrared radiation about 50% and ultraviolet light about 10%.

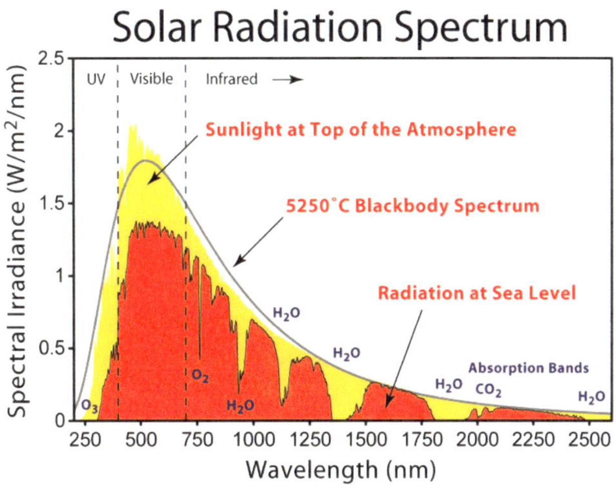

Fig. 2.1. Solar irradiance spectrum above atmosphere and at surface
Image created by Robert A. Rohde / Global Warming Art
(http://www.globalwarmingart.com/wiki/File:Solar_Spectrum_png)

2.2 Air mass

When the sunlight passes through the atmosphere, some of light will be absorbed by the air and dust. This means part of the light power will be lost. The Air Mass quantifies the power reduction when the light passes through the atmosphere. It is defined as:

$$AM=\frac{1}{\cos(\theta)} \tag{2-3}$$

where θ is the angle from the vertical direction (Figure 2.2). When the sun is directly overhead, the Air Mass is 1 (θ=90°).

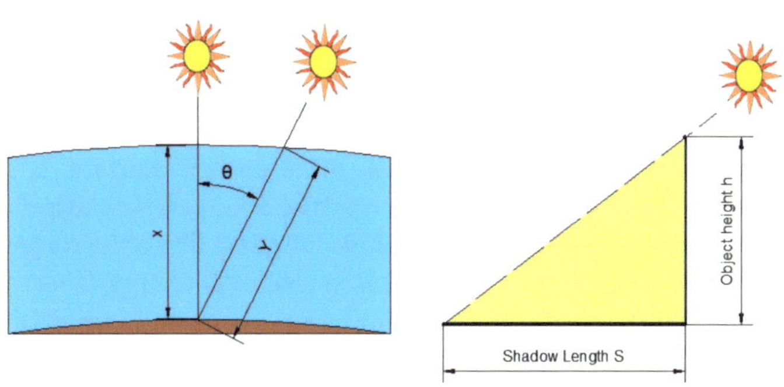

Fig. 2.2. AM measurement

The measurement of AM is in Figure 2.2. Here h is the object height, s is the shadow length. From the geometry, the AM can be calculated:

$$AM=\frac{1}{\cos\theta}=\sqrt{1+\left(\frac{s}{h}\right)^2} \tag{2-4}$$

The above formula is an ideal condition that assumes the atmosphere is a flat horizontal layer, but the Earth is a sphere and the atmosphere is like sphere shell. An equation which incorporates the curvature of the earth is:

$$AM=\frac{1}{\cos\theta+0.50572(96.07995-\theta)^{-1.6364}} \tag{2-5}$$

2.3 Standardised solar spectrum and solar irradiation

The power and the spectrum of the incident light can affect the efficiency of a solar cell. For measuring the solar cell efficiency, a standard spectrum and power density has been defined for both radiations outside the Earth's atmosphere and at the Earth's surface. AM0 is the standard spectrum outside the Earth's atmosphere, it is used for space cell. AM1.5G and AM1.5D are the standard spectrums at the Earth's surface. AM1.5G is for global and includes both direct and diffuse radiation. AM1.5D only includes direct radiation. Due to absorption (18%) and to scattering (10%) the intensity of AM1.5D radiation can be approximated by reducing the AM0 spectrum by 28%. The global spectrum is 10% higher than the direct spectrum. For the convenience of calculation, AM1.5G spectrum has been normalized to $1kW/m^2$.

2.4 Photovoltaic energy conversion

The "photovoltaic effect" is the basic physical process through which a solar cell converts solar energy into electricity. Solar cell is actually a semiconductor diode. Sunlight is composed of photons, or "packets" of energy. These photons contain various amounts of energy corresponding to the different wavelengths of light. When photons strike a solar cell, they may be reflected or absorbed. When a photon is absorbed, the energy of the photon is transferred to thermal energy or electrical energy. The thermal energy can heat up the solar cell to a very high temperature and damage the solar cell. The absorbed energy can excite an electron in an atom of the cell. With its newfound energy, the electron is able to escape from its normal position associated with that atom to become part of the current in an electrical circuit. The absorbed energy of a photon is changed to thermal energy or electrical energy is related to the band-gap of the semiconductor material of the PV cell. A photon with energy equal to or greater than the band-gap of the materials is able to excite one electron when absorbed. If a photon with energy less than the band-gap all its energy will become thermal energy when absorbed. For the photons with more energy than the band-gap, the excess energy above the band-gap will also become thermal energy when absorbed. Because of this constraint, the efficiency of solar cell is very low. A good designed solar cell tries to absorb all the sunlight and change all the solar energy to electrical energy. This means the solar cell materials can be composites with different band-gap according to different sunlight photons energy.

Figure 2.3 shows photovoltaic current which is produced from semiconductor diodes. The diode connected with the current source represents the solar cell. I is the output current, and Ip is the photocurrent. Figure 2.3(a) shows the simplest solar cell model, where a diode and a current source are connected in parallel. It is noted that the source current (photocurrent) is directly proportional to the solar radiation. The diode represents the p-n junction of a solar cell. The current (I) passes through the load is the difference between the photocurrent (Ip) and the current through the diode (I_D), and can be expressed by:

$$I=I_p-I_D=I_p-I_S\left[\exp\left(\frac{V}{mV_T}\right)-1\right] \quad (2\text{-}6)$$

In formula (2-6), m is the diode ideality factor, in the simplest cell m is represented by the ideal diode model, m=1. I_S is the reverse saturation current of a diode. I_S is in the order of 10^{-8} A/m^2. V_T is the thermal voltage of a diode. The relationship of the V_T and the temperature of the solar cell (T) can be shown in formula (2-7):

$$V_T=\frac{kT}{q} \quad (2\text{-}7)$$

where κ is the Boltzmann constant, and $k=1.38\times10^{-23}\left(\frac{J}{K}\right)$. q is the charge of electron, and $q=1.6\times10^{-19}(C)$.

Figure 2.3(b) shows a real solar cell circuit. There are serial (Rs) and parallel or shunt (R_{SH}) resistance in the equivalent circuit. The output current is

$$I=I_p-I_D-I_{SH}=I_p-I_S\left[\exp\left(\frac{V}{mV_T}\right)-1\right]-\frac{V+IRs}{R_{SH}} \quad (2\text{-}8)$$

where I_{SH} is the current passing through the shunt resistance and R_{SH} is the shunt resistance. Formula (2-8) is a transcendental equation, there is no analytical solution. But for an operating voltage V, the output current I can be determined numerically. In addition, the open circuit voltage, V_{OC}, can be calculated by setting I=0, i.e.

$$VOC\approx\frac{kT}{q}\ln\left(\frac{I_P}{I_S}-1\right) \quad (2\text{-}9)$$

In figure 2.3, short circuit current (I_{sc}) is the current when the terminals are connected to each other (zero load resistance). For a high-quality solar cell (low R_S, and high R_{SH}) the short-circuit current I_{SC} is

$$I_{SC}\approx I_p$$

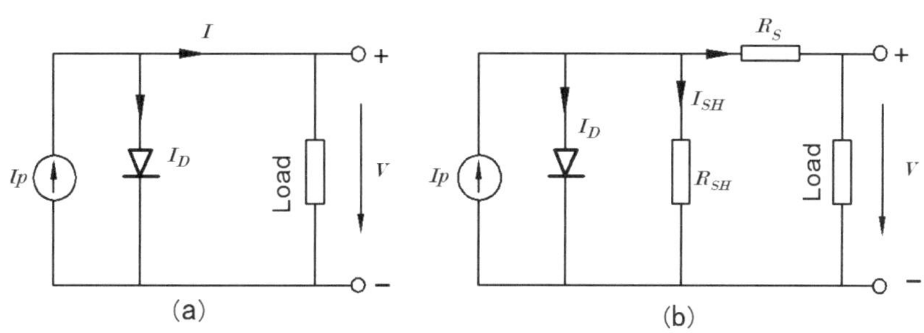

Fig. 2.3. The equivalent circuits of solar cell (a) Ideal model; (b) Practical model.

Current density

The values of I_S, R_S, and R_{SH} are dependent upon the physical size of the solar cell. In order to specify the properties of solar cell clearly, current density is introduced. It is the current produced per unit cell area:

$$J=J_p-J_S\left[\exp\left(\frac{V}{mV_T}\right)-1\right]-\frac{V+Jr_S}{r_{SH}} \quad (2\text{-}10)$$

where
J = current density (amperes/cm^2)
J_P = photogenerated current density (amperes/cm^2)
J_S = reverse saturation current density (amperes/cm^2)
r_S = specific series resistance (Ω-cm^2)
r_{SH} = specific shunt resistance (Ω-cm^2)

Efficiency η:

The power density is the product of voltage and current density.

$$P=J\,V \quad (2\text{-}11)$$

The maximum power density occurs somewhere between $V = 0$ (short circuit) and $V = V_{oc}$ (open circuit) at a voltage V_m. The corresponding current density is J_m, and thus the maximum power density is $P_{d,m} = J_m V_m$. (Figure 2.4).
The efficiency of a solar cell is defined as the input power divided by the output power. If the incoming light has a power Ps, the efficiency will be:

$$\eta=\frac{P_{d,m}}{P_s} \quad (2\text{-}12)$$

FF is the solar cell fill factor.
It is defined as

$$FF=\frac{J_m V_m}{J_{SC} V_{OC}} \quad (2\text{-}13)$$

It gives a measure of how much of the open circuit voltage and short circuit current is "utilized" at maximum power (Figure 2.4). Using FF we can express the efficiency as:

$$\eta=\frac{J_{SC} V_{OC} FF}{P_S} \quad (2\text{-}14)$$

The four quantities J_{sc} (short circuit current density), V_{oc}, FF and η are frequently used to characterize the performance of a solar cell. They are often measured under standard lighting conditions, which implies Air Mass 1.5 spectrum, light flux of 1000W/m^2 and temperature of 25°C.
If shadows and defects present, they cause power loss because these areas become loads instead of generating power. The DC power may be converted to AC power by an inverter. The conventional energy causes pollution and greenhouse effect. Energy shortage is critical for most of the countries. These problems cause the demand for clean and renewable energy and lead to the significant expansion of the manufacture of solar cells. Currently available

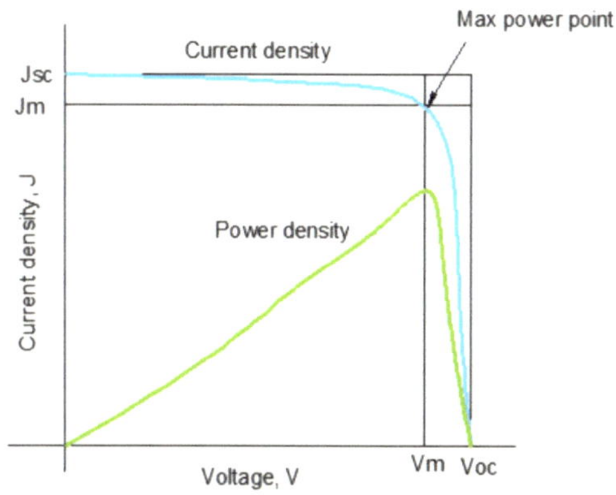

Fig. 2.4. The current-voltage and power-voltage characteristics of an ideal solar cell.

solar cells have the efficiencies about 6% (for the first generation silicon cells) to 40% (for the second and third generation multiple junction cells, Figure 2.5). It is estimated that over 8000 MW useful energy can be harvested over the world from sunlight each year with the 10% efficiency. But even now, the cost of photovoltaics is around \$0.30/kWh, which is much higher than conventional power. The solar cell projects still need government support for the high cost. Only when the cost is close to that of the traditional power, the solar energy era will come.

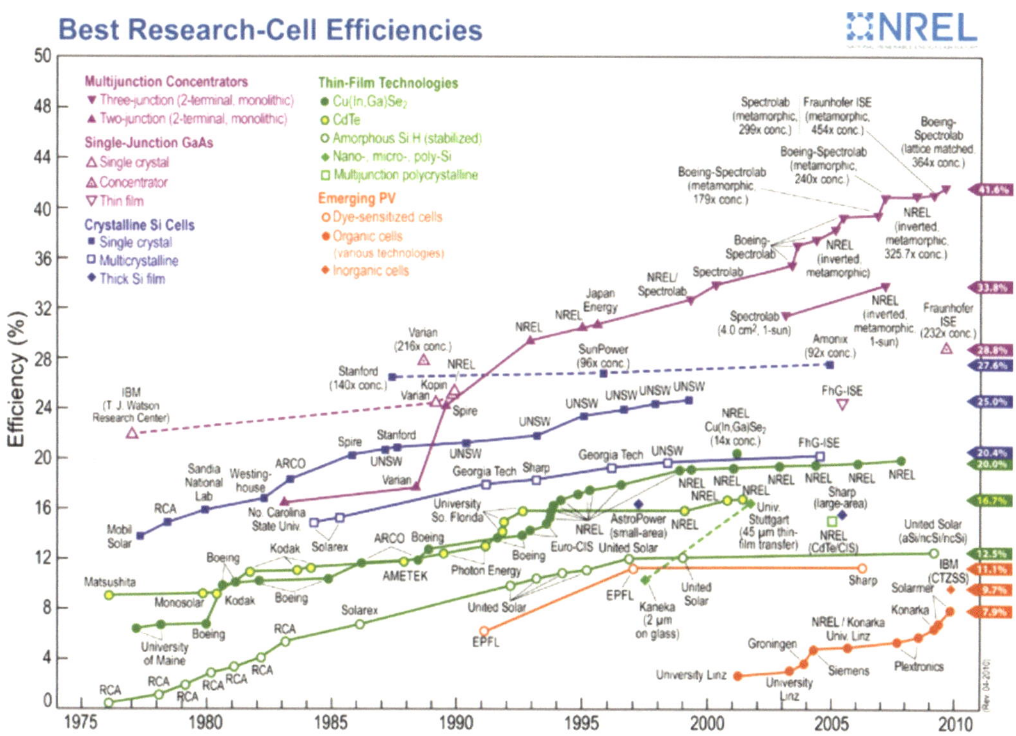

Fig. 2.5. NREL compilation of best research solar cell efficiencies
Chart courtesy of Larry Kazmerski, NREL

The National Renewable energy Lab (NREL) gives solar cell efficiencies based on different solar cell materials. Figure 2.5 shows the efficiencies of all kinds of solar cells and displays a very good prospect for all kinds of solar cells. Spectrolab (subsidiary of The Boeing Company) has set a new world record for terrestrial concentrator solar cell efficiency. The cell can convert 41.6% of concentrated sunlight into electricity. This is a multi-junction solar cell. This type of cell achieves a higher efficiency by capturing more of the solar spectrum. In a multi-junction cell, individual cells are made of layers, where each layer captures part of the sunlight passing through the cell. So this solar cell can absorb more energy from the sun's light (King, 2009). Table 2.1 gives the efficiencies description of the different solar cells in Figure 2.5.

Classification	Area cm^2	V_{OC} (V)	J_{SC} (mA/cm^2)	FF (%)	Efficiency (%)	Reference
GaInP/ GaInAs/ Ge	0.3174	3.192	1.696	88.74	41.6	(King, 2009)
Organic	0.0441	0.756	14.7	70.9	7.9	(Green et al., 2010)
CdTe	1.032	0.845	26.1	75.5	16.7	(Wu et al., 2001; Wu, 2004)
CIGS	0.419	0.69	35.5	81.2	19.9	(Repins et al., 2008)
Dye-Sensitized	0.219	0.736	20.9	72.2	11.1	(Chiba et al., 2006)

Table 2.1. Best efficiencies from Figure 2.5

First generation solar cell: single crystal silicon wafers (c-Si).

Because of broad spectral absorption range, Single crystal silicon solar cell can get high efficiency as 27.6% (Fig 2.5). The first generation photovoltaic cells are the dominant commercial products of solar cells taking more than 80% of the solar cell market. But the material processing and manufacturing are expensive.

Second generation: thin film

Based on thin film technology, the second generation solar cells also become commercial products. Thin film technology can reduce mass of material required for cell design, so it can reduce the cost for manufacturing solar cell. Normally, the efficiencies of thin-film solar cells are lower than the first generation solar cells. But thin film technology can make the solar cell panels on such substrates as light or flexible materials, even textiles. The second generation solar cells mainly include amorphous silicon (a-Si), polycrystalline silicon (poly-Si, not thin film), cadmium telluride (CdTe), copper indium gallium diselenide (CIGS) alloy etc. CdTe and CIGS are the mainly thin film solar cells on the market. The highest efficiencies of CdTe and CIGS can reach 16.5% and 19.9% respectively (Figure 2.5). Thin film solar cells are developing very fast. The market share is from about 10% (2007) to about 19% (2009).

Third generation and fourth generations:

The third generation solar cells include nanocrystal solar cells, photoelectrochemical (PEC) cells, polymer solar cells (thin film), dye sensitized solar cell (DSSC, thin film). The efficiency of the third generation is about 10%. The fourth generation cells are considered to be hybrid-inorganic crystals within a polymer matrix.

3. Nanocomposite manufacturing

In thin film soar cell, the basic manufacturing processing is the ability to deposit thin films of material. In this section, the main nano-manufacturing techniques for thin film solar cell will be introduced. These techniques include chemical bath deposition (CBD), close space sublimation (CSS), chemical vapor deposition (CVD), Magnetron sputtering and Electrodeposition (ED). There are many aspects should be considered for chose which process, such as cost, area size, in lab or in industry etc. The most important is the solar cell manufacturing technique includes a serial deposition processes. Each process has many possible choices. For example, making a CdS/CdTe solar cell includes transparent conductive oxides (TCO) deposition, buffer layer (CdS) deposition, absorber layer (CdTe) deposition and back contact deposition. CdTe layer can be deposited through one of CBD, CVD, CSS, mangnetron sputtering and electrodeposition. When one process is chosen, this process also affects the former (CdS) process and latter process (back contact), even the annealing process. The current status of technique for nanomanufacturing is discussed below.

3.1 Chemical bath deposition (CBD)

In 1933 Bruckman deposited PbS thin film by chemical bath deposition (CBD) or solution grown method. From then chemical bath deposition (CBD) has been used for decades to deposit thin films of various semiconductors (Hodes, 2003). The chemical bath deposition method is one of the cheapest methods to deposit thin films and nanomaterials. The CBD is very sample. It does not depend on expensive equipment and requires only solution containers and substrate mounting devices. CBD is a scalable technique that can be employed for large area batch processing or continuous deposition. The key advantages are low cost, large area, low temperature and atmospheric processing. With the development of thin film solar cell technology, CBD as the economical method is widely used in depositing buffer layers of CdS (and similar materials) in thin-film photovoltaic cells based on CdTe and $CuInSe_2$ (Mendoza-Perez et al., 2009; Ochoa-Landın et al., 2009; Lee, 2005; Castillo-Alvarado et al., 2010; Rose et al, 1999; Guillemoles et al., 2000; Hariskos et al., 2005; Romeo et al., 2004; Kylner, 1999; Shirakata et al., 2009; Noufi, 2006; Ennaoui et al., 2001). It takes advantage of the use of a reaction from a solution were different precursors can be dissolved easily. Depending on the deposition condition, the film growth can take place by ion–ion condensation or by adsorption of colloidal particles (cluster by cluster) from the solution onto substrates. Films deposited by this technique are proved to be of comparable quality with those produced by other sophisticated and expensive methods. There are some drawbacks of this method, such as the low material yield for film formation and the wastage of solution after every deposition which leads to treatment costs. Figure 3.1 shows the typical CBD devices.

Figure 3.1 shows the typical CBD-CdS devices in lab. The container is a beaker. Hot plate can heat the solution up to 90°C. The chemical solution contains Cadmium Acetate, Ammonium Acetate, Ammonium Hydroxide, and Thiourea. The chemical reaction results in the deposition of CdS onto all surfaces in the bath, including the surfaces of the ITO coated glass. Typically, for a 20-minute run, a thickness of 800-1000 Å can be achieved. The bath parameters such as concentration, bath temperature and deposition time greatly affect the morphology, thickness, refractive index, electrical resistivity, even the band gap of the deposited film (Lee, 2005; Contreras et al., 2002; Eze & Okeke, 1997; Marayanan et al., 1997). These properties of the buffer layer affect the performance of CdTe and CIGS solar cells.

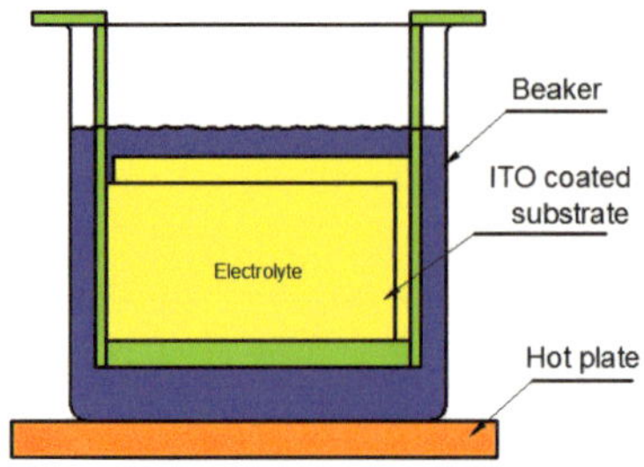

Fig. 3.1. Chemical bath deposition (CBD) devices for CdS in lab

3.2 Close space sublimation (CSS)

The close space sublimation (CSS) is widely used in CdS (and similar semiconductors) and CdTe thin film deposition in CdTe solar cells (Matsune et al., 2006; Kumazawa et al., 1997; Aramoto et al., 2003; Tsuji et al., 2000; Britta & Ferekides., 1993; Dzhafarov et al., 2005; Dobson et al., 2000; Enrıquez et al., 2007; Romeo et al., 2004; Khrypunov et al., 2006; Okamotoet al., 2000; Dhere et al., 1997; Ferekides et al., 2000). CSS is an attractive process and has many advantages. Compared with other deposition techniques, it can offer high deposition rate, it is able to produce films with large grains and can be easily scaled up for manufacturing purposes. CdTe CSS process is about 550-620°C, and CdS CSS process is about 500-600°C. The temperature is over CdTe and CdS recrystallization temperature, So, the grains size is larger than that with PVD or CVD process. CSS-CdTe grains size is about 290nm, PVD-CdTe giains size is 13nm (Moutinho et al., 2008).

Unlike other thin film deposition techniques, the distance between the source and the substrate is very short with millimeters or less, the source chosen is very important, because the source strongly affects the control of the deposition parameters, especially the deposition rate (Pinheiro et al., 2006). The close space sublimation (CSS) technique is characterized by the short distance between the source and the substrate, which is usually less than 1 mm. Figure 3.2 shows the CSS system in lab.

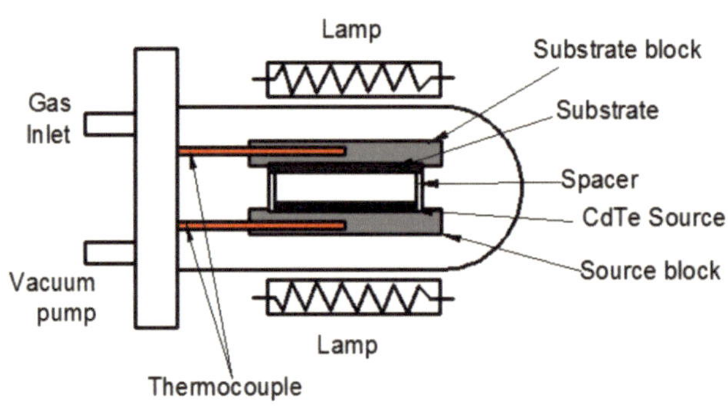

Fig. 3.2. CSS-CdTe deposition system

The CSS system is set up in a closed quartz chamber. When depositing films, the chamber can be filled a flow of gas from gas inlet. The gas can be hydrogen or inert gas such as Argon or Helium. O_2 is also found in CSS system. For example, CSS-CdS and CSS-CdTe deposition is in the ambient with O_2 (Ferekides et al., 2000). This was due to the fact that solar cells fabricated with CdS films prepared in O_2 ambient exhibited higher efficiencies than devices fabricated with CdS films deposited in inert ambient. Some CSS systems also work under vacuum conditions of about 1 Torr.

The reactor consists of a source (CdS or CdTe) and a substrate, separated by quartz spacers. The source is in underside and the substrate is above. They are held by two graphite susceptors (blocks) respectively. Each graphite block is set one thermocouple to monitor the temperature of the source and the substrate. The aim of the CSS technique is to provide a temperature difference between the source and the substrate. This temperature difference enables a diffusion controlled transport mechanism.

Radiant heat from outside lamps can heat the graphite blocks and this heat then is transmitted to the source and the substrate. The heat can also be provided by the electrical current (resistance heating), using the two graphite blocks as resistances. A combination of the two can also be adopted.

In CSS system, the main parameters acting on the deposition rates are the spacing, the source and the substrate temperature, pressure and the ambient gas.

3.3 Chemical vapor deposition (CVD)

Chemical vapor deposition (CVD) is a chemical process used to produce high-purity, high-performance solid materials. In solar cell industry, CVD process is often used to produce semiconductor thin films such as SnO_2 (Wu et al., 2006), ZnO, CdS (Kumazawa et al., 1997) and CdTe (or similar semiconductors). In a typical CVD process, the sources (precursors) are volatile. When heated, they react/or decompose on the substrate surface to produce the desired deposit. Frequently, volatile by-products are also produced, which are removed by gas flow through the reaction chamber. Figure 3.3 shows the CVD system and process.

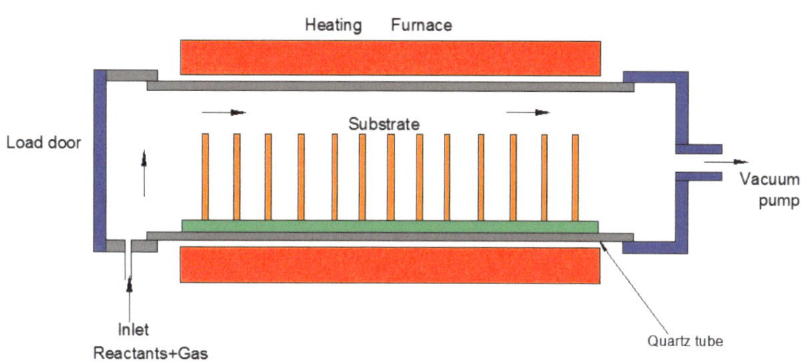

Fig. 3.3. CVD system and process

In this process, the substrate is placed inside a reactor to which reactants and inert dilute gases are supplied. The gaseous species move to the substrate and reactants are adsorbed onto the substrate with condenses. The gaseous by-products of the reactions are desorbed from the surface of the specimens and removed from the reaction chamber. The exhaust is subsequently trapped in a recycle tank through a vacuum pump. To precisely control the temperature at different locations in the chamber, multi-zone resistance heaters are used.

The deposition rate in a CVD process is controlled by the temperature. In high temperature range, mass-transfer controls the whole process of deposition. Nevertheless, in the low temperature range, the chemical reaction determines the rate of deposition. Transition between the two mechanisms is also found.

In thin film solar cell application, a number of forms of CVD are in wide use and are frequently referenced in the literature. The different types of CVD processes as follows:
According to pressure:

a. Atmospheric pressure CVD (APCVD) - CVD processes at atmospheric pressure
b. Low-pressure CVD (LPCVD) (Kumazawa et al., 1997; Rose et al., 1999)- CVD processes at subatmospheric pressures. Most modern CVD processes are either LPCVD or UHVCVD. The LPCVD process produces layers with excellent uniformity of thickness and material characteristics. The main problems with the process are the high deposition temperature (higher than 550°C) and the relatively slow deposition rate.
c. Ultrahigh vacuum CVD (UHVCVD) - CVD processes at a very low pressure, typically below 10^{-6} Pa (~10^{-8} torr).

Other methods:

PECVD—Plasma-Enhanced CVD; CVD processes that utilize plasma to enhance chemical reaction rates of the precursors. PECVD processing allows deposition at lower temperatures, which is often critical in the manufacture of semiconductors.
ALCVD—Atomic layer CVD (Naghavi et al., 2003); It is based on the sequential use of a gas phase chemical process. The majority of ALD reactions use two chemicals (precursors). These precursors react with a surface one-at-a-time in a sequential manner. By exposing the precursors to the growth surface repeatedly, a thin film is deposited.
MOCVD—Metalorganic chemical vapor deposition; CVD processes based on metal organic precursors. MOCVD is a common process for thin film solar cell deposition. It can be used on TCO, buffer layer and absorber layer, such as SnO_2 (Ferekides et al., 2000), CdS (Matsune et al., 2006; Kato et al., 2001; Tsuji et al., 2000), ZnO (Hariskos et al., 2005; Ennaoui et al., 2001) and CdTe (Sharma et al., 2003) etc.
There are also other type of CVD for different materials. The quality of the material varies from process to process, but in general, higher process temperature yields a material with higher quality and less defects.

3.4 Magnetron sputtering

Sputter deposition is a physical vapor deposition (PVD) method of depositing thin films by sputtering. In sputtering process, the material is released from the source at much lower temperature than evaporation. Figure 3.4 shows the sputter deposition process.

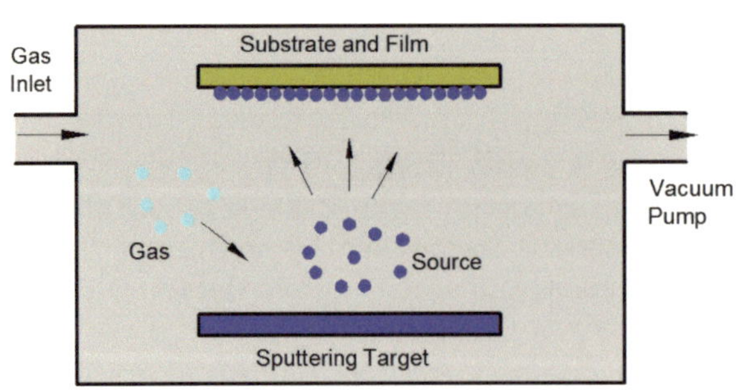

Fig. 3.4. Typical RF sputtering system.

The sputtering is ejecting material from a "target" (source), and then is deposited onto a "substrate". In figure 3.4, the substrate is placed the upside of a vacuum chamber with the source material, named a target, underside. An inert gas (such as argon) is introduced at low pressure from gas inlet. The radio frequency (RF) power source causes the gas to become ionized. The ions are accelerated towards the target to bombard the atoms of the source material. The surface atoms are broken off from the target in vapor form and condense on the substrate surface to form thin film. The basic principle of sputtering is the same for all sputtering technologies. The sputtering is considered as a low temperature evaporation, which is realized by the ion bombardment of the target.

Normally the sputtering inert gas is argon. The principle for inert chosen is that the atomic weight of the sputtering gas should be close to the atomic weight of the target as for efficient momentum transfer. Neon is for sputtering light elements and krypton or xenon is for heavy elements. Reactive gases such as F_2, O_2, H_2 and N_2 etc can also be used to sputter compounds. The compound can be formed on the target surface, in-flight or on the substrate depending on the process parameters. The compounds have very good properties for application.

For example ZnTe:N is nitrogen-doped ZnTe. It can be the transparent back contact of a CdS/CdTe solar (Compaan et al., 2004). ZnTe:N films with ~750 nm thickness were deposited by reactive RF sputtering at ~350°C with 3% N_2 in the Ar sputter gas on aluminosilicate glass (Corning 1737). CdS:O (Wu, 2004) is CdS sputtered at room temperature in an oxygen/argon gas mixture. The CdS:O film has a higher optical bandgap (2.5-3.1 eV) than the poly-CdS film. The bandgap increases with an increase of oxygen content. A CdS thin film sputtered in a 2% oxygen/argon environment is still polycrystalline, but average grain size is around 3-5 nm, much smaller than the CdS film sputtered in an oxygen-free environment (~25 nm).

Sputtering is used extensively in the semiconductor industry to deposit thin films of various materials. For making efficient thin film photovoltaic (CdTe and CIGS) solar cells, sputtering is very useful technique.

In thin film CdTe solar cell, every layer can be deposited by sputtering including front contact transparent conductive oxide (TCO) (Wu et al., 2006; Wu et al., 2005; Romeo et al., 2010; Gupta et al., 2004; Santos-Cruz et al., 2006; Romeo et al., 2007; Tiwari et al., 2004; Lee, 2005; Romeo et al., 2003; Singh & McClure, 2003; Rose et al., 1999; Mitchell et al., 1977; Compaan et al., 2004; Wu, 2004; Colombo et al., 2009; Romeo et al., 2006; Minami, 2005; Singh et al., 1999;], buffer layer CdS (Wu et al., 2006; Wu et al., 2005; Romeo et al., 2010; Gupta et al., 2004; Lee, 2005; Romeo et al., 2003; Krishnan et al., 2009; Gupta et al., 2006; Compaan et al., 2004; Wu, 2004; Colombo et al., 2009; Singh et al., 1999), CdTe (Compaan et al., 2004; Gupta et al., 2004; Santos-Cruz et al., 2006; Mathew et al., 2004; Krishnan et al., 2009; Gupta et al., 2006; Compaan et al., 2004) and back contact (Compaan et al., 2004; Romeo et al., 2003; Fahrenbruch, 2007).

Xuanzhi Wu (Wu, 2004) introduced a kind of CdTe solar cell with a new structure and a new deposition process. In this structure, three layer of TCO film CTO, buffer layer ZTO, and window layer nano-CdS:O are deposited by RF sputtering in room temperature. RF sputtering is a mature technology with demonstrated production scalability. The new process has only one heat-up segment in the entire device fabrication process. The recrystallization of the first three layers and the interdiffusion at the three interfaces (CTO/ZTO, ZTO/CdS, and CdS/CdTe interfaces) are completed during CdTe deposition by the CSS technique. This process provides attractive alternatives for producing CdTe modules with a potential of high throughput and low cost by (1) increasing device efficiency up to 16.5%, (2) improving device yield,and (3) simplifying the device fabrication process.

Some researches made CdTe solar cell with all sputtering process. Gupta & Compaan (Gupta & Compaan, 2004) developed ZnO:Al/CdS/CdTe solar cell with all sputtering process. High temperature will make ZnO:Al degraded. This cell got efficiency as 14%. This result confirms that the moderate temperatures possible with magnetron sputtering can provide important advantages in cell fabrication and expand the range of materials available for thin- film polycrystalline solar cells. Marsillac et al (Marsillac et al., 2007) developed an Ultra-thin bifacial CdTe solar cell SnO2:F/CdS-/CdTe/ZnTe:N/ITO with all sputtering process. The efficiency can get over 5%.

In thin film CIGS solar cell, sputter is also a very common deposition process for thin film. Buffer layer (CdS or ZnO) and transparent conductive oxide (TCO, such as ITO, ZnO, ZnMgO and In_2S_3 etc) can be deposited by sputtering (Hariskos et al., 2005; Nakada & Mizutani, 2002;

Wuerz et al., 2009; Kylner, 1999; Romeo et al., 2004; Bhattacharya et al., 2001; Ennaoui et al., 2001; Ohashi et al., 2001; Delahoy et al., 2004; AbuShama et al., 2004; Contreras et al., 1999; Romeo et al., 2003). CIGS absorber layer also can be deposited by sputter. Zhang et al (Zhang et al., 2010) introduced a magnetron-sputtering process for CIGS absorber layer deposition. The deposition of CIGS films using CIGS quaternary-alloyed target is fabricated through sintering process. Cu2Se, In2Se3, and Ga2Se3 powders are used as raw materials. Selenium can be directly incorporated into absorbers from the sputtering of CIGS target. The results indicated that it is feasible to prepare CIGS absorbers with this method. Therefore, the selenization process that is difficult to control and perhaps use toxic H_2Se gas may be unnecessary.

Alan E. Delahoy et al (Delahoy et al., 2004) used hybrid process to form CIGS thin film. This hybrid process includes both thermal evaporation and sputtering. The linear source thermal evaporation supplied the In, Ga and Se, and the sputtering provided Cu. This hybrid process is applied to both stationary and moving glass substrates. The motivation for this work was the need to improve the ease and uniformity of Cu delivery. The hybrid process for CIGS formation shows good reproducibility, and shows promise for being manufacturable.

In the method of sputtering CIGS absorbers, a metal film of Cu, In, and Ga is sputtered at or near room temperature and reacted in a Se atmosphere at high temperature. This process has higher throughput than coevaporation and compositional uniformity can be more easily achieved. Dhere (Dhere, 2006) reviewed the CIGS layer deposition technique and introduced the reactive co-sputtering CIGS layer process. Co-sputtering Cu, In, Ga and Se together has the problem that Se can make the target insulated and prevents the accelerating ions to bombard the targets. The new reactive co-sputtering process is dual cylindrical magnetron sputtering from Cu–In–Ga targets in the presence of selenium vapor. In Cu-In-Ga sputtered cycle, each of the two targets becomes negatively biased alternatively. The sputtering takes place for a part of the cycle from one target and for the remainder of the cycle from the other target. This deposition process is efficient and can get appropriate Cu:In:Ga composition and allows the deposition of slightly Cu-poor, near-stoichiometric CIGS thin films.

Sputter deposition is a low temperature PVD process and evaporation is a high temperature process. Even the materials have very high melting points, they also can be easily deposited by sputtering. But for these materials with high melting points, evaporation is problematic or impossible. Sputter deposited films have a composition close to that of the source material. These films typically have a better adhesion on the substrate than evaporated

films. Sputtering process is in low temperature, this is compatible with reactive gases such as oxygen.
In sputtering process, the atoms can go anywhere after bombarded, which can lead to contamination problems. Also, active control for layer-by-layer growth is difficult compared to pulsed laser deposition and inert sputtering gases are built into the growing film as impurities.

3.5 Electrodeposition (ED)

Electrodeposition is also called electroplating. It is a plating process that uses electrical current to reduce cations of a desired material from a solution (electrolyte) and coat a conductive object with a thin layer of the material. Figure 3.5 shows a very simple electrodepostion device. The main components include DC voltage source, Anode, Cathode and electrolyte. There are additional devices such as heater and stirrer not shown here. In the electroplating process, the substrate is setup as cathode in the liquid solution (electrolyte). The electrolyte contains one or more dissolved metal salts as well as other ions that permit the flow of electricity. The Anode is a kind of metal. After added positive potential on the anode, the Anode metal atoms will be oxidized and dissolve in the solution. At the cathode, the dissolved metal ions in the electrolyte solution are reduced to atom and form a metal thin film on the cathode surface. The ions in the electrolyte bath are continuously replenished by the anode metal. This process is also called sacrificing anode method.

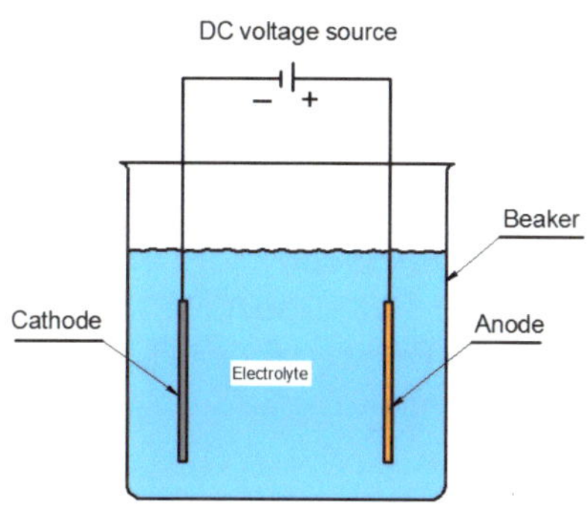

Fig. 3.5. Electrodeposition device in lab.

Metal thin films such as copper, molybdenum, gold and nickel etc can be deposited by electrodeposition process. The film thickness is in variations from nanosize to 100μm, which is controlled by the external electrical potential. In any process, the surface of the substrate must have an electrically conducting coating before the deposition can be done.
In thin film solar cells, such as CdTe, CIGS and DSSC solar cells, most of the thin film solar cell layers can be deposited by electrodeposition. For example, in a CdTe/PMeT structure solar cell (Gamboa et al., 1999), the CdTe absorber layer was deposited by electrodeposition process. In CdTe/ZnO structure solar cell, buffer layer ZnO was deposited on a conductive glass substrate by electrodeposition process (Levy-Clement et al., 2002). The glass was covered by a fluorine -doped tin dioxide ($F:SnO_2$) thin film.
Electroposition is quite slow process. Normally in CdS/CdTe solar cell, depositing a 2μm thin film of CdTe by electrodeposition process needs 2-3 hours. With a channel flow cell, this process was grown rapidly and the time can be cut to 20 min. The as-deposited films are

structurally more disordered, but after annealing and type-conversion they display superior electronic properties. The solar cell fabricated using a film prepared in the channel cell achieved efficiency 6% (Peter & Wang, 1999).

Electrodeposition is an economic, simple and successful method for preparing large area thin films. It is also a very convenient method for preparation of thin film CdTe on metallic (stainless steel) substrates (Mathew et al., 1999).

As a low cost process, electrodeposition is also widely used on CIGS solar cell thin film deposition. Kois et al (Kois et al., 2008) introduced a CIGS deposition process. Cu–In–Ga (CIG) layers of graded composition have been grown by one-step electrodeposition from thiocyanate complex electrolytes. Then the as-prepared CIG precursors are selenizated at Se-vapour to form CIGS thin film. In this process, the Ga-content in CIG films can be tailored by stirring in the process of simultaneous electrodeposition of Cu-In-Ga thin films.

Electrodeposition can be used in DSSC solar cell thin film deposition. In a ZnO/DSSC structure solar cell, electrodeposited nanoporous ZnO films exhibiting enhanced performance in dye-sensitized solar cells. Nanoporous ZnO films with grain size of 20–40 nm were grown by cathodic electrodeposition from an aqueous zinc nitrate solution. DSSCs were fabricated from nanoporous ZnO films and the cell performance could be greatly improved with the increase of ZnO film thickness. This kind of DSSC with double-layer ZnO films (8 μm thick nanoporous ZnO films on a 200 nm thick compact nanocrystalline ZnO film) achieved efficiency as high as 5.08% (Chen et al., 2006).

A counter electrode was prepared for a dye-sensitized solar cell (DSSC) through electrochemical deposition of mesoporous platinum can improve the performance of the solar cell. This kind of DSSC rendered a higher solar-to-electricity conversion efficiency of 7.6%, which is much higher than that of the sputter-deposited or most commonly-employed thermal deposited Pt counter electrodes (about 6.4%) (Yoon et al., 2008).

3.6 Other manufacture for thin film solar cell

There are still many process for thin film deposition such as close space vapor transport (CSVT) for CdTe thin film (Mendoza-Perez et al., 2009), physical vapor deposition (PVD)for CIGS thin film (Hermann et al., 2001) etc.

4. Nanocomposite application in thin film solar cell (Photovoltaic Cell)

A thin-film solar cell is made by depositing one or more thin layers (thin films) of photovoltaic materials on a substrate. The thickness of the thin-films is from a few nanometers to tens of micrometers. Thin film nano-composite solar cell materials include possible combinations of CdTe, TiO_2, CdS, $CuInS_2$, $CuInGaSe_2$ (CIGS), ZnS, ZnO, ZnTe, HgS, CuSCN, Ge, PbS etc. Some of them become commercial products, for example CdTe, $CuInGaSe_2$ (CIGS). However, most of them are still in lab research.

4.1 Solar cell structures

There are four basic structures for the photovoltaic solar cells, homojunction, heterojunction, p-i-n and n-i-p, multijunction.

Homojunction

Crystalline silicon solar cell is homojunction device. It is a simple P/N junction semiconductor. The free electrons and holes generated by light deep in the silicon diffuse to the P/N junction, then separate to produce a current if the silicon is of sufficient high quality.

Heterojunction

Thin film solar cell such as CdTe and CIGS are heterojunction device. This device has two different semiconductors. For example, CIGS has two semiconductors as CdS and Cu(In, Ge)Se_2. The solar cell with this structure can absorb light much better than silicon. The two semiconductors layers have different roles. The top layer (CdS) is a "window" layer. This layer is transparent to light and allows almost all incident light to reach the bottom layer. The bottom layer is absorber which can generate electrons and holes when absorbs the sunlight. Figure 4.1 gives two kinds of heterojunction structure CdTe and CIGS solar cells (Romeo et al., 2004).

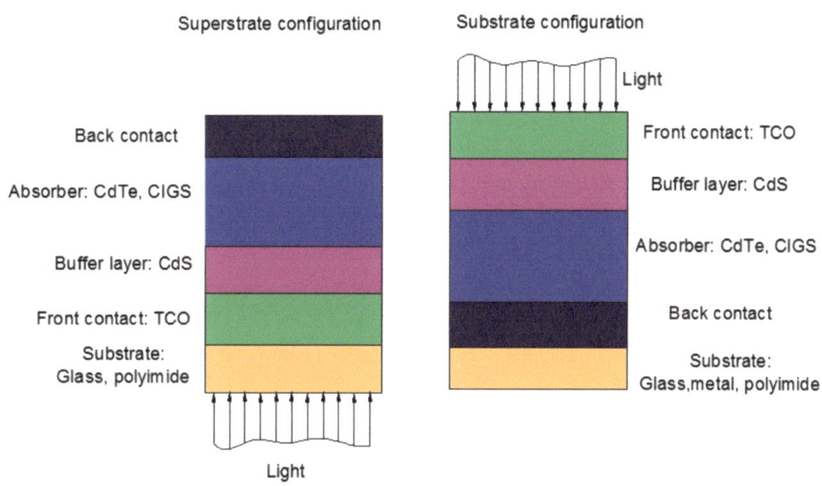

Fig. 4.1. Schematic cross-section of 'superstrate' and 'substrate' configurations for CdTe and CIGS solar cells

CdTe and CIGS have the same structures and compose 5 parts: substrate; front contact; buffer layer; absorber layer and back contact.

p-i-n /n-i-p Devices

Amorphous silicon thin-film cells are p-i-n structures. This structure has sandwiched three layers, p-type, i-type (intrinsic, undoped) and n-type layers. An electric field between the p- and n-type regions is set up which stretches across the middle intrinsic resistive region. The electric field can separate the free electrons and holes which generated by the incident light in the intrinsic region.

Multijunction Devices

Multijunction solar cells are called tandem cells. Normally, 2-junction or 3-junction solar cell is researched. This structure can achieve very high total conversion efficiency by capturing a larger portion of the solar spectrum. For example a 3-junction solar cell can achieve 41.6% efficiency (Figure 2.5). This cell is formed with individual cells and different band-gaps cells are stacked on top of one another. Each individual cell can capture one part of solar radiation. The band-gaps order is from top to bottom position with large to small band-gap. Most research in multijunction cells focuses on gallium arsenide cells.

4.2 CdS/CdTe thin film solar cell

CdS/CdTe thin-film solar cells have the potential to be mass-produced at low cost. Cadmium telluride (CdTe) has a band-gap of ~1.5 eV and the absorption spectrum is about 850nm. It is nearly ideal for sunlight absorption. So it has been recognized as a strong candidate for thin film solar cell applications. Cadmium telluride (CdTe) thin film solar cell is based on the use of cadmium telluride thin film, a semiconductor layer designed to absorb and convert sunlight into electricity. CdTe is a heterojunction p-type semiconductor. The cell was completed by adding top and bottom contacts. CdTe solar cells have very good optical property that one or two microns thickness thin film can absorb 98 percent of the sunlight. Although CdTe is most often used in PV devices without being alloyed, it is easily alloyed with zinc, mercury, and a few other elements to vary its properties. Many methods have been used for the fabrication of CdTe layers, for example close-spaced sublimation (CSS) (Chu et al., 1991; Kumazawa et al., 1997; Aramoto et al., 2003; Ferekides et al., 2000), electrodeposition (Gamboa et al., 1999), magnetron sputtering (Compaan et al., 2004; Gupta & Compaan, 2004), chemical vapor deposition (CVD) (Meyer & Saura., 1992), and metal-organic chemical vapor deposition (MOCVD) (Zoppi et al., 2006) and vapor phase epitaxy (VPE) (Levy-Clement et al., 2002).
CdS is one of the most crucial films, which serves as the window layer. Cell performance depends primarily on the electrical and optical properties of CdS film. The deep research shows the conversion efficiency of CdS/CdTe solar cells strongly depends on the ruggedness of the CdS surface (Tsuji et al., 2000). CdS is an n-type semiconductor. The optimum band gap is ~2.4 eV for solar spectrum and the direct band gap yields high optical absorption coefficient. CdS can be deposited by many methods such as CSS (Luschitz et al., 2009), MOCVD (Matsune et al., 2006; Tsuji et al., 2000), CVD (Kumazawa et al., 1997), magnetron sputtering (Gupta & Compaan., 2004), chemical bath deposition (CBD) and vacuum evaporation (Lee, 2005).
Much research has been performed on CdS/CdTe solar cells and very high energy conversion efficiencies have been achieved (Table 4.1).

J_{SC} (mA/cm^2)	V_{OC}(V)	F.F. (%)	Efficiency %	reference
23.6	0.814	73.25	14	(Compaan et al., 2004)
25.5	0.82	72	15.1	(Matsune et al., 2006)
25.36	0.826	72.2	15.12	(Kumazawa et al., 1997)
25.1	0.843	74.5	15.8	(Britta & Ferekides, 1993)
25.88	0.845	75.51	16.5	(Wu, 2004)

Table 4.1 High efficiencies CdS/CdTe solar cell parameters
(Measured under the standard condition: AM1.5, 100 mW/cm^2)

Morales-Acevedo (Morales-Acevedo, 2006) has made a physical analysis of the typical CdS-/CdTe superstrate solar cell. It shows that present record efficiencies are very close to the practical efficiency limit for a CdS/CdTe hetero-junction cell. The estimation of the maximum efficiency of hetero-junction CdS/CdTe solar cells is around 17.5%. The recorded highest efficiency is 16.5% (Wu, 2004), only 1% lower than the efficiency limitation. His work explains why the record efficiency for this kind of cells has been stable for the last 10 years, going up by less than 1% from 15.8% (Britta & Ferekides, 1993) to only 16.5%.

4.2.1 CdS/CdTe solar cell structure

Figure 4.2 gives the conventional structures of CdS/CdTe thin film solar cell. This structure has been developed over 30 years. Normally it has 5 parts:

1. Substrate-Glass or metal foil;
2. Front contact-Transparent conducting oxide (TCO);
3. CdS window layer;
4. CdTe absorber;
5. Back contact.

A transparent conducting oxide (TCO) is used as an antireflection coating. It can provide a low resistance contact to CdS layer. SnO_2 is traditional TCO. SnO_2 has two thin film layers. The first thin film is the fluorine doped tin oxide layer (SnO_2:F). This layer also called conductive tin oxide or c-SnO_2. It is the transparent contact that provides current collection from the front of the device. The second thin film is the undoped SnO_2 layer (also called i-SnO2 for 'intrinsic' or `insulating' SnO_2). It can help protect the open-circuit voltage of the device in some situations (Rose et al., 1999). Cd_2SnO_4 (Cadmium tin oxide, CTO) is the new development TCO. CTO has much more advantages than SnO_2, such as high conductivity and better optical property. The n-type CdS layer serves as the window layer and the CdTe layer serves as the absorber layer for the incident light. P-type CdTe films will lead to large internal resistance losses. So, in this structure, the CdTe layer is intrinsic (that is, neither p-type nor n-type, but natural), and add a layer of p-type zinc telluride (ZnTe) between the CdTe and the back electrical contact. This is n-i-p structure. The electrical field is formed between the n-type CdS and the p-type ZnTe which extends right through the intrinsic CdTe. The buffer layer ZnTe:Cu–doped graphite with high conductivity can improve the Ohmic conductance. The HgTe:Cu-doped graphite layer also can be used as buffer layer which produces an ohmic contact to the CdTe, then the silver layer is used to decrease the lateral resistivity of the back contact (Rose et al., 1999).

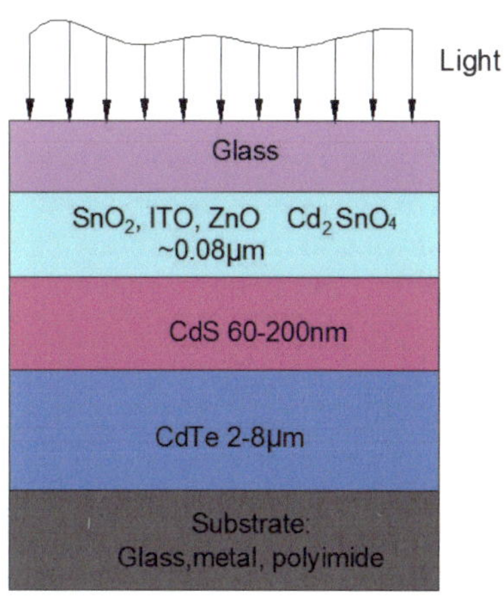

Fig. 4.2. Conventional CdS/CdTe structure

4.2.2 Unconventional structure

Base on the conventional CdS/CdTe solar cell structure, Wu. (Wu, 2004) introduced a new structure CTO/ZTO/CdS/CdTe solar cell. This structure has been developed since the year of 2001 with total-area efficiency of 16.5% (V_{oc}=845.0 mV, J_{sc}=25.88 mA/cm^2, FF=75.51%, and area=1.032 cm^2). This is the highest efficiency ever reported for CdTe solar cells.

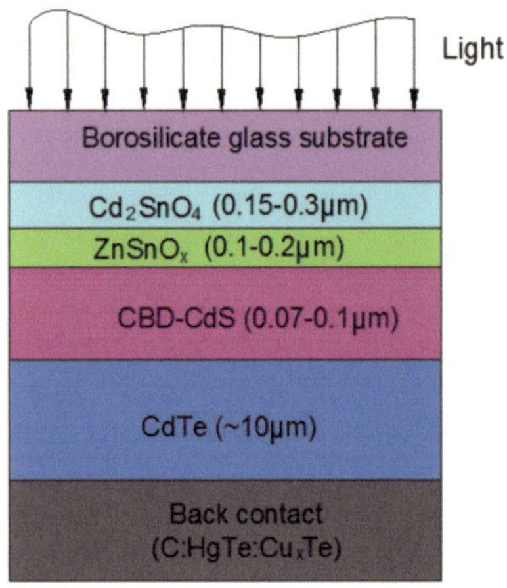

Fig. 4.3. Modified CTO/ZTO/CdS/CdTe device structure (Wu, 2004)

Figure 4.3 shows the CTO/ZTO/CdS/CdTe solar cell structure. The cadmium stannate (Cd_2SnO_4, or CTO) transparent conductive oxide (TCO) films have lower resistivity, higher transmittance and smoother surfaces than conventional SnO_2 TCO films. They can improve the Jsc and FF of a solar cell. Thin film $ZnSnO_x$ (ZTO) is a buffer layer, it can improve device performance and reproducibility. The ZTO film has a high optical bandgap (~3.6 eV) and near-zero absorbance. This device performance can be enhanced and transmission loss due to the TCO front-contact can be reduced by use of the CTO and ZTO bilayers (Wu et al., 1998; Wu et al., 2006). ZTO buffer layer can significantly reduce resistivity between the CTO and CdS layers in two reasons. First it can reduce the probability of forming a localized TCO/CdTe junction with low Voc and FF when the CdS film is thinned. Second, it can greatly reduce shunting problems (Wu et al., 2001). A buffer layer could help relieve stresses between these layers, thereby improving adhesion during the CdCl2 treatment.

Zinc stannate films have the properties of high bandgap, high transmittance, low absorbance, and low surface roughness. These films are chemically stable and exhibit higher resistivity. They can match well with CdS window layer. In fact ZTO buffer layer can significantly enhance the performance and reproducibility of both SnO_2-based and Cd_2SnO_4 (CTO)-based CdS/CdTe devices (Wu et al., 1998; Gayam et al., 2007).

4.2.3 Substrate

Soda-lime glass

Soda-lime glass is also called soda-lime-silica glass. It is the most common glass and widely used for windowpanes and glass containers (bottles and jars) for beverages, food, and some commodity items. Soda-lime glass is the basic substrate materials for CdS/CdTe thin film solar cell (Matsune et al., 2006). But soda-lime (SL) glass has poorer properties than borosilicate glass, such as a higher thermal expansion coefficient, higher Na and Fe content, higher absorbance, and lower softening temperature.

Borosilicate glass

Borosilicate glass is noted for its low thermal expansion and chemical resistance. Therefore it is widely used in laboratory equipment and near heat sources such as lamps. Borosilicate glass is very good for thin film substrate. Corning glass is one kind of alkali free borosilicate glass. It was originally developed for thin film electronic circuits, which require an extremely smooth substrate with special electrical properties. Corning glass has very good surface flatness, surface smoothness and very low thermal expansion. The glass has an alkali level under 0.3%. This is very important since alkali ions are known to be detrimental to performance, reliability, and longevity of thin film devices. Corning glass 7059 (Wu et al., 2006; Compaan et al., 2004; Kumazawa et al., 1997; Britt & Ferekides, 1993) and 1737 (Matsune et al., 2006) are ideal substrate glass for CdS/CdTe thin film solar cell.

4.2.4 Front contact-transparent conducting oxide (TCO)

The most important characteristics that a transparent conductive oxides (TCO) front contact must exhibit are a low sheet resistance and a high transparency in the visible region. TCOs are important semiconductors thin films used on solar cells. Most of these films are fabricated with polycrystalline or amorphous microstructures. They are nanocomposites. The important TCOs are impurity-doped ZnO, In_2O_3 , SnO_2 and CdO as well as multi-component oxides consisting of combinations of ZnO, In_2O_3 and SnO_2, including some ternary compounds existing in their systems (Minami, 2005). The ternary compounds include Cd_2SnO_4, $CdSnO_3$, $CdIn_2O_4$, Zn_2SnO_4, $MgIn_2O_4$, $CdSb_2O_6$ and $In_4Sn_3O_{12}$, $ZnSnO_3$, $Zn_2In_2O_5$, $Zn_3In_2O_6$, In_2SnO_4. Sn doped In_2O_3 (ITO) and F doped SnO_2 TCO thin films are the preferable materials for most present applications. In solar cells, TCOs need meet some requirements such as low resistivity (below 10^{-3} $\Omega\cdot$cm) and good transmittance of incident light (over 80%). The industry standard in TCO is ITO, or tin-doped indium-oxide. This material has a low resistivity of ~10^{-4} $\Omega\cdot$cm and a transmittance of greater than 80%. But ITO is very expensive. Alternative materials such as ZnO:Al (AZO) and ZnO:Ga (GZO) can be the solution. Figure 4.4 gives the practical TCO used on thin film transparent electrodes. Cd containing TCOs are not listed here because of the toxicity of Cd.

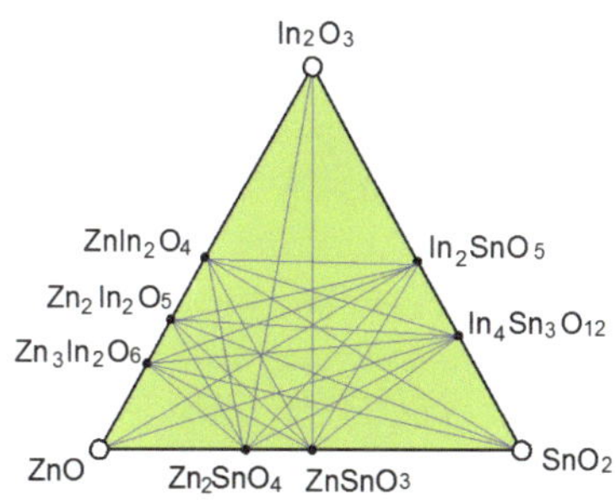

Fig. 4.4. Practical TCO semiconductors for thin-film transparent electrodes.

The study on impurity-doped ZnO, In_2O_3 and SnO_2 thin film TCO shows that the obtained minimum resistivities of impurity-doped ZnO are still decreasing and close to those of In_2O_3, whereas those of impurity-doped SnO_2 and In_2O_3 films have essentially remained unchanged for more than the past twenty years (Minami, 2005). So, AZO is the best alternative for ITO.

In the tandem thin-film solar cells, the requirements for TCO front-contact are high band-gap, high transmission and high conductivity (low resistivity). This TCO can be a transparent contact in the top cell. It must also have high transmission in the sunlight wavelength. But the conventional TCOs (such as ITO, SnO_2, and ZnO) cannot completely meet these requirements. Wu. developed a high-quality CTO TCO film (Fig 4.3), which has transmission of 80-90% in the region 400–1300 nm and bulk resistivity of $1.8x10^{-4}\Omega\cdot cm$ (Wu, 2004). The problem of this TCO contains Cd element, it is toxic.

SnO_2:F TCO

SnO_2:F is the SnO_2 films doped with fluorine. SnO_2:F is the conventional TCO and used for more than 30 years. The SnO_2:F TCO has good absorbance (A) and transmittance (T) in the visible region (Chu et al., 1991; Wu et al., 2006; Compaan et al., 2004; Britt & Ferekides, 1993). It is a quite stable material but it exhibits a high resistance and not good for this kind of solar cells.

Indium tin oxide (ITO) TCO

Indium tin oxide (ITO) (Matsune et al., 2006; Minami, 2005) is one of the most widely used transparent conducting oxides (TCO) on thin film solar cell because of its very good electrical conductivity and optical transparency. Normally, ITO is In_2O_3, but some In can diffuse into CdS and/or CdTe when this material is used as a front contact. In this case a SnO_2 buffer layer can be used as a diffusion barrier. ITO thin film is a composite with indium (III) oxide (In_2O_3) and tin (IV) oxide (SnO_2), typically 90% In_2O_3, 10% SnO_2 by weight. It is transparent and colorless. In the infrared region of the spectrum it is a metal-like mirror. Like other transparent conducting oxides, when a compromise (In_2O_3 and SnO_2) has to be reached during its film deposition, the high concentration of charge carriers will increase the material's conductivity, but decrease its transparency. ITO thin films are most commonly deposited on surfaces by electron beam evaporation (EBE), physical vapor deposition (PVD), or a range of sputter deposition (SD) techniques.

A new TCO that is fluorine doped In_2O_3 was developed. It exhibits very good properties that the resistivity is $2x10^{-2}\Omega m$, the transparency better than 90% between 400 and 800 nm and this materials is quite stable at a temperature of 500^0C. The best CdS/CdTe solar cells with an efficiency of approximately 14% are obtained by using 0.4 μm of fluorine doped In_2O_3 as a TCO (Romeo et al., 2003).

Nano-scale ITO thin films can provide a path to a new generation of solar cells. These materials applied on solar cells can make the cells to be low-cost, ultra-lightweight, and flexible. Because of the nanoscale dimensions of the nanorods, quantum-size effects influence their optical properties. But a stable supply of indium-tin-oxide (ITO) cannot be assured because indium is a very expensive and scarce material.

The main problem about ITO is the cost. ITO's price is several times that of aluminum zinc oxide (AZO). But ITO is much better than AZO in almost every performance. For example, ITO has very good chemical resistance to moisture and it can survive in a CIGS cell for 25–30 years on a rooftop.

ZnO:Al (AZO) TCO

ITO is very expensive TCO. AZO is one of alternatives to ITO. AZO thin films, with a low resistivity of the order of 10^{-5} Ω·cm and source materials that are inexpensive and non-toxic, are the best candidates (Minami, 2005; Minami, 2008). AZO has relative good optical transmission performance in the solar spectrum. The ZnO:Al film is more transparent than SnO_2:F over the whole spectrum due to higher electron mobility. But the SnO_2:F has more better absorbance than ZnO:Al (Figure 4.5). ZnO:Al has high transparency well into the infrared and excellent sheet resistance with high mobility. Thus it is an attractive candidate as a TCO. AZO has been used on CdS/CdTe solar cell and achieve a CdTe solar cell with 14.0% efficiency at one sun for an air-mass-1.5 global spectrum (Compaan et al., 2004).

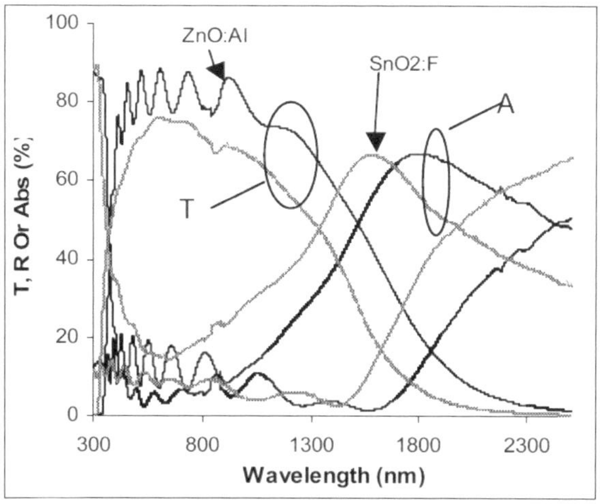

Fig. 4.5. ZnO:Al & SnO2:F optical properties (Compaan et al., 2004).
A- absorbance and T- transmittance

But In comparison to sputtered cells on commercial SnO_2:F, the stability of ZnO:Al-based cells is poorer. This may be due to interdiffusion across the ZnO:Al/CdS interface (Gupta & Compaan., 2004).

4.2.5 Back contact

The back contacts on CdS/CdTe solar cells often show a non-Ohmic behavior because of the semiconductor Schottky barrier. This Schottky barrier acts as a diode reverse biased to the CdS/CdTe junction diode and increases the contact resistance, thereby reducing the solar cell performance [Te901]. The main function of back contacts is to eliminate the Schottky barrier to reduce the contact resistance. It is well known that the addition of Cu to back-contacts is commonly used to improve the performance of CdS/CdTe solar cells. This may be due to the reaction with a Te-rich CdTe layer and the formation of a Cu_xTe/CdTe back-contact (Wu, 2006). But copper is a fast diffuser and it influences the long-term stability of such cells. Cu should be avoided in the back contact to obtain a long term stable CdTe/CdS solar cell

Conventional back contacts on CdS/CdTe solar cells are commonly made with Cu/Au or Hg (Compaan et al., 2004) or Pb and Cu/graphite (Aramoto et al., 2003). Cu-containing back contacts can influence the solar cells efficiency and the performance degrades because of Cu diffusion to the junction. Cu can easily combine with Te to form Cu_2Te. This material is unstable. In order to get stable back contacts Sb has been applied [Te901]. Sb/Au back contact also shows typical diffusion which is causing degradation just like Cu/Au back contact. But the degradation for the Sb/Au back contact is not as strong as that for the Cu/Au contact. Different back contact materials have been investigated so far, such as Cu_xTe (Wu et al, 2006), ZnTe:Cu (Wu et al, 2006), As_2Te_3:Cu (Romeo et al., 2010; Romeo et al., 2007), Bi (Vigil-Galán et al., 2007), Sb_2Te_3 (Romeo et al., 2007), even ITO (Romeo et al., 2007; Tiwari et al., 2004).

ZnTe:N/ITO (Marsillac et al., 2007) is one kind back contact of CdS/CdTe solar cell. The cell structure is SnO_2:F/CdS/CdTe/ZnTe:N/ITO. After $CdCl_2$ treatment, ZnTe:N was achieved through sputtering an undoped ZnTe in a 5% nitrogen–95% argon environment. Then ITO thin film was deposited. Because the ZnTe:N layer is too thin and the conductivity of the ZnTe:N film is not very high, a secondary transparent electrode was used to collect the photogenerated current. ITO TCO film can be used as the secondary transparent electrode. This back contact is the same as Cu_xTe/ITO (Wu et al, 2006). It is for ultra-thin absorber (eta) soalr cell. This back contact is transparent just like transparent front contact. It can be used in polycrystalline thin-film tandem cells because there is enough light transmitted to the bottom cell (Romeo et al., 2007; Tiwari et al., 2004). TCO back contact on CdTe provides superior cell stability, simplified processing and a potential for low-cost production. The average efficiency of reference solar cells with standard Cu/Au back contacts on CdTe is in the range of 10-11%. It was observed that most of the solar cells with ITO back contact were in the efficiency range of 7-8%. A. N. Tiwari et al developed a kind of CdTe solar cell with SnO_2:F as a front contact for CdS and ITO as a back contact on CdTe. The efficieny achieved 7.9% (V_{oc}=702mV, J_{sc}=18.2 mA/cm^2, FF=0.62) (Tiwari et al., 2004).

4.2.7 Fabrication

Fabricating CdTe solar cell includes a serial deposition processes such as TCO layer, CdS layer, CdTe layer, and back contact layer etc deposition processes. Each individual process has many deposition methods such as CBD, CSS, CVD, Ed and sputtering process etc. Figure 4.6 gives the deposition processes of one example of conventional structure and one

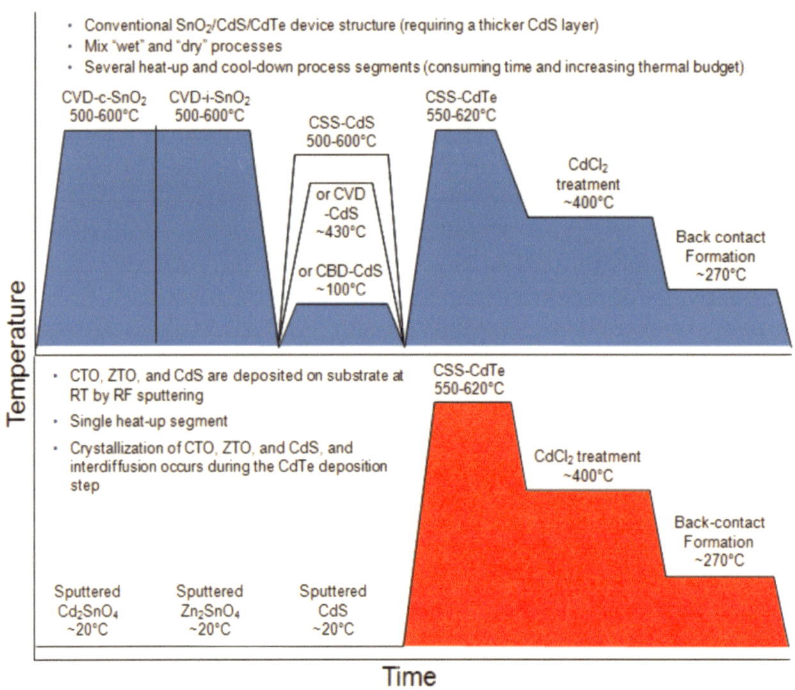

Fig. 4.6. The conventional solar cell and CTO/ZTO/CdS/CdTe solar cell deposition process (Noufi, 2006).

example of CTO/ZTO/CdS/CdTe device structure (Noufi, 2006). For conventional structure, both SnO_2 (c-Sno_2 and i-SnO_2) can be deposited on glass by CVD; thin film CdS is formed on SnO_2 layer by CBD, CVD or CSS; CdTe film is deposited by CSS. CSS has received the most attention for CdTe deposition recently because it is well-suited to large-scale manufacturing (Rose et al., 1999).

For the CTO/ZTO/CdS/CdTe structure, the first three layers Cd_2SnO_4 (CTO), Zn_2SnO_4 (ZTO) buffer layer and CdS window layer are prepared by the same deposition technique-RF magnetron sputtering at room temperature. RF sputtering is a mature technology with many advantages such as thickness control and easy operation. This deposition process has only one heat-up segment (CdTe CSS process) in the entire device fabrication process. In the CdTe CSS process, the recrystallization of the first three layers and the elements interdiffusion at the three interfaces (CTO/ZTO, ZTO/CdS and CdS/CdTe interfaces) are completed. This kind of solar cell can reach efficiency of 14.7% (Voc = 833.8 mV, Jsc = 24.06 mA/cm2, FF = 73.29%, and area=1.159cm^2) (Wu, 2004).

After CdTe CSS deposition, there is a $CdCl_2$ annealing in both processes. The main effect of the $CdCl_2$ heat treatment on the physical properties of CdTe thin films is to promote recrystallization and grain growth. But here the CSS process is in a higher temperature and has large grain size. The recrystallization process has finished in CdTe CSS process. $CdCl_2$ treatment has several substantial benefits such as: grain-boundary passivation, increased CdS/CdTe interface alloying, and reduced lattice mismatch between the CdS and CdTe layers. The ZTO ($ZnSnO_x$) films were deposited by RF sputtering at room temperature which has a very high resistivity. After $CdCl_2$ annealing at a higher temperature, the film resistivity is reduced greatly. The ZTO band-gap (Eg) remains the same (~3.6 eV), but its optical transmission is slightly improved. The reason is the inter-diffusion at the interfaces. The inter-diffusion of the CdS and ZTO layers improved the quantum efficiency of a CdTe cell over the entire active wavelength region (400-860 nm). Even in the conventional device structure, the inter-diffusion also can improve device performance and reproducibility. In spite of the 9.7% lattice mismatch between hexagonal CdS and cubic CdTe, this structure still can get very high efficiency due to the interdiffusion. After anneal, the V_{oc} and FF increase, so does the efficiency. Cells made without the anneal generally have efficiencies between 6% and 10%, whereas cells made with the anneal are generally more than 12% efficient. (Wu et al., 2005; Rose et al., 1999; Wu et al., 2004).

4.2.8 Cu in CdS/CdTe solar cell

K. Barri et al has studied the role of copper in CdTe solar cell. Copper is typically introduced in CdTe during the application of the back electrode, to enhance device performance by facilitating the formation of an ohmic back contact. But Cu and Te can form unstable component Cu_2Te, which is associated with long time stability (Barri et al, 2005, Dobson et al., 2000). Here, Cu was introduced in the CdS film prior to the deposition of the CdTe and both plain graphite and Sb_2Te_3/Mo were used as back contacts (Fig.4.7). The high performance of the solar cell achieved (V_{OC} of 830 mV and FF's in the high 60's) clearly indicates that Cu enhances device performance, even when intentionally introduced in CdS. But excessive amounts of Cu can lead to shunting and poor collection. The formation of Cu_2Te may not be necessary to achieve effective back contacts to CdTe (Barri et al, 2005).

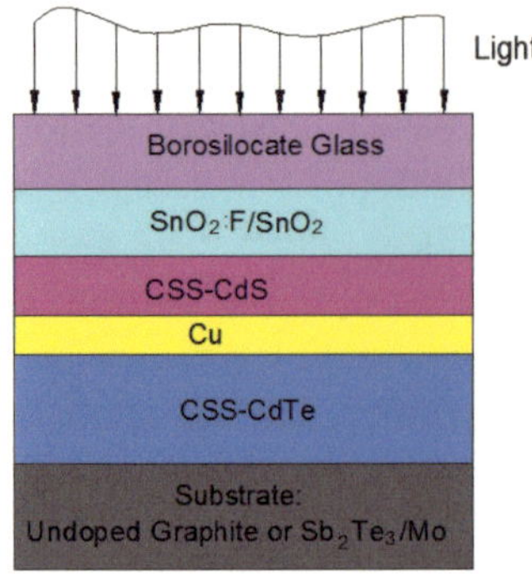

Fig. 4.7. Cu was incorporated into the device prior to the CdTe deposition (Barri et al, 2005)

4.2.9 Thickness influence

Large scale manufacturing of CdS/CdTe solar cells is constrained by the cadmium and tellurium materials. Because the tellurium is rare element and limited availability and the cadmium is hazardous to human health. Gupta et al (Gupta et al., 2006) found that it is possible to reduce the CdTe layer thickness without much compromise in efficiency. Normally the thickness of CdTe is about 2-8µm. But in the ultra-thin films solar cells, the thickness is 0.7-1.28µm. The CdS/CdTe solar cells were fabricated using magnetron sputtering method. The best thin CdTe cell was obtained with 1µm CdTe and had efficiency of 11.9%. Cells with 0.7 µm CdTe show efficiency 11.2%, this is the thinnest CdTe cell ever reported with efficiency above 10%. The standard CdTe cells (2.3 µm CdTe thickness) efficiency is 13%.

The thickness of CdS film is a critical factor affecting cell performance. Kumazawa et al (Kumazawa et al., 1997) made this study. As the thickness of CdS film decreases, the open-circuit voltage (V_{oc}) becomes low and when the CdS film thickness is less than 60nm the V_{oc} decreases dramatically. The short-circuit photocurrent density (J_{sc}) becomes a little higher when the thickness of CdS decreases (figure 4.8). It shows that CdS thickness best range is ~60 nm.

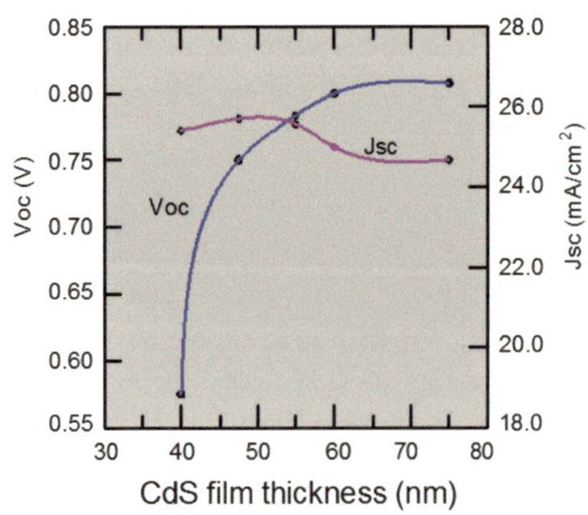

Fig. 4.8. The V_{oc} and J_{sc} as a function of CdS film thickness
Data comes from (Kumazawa et al., 1997)

4.2.10 Other type CdTe structure

1. ZnO/CdTe/CuSCN heterostructure (Tena-Zaera et al., 2005)

This structure can be made as eta (extremely thin absorber) solar cell. It uses SnO_2:F as the TCO. ZnO/CdTe is an n-type window layer semiconductor and CuSCN is a p-type absorber layer semiconductor. The energy gap of ZnO is 3.31 eV and that of CdTe is 1.54 eV. In the 400-800 nm (AM1.5) solar spectrum range, the effective absorption (AE) is about 87% and the effective reflectance (RE) is only 10%. It is very favorable for the use of extremely thin absorber (eta) solar cell. TiO_2/CdTe/ZnTe and $TiO_2/CuInS_2$/CuSCN have the similar heterostructures as that of ZnO/CdTe/CuSCN.

2. Au-Cu/p-CdTe/n-CdO/glass-type solar cells (Santos-Cruz et al., 2006)

The CdO:F films were grown by the sol-gel method, The resistivity is $4.5X10^{-4}\Omega\cdot$cm and the optical transmission is higher than 85%. The CdTe:Sb films were prepared by the RF sputtering technique, the resistivity value is $10^{6}\Omega\cdot$cm.The Au-Cu contacts were thermally evaporated. This kind of heterostructure PV solar cell can achieve the highest energy conversion efficiency 5.48%.

3. ZnS and ZnCd as window layer (Contreras-Puente et al., 2000)

G. Contreras-Puente et al developed $SnO_2/Zn_{0.9}Cd_{0.1}$/CdTe and SnO_2/ZnS/CdTe two types semiconductor thin films solar cells. The efficiencies are 1.26% (V_{oc} = 489 mV, J_{sc} = 8.9 mA/cm^2, FF = 29%) and 3.12%((V_{oc} = 324 mV, J_{sc} = 22 mA/cm^2, FF = 42%) respectively. The efficiencies are very low.

4.2.11 CdTe/CdS Solar cells on flexible substrates

Normally, the substrates of thin film CdS/CdTe solar cells are glass. But glass are hard, weight and fragile. The post deposition annealing of the films needs a high temperature (420°C). This temperature can cause rupture of the glass substrate. On the other hand, solar cells on flexible metallic substrates are light weight, free of damage and are suitable for storage, transportation and installation. CdTe has been successfully electrodeposited on various foils such as stainless steel (SS), Mo, Ni and Cu. Molybdenum is considered as the suitable substrate material from the point of view of the matching thermal expansion coefficient with CdTe. Pantoja Enriquez et al (Pantoja Enriquez et al., 2004) developed a CdTe/CdS device on flexible molybdenum (Mo) substrate with the efficiency of 3.5% (Voc=0.5 V, Jsc=10.6mAcm^{-2}, FF=0.4).

Figure 4.9 gives another unconventional CdS/CdTe solar cell structure (Singh et al., 1999)

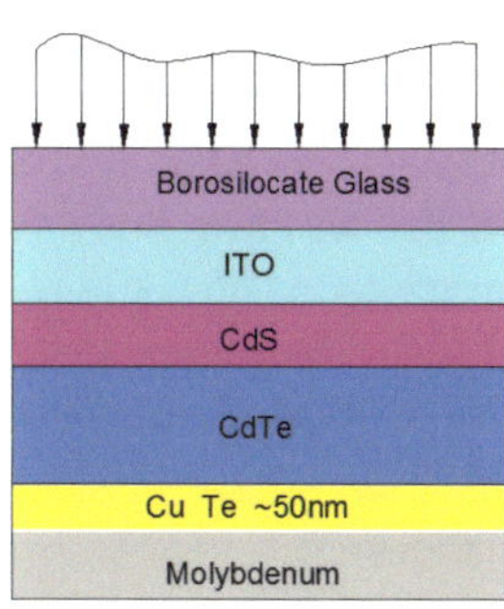

Fig. 4.9. Schematic of device configuration on Mo foil substrate (Singh et al., 1999)

The substrate is molybdenum foil. CdS is deposited on top of CdTe. Thin film inter-layers (approximately 50 nm thick) of Cu and Te are used to improve the conductivity between Mo and CdTe. Thin film Cu and Te has the same role of ZnTe as in fig. 4.1. Cu and Te can dope CdTe and make it heavily p-type, which facilitates tunneling. Also Cu and Te can form compound $Cu_{2\sim x}Te$ between Mo and CdTe and make tunneling more effective. This structure is made as follows: Inter-layers of Cu and Te are evaporated onto the Mo substrate; CdTe is deposited by thermal evaporation and then treated with a $CdCl_2$ solution and annealed; CdS film is deposited by thermal evaporation and then treated with $CdCl_2$, annealed and indium doped. The thermal evaporation processes are in low temperature (220°C), so $CdCl_2$ treatment has two functions: one is to promote the diffusion of Te and Cu into CdTe to create a p-type region and improve the conductivity between CdTe film and molybdenum; the other is to promote crystal growth of CdTe and CdS through recrystalization. The top contacting material is made by sputtering ZnO, ITO or a combination of ZnO and ITO. This molybdenum foil substrate solar cell cannot get such high efficiency as that of glass substrate solar cell. One reason is the high series resistance of the device, The other reason is that the high defect density associated with rough CdTe/CdS interface, which results in low shunt resistance.

Matulionis et al (Matulionis et al., 2001) made the same structure of Mo/CdTe/CdS/ITO thin-film solar cells by radio-frequency magnetron sputtering. The conversion efficiency is 7.8 percent on 0.05 cm^2 area device. The high efficiency is due to the back contact. Here, the back contact between the molybdenum and the CdTe is ZnTe:N, it can improve the conductivity.

Au/Pd alloy layer also can be the interlayer between Mo substrate and CdTe layer (Pantoja Enriquez et al., 2004).

The polymer also can be the substrate of CdTe/CdS solar cell. The problems of this substrate are low light absorption and high temperature stability. The highest reported efficiency of a flexible CdTe/CdS solar cell on polymer substrate is 11.3% (Romeo et al., 2006). This efficiency can compare well with the efficiency of CIGS or a-Si solar cells developed on polymer foils.

4.3 Cu(In, Ga)Se_2 (CIGS) thin film solar cell

Copper indium gallium (di)selenide (CIGS), is a compound semiconductor material composed of copper, indium, gallium, and selenium. It is used as light absorber material for thin-film solar cells. The chemical formula of this material is $CuIn_{(1-\chi)}Ga_{\chi}Se_2$, where the value of χ can vary from 0 to 1. When χ is 0, it is copper indium selenide (CIS) and when χ is 1, it is copper gallium selenide (CGS). In laboratory research, the efficiency of CIGS solar cell is approaching 20%. CIGS solar cell can be deposited on a variety of cheap substrates (e.g., glass, plastic, foil) and has acceptable environmental stability characteristics. There is no toxic Cd in the absorber layer, even the window layer (CdS) can be replaced by Cd-free materials. These advantages attract much focus on CIGS-based devices.

Depending on the Ga/(In+Ga) ratio, the bandgap of CIGS can be varied continuously between 1.04 (CuInSe2) and 1.68 eV (CuGaSe2). The current high-efficiency devices are prepared with bandgaps in the range 1.20–1.25eV, this corresponds to a Ga/(In+Ga) ratio between 25 and 30% (Romeo et al., 2004).

Compared with CdTe thin film solar cell, CIGS solar cell absorber layer is complex. It has four elements as Cu, In, Ga and Se. The properties of the device are significantly impacted by the detailed compositional profile in the absorber. CuInGaSe2 has some drawbacks due to limited availability and increasing cost of indium and gallium elements.

CIS, CIGS,CGS and Ga grading

CIS is an abbreviation of copper indium selenide ($CuInSe_2$). If added gallium, it will become CIGS. CGS is an abbreviation of copper gallium selenide (CuGaSe2). If added indium, it change to CIGS. So, CIGS is a variation of CIS or CGS. CuInSe2 and CuGaSe2 have the chalcopyrite lattice structure; it is a diamond-like structure. CIS and CGS have high optical absorption coefficients and versatile optical and electrical characteristics and is especially attractive for thin film solar cell application. CIS (no Ga) and CGS (no In) solar cells have achieved 15% and 10.2% efficiencies respectively (NERL) (AbuShama et al., 2004). At present the manufacture of CIS or CGS solar cells is high when compared with amorphous silicon solar cells. The use of gallium increases the optical band-gap of the CIGS layer as compared to pure CIS, thus increasing the open-circuit voltage, but decreasing the short circuit current. Figure 4.10 gives the CIS Ternary phase diagram.

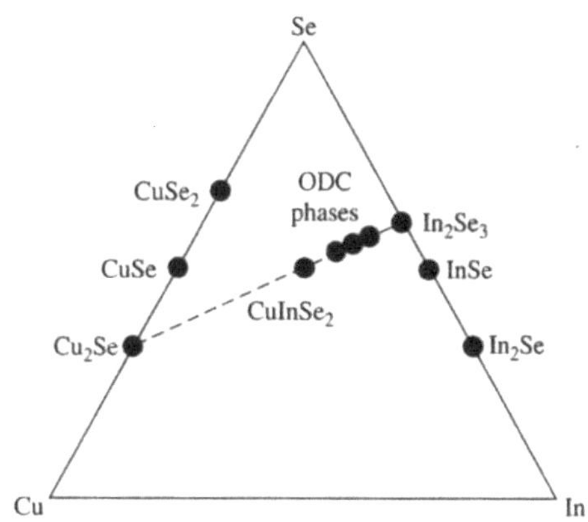

Fig. 4.10. Ternary phase diagram of the Cu-In-Se system (Anderson et al., 2003)

The binary phase Cu_2Se and In_2Se_3 can be alloyed to form $CuInSe_2$; similarly the binary phase Cu_2Se and Ga_2Se_3 can be alloyed to form $CuGaSe_2$. Ternary phase $CuInSe_2$ can be alloyed in any proportion (χ) with ternary $CuGaSe_2$ to form $Cu(In,Ga)Se_2$. The chemical formulas as follows:

$Cu_2Se+In_2Se_3 \rightarrow 2CuInSe_2$

$Cu_2Se+Ga_2Se_3 \rightarrow 2CuGaSe_2$

$(1-\chi)CuInSe_2+\chi CuGaSe_2 \rightarrow CuIn_{\chi}Ga_{1-\chi}Se_2$

So, $Cu(In,Ga)Se_2$ can accomodate large variations in composition. The compositional ratios of Cu/(In+Ga) and Ga/(In+Ga) play critical roles on the characteristics of CIGS solar cells. The Cu/(In+Ga) ratio largely affects the morphology of the CIGS absorber, while the Ga/(In+Ga) ratio determines the band-gap, both of which are critical for cell performance (Sakurai et al., 2003).

The ratio (χ) of Ga/(In+Ga) can change the band-gap of the CIGS absorber layer. According to different Ga/(In+Ga) ratio, the CIGS solar cells with different band-gap can be manufactured. The most important is this ratio not only to optimize the general band gap level, but also to obtain different band gaps at different depths in the CIGS thin film. This is called band gap profiling. In CIGS thin film solar cells an in-depth band gap variation due to changes in the Ga/(In+Ga) ratio is commonly referred to as Ga-grading (Lundberg et al., 2005). By increasing the Ga/(Ga+In) ratio (χ) from 0 to 1 the band-gap varies continuously from about 1.02 eV (CIS) to about 1.66 eV (CGS). The various band gaps are suitable to the solar spectrum to improve light absorption and thereby further improve the solar cell performance and increase the efficiency. For high performance devices, the ratios are Ga/(In+Ga)=0.2-0.3 and Cu/(In+Ga)=0.7-1.0. In the recorded high efficiency (19.9%) CIGS solar cell, the ratios are Ga/(In+Ga)=0.3 and Cu/(In+Ga)= 0.81 (Repins et al., 2008).

4.3.1 Structure of a CIGS thin-film solar cell

CIGS thin film solar cell has the same structure as CdTe solar cell (Figure 4.2). It has 5 different thin film layers as:

1. Substrate;
2. Front contact-Transparent conducting oxide (TCO);
3. CdS window layer;
4. CIGS absorber;
5. Back contact.

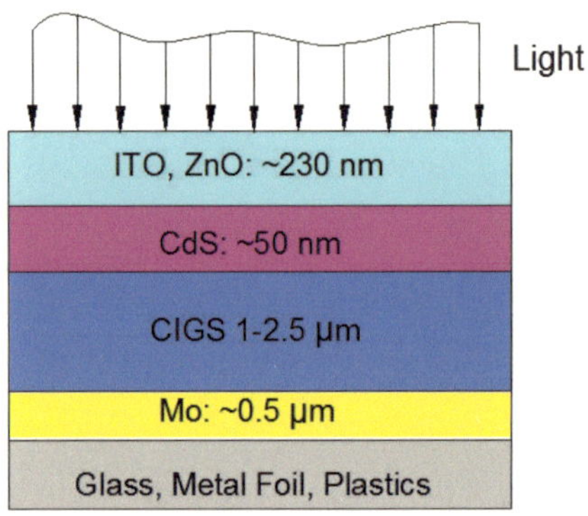

Fig. 4.11. CIGS thin-film solar cell structure.

Basic CIGS thin film solar cell structure is depicted in Figure 4.11. The Individual layer thicknesses are approximate and may differ somewhat among laboratories. The substrates can be glass (NAKADA & MIZUTANI, 2002), metal foils (Wuerz et al., 2009) or plastics (Huang et al., 2004). The most common substrate is soda-lime glass, which is about 1–3 mm thickness. Metal molybdenum (Mo) is coated on the substrate as back contact pole (NAKADA & MIZUTANI, 2002). The semiconductors CIGS, CdS and ZnO are shaped into the heterojunction. Here CIGS is p-type semiconductor and ZnO is n-type semiconductor. CdS is buffer layer. In this asymmetric structure, the CIGS layer serves as an absorber, ZnO and CdS layers serve as window layers and the band-gaps: $E_{g,ZnO}$=3.2 eV and $E_{g,CdS}$=2.4 eV. ZnO layer also serves as front contact pole for current collection.

CIGS absorber layer is complex, according to detailed thin film composition. It can be CIS, CIGS or CGS.

CdS window layer can be replaced by ZnS etc thin film to realize Cd-free thin films CIGS solar cells. Front contact TCOs can be ZnO (Repins et al., 2008), ZnO:Al (NAKADA & MIZUTANI, 2002), ITO etc. Back contact can be Mo (formation of $MoSe_2$, back surface field), Cr (Na barrier), Cu (CIS Cut process) or TCOs (superstrate cells, reversed configuration).

The standard thickness of the Cu(In,Ga)Se_2 (CIGS) absorber thin film layer is 1-2.5 µm (Figure 4.11). Reducing the thickness can reduce the materials cost and lower the production cost. Lundberg et al (Lundberg et al., 2003) research shows that when the thickness of the CIGS absorber layer is down to 0.8-1.0 µm, this structure can maintain a very high performance. When the CIGS layer was further reduced in thickness the loss in performance increased. When the thickness is 1.8 µm, the efficiency is 16.1%; when the thickness is 1.0µm, the efficiency is 15%; but when the thickness is 0.6 µm, the efficiency is 12.1%. A pronounced loss in the efficiency is shown.

4.3.2 Manufacturing

The CIGS layer can be deposited in a polycrystalline form directly onto molybdenum coated glass sheets or steel bands. Compared to large crystal, it can save energy. CIGS films can be manufactured by several different methods. A vacuum-based process (NAKADA & MIZUTANI, 2002; Repins et al., 2008; Delahey et al., 2004) is very common. Co-evaporating or co-sputtering copper, gallium, and indium can form a CuInGa thin film. Then the film is annealed in a selenide vapor to form the CIGS structure. The CIGS also can be formed with directly co-evaporating copper, gallium, indium and selenium onto a heated substrate. A non-vacuum-based process deposits nanoparticles of the precursor materials on the substrate and then sinters them in situ. CIGS layer also can be achieved through electroplating, which is the low cost way. Vacuum processes are expensive and can achieve very high efficiency of almost 20% (NAKADA & MIZUTANI, 2002; Repins et al., 2008). Non-vacuum solution processes progressed quickly and can get efficiencies of 10%-15% (Bhattacharya et al., 2001; Kapur et al., 2003), such companies as ISET, Nanosolar and IBM have this technology.

The recorded high efficiency (19.9%) CIGS solar cell deposition process is so-called three-stage process introduced by NREL (Gabor et al., 1994). It is a vacuum process. In first stage, In, Ga and Se are co-deposited to form $(In_x,Ga_{1-x})_2Se_3$, followed by the co-deposition of Cu and Se until Cu-rich composition CIGS is reached, and finally the overall Cu concentration is readjusted by subsequent deposition of In, Ga and Se. The process yields smooth films and can result in compositional profiles contributing to both high currents and voltages in the devices. This smoother surface facilitates the uniform conformal deposition of a thin buffer layer and prevents ion damage in CIGS during sputter deposition of ZnO/ZnO:Al (Gabor et al., 1994; Pomeo et al., 2004). The overall solar cell is made as follows: Soda-lime glass (SLG) substrate, sputtered Mo back contact, three stage co-evaporated CIGS, chemical-bath-deposited (CBD) CdS, sputtered resistive/conductive ZnO bi-layer, e-beam-evaporated Ni/Al grids, MgF2 antireflective coating, and photolithographic device isolation (Repins et al., 2008; Contreras et al., 1999; Ramanathan et al.,2003).

There are many deposition processes of CIGS thin film. Co-evaporation and two-step are the main processes (Shafarman & Stolt., 2003). Figure 4.12 shows the schematic of co-evaporation process.

The process uses line-of-sight delivery of the Cu, In, Ga, and Se from open-boat sources to the heated substrate. The source evaporation temperature is under control. Typical ranges are 1300 to 1400°C for Cu, 1000 to 1100°C for In, 1150 to 1250°C for Ga, and 300 to 350°C for Se evaporation. The advantage of the coevaporation process to deposit CIGS thin film is its considerable flexibility to choose the process specifics and to control film composition and band gap. The disadvantage is the difficulty in control, particularly the control of the Cu-evaporation source. The deposition, diagnostic, and control technology need be improved.

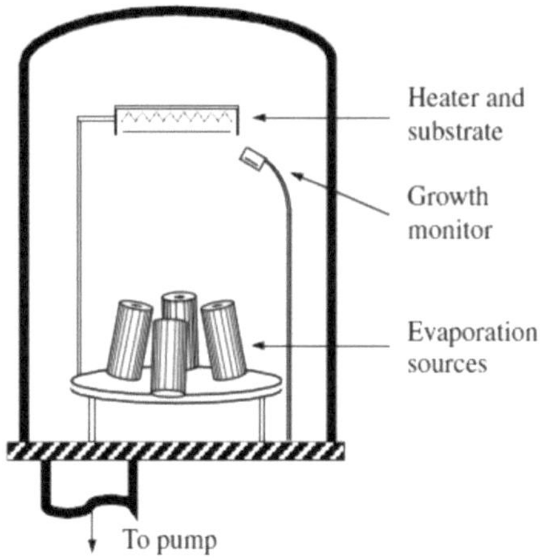

Fig. 4.12. Configuration for multisource elemental coevaporation.
(Shafarman & Stolt., 2003).

Two-step process is also called selenization. In selenization, copper, indium and gallium source atoms are ejected from solid targets by high-energy ions (also called "sputtering"). The selenium is added in a second step through a "selenization" process using a high temperature hydrogen selenide gas or solid source selenization. Sputtering is a mature technique. This is the primary advantage of selenization. But this process has the limited ability to control composition and increase band gap, which may limit device and module performance. Other difficulties that must be overcome include poor adhesion and the use of hydrogen selenide, which is hazardous and costly to handle.

Other Deposition includes hybrid sputtering in which Cu, In, and Ga are sputtered while Se is evaporated, closed space sublimation (CSS), chemical bath deposition (CBD) etc. These methods are reviewed in Reference (Shafarman & Stolt., 2003).

CIGS thin film solar cells have become commercial products successfully. A lower-cost process should feature high deposition rates, high material utilization, and simpler equipment capable of processing very large substrates. In order to lower the cost, there are still some critical issues for CIGS application (Ullal, H.S. & Roedern, B. von. 2007):

1. Standardization of equipment and technology for the growth of the CIGS absorber films;
2. Higher module efficiencies (over 20%);
3. Prevention of moisture ingress for flexible CIGS modules;
4. Thinner absorber layers of less than 1 micrometer or less;
5. CIGS absorber film stoichiometry and uniformity over large areas.

4.3.3 Efficiency

CIGS ($CuIn_xGa_{(1-x)}Se_2$) is one of the most prospective absorber materials for low cost polycrystalline thin film solar cells. Unlike the homojunction silicon cells, the structure of CIGS cells is a more complex heterojunction system. CIGS solar cell can achieve very high efficiencies such as 19.5% (x~0.3) (Contreras et al., 2005) ,19.9% (x~0.3) (Repins et al., 2008) etc. In National Renewable Energy Laboratory (NREL), the new world record total area efficiencies of 15.0% for CIS and 10.2% for surface modified CGS solar cells had been achieved. The use of gallium increases the optical band gap of the CIGS layer as compared to pure CIS, thus increasing the open-circuit voltage. Crystalline silicon solar cells can get efficiency as high as 24.7%, and CIGS efficiency is lower than 20%. But CIGS is much cheaper due to the much lower material cost and potentially lower fabrication cost. CIGS has very strong light absorption as a direct band-gap material. Most of the sunlight can be absorbed with 1-2um thickness CIGS layer. Compared to CdTe solar cells, CIGS are more pro-environmental solar cells. Because CIGS can consume smaller cadmium than CdTe, and cadmium is toxic material. Table 4.2 lists some conventional CIGS solar cell efficiencies.

CIGS cell	Efficiency (%)	Voc (mv)	Jsc (mA/cm2)	FF (%)	Area (cm2)	Reference
Thin-film ZnO/CdS/CIGS	19.9	690	35.5	81.2	0.419	Repins et al., 2008
Thin-film ZnO/CdS/CIGS	19.5	690	35.22	79.9	0.409	Contreras et al., 2005
Thin-film ZnO/CdS/CGS	10.2	823	18.61	66.77	0.419	AbuShama et al., 2005
Thin-film ZnO/CdS/CIS	15	490	40.5	75.2	0.403	AbuShama et al., 2005
Thin-film ZnO/CIGS	18.5	670	35.11	78.8	0.402	Kazmerski, 2004

Table 4.2. CIGS solar cell efficiencies

4.3.4 Cd-Free Cu(In, Ga)Se2 thin-film solar cells

The recent trend in buffer layers is to substitute CdS with 'Cd-free' wide-bandgap semiconductors and to replace the CBD technique with in-line-compatible processes. The standard device structure of $Cu(In,Ga)Se_2$ (CIGS)-based solar cells use a very thin chemical-bath-deposited (CBD) CdS buffer layer as window layer. But Cd is a kind of toxic materials. In the last decade, serious efforts to substitute the CdS buffer layer by other nontoxic low-absorbing materials have been made and the results are encouraging. For example, M.A. Contreras et al (Contreras et al., 2003) developed a Cd-free $Cu(In,Ga)Se_2$ thin-film solar cell. Thin film ZnS is used as window layer, which is a wider band gap material than CdS and can improve the quantum efficiency at short wavelengths. This structure solar cell can get energy conversion efficiency as high as 18.6%. This result suggests that CIGS thin film solar cells with efficiencies as high as those fabricated using CdS buffer can be achieved even if toxic Cd compounds are not utilized.

As an alternative to CdS, various materials show promising results. CBD-ZnS, MOCVD-ZnSe, ALD-ZnSe, CBD-ZnSe, CBD-ZnO, co-sputtered (Zn,Mg)O, CBD-$In(OH)_3$, ALCVD-In_2S_3, Co-evap-In_2Se_3, Co-evap-$InZnSe_x$, CBD-SnO_2 etc can be alternative buffer layers to replace the traditional CdS window layer and realize the Cd-free CIGS solar cells. The deposition methods can be: chemical bath deposition (CBD), atomic layer chemical vapour deposition (ALCVD), metal organic chemical vapour deposition (MOCVD), ion layer gas reaction (ILGAR), sputtering, thermal evaporation, and electrodeposition (ED) (Hariskos et al., 2005). Some processes can get very high efficiencies. CBD ZnS based buffer layer CIGS solar cells have 18.6% efficiency which is comparable with the CBD CdS. However, Zn-based compounds tend to form a blocking barrier due to the band alignment with CIGS. If the layer thickness is less than 50 nm, and the deposition quality is high, and the CIGS surface is uniform, the barrier can be reduced (Remeo et al., 2004). Hariskos et al has over-viewed the development of Cd-free materials and manufacture for $Cu(In,Ga)Se_2$-based thin-film solar cells and modules. Table 4.3 lists some CIGS solar cells condition with different buffer layer materials.

Material	Method	Efficiency (%)	Reference
CdS	CBD	19.9	Repins et al., 2008
ZnS-based	CBD	18.6	Hariskos et al., 2005
$In(OH)_3$:Zn-based	CBD	14	Tokita et al., 2003
In_2S_3-based	ALCVD	16.4	Naghavi et al., 2003
ZnSe-based	CBD	15.7	Ennaoui et al., 2001
ZnInSex	Coevap.	15.3	Hariskos et al., 2005
InxSey	Coevap.	13	Hariskos et al., 2005
ZnMgO	Sputtering	16.2	Negami et al., 2002
ZnO	CBD	15.7	Hubert et al., 2009
SnO_2	CBD	12.2	Hariskos et al., 2005

Table 4.3. CIGS efficiencies and manufacture with different window layers (Hariskos et al., 2005).

4.4 Nanocomposite application in dye-sensitized solar cells

A dye-sensitized solar cell (DSSC or DSC) is a low-cost solar cell like CdTe and CIGS thin film solar cells. DSSC has developed many years, but for a very long time the energy conversion efficiency is less than 1%. The high efficiency DSSC solar cell (over 7%) was invented by Michael Grätzel and Brian O'Regan at the École Polytechnique Fédérale de Lausanne (one of the two Swiss Federal Institutes of Technology) in 1991 (O'Regan & Grätzel., 1991). For this significant discovery, DSSC solar cell attracted considerable attention as a potential alternative to conventional inorganic photovoltaics and developed rapidly during the 1990s. Dr. Grätzel created DSSC solar cell from low to medium-purity materials through low-cost process, which exhibits a commercially realistic energy-conversion efficiency. So, DSSC solar cell is called as Grätzel cell also. Up to date, the DSSC solar cell can reach 11.1% efficiency (CHIBA et al., 2006; Han et al., 2006).

DSSC can be classified into the group of nanocomposite thin film solar cells. It is based on a semiconductor formed between a photo-sensitized anode and an electrolyte, a photoelectrochemical material. Because of the high surface area of the semiconductor film and the ideal spectral characteristics of the dye, the device harvests a high proportion of the incident solar energy flux (46%) and shows exceptionally high efficiencies for the conversion of incident photons to electrical current (more than 80%) (O'Regan & Grätzel., 1991). DSSC solar cell is insensitive to temperature change. When the temperature rises from 20 to 60°C, which is the normal natural condition, the power conversion efficiency has no change. In contrast, conventional silicon cells exhibit a significant decline over the same temperature range amounting to about 20%. DSSC solar cell has the lower sensitivity to angle of light incidence as compared to silicon-based cells. These advantages made DSSCs attractive for practical applications (Grätzel, 2004). Unlike solid semiconductor solar cell, the dye-sensitized solar cell uses liquid electrolyte to transport electron excited by sunlight. The liquid electrolyte dye-sensitized solar cell has some problems such as leaking and degradation. It is not stable. So colloids are developed and used in dye-sensitized solar cells. Normally a dye-sensitized solar cell uses nano-crystalline TiO_2 as the semiconductor material (O'Regan & Grätzel., 1991). Here the semiconductor is solely used for separating charge. Photoelectrons are generated from the photosensitive dye. Photovoltaic performance of the solar cell depends remarkably on the semiconductor materials.

4.4.1 DSSC solar cell structure and working cycle

Figure 4.13 is a schematic diagram of a DSSC solar cell. A typical DSSC includes back contact transparent conducting oxide (TCO), semiconductor titanium dioxide (TiO_2) thin film, electrolyte (usually a ruthenium bipyridyl complex), Pt catalyst and front contact TCO. The two transparent conducting oxide (TCO) glass slides are coated with fluorine doped tin oxide (FTO). TCO can be ZnO, ITO and SnO_2 etc. TCO in DSSC solar cell has the same functions with that in CdTe and CIGS solar cells. The n-type semiconductor TiO_2 thin film is attached to the substrate serves as a back electrode. The electrolyte contains a reduction-oxidation (redox) couple (I^- / I_3^-). The counter electrode TCO glass is covered with a very small amount of platinum (5-10 μg/cm²), which is responsible for catalytic cathodic reduction of triiodide to iodide. The electrolyte fully fills the space between the two electrodes.

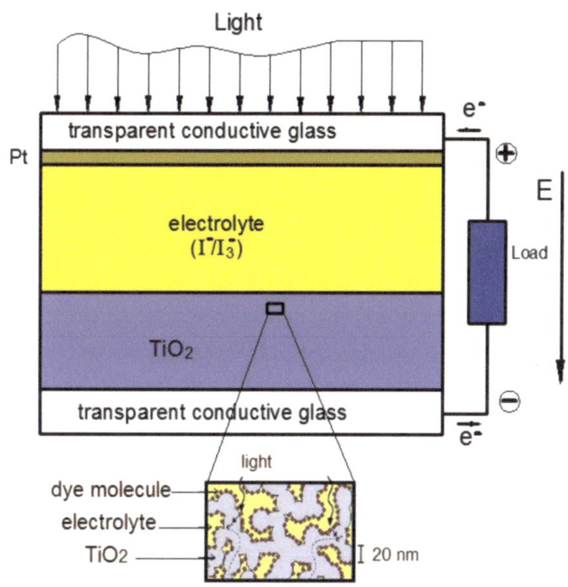

Fig. 4.13. A schematic diagram of structure and components of a DSSC solar cell
http://www.umk.fi/en/newsletter_newsletter_0108_Aitola_more.html

The semiconductor TiO_2 thin film is about 10μm thickness with the particles in nano-size about 20nm (O'Regan & Grätzel., 1991). Dye molecules are attached to the TiO_2 surface. TiO_2 is porous nano-crystalline structure and the internal surface area is thousands of times greater than the dimension of the cell (Yongbai, 2007). Under solar radiation, the dye molecules absorb the photons and excite electrons. The charge separation from the dye to the TiO_2 happens at the surface between the semiconductor and electrolyte surface. The application of porous nanocrystalline TiO_2 semiconductor and organic electrolytes (ruthenium complex) was Dr. Grätzel's great invention. The smooth surface between electrolyte and non-porous crystalline TiO_2 can absorbs incident light less than 1%. But the interface between electrolyte and porous nanocrystalline TiO_2 can absorb incident light over 80%.

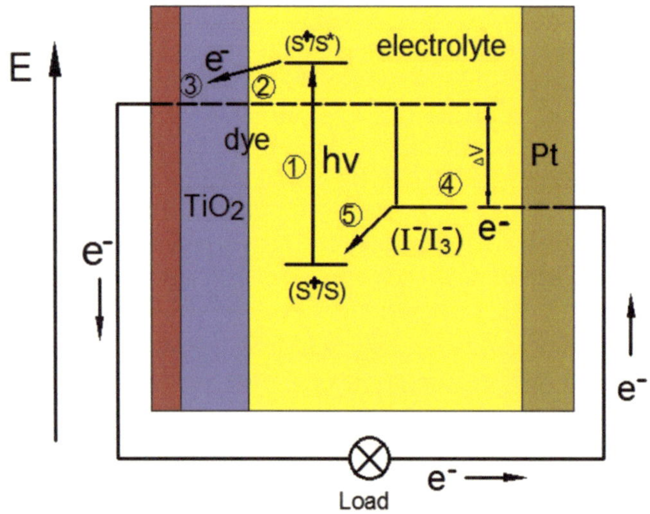

Fig. 4.14. Principle of operation and energy level scheme of the dye-sensitized nanocrystalline solar cell. Potential are referred to the standard calomel electrode (SCE) (Hagfeldtt & Grätzel, 1995). The redox system is iodide/triiodide-based redox eletrolyte. S stands for sensitizer; S*, electronically excited sensitizer; S^+, oxidized sensitizer

The DSSC is different with any other solar cells. This photovoltaic device uses molecules to absorb photons and convert them to electric charges without the need of intermolecular transport of electronic excitation, whereas others use atoms. DSSC solar cell separates the two functions of light harvesting and charge-carrier transport, whereas conventional and all of the other known photovoltaic devices perform both operations simultaneously (Grätzel, 2009). Figure 4.14 shows the DSSC solar cell working cycle and the relative energy levels of the cell. Incident sunlight passes through the transparent electrode into the dye layer where it can excite electrons that then flow into the semiconductor TiO_2 layer. The electrons flow toward the transparent electrode where they are collected for powering a load. After flowing through the external circuit, they are re-introduced into the cell on a metal electrode on the back, flowing into the electrolyte. The electrolyte then transports the electrons back to the dye molecules. This cycle includes 5 steps of photoelectronic chemistry process (Figure 4.14) (Yongbai, 2007):

1. An incident photon is absorbed by the dye molecule and an electron from a molecular ground state S is then excited to a higher energy state S* (at anode);

$$S + h\nu \rightarrow S^+ \tag{1}$$

2. the excited electron is injected to an oxidized state S^+ (at anode);

$$S^* \rightarrow S^+ + e^- \tag{2}$$

$$2S^{+}+3I^{-} \rightarrow 2S+I_3^{-} \quad (3)$$

3. the injected electron passes through the porous nanocrystalline material and reaches the transparent conducting oxide layer;
4. the electron is transferred to the triiodide in the electrolyte to yield iodine (at cathode);

$$I_3^{-}+2e^{-}(Pt)=3I^{-} \quad (4)$$

5. the reduction of the oxidized dye by the iodine in the electrolyte.

$$S^{+}+e^{-} \rightarrow S \quad (5)$$

hv is the photon energy; S stands for sensitizer; S*, electronically excited sensitizer; S^{+}, oxidized sensitizer.

4.4.2 Dye and efficiency

Different sensitizers have different transparency and absorbance on the sunlight spectral region. In dye solar cells, the dye is one of the key components for high power conversion efficiencies. Since highly efficient dye-sensitized solar cells (DSSCs) were first reported by Grätzel, substantial research has been carried out in the engineering of novel dye structures in order to enhance the performance of the system. At least 80 groups attended this research. The energy conversion efficiencies have been achieved over 10%.

Scientifically, a DSSC can achieve efficiency as high as 15% or more (Grätzel, 2003; Kroon et al., 2007). But till now, the highest efficiency is around 12%. This means there is a high potential for improvement in efficiency. The efficiency (η) is the function of the open circuit voltage (V_{oc}), the short circuit current (J_{sc}) and Fill factor (FF). The limitation of efficiency is on the three parts (Kroon et al., 2007):

a. Inefficient light absorption by existing sensitizer dyes in whole sunlight spectrum. Some are sensitized on UV-visible region but insensitive on IR region; some are sensitized on IR region, but insensitive on UV-visible region.
b. Sub-optimum photovoltage output. The improvements in the photovoltage could substantially increase device efficiencies by up to 50%.
c. Fill factors (FF) are limited in general by series resistance losses, light-intensity dependent recombination, non-ideal dark diode currents and, in some cases, shunt resistance losses.

Table 4.4 provides a summary of the best performance data obtained (state of the art) to date in solar light conversion, as well as open circuit voltages (V_{OC}) at various fill factors (FF) using various dyes for a number of surface areas.

In the new reports from M. Graetzel, EPFL has got very high efficiencies about sensitizers of C101 (11.3%, Thampi et al., 2008) and Z991 (11.91% (Grätzel, 2008),12.3% (Grätzel, 2009). There are still a lot of work to do to achieve the theory efficiency 15% (Kroon et al., 2007).

Many different dyes with a variety of chromophoric ligands have been synthesized and studied in DSSCs: polypyridine complexes of transition metals, metalloporphyrins, and metallo-phathalocyanines as well as different metal-free, donor-acceptor type dyes (Kalyanasundaram & Grätzel 2009). In most of the experiments, Ruthenium-based dyes are the main sensitizers. They are the best choices for DSSCs. They includesN3, N719, N749 (Black dye), N621 (η=9.57%, (Nazeeruddin et al., 2005), C104, Z907 (η=9.5%, Kalyanasundaram & Grätzel 2009) etc.

Sensitizer	Efficiency (%)	Voc (mv)	Jsc (mA/cm2)	FF (%)	Area (cm2)	research institution	Reference
N3	7.12-7.9				0.5	EPFL	O'Regan & Grätzel 1991
N3	10	720	18.2	73	0.31	EPFL	Nazeeruddin et al., 1993
N749	10.4	720	20.5	70	0.18	EPFL	Nazeeruddin et al., 2001
N3	9.3	770	16.8	72	1	EPFL	Nazeeruddin et al., 2003
N3	11.04	840	16.8	79		EPFL	Grätzel, 2004
Z910	10.2	777	17.2	76.4		EPFL	Wang et al., 2004
N719	11.18	846	17.7	75	0.16	EPFL	Nazeeruddin et al., 2005
N749	11.1	736	20.9	72.2	0.22	Sharp	CHIBA et al., 2006
N719	10.8	760	17.3	76	0.25	AIST	Chiba et al., 2006
N719	10.1	826	17	72	1.31	EPFL	Kroon et al., 2007
K77	10.5	780	19.2	72.5	0.158	EPFL	Kuang et al., 2007
C104	10.53	760	17.87	77.6	0.158		Gao et al., 2008

EPFL: Ecole Polytechnique Federale de Lausanne
AIST: the National Institute of Advanced Industrial Science and Technology (Japan)

Table 4.4 High efficiencies dye-sensitized solar cells development

N3 dye is one of the most commonly used sensitizing dyes for high-performance DSSCs. N719 dye is one of its derivatives,in which two protons are replaced by tetrabutylammonium cations (TBA) (figure 4.14). Both N3 and N719 have very good absorbance in UV-visible spectrum region, and can reach very high solar-energy-to-electricity-conversion efficiency (table 4.4). N749 (black dye) is sensitive in the low-frequency range of red and IR light.
Figure 4.15 and Table 4.5 show the structures and compositions of some of these high performance dyes respectively.
Recently some organic dyes have been developed. For example the D205 indoline dye (Ito et al., 2008) and C217 (Zhang et al., 2009) achieved efficiencies 9.5% and 9.8%, respectively.

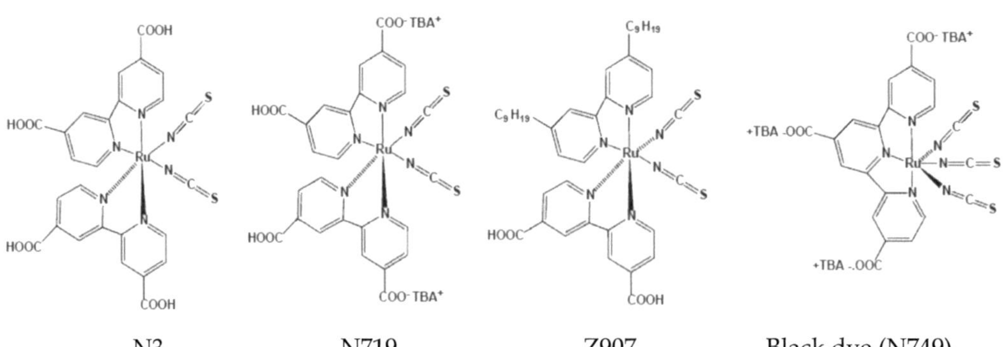

N3	N719	Z907	Black dye (N749)
η= 11.04%	η= 11.18%	η= 9.5%	η= 11.1%

Fig. 4.15. Structures of some of the most efficient Ru-based DSSCs

Dye	Name
N3	cis-bis(isothiocyanato)bis(2,2′-bipyridil-4,4′-dicarboxylate) ruthenium(II)
N719	cis-bis(isothiocyanato)bis(2,2′-bipyridyl-4,4′-dicarboxylate) ruthenium(II) bis-tetra-n-butylammonium
N621	cis-bis(isothiocyanato)(2,2′-bipyridyl-4,4′-dicarboxylate)(4,4′-ditridecyl-2,2′-bipyridine)ruthenium(II)
Z907	cis-bis(isothiocyanato)(2,2′-bipyridyl-4,4′-dicarboxylate)(4,4′-dinonyl-2,2′-bipyridine)ruthenium(II)
N749	tris(cyanato)-2,2',2″-terpyridyl-4,4',4″-tricarboxylate)Ru(II)

Table 4.5. Some dyes' compositions

4.4.3 Metal oxides

Semiconductor oxides used in dye-sensitized solar cell include TiO_2, ZnO, SnO_2, Nb_2O_5 etc., which serve as the carrier for the monolayers of the sensitizer using their huge surface and the medium of electron transfer to the conducting substrate. Anders Hagfeldtt et al (Hagfeldtt & Grätzel., 1995) even introduced nano-crystalline CdSe, CdS, WO_3, Fe_2O_3, In_2O_3 and Ta_2O_5 etc as the semiconductor oxides on the DSSC applications. Conventional high efficiency DSSC uses nano-crystalline TiO_2 as the semiconductor because TiO_2 has the properties such as low-cost price, abundance in the market, nontoxicity etc. ZnO has an energy gap of 3.37 eV, nearly identical to that of TiO_2. DSSCs built from ZnO nanoparticles show the second highest efficiencies after TiO_2. Karin Keis et al (Keis et al., 2002) developed a DSSC solar cell with nanoporous ZnO as the semiconductor to replace TiO_2 and N3 as the sensitizer. By improving the interfacial contact between dyes and ZnO particles in the film, overall solar-to-electric energy conversion efficiencies of up to 5% were obtained. Ming-Hong Lai et al (Lai et al., 2010) developed a new structure DSSC solar cell using ZnO to replace TiO_2. In this structure, a ZnO thin film is used as the TCO and a ZnO nanorod layer as semiconductor. There are also a lot of research about ZnO-based DSSC (Hongsith & Choopun., 2010; Martinson et al., 2007). Unfortunately, the ZnO-based DSSC efficiency is much lower compared to that based on TiO_2. One of the reasons for the low efficiency of ZnO-based DSSCs is the excessive dye aggregation on the ZnO surface, which in turn causes slower electron injection from the dye to ZnO (Zhang et al., 2008). The conditions as SnO_2 and other semiconductors are same low efficiencies. Up to date Nano-porous TiO_2 is still the best choice in semiconductor.

4.4.4 Electrolytes

The electrode in DSSC is to transport the charge carrier between photoanode and counter electrode. In the DSSC working cycle, the dye releases the electrons excited by the incident light into the conduction band of TiO_2. The dye loses electrons and must be reduced to its ground state rapidly. The ionic electrolyte must send electrons to the dye quickly. So, the electrode should have very good conductivity and good interfacial contact with the porous nanocrystalline layer and the counter electrode.

The electrolyte mixtures used in high-efficiency DSSC cells include liquid, solid-state and quasi-solid-state electrolytes.
Liquid electrolyte is widely used in DSSC. Most experiments of DSSC use liquid electrolyte including I^- / I_3^- redox couple. The light-to-electricity conversion efficiency based on I^- / I_3^- redox couple liquid electrolyte has been achieved to over 12%. This is the most efficient electrolyte. But the problem about this electrolyte is the corrosion for metals, which can cause sealing and long-time stability. Other kind redox couples such as Br^-/Br_2, SCN^-/SCN_2, $SeCN^-/SeCN_2$ also can be used in liquid electrolyte, but the light-to-electricity conversion efficiency is low. The limitations of liquid electrolyte are leakage of device, volatilization of organic solvents and long term stability. Solid-state electrolytes with no liquid materials can overcome the disadvantage of fluidity and volatility for liquid electrolytes and has long term stability. But the light-to-electricity conversion efficiency of DSSC with solid-state electrolytes is very low because of poor interface contact property and lower conductivity (Wu et al., 2008). Quasi-solid-state electrolyte is between liquid electrolyte and solid-state electrolyte. It has better long-term stability than liquid electrolytes but less than that of solid-state electrolyte. It has better ionic conductivity and interfacial contact property than that of solid-state electrolyte, but not as good as that of liquid electrolyte. Now DSSCs based on quasi-solid-state electrolyte are attracting more interest. Some researches got good results and the light-to-electricity conversion efficiencies are very high, such as 6.9% (Stathatos & Lianos., 2007) and 7.3% (Sathiya Priya et al., 2008). Compared with liquid electrolyte based DSSCs, the efficiencies of quasi-solid-state based DSSCs are very low.

4.4.5 Tandem-structure and dye-bilayer structure DSSC

Up to date, no sensitizer has good incident absorbance along the whole sunlight spectrum. Like conventional solar cell, the DSSC also can be synthesized as tandem-structure. But DSSC has special property that the nanocrystalline dye-sensitized solar cell DSSC can be designed by appropriate choice of the sensitizer to absorb incident photons in selective spectral regions of the solar spectrum while maintaining high transparency in the remaining wavelength range (Liska et al., 2006). So, in the tandem-structure DSSC, just choosing different sensitizers which have different transparency and absorbance on the sunlight spectral region to make top and bottom cells can get high conversion efficiencies. For example, Yanagida et al in AIST group developed a kind of DSSC tandem-structure dye-sensitized solar cell (Yanagida et al., 2010). N719 is sensitized on UV-visible region and N749 (black dye) is sensitized near infrared region (figure 4.16). The spectral response of the two systems is complementary to each other and reveals the advantage of employing them in a tandem mode. Tandem-structure DSSC with N719 as top cell and N749 as bottom cell sensitized on incident light from UV to infrared region can reach the highest photovoltaic conversion efficiency of 10.6% (parallel structure) under solar-simulating light conditions (AM1.5, $100mWcm^{-2}$).
A DSSC cell can combine a CIGS cell to buildup a tandem-structure cell. Liska et al developed a photovoltaic tandem cell comprising a nanocrystalline dye-sensitized solar cell (N719 based) as a top cell for high-energy photons and a copper indium gallium selenide (CIGS) thin-film bottom cell for lower-energy photons (Liska et al., 2006). Figure 4.17 shows photon-to-current conversion efficiency (IPCE) of this DSSC/CIGS tandem-structure solar cell. The DSSC top cell shows a strong response in the UV, blue, and green wavelength domains and the CIGS bottom

cell exhibits high external quantum efficiencies in the red and near IR parts of the spectrum extending from 700 to 1150 nm where the DSSC is insensitive to light. This DSSC/CIGS tandem-structure is sensitized on the spectrum from UV to IR region.

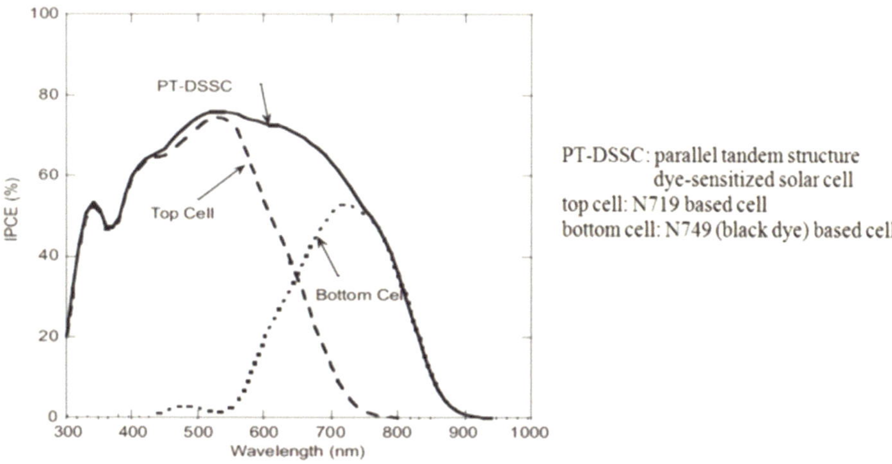

Fig. 4.16. The incident monochromatic photon-to-current conversion efficiency (IPCE) spectra of a PT-DSSC (solid line) and its top and bottom cells (dashed and dotted lines, respectively).

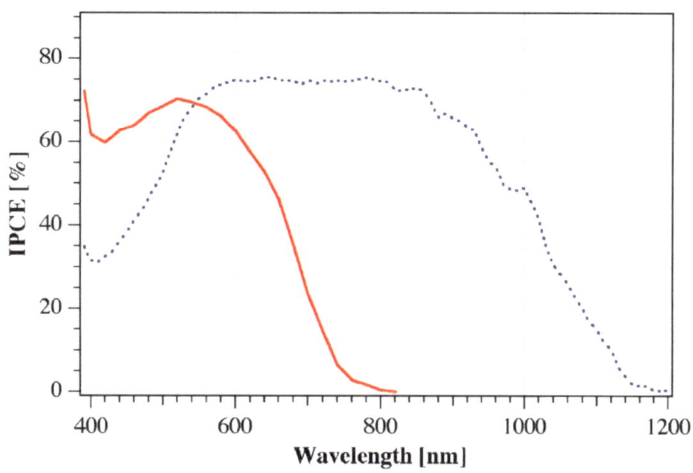

Fig. 4.17. Spectral response curves of the photon-to-current conversion efficiency (IPCE) for a DSSC top cell (bold line) and a CIGS bottom cell (dotted line) (Liska et al., 2006)

This cell can achieve photovoltaic conversion efficiencies as high as 15.75% which was significantly higher than that of the individual cells. Here the N719 based DSSC top cell efficiency is 7.85% and the CIGS bottom cell efficiency is 8.17%.

Inakazu et al studied a dye-sensitized solar cells (DSSC) containing dye-bilayer structure of black dye and NK3705. This structure was fabricated by staining one TiO_2 layer with these two dyes, step by step. The short circuit current (J_{sc}) and the incident photon to current efficiency (IPCE) of this cell are almost the sum of those of DSSC stained with black dye only and DSSC stained with NK3705 only. The efficiency of this dye-bilayer solar cell is 9.16%, which is higher than that of black-dye based solar cell (7.28%) and NK3705 based cell (1.85%) (Inakazu et al., 2008).

4.5 Polymer solar cells

A polymer solar cell is also called an organic solar cell or a plastic solar cell. The polymer as substrate can make the cell flexible. The polymer solar cell is typically made from a nanocomposite thin film material. Such a solar cell produces electricity from sunlight by the polymer. Compared with silicon-based solar cells, polymer solar cells can greatly lower the cost. For a polymer solar cell, there are three major ways to lower the cost. First, it uses low cost plastic as the active materials to convert solar energy into electricity; second, the active plastic layer thickness is in nano-scale, this means the raw materials consumption will be significantly decreased; the third aspect is the low manufacturing cost, the solution processing is low easy and low cost. The ink-jet printing, micro-contact printing, and other soft lithography techniques have further improved the potential of conjugated polymers for low-cost fabrication of large-area integrated devices on both rigid and flexible substrates. These techniques are similar to printing on newspaper and the cost is very low. Polymer solar cells is cheaper to make than thin-film cadmium-telluride (CdTe) or copper indium gallium selenide (CIGS) ones because they use low-cost materials and are easy to print. The main reason for the extensive interest in polymer semiconducting materials is their potential for the realization of a low cost, easily processed, light-weight and mechanical flexible device. The main problems of polymer solar cell are photochemical degradation and low light-electric conversion efficiency. The research of polymer solar cells concentrates on increasing photochemical stability and energy conversion efficiency.

4.5.1 Structure of polymer solar cell

Figure 4.18 shows the polymer solar cell structure. It is similar with CdTe, CIGS and DSSC. The difference is the active layer. It is polymer blend, including a p-type semiconductor (donor) and an n-type semiconductor (acceptor). The basic process includes absorption of light, charge transfer and separation of the opposite charges, charge transport and charge collection. The process is the same with DSSC. So, it can also be classified as dye-sensitized solar cell.

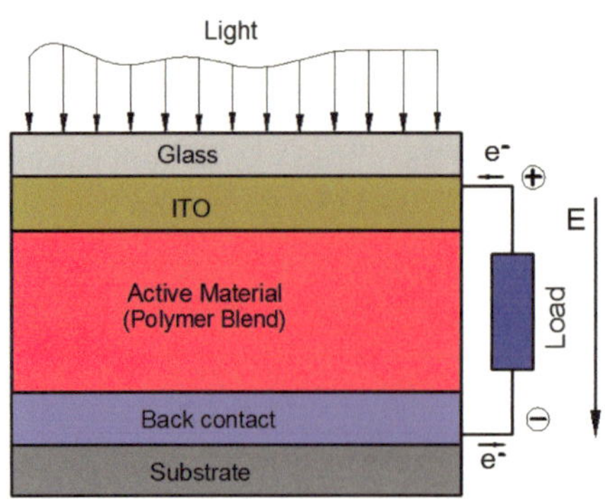

4.18. Typical polymer solar cell structure

Donors:
MDMO-PPV = poly[2-methoxy-5-(3´,7´-dimethyloctyloxy)-p-phenylene vinylene];
P3HT= poly(3-hexylthiophene);
PFDTBT: poly[2,7-[9-(2´-ethylhexyl)-9- hexylfluorene]-alt-5,5-(4´,7´-di-2-thienyl-2´,11´,3´-benzothiadiazole)].
Acceptors:
PCBM: 3´-phenyl-3´H-cyclopropa[1,9][5,6]fullerene-C60-Ih-3´-butanoic acid methyl ester;
PCBM: 3´-phenyl-3´H-cyclopropa[8,25][5,6]fullerene-C70-D5h(6)-3´-butanoic acid methyl ester.

4.5.2 Efficiency

Due to the inferior charge-transport properties and limited spectral absorption range of the polymer active layer, the overall power conversion efficiencies (PCE) reported are still low. Green et al have reported the high efficiencies polyer solar cells. For single solar cell, the efficiency is 5.15% (Voc, 876mV; Jsc, 9.39mA/cm2; FF, 62.5%). For 2-cell tandem solar cell, the efficiency is 6.1% (Voc, 1589mV; Jsc, 6.18mA/cm2; FF, 61.9%) (Green et al., 2010). This low efficiency limits the application of commercialization. Solarmer Energy, Inc is the leading developer company in polymer solar cell technology. In 2009 alone, Solarmer produced three certified world records: 6.8%, 7.6% and 7.9%. Recently a new world record of 8.13% for their organic photovoltaic (OPV) cell efficiency was achieved, which was certified by the National Renewable Energy Laboratory (NREL). (http://www.solarmer.com). This new development made the commercialization of polymer solar cells is possible. This cutting-edge technology has the potential to drive energy production cost down to 12-15 cents/kWh. But for polymer solar cell, the efficiency over 15% can have a major impact on the solar cell power market. In theory, the efficiency with 15-20% is possible. This means there is still much work to do to improve the efficiency.

5. Summery and conclusions

Based on the review of thin film typed nanocomposites for solar energy conversion, the following conclusions may be made. Various nanocomposite thin films for solar cell fabrication are still under development. The principle of photovoltaic energy conversion of nanocomposites is different from that of the first generation of solar cells, i.e. silicon cells because of the structure complexity. The efficiency of solar cells could potentially be improved by using two dimensional nanocomposite cells such as $CuInGaSe_2$(CIGS), dye-sensitized solar cell (DSSC). There are many challenges in polymer based solar cells research because they have very low efficiencies.

6. Reference

AbuShama, J. A., Johnston, S., Moriarty, T., Teeter, G., Ramanathan, K., & Noufi, R. (2004). Properties of ZnO/CdS/$CuInSe_2$ Solar Cells with Improved Performance. *Prog. Photovolt: Res. Appl.*, *12*, pp.39–45.

AbuShama, J., Noufi, R., & Johnston, S. (2004). Improved Performance in $CuInSe_2$ and Surface-Modified $CuGaSe_2$ Solar Cells. *the 2004 DOE Solar Energy Technologies, Program Review Meeting*. Denver, Colorado.

Anderson, T., Crisalle, O., Li, S., & Holloway, P. (2003). *Future CIS Manufacturing Technology Development.* Gainesville, Florida.

Aramoto, T., Adurodija, F., Nishiyama, Y., Arita, T., Hanafusa, A., Omura, K.,& Morita, A. (2003). A new technique for large-area thin film CdS/CdTe solar cells. *Solar Energy Materials & Solar Cells , 75*, pp. 211–217.

Barri, K. J., & Ferekides, C. (2005). INTRODUCTION OF Cu IN CdS AND ITS EFFECT ON CdTe SOLAR CELLS. *31st IEEE Photovoltaic Specialists Conference.*

Beecroft, L. L., & Ober., C. K. (1997). Nanocomposite Materials for Optical Applications. *Chem. Mater. , 9*, 1302-1317.

Bhattacharya, R., Balcioglu, A., & Ramanathan, K. (2001). Deep-level transient spectroscopy (DLTS) of CdS/$CuIn_{(1-x)}Ga_xSe_2$-based solar cells prepared from electroplated and autoplated precursors, and by physical vapor deposition. *Thin Solid Films , 384*, pp. 65-68.

Britta, J., & Ferekides, C. (1993). Thin-film Cd/CdTe solar cell with 15.8% efficiency. *Appt. Phys. Lett. , 62* (22), 2851-2852.

Castillo-Alvarado, F., Inoue-Chávez, J., Vigil-Galán, O., Sánchez-Meza, E., López-Chávez, E., & Contreras-Puente, G. (2010). C–V calculations in CdS/CdTe thin films solar cells. *Thin Solid Films , 518*, 1796–1798.

Chen, Z., Tang, Y., Zhang, L., & Luo, L. (2006). Electrodeposited nanoporous ZnO films exhibiting enhanced performance in dye-sensitized solar cells. *Electrochimica Acta , 51*, 5870–5875.

Chiba, Y., Islam, A., Komiya, R., Koide, N., & Han, L. (2006). Conversion efficiency of 10.8% by a dye-sensitized solar cell using a TiO_2 electrode with high haze. *Appl. Phys. Lett. , 88*, 223505 1-3.

CHIBA, Y., ISLAM, A., WATANABE, Y., KOMIYA, R., KOIDE, N., & HAN, L. (2006). Dye-Sensitized Solar Cells with Conversion Efficiency of 11.1%. *Japanese Journal of Applied Physics , 45* (25), L638–L640.

Chu, T. L., Chu, S. S., Ferekides, C., Wu, C. Q., Britt, J., & Wang, C. (1991). 13.4% efficient thin-film CdS/CdTe solar cells. *J. Appl. Phys. , 70* (12), pp. 7608-7612.

Colombo, E. B., Calusi, S., Giuntini, L., Lo Giudice, A., Manfredotti, C., Massi, M., Olivero, P., Romeo, A., Romeo, N. & Vittone E. (2009). IBIC analysis of CdTe/CdS solar cells. *Nuclear Instruments and Methods in Physics Research B , 267*, pp. 2181–2184.

Compaan, A. D., Gupta, A., Drayton, J., Lee, S.-H., & Wang, S. (2004). 14% sputtered thin-film solar cells based on CdTe. *phys. stat. sol. (b) , 241* (3), pp. 779–782.

Compaan, A. D., Gupta, A., Lee, S., Wang, S., & Drayton, J. (2004). High efficiency, magnetron sputtered CdS/CdTe solar cells. *Solar Energy , 77*, pp. 815–822.

Contreras Puente, G., Vigil, O., Ortega-LoÂpez, M., Morales-Acevedo, A., Vidal, J., & Albor-Aguilera, M. (2000). New window materials used as heterojunction partners on CdTe solar cells. *Thin Solid Films , 361-362*, 378-382.

Contreras, M. A., Egaas, B., Ramanathan, K., Hiltner, J., Swartzlander, A., Hasoon, F., & Noufi, R. (1999). Progress Toward 20% Efficiency in Cu(In,Ga)Se_2 Polycrystalline Thin-film Solar Cells. *Prog. Photovolt: Res. Appl. , 7*, pp.311-316.

Contreras, M. A., Ramanathan, K., AbuShama, J., Hasoon, F., Young, D. L., Egaas, B., & Noufi, R. (2005). Diode Characteristics in State-of-the-Art ZnO/CdS/Cu($In_{1-x}Ga_x$)Se_2 Solar Cells. *Prog. Photovolt: Res. Appl.*, *13*, 209–216.

Contreras, M. A., Romero, M. J., To, B., Hasoon, F., Noufi, R., Ward, S., & Ramanathan, K. (2002). Optimization of CBD CdS process in high-efficiency Cu(In,Ga)Se_2-based solar cells. *Thin Solid Films*, *403-404*, 204-211.

Contreras, M., Nakada, T., Hongo, M., Pudov, A., & Sites, J. (2003). ZnO/ZnS(O,OH)-/Cu(In,Ga)Se_2/Mo SOLAR CELL WITH 18.6% EFFICIENCY. *Proceedings 3rd World Conference of Photovoltaic Energy Conversion*, (pp. 570-573). Osaka, Japan.

Delahoy, A. E., Chen, L., Akhtar, M., Sang, B., & Guo, S. (2004). New technologies for CIGS photovoltaics. *Solar Energy*, *77*, pp. 785–793.

Dhere, N. G. (2006). Present status and future prospects of CIGSS thin film solar cells. *Solar Energy Materials & Solar Cells*, *90*, pp.2181–2190.

Dhere, N. G. (2006). Present status and future prospects of CIGSS thin film solar cells. *Solar Energy Materials & Solar Cells*, *90*, 2181-2190.

Dhere, R., Rose, D., Albin, D., Asher, S., Al-Jassim, M., Cheong, H., Swartzlander, A., Moutinho, H., Coutts, T. & Sheldon, P. (1997). Influence of CdS/CdTe Interface Properties on the Device Properties. *the 26th IEEE Photovoltaic Specialists Conference.* Anaheim, California.

Dobson, K. D., Visoly-Fisher, I., Hodes, G., & Cahen, D. (2000). Stability of CdTe/CdS thin-film solar cells. *Solar Energy Materials & Solar Cells*, *62*, pp. 295-325.

Dobson, K. D., Visoly-Fisher, I., Hodes, G., & Cahen, D. (2000). Stability of CdTe/CdS thin-film solar cells. *Solar Energy Materials & Solar Cells*, *62*, 295-325.

Dzhafarov, T., Yesilkaya, S., Yilmaz Canli, N., & Caliskan, M. (2005). Diffusion and influence of Cu on properties of CdTe thin films and CdTe/CdS cells. *Solar Energy Materials & Solar Cells*, *85*, 371–383.

Ennaoui, A., Siebentritt, S., Lux-Steiner, M., Riedl, W., & Karg, F. (2001). High-effciency Cd-free CIGSS thin-film solar cells with solution grown zinc compound buffer layers. *Solar Energy Materials & Solar Cells*, *67*, 31-40.

Enrıquez, J., Pantoja, E., Barojas, G., Silva Gonzalez, R., & Pal, U. (2007). S and Te inter-diffusion in CdTe/CdS hetero junction. *Solar Energy Materials & Solar Cells*, *91*, pp. 1392–1397.

Eze, F., & Okeke, C. (1997). Chemical-bath-deposited cobalt sulphide film: preparation effects. *Materials Chemistry and physics*, *47*, 31-36.

Fahrenbruch, A. L. (2007). Exploring Back Contact Technology to Increase CdS/CdTe Solar Cell Efficiency. *the Materials Research Society Meeting*, (pp. Y7-5). San Francisco, CA.

Ferekides, C., Marinskiy, D., Viswanathan, V., Tetali, B., Palekis, V., Selvaraj, P., & Morel, D. L. (2000). High efficiency CSS CdTe solar cells. *Thin Solid Films*, *361-362*, pp. 520-526.

Gabor, A. M., Tuttle, J. R., Albin, D. S., Contreras, M. A., Noufi, R., & Hermann, A. M. (1994). High-efficiency $CuIn_xGa_{1-x}Se_2$ solar cells made from $(In_x,Ga_{1-x})_2Se_3$ precursor films. *Appl. Phys. Lett.*, *65* (2), 198-200.

Gamboa, S., Sebastian, P., Mathew, X., Nguyen-Cong, H., & Chartier, P. (1999). A CdTe/PMeT photovoltaic structure formed by electrodeposition and processing. *Solar Energy Materials & Solar Cells , 59,* 115-124.

Gao, F., Wang, Y., Zhang, J., Shi, D., Wang, M., Humphry Baker, R., Wang, P., Zakeeruddin, S. M. & Grätzel, M. (2008). A new heteroleptic ruthenium sensitizer enhances the absorptivity of mesoporous titania film for a high efficiency dye-sensitized solar cell, Chem. Commun., 2008, 2635–2637.

Gayam, S., Bapanapalli, S., Zhao, H., Nemani, L., Morel, D., & Ferekides, C. (2007). The structural and electrical properties of Zn–Sn–O buffer layers and their effect on CdTe solar cell performance. *Thin Solid Films , 515,* 6060-6063.

Grätzel, M. (2009). The Advent of Mesoscopic Solar Cells. *International Symposium on Innovative Solar Cells 2009.* Tokyo, Japan.

Grätzel, M. (2004). Conversion of sunlight to electric power by nanocrystalline dye-sensitized solar cells. *Journal of Photochemistry and Photobiology A: Chemistry , 164,* 3-14.

Grätzel, M. (2003). Dye-sensitized solar cells. *Journal of Photochemistry and Photobiology C: Photochemistry Reviews , 4,* 145-153.

Grätzel, M. (2008). Power from the Sun, The advent of mesoscopic solar cells. *International Symposium on Solar Cells and Solar Fuels.* Dalian, China.

Grätzel, M. (2009). Recent Advances in Sensitized Mesoscopic Solar Cells. *ACCOUNTS OF CHEMICAL RESEARCH , 42* (11), 1788-1798.

Green, M. A., Emery, K., Hishikawa, Y., & Warta, W. (2010). Solar cell efficiency tables (version 35). *PROGRESS IN PHOTOVOLTAICS: RESEARCH AND APPLICATIONS, 18,* 144-150.

Guillemoles, J.-F., Kronik, L., Cahen, D., Rau, U., Jasenek, A., & Schock, H.-W. (2000). Stability Issues of Cu(In,Ga)Se2-Based Solar Cells. *J. Phys. Chem. , B* (104), 4849-4862.

Gupta, A., & Compaan, A. D. (2004). All-sputtered 14% CdS/CdTe thin-film solar cell with ZnO:Al transparent conducting oxide. *Appl. Phys. Lett. , 85* (4), pp. 684-686.

Gupta, A., Parikh, V., & Compaan, A. D. (2006). High efficiency ultra-thin sputtered CdTe solar cells. *Solar Energy Materials & Solar Cells , 90,* pp. 2263–2271.

Hagfeldtt, A., & Grätzel, M. (1995). Light-Induced Redox Reactions in Nanocrystalline Systems. *Chem, Rev. , 95,* 49-68.

Han, L., Fukui, A., Fuke, N., Koide, N., & Yamanaka, R. (2006). High Efficiency of Dye-Sensitized Solar Cell and Module. *Photovoltaic Energy Conversion, Conference Record of the 2006 IEEE 4th World Conference on, 1,* pp. 179-182. Waikoloa, HI.

Hariskos, D., Spiering, S., & Powalla, M. (2005). Buffer layers in Cu(In,Ga)Se_2 solar cells and modules. *Thin Solid Films 480–481,* pp. 99– 109.

Hermanna, A., Gonzalez, C., Ramakrishnan, P., Balzar, D., Popa, N., Rice, P., Marshall, C. H., Hilfiker, J. N., Tiwald, T., Sebastian, P. J., Calixto, M. E. & Bhattacharya, R. N. (2001). Fundamental studies on large area Cu(In,Ga)Se_2 films for high efficiency solar cells. *Solar Energy Materials & Solar Cells , 70,* 345–361.

Hodes, G. (2003). *Chemical Solution Deposition of Semiconductor Films.* New York: Marcel Dekker, Inc.

Hongsith, N., & Choopun, S. (2010). ZnO Nanobelts as a Photoelectrode for Dye-Sensitized Solar Cell. *Chiang Mai J. Sci.*, *37* (1), 48-54.

Huang, C., Meen, T., Lai, M., & Chen, W. (2004). Formation of $CuInSe_2$ thin films on flexible substrates by electrodeposition (ED) technique. *Solar Energy Materials & Solar Cells*, 553-565.

Hubert, C., Naghavi, N., Roussel, O., Etcheberry, A., Hariskos, D., Menner, R., Powalla, M., Kerrec, O. & Lincot, D. (2009). The Zn(S,O,OH)/ZnMgO Buffer in Thin Film $Cu(In,Ga)(S,Se)_2$ -Based Solar Cells Part I: Fast Chemical Bath Deposition of Zn(S,O,OH) Buffer Layers for Industrial Application on Co-evaporated $Cu(In,Ga)Se_2$ and Electrodeposited $CuIn(S,Se)_2$ Solar Cells. *Prog. Photovolt: Res. Appl.*, *17*, 470-478.

Inakazu, F., Noma, Y., Ogomi, Y., & Hayase, S. (2008). Dye-sensitized solar cells consisting of dye-bilayer structure stained with two dyes for harvesting light of wide range of wavelength. *APPLIED PHYSICS LETTERS*, *93*, 093304 1-3.

Ito, S., Miura, H., Uchida, S., Takata, M., Sumioka, K., Liska, P., Comte, P., Pechy, P. & Grätzel, M. (2008). High-conversion-efficiency organic dye-sensitized solar cells with a novel indoline dye. *Chem. Commun.*, 5194-5196.

Ivanov, V., Balakai, V., Kurnakova, N., Arzumanova, A., & Balakai, I. (2008). Synergistic Effect in Nickel–Teflon Composite Electrolytic Coatings. *Russian Journal of Applied Chemistry*, *81* (12), 2169–2171.

Kalyanasundaram, K., & Grätzel, M. (2009). Efficient Dye-Sensitized Solar Cells for Direct Conversion of Sunlight to Electricity. *Material Matters*, *4* (4), 88-90.

Kapur, V. K., Bansal, A., Le, P., & Asensio, O. I. (2003). Non-vacuum processing of $CuIn_{(1-x)}Ga_xSe_2$ solar cells on rigid and flexible substrates using nanoparticle precursor inks. *Thin Solid Films*, *431–432*, 53-57.

Kato, K., Hibino, T., Komoto, K., Ihara, S., Yamamoto, S., & Fujihara, H. (2001). A life-cycle analysis on thin-film CdS/CdTe PV modules. *Solar Energy Materials & Solar Cells*, *67*, pp. 279-287.

Kazmerski, L. (2004). Photovoltaics R&D: At the Tipping Point. *the 2004 DOE Solar Energy Technologies Program Review Meeting.* Denver, Colorado.

Keis, K., Magnusson, E., Lindstrom, H., Lindquist, S. E., & Hagfeldt, A. (2002). A 5% efficient photoelectrochemical solar cell based on nanostructured ZnO electrodes. *Solar Energy Materials & Solar Cells*, *73*, 51-58.

Khrypunov, G., Romeo, A., Kurdesau, F., Batzner, D., Zogg, H., & Tiwari, A. (2006). Recent developments in evaporated CdTe solar cells. *Solar Energy Materials & Solar Cells*, *90*, pp. 664–677.

King, R. (2009). Raising the Efficiency Ceiling in Multijunction Solar Cells. *Stanford Photonics Research Center Symposium.*

Kois, J., Ganchev, M., Kaelin, M., Bereznev, S., Tzvetkova, E., Volobujeva, O., Stratieva, N. & Tiwari, A. N. (2008). Electrodeposition of Cu–In–Ga thin metal films for Cu(In, Ga)Se2 based solar cells. *Thin Solid Films*, *516*, 5948-5952.

Krishnan, S., Sanjeev, G., Pattabi, M., & Mathew, X. (2009). Effect of electron irradiation on the properties of CdTe/CdS solar cells. *Solar Energy Materials & Solar Cells*, *93*, pp. 2–5.

Kroon, J. M., Bakker, N. J., Smit, H. J., Liska, P., Thampi, K. R., Wang, P., Zakeeruddin, S. M., Grätzel, M., Hinsch, A., Hore, S., Wu¨rfel, U., Sastrawan, R., Durrant, J. R., Palomares, E., Pettersson, H., Gruszecki, T., Walter, J., Skupien, K. & Tulloch, G. E. (2007). Nanocrystalline Dye-sensitized Solar Cells Having Maximum Performance. *Prog. Photovolt: Res. Appl.*, *15*, 1–18.

Kuang, D., Klein, C., Ito, S., Moser, J.-E., Humphry-Baker, R., Evans, N., Duriaux, F., Grätzel, C., Zakeeruddin, S. M. & Grätzel, M. (2007). High-Efficiency and Stable Mesoscopic Dye-Sensitized Solar Cells Based on a High Molar Extinction Coefficient Ruthenium Sensitizer and Nonvolatile Electrolyte. *Adv. Mater.*, *19*, 1133-1137.

Kumazawa, K., Shibutani, S., Nishio, T., Aramoto, T., Higuchi, H., Arita, T., Hanafusa, A., Omura, K., Murozono, M. & Takakura, H. 15.1% Highly efficient thin film CdS/CdTe Solar cell. *Solar Energy Materials and Solar*, *49*, pp. 205-212.

Kylner, A. (1999). Effect of impurities in the CdS buffer layer on the performance of the Cu(In,Ga)Se_2 thin film solar cell. *J. Appl. Phys.*, *85* (9), 6858-6865.

Lai, M. H., Tubtimtae, A., Lee, M. W., & Wang, G. J. (2010). ZnO-Nanorod Dye-Sensitized Solar Cells: New Structure without a Transparent Conducting Oxide Layer. *ZnO-International Journal of Photoenergy*, *2010*.

Lee, J. (2005). Comparison of CdS films deposited by different techniques: Effects on CdTe solar cell. *Applied Surface Science*, *252*, 1398–1403.

Levy-Clement, C., Katty, A., Bastide, S., Zenia, F., Mora, I., & Munoz-Sanjose, V. (2002). A new CdTe/ZnO columnar composite film for Eta-solar cells. *Physica*, *E* (14), 229-232.

Liska, P., Thampi, K. R., & Grätzel, M. (2006). Nanocrystalline dye-sensitized solar cell/copper indium gallium selenide thin-film tandem showing greater than 15% conversion efficiency. *APPLIED PHYSICS LETTERS*, *88*, 203103 1-3.

Lu, C., Cui, Z., Wang, Y., Li, Z., Guan, C., Yang, B., & Shen, J. (2003). Preparation and characterization of ZnS–polymer nanocomposite films with high refractive index. *Journal of Materials Chemistry*, *13*, 2189–2195.

Lu, H., Liu, Y., Gou, J., Leng, J., & Du, S. (2010). Synergistic effect of carbon nanofiber and carbon nanopaper on shape memory polymer composite. *APPLIED PHYSICS LETTERS*, *96*, 084102 1-3.

Lundberg, O., Bodegard, M., Malmstrom, J., & Stolt, L. (2003). Influence of the Cu(In,Ga)Se_2 Thickness and Ga Grading on Solar Cell Performance. *Prog. Photovolt: Res. Appl., 11*, 77-88.

Lundberg, O., Edoff, M., & Stolt, L. (2005). The effect of Ga-grading in CIGS thin film solar cells. *Thin Solid Films*, *480-481*, 520-525.

Luschitz, J., Siepchen, B., Schaffner, J., Lakus-Wollny, K., Haindl, G., Klein, A., et al. (2009). CdTe thin film solar cells: Interrelation of nucleation, structure and performance. *Thin Solid Films*, *517*, 2125–2131.

Marayanan, K., Vijayakumar, K., Mnair., K., & Nrao, G. (1997). Chemical bath deposition of CdS thin films and their partial conversion to CdO on annealing. *Bull. Mater. Sci.*, *20* (3), pp. 287-295.

Marsillac, S., Parikh, V., & Compaan, A. (2007). Ultra-thin bifacial CdTe solar cell. *Solar Energy Materials & Solar Cells*, *91*, pp. 1398–1402.

Martinson, A. B., Elam, J. W., Hupp, J. T., & Pellin, M. J. (2007). ZnO Nanotube Based Dye-Sensitized Solar Cells. *Nano Lett.*, *7* (8), 2183-2187.

Mathew, X., J. Enriquez, P., Romeo, A., & Tiwari, A. N. (2004). CdTe/CdS solar cells on flexible substrates. *Solar Energy*, *77*, pp. 831–838.

Mathew, X., Sebastian, P., Sanchez, A., & Campos, J. (1999). Structural and optoelectronic properties of electrodeposited CdTe on stainless steel foil. *Solar Energy Materials & Solar Cells*, *59*, 99-114.

Matsune, K., Oda, H., Toyama, T., Okamoto, H., Kudriavysevand, Y., & Asomoza, R. (2006). 15% Efficiency CdS/CdTe thin film solar cells using CdS layers doped with metal organic compounds. *Solar Energy Materials & Solar Cells*, *90*, pp. 3108-3114.

Matulionis, I., Sijin, H., Drayton, J., Price, K., & Compaan, A. (2001). Cadmium telluride solar cells on molybdenum substrates. *In Proceedings of the Materials Research Society symposium on II–VI compound semiconductor photovoltaic materials.* Warrenda, PA.

Mendoza-Perez, R., Sastre-Hernandez, J., Contreras-Puente, G., & Vigil-Galan, O. (2009). CdTe solar cell degradation studies with the use of CdS as the window material. *Solar Energy Materials & Solar Cells*, *93*, 79-84.

Meyer, G., & Saura, J. (1992). CdTe crystals grown by chemical vapour deposition. *Journal of materials science letters*, *11*, 143-144.

Minami, T. (2008). Present status of transparent conducting oxide thin-film development for Indium-Tin-Oxide (ITO) substitutes. *Thin Solid Films*, *516*, 5822–5828.

Minami, T. (2005). Transparent conducting oxide semiconductors for transparent electrodes. *Semicond. Sci. Technol.*, *20*, pp. S35–S44.

Mitchell, K. W., Fahrenbruch, A. L., & Bube, R. H. (1977). Evaluation of the CdS/CdTe heterojunction solar cell. *J. Appl. Phys.*, *48* (10), pp. 4365-4371.

Morales-Acevedo, A. (2006). Can we improve the record efficiency of CdS/CdTe solar cells? *Solar Energy Materials & Solar Cells*, *90*, 2213–2220.

Moutinho, H., Dhere, R., Romero, M., Jiang, C.-S., To, B., & Al-Jassim, M. (2008). RECRYSTALLIZATION OF PVD CdTe THIN FILMS INDUCED BY $CdCl_2$ TREATMENT-A COMPARISON BETWEEN VAPOR AND SOLUTION PROCESSES. *the 33rd IEEE Photovoltaic Specialists Conference*. San Diego, California.

Naghavi, N., Spiering, S., Powalla, M., Cavana, B., & Lincot, D. (2003). High-Efficiency Copper Indium Gallium Diselenide (CIGS) Solar Cells with Indium Sulfide Buffer Layers Deposited by Atomic Layer Chemical Vapor Deposition (ALCVD). *Prog. Photovolt:Res. Appl.*, *11*, pp. 437–443.

NAKADA, T., & MIZUTANI, M. (2002). 18% Efficiency Cd-Free Cu(In, Ga)Se_2 Thin-Film Solar Cells Fabricated Using Chemical Bath Deposition (CBD)-ZnS Buffer Layers. *Jpn. J. Appl. Phys.*, *41*, pp. L 165–L 167.

Nazeeruddin, M. K., Angelis, F. D., Fantacci, S., Selloni, A., Viscardi, G., Liska, P., Ito, S., Takeru, B. & Grätzel, M. (2005). Combined Experimental and DFT-TDDFT Computational Study of Photoelectrochemical Cell Ruthenium Sensitizers. *J. AM. CHEM. SOC.*, *127*, 16835-16847.

Nazeeruddin, M. K., Humphry Baker, R., Liska, P., & Grätzel, M. (2003). Nazeeruddin, Md. K.; HumphrInvestigation of Sensitizer Adsorption and the Influence of Protons on Current and Voltage of a Dye-Sensitized Nanocrystalline TiO_2 Solar Cell.

Nazeeruddin, Md. K.; Humphry-Baker, R.; Liska, P. ; Grätzel, M. (2003). Investigation of Sensitizer Adsorption and the InfluencJ. Phys. Chem. B , 8981-8987.

Nazeeruddin, M. K., Kay, A., Rodicio, l., Humpbry-Baker, R., Miiller, E., Liska, P., Vlachopoulos, N. & Grätzel, M. (1993). Conversion of Light to Electricity by cis-XzBis (2,2'-bipyridyl-4,4'-dicarboxylate)ruthenium(11), Charge-Transfer Sensitizers ($X=Cl^-$, Br^-,I^-,CN^-, and SCN^-) on Nanocrystalline TiO_2 Electrodes. *J. Am. Chem. SOC. , 115*, 6382-6390.

Nazeeruddin, M. K., Pechy, P., Renouard, T., Zakeeruddin, S. M., Humphry-Baker, R., Comte, P., Liska, P., Cevey, L., Costa,E., Shklover, V., Spiccia, L., Deacon, G. B., Bignozzi, C. A. & Grätzel, M. (2001). Engineering of Efficient Panchromatic Sensitizers for Nanocrystalline TiO2-Based Solar Cells. *J. Am. Chem. Soc. , 123*, 1613-1624.

Negami, T., Aoyagi, T., Satoh, T., Shimakawa, S. i., Hayashi, S., & Haskimoto, Y. (2002). Cd free CIGS solar cells fabricated by dry processes. *Photovoltaic Specialists Conference, 2002. Conference Record of the Twenty-Ninth IEEE*, (pp. 656-659).

Noufi, R. (2006). High Efficiency CdTe and CIGS Thin Film Solar Cells: Highlights of the Technologies Challenges. *the 2006 IEEE 4th World Conference on Photovoltaic Energy Conversion (WCPEC-4)*. Waikoloa, Hawaii.

Ochoa-Landın, R., Vigil-Galan, O., Vorobiev, Y., & Ramırez-Bon, R. (2009). Chemically deposited Te layers improving the parameters of back contacts for CdTe solar cells. *Solar Energy , 83*, 134-138.

Ohashi, D., Nakada, T., & Kunioka, A. (2001). Improved CIGS thin-film solar cells by surface sulfurization using In_2S_3 and sulfur vapor. *Solar Energy Materials & Solar Cells , 67*, pp. 261-265.

Okamoto, T., Yamada, A., & Konagai, M. (2000). Optical and electrical characterizations of highly efficient CdTe thin film solar cells prepared by close-spaced sublimation. *Journal of Crystal Growth , 214/215*, pp. 1148-1151.

O'Regan, B., & Gratzel, M. (1991). A low-cost, high-efficiency solar cell based on dye-sensitized colloidal TiO2 films. *Nature , 353*, 737-740.

Pantoja Enriquez, J., Mathew, X., Hernandez, G., Pal, U., Magana, C., Acosta, D. R., Guardian, R., Toledo, J. A., Contreras Puente, G. & Chavez Carvayar, J. A. (2004). CdTe/CdS Solar cells on flexible molybdenum substrates. *Solar Energy Materials & Solar Cell , 82*, 307-314.

Peter, L., & Wang, R. L. (1999). Channel flow cell electrodeposition of CdTe for solar cells. *Electrochemistry Communications , 1*, 554–558.

Pinheiro, W. A., Falcão, V. D., Cruz, L. R., & Ferreira, C. L. (2006). Comparative Study of CdTe Sources Used for Deposition of CdTe Thin Films by Close Spaced Sublimation Technique. *Materials Research , 9* (1), pp. 47-49.

Ramanathan, K., Contreras, M. A., Perkins, C. L., Asher, S., Hasoon, F. S., Young, D., Romero, M., Metzger, W., Noufi, R., Ward, J. & Duda, A. (2005). Properties of 19.2% Efficiency ZnO/CdS/$CuInGaSe_2$ Thin-film Solar Cells. *Prog. Photovolt: Res. Appl. , 11*, 225-230.

Repins, I., Contreras, M. A., Egaas, B., DeHart1, C., Scharf, J., Perkins, C. L., To, B. & Noufi, R. (2008). 19.9%-efficient ZnO/CdS/CuInGaSe_2 Solar Cell with 81.2% Fill Factor. *PROGRESS IN PHOTOVOLTAICS: RESEARCH AND APPLICATIONS*, *16*, 235-239.

Rogacheva, E. I., Tavrina, T. V., Nashchekina, O. N., Grigorov, S. N., Nasedkin, K. A., Dresselhaus, M. S. & Cronin, S. B. (2002). Quantum size effects in PbSe quantum wells. *APPLIED PHYSICS LETTERS*, *15*, 2690-2691.

Romeo, A., Khrypunov, G., Galassini, S., Zogg, H., & Tiwari, A. (2007). Bifacial configurations for CdTe solar cells. *Solar Energy Materials & Solar Cells*, *91*, pp.1388–1391.

Romeo, A., Khrypunov, G., Kurdesau, F., Arnold, M., Batzner, D., Zogg, H., & Tiwari, A. N. (2006). High-efficiency flexible CdTe solar cells on polymer substrates. *Solar Energy Materials & Solar Cells*, *90*, pp. 3407–3415.

Romeo, A., Terheggen, M., Abou-Ras, D., Ba¨tzner, L. D., Haug, F.-J., Ka¨lin, M., Rudmann, D. & Tiwari, A. N. (2004). Development of Thin-film Cu(In,Ga)Se_2 and CdTe Solar Cells. *Prog. Photovolt: Res. Appl.*, *12*, 93-111.

Romeo, N., Bosio, A., & Romeo, A. (2010). An innovative process suitable to produce high-efficiency CdTe/CdS thin-film modules. *Solar Energy Materials & Solar Cells*, *94*, pp. 2–7.

Romeo, N., Bosio, A., Canevari, V., & Podesta, A. (2004). Recent progress on CdTe/CdS thin film solar cells. *Solar Energy*, *77*, pp. 795–801.

Romeo, N., Bosio, A., Canevari, V., Terheggen, M., & Vaillant Roca, L. (2003). Comparison of different conducting oxides as substrates for CdS/CdTe thin film solar cells. *Thin Solid Films*, *431-432*, pp.364–368.

Romeo, N., Bosio, A., Mazzamuto, S., Romeo, A., & Vaillant-Roca, L. (2007). A HIGH EFFICIENCY CdTe/CdS THIN FILM SOLAR CELLS WITH A NOVEL BACK-CONTACT. *22nd European Photovoltaic Solar Energy Conference.* Milan, Italy.

Rose, D. H., Hasoon, F. S., Dhere, R. G., Albin, D. S., Ribelin, R. M., Li, X., Yoxa Mahathongdy, S., Gessert, T. A. & Sheldon, P. (1999). Fabrication Procedures and Process Sensitivities for CdS/CdTe Solar Cells. *PROGRESS IN PHOTOVOLTAICS: RESEARCH AND APPLICATIONS*, *7*, 331-340.

Sakurai, K., Hunger, R., Tsuchimochi, N., Baba, T., Matsubara, K., Fons, P., Yamada, A., Kojima, T., Deguchi, T., Nakanishi, H. & Niki, S. (2003). Properties of CuInGaSe_2 solar cells based upon an improved three-stage process. *Thin Solid Films*, *431-432*, 6-10.

Santos-Cruz, J., Torres-Delgado, G., Castanedo-Perez, R., Jimenez-Sandoval, S., Marquez-Marın, J., & Zelaya-Angel, O. (2006). Au–Cu/p-CdTe/n-CdO/glass-type solar cells. *Solar Energy Materials & Solar Cells*, *90*, pp. 2272–2279.

Sathiya Priya, A. R., Subramania, A., Jung, Y. S., & Kim, K. J. (2008). High-Performance Quasi-Solid-State Dye-Sensitized Solar Cell Based on an Electro spun PVdF-HFP Membrane Electrolyte. *Langmuir*, *24*, 9816-9819.

Shafarman, W. N., & Stolt, L. (2003). Cu(InGa)Se_2 Solar Cells. In A. Luque, & S. Hegedus, *Handbook of Photovoltaic Science and Engineering.* John Wiley & Sons, Ltd.

Sharma, R. K., Jain, K., & Rastogi, A. C. (2003). Growth of CdS and CdTe thin films for the fabrication of n-CdS/p-CdTe solar cell. *Current Applied Physics*, *3*, pp. 199–204.

Shirakata, S., Ohkubo, K., Ishii, Y., & Nakada, T. (2009). Effects of CdS buffer layers on photoluminescence properties of Cu(In,Ga)Se_2 solar cells. *Solar Energy Materials & Solar Cells*, *93*, 988–992.

Singh, V. P., & McClure, J. C. (2003). Design issues in the fabrication of CdS–CdTe solar cells on molybdenum foil substrates. *Solar Energy Materials & Solar Cells*, *76*, pp. 369–385.

Singh, V. P., McClure, J. C., Lush, G., Wang, W., Wang, X., Thompson, G., & Clark, E. (1999). Thin film CdTe-CdS heterojunction solar cells on lightweight metal substrates. *Solar Energy Materials & Solar Cells*, *59*, pp. 145-161.

Stathatos, E., & Lianos, P. (2007). Increase of the Efficiency of Quasi-Solid State Dye-Sensitized Solar Cells by a Synergy between Titania Nanocrystallites of Two Distinct Nanoparticle Sizes. *Adv. Mater.*, *19*, 3338-3341.

Sternitzke, M. (1997). Review: Structual Ceramic Nanocomposites. *Journal of the Europ-ean Ceramic*, *17*, 1061-1082.

Tena-Zaera, R., Katty, A., Bastide, S., Levy-Clement, C., ORegan, B., & Munoz-Sanjose, V. (2005). ZnO/CdTe/CuSCN, a promising heterostructure to act as inorganic eta-solar cell. *Thin Solid Films*, *483*, 372-377.

Thampi, K., Bessho, T., Gao, F., Zakeeruddin, S., Wan, P. & Grätzel, M. (2008). New materials improve dye sensitized solar cell performance. *23rd European Photovoltaic Solar Energy Conference and Exhibition*, (pp. 153-155). Valencia, Spain.

Tiwari, A. N., Khrypunov, G., Kurdzesau, F., Batzner, D. L., Romeo, A., & Zogg, H. (2004). CdTe Solar Cell in a Novel Configuration. *Prog. Photovolt: Res. Appl.*, *12*, pp. 33–38.

Tjong, S. (2006). Structural and mechanical properties of polymer nanocomposites. *Materials Science and Engineering*, *R* (53), 73-197.

Tokita, Y., Chaisitsak, S., Yamada, A. & Konagai, M. (2003). High-efficiency Cu(In,Ga)Se_2 thin-film solar cells with a novel $In(OH)_3:Zn^{2+}$ buffer layer. *Solar Energy Materials & Solar Cells*, *75*, 9-15.

Tsuji, M., Aramoto, T., Ohyama, H., Hibino, T. & Omura, K. (2000). Characterization of CdS thin film in high efficient CdS/CdTe solar cells. *Journal of Crystal Growth*, *214/215*, pp. 1142-1147.

Ullal, H. & Roedern, B. v. (2007). Thin Film CIGS and CdTe Photovoltaic Technologies: Commercialization, Critical Issues, and Applications. *the 22nd European Photovoltaic Solar Energy Conference (PVSEC) and Exhibition*. Milan, Italy.

Vigil-Galán, O., Sánchez-Meza, E., Ruiz, C., Sastré-Hernández, J., Morales-Acevedo, A., Cruz- Gandarilla, F., Aguilar-Hernández, J., Saucedo, E., Contreras-Puente, G. & Bermúdez, V. (2007). Physical properties of Bi doped CdTe thin films grown by CSVT and their influence on the CdS/CdTe solar cells PV-properties. *Thin Solid Films*, *515*, 5819–5823.

Wang, P., Zakeeruddin, S. M., Moser, J. E., Humphry-Baker, R., Comte, P., Aranyos, V., Hagfeldt, A., Nazeeruddin, M. K. & Grätzel, M. (2004). Stable New Sensitizer with Improved Light Harvesting for Nanocrystalline Dye-Sensitized Solar Cells. *Adv. mater.*, *16* (20), 1806-1811.

Wu, J., Lan, Z., Hao, S., Li, P., Lin, J., Huang, M., Fang,L & Huang, Y. (2008). Progress on the electrolytes for dye-sensitized solar cells. *Pure Appl. Chem.*, *80* (11), 2241-2258.

Wu, X. (2004). High-efficiency polycrystalline CdTe thin-film solar cells. *Solar Energy , 77,* pp. 803–814.

Wu, X., Dhere, R. G., Albin, D. S., Gessert, T. A., DeHart, C., Keane, J. C., Duda, A., Coutts,T.J., Asher, S., Levi, D. H., Moutinho, H. R., Yan, Y., Moriarty, T., Johnston, S., Emery, K. & Sheldon, P. (2001). High-Efficiency CTO/ZTO/CdS/CdTe Polycrystalline Thin-Film Solar Cells. *NCPV Program Review Meeting.* Lakewood, Colorado.

Wu, X., Sheldon, P., Mahathongdy, Y., Ribelin, R., Mason, A., Moutinho, H.R. & Coutts, T. J. (1998). CdS/CdTe Thin-Film Solar Cell with a Zinc Stannate Buffer Layer. *National Renewable Energy Laboratory Presented at the National Center for Photovolt Program Review meeting.* Denver, Colorado.

Wu, X., Zhou, J., Duda, A., Keane, J. C., Gessert, T. A., Yan, Y. & Noufi, R. (2006). 13.9%-efficient CdTe Polycrystalline Thin-film Solar Cells with an Infrared Transmission of ~50%. *Prog. Photovolt: Res. Appl. , 14,* pp. 471–483.

Wu, X., Zhou, J., Keane, J., Dhere, R., Albin, D. S., Gessert, T.A., DeHart, C., Duda, A.,Ward, J. J., Yan, Y., Teeter, G., Levi, D. H., Asher, S., Perkins, C., Moutinno, H. R., To, B., Emery, K., Moriarty, T., Zhang, Y., Wei, S., Coutts, T.& Noufi, R. (2005). Advances in CdTe R&D at NREL. *the 2005 DOE Solar Energy Technologies Program Review Meeting.* Denver, Colorado.

Wuerz, R., Eicke, A., Frankenfeld, M., Kessler, F., Powalla, M., Rogin, P. & Yazdani-Assl, O (2009). CIGS thin-film solar cells on steel substrates. *Thin Solid Films , 517,* pp. 2415–2418.

Yanagida, M., Onozawa-Komatsuzaki, N., Kurashige, M., Sayama, K., & Sugihara, H. (2010). Optimization of tandem-structured dye-sensitized solar cell. *Solar Energy Materials & Solar Cells , 94,* 297-302.

Ying, D., & Zhang, D. (2000). Processing of Cu–Al_2O_3 metal matrix nanocomposite materials by using high energy ball milling. *Materials Science and Engineering , A286,* 152–156.

Yongbai, Y. (2007). Nanocomposite thin films for solar energy conversion. In S. Zhang, & N. Ali, *Nanocomposite thin films and coatings Processing, Properties and Performance* (pp. 381-417). Imperial College Press.

Yoon, C. H., Vittal, R., Lee, J., Chaec, W.-S., & Kima, K.-J. (2008). Enhanced performance of a dye-sensitized solar cell with an electrodeposited-platinum counter electrode. *Electrochimica Acta , 53,* 2890–2896.

Zhang, G., Bala, H., Cheng, Y., Shi, D., Lv, X., & Yu, Q. &. (2009). High efficiency and stable dye-sensitized solar cells with an organic chromophore featuring a binary π-conjugated spacer. *Chem. Commun. ,* 2198-2200.

Zhang, N., Zhuang, D.-M., & Zhang, G. (2010). An investigation on preparation of CIGS targets by sintering process. *Materials Science and Engineering B , 166,* pp. 34–40.

Zhang, R., Pan, J., Briggs, E. P., Thrash, M., & Kerr, L. L. (2008). Studies on the adsorption of RuN3 dye on sheet-like nanostructured porous ZnO films. *Solar Energy Materials & Solar Cells , 92,* 425-431.

Zhao, Q., Liu, C., Liu, H., Zhang, L., & Yao, X. (1994). Quantum size effects and nonlinear optical properties of ZnS and CdS semiconductor-doped silica glasses prepared by sol-gel process. *Applications of Ferroelectrics, 1994 .ISAF '94., Proceedings of the Ninth IEEE International Symposium on,* (pp. 797-800). University Park, PA.

Zoppi, G., Durose, K., Irvine, S. J., & Barrioz, V. (2006). Grain and crystal texture properties of absorber layers in MOCVD-grown CdTe/CdS solar cells. *Semicond. Sci. Technol.*, *21*, 763-770.

9

Composite Nanomaterials for Hydrogen Technologies

Liga Grinberga and Janis Kleperis
Institute of Solid State Physics, University of Latvia
Latvia

1. Introduction

Analyzing today's situation and tracing tendencies, it is clear that the primary energy consumption is increasing but reserves are running out very rapidly. Meanwhile, global utilization of fossils is causing environmental problems throughout the world.

As a consequence, investigations of alternative energy strategies have recently become important, particularly for future world stability. The most important property of alternative energy sources is their environmental compatibility.

One such new energy carrier currently being investigated is hydrogen. Many hold the hopes that it could maintain mankind's growing need for energy. However, the hydrogen alternative has both positive and negative aspects.

The main advantage of hydrogen as a fuel is sustainable development. It is a non-toxic energy carrier and holds a higher energy content: 9.5 kg of hydrogen is equivalent to that of 25 kg of gasoline (Midilli, 2005). Hydrogen can be produced in many ways, however the most environmentally friendly and the less fossils consuming are processes driven directly by sunlight - photocatalysis and photo-biological methods.

Studies demonstrate that solid materials can be utilized to solve the storage problem by reversible absorption and desorption of large amounts of hydrogen (Schlapbach & Zuttel, 2001). Although several metal hydrides and composite materials are capable of meeting this target, the high desorption temperatures, slow absorption/desorption rates and small cycling capacity limit the widespread application of current metal hydrides.

Nanostructuring of materials and enhancement of surface absorption capability are two main factors to increase the amount of sorbed hydrogen. One way to combine the effectiveness of hydrogen absorption in metal hydrides and the desirable weight/volume proportion is to make composite material from alloy forming hydride and appropriate support material.

In this chapter a short description of water photocatalysis and photosynthetic hydrogen production and work of our laboratory on these subjects are given. There are described various composite materials for solid hydrogen storage and compared their characteristics. The chapter is concluded with the main results of our work on hydrogen storage in the modified AB_5 type metal hydride where the idea of the possible gain of using the spill-over effect to enhance catalytic activity and the amount of absorbed hydrogen was explored.

2. Hydrogen production

Hydrogen production is the first step toward the transition to a hydrogen economy. Fossil fuel systems for hydrogen production are the oldest technologies and tend to be the cheapest. Because fossil fuels are carbon-based, carbon dioxide is produced as a by-product when they are decomposed to release energy. Biomass pyrolysis and gasification processes are very similar to fossil fuel reforming and gasification processes.
The water splitting methods can use nuclear heat or alternative energy sources do not produce harmful emissions but they are more expensive than fossil fuel processes. Photocatalytic and photobiological processes use solar energy, and the sulphur iodine process uses nuclear heat. In the case of electrolysis, electricity supplies the energy required and there could be used alternative energy sources like wind, solar, or water power. These processes are attractive because the water feedstock contains only hydrogen and oxygen, so no carbon dioxide is released in its decomposition.

2.1 Photocatalysis

Photocatalytic water splitting using solar energy could be one of solutions of environmentally friendly and clean ways of hydrogen production. The basic material for the production of 'solar hydrogen' is water that is a renewable resource and on the earth it is enough and easy to access.
However, there are many problems that must be solved before this technology become economically feasible. One of tasks is the development of efficient nano sized photocatalyst that works in the visible light. The other task is to increase the efficiency of solar energy utilization.
The photochemical water splitting concept lies on the materials (photocatalysts) that can produce chemical reactions by absorption a quantum of light. The photocatalysts mainly are semiconductors that use photons to excite an electron from the valence band to the conduction band. The excited electrons by 'moving' to the conduction band and 'leaving' holes in the valence band cause reduction - oxidation reactions similarly to electrolysis (1). Electron is creating hydrogen by water molecule reduction (2) while holes form oxygen by oxidation (3) (Fig.1.).
Energy of absorbed photons must be greater than band-gap energy of a semiconductor, although the band-gap should be higher than 1.23 eV, which is the energy needed for to split water. Due to the orbital configuration of the oxide semiconductor metal cations the energy levels are more positive than oxidation potential of water and, consequently, the band-gap become wider. For ultra violet light absorption the band-gap exceeds 3 eV, however for visible light absorption the band-gap should be around 2 – 2,2 eV (Navarro et. al., 2009).
Photocatalysis reactions are considered as 'up-hill' reactions because back reactions can proceed very easy. There are several important requirements that must be fulfilled for successful photocatalytic water decomposition.
Generated electrons and holes have to be separated and before recombination reaction takes place they should migrate to the surface. In this process the bulk properties of semiconductors are determinative; tough the reduction and oxidation reactions are dependent of a surface area and active sites on the surface of a photocatalyst. Summarizing, functional photocatalyst is a complex material that should provide the right band structure and suitable bulk and surface properties.

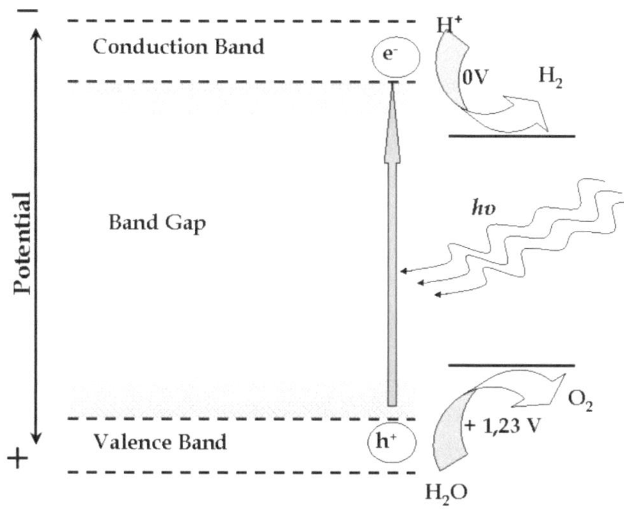

Overall reaction:

$$H_2O \rightarrow H_2 + {}^1/{}_2O_2 \quad (1)$$

Reduction:

$$2H^+ + 2e \rightarrow H_2 \quad (2)$$

Oxidation:

$$H_2O + 2h^+ \rightarrow 2H^+ + {}^1/{}_2O_2 \quad (3)$$

Fig. 1. Schematic description of water decomposition on semiconductor photocatalysts

Photocatalysts	**Band-gap energy (eV)**	**Cocatalyst**	**Sacrificial reagent**	**Reference**
TiO_2-Cr-Sb	2,2	-	$AgNO_3$	Kato &Kudo, 2002
$SrTiO_3$ - Cr - Ta	2,3	Pt	CH_3OH	Ishii et al., 2004
$SrTiO_3$ - Cr - Sb	2,4	Pt	CH_3OH	Kato &Kudo, 2002
$La_2Ti_2O_7$ - Cr	1,8 – 2,3	Pt	CH_3OH	Hwang et al., 2005
TaON	2,5	Pt	CH_3OH	Hitoki et al., 2002
		-	$AgNO_3$	
$CaTaO_2N$	2,4	Pt	CH_3OH	Yamashita et al., 2004
		-	$AgNO_3$	
$SrTaO_2N$	2,1	Pt	CH_3OH	Yamashita et al., 2004
		-	$AgNO_3$	
$Sr_2Nb_2O_{7-x}N_x$	2,1	Pt	CH_3OH	Ji et al., 2005
		-	$AgNO_3$	
$BiVO_4$	2,4	-	CH_3OH	Kudo et al., 1998
		-	$AgNO_3$	
$(Ga_{1-x}Zn_x)(N_{1-x}O_x)$	2,4 – 2,8	Cr/Rh	-	Maeda et al., 2006
$(Zn_{1+x}Ge)(N_2O_x)$	2,7	RuO_2	-	Lee et al., 2007
CdS	2,4	-	S^{2-}/SO^{2-}_3	Navarro et al., 2008
CdS - CdO - ZnO	2,3	-	S^{2-}/SO^{2-}_3	Navarro et al., 2008
$Cd_{0.7}Zn_{0.3}S$	2,68	-	S^{2-}/SO^{2-}_3	del Valle et al., 2008
ZnS - Cu	2,5	-	SO^{2-}_3	Kudo and Sekizawa, 1999
$(AgIn)_xZn_2(1-x)S_2$	2,4	Pt	S^{2-}/SO^{2-}_3	Kudo et al., 2002
$(CuAg\ In)_xZn_2(1-x)S_2$	2,4	Ru	S^{2-}	Tsuji et al., 2005
$Na_{14}In_{17}Cu_3S_{35}$	2,0	-	-	Zheng et al., 2005

Table 1. Selection of photocatalysts developed for water splitting reaction under visible light (Navarro et al., 2008)

To develop suitable visible light photocatalysts the band-gap tuning by doping of transition-metal cations has often been used. Unfortunately photocatalytic activity significantly decreases because doped cations inducing a formation of recombination centres between photogenerated electrons and holes and impend a migration of holes. Making a solid solution is other way of preparation of photocatalysts. There the different width band-gap semiconductor ratio can change energy levels in the composite material in total. Table 1 shows several photocatalyst materials that are developed and investigated lately.

Our group's investigations thanks to European Social Fund project 2009/0202/1DP/ 1.1.1.2.0/09/APIA/VIAA/141 lies on a developing of a new photocatalysts where the method of doping and solid solution preparation is combined.

2.2 Bio-hydrogen

A number of technologies for biological H_2 production are available, but they are not established for significant amount of hydrogen production yet. Methods for engineering and manufacturing these systems have not been fully evaluated.

Biological processes of hydrogen recovery and collection from organic resources such as municipal wastewater and sludge facilitate recycling of sewage are environmentally benign and necessary for alternative independent household support. Nowadays, many institutions and universities worldwide are involved in the research of hydrogen production by microorganisms and algae (Das & Veziroglu, 2008, Kotay Meher & Das, 2008, Donohue & Cogdell, 2006).

Microorganisms are capable of producing H_2 via two main pathways: fermentation and photosynthesis. The processes of bio-hydrogen production include:

1. direct biophotolysis by green algae – the photosynthetic production of hydrogen by splitting water into molecular hydrogen and oxygen using sunlight under specific conditions;
2. indirect biophotolysis by cyanobacteria with specialized cells (heterocysts) that perform nitrogen fixation and contain enzymes (nitrogenase and hydrogenase) directly involved in hydrogen metabolism and synthesis of molecular H_2;
3. photo-fermentation by purple non-sulfur bacteria that evolve molecular H_2 catalyzed by nitrogenase enzyme under nitrogen-deficient conditions using the energy of light and organic acids;
4. dark-fermentation by anaerobic bacteria grown in the dark on carbohydrate-rich substrates (Holladay et. Al., 2009, Das & Veziroglu, 2001, Levin, 2004).

Bacterial hydrogen production by fermentation of carbohydrate-containing substrates (glucose, cellulose, starch and organic waste materials) is frequently preferred to photolysis, because it does not rely on the availability of light sources. In the fermentation of glucose by enterobacteria, e.g. Escherichia coli, one of the pyruvate oxidation products, alongside with acetyl-CoA, is formate, which is produced by pyruvate formate lyase and is the sole source of hydrogen in these bacteria. The formate is split into CO_2 and H_2 by formate hydrogen lyase (FHL) complex, which comprises seven proteins, six of them being encoded hyc operon. Five hyc operon encoded proteins are membrane-embedded electron transporters. The hycE protein is one of the three E.coli NiFe hydrogenases. The hycE and FDH-H components of FHL complex are soluble peri¬plasmic proteins. The hydrogen evolved from FHL is consumed by E.coli uptake hydrogenases Hyd-1 and Hyd-2. In contrast to enterobacteria, strictly anaerobic fermenters, e.g. Clostridia, use a reduced ferredoxin (required to oxidize pyruvate to acetyl-CoA) for H_2 production by the hydrogenase that generates ferredoxin in the oxidized form and releases electrons as molecular hydrogen (Nath & Das, 2004, Hallenbeck & Benemann, 2002, Maeda et. al., 2008).

Glucose fermentation by enteric bacteria yields the maximum of 2 mol H2/ mol glucose (Wang & Wan, 2009). To enhance the hydrogen production and utilize the substrate in full measure for complete conversion, the synergy of biological processes (two-stage/hybrid ones) should be applied.

Gaseous hydrogen is formed in liquid media during bacterial fermentation. Hydrogen gas hardly dissolves in aquatic solutions, but special methods are required to discharge hydrogen into atmosphere and in so doing to escape oversaturation (Mandelis & Christofides, 1993). In the atmosphere the parameters of hydrogen gas are measured by classical volumetric, mass-spectrometric and chromatography methods, or using chemical gas sensors. Wilkins (Wilkins et. al., 1974) described a method for measuring gas production by microorganisms using a platinum electrode and a reference Calomel ($Hg–Hg_2Cl_2$) electrode. To measure hydrogen gas concentration in liquid a hydrogen electrode is usually used (Pt or another noble metal – gold, rhodium, palladium, etc.).

The hydrogen H+ ions and the molecular hydrogen H2 set the equilibrium potential in compliance with the reaction: $H2 \Leftrightarrow 2H^+ + 2e^-$. This reaction proceeds very fast, so in its course the equilibrium state remains stable; in electrochemistry this electrode is adopted as zero reference (with zero potential). In microbiology, to measure dissolved oxygen and hydrogen gases the micro-respiration Clark electrodes are used (Ghirardi e.al., 1994).

In a Clark's electrode the cathode polarized versus an internal Ag/AgCl anode is placed behind an electrically insulating silicone rubber membrane, which is extremely permeable to oxygen. The flow of electrons from the anode to the oxygen-reducing cathode reflects linearly the partial oxygen pressure around the sensor tip and is in the pA range.

The same principle holds for a hydrogen Clark-type sensor: the environmental hydrogen is driven by the external partial pressure and penetrates through the sensor tip membrane to be oxidized at the platinum anode surface. Flynn et al. [Flynn et. al., 2002) used chemochromic sensors for screening in order to identify positive (i.e. hydrogen-producing) algal colonies. A chemochromic sensor film, which is normally transparent, turns blue in the presence of hydrogen gas.

Hydrogen gas is produced during the bacterial fermentation process in anaerobic conditions. In practice, hydrogen is collected in the gaseous state, since dissolved hydrogen tends to become gas. To optimize the hydrogen collection methods it is necessary to study properly the hydrogen production kinetics in liquid phase during the fermentation process. The experimental test system for bacterial hydrogen production and micro-sensors were used to determine the hydrogen gas concentrations in liquid; the mass-spectrometry method was employed for measurements in the hydrogen-containing head space.

In our experiments Escherichia coli strain MSCL 332 (i.e. from Microbial Strain Collection of Latvia) was grown on Luria-Bertani (LB) nutrient agar plates (5 g/l yeast extract, 10 g/l tryptone, 10 g/l sodium chloride, 15 g/l Bacto agar). E.coli from single colonies on the agar plates were inoculated in 2 x 150 ml flasks containing LB liquid medium. The flasks were aerobically shaken at 37°C for 12 hours at 120 rpm using a multi-shaker PSU-20.

The bacteria cell number in the overnight culture was titrated at 10-6 dilution. The amount of bacterial cell protein was calculated assuming that one E.coli cell contains 1.54×10^{-13} g of protein. The overnight culture in LB liquid medium was mixed (1:1) with phosphate buffer saline (PBS) pH 7.3 (0.8 g/l NaCl, 0.2 g/l KCl, 1.43 g/l Na_2HPO_4, 0.2 g/l KH_2PO_4) in a vessel sterilized for measurements. The PBS contained a complex trace element medium pH 6.5 (0.039g/l $Fe(NH_4)_2{\cdot}SO_4{\cdot}6H_2O$, 0.172 mg/l Na_2SeO_3, 0.02 mg/l $NiCl_2$, 0.4 mg/l $(NH_4)6Mo_7O_{24}$). Glucose (3.3mM, final concentration, sterilized through membrane 0.2µm filter) was added at the start of experiment.

The hydrogen and oxygen concentrations were measured with Clark-type microsensors in the sample liquid phase. The microsensors were connected with the signal amplifier – a pico-ammeter and an A/D current converter connected to PC using USB port.

Before the measurements, both oxygen and hydrogen microsensors were calibrated in a liquid culture medium (similar to the sample measured by 15 min bubbling Ar) for zero concentrations and hydrogen gas and clean air for 100% dissolved H_2 and O_2 concentrations (730 and 760 µmol/l, accordingly). The system is able to work independently when measurements are made in one sample. If there are several samples at a time, it is necessary to move microsensors manually and to sterilize the sensor tip using 96% ethanol, 0.1 M NaOH and distilled water every time when it is taken out from the sample.

The gas from the headspace of liquid bacterial culture in the test vessel was taken to an RGAPro-100 mass-spectrometer to analyze its components. The gas from an argon balloon through a diffuser was let in the test vessel with bacteria culture to sustain the anaerobic environment (see Fig.2) and put in a water bath to maintain a temperature of 37±2 °C.

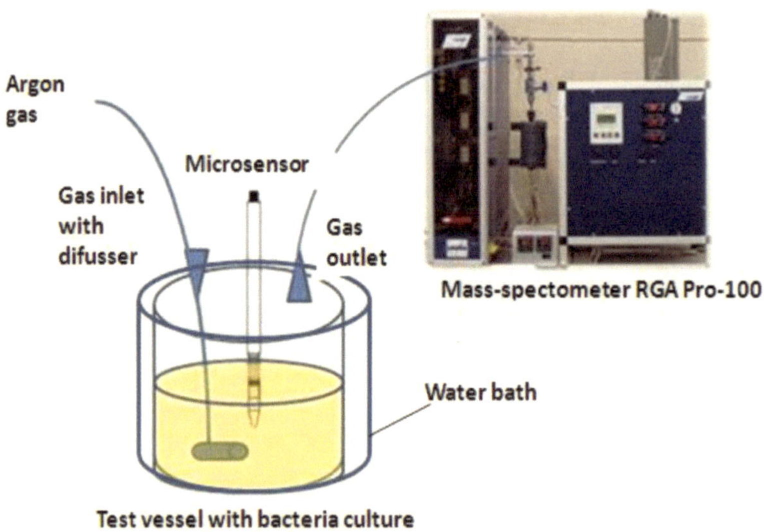

Fig. 2. Experimental test system for H2 concentration measurements with mass-spectrometer

Argon gas was bubbled through the liquid for 15 min (flow 13 l/h) and gas measurements were made with 30 min and an hour intervals. The total time of measurements was six hours. The gas volume taken for each analysis was 20 cm3. During the mass-spectrometric analysis, simultaneous measurements with a hydrogen microsensor were taken in order to make unbiased comparison of mass-spectrometric and hydrogen microsensor analyses. The concentration of dissolved hydrogen gas was measured with a microsensor; the mass-spectrometric analyses were made for the atmospheric composition in the headspace of the sample bacteria culture and nutrients.

The experimental results of microsensoric measurements were analyzed using Sensor Trace Basic and MicOX (A/S Unisense) programs, and processed by Microsoft Office Excel 2007. The mass-spectrometric data were analyzed by RGA 3.0 Software for SR Residual Gas Analyzers program.

Summarizing experiments: The hydrogen output was measured for seven hours after the beginning of fermentation process; increase in the hydrogen concentration was observed starting from the second hour after adding glucose. The constancy of oxygen concentration in the measurements evidences that the system had reliable anaerobic conditions (Fig. 3.). As is seen from this figure, the concentration of dissolved hydrogen stopped increasing after 5-10 h as glucose exhausted.

The maximum rate of hydrogen formation in the test system was 612 μmol/ l/20 min or 1.4 mmol [2.4 mg] /l per h for 43 mg protein mass (i.e. 32.6 μmol/mg protein mass). The maximum concentration of dissolved hydrogen (2481 μmol/l or 2.5 mmol/l) is reached in the fourth hour of fermentation as is seen in Fig. 2. This concentration at least three times exceeds the maximum thermodynamically allowed concentration of dissolved hydrogen in water.

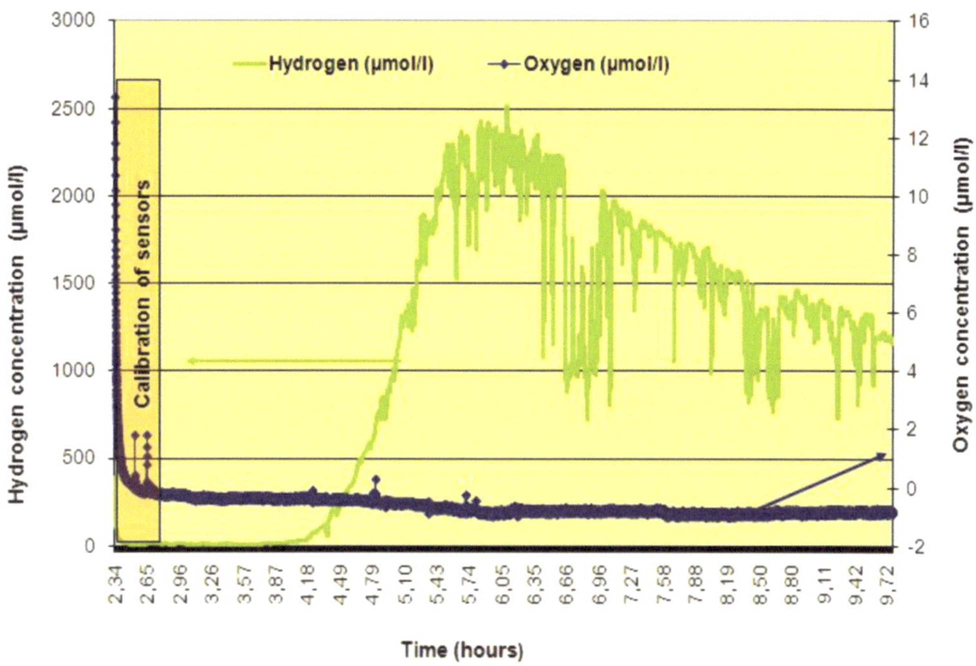

Fig. 3. Microsensoric fermentation measurements on the sample with E.coli

To demonstrate that the hydrogen production began only after glucose had been added various glucose concentrations were tested. The correlation between the glucose concentration and the hydrogen output is shown in Fig. 4.

To calculate the partial pressure of hydrogen in the headspace Henry's law was used. The calculations were done using the measured dissolved hydrogen concentrations in the test system (2481 μmol/l after 4 h fermentation). At room temperature and normal atmospheric pressure the Henry constant is k_H = 1282,05 (l/atm·mol), therefore pH = 1282.05·cH (atm) and in our case pH = 1282.05·2481·10^{-6} = 3.18 atm, which obviously does not fit the experimental results obtained in the mass-spectrometric analysis.

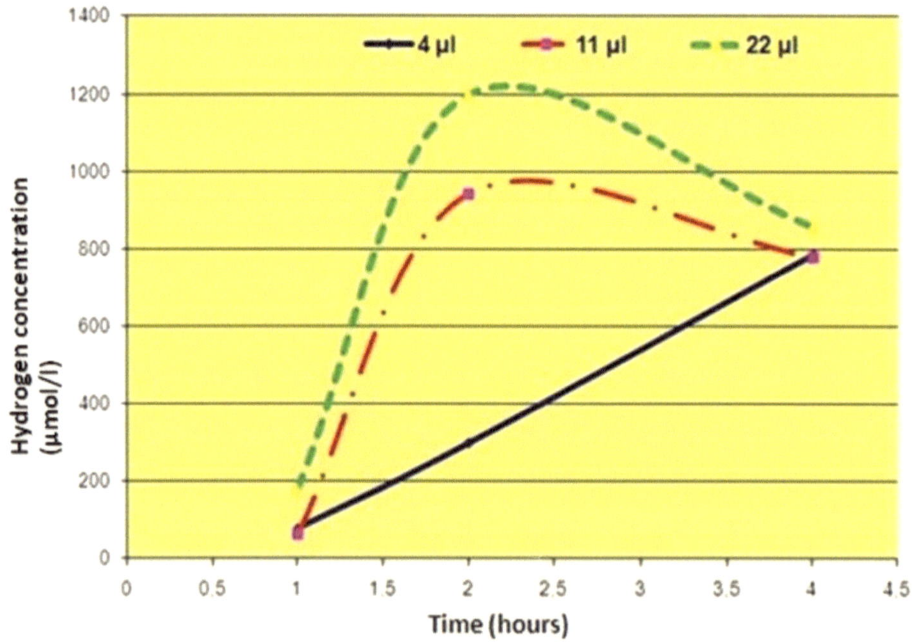

Fig. 4. Time dependence of hydrogen output at different glucose concentrations

To calculate the partial pressure of hydrogen in the headspace Henry's law was used. The calculations were done using the measured dissolved hydrogen concentrations in the test system (2481 µmol/l after 4 h fermentation). At room temperature and normal atmospheric pressure the Henry constant is k_H = 1282,05 (l/atm·mol), therefore pH = 1282,05·cH (atm) and in our case pH = 1282,05·2481·10^{-6} = 3.18 atm, which obviously does not fit the experimental results obtained in the mass-spectrometric analysis.

As is seen from Fig. 5, the partial pressure of hydrogen of 3.18 atm above the test system's headspace is inadequate to the concentrations determined by mass-spectrometric analysis – only 6·10^{-3} atm or 0.6% vol.

The mass-spectrometric analysis has revealed the presence of different volatile substances – the end products of bacterial formation: acetate, carbon dioxide, ethanol, acteone, and hydrogen gas. In three separate measurements (without Ar bubbling and liquid mixing, with argon bubbling only, and with argon bubbling & liquid mixing) the mass-spectrometric analysis showed a hydrogen concentration increase from 0% to 0.4% after 6 h fermentation only in one measurement when no argon bubbling and mixing was applied. Such an increase in the hydrogen concentration (only 3.6·10–3 atm partial pressure) is not convincing as compared with the concentrations measured in liquid phase. A reason for that could be the limited hydrogen migration from liquid to gaseous phase (dissolved hydrogen oversaturation); besides, the mass-spectrometry measurement method was not perfect (it is to be optimized in the future experiments).

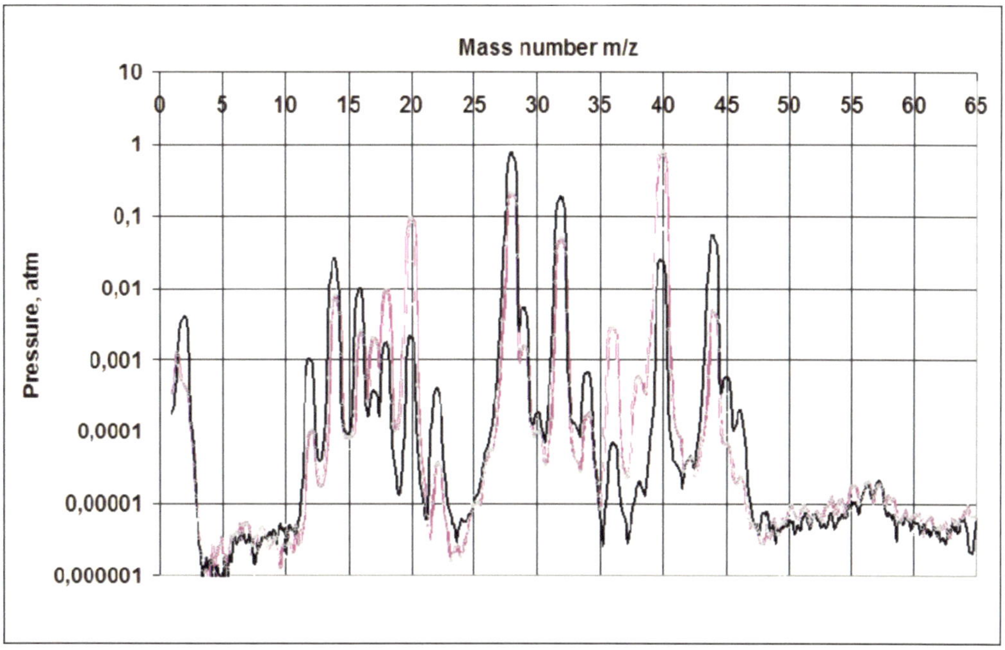

Fig. 5. Mass-spectometric analysis of the sample with bacteria E.coli before fermentation (light gray curve) and after 6h (black curve)

The mass-spectrometric analysis has revealed the presence of different volatile substances – the end products of bacterial formation: acetate, carbon dioxide, ethanol, acteone, and hydrogen gas. In three separate measurements (without Ar bubbling and liquid mixing, with argon bubbling only, and with argon bubbling & liquid mixing) the mass-spectrometric analysis showed a hydrogen concentration increase from 0% to 0.4% after 6 h fermentation only in one measurement when no argon bubbling and mixing was applied. Such an increase in the hydrogen concentration (only 3.6·10–3 atm partial pressure) is not convincing as compared with the concentrations measured in liquid phase.

A reason for that could be the limited hydrogen migration from liquid to gaseous phase (dissolved hydrogen oversaturation); besides, the mass-spectrometry measurement method was not perfect (it is to be optimized in the future experiments). Concentrations of dissolved gas are higher than theoretically possible in the anaerobic processes where gases are formed in the liquid phase and tend to reach the gas phase. Such over-saturation could be associated with biological processes: lower pH due to the formation of gases (e.g. CO_2, H_2S) in anaerobic processes; besides, a negative thermodynamic effect is caused by inhibator gases – e.g. H_2, since the hydrogen synthesising enzymes are sensitive to H2 concentrations and are subject to the end-product inhibition. As concentrations of hydrogen increase, its synthesis rate decreases: the evolved H_2 is consumed by E.coli uptake hydrogenases Hyd-1 and Hyd-2.

As mentioned, the hydrogen synthesising enzymes are sensitive to the end product – the hydrogen gas concentration. As this concentration increases the synthesis rate decreases, with formation of mixed-acid hydrogen-containing fermentation products (ethanol, acetate, butane). To enhance the hydrogen gas output, the bacterial metabolism has to be switched from alchocol and acid formation to volatile fatty acids. This can be facilitated by the system's optimization, for example, using continuous bubbling with inert gas to reduce the partial pressure of hydrogen in the liquid phase thereby increasing its formation in the gaseous phase.

To enhance hydrogen formation, very delicate bubbling/mixing procedures should be applied, since in our measurements, with intense bubbling and mixing by a magnetic stirrer, the least hydrogen increase in the headspace was observed.

2.3 Conclusions

Escherichia coli wild-type strain MSCL 332 from Microbial Strain Collection of Latvia was successfully used. The hydrogen concentration in the headspace was analyzed by mass spectrometry. Due to incompleteness of the test system the hydrogen concentration in the gaseous phase was not detected, which therefore remains to be a subject of future activities. Also, the gas measuring system should be improved based on the mass-spectrometry. To enhance the transfer of dissolved hydrogen into the headspace, continuous gas bubbling and mixing are required.

The dissolved hydrogen and oxygen concentrations in liquid phase during the fermentation process were measured using microsensors. The maximum of dissolved hydrogen concentration (2481 μmol/l) was reached by the fourth hour of fermentation, which is markedly higher than predicted by Henry`s law (730 μmol/l). Also, alternative methods should be employed for the hydrogen collection directly from the nutritional broth thus making it possible to develop commercial hydrogen production.

3. Hydrogen storage

Hydrogen storage is clearly one of the key challenges in developing hydrogen economy and hydrogen storage for vehicle applications is one of the most important challenges.

Hydrogen storage basically implies the reduction of the huge volume of the hydrogen gas because a 1kg of hydrogen at ambient temperature and atmospheric pressure takes a volume of 11 m^3. To increase a hydrogen density the compressing of hydrogen or decreasing of temperature below critical must be performed.

Available technologies permit to store hydrogen directly by modifying its physical state in gaseous or liquid form in pressurized or in cryogenic tanks. Storage by absorption as chemical compounds or by adsorption on carbon materials have definite advantages from the safety perspective such that some form of conversion or energy input is required to release the hydrogen for use. A great deal of effort has been made on new hydrogen-storage systems, including metal, chemical or complex hydrides and carbon nanostructures.

Hydrogen interaction with other elements depends from material, it occurs as anion (H^-) or cation (H^+) in ionic compounds, it participates with its electron to form covalent bonds, and it can even behave like a metal and form alloys at ambient temperature. The hydrogen molecule H_2 can be found in various forms depending on the temperature and the pressure which are shown in the phase diagram (Fig. 5.).

The phase diagram shows that the liquid hydrogen with a density of 70.8 kg m^{-3} at -253°C only exists between the solid line and the line from the triple point at 21.2K and the critical point at 32K. At low temperature hydrogen is a solid with a density of 70.6 kg m^{-3} at -262°C and is a gas at higher temperatures with a density of 0.089886 kg m^{-3} at 0°C and a pressure of 1 bar (Zuttel, 2004).

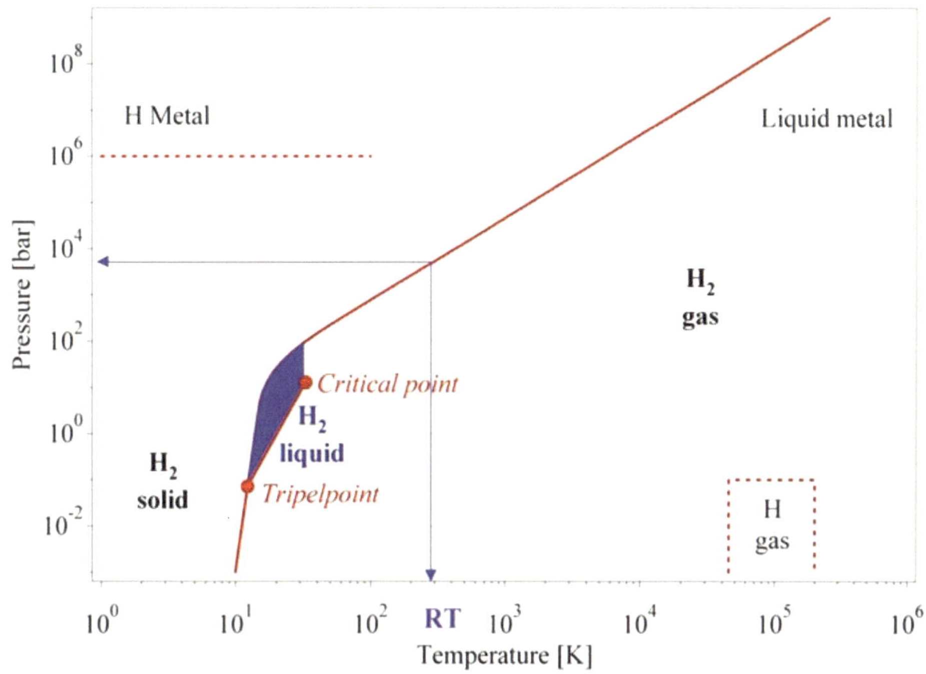

Fig. 5. Primitive phase diagram for hydrogen (Zuttel, 2004)

3.1 Metal hydrides

Hydrogen is a highly reactive element and forms hydrides and solid solutions with thousands of metals and alloys that will release hydrogen at elevated temperatures. Metal hydrides are arranged of metal atoms that form a host lattice, and hydrogen atoms that are installed in the interstitial sites.

The absorption process of hydrogen in the metal can be described using one-dimensional Lennard-Jones potential scheme for hydrogen molecule and 2 hydrogen atoms (Fig. 6) (Zuttel, 2004). The potential of molecule and both atoms are separated by the heat of dissociation energy E_D=435.99 kJ/mol in some distance from the metal surface. Approximately one hydrogen molecule radius from the metal surface the hydrogen molecule interacts with metal surface and due to the Van der Waals forces physisorption happens that is illustrated as the flat minimum in the H_2+M curve. Closer to the surface the hydrogen has to overcome an activation barrier for dissociation and formation of the hydrogen metal bond and hydrogen becomes chemisorbed that is showed as deep minimum of the 2H+M curve. If the both curves cross above zero energy level for the chemisorption activation energy is needed and the kinetics of adsorption is getting more slowly.

Depending on the surface elements the activation barrier height can be changed and chemisorbed hydrogen atoms may have a high surface mobility, interact with each other and form surface phases. Furthermore the chemisorbed hydrogen atom can jump in the subsurface layer and finally diffuse on the interstitial sites through the host metal lattice (Schlapbach, 1992). Hydride formation from gaseous phase can be described by pressure-composition isotherms (Fig. 7.). The host metal dissolves some hydrogen and solid solution phase or α-phase is formed. The metal lattice expands proportionally to the hydrogen concentration by approximately 2 – 3Å³ per hydrogen atom.

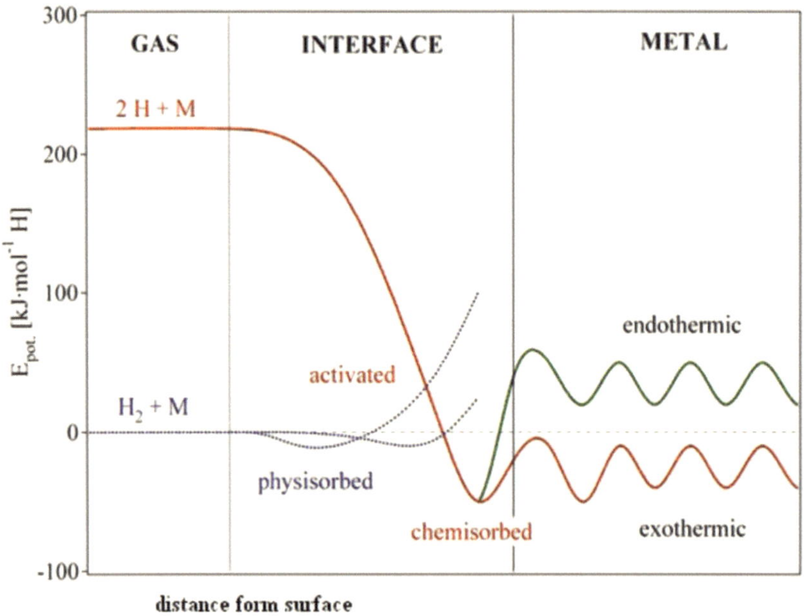

Fig. 6. Potential energy curves for activated and non-activated chemisorption of hydrogen on metal surface (Zuttel, 2003)

If the hydrogen pressure and concentration Hydrogen/Metal exceeds ratio 0,1 a H–H interaction becomes significant and the β-phase nucleates and grows. While the α-phase and β-phase coexists, the isotherms show a flat plateau, the length of which determines how much H_2 can be reversibly stored.

When the α-phase completely transfers to the β-phase, the H_2 pressure rises steeply with the concentration. Further enlargements of hydrogen pressure can cause formation of other plateaux and hydride phases. The two-phase region ends in a critical point T_C, above which the transition from α- to β-phase, is continuous (Fig. 7.).

The hydrogen concentration in the hydride phase is often found to be H/M=1. The volume expansion between the coexisting α- and the β-phase corresponds in many cases to 10–20% of the metal lattice. Therefore, at the phase boundary a large amount of stress is built up and often leads to a decrepitation of brittle host metals such as intermetallic compounds. The final hydride is a powder with a typical particle size of 10–100 μm (Zuttel, 2004; Schlapbach, 1998; Schlapbach, 1992).

The plateau pressure strongly depends on temperature that is related to the changes of enthalpy and entropy (4). Solving the Van't Hoff equation and from the gained slope where the pressure is a function of temperature, the heat of hydride formation can be evaluated.

$$\ln\left(\frac{p_{eq}}{p_{eq}^{0}}\right) = \frac{\Delta H}{R} \cdot \frac{1}{T} \cdot \frac{\Delta S}{R}, \qquad (4)$$

where p_{eq} is plateau pressure at equilibrium state, p^0_{eq} stands for the plateau pressure at standard conditions, R is the universal gas constant, T is the temperature at p_{eq} and ΔH is the enthalpy change.

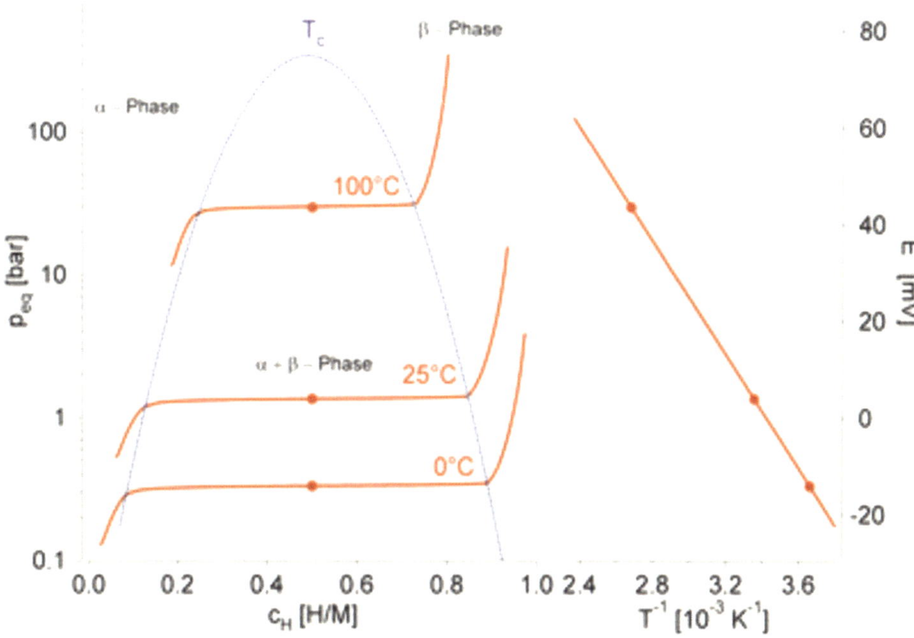

Fig. 7. Pressure composition isotherms for typical intermetallic compound is shown on the left side. The construction of the Van't Hoff plot is shown on the right hand side (Zuttel, 2004)

The enthalpy term characterizes the stability of the metal hydrogen bond. A hydride forming entropy changes leads to great heat generation during the hydrogen absorption. The same heat has to be provided to the metal hydride to desorb the hydrogen.

Hydrogen absorption is very much involved with a phase transition. Pressure does not increase with the amount of absorbed hydrogen as long as the phase transition takes place therefore a, metal hydrides can absorb large amounts of hydrogen at a constant pressure. Hydrogen sorption characteristics can be changed by partial substitution of the hydride forming elements, thereby it is possible to form some metal hydrides that works at ambient temperature and close to atmospheric pressure (Schlapbach, 1992)

A particular interest is about intermetallic hydrides because the variations of the elements allow modification of the properties of the hydrides (Table 2). The element A usually is a rare earth or an alkaline earth metal and tends to form a stable hydride or has a high affinity to hydrogen. The B element is often a transition metal and forms unstable hydrides or has a low affinity to hydrogen. The combination of the both elements A and B gives as alloys suitable for practical applications.

3.2 AB_5 intermetallic compounds

The on of the classical alloys is a combination of the La and Ni, where La individually forms LaH_2 at 25°C, $p=3\cdot10^{-29}$ atm, $\Delta H_f=-208$ kJ/mol H_2, with Ni that alone forms NiH at 25°C, p=3400atm, ΔH_f=-8,8 kJ/mol H_2. Although the alloy $LaNi_5$ forms a hydride at 25°C, p=1,6 atm, ΔH_f=-30,8 kJ/mol H_2, that can be considered as a interpolation between boundaries of elemental hydride forming activities (Sandrock, 1999).

Intermetallic compound	Prototype	Hydrides	Structure
AB_5	$LaNi_5$	$LaNi_5H_6$	Hexagonal
AB_2	ZrV_2, $ZrMn_2$, $TiMn_2$	$ZrV_2H_{5,5}$	Hexagonal or cubic
AB_3	$CeNi_3$, YFe_3	$CeNi_3H_4$	Hexagonal
A_2B_7	Y_2Ni_7, Th_2Fe_7	$Y_2Ni_7H_3$	Hexagonal
A_6B_{23}	Y_6F_{23}, Ho_6Fe_{23}	$Ho_6Fe_{23}H_{12}$	Cubic
AB	TiFe, ZrNi	$TiFeH_2$	Cubic
A_2B	Mg_2Ni, Ti_2Ni	Mg_2NiH_4	Cubic

Table 2. The most important hydride forming intermetallic compounds (Zuttel, 2004)

$LaNi_5$ has a $CaCu_5$-type, structure containing three octahedral and three tetragonal sites per elemental cell unit. The alloy forms at least two hydrides: α phase - $LaNi_5H_{0.3}$ - with low hydrogen content and β phase - $LaNi_5H_{5.5}$ - with high hydrogen–content. Both hydrides differ significantly in the specific lattice volume; β phase has a 25% larger lattice expansion as α phase that causes a crumbling of the alloy particles on hydriding - dehydriding cycles.

The alloy of rare earth elements in composition with nickel was used for the first time by Lindholm (Lindholm, 1996) as the electrode in the fuel cells in 1966 and then by Dilworth and Wunderlin (Dilworth & Wunderlin, 1968) in 1968. The term $MmNi_5$ was used where Mm (mischmetal)-represents a natural mixture of rare earth elements, mostly consisting of Ce (30-52 wt%), La (13 to 25 wt%), Nd, Pr and Sm (13-57 wt%) where an amount and elements of additives depends on the place of origin.

Though, the element substitutions in the AB_5-type alloys have been made also artificially to get better alloys for practical use. La can be replaced with Mm, Ce, Pr, Nd, Zr, Hf and Ni can be exchanged with Al, Mn, Si, Zn, Cr, Fe, Cu, Co; thereby altering the hydrogen storage capacity, the stability of the hydride phase or the corrosion resistance. For example, a partial replacement of the A and B components significantly changes macrostructure of an alloy and other properties (Table 3).

The stoichiometry of an alloy influences its durability in the long-term hydriding - dehydriding cycles, and typical commercial AB5-type alloys consists of at least 5-6 different metals, for example La0.64Ce0.36Nd0.46Ni0.95Cr0.19Mn0.41Co0.15 (Bernd, 1992).

Commonly metal hydrides are very effective for storing large amounts of hydrogen in a safe and compact way but they are mostly heavy or working in the not suitable conditions for vehicle applications. The transition hydrides are reversible and works around ambient temperature and atmospheric pressure but the gravimetric hydrogen density is limited to <3 mass%. It is still a challenge to explore the properties of the lightweight metal hydrides or investigate new hydride composite materials.

Composition	Elements and their role
Substitutions of A in AB_5 $La_{1-y}M_yB_5$	Zr, Ce, Pr, Nd decrease the unit cell volume, improve activation, high-rate discharge and cycle life, but increase the self-discharge due to a higher dissociation pressure of the metal hydride. The use of Mm instead of La reduces the alloy costs.
Substitutions of B in AB_5 $A(Ni_{1-z}M_z)_5$	A= La, Mm;M= Co, Cu, Fe, Mn, Al; $0<z<0.24$ **Ni** (1-z)>2.2 is indispensable to prevent the decrease of the amount of absorbed hydrogen and the electrode capacity **Co** decreases the volume expansion upon hydriding, retards an increase of the internal cell pressure, decreases the corrosion rate and improves the cycle life of the electrode, especially at elevated temperature (40 C), but increases the alloy costs **Substitution of Co by Fe** allows cost reduction without affecting cell performance, decreases decrepitation of alloy during hydriding. **Al**. increases hydride formation energy, prolongs cyclic life. **Mn** decreases equilibrium pressure without decreasing the amount of stored hydrogen. **V** increases the lattice volume and enhances the hydrogen diffusion. **Cu** increases high rate discharge performance.
Special additions to B in AB_5 $A(Ni,M)_{5-x}B_x$	A= La, Mm; M = Co, Cu, Fe, Mn, Al; B= Al, Si, Sn, Ge, In, Tl, **Al, Si, Sn and Ge** - minimise corrosion of the hydride electrode. **Ge**-substituted alloys exhibit facilitated kinetics of hydrogen absorption/ desorption in comparison with Sn-containing alloys. **In, Tl, Ga** increase overvoltage of hydrogen evolution (prevent generation of gaseous hydrogen).
Nonstoichiometric alloys $AB_{5\pm x}$	A= La, Mm; B=(Ni,Mn,Al,Co,V,Cu) Additional **Ni** forms separate finely dispersed phase. In **$MmB_{5.12}$** the **Ni_3Al**-type second phase with high electrocatalytic activity is formed. **Alloys poor in Mm** are destabilised and the attractive interaction between the dissolved hydrogen atoms increases. Second phase (**Ce_2Ni_7**), which forms very stable hydride is present in $MmB_{4.88}$,. When (5-x)<4.8, the hydrogen gas evolution during overcharge decreases.
Addition of alloys with increased catalytic activity $AB_5 + DE_3$	D= Mo, W, Ir; E= Ni, Co DE_3 is a catalyst for hydrogen sorption-desorptionreactions.
Mixture of two alloys $A^1B^1{}_5 + A^2B^2{}_5$	Mixing of two alloys characterised by various hydrogen equilibrium absorption pressures increases the electrode performance.

Table 3. Effect on composition on properties of AB_5-type alloys

Our laboratory at the Institute of Solid State Physics works on the classical AB_5 hydride type material $LaNi_5$ modification. As hydrogen absorber, $LaNi_5$ has been one of the most investigated intermetallic compounds during the last decades. Despite of its high hydrogen capacity of a one hydrogen atom to the each metal atom and easy activation, the binary compound is not suitable for applications due to its high plateau pressure and short lifecycle (Bittner & Badcock, 1983). However, modification of the physical and chemical properties of $LaNi_5$ can be achieved by substituting lanthanum atom with a rare earth metal (e.g. Ce, Pr, Nd, Er) or nickel with a transition metal (Al, Mn, Co, Cr). Specific sample preparation methods, such as melt-spinning, sputtering and mechanical milling have been used to improve the hydrogenation kinetics of intermetallic compounds. The resulting alloys exhibit particular structural characteristics as nano-crystalline grains with a high density of grain boundaries and lack a long-range order (similar to an amorphous state). These microstructures currently provide fast hydrogenation kinetics and better lifecycle behaviour. Mechanical milling has become a popular technique because of its simplicity, relative inexpensive equipment and applicability to most intermetallic compounds. This technique has been used for several hydrogen storage alloys there was observed a good improvement in hydrogen activation and kinetics (Ares et. al., 2004, Zaluska et. al., 2001).

The diffusion of active species on the surface may play an important role in reactions on multifunctional catalyst transport phenomena. Especially, migration of hydrogen atoms from a metal to an oxide or carbon surface that by itself has no activity for dissociate hydrogen adsorption is important. It is well known that noble metals, like Pt and Pd, can adsorb and diffuse hydrogen in reactive forms over relatively large distances. This property named as spill-over effect, is widely exploited in catalysis (Scarano et. al., 2006).

The spill-over of hydrogen involves a transfer of electrons to acceptors within the support; this process modifies the chemical nature of the support and can also activate a previously inactive material and/or induce subsequent hydrogen physisorption (Roland et. al., 1997). Dissociation of hydrogen molecule on a metal and subsequent spill-over of atomic hydrogen to its support is highly dependent upon the chemical bridges formed at the interface. Hydrogen spill-over can be assessed in a number of ways, but perhaps the most common is simple calculation of the hydrogen to metal ratio, either the surface metal or total metal content. When spill-over occurs, the relation $H:M_{surface}$ will typically exceed unity. In the case of materials that form hydrides, this relation will exceed the stoichiometric ratio of the hydride.

An AB_5 type alloy with a trade name 7-10 produced by the company Metal Rare Earth Limited of China was chosen for experiments. To study this material the measurements were carried out using Scanning Electron Microscope (SEM) of Carl Zeiss brand, model EVO 50 XVP located at the Institute of Solid State Physics. The SEM images were taken in secondary electrons (SE); the acceleration voltage was equal to 30 kV, and the emission current was between 0,5pA and 500nA. The energy dispersive detector for X-rays (EDX) was used for composite determination in the alloy 7-10.

Structural properties of the samples were studied by X-ray Diffractometer System X-STOE Theta/theta, using Kα Cu radiation but the diffraction patterns were analyzed by appropriate software of STOE system at the DTU RISOE National Laboratory (RNL).

Tungsten - carbide crucibles with 2 balls from the same material and a high energetic ball mill Retsch® MM200 was used for grinding the raw material 7-10 for 30 minutes at frequency 25 Hz per min as well as for preparing a composite with a glass. The composite consists of 3,7 weight parts of an alloy 7-10 and of a 1 part of the Pyrex glass.

A thermogravimetric technique has been used to study the hydrogen sorption on prepared samples. The measurements were performed by equipment based on the Sartorius high pressure balance (HPB) combined with pressure, temperature and gas supplying systems. The sample is placed in the steel container that can be sealed to provide a vacuum or gas atmosphere and the pressure and temperature ranges of 10^{-3} - 30 bar and from room temperature to 300°C, respectively. The studied sample 340±2 mg was initially degassed under vacuum down to 10^{-3} bars at the ambient temperature and flushed with helium gas. That was prolonged by heating up the system until 260°C with followed vacuuming, activating at 10^{-3} bar hydrogen pressure and cooling down. Subsequently the previous actions were repeated two times and finished with the vacuuming of system. Afterwards, stepwise changes of the pressure inside the measuring device were applied. At a constant temperature the increase of the sample weight, as a function of time was measured for each pressure step. The same weight and treatment procedures was chosen for a composite material, though, for calculations one have to remember that in composite an amount of hydride forming alloy is less than that for pure alloy.

SEM and EDX results (Fig. 8, Table 4) shows that the composition of the alloy 7-10 corresponds to formula $A_{0,96}B_{5,04}$ (A=La, Ce, Nd, Pr; B=Ni, Co, Mn, Al, Cr) that is close to AB_5 stoichiometry and the molecular mass of this sample becomes 435,74 g/mol. For determination of the molecular mass of the alloy and further calculations there was assumed that after the treatment a metalhydride material contains a diminutive amount of oxygen.

The XRD pattern shows that alloy 7-10 belongs to a single phase $LaNi_5$ hexagonal $CaCu_5$-type structure in the space group P6/mmm.

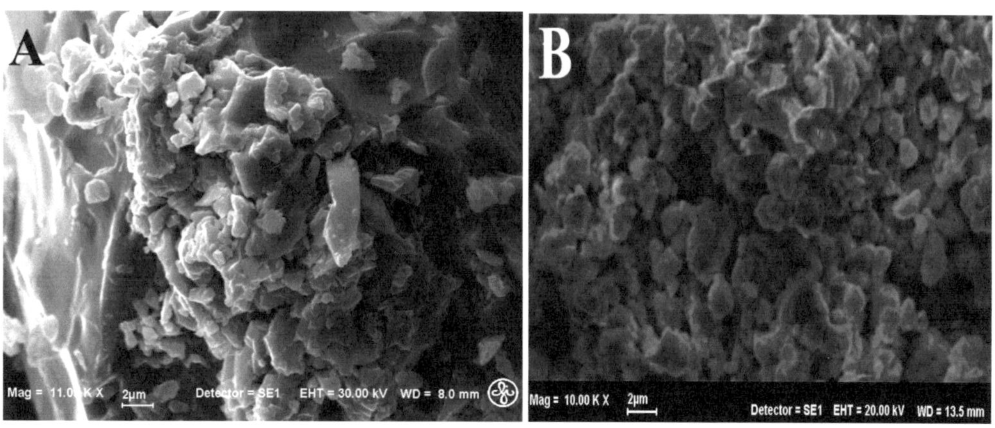

Fig. 8. SEM images of A - an alloy 7-10, grain size 50 µm and smaller, magnification 11 000; B - ball milled alloy 7-10, grain size 2 µm and smaller but agglomerated in clusters, magnification 10 000

Analysing a treated data it was figured out that a raw hydride material 7-10 absorbing more hydrogen than a classical $LaNi_5$ by itself. The HPB plots clearly shows that a pure $LaNi_5$ absorbing and desorbing hydrogen when the pressure is changed. There just a 0.26 w% of hydrogen is remaining in the pure $LaNi_5$ sample at the one atmosphere. Though, the data plot of a sample 7-10 displays opposite tendency - the absorbed hydrogen amount just slightly decreasing after lowering the pressure to the one atmosphere (Fig. 9.).

Element	Weight %	Atomic %
La	17,51	8,24
Ce	9,88	4,61
Nd	3,07	1,39
Pr	0,98	0,45
Ni	52,89	58,89
Co	6,23	6,91
Mn	5,08	6,05
Al	1,93	4,67
Cr	0,39	0,49
O	2,03	8,30

Table 4. Quantitative EDX data for the total surface of the sample 7-10

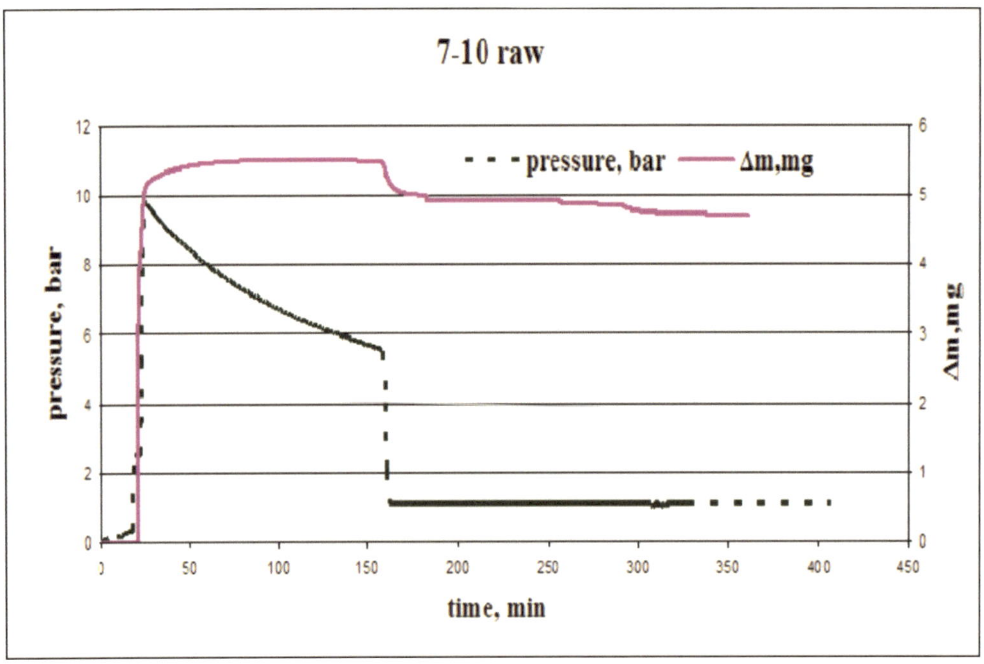

Fig. 9. HPB data plot of the raw sample 7-10 at the room temperature

Comparison of HPB results of both materials with the glass phase additives confirmed the same tendency – the composite of the alloy 7-10 and a glass absorbed more hydrogen than a composite $LaNi_5$ with a glass (Fig.10.).

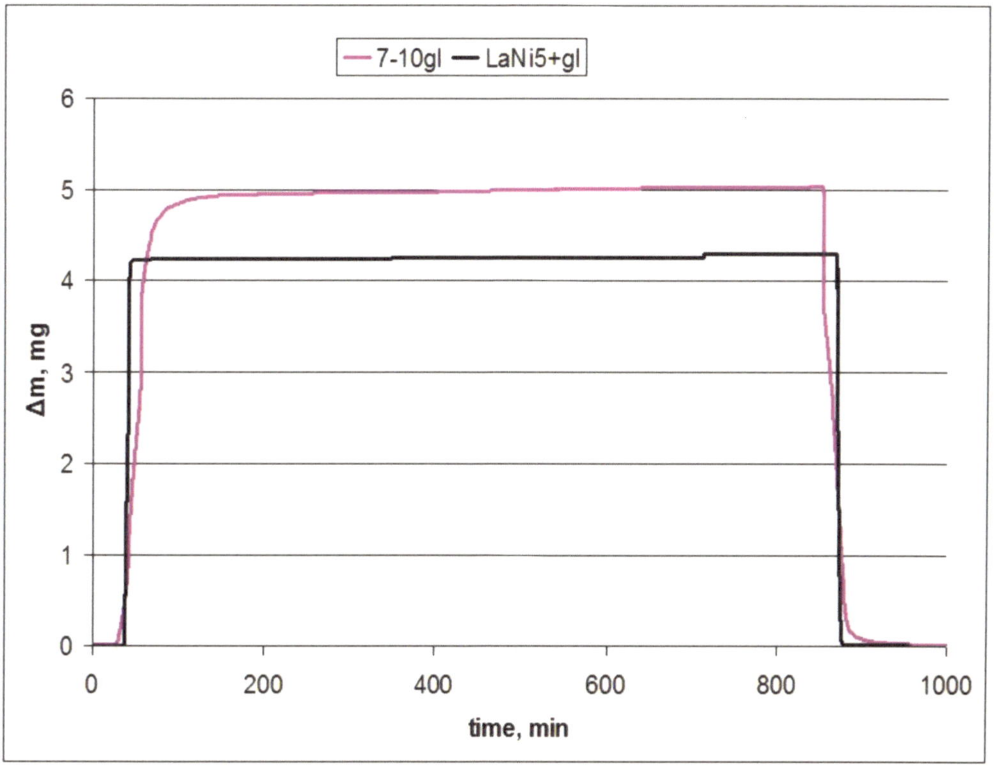

Fig. 10. HPB data of the weight change in time of the composite alloy 7-10 +glass (dashed line) and the composite $LaNi_5$ + glass (solid line)

7-10	**Δm, mg**	**w%**	**x**
Δm_{total}	5,507E-03	1,582	6,966
Δm_{av}	5,138E-03	1,477	6,499
Δm(1atm)	4,883E-03	1,405	6,177
7-10+glass	**Δm, mg**	**w%**	**x**
Δm_{total}	5,048E-03	1,643	7,242
Δm_{av}	4,616E-03	1,504	6,621
Δm(1atm)	4,393E-03	1,433	6,302

Table 5. Calculations of HPB data for raw alloy (7-10) and composite (7-10+glass)

The calculations of amount of the absorbed hydrogen in a raw alloy 7-10 and in the composite proved that the composite of 7-10 and glass have absorbed more than pure alloy 7-10 and are showed in the table 2, where Δm total – total change of the weight, Δmav – average change of the weight during cycling, Δm (1 atm) weight remaining at the 1 atm of pressure, w% - weight percents of hydrogen in alloy, x = a value from stoichiometric formula of hydride AB_5H_x

The XRD analysis of hydrogenated samples showed a good agreement with calculations of HPB data (Fig. 11, Table 6). From the XRD plot is well observable that the diffraction peaks of the hydrogenated alloy 7-10 are largely shifted to the smaller angles than that of the starting alloy, indicating that the α-phase of hydride is changed into the β-phase and the lattice parameters and cell volume of the hydride is larger than that of the starting alloy. An observed shift of XRD peaks after hydrogenation for the composite is even larger than that for the pure 7-10 alloy. Also corresponding lattice parameters and a cell volume for the hexagonal P6/mmm symmetry accordingly is larger of hydrogenated composite sample as for fully hydrogenated alloy 7-10 (Table 6).

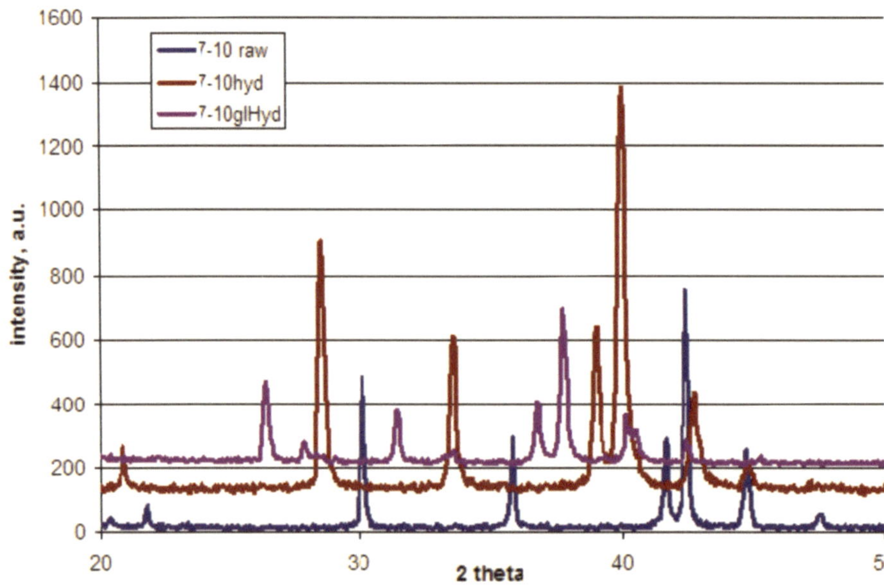

Fig. 11. XRD plot of raw (7-10raw), hydrogenated pure 7-10 (7-10H) and hydrogenated alloy 7-10 with glass phase (7-10gl-H)

Cell size / Sample	a, Å	c, Å	V, Å3
7-10	5,0083	4,0567	88,12
7-10hydrogenated	5,326	4,234	104,0
7-10+glass hydrogenated	5,369	4,2754	106,78

Table 6. Structural parameters of raw and hydrogenated samples of alloy 7-10 and composite

3.3 Conclusions

The HPB data treatment and calculations showed that hydrogen uptake in the composite alloy 7-10 with glass exceeds the pure alloy that can be explained as the spillover from the

AB_5 catalyst. The following mechanism is deduced - the hydrogen chemisorbs at the surface sites found on the AB_5 (mostly Ni sites). Bridges between the catalyst and glass particles allow the chemisorbed hydrogen to migrate onto the glass surface. Desorption occurs directly from the relatively lower energy glass sites without migration back to the catalyst. Hydrogen spillover depends upon the glass-catalyst contact. The contact changes with the quality of the mixing and milling, as well as the position of alloy 7-10 grains in the mixture. It was observed from the X-ray diffraction patterns, that the beta phase of the ball milled composite sample occurred faster than in the pure alloy sample and the peak shift to the smaller angles is noticeable larger. It is possible to assume, that the gamma hydride phase (γ) is forming when the alloy 7-10 is mixed in composite with Pyrex glass that isn't observable for pure alloy. Large lattice distortions in the γ phase are caused by hydrogen atom location sites close to Ni atoms in the elementary cell that produces inhomogeneous distribution of atoms but in the same time allows include more hydrogen atoms in the cell volume than the β phase.

4. References

Ares J. R., Cuevas F., Percheron-Guegan A., (2004). Influence of thermal annealing on the hydrogenation properties of mechanically milled AB5-type alloys, *Materials Science and Engineering B,*vol. 108, No. 1-2, (Apr., 2004) 76-80, ISSN 0921-5107

Bittner H. F. & Badcock C. C., (1983). Electrochemical Utilization of Metal Hydrides, *Journal of Electrochemical Society* Vol. 130, No. 5 (May, 1983) pp. 193C-198C, ISSN 0013-4651

Das.D, & Veziroglu, T.N. (2001). Hydrogen production by biological processes: a survey of literature. *Intern. J. Hydrogen Energy,* Vol. 26, No. 1, (Jan., 2001) pp. 13–28, ISSN 0360-3199

Das D., & Veziroglu N.T. (2008). Advances in biological hydrogen production processes. *Intern. J. Hydrogen Energy,* Vol. 33, No. 21, (Nov., 2008) pp. 6046-6057, ISSN 0360-3199

Dilworth L.R. & W.J. Wunderlin (1968). Allis-Chalmers manufacturing Company (US) Patent US- 3,405,008.

Donohue T.J., & Cogdell R.J. (2006). Microorganisms and clean energy. *Nature Reviews Microbiology* Vol. 4, 800. doi:10.1038/nrmicro1534 (Nov. 2006), ISSN : 1740-1526

Flynn T., Ghirardi M.L., & Seibert M. (2002). Accumulation of O2-tolerant phenoltypes in H2-producing strains of Chlamydomonas reinhardtii by sequential applications of chemical mutagenesis and selection. *Intern. J. Hydrogen Energy,* Vol. 27, No. 11-12, (Nov.-Dec., 2002) pp. 1421-1430, ISSN 0360-3199

Ghirardi M, Togasaki R.K., & Seibert M. (1997). Oxygen sensitivity of algal H2-production. *Applied Biochem Biotechnol.* Vol. 63-65, pp.141–151, ISSN: 0273-2289

Hallenbeck P.C., & Benemann J.R. (2002). Biological hydrogen production; fundamentals and limiting processes. *Intern. J. Hydrogen Energy,* Vol. 27, No. 11-12, (Nov. -Dec. 2002) pp. 1185-1193, ISSN 0360-3199

Holladay J.D., Hu J., King D.L., & Wang Y. (2009). An overview of hydrogen production technologies. *Catalysis Today,* Vol. 139, No. 4, (Jan., 2009) pp 244-260, ISSN 0920-5861

Kotay Meher S., & Das D. (2008). Biohydrogen as a renewable energy resource – prospects and potentials. Intern. J. Hydrogen Energy, Volume 33, Issue 1, January 2008, Pages 258-263 ISSN 0360-3199

Levin D.B., Pitt L., & Love M. (2004). Biohydrogen production: prospects and limitations to practical application. *Intern. J. Hydrogen Energy*, Vol. 29, No. 2, (Feb. 2004) pp. 173-185, ISSN 0360-3199

Lindholm I., (1966) Allmanna Svenska Elektriska Aktiebolaget (SE) Patent US- 3,262,816

Maeda T., Sanchez-Torres V. & Wood T.K. (2008). Enhanced hydrogen production from glucose by metabolically engineered Escherichia coli. *Applied Microbiology and Biotechnology*, Vol. 77, Nr. 4, pp. 879-890, DOI: 10.1007/s00253-007-1217-0, ISSN 1432-0614

Midilli A., Ay M., Dincer I., Rosen M. A., On hydrogen and hydrogen energy strategies: I: current status and needs., *Renewable and Sustainable Energy Reviews* Vol. 9, No. 3, (Jun., 2005) pp. 255-271, ISSN 1364-0321

Nath K., & Das D. (2004). Improvement of fermentative hydrogen production: various approaches. *Appl. Microbiol. Biotechnol.*, Vol. 65 (Oct., 2004) pp. 520–529, ISSN: 1432-0614

Navarro R.M., del Valle F., Villoria de la Mano J.A., Álvarez-Galván M.C., Fierro J.L.G. (2009) Photocatalytic Water Splitting Under Visible Light: Concept and Catalysts Development Advances in Chemical Engineering, *Advances in Chemical Engineering - Photocatalytic Technologies*, Vol.36, (Jul, 2009) pp. 111-143, ISSN: 0065-2377

Roland U., Braunschweig T., Roessner F., (1997) On the nature of spilt-over hydrogen, *Journal of Molecular Catalysis A: Chemical*, Vol. 127, No. 1-3, (Dec. 1997) pp 61-84, ISSN 0733-9372

Sandrock, G., A panoramic overview of hydrogen storage alloys from a gas reaction point of view. *Journal of Alloys and Compounds*, Vol. 293-295, (Dec., 1999) pp. 877-888, ISSN 0925-8388

Scarano D., Bordiga S., Lamberti C., Ricchiardi G., Bertarione S., Spoto D G., *Applied Catalysis A: General,*, Vol. 307, No. 1, (Jun., 2006) pp 3-12, ISSN 0926-860X

Schlapbach L., (1992) Surface properties and activation. In: *Hydrogen in intermetallic compounds II*, Vol. 67, ed. L. Schlapbach. Pp. 15-95, Springer: Berlin Heidelberg, ISBN 9780387546681, New York

Schlapbach L. and Zuttel A., (2001) Hydrogen-storage materials for mobile applications *Nature*, Vol. 414, (Nov., 2001), pp. 353-358, ISSN 0028-0836

Wang J., & Wan W. (2009). Factors influencing fermentative hydrogen production: A review. *Intern. J. Hydrogen Energy*, Vol. 34, No. 2, (Jan., 2009) pp. 799-81, ISSN 0360-3199

Wilkins J. R., Stoner G.E., & Boykin E.H. (1974). Microbial Detection Method Based on Sensing Molecular Hydrogen. *Applied Microbiology*, Vol. 5 (May 1974), pp. 949–952, ISSN 1365-2672

Zaluska A., Zaluski L., & Ström-Olsen J. O. (2001) Structure, catalysis and atomic reactions on the nano-scale: a systematic approach to metal hydrides for hydrogen storage, *Applied Physics A: Materials Science & Processing*, Vol. 72, No 2, pp. 157 - 165, ISSN 0947-8396

Züttel A., (2003) Materials for hydrogen storage, *Materials Today*, Vol. 6, No 9 (Sept., 2003) p. 24-33, ISSN 1369-7021

Züttel, A., (2004) Hydrogen storage methods. *Naturwissenschaften*, Vol. 91. No. 4, (Apr. 2004) p. 157-172, ISSN 0028-1042

10

Development of Tunable Nanocomposites Made from Carbon Nanotubes for Electrochemical Applications

María José Esplandiu[1], Mireia Baeza[2], Rosa Olivé-Monllau[2], Francisco Céspedes[2] and Jordi Bartrolí[2]

[1]*Centro de Investigación en Nanociencia y Nanotecnología, CIN2 (CSIC-ICN), Campus UAB, 08193 Bellaterra (Barcelona)*

[2]*Grup de Sensors i Biosensors, Departament de Química, Facultat de Ciències, Edifici C-Nord, Universitat Autònoma de Barcelona, 08193 Bellaterra (Barcelona)*

Spain

1. Introduction

The advent of the nanotechnology field has brought about novel devices and materials at the nanometer scale triggered by the demand of miniaturizing electronic, optical, actuating and sensing systems. In this nanotechnology era, the concept of nanocomposite has been brought to light though it has been present in nature and used from historical times. A nanocomposite can be defined as a material made of more than one solid phase where at least one of the constituent parts has a nanometer scale dimension. Building blocks with dimensions in the nanosize range make possible to design and create new materials with unprecedented versatility and improvement in their physical and chemical properties. The promise of nanocomposites lies in their multifunctionality for different applications and the possibility of unique combinations of properties unachievable with conventional materials.

As mentioned before, the concept of enhancing properties and improving characteristics of materials through the creation of multiple phase nanocomposites is not recent. Mother Nature has a lot of examples of nanocomposites such as the structure of seashells and the bones. The idea has also been practiced since civilization started and humanity began producing more efficient materials for functional purposes. The Maya blue pigment, an ancient nanostructured material which was found to be very resistive to acids or (bio)corrosion phenomena, is a composite of organic and inorganic constituents, primarily leave dyes combined with a natural clay (Jose-Yacaman et al., 1996).

Nanocomposites comprise a wide scope ranging from metal/ceramic nanocomposites, polymer-based nanocomposites to natural or biomimetic nanobiocomposites. They still share tremendous challenges, especially in the control over the distribution in size and dispersion of the nanosize constituents and in the tailoring and understanding of the role of the interfaces (Ajayan et al., 2003; Harris, 2004; Moniruzzaman et al., 2006; Spitalsky et al., 2010).

Among the broad world of nanocomposites, the conducting ones attract special attention because of their special applications in electronics and electrochemistry (Grossiord et al.,

2006; Céspedes et al., 1996; Xiao & Li., 2008; Li et al., 2008; Li et al., 2009; Rajesh et al., 2009; Navratil & Barek, 2009). Nanocomposite for electrode applications are formed by at least one conducting and one insulating phase. In this context, carbon nanotubes arise as ideal conducting phase due to their unique chemical and physical properties. At this moment carbon nanotube composites comprise a very dynamical emerging field and it turns out very difficult to cover the entire spectrum of such topic (Ajayan et al., 2003; Rubianes & Rivas, 2003; Wang & Musameh, 2003; Barrera et al., 2005, Harris, 2004; Moniruzzaman et al., 2006; Spitalsky et al., 2010; Li et al. 2008; Xiao & Li, 2008). Therefore in this chapter we will focus on conducting nanocomposites based on carbon nanotubes/insulating polymers and special attention will be stressed on CNT epoxy composites as robust, easy-processing and cheap platforms for electrochemical applications (Céspedes et al., 1996; Pumera et al., 2006; Pacios et al, 2008; Olivé-Monllau et al., 2010). We will introduce an electrochemical characterization approach based mainly on impedance spectroscopy which can allow us to find the optimized CNT composition for electroanalytical purposes and which can be extended to other electrochemical applications. Additionally, we will briefly describe basic concepts and provide generic examples that can give important insights in this field.

2. Basic considerations on the structure and properties of carbon nanotubes

Carbon nanotubes have captured the interest as nanoscale materials due to their structure (as a model of a 1D system) and their impressive list of superlative and outstanding properties such as high Young´s modulus and strength, unique richness in electrical properties, high thermal conductivity, chemical and electrochemical anisotropy and biocompatibility.

The graphene layers have become the starting point to explain the structure of carbon nanotubes. The way the graphene sheet is wrapped up can be described by a pair of indices (*n,m*) that define the chiral vector, $\vec{C} = n\vec{a}_1 + m\vec{a}_2$, in which $\vec{a}_1$ and $\vec{a}_2$ are the basis vectors of the hexagonal graphene lattice (Fig. 1). Three different types of nanotube structures can be generated by rolling up the graphene layer: zigzag (*m=0*), armchair (*n=m*) and chiral nanotubes (the rest of vectors) (Saito et al., 1998; Louie, 2001). Although CNT are closely related to 2D graphene, the tube curvature and the quantum confinement in the circumferential direction of the nanotube bring about the unique properties that make CNT different from graphene. One of these unusual properties is the electronic conductivity which strongly depends on the chirality and CNT diameter. CNTs can exhibit singular electronic band structures and can show metallic and semiconduncting behavior. As a general rule, *n,m* tubes with *n-m* being an integer multiple of 3 are metallic while the remaining tubes are semiconducting. The band gap of semiconducting tubes can be approximated by the relation $E_g = 0.8eV / d$, with *d* being the diameter of the nanotube (Wildoer et al., 1998). Therefore the bigger the diameter, the more metallic behavior is found. Moreover, carbon nanotubes can behave like quantum wires due to the confinement effects on the tube circumferences. They can exhibit a ballistic flow of electrons with electron mean-free paths of several micrometers, and are capable of conducting very large current densities (Javey et al., 2004). For instance, carbon nanotubes have been shown to withstand current densities up to 10^9 A/cm^2, a current density which is about 2-3 orders of magnitude larger than in metals such as Al or Cu.

So far we have considered CNTs formed by a single graphene wall (SWNT) with diameters ranging generally from 0.4 nm up to 1.5 nm and with an important diversity in electronic conductivity, from semiconducting to metallic behavior. However CNTs can also be made by multiple graphene layers disposed in different arrangements. Such structures, known as multi-walled carbon nanotubes (MWCNTs), have typically diameters in the range of 2-100 nm and metallic behavior. MWCNTs can appear as several concentric tubes fitted one inside the other (hollow MWCNTs) or with the graphene planes forming an angle with respect to the axis of the tube (herringbone or bamboo MWCNTs), see schemes in Fig. 1. The bamboo-like tubes differ from the herringbone ones in that some of the graphene layers are periodically closed along the length of the tube forming compartments. The main difference between the hollow morphology and the herringbone and bamboo ones lies on the high density of terminating edge planes that contain the last two morphological variations of MWCNTs.

The bonding in CNT is basically sp^2, however the circular curvature, apart form yielding quantum confinement and quantized conductance, also causes σ-π rehybridization, an effect that is stronger as the CNT diameter decreases (Han, 2005; Srivastava, 2005). In such case three σ bonds are slightly out of plane and for compensation the π orbitals are more delocalized outside the tube (Fig. 1). The electron cloud distortion induced by the curvature, which yields a rich π-electron conjugation outside the tube, can make CNT more electrochemically active and electrically and thermally more conductive. Moreover, the latter fact together with the combination of size, structure, topology and light weight confer nanotubes remarkable mechanical properties such as high stability, high strength and stiffness together with low density and elastic deformability. That is a big difference with respect to graphite or graphene in which the sp^2 hybrid orbital form three in-plane σ bonds with an out-of-plane π orbital.

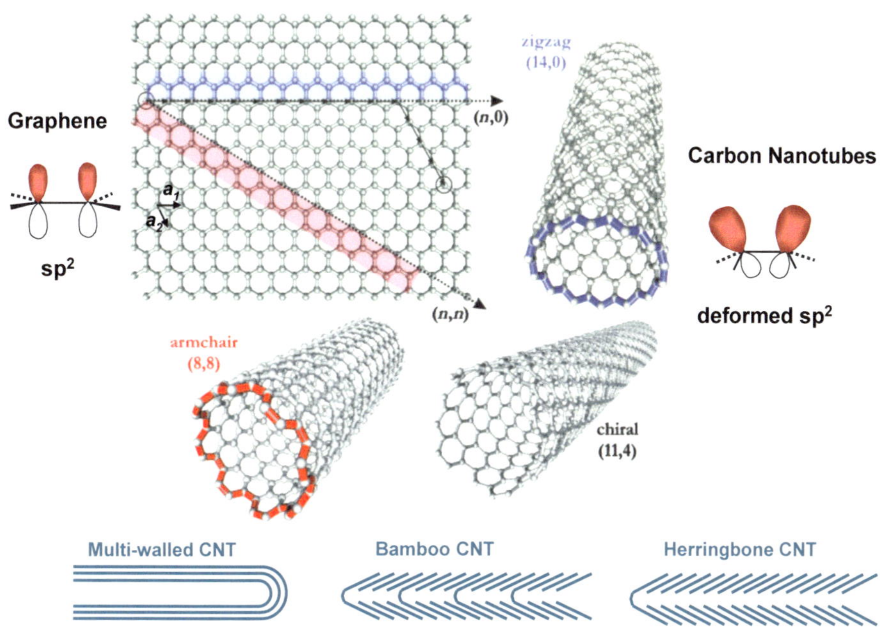

Fig. 1. Roll-up of a graphene sheet leading to three different types of SWNTs together with their bonding structures. Different structures of multi-walled CNT. Adapted from Ref. (Balasubramanian, 2005).

CNTs provide a high specific surface system which together with the σ-π rehybridization and the presence of structural defects facilitate different chemical processes such as (bio)chemical derivatization, intercalation, molecular adsorption, doping, charge transfer, etc (Han, 2005; Niyogi, 2002; Chen, 2003; Hirsch, 2002; Burghard, 2005; Tasis, 2006). These characteristics are being considerably exploited for different applications such as biochemical and chemical sensing, energy storage or separation techniques. A good understanding of the chemical properties of CNTs is mandatory for enhancing the efficiency of practical devices and also for comprehending related fundamental processes such as their electrochemical properties.

Although graphite, fullerene and carbon nanotubes are built from the same basic element, the chemical reactivity is substantially different among them. The chemical reactivity is higher for a CNT than for a graphene layer but lower if compared with fullerene (Niyogi, 2002). Such behavior can be closely related to the surface curvature of the carbon structure (Niyogi, 2002; Han, 2005; Chen, 2003; Burghard, 2005) and the higher strain that is generated by such spherical geometry. A higher surface reactivity with decreasing nanotube diameter is thus expected (Chen, 2003).

The reactivity of different graphitic carbon allotropes is also characterized by having chemical anisotropy. Taking as a reference, for instance, the layered structure of Highly Oriented Pyrolytic Graphite (HOPG), it is known the its basal plane is more inert than the highly reactive edges (edge HOPG) which contain unsatisfied valences or dangling bonds prone to reaction with oxygen or water (McCreery, 2008). Such anisotropy is also observed in adsorption processes. The same anisotropy in the chemical reactivity can be considered on CNT, since the walls behave differently than the ends which can be either capped or just finishing in terminal edges (Balasubramanian, 2005; Burghard, 2005; Hirsch, 2002; Katz, 2004). The chemical reactivity is increased at the capped ends (due to the presence for instance of pentagons) or at the edges of an open nanotube, which become the more reactive sites of the CNT irrespective of its diameter. Such anisotropy in the chemical reactivity will also have strong implications from the point of view of electrochemistry.

Although, the CNT ends exhibit a higher chemical reactivity than the walls, the presence of local defects on the walls constitute also efficient reactive sites. For instance, when CNTs are subjected to oxidative acid treatment, the local generation of carboxylic, cetones, alcohols and ester groups leads to active sites that can be profited for (bio)molecular anchoring (Hirsch, 2002; Katz, 2004; Tasis, 2006; Esplandiu, 2009a). Such acid treatments also open the CNT capped ends, bringing about an increase of oxygen rich terminal groups. Such defects together with the reactivity associated with the generated strain from the curvature are very important for the attachment of a wide diversity of molecules (complexing agents, fluorescent and electroactive groups, (bio)catalysts, or biomolecules, (proteins, carbohydrates, nucleic acids, etc). Chemical modification turns out to be one of the key issues for the development of selective electrochemical (bio)sensors for (bio)recognition or catalyst materials to be used in energy-storage devices. Moreover, chemical modification also aids to overcome one of the drawbacks for CNT applications which is the difficulty of dispersing them in solvents, especially in aqueous media.

In general, the electrochemical performance of carbon materials is basically determined by the electronic properties (specially their band structure in terms of density of states), and given its interfacial character, by the surface structure and surface chemistry (i.e. surface terminal functional groups or adsorption processes) (McCreery, 2008). Such features will affect the electrode kinetics, potential limits, background currents and the interaction with molecules in solution.

We have mentioned the chemical anisotropy characteristic of graphitic materials, which is also present in their electronic properties. For instance, it is well known that in the layered structure of HOPG, the electronic conductivity in the plane parallel to the graphene layers (basal plane) is about four orders of magnitude lower than in the perpendicular one (edge plane) (McCreery, 1990). The basal plane of HOPG is atomically ordered, with higher electronic resistance and chemically more inert than edge plane HOPG. Such anisotropy also influences the electrochemical behavior inducing an electrochemical anisotropy as well. It has been demonstrated that in graphite, the basal plane exhibits slow electron transfer kinetics whereas the reactive edge sites can increase the electron transfer rate in almost five orders of magnitude. The same electrochemical anisotropy can also be expected in CNTs. The open ends of carbon nanotubes have been likened to the edge planes of HOPG whereas the tube walls are suggested to have similar electrochemical properties to those of basal HOPG (Moore, 2004; Banks, 2005).

Electrochemical experiments with carbon nanotube electrodes showed enhanced electron transfer reactivity (increased redox currents and reduced peak separations in the voltammetry which enables lower detection limits and enhanced sensitivity) for a variety of systems as compared to other electrode materials (Wang, 2005; Gooding, 2005; Banks, 2006). These features have impelled some research groups to label CNTs as unique electrocatalytical materials. However this issue has open a big debate since all these claims have been concluded from analyzing CNT systems that were little characterized in terms of CNT purity (amorphous carbon, metal catalyst, etc.). Indeed metal impurities from the CNT growth procedure can produce electrocatalytical effects and might bring about misinterpretations. So far there is no an absolute assessment of the relative contribution of metal catalyst and the CNT itself in the electrochemical process. Caution has to be taken in attributing enhanced electrocatalytical properties of CNTs on different redox systems as compared to other graphitic carbon electrodes. It is quite probable that CNTs themselves exhibit electrocatalytic effects but in the same way as what was found on edge pyrolytic graphite systems (Moore, 2004). CNT defects such as edge-plane like sites (like the ones at the open ends of nanotubes) can have an important role in such electrocatalytic behavior. However, the possible electrochemical enhancement at the defect sites should not induce us to completely disregard the electrochemical behavior of the sidewalls. Although they can be considered less electrochemically active some groups have already shown important electrochemical performance from the sidewall sites (Dumitrescu et al., 2009).

Moreover, it has also been suggested that a thin layer diffusion within the CNT porous layer can also contribute to explain the electron kinetics observed with CNTs specially when the CNT electrode is composed of CNT multilayers (Streeter et al., 2008).

From the discussion above one can conceive that the electrochemical behavior of CNTs is rather complex for an unequivocally assessment. There are many concerns that have to be taken into account which arise from the different growth techniques, the quality of the CNTs, the chemical processing steps employed, incomplete characterization, CNT electrode configuration, etc (Dumitrescu et al, 2009). However unique advantages of the CNTs are coming from their nanometer size which can be manipulated to design novel electrode architectures (Gooding, 2005; Esplandiu, 2009a) or for signal amplification purposes (Wang, 2005; Merkoci, 2006; Kim, 2007). Related with the latter issue, CNT systems exhibit a high effective surface area which is beneficial not only for enhancing electrochemical currents of diffusing electroactive species but also for allowing the load of high density electrochemical active (bio)molecules which can improve electrochemical signals. Another example related with the fact that CNTs are versatile materials for the design of unusual electrode configurations can be provided by the CNT modified electrodes with the nanotubes disposed in an upright configuration, which due to the nanometer size they can establish excellent wiring to the redox center of metalloproteins or enzymes favoring the direct electron transfer process (Yu, 2003; Patolsky, 2004; Liu, 2005; Esplandiu, 2009b).

3. CNT-polymer nanocomposite electrodes

3.1 General considerations and fabrication methods

As mentioned before, the nanometer dimensions of CNTs, their remarkable high specific surface area, high aspect ratio, low-weight and their additional extraordinary physical and chemical properties make them ideal fillers for polymer matrices and thus for imparting new functionalities to polymer nanocomposites. The use of carbon nanotubes has attracted a great deal of attention not only to develop better structural-reinforced materials thanks to the advantageous CNT mechanical properties (high stiffness, tensile strength) but also for conferring remarkable conductivity to polymer matrices (Ajayan et al., 2003; Wang & Musameh, 2003; Barrera et al., 2005, Harris, 2004; Moniruzzaman et al., 2006; Coleman, 2006; Spitalsky et al., 2010; Li et al. 2008; Xiao & Li, 2008; Martone et al., 2010).

However the fabrication of high performance CNT composites still results challenging and several important requirements should be fulfilled for an effective improvement of CNT-based composite properties. One of the crucial ones is to get a very homogenous dispersion of CNTs in the solvent or host matrix and an optimized concentration of the CNT load (Barrera et al., 2005; Olek, 2006; Grossiord et al., 2006; Sahoo et al. 2010; Grady, 2010). High CNT purification, good interfacial bonding or good adhesion between nanotubes and polymers are also necessary conditions for improving mechanical/electrical properties of the composites. Moreover, alignment of the CNTs in composites which provides enhanced anisotropic characteristics has also been observed to improve in some cases the mechanical properties of the composites (Olek, 2006).

CNTs are difficult to disperse and dissolve in any organic and aqueous medium due to the strong Van der Waals interactions which drive their aggregation. Dispersion can be achieved by mechanical and chemical methods. The mechanical techniques involve physically separating the tubes from each other by mechanical agitation (shear mixing, ultrasonication, magnetic stirring, etc). The chemical methods often use surfactants or chemical treatment of the tube surface which enhance the interactions between the filler and the matrix.

Several processing methods are available for fabricating the CNT/polymer composites based on the nature of the polymer being used (Grossiord et al., 2006; Olek, 2006; Spitalsky et al., 2010; Sahoo et al., 2010; Sánchez et al. 2009). The most common method is based on the mixing of the CNTs and a polymer in a suitable solvent before evaporating the solvent to form the composite films. For thermoplastic polymers which are insoluble in any solvent the melting processing is the common alternative (e.g. melting of thermoplastic polymers to form a viscous liquid and then mixing with CNTs). For thermosetting polymers such as epoxy, the CNTs are mixed with monomers (usually liquid) and then they are thermally cured with crosslinking/catalyzing agents. Other approaches are the layer by layer assembly of polymers on CNTs (normally with oppositely charged polyelectrolytes), in situ polymerization on the CNT surface, coagulation of CNT/polymer by phase inversion such as in the case of polysulfone/CNT composites, etc.

As mentioned in the introduction, we will mainly focus on the fabrication and properties of rigid CNT-polymer composite electrodes based on an epoxy matrix. Epoxy resins are a family of polymers widely used for their excellent chemical properties, their good adhesion to other materials and their excellent insulating characteristics (Céspedes et al., 1996). Furthermore, epoxy CNT composites are easily prepared, inexpensive, widely available, ease of machining and malleable before curing and exhibit the advantage of generating a fresh surface by only polishing every time that is needed for successive experiments. CNTs have atomically smooth and quite inert surfaces and as such there is a lack of interfacial bonding or adhesion between the CNT and the polymer chain that limits the composite performance. Chemical modification can provide optimum surface sites that enhance the interaction with the host matrix. Accordingly, our typical procedure consisted in CNT treatment with concentrated HNO_3 acid to increase the CNT purity and to generate oxygen functionalities which provide better interaction or adhesion with the polymeric resin. After acid treatment, the CNTs were dispersed in the resin and hardener agents by mechanical agitation and allowed to cure during 24 h at 80ºC. Electrode surface was then polished with different sandpapers of decreasing grain size (Pacios et al., 2009; Olivé-Monllau et al., 2009). Figure 2 shows a schematic approach of the composite electrode preparation.

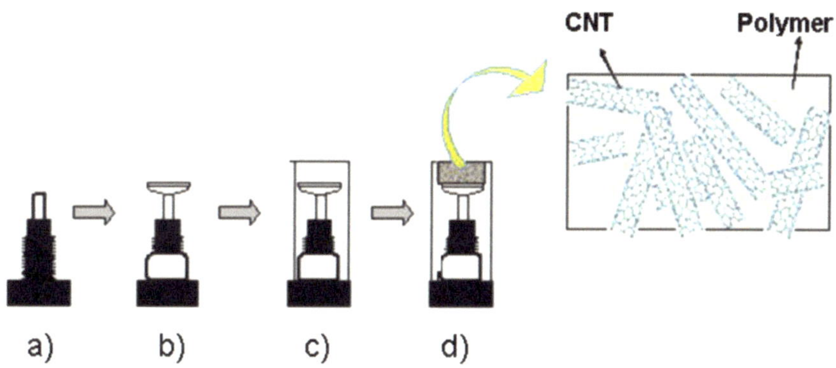

Fig. 2. Electrode assembly. Composite electrodes were prepared by using a PVC tube body small copper disk soldered at the end of an electrical connector. The composite was prepared by first mixing manually the epoxy resin and the hardener in a 20:3 ratio (w:w), respectively. Then different weight proportions of CNT were added to the epoxy mixture. The carbon paste was well mixed during one hour and put in the cavity of the PVC body and thermally cured. A) Electrical connector; B) copper disk fixture; C) mount of PVC body; D) introduction of carbon-epoxy paste.

3.2 Electrical properties of composites: percolation curves

Many studies have been performed to incorporate and adjust the CNT amount in polymer matrices to tailor electrical properties according to the different purposes since many applications (electromagnetic interference shielding, antistatic coatings, etc.) normally require specific levels of conductivity (Mylvaganam & Zhang, 2007). Following our main interest, that is the use of CNT composites for electrochemical applications, one has to assure, as a first step, the amount of CNT necessary in the matrix for a good electrical conductivity. One way to get a wide overview of the electrical properties of the nanocomposites is by performing percolation curves. Systems composed of polymeric matrix and conductive filler experience the percolation transition which refers to the critical concentration of the filler at which the electrical properties of the composite are significantly changed (Céspedes et al., 1996; Martin et al., 2004; Olek, 2006; Singh et al., 2009; Faiella et al., 2009; Olivé-Monllau et al., 2009; Kara et al., 2010). The CNT composite becomes conductive above a critical value which is called percolation threshold and defines the insulator - conductor transition. At this point the first conductive network is formed through the matrix. For reaching the electrical percolation threshold and therefore be electrically conductive, it is not necessary a direct physical connection or overlapping of the CNTs. Nanotubes can just be close enough to allow for a hopping/tunneling electron effect.

The electrical percolation threshold depends on many factors including the size and the shape of the filler, the matrix properties, preparation method, filler properties, dispersion of the filler in the matrix, interaction between compounds, etc. One of the interesting features of CNT composites is that they can reach percolation thresholds at a very low weight fraction of nanotubes due to their high aspect ratio (Sandler, J.K.W. et al., 2003). Numerous studies have been performed on the evaluation of the electrical percolation threshold of CNT/polymer systems (Martin et al., 2004; Olek, 2006; Singh et al., 2009; Faiella et al., 2009; Kara et al., 2010). Depending on the functionalization, exfoliation and dispersion of CNTs, composite processing and properties of the components, different values of percolation thresholds were obtained ranging from 0.005 up to 5 wt%.

As a matter of illustration, let´s concentrate on the percolation curves for CNT epoxy composite in comparison with the one corresponding to graphite epoxy composite (fig. 3) (Olivé-Monllau et al., 2009). The figure shows the variation of resistivity as a function of the percentage of carbon nanotubes or graphite. In the case of CNT composite, the percolation threshold is achieved around 1% of CNT content whereas in the case of the graphite composite the percolation threshold was about 12%. Below that critical value the carbon composites behave as insulators and above that point the resistivity of the composites is very low and their variations are almost negligible. It is worth noting the smaller percolation threshold of the CNT composite as compared to its graphite analog.

Another way to characterize the electrical properties of the carbon composites is by using conductive Atomic Force Microscopy (C-AFM) (O'Hare et al., 2002; Pacios et al., 2008). This technique consists in the use of a conductive tip operating in standard contact mode with the sample. As a voltage is applied between tip and sample, a current is generated, the intensity of which will depend on the sample nature. Thus, a map of current distribution along the sample is measured simultaneously with the surface topography. The C-AFM mapping images allowed us to observe the conductive microzones where the electrochemical charge transfer can take place (light regions), and the non conductive insulating microzones where such processes are suppressed (dark regions). As seen in Fig. 4 the composite with 20% of MWCNT loading showed slightly more conductive areas than in the case of the composite with 10% of MWCNT loading, as expected. Moreover, from the topographic AFM images one can estimate a value of roughness (R). Indeed, as the MWCNT loading of the composite increases, the roughness of the surface also increases. Such increase in the conducting areas and in the surface roughness with the CNT load also greatly influence the double-layer capacitance values as will be discussed later on with the electrochemical characterization.

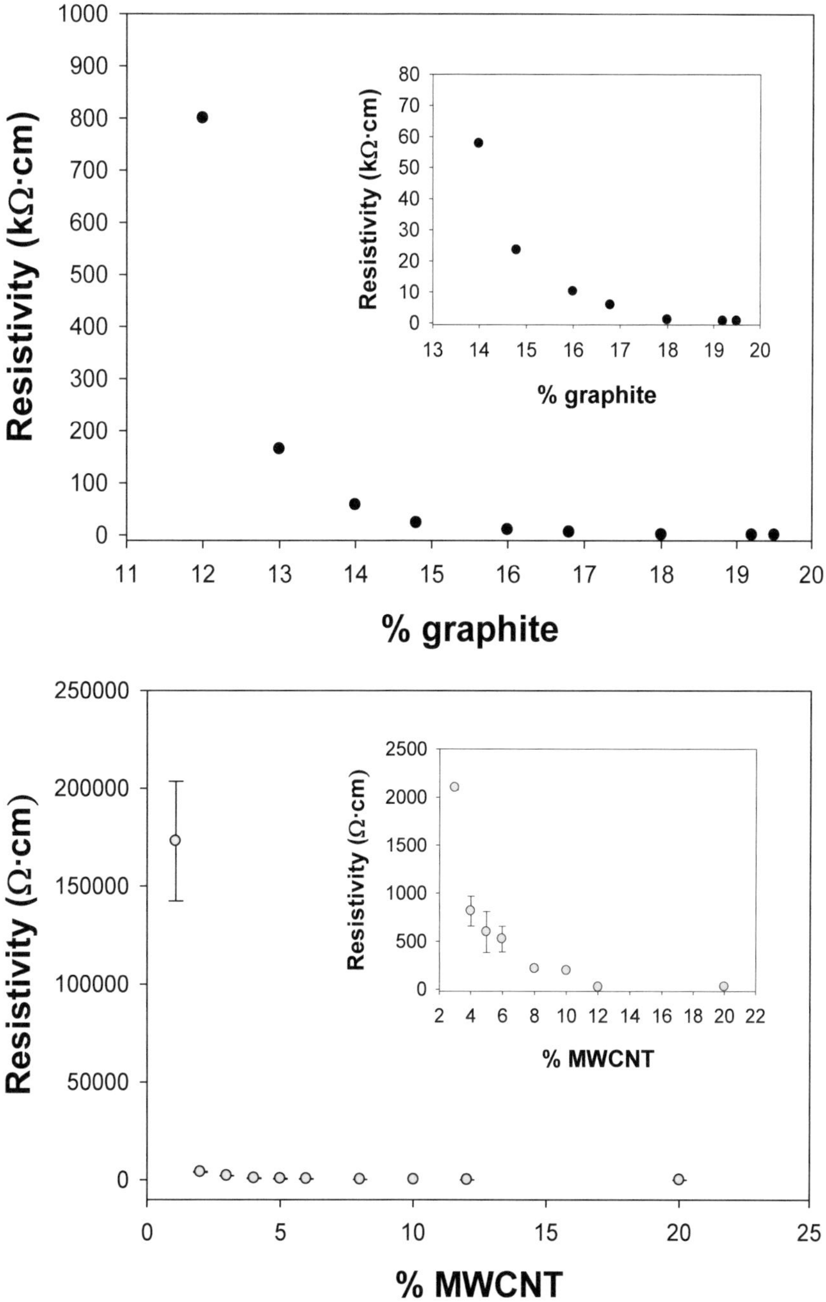

Fig. 3. Percolation curve obtained for graphite and CNT epoxy composites. The trend follows the percolation theory. The insets show the results in a zoom scale. Graphite composite percolation curve was adapted from Ref. (Olivé-Monllau et al., 2009) and the carbon nanotube composite curve was reproduced from Ref. (Olivé-Monllau et al., 2010) with permission. Copyright, 2010, Elsevier.

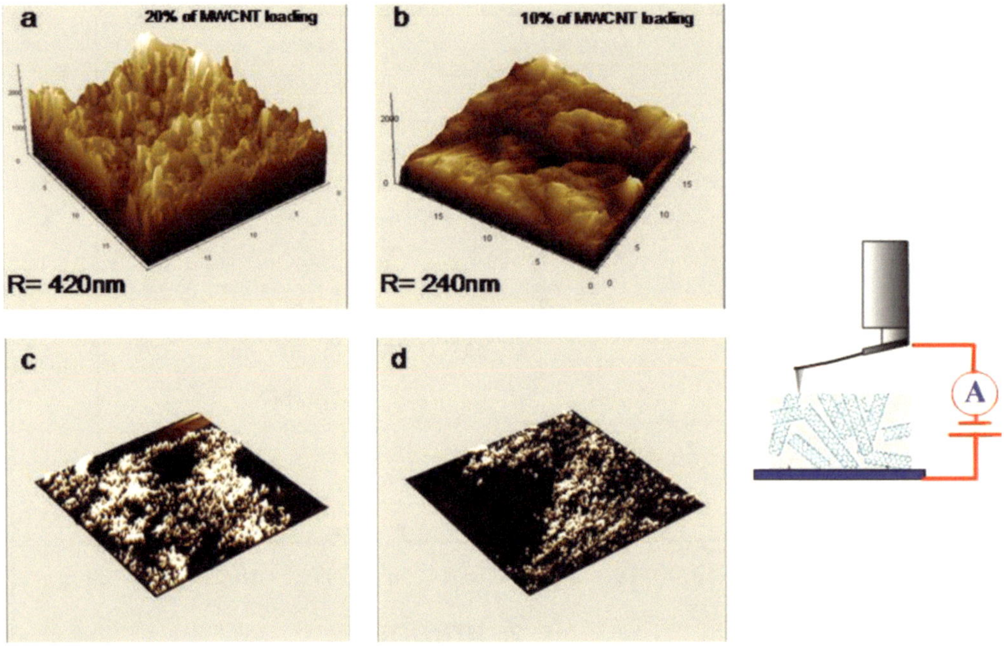

Fig. 4. Topographical AFM images with their corresponding conductivity mappings for 20% of CNT loading (a,c) and 10% of CNT loading (b,d). The x-axis, y-axis and z-axis units are µm, µm and nm respectively. Scheme of a C-AFM set-up. Adapted from Ref. (Olivé-Monllau et al., 2010) with permission. Copyright, 2010, Elsevier.

3.3 Electrochemical properties

Carbon composites for electrochemical applications were introduced by Adams in the fifties by using carbon paste electrodes (Adams, 1958), a mixture of a polycrystalline graphite powder with water-immiscible insulating organic liquid (Nujol, hexadecane, etc.). After that, other polymer alternatives were used such as epoxy, Nafion or conducting polymers (Navarro-Laboulais et al., 1994; Céspedes et al, 1996; Zhao & O' Hare, 2008; Olivé-Monllau et al., 2009; McCreery, 2008; Navratil & Barek, 2009). Since then the electrochemical properties of different carbon composite electrodes were studied in detail covering a wide range of issues from electroanalysis, pharmaceuticals, biological, redox processes and mechanistic electrochemistry. However with the advent of CNTs, the use of composite electrodes has been boosted. CNT composite electrodes are a suitable material for ion intercalation, catalyst entrapment/deposition for applications in electromechanical systems (artificial muscles or motors), energy applications (lithium batteries, supercapacitors, solar cells, (bio)fuel cells, etc.) and for (bio)chemical sensing (diagnostic medicine, genetics, drug, food and environmental sensing). Many groups have reported the use of CNT (bio)composites based on Teflon, epoxy, chitosan, poplystyrene or polysulfone for a wide variety of electroanalytical assays (Wang, 2005; Pumera et al., 2006; Xiao et al., 2008; Yogeswaran & Chen, 2008; Pacios et al., 2008; Sánchez et al., 2009; Kara et al., 2010; Olivé-Monllau et al., 2010).

In general carbon composite materials present improvements in the electrochemical response over conventional solid carbon electrodes such as the typical glassy carbon electrodes or other metal electrodes (McCreery, 2008). First, the composites exhibit the interesting advantage of easy surface renewal without extensive polishing. Second, the graphitic particles or nanotubes exhibit randomly exposed edges and are usually quite electrochemically reactive. Third, the background current (I_c) responsible of the noise in the electrochemical signal is considerable small. Such background current is related with the double layer charging capacitance of the electrode which is mainly determined by the active area, that is, the fractional area of carbon exposed to the solution (A_{act}). The active area is much smaller than the geometric area (A_{geom}) of the electrode since the remaining part is coated with the host polymer. On the other side the faradaic current (I_f) related to the electrochemical electron transfer process is more proportional to the geometric area of the electrode if the active areas are not so far away from each other. That makes that the electrochemical signal(I_f)/noise (I_c) ratio be proportional to A_{geom}/A_{act} and hence bigger than in the case of conventional metal electrodes of the same geometric area. Fourth, such nanocomposite electrodes can behave, under certain conditions, as an array of microelectrode with all the advantages that they comprise for electroanalytical applications (fig.5) (Arrigan, 2004). Apart from exhibiting low noise current which leads to lower limits of detection than macroelectrodes, they can exhibit increased electrochemical signal and fast

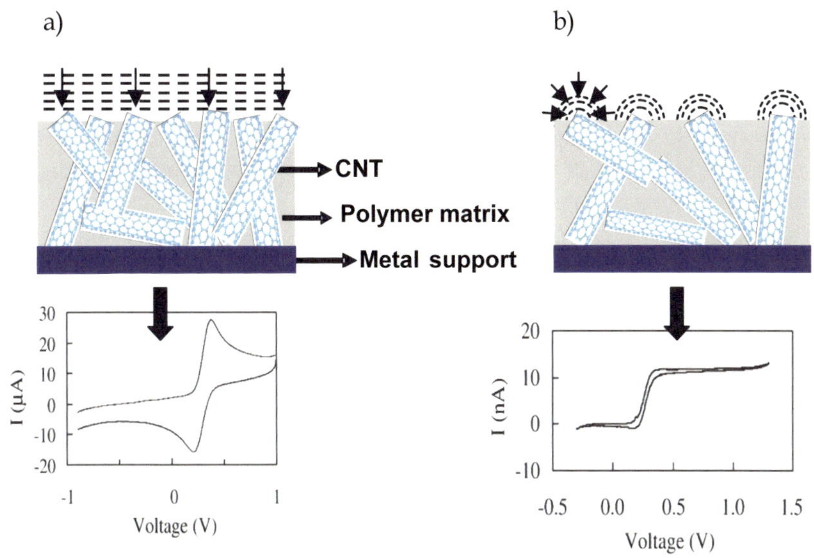

Fig. 5. Schematics showing the diffusion profile of the electroactive species for different composites and their corresponding cyclic voltammogram (CV) response. In case a) the active exposed areas of the electrodes are closer, so the diffusion layer of the different electrochemical active spots overlap generating a linear diffusion of the electroactive species and the CV response is in a peak-shaped profile. In case b) the composite behaves as a microelectrode array, the electrochemical active spot of the electrode are sufficiently far away that the diffusion profiles are hemispherical and very dependent on the perimeter of the active surface. That brings about an enhanced and fast mass transport of the electroactive species and consequently fast response and higher faradaic signal with respect to the background noise. The CV profile changes to a sigmoidal shape in which a steady state current is rapidly reached.

response due to the enhanced mass transport to the interfacial reaction zone. As the electrochemically active zones electrode dimension becomes smaller than the diffusion length, relative to the time scale of the measurement, the flux of reactant to the electrode is enhanced because it becomes dominated by hemispherical diffusion causing the electrode to approach a steady state limiting current. Both of these factors, the low background current and the enhanced mass transport lead to enhanced signal/background ratios, high sensitivity, fast electrode response and low detection limits. Fifth, carbon composites retain the wide potential window of carbon materials compared with conventional metals. Sixth, a composite permits addition of a variety of reagents to the host material, including electrocatalysts, enzymes, chemical recognition agents, to confer selectivity (Fig.6). Seventh, the host polymer provides big robustness to the electroactive material, a feature that can only be limited in presence of certain organic solvents.

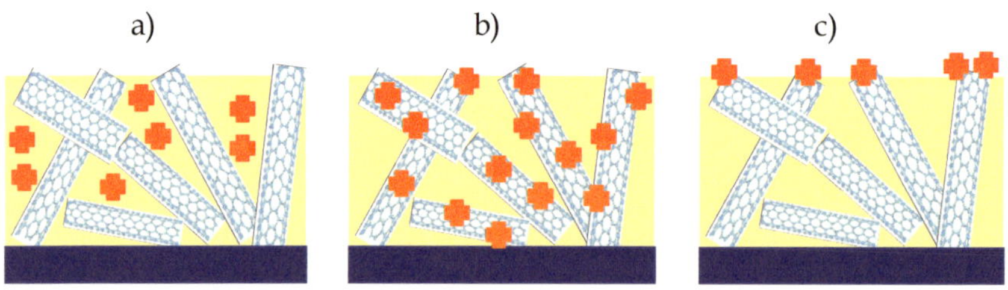

Fig. 6. Different strategies for (bio)chemical modification of the composite matrix that can confer selectivity for (bio)sensing or increase electrocatalytic effects. Composites can be modified with a wide variety of species (catalyst nanoparticles, redox mediators, enzymes, proteins, biomolecular receptors or ligands, etc.) in bulk or at the surface. In case a) the modifiers are introduced when preparing the carbon + polymer mixture and remain embedded or trapped in the composite. In case b) the carbon material is previously covalently or non-covalently functionalized with the species and then mixed with the polymer. In case c) the (bio)species are immobilized on the surface of the already prepared composite.

Carbon nanotube composites apart from sharing the electrochemical advantages of the carbon composites outlined before, they have added features coming from their nanoscale size, structure and high aspect ratio. Such distinct characteristics can contribute to innovative electrochemical applications and to establish the differences with respect to other carbon materials (McCreery, 2008; Esplandiu, 2009a). For instance, novel composite electrode architectures can be accomplished and tailored according to electrochemical applications, a field that is still far from being fully exploited, and which can also provide important insights for a more fundamental understanding of CNT electrochemistry. Thus, the high aspect ratio of carbon nanotubes can be used to form controllable CNT microelectrode array composites in similar direction than the CNT composites proposed by Meyyappan and coworkers (Fig. 7) (Koehne et al., 2004). Taking advantage of the electrochemical anisotropy of the CNTs, one could tailor the tubes in an upright configuration with the more electrochemical reactive sites or tube endings facing to the analytes. That could greatly improve the sensitivity and detection limits to the range of few molecules. At the same time the CNT edges can promote (bio)functionalization in a easier way than other forms of carbon and act as molecular wires where electroactive biomolecules can be plugged at their ends facilitating the direct electron transfer. Alternatively, and following the continuous search of ultrasensitive electrochemical detectors, CNT composites can be used as specific biorecognition indicators profiting from their high surface area available for loading electroactive markers which can amplify biorecognition events. On the other hand, the design of 3D CNT ensembles with high surface area and mesoporous character can yield large double layer capacitances, or provide suitable electrode frameworks for ion intercalation or catalyst entrapment/deposition, all being important for the development of electromechanical systems or electrochemical energy storage systems (Esplandiu, 2009a).

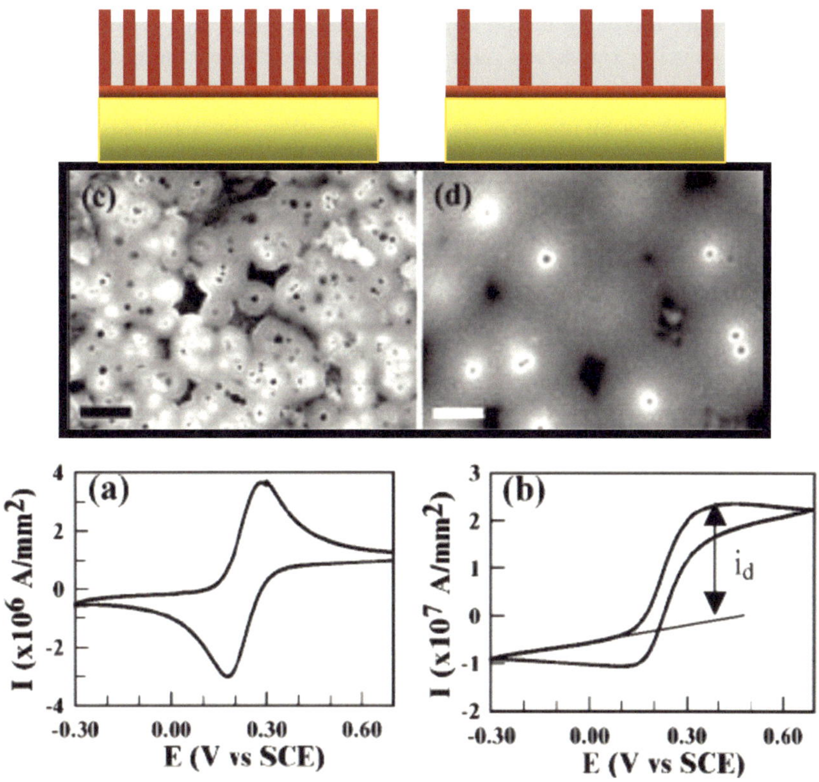

Fig. 7. CNT nanocomposite formed from vertically aligned CNTs grown by plasma CVD deposition and then encapsulated in a dielectric (SiO_2). After that, a planarization was performed by using chemical mechanical polishing which allowed the ends of the carbon nanotubes to be exposed to the solution. The figure shows two composite schemes, with higher and lower density of vertically aligned CNTs. a) and b) shows the CV measurements in presence of a redox couple with the high-density MWCNT nanoelectrode array ($2.\ 10^9$ electrodes/cm^2) and low density one (with ca. 7.10^7 electrodes/cm^2) respectively. C) and d) show the scanning electron images for the high and low density arrays. Adapted from Ref. (Koehne et al., 2004).

However, CNT composites still exhibit difficulties that can limit the desired electrochemical performance, some of them mentioned in sections before like difficulties in the dispersion and adhesion with the polymer host together with the lack of homogeneity of the different commercial CNT lots, different amounts of catalyst or amorphous carbon impurities in the nanotubes as well as big dispersion in their diameter/length. These variations are difficult to quantify and make mandatory not only an electrical characterization but also an electrochemical characterization of the composite before being used in the desired application together with the optimization of the CNT load in the composite. Many electrochemical groups have done the optimization of the composite proportions under the criteria of maximizing the particle loading without losing its physical and mechanical stability (Céspedes et al., 1996; Pumera et al., 2004; Pacios et al., 2008). The type of carbon nanotubes can affect the maximum attainable carbon loading and the goal of those approaches was to achieve the maximum conductivity value. However and taking into account the low reproducibility of the composite electrochemical performance due to the mentioned limitations, it is important not only to consider the conductivity of the composite material (which can be extracted from the electrical characterization through the percolation curves) but also the reproducibility regarding their electrochemical properties.

For instance, from the electroanalytical point of view, the requirements to achieve a good electrode performance are high sensitivity, rapid response time and low limit of detection. These features are related with some physical parameter such as material resistivity, heterogeneous electron transfer and double layer capacitance of the composite electrode. Accordingly, it is important to perform the electrochemical characterization of such parameter for each electrode composition to determine the optimum CNT composite which can bring about the best electroanalytical properties (Olivé-Monllau et al., 2010).

Electrochemical impedance spectroscopy (EIS) can be used to characterize each resulting conductive composite. This technique provides, in an easy way, information about the electron transfer rate, double-layer capacitance, contact resistance and resistance of the solution (Pacios et al., 2008; Esplandiu et al., 2009b; Olivé-Monllau et al., 2010). Thereby, we can determine the composite that exhibits a high electron transfer rate, the lowest double layer capacitance and ohmic resistance to guarantee a high signal/noise ratio, high sensitivity and low detection limits. These results can also be contrasted with voltammetric measurements. For illustration of the electrochemical characterization of CNT composites for electroanalytical purposes, we will continue with the CNT epoxy system characterized by electrical means in the section before.

3.3.1 Electrochemical impedance spectroscopy measurements

Impedance spectroscopy is a technique which is used to characterize electrode processes and complex interfaces. This method studies the system response to the application of a periodic small-amplitude AC signal (around 10 mV). An AC potential is applied ($E(t) = E_o \cos(\omega + t)$) and a AC current is obtained $I(t) = I_0 \cos(\omega t - \varphi)$ (Fig. 8 a). From the relation of both signals the impedance (Z) is obtained ($Z = E(t)/I(t)$). The measurements are carried out at different frequencies and thus the name impedance spectroscopy. In our case it was about (0.1Hz to 100 kHz). Impedance methods allows characterizing the double layer interface at the electrodes and the physicochemical processes of widely differing time constants, sampling electron transfer at high frequency and mass transfer at low frequency. Impedance results are commonly fitted to equivalent electrical circuits of resistors and capacitors, such as the circuit shown in Fig 8 b, which is often used to interpret simple electrochemical systems. There are many ways to plot impedance data. Since the function impedance is an AC signal, one can express it in terms of complex number ($Z = Z_{real} + j Z_{imag}$) and plot the Z_{real} vs Z_{imag}. Such kind of plot is called Nyquist plot as the one shown in Fig. 8 c, and provides visual insight into the system dynamics at the electrochemical interface. Normally such plot exhibits a kind of semicircle profile plus a linear region (the points at which the impedance data cut the real impedance axis represent resistances). In the typical plot of Fig. 8 b R_{ct} is the charge-transfer resistance, which is inversely proportional to the rate of electron transfer (a rough estimation of R_{ct} is related to the width of the semicircle); C_d is the double-layer capacitance, R_s is the electrolyte resistance and, Z_w is the Warburg impedance, which arises from mass-transfer limitations and can be used to measure effective diffusion coefficients. If in an electrochemical system the electron transfer rate is the limiting step, the charge resistance is quite large and the impedance spectrum is dominated by the semicircle feature. Then it is said that the electrochemical process is kinetically controlled. On the other side, if the electron transfer is fast and the diffusion of the electroactive species is the limiting factor, the linear part dominates the impedance spectra and it is said that the process is diffusion controlled.

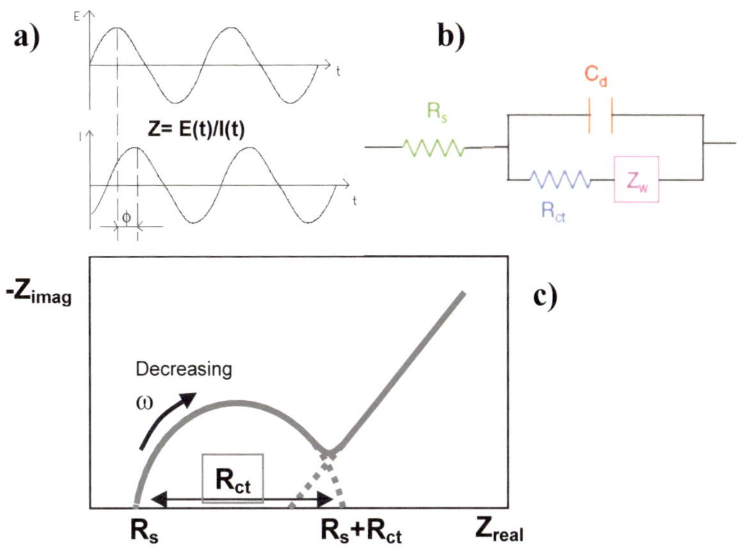

Fig. 8. Impedance function and typical equivalent circuit at the electrochemical interface. Representations of the impedance data (Nyquist plot).

Thus the impedance spectra can give us a broad overview of the different processes taking place at the electrochemical interface (capacitive, resistive, diffusion effects) and which one is dominating more at specific range of frequencies.

By means of electrochemical impedance spectroscopy we could obtain general trends in the electrochemical parameters of our CNT epoxy composites such as the solution resistance or the ohmic resistance ((R_Ω) which comprises the resistance of the solution (Rs) plus any contact resistance (Rc)), charge transfer resistance (R_{ct}) and the double-layer capacitance (C_{dl}) for composites with different MWCNT loading. Such electrochemical characterization was performed in presence of benchmark redox species such as $Fe(CN)_6^{3-/4-}$ which is very sensitive to the electrode surface characteristics. These parameters were obtained by fitting the impedance spectra to a simple equivalent circuit (inset Fig. 9). This circuit was sufficiently suitable to interpret the R_Ω, C_{dl} and R_{ct} values in terms of the interfacial phenomena that occur at the electrochemical cell (Olivé-Monllau et al., 2010).

We studied different compositions of MWCNT-epoxy composite electrodes by varying the MWCNT loading from 3% to 20%. Fig. 9 shows, on one hand, the impedance response of composites with low resistivity (from 20% to 10% of MWCNT loading) (Olivé-Monllau et al. 2010). In such cases the impedance behavior is dominated by a small diameter semicircle representing kinetic-controlled electrode process, though in some cases the diffusion-controlled process starts to be discerned at low frequencies (inset of the figure). On the other hand, the impedance plot for composites with higher resistivity (from 7% to 3% of MWCNT loading) appears to be dominated by a big diameter semicircle and only the kinetic-controlled electrode process is present in the recorded frequency range.

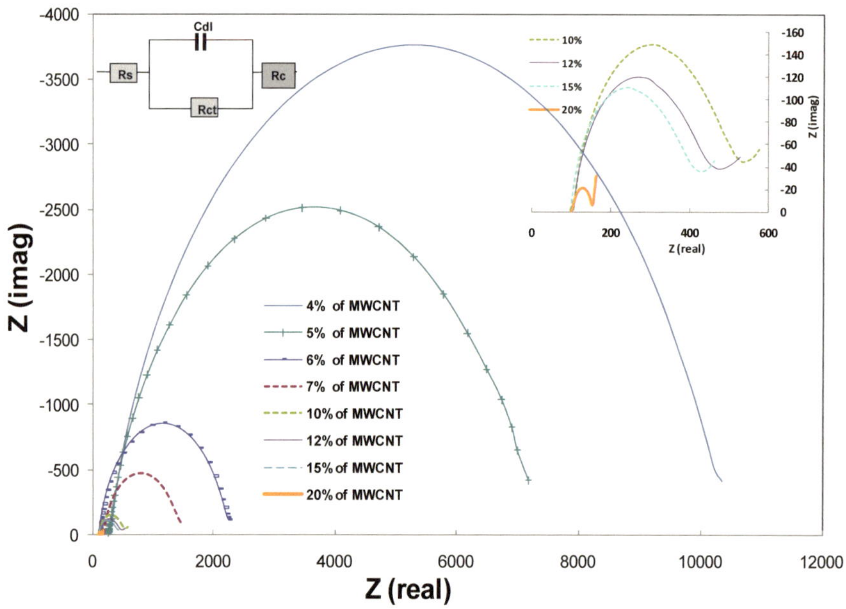

Fig. 9. Impedance spectra for different MWCNT loading electrodes in the presence of $Fe(CN)_6^{3-/4-}$. The insets in the figure show the equivalent circuit used for the impedance spectra fitting and a zoom of the low impedance composite electrodes. Reproduced from Ref. (Olivé-Monllau et al., 2010) with permission. Copyright, 2010, Elsevier.

The ohmic resistance parameter consists of the solution resistance (which is dependent on the ionic concentration, the type of ions and also the electrode area) in series with the contact or the ohmic composite resistance. The latter resistance is the one that has a direct relation with the dry resistance taken for the percolation plot of Fig. 1. Since the solution and contact resistance appear in series, they cannot be independently resolved in the impedance data. Fig. 10 A shows the variations of the ohmic resistance as a function of the carbon nanotube composition. At low carbon loads (< 6% of MWCNT loading), the ohmic resistance is dominated by the contact or composite resistance whereas at higher carbon loads (≥ 6% of MWCNT loading), the ohmic resistance is more dominated by the solution resistance reaching an averaged value of about 100 ohms. This behavior is in agreement with the obtained results in the percolation curve. Low bulk resistance is suitable for electroanalytical materials since it favorably affect the response time and sensitivity of the electrode.

The quantitative values of R_{ct} are depicted in Fig. 10B (Olivé-Monllau et al., 2010). The R_{ct} parameter is inversely proportional to the heterogeneous charge transfer rate and also affects the sensitivity and response time of the electrode. From Fig. 10 B, it is clearly observed the decrease of R_{ct} with the increasing of CNT load. This parameter normalized by the electrochemical active area should be constant for conventional metal electrode interfaces. However in the case of CNT materials that might not be expected. As mentioned before, it is quite accepted that CNTs exhibit electrochemical anisotropy with higher electrode kinetics on edges and lower electron transfer rates on the walls. That makes the electrode kinetics of CNT electrodes to be very dependent on the carbon nanotubes nature and structure. Indeed, we have normalized R_{ct} with respect to the electrochemical active area (see section next section for electroactive area evaluation) and listed its variation with the CNT composition in Table 1. Again a decrease of the R_{ct} parameter with CNT proportion has been obtained which probably speaks of the strong relation between electrochemical reactivity and the surface characteristics of the conducting material. As the carbon load increases, the probability of having more electroactive sites increases and hence the electrode kinetics.

Therefore, composites with low charge transfer resistances are appropriated to be used in electrochemical measurements. According to the results, composites between 20% and 9% of MWCNT loading presented lower charge transfer resistance values. The use of composites in such range of proportions will guarantee fast electron exchange (Olivé-Monllau et al., 2010). However and in spite of the enhanced kinetics, high load of conducting material can increase the background current and smear the faradaic signal response, especially when the electroactive species are present in low concentration.

Therefore it is important to consider the remaining impedance parameter represented by the double layer capacitance which is directly related to the charging or background current. This parameter exhibits increased values with electrodes comprising high surface area of conducting material. In general composites contain only a fraction of conductive area exposed to the solution with the remainder occupied by the insulating polymer. The electrode capacitance, which is determined nearly exclusively by the exposed carbon becomes low, and in turn decreases the background current. That enhances the signal to noise ratio and consequently decreases the analyte detection limits.

Fig. 10C depicts the decrease of the double-layer capacitance values with the decrease of the MWCNT loadings. Note the remarkable lower values of the capacitance double layer between 3% and 11% of MWCNT loading. Higher carbon loads, not only increase the conducting areas but also the surface roughness and porosity of the composites and consequently the background current increments remarkably.

According to the impedance results and taking into account the properties required by an electrode for electroanalytical purposes, such as rapid response time, low limit of detection and high sensitivity, the interval between 9% of MWCNT loading and 11% of MWCNT loading seems to be optimal to fulfill all these requirements. It is important to stress that the fabricated composites in such interval present similar electrochemical performance. The benefit of using composites in such range is that small variations in the composite composition (due to, for example, limitations in the CNT dispersion, purification, etc.) produce a small change in the electrochemical composite behavior. Indeed, under these conditions the electrochemical reproducibility is increased (Olivé-Monllau et al., 2010).

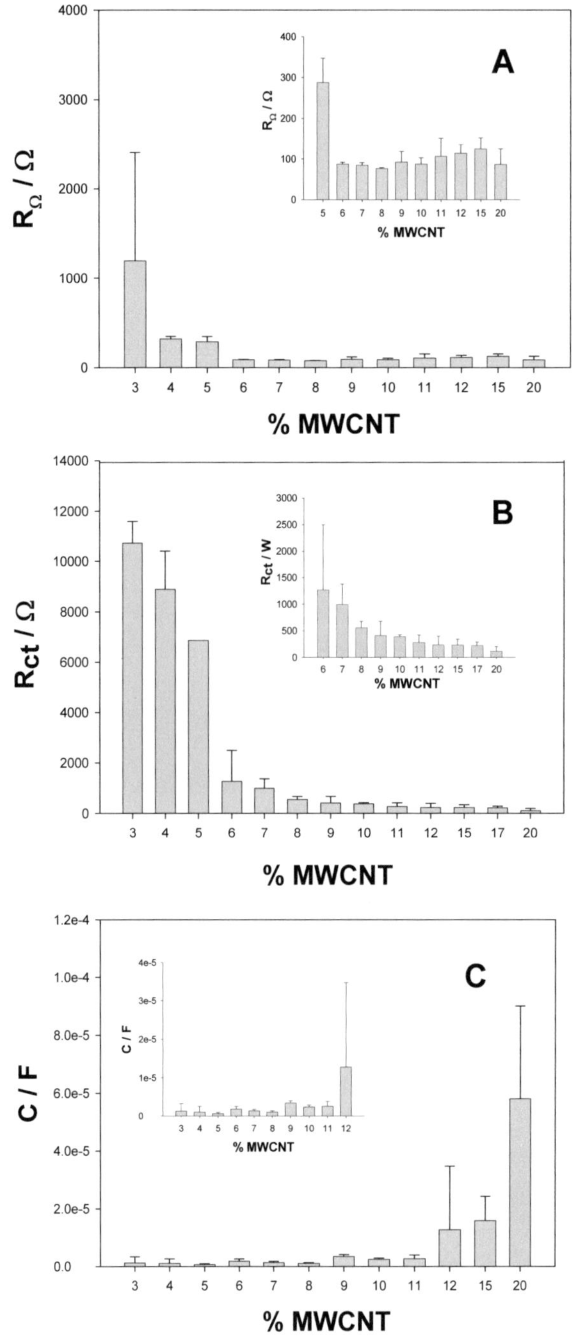

Fig. 10. A) Values of the ohmic resistance, B) charge transfer resistance and C) double layer capacitances extracted from the fitting of the impedance spectra together with their standard deviation for the different CNT loading electrodes. The insets show the values at the lower scale for each evaluated physical parameter. Readapted from Ref. (Olivé-Monllau et al., 2010) with permission. Copyright, 2010, Elsevier.

3.3.2 Cyclic voltammetry characterization

We can also perform cyclic voltammetry in order to and compare the results obtained with the impedance technique. Cyclic voltammograms were taken for the different composite composition electrodes in presence of the benchmark $Fe(CN)_6^{3-}$/ $Fe(CN)_6^{4-}$redox couple (Olivé-Monllau et al. 2010). First of all, it is important to point out the abrupt change in the current/potential shape between 5% and 6% of the MWCNT proportion. At 5% MWCNT proportion, the current potential profile has a more sigmoidal shape as can be appreciated in more detail in the inset of Fig.11, which can be ascribed to an electrochemical behavior more related with a microelectrode array. Beyond 6% of MWCNT proportion, the composite electrodes exhibit the typical peak-shaped profile corresponding to more massive electrodes with planar diffusion characteristics. Different parameters were extracted from the cyclic

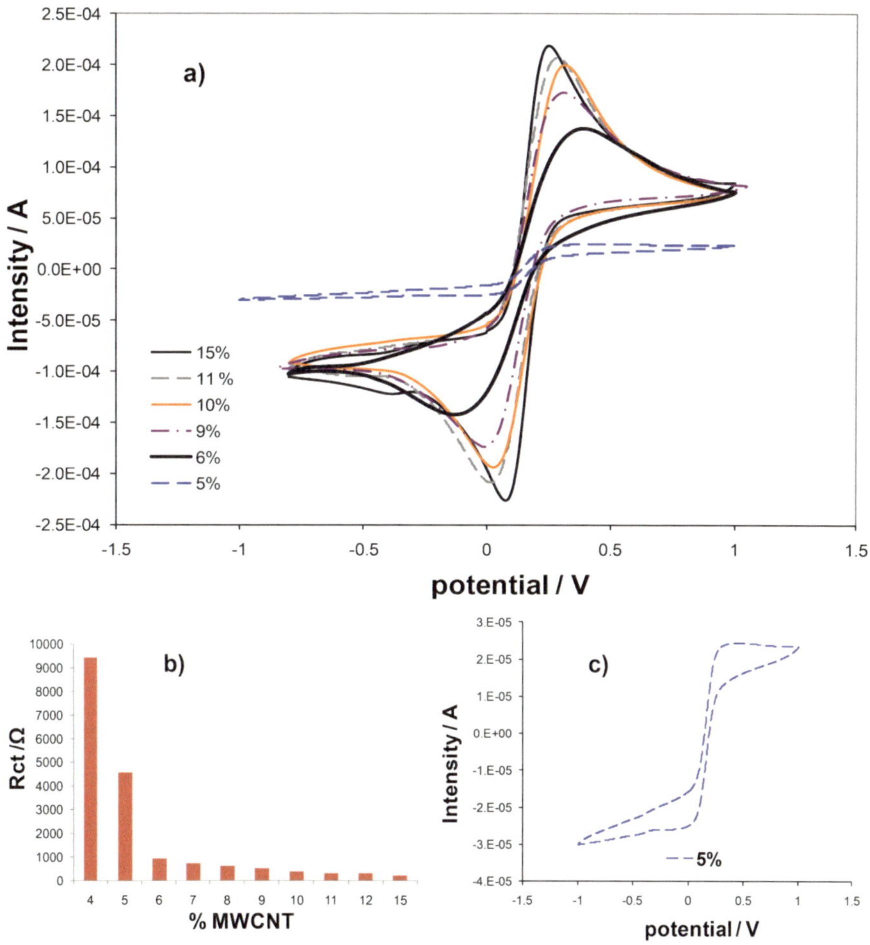

Fig. 11. a) Cyclic voltammograms for the different CNT composites in presence of the ferrocyanide/ferrycianide couple and compensated from any ohmic resistance. B) Trend of the charge transfer resistance for different composite compositions. C) A zoom of the cyclic voltammogram obtained by a composite with 5 % of CNT loading. Reproduced from Ref. (Olivé-Monllau et al., 2010) with permission. Copyright, 2010

voltammograms such as the peak separation potential (ΔE) and peak current (Ip) as shown in table 1. One can observe an increase of the peak current with the MWCNT loading, due to an increase of the electroactive area, together with a decrease of peak separation related to an enhancement of the electron transfer rate. The relative electroactive area (Table 1) was estimated from the peak-shaped voltammograms (up to a proportion of 5%) by quantifying the peak current with the use of this relationship, $I_p=3.10^5 \cdot n^{3/2}(\alpha D_{red}\upsilon)^{1/2}AC^*_{red}$ (Pacios et al, 2008; Olivé-Monllau et al., 2010), which is appropriate for electron transfer-controlled processes. In this equation α represents the transfer coefficient which was considered to be approximately 0.5, D_{red}= 6.32 x 10^{-6} $cm^2 \cdot s^{-1}$ corresponds to the diffusion coefficient of the reduced species, υ =0.01 $V \cdot s^{-1}$ represents the scan rate, A is the electroactive area and C^*_{red} = 0.01M is the bulk concentration of the electroactive species. We also evaluated the exchange current (i_o) from Tafel plots (log current vs. potential), a parameter which provides information about the reversibility of the process (Sánchez et al., 2009; Olivé-Monllau et al., 2010). From the value of the exchange current we can also evaluate the charge transfer resistance through the relation ($i_o=RT/nFR_{ct}$). The trend observed for this parameter resembles very closely the results obtained from the impedance measurements. From the comparison of the R_{ct} values extracted from voltammetry with the ones obtained by impedance measurements (Table 1), it can be observed that their values agree quite well for higher MWCNT loads (greater than 7%). By normalizing R_{ct} with respect to the electroactive area, one can observe that this parameter is also decreasing with the increase of the active area, similar to what was observed in the impedance section. That again indicates the influence of the electrochemical anisotropy of carbon nanotubes which can be more noticeable as the MWCNT loading is increased.

Electrodes	**i_o (A)**	**R_{ct} (Ω)**	**i_p (A)**	**A (cm^2)**	**ΔE (V)**	**R_{ct} A (Ω cm^2)**	**R_{ct} ^{EIS} A (Ω cm^2)**
4%	$2.68 \cdot 10^{-6}$	9418					
5%	$5.52 \cdot 10^{-6}$	4572					
6%	$2.70 \cdot 10^{-5}$	936	$1.61 \cdot 10^{-4}$	0.30	0.476	280.8	641.4
7%	$3.36 \cdot 10^{-5}$	751	$1.96 \cdot 10^{-4}$	0.37	0.427	277.9	468.4
8%	$4.04 \cdot 10^{-5}$	625	$2.07 \cdot 10^{-4}$	0.39	0.335	243.8	250.6
9%	$4.63 \cdot 10^{-5}$	545	$2.33 \cdot 10^{-4}$	0.43	0.299	234.3	257.6
10%	$6.09 \cdot 10^{-5}$	415	$2.40 \cdot 10^{-4}$	0.45	0.283	186.8	186.3
11%	$7.24 \cdot 10^{-5}$	348	$2.53 \cdot 10^{-4}$	0.47	0.261	163.6	178.1
12%	$7.45 \cdot 10^{-5}$	339	$2.71 \cdot 10^{-4}$	0.50	0.261	169.5	175.3
15%	$1.04 \cdot 10^{-4}$	243	$2.72 \cdot 10^{-4}$	0.51	0.169	123.9	117.9

Table 1. Cyclic voltammetry parameter for the different composite compositions, i_o corresponds to the exchange current, R_{ct} to the charge transfer resistance, I_p to peak current, A to the active area and ΔE to the peak separation potential, R_{ct} A^{-1} and R_{ct}^{EIS} A^{-1} represent the normalized R_{ct} with respect to the active area obtained by voltammetry and EIS respectively. Adapted from Ref. (Olivé-Monllau et al., 2010).

3.3.3 Electroanalytical performance

In this part we will illustrate the impact of the CNT composite loading on the electroanalytical detection. Ascorbic acid was used as an analyte for evaluating the electroanalytical characteristics. Measurements were carried out at a fixed potential of 0.6 V. The MWCNT composite response to changes in concentration of ascorbic acid was evaluated by chronoamperometric measurements. The analytical parameters as the detection limit, sensitivity and the linear range were evaluated for 10% of MWCNT composite and compared to those obtained with 20% of MWCNT loading (composition used in the vast majority of previous studies reported by our research group (Pumera et al, 2006; Pacios et al., 2008). For each composition three electrodes were evaluated. Table 2 shows the calibration plot parameters for 20% and 10% of MWCNT loading composite electrodes using ascorbic acid as analyte. The experimental results show that when the MWCNT loading decreases from 20% up to 10%, the sensitivity slightly decreases, but the linear range remarkably increases and the limit of detection (LOD) becomes one order of magnitude lower (Olivé-Monllau et al., 2010). These results suggest that MWCNT loading of 10% seems to be an optimal proportion, since it can achieve a quite good sensitivity but what is more importantly very low limits of detection and wide range of linearity. That agrees very well with the results obtained by impedance spectroscopy which showed that the range between 9-11% exhibits a good behavior of the impedance parameters for the electroanalytical purposes with low charge transfer resistance, ohmic resistance and double layer capacitance. Although a microelectrode array response was obtained at 5% of CNT loading that could be quite beneficial for electroanalytical purposes, we decided to take a range with slightly higher CNT percentage. We found that the ohmic resistance for 5% CNT composite is rather bigger than in the case of CNT composites with a CNT loading between 9-11% and that could smear the electroanalytical response.

Electrodes	Sensitivity/μA L mg^{-1} (%RSD$^{95\%}_{n=3}$)	LOD/mg L^{-1}	Lineal Range /mg L^{-1}
20% of MWCNT	0.29 (9%)	0.40	0.40 - 330
10% of MWCNT	0.20 (7%)	0.04	0.04 - 700

Table 2. Calibration parameters for 20% and 10 % of MWCNT composite electrode extracted from chronoamperometric measurements with ascorbic acid as analyte and 0.01M KNO_3/HNO_3 as background electrolyte. Reproduced from Ref. (Olivé-Monllau et al., 2010) with permission. Copyright, 2010, Elsevier.

The results, above and in the previous sections, show that the adjustment of the CNT composition, according to the electrical and electrochemical parameters, is very important to further get improved electroanalytical features. These studies can be extended to more complex electrochemical systems such as the development of (bio)sensors. That issue comprises one of the cornerstones of CNT composite applications since a frenetic research activity has burst around the field of enzymatic and protein sensors, DNA and genosensors and immunosensors. Thus, tailoring the CNT amount of the composites together with a proper (bio)functionalization (either in bulk or at the surface) of (bio)receptors, redox mediators and catalysts one can get efficient (bio)chemical sensors that can be even integrated in microfluidics systems. Further improvements can be achieved by a controlled geometrical rearrangement of the CNTs in the polymer matrix to get an enhanced electron transfer process. However, the applications of CNT nanocomposites are not only limited to the electroanalysis field with the development of (bio)chemical sensors. Polymer CNT nanocomposites as actuators (artificial muscles) or as novel electrochemical energy harvesting devices comprise another promising though changeling field. For energy generation and storage, CNT composites hold promise as supercapacitors, Li-ion batteries, solar cells and fuel cells due to their high specific surface area and mesoporous character. It is quite expected that for most of the latter applications and in contrast to the electrochemical (bio)sensing devices, the capacitance parameter has to be maximized to get the intended purposes. Again, the impedance studies as a function of the CNT loading can result a very useful tool for tuning the CNT proportion amount to achieve the best performance of those applications.

4. Conclusions

The advent of CNT nanocomposites has opened innovative perspectives for research, development and applications and nowadays these composite comprise a broad world difficult to embrace.

CNT composites share the advantages of other graphitic carbon composites such as high signal to noise ratio, broad potential window, easy surface renewal, (electro)chemical anisotropy with higher electroactivity at edges, easily machinable and excellent platform for (bio)chemical modification. Apart from that, the main advantage of CNT nanocomposites in electrochemistry comes from their nanometer size, structure and high aspect ratio. Such distinct features provide versatility to tailor the CNT composites in different electrode arrangements which can allow the controlled fabrication of microelectrode arrays, the exposition of the more electroactive sites, the enhancement of the electron exchange or the promotion of better ways of (bio)functionalization. Moreover, controlled and novel electrode arrangements together with a proper CNT chemical processing and complete characterization could also help to unveil the uncertainties related to the electrochemical process itself in CNTs. Another advantage of CNT composites comes from their high surface area which can be used to load high amount of (bio)recognition indicators and amplify the electrode response. Moreover, the design of 3D CNT composites with high surface area and mesoporous character can yield large double layer capacitances, or provide suitable electrode frameworks for ion intercalation or catalyst entrapment/deposition, all being of big impact for the development of electromechanical systems or electrochemical energy storage systems. Though the CNT composites exhibit many benefits for the electrochemical purposes, they also suffer from some limitations that can smear the electrochemical performance. These limitations are coming from difficulties in the dispersion and adhesion with the polymeric matrix together with a lack of homogeneity in the CNT production or different amounts of the catalyst impurities. That makes necessary a previous electrochemical characterization of the composite that can help to optimize the CNT load before being used in the desired application.

Therefore in this chapter, we have attempted to highlight CNT epoxy composites as robust, easy-processing, and cheap platforms for basically electroanalysis applications and an electrochemical procedure to evaluate the physical parameters for optimizing the nanocomposite performance according to the intended purpose. Among the electrochemical techniques, EIS provides a versatile tool to optimize the composite material. On one side, it

is possible to extract the ohmic resistance which can be related to the percolation resistivity. On the other side, other useful parameters can be extracted such as the charge transfer resistance, which is related with the heterogeneous electron transfer rate and which depends on the surface electrochemical reactivity. Moreover, EIS also allows the extraction of the electrode capacitance which can be correlated to the background current, an important parameter to minimize in order to enhance the signal/noise ratio. Consequently all these parameters are relevant for the composite response and by a proper evaluation of them as a function of the CNT amount in the composite, one can choose the proportion which fulfills the electroanalytical requirements of high sensitivity, fast response and low limits of detection. In the present work such composition ranges between 9% and 11% of MWCNT loading. Moreover, small variations in composition around this optimal range do not produce high changes in the electrochemical composite behavior. To confirm these predictions, we showed that the electroanalytical detection of ascorbic acid was easily improved using the optimal 10% of MWCNT loading electrode. We achieved at that CNT proportion remarkable lower LOD and wider lineal range than using the 20% of MWCNT loading electrode. Such optimal CNT loading values can allow us to fabricate extremely attractive and robust composite electrodes with very interesting application as amperometric (bio)sensors at low electroanalyte concentration.
The electrochemical characterization here proposed can be extended to more complex systems like (bio)chemical CNT composite sensors or CNT nanocomposites for being used as actuators or oriented energy applications.

5. References

Adams, R.N. (1958). Carbon Paste Electrode. *Analytical Chemistry*, 30, (9), pp. 1576-1586.

Ajayan, P.M.; Schadler, L.S.; Braun, P.V (2003). *Nanocomposite Science and Technology*, Ajayan, P.M.; Schadler, L.S.; Braun, P.V (Ed.), 1-207, Wiley-VCH Verlag, ISBN 527-30359-6, Weinheim, Germany.

Arrigan, D.W.M. (2004). Nanoelectrodes, nanoelectrode arrays and their applications. Analyst 124 (12) 1157-1165.

Balasubramanian, K; Burghard, M. (2005). Chemically functionalized carbon nanotubes. *Small*, 1, (2), pp.180-192.

Banks, C.E.; Compton, R.G. (2006). New electrodes for old: from carbon nanotubes to edge plane pyrolytic graphite. *The Analyst*, 131, (1), pp. 15-21.

Banks, C.E.; Davies, T.J.; Wildgoose, G.G., Compton, R.G. (2005). Electrocatalysis at graphite and carbon nanotube modified electrodes: edge-plane sites and tube ends are the reactive sites. *Chemical Communications* (7), pp. 829-841.

Barrera, E.V.; Shofner, M.L.; Corral, E.L (2005). Applications: Composites, in: *Carbon Nanotubes: Science and Applications*, M. Meyyappan (Ed.), 253-273, CRC Press LLC, ISBN 0-8493-2111-5, Florida, USA.

Burghard, M. (2005). Electronic and vibrational properties of chemically modified single-walled carbon nanotubes. *Surface Science Reports*, 58, (1-4), pp. 1-109.

Céspedes, F.; Martínez-Fàbregas, E; Alegret, S. (1996). New materials for electrochemical sensing I. Rigid conducting composites. *Trends in Analytical Chemistry*, 15, (7), pp. 296-304.

Chen, Z.; Thiel, W.; Hirsch, A. (2003). Reactivity of the convex and concave surfaces of single-walled carbon nanotubes (SWCNTs) towards addition reactions: dependence on the carbon-atom pyramidalization. *ChemPhysChem*, 4, (1), pp. 93-97.

Coleman, J.; Khan, U.; Gun'ko, Y.K. (2006). Mechanical reinforcement of polymers using carbon nanotubes. *Advanced Materials*, 18, (26), pp. 689-709.

Dumitrescu, I; Unwin, P.R.; Macpherson, J.V. (2009) Electrochemistry at Carbon Nanotubes: Perspective and Issues. *Chemical Communications*, (45), pp. 6886-6901.

Esplandiu, M.J. (2009a). Electrochemistry on Carbon Nanotube modified surfaces, in: *Chemically Modified Electrodes*, R.C. Alkire, D.M. Kolb, J. Lipkowski, P.N. Ross (Eds.), 57-209, Wiley-VCH Verlag, ISBN 978-3-527-31420-1, Weinheim, Germany.

Esplandiu, M.J.; Pacios, M.; Cyganek, L.; Bartroli, J.; del Valle, M. (2009b).

Enhancing the electrochemical response of myoglobin with carbon nanotube electrodes. *Nanotechnology*, 20, (35), 355502 (8pp).

Faiella, G.; Piscitelli, F.; Lavorgna, M.; Antonucci, V.; Giordano, M. (2009). Tuning the insulator to conductor transition in a multiwalled carbon nanotube/epoxy composite at substatistical percolation threshold. *Applied Physics Letters*, 95, (15), 153106 (2 pp.).

Gooding, J.J. (2005). Nanostructuring Electrodes with Carbon Nanotubes: A Review on Electrochemistry and Applications for Sensing. *Electrochimica Acta*, 50, (15), pp.3049-3060.

Grady, B. (2010). Recent developments concerning the dispersion of carbon nanotubes in polymers. *Macromolecular Rapid Communications*, 31, (3), pp. 247-257.

Grossiord, N.; Loos, J.; Regev O.; Koning, C.E. (2006). Toolbox for dispersing Carbon Nanotubes into polymers to get conductive nanocomposites. *Chemistry of Materials*, 18 (5), pp.1089-1099.

Han, J. (2005). Structures and properties of carbon nanotubes, in: *Carbon Nanotubes: Science and Applications*, M. Meyyappan (Ed.), pp.2-21, CRC Press LLC, ISBN 0-8493-2111-5, Florida, USA.

Harris, P.J.F. (2004). Carbon nanotube composites. *International Materials Reviews*, 4, (1), pp.31-43.

Hirsch, A. (2002). Functionalization of single-walled carbon nanotubes. *Angewandte Chemie International Edition*, 41, (11), pp. 1853-1859.

Javey, A.; Qi, P.; Wang, Q; Dai, H. (2004). Ten-to 50-nm-long quasi-ballistic carbon nanotube devices obtained without complex lithography. *Proceedings of the National Academy of Sciences of the United States of America*, 101, (37), pp.13408-13410.

Jose-Yacaman, M.; Rendon, L.; Arenas, J.; Serra Puche, M. C. (1996). Maya Blue Paint: An Ancient Nanostructured Material. *Science*, 273, (5272) pp. 223-225.

Kara, S.; Arda, E.; Dolastir, F.; Pekcan, O. (2010). Electrical and optical percolations of polystyrene latex-multiwalled carbon nanotube composites. *Journal of Colloid and Interface Science*, 344, (2), pp. 395-401.

Katz, E.; Willner, I. (2004). Biomolecule-functionalized carbon nanotubes: applications in nanobioelectronics. *ChemPhysChem*, 5, (8), pp.1084-1104.

Kim, S.N.; Rusling, J.F.; Papadimitrakopoulos. (2007). Carbon Nanotubes for electronic and electrochemical detection of biomolecules. *Advanced Materials*, 19, (20), pp. 3214-3228.

Koehne, J.; Li, J.; Cassell, A.M.; Chen, H.; Ye, Q.; Ng, H.T.; Han, J.; Meyyappan,

M. (2004). The fabrication and electrochemical characterization of carbon nanotube nanoelectrode arrays. *Journal of Materials Chemistry*, 14, pp. 676-684.

Li C.; Bai, H; Shi, G.Q. (2009). Conducting polymer nanomaterials: electrosynthesis and applications. *Chemical Society Reviews*, 38, (8), pp. 2397-2409.

Li, C; Thostenson E.T.; Chou, T. (2008). Sensors and actuators based on carbon nanotubes and their composites: a review. *Composites Science and Technology*, 68, (6), pp. 1227-1249.

Liu, J.; Chou, A.; Rahmat, W.; Paddon-Row, M.N.; Gooding, J.J. (2005). Achieving direct electrical connection to glucose oxidase using single walled carbon nanotube arrays. *Electroanalysis*, 17, (1), pp. 38-45.

Louie, S.G. (2001). Electronic properties, junctions and defects of carbon nanotubes in: *Topics in Appl. Physics Carbon Nanotubes, Synthesis, Structure, Properties and Applications*, M.S. Dresselhaus, G. Dresselhaus, Ph. Avouris (Eds.), Springer-Verlag, pp. 113-142, ISBN 3-540-41086-4, Berlin Heidelberg.

Martin, C.A.; Sandler, J.K.W.; Shaffer, M.S.P.; Schwarz, M.K.; Bauhofer, W.; Schulte, K.; Windle, A.H. (2004). Formation of percolating networks in multi-wall carbon nanotube epoxy composites. *Composites Science and Technology*, 64, (15), pp. 2309-2316.

Martone, A.; Formicola, C.; Giordano, M.; Zarrelli, M. (2010). Reinforcement efficiency of multi-walled carbon nanotube/epoxy nanocomposites. *Composites science and technology*, 70, (7), pp. 1154-1160.

McCreery, R.L. (1991) in Carbon electrodes: Structural effects on electron transfer kinetics, in: *Electroanalytical Chemistry*, A.J. Bard (Ed.), vol 17, pp. 221-374, Marcel Dekker, Inc., ISBN 0-8247-8409-X, New York.

McCreery, R.L. (2008). Advanced Carbon electrode materials for molecular electrochemistry. *Chemical Reviews*, 108 (7) pp. 2646-2687.

Merkoci, A. (2006). Carbon Nanotubes in Analytical Sciences. *Microchimica Acta*, 152 (3-4) pp. 157-174.

Moniruzzaman, M.; Winey, K.I. (2006). Polymer Nanocomposites containing carbon nanotubes. *Macromolecules*, 39, (16), pp. 5194-5205.

Moore, R.R.; Banks, C.E; Compton, R.G. (2004). Basal plane pyrolytic graphite modified electrodes: comparison of carbon nanotubes and graphite powder as electrocatalysts. *Analytical chemistry*, 76, (10), pp. 2677-2682.

Mylvaganam, K.; Zhang, L.C. (2007). Fabrication and application of polymer
composites comprising carbon nanotubes. *Recent Patents on Nanotechnology*, 1, (1), pp. 59-65.

Navarro-Laboulais, J.; Trijueque, J.; Vicente, F.; Scholl, H. (1994).
Voltammetric determination of optimal conductive load proportion in graphite-epoxy composite electrodes. *Journal of Electroanalytical Chemistry*, 379, (1-2), pp. 159-163.

Navratil, T.; Barek, J. (2009). Analytical Applications of composite solid electrodes. *Critical Reviews in Analytical Chemistry*, 39, (3), pp. 131-147.

Niyogi, S; Hamon, M.A.; Hu, H.; Zhao, B; Bhowmik, P.; Sen, R.; Itkis, M.E.; Haddon, R.C. (2002). Chemistry of single-walled carbon nanotubes. *Accounts of Chemical Research*, 35, (12), pp.1105-1113.

O'Hare, D; Macpherson, J.V.; Willows, A. (2202). On the microelectrode behavior of graphite epoxy composite electrodes. *Electrochemistry Communications*, 4, (3), pp. 245-250.

Olek, M. (2006). Carbon Nanotube Composites- Mechanical, electrical and optical properties. PhD Thesis, Bonn, Germany.

Olivé-Monllau, R.; Baeza, Mireia; Bartrolí, J.; Céspedes, F. (2009). Novel amperometric sensor based on rigid near-percolation composite. *Electroanalysis*, 21, (8), pp. 931-938.

Olivé-Monllau, R; Esplandiu, M.J.; Bartrolí, J.; Baeza, M.; Céspedes, F. (2010). Strategies for the optimization of carbon nanotube/polymer ratio in composite materials: applications as voltammetric sensors. *Sensors and Actuators B-Chemical*, 146, (1), pp. 353-360.

Pacios, M. ; del Valle, M. ; Bartrolí, J. ; Esplandiu, M.J. (2008). Electrochemical behavior of rigid carbon nanotube composite electrodes. *Journal of Electroanalytical Chemistry*, 619, pp. 117-124.

Patolsky, F.; Weizmann, Willner, I. (2004). Long-range electrical contacting of redox enzymes by SWCNT connectors. *Angewandte Chemie International Edition*, 43, (16), pp. 2113-2117.

Pumera, M.; Merkoci, A.; Alegret, S. (2006). Carbon nanotube-epoxy composites for electrochemical sensing. *Sensors and Actuators B-Chemical*, 113, (2), pp. 617-622.

Rajesh; Ahuja, T., Kumar D. (2009). Recent progress in the development of nano-structured conducting polymers/nanocomposites for sensor applications. *Sensors and Actuators B-Chemical,* 136, (1), pp. 275-286.

Rubianes, M.D.; Rivas, G.A. (2003). Carbon nanotubes paste electrode. *Electrochemical Communications,* 5, (8), pp. 689-694.

Sahoo, N.G.; Rana, S.; Cho, J.W.; Chan, S.H. (2010). Polymer nanocomposites based on functionalized carbon nanotubes. *Progress in Polymer Science,* 35, (7), pp. 837-867.

Saito, R.; Dresselhaus, G.; Dresselhaus M.S. (1998). *Physical Properties of carbon nanotubes,* Imperial College Press, pp. 1- 272, ISBN 1-86094-093-5, London.

Sánchez, S.; Pumera, M.; Fábregas, E.; Bartrolí, J.; Esplandiu, M.J. (2009). Carbon nanotube/polysufone soft composites: preparation, characterization and application for electrochemical sensing of biomarkers. *Physical Chemistry Chemical Physics,* 11, (35), pp. 7721-7728.

Sandler, J.K.W.; Kirk, J.E.; Kinloch, I.A.; Shaffer, M.S.P.; Windle, A.H. (2003). Ultra-low electrical percolation threshold in carbon-nanotube –epoxy composites. *Polymer,* 44, (19), pp. 5893-5899.

Singh, I.; Verma, A.; Kaur, I.; Bharadwaj, L.M.; Bhatia, V.; Jain, V.K.; Bhatia, C.S.; Bhatnagar, P.K.; Mathur, P.C. (2010). The effect of length of single-walled carbon nanotubes (SWNTs) on electrical properties of conducting polymer-SWNT composites. *Journal of Polymer Science: Part B: Polymer Physics,* 48, (1), pp. 89-95.

Spitalsky, Z.; Tasis, D.; Papagelis, K.; Galiotis, C. (2010). Carbon nanotube-polymer composites: chemistry, processing, mechanical and electrical properties. *Progress in Polymer Science,* 35, (3), pp. 357-401.

Srivastava, D. (2005). Computational nanotechnology of carbon nanotubes, in: *Carbon Nanotubes: Science and Applications,* M. Meyyappan (Ed.), pp. 25-59, CRC Press LLC, ISBN 0-8493-2111-5, Florida, USA.

Streeter, I.; Wildgoose, G.G.; Shao, K., Compton, R.G. (2008). Cyclic voltammetry on electrode surfaces covered with porous layers: An analysis of electron transfer kinetics at single-walled carbon nanotube modified electrodes. *Sensors and Actuators B,* 133, (2), pp. 462-466.

Tasis, D.; Tagmatarchs, N.; Bianco, A.; Prato, M. (2006). Chemistry of Carbon Nanotubes. *Chemical Reviews,* 106, (3), pp. 1105-1136.

Wang, J. (2005). Carbon-nanotube based electrochemical biosensors: a review. *Electroanalysis,* 17, (1), pp. 7-14.

Wang. J.; Musameh, M. (2003). Carbon nanotube/Teflon composite electrochemical sensors and biosensors. *Analytical Chemistry,* 75, (9), pp. 2075-2079.

Wildoer, J.W.G.; Venema, L.C.; Rinzler, A.G.; Smalley, R.E.; Dekker, C. (1998). Electronic structure of atomically resolved carbon nanotubes. *Nature,* 391, pp. 59-62.

Xiao, Y.; Li, C (2008). Nanocomposites: from fabrications to electrochemical applications. *Electroanalysis,* 20, (6), pp. 649-662.

Yogeswaran, U.; Chen, S. (2008). Recent trends in the application of carbon nanotubes-polymer composite modified electrodes for biosensors: a review. *Analytical Letters,* 41, (2), pp. 210-243.

Yu, X.; Chattopadhyay, D.; Galeska, I.; Papadimitrakopoulos, F. ; Rusling, J.F. (2003). Peroxidase activity of enzymes bound to the ends of single-wall carbon nanotube forest electrodes. *Electrochemistry Communications,* 5, (5), pp. 408-411.

Zhao, H.; O'Hare, D. (2008). Characterisation and modeling of conducting composite electrodes. *Journal of Physical Chemistry C,* 112, (25), pp. 9351-9357.

11

Synthesis, Sintering Behaviour and Mechanical Properties of Lead-Ceramic (Nano-)Composites for Acid Battery Grids

Alexandre Maître[1] and Michel Vilasi[2]
[1]*Laboratoire Science des Procédés Céramiques et de Traitements de Surface – UMR CNRS 6638 – Centre Européen de la Céramique – 12 rue Atlantis F-87068 Limoges Cedex*
[2]*Institut Jean Lamour – UMR CNRS 7555 – Faculté des Sciences et Techniques – Boulevard des Aiguillettes – F-54506 Vandœuvre-lès-Nancy Cedex*
France

1. Introduction

In recent years, lead-calcium-tin alloys have been widely used for producing lead/acid battery grids. In particular, lead-calcium-tin grids are usually employed for the positive grids of Valve-Regulated Lead Acid batteries (VRLA) or maintenance free storage lead/acid batteries (Giess, 1995 ; Bagshaw, 1995). Indeed, Pb-Ca-Sn alloys present better mechanical and electrochemical properties considering Pb-Ca or Pb-Sb systems. In particular, the addition of tin to lead-calcium alloys dramatically improves the conductivity of corrosion products on the grid surface, allowing improved corrosion resistance and inhibition of the formation of non conductive layers at the grid-active material interface (Miraglio et al., 1995; Takahashi et al., 1995). Furthermore, tin increases the resistance corrosion in overcharge conditions because it leads to the rising of the oxygen overvoltage (Bui et al., 1997).
From a metallurgical point of view, the age-hardening process of the most commonly used ternary alloys Pb-Ca(0.08wt.%)Sn(2wt.%) proceeds by discontinuous or continuous precipitation of $L1_2$ phase (*i.e.* $(Pb_{1-x}Sn_x)_3Ca$) from supersaturated α solid solution (Maître et al., 2003 ; Bouirden et al., 1991). Moreover, at room temperature, the residual tin and calcium supersaturation of aged ternary alloys can lead to the overageing appearance. The overageing process of Pb-Sn alloys takes place in two stages (Maître et al., 2003): first, a discontinuous precipitation of a coarse $L1_2$ phase, second, the lamellar coalescence of these latter precipitates. Globally, the beginning of the overageing can be associated to the rapid decrease of mechanical and electrochemical properties of PbCaSn alloys (Bourguignon et al., 2003).
To delay the overageing phenomenon, annealing treatments are usually performed. These treatments are issued from Transformation-Time-Temperature (TTT) diagrams of Pb-Ca-Sn ternary alloys (Hilger et al., 1996). So, a TTT diagram must be established for each alloy composition. Since the common PbCaSn alloys contain numerous impurities such as Ag, Al, Bi, Sb, Sn that stem from secondary lead used for grid manufacturing, the determination of corresponding TTT curves can become a very tiresome work.

Consequently, in order to avoid the building of several TTT diagrams or the using of secondary elements, it is required to set up new manufacturing processes of lead-based materials which inhibit the recrystallization phenomenon. In this context, it becomes pertinent to introduce a secondary phase at the grain boundaries which exhibits the properties listed below:

i. its mechanical properties (*e.g.* hardness) must be much more higher than those of PbCaSn alloys;
ii. its solubility should be negligible in the lead matrix ;
iii. a thermodynamical equilibrium must be reached between this secondary phase and the matrix.

These properties appear to be quite similar to those encountered in the spatial or aeronautics applications areas for which materials are submitted to high temperature ($T > 0.5\ T_m$) for long working time. For example, the fine dispersion of ceramic particles (*e.g.* yttria, zirconia) in platinum matrix leads to the slow down of the recrystallization. As a general rule, this mechanical reinforcement namely called ODS (Oxide Dispersion Strengthening) was widely dedicated to high temperature alloys (Reppich, 2002). Among the benefits effects, the fine dispersion of ceramic particles through the alloy specimen allows anchoring the grain boundaries, prohibiting the discontinuous transformations (*i.e.* recrystallisation or overageing) and standing in the way of the dislocation movement.

Several manufacturing methods of ODS lead-based alloys have been reported in the literature. In particular, previous studies deal with the interest of physical methods. Jiang et al. (Jiang et al., 1998) were able to elaborate nanosized lead powders by using a pulsed wire discharge. This discharge is generated from a thin lead film and leads to its vaporization under the form of fine lead particles. Therefore, no study displays the ability to produce *in situ* composites lead-based powders. Nevertheless, an uncommon work (Devaux et al., 1997) succeeds the elaboration of nanocomposites bismuth-antimony-silica powders thanks to an arc plasma assisted melting method. This latter allows synthesizing fine, pure metallic and ODS powders from a versatile process.

Otherwise, the chemical way has been also investigated to prepare lead-based composite powders (Tilman et al., 1971). So, the $Pb\text{-}Al_2O_3$ composite has been elaborated in high amounts from the precipitation of aqueous solution of $Pb(NO_3)_2$ and $Al(NO_3)_3{,}9H_2O$ in presence of $(NH_4)_2CO_3$ and NH_4OH at controlled value of pH. After washing, drying at 120°C then at 450°C and 600°C, mixtures of alumina and lead monoxide powders have been obtained. Finally, the $Pb\text{-}Al_2O_3$ composite was synthesized by heating the starting mixture under a reducing atmosphere (*i.e.* hydrogen) at 300°C for 18 h. According to these authors, the sintering must be conducted under flowing hydrogen between 300 and 315°C for 24h. Another method has been tried some years earlier by Roberts (Roberts, 1964). This latter consists in mixing a lead powder for 6h at room temperature under air to form thin oxide layer on the grain surface. This layer is crushed and then dispersed within the powder during its shaping by extrusion. Therefore, a recent work (Balaz et al., 2004) reported that the synthesis of lead-based powder during milling of a lead sulphide powder could be considered. In fact, the predominant reaction is the reduction of the lead sulphide in presence of iron issued from the wearing of steel balls:

$$PbS_{(s)} + Fe_{(s)} \rightarrow Pb_{(s)} + FeS_{(s)} \qquad (1)$$

The conversion rate can reach 84% after an hour of milling. In these conditions, it can be noticed that the average diameter of particles tends to 15 nm whereas a rising of the milling

time is related to a significant grain growth. For example, the mean grain size could range between 60 and 150 nm for long milling time. Nevertheless, a promising synthesis method has been developed by Dailly (Dailly et al., 2003) and Balan (Balan et al., 2004) for antimony and tin, respectively. This way that relies on the alkaline hydride reduction of metal chloride in the THF solvent provides nanosized metallic particles. These particles appeared to be covered by an alkoxide layer which inhibits both the oxidation and the pre-sintering of powders. Consequently, this method could be easily applied to the elaboration in large amount of ultrafine, pure lead-based powder or lead-lead oxide composite and will be retained afterwards.
The first part of the present chapter is focused on the investigation of the original routes to synthesize lead-ceramic nanocomposites: in particular, the physical route using a plasma arc discharge or the chemical way requiring the reduction of lead-based salts has been selected. In a second part, the potentialities in terms of shaping or sintering of different batches of powders will be examined in comparison with the usual powder metallurgy route (*i.e.* from commercial lead and ceramic powders). In the next section, the mechanical and/or the electrochemical properties of these materials have been determined under the working conditions encountered by lead acid battery grids and have been compared to those obtained for usual PbCaSn alloys. Concerning the electrochemical behaviour, the role played by ceramic films deposited on lead-based substrates has been investigated.

2. Elaboration of lead-based composites

2.1 Synthesis of the pure lead and lead-ceramic powders

The elaboration process differs from the synthesis of the starting pure lead and lead-ceramic composite powders. Before shaping and sintering, the first part of this section will be focused on the description of the different synthesis methods.

2.1.1 Conventional process using commercial powders

This method which is described in details in a previous work (Cartigny et al., 2007) leads to the preparation of lead-ceramic composites which will play the role of "reference" for this study. The lead and ceramic powders (Table 1) were mixed in Turbula apparatus then were

Powder (supplier)	density	Granulometry (μm) furnished by			
		Supplier	Image analysis	Laser granulometry	BET method
Lead (Aldrich)	11.34	40	4.3	-	-
Y_2O_3 (S.E.P.R.)	5.01	<10	1×2 (cubic shaped)	-	0.08
TiO_2 (S.E.P.R.)	4.26	0,2	0.1	0.2	0.09
SiC (Cerac)	3.2	<1	1	-	0.06
SiO_2 (Cerac)	2.2	<5	1	-	0.63

Table 1. Main characteristics of the starting powders

crushed in agate mortar. It must be noticed that ceramic powders such as SiC, SiO_2, TiO_2, Y_2O_3 have been retained because of their chemical inertia in a 5M concentrated sulphuric acid solution. Since the lead grains are partially oxidized as shown by TEM observation and EDXS analysis (Fig.1), the commercial lead powder was previously purified by introducing it in an acid acetic and ethanol mix. Then, the powder was dried under vacuum to remove the solvent traces.

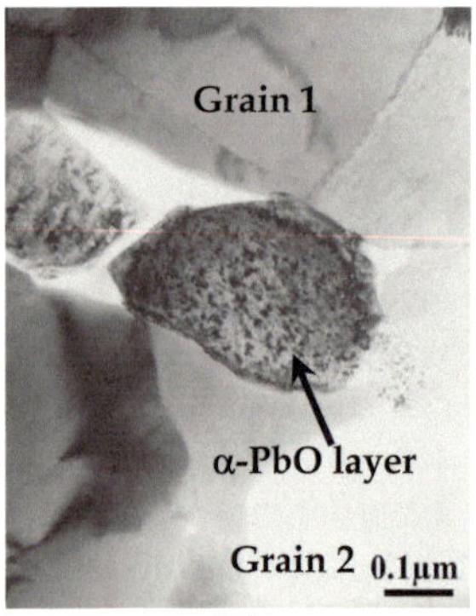

(a)

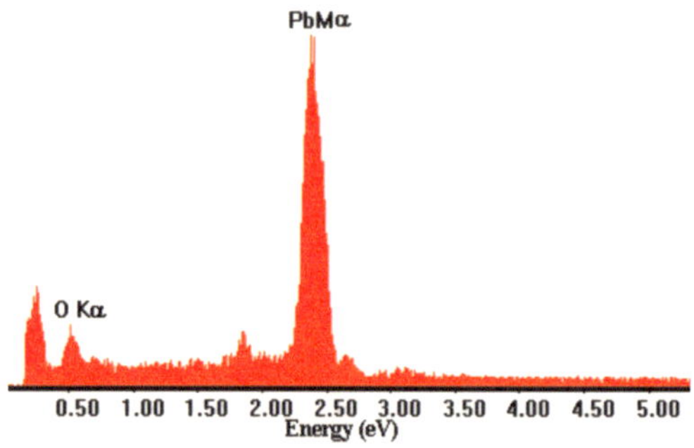

(b)

Fig. 1. TEM micrograph of oxidized lead grains (a) and corresponding analysis of lead oxide by Energy Dispersive X-ray Spectrometry (b)

2.1.2 Synthesis of composite powders assisted by arc plasma process

This method consists in the plasma arc evaporation or co-evaporation of a pure lead anode recovered by silica. This apparatus has been previously described in the literature (Brochin et al., 1999) for the synthesis of bismuth-based composites. The main schema of this device is reported in Fig.2. An electrical arc was produced between a tungsten cathode and an anode-target which is made of the metal to vaporize. For the synthesis of pure lead powder, the molecular hydrogen (H_2) which is dissociated in the arc due to the higher temperature (>10 000K) is dissolved mainly in the metal melt. Its solubility decreases as a function of the distance from the target centre. In the outer part of the metal target, the hydrogen radicals react to form hydrogen molecules. This latter reaction is accompanied by heat emission which enhances the kinetic of the metal vaporization. When a silica target is used to prepare lead-ceramic composite powders, the reaction sequence is quite similar. Nevertheless, further reactions related to silica could occur. First, the silica dissociation leading to the formation of silicon monoxide and molecular oxygen as follows for temperature higher than 3190K:

$$SiO_{2(s)} \rightarrow SiO_{(g)} + \frac{1}{2} O_{2(g)} \quad (2)$$

Then, the oxygen dissociation occurs when the temperature exceeds 3810K according to equation:

$$\frac{1}{2} O_{2(g)} \rightarrow 2O_{(g)} \quad (3)$$

Finally, the silicon monoxide transforms into gaseous silicon for temperature higher than 4730K:

$$SiO_{(g)} \rightarrow Si_{(g)} + \frac{1}{2} O_{2(g)} \quad (4)$$

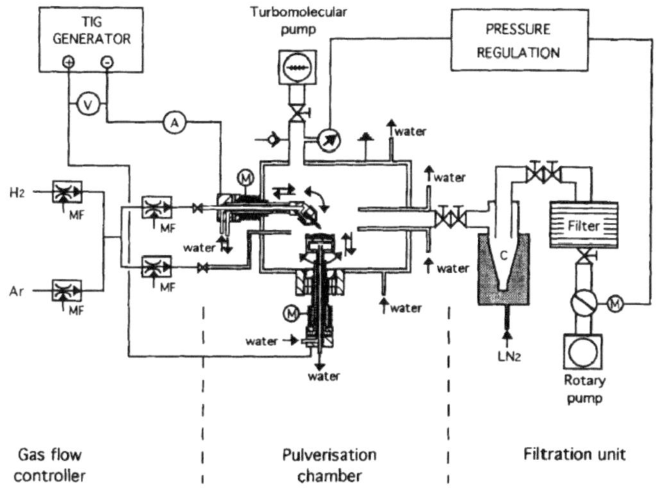

Fig. 2. Device for arc plasma powder production (Brochin et al., 1999)

The first step of the experimental protocol consisted in preparing the metallic anode. So, shavings of lead (99.999%, GoodFellow, UK) which were previously washed in a concentrated acid acetic solution and then in water, ethanol were put in a graphite crucible. The melting of metal was performed under argon (P_{Ar} = 0.13bar) by creating an electrical arc with the cathode. To synthesize a powder of lead-ceramic composite, a silica sheet (99.5%, Heraus, Germany) was put over the metal. A hole was drilled in the centre of the silica sheet to allow the formation of an electrical arc during all the experiment.

The main parameters for the production of pure lead powders have been chosen taking into account the results obtained for the synthesis of bismuth-based powders (Brochin et al., 1999):

i. the working distance between anode and cathode was fixed to 5mm ; the geometry of device coupled with the anode-cathode distance and the residual pressure in the reactor provided the voltage ;
ii. the argon flow ranges between 10 and 12L.min^{-1} to master the powder production. The residual pressure was fixed to 100 torr.

Further experiments have been carried out to find optimised parameters (*i.e.* the anode-cathode distance, the gas flow, the arc intensity) for pure lead powder production. So, for a voltage equal to 16.5V, it appeared that the yield per hour increased when the current intensity was around 60A (Fig.3a). Indeed, a higher intensity inferred a feed powder toward the collector less efficient: some particles agglomerates form a deposit just around the crucible or on the reactor walls. In the same manner, when the current intensity was fixed to 60A, it can be noticed that the yield remains much higher for a voltage near 11-12V (*i.e.* for an anode-cathode distance of 5mm) (see Fig.3b). Finally, under these optimised conditions, a pure lead powder flow of 1g/min can be expected.

In a second step, the synthesis parameters of the nanocomposite Pb-SiO_2 powder have been determined. An argon flow of 10L.min^{-1} has been maintained in the reactor for all the experiments. As revealed in Fig.4a, a good compromise between the yield and the amount of collected powder can be reached for a current intensity of 120A and a voltage initially fixed to 16.5V. In the same way, Fig.4b clearly demonstrates that the voltage between anode and cathode of device should be equal to 16.5V to target the higher powder yield for a current intensity of 120A. These optimised process parameters have been chosen to obtain a silica content of 2.5 vol.% in nanocomposite powders with a production flow of 3g.min^{-1}.

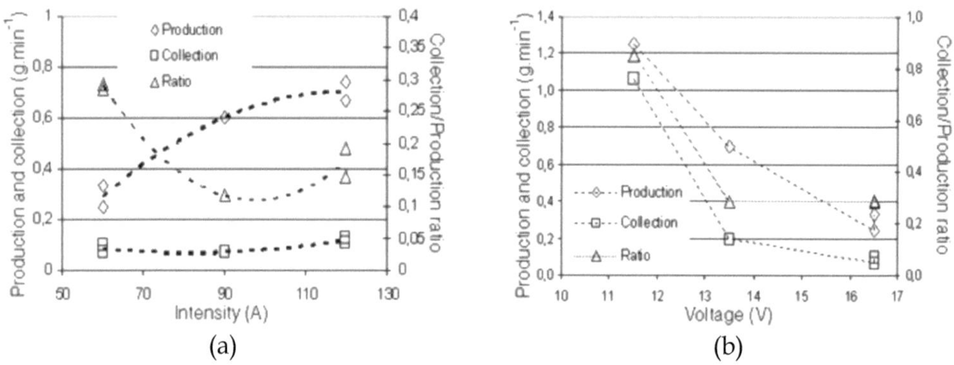

Fig. 3. Influence on the production, collection and ratio of pure lead powder (a) of electric arc intensity for a given voltage (16V) ; (b) of electric arc voltage for a given intensity (60A)

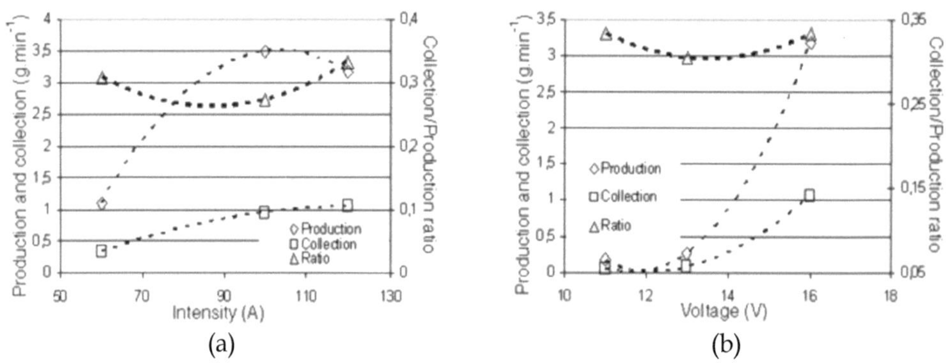

Fig. 4. Influence on production, collection and ratio of Pb/SiO_2 composite powder (a) of electric arc intensity for a given voltage (16V); (b) of electric arc voltage for a given intensity (120A)

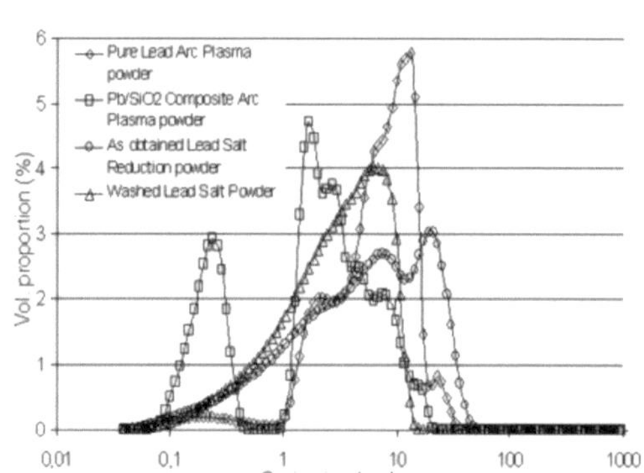

Fig. 5. Grain size distribution for plasma arc powders (pure lead and Pb/SiO_2 composite) and lead salt reduction powders (as obtained and ethanol/water/acetic acid washed)

Pure lead powder synthesized by the plasma arc process does not contain any oxide amount. Nevertheless, this powder is characterized by a wide grain size distribution (Fig.5) which exhibits at least two populations of grains. The first population with an average grain size of around 10µm is associated to the presence of agglomerates which results from the pre-sintering of the elementary particles. Indeed, it can be observed the appearance of necks between particles (Fig.6a) which could be attributed to the early stages of the powder sintering at room temperature. The second population can be associated to a mean grain size of elementary spherical particles (*i.e.* <0.1µm).

In the same manner, the Pb-SiO_2 grain size distribution seems to be bimodal ($\varnothing_{m1}$ = 250nm, $\varnothing_{m2}$ = 2µm) (Fig. 5). The amount of agglomerates which average size is higher than 10µm remains weak. Otherwise, few nanosized particles can be detected from SEM micrographs (Fig.6b). These results are similar to those reported in literature for the synthesis by the same process of Bi-SiO_2 powders (Brochin et al., 1999). Furthermore, the corresponding XRD pattern (Fig. 7) does not reveal silica. That means that the silica phase is not crystallized and/or its contents remains not sufficient to be detected. The appearance of low intensity peaks of the lead monoxide phase can be interpreted considering the characterization conditions (*e.g.* long exposure time in air during XRD experiments).

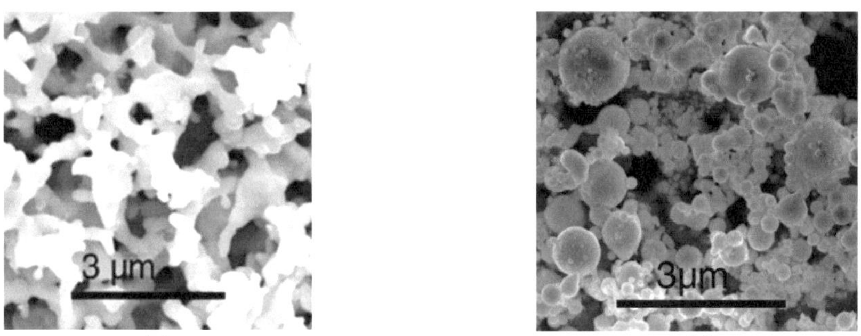

(a) (b)

Fig. 6. SEM micrographs of the plasma arc powders: (a) pure lead, (b) Pb/SiO_2 composite

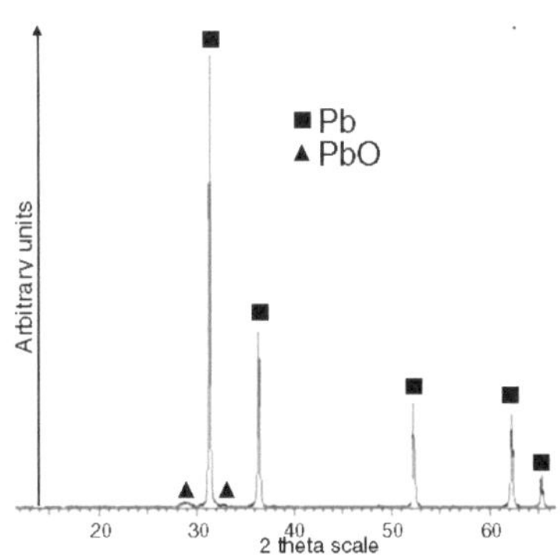

Fig. 7. XRD patterns of Pb-SiO_2 composite obtained by a plasma arc process (JCPDS for lead: 00-004-0686, JCPDS for PbO: 00-035-1483)

The synthesis of pure lead powder by arc plasma process leads to non oxidized powders but with a large grain size distribution for which the higher grain diameter does not exceed few micrometers. This morphology of so-obtained powders would result from the pre-sintering of pure lead particles at room temperature. Conversely, the elaboration of lead-silica composite powders provides low amounts of agglomerates because of the presence of ceramic particles that inhibits the neck formation between metal particles. The composite powder looks like well dispersed with a bimodal distribution.

2.1.3 Elaboration of composite powders by the route of lead salt reduction

The synthesis of the lead powder rested on the lead salt reduction ($PbBr_2$) by a reducing agent (NaH) according to:

$$PbBr_2 + 2NaH/2tBuONa \rightarrow Pb/2\ tBuONa + H_2 + 2\ NaBr \tag{5}$$

After reaction, the presence de alkoxide in the powder inhibits both the oxidation of lead particles and their agglomeration by pre-sintering (Balan et al., 2004). The stoichiometric proportions of reagents for the reaction (5) were: 2 moles of BuOH (2-methyl-2-propanol, >99%, Aldrich), 4 moles of NaH (55-65%, Fluka) and 1 mole of $PbBr_2$ (>98%, Aldrich). NaH component was beforehand introduced in excess (+5%) so that the reaction (5) can be considered as complete. The quantity of solvent (Tetrahydrofurane (THF)) was equal to 30mL per millimole of lead bromide.

A view of the device used to synthesize the lead powder is proposed in Fig.8. The first step of the experimental protocol consisted in introducing drop by drop the alkoxide (4.4g) in a THF bath (100mL) containing NaH and heated at 100°C. This step was accompanied by the hydrogen release. In a second step, 11g of lead bromide was added to the bath which was then heated at 100°C for 3h. After cooling, the solvent was removed thanks to a rotary evaporator and the powder so obtained was dried under secondary vacuum. Since this powder should be composed (theoretical proportions) of 31.4wt.% Pb, 33.9wt.% NaBr, 31.7wt.% alkoxide, 0.3wt.% NaH, a second washing step in polar solvent was required to remove the residuals salts. Acetonitrile, in spite of its polar momentum (3.92D) was not sufficiently efficient to remove NaBr salt. It was the reason why ethanol (1.62D) has been used first then water (1.85D). In a final step of washing, acetic acid diluted in ethanol to eliminate residual lead oxides has been retained.

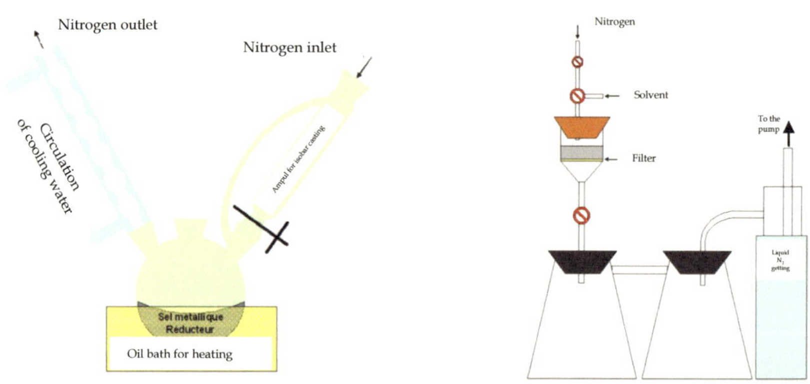

(a) (b)

Fig. 8. Devices for the powder synthesis by lead salt reduction (a) and for washing it (b)

The powders obtained by lead salt reduction and after washing/drying exhibited a small amount of $PbO/PbBr_2$ phases as shown in Fig.9. These observations have been confirmed by TEM characterization of this powder. Indeed, TEM observations (Fig.10a) coupled with electron microdiffraction characterizations (Fig.10b) attested the presence of small oxide particles like plates whereas the lead grains seemed to be spherical with an average grain size of around one hundred micrometers. In particular, the intense diffraction ring detected at 0.299nm could be indexed considering a PbO-$PbBr_2$ mixed compound (*e.g.* $6PbO$-$PbBr_2$) as suggested in the litterature (Knowles, 1951). Otherwise, the grain size distribution spreads out two orders of magnitude between 100nm and 15μm.

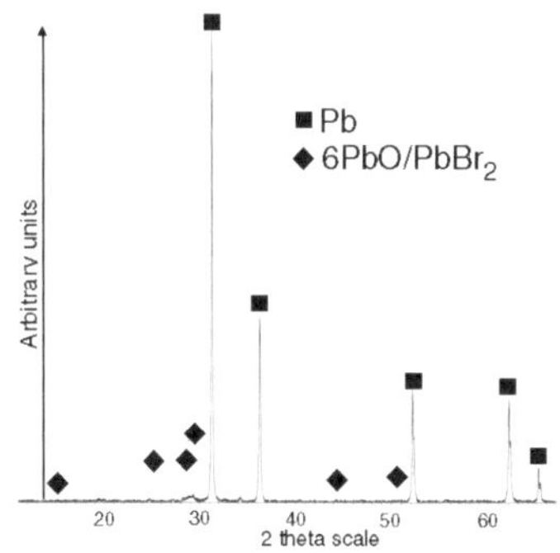

Fig. 9. XRD pattern for an ethanol/water washed powder (JCPDS file for lead: 00-004-0686, JCPDS file for $6PbO/PbBr_2$: 00-006-0384)

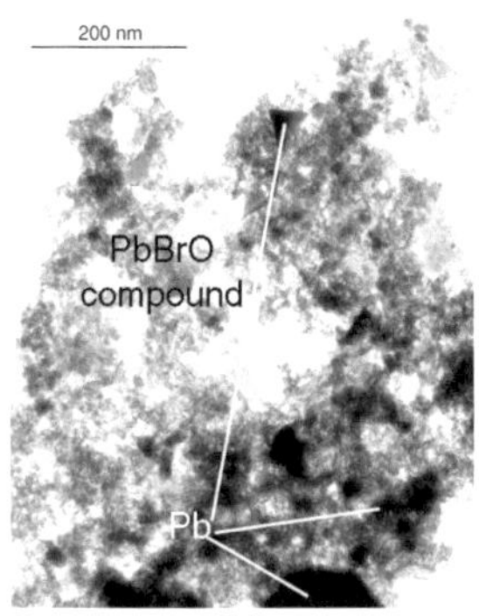

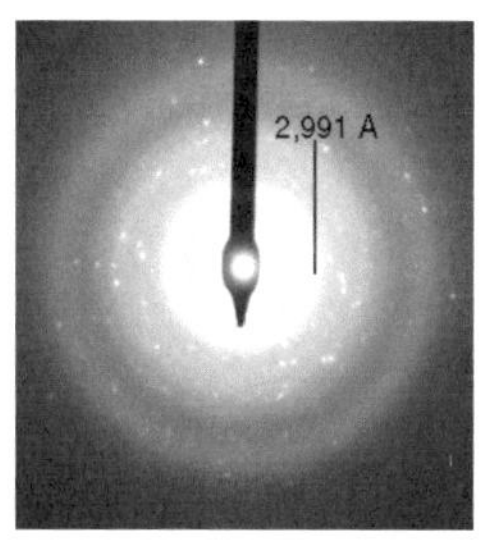

(a) (b)

Fig. 10. (a) TEM observation of washed lead salt reduction powder, (b) corresponding electron microdiffraction pattern of the $Pb_xBr_yO_z$ compound

2.2 Thermal behaviour and shaping

2.2.1 Calcination of powders

The thermal behaviour of powders obtained from the physical and chemical routes has been analysed from Differential Scanning Calorimetry (DSC) runs. The DSC thermograms of lead-based powders appear to be different as a function of the synthesis method. So, the thermograms of plasma arc powders only display an endothermic peak which can be associated to the lead melting point (Fig.11a). The corresponding peak shifts toward the lower temperature what would be due to the small particles size of the starting powder (Ben David et al, 1995).

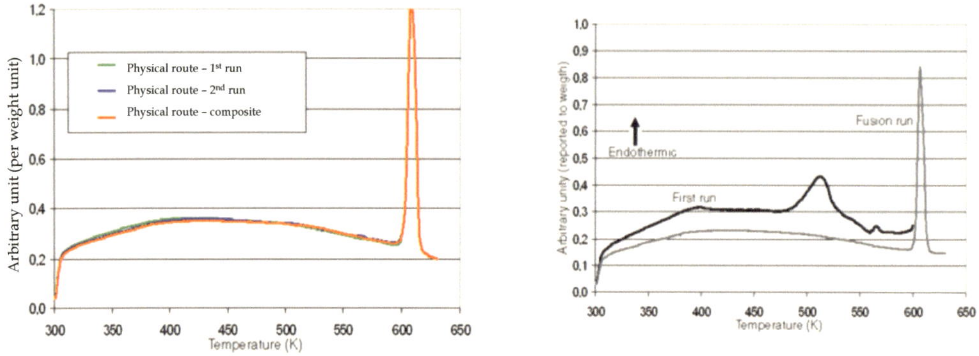

(a) (b)

Fig. 11. DSC thermograms of the lead-SiO_2 plasma arc powder (a) and of the ethanol/water/acetic acid washed lead salt reduction powder (heating rate = 3K.min^{-1}) (b)

The DSC thermogram of the lead salt reduction powder after washing with ethanol, water and acetic acid (Fig.11b) shows a weak endothermic peak near 400K. This thermal effect could be attributed to the vaporization of ethanol or acid acetic traces. Moreover, a second endothermic peak is detected at 480K and is associated to the lead (II) acetate removing. Finally, the displacement toward the lower temperatures of the lead melting peak remains characteristic of the presence of fine lead particles.

2.2.2 Shaping and sintering

Whatever the starting powders, discs of 20mm diameter have been prepared at room temperature by uniaxial pressing with an applied load of 220MPa. Then, the samples have been hot pressed under vacuum ($PO_2 < 10^{-6}$ Pa) thanks to a specific device developed in LCSM laboratory (Cartigny et al., 2007). The choice of the sintering parameters (*e.g.* temperature, soaking time) resulted from dilatometric experiments carried out for commercial pure lead and lead-ceramic mixtures. These parameters have been applied to the whole of compacts obtained from the arc plasma and lead salt reduction powders. As an example, Fig.12 shows dilatometric curves of pure lead and Pb-0.5vol.%TiO_2 compacts which appear similar until 225°C. The average thermal expansion coefficient (α), calculated between 20 and 225°C, was approximately equal to 22.7×10^{-6} K^{-1}. Between 215 and 235°C, a rapid changing of the dilatation rate was observed. For upper temperatures, linear dilatation behaviour can be noticed. Moreover, these observations gave rise to the following comments: i) no shrinkage was observed during thermal treatment, ii) the dilatation effect was more important for pure lead than for composite.
To understand this "anomalous" dilatation for pure lead, several samples were pulled out the dilatometer at 215 and 260°C to be air quenched. After cooling, SEM micrograph allowed to establish the grain size distribution (Fig.13) which showed the dependence between the grain size distribution and temperature. So, for the higher temperature, new particle ranks appeared for the greater diameters, *i.e.* between 12 and 20µm. That could be related to the appearance of an abnormal grain growth mechanism during the thermal treatment. Moreover, Fig.14b confirms the presence of coarse grains ($\varnothing_m$>8µm) only when the pure lead sample is treated at the higher temperatures. The coarse grains are well detected by comparison of heated sample microstructures with those of non-treated samples (Fig.14a). This abnormal grain growth would be probably due to the secondary

recrystallisation process that could occur for the cold-pressed lead samples. Conversely, lead-TiO_2 composite keeps a fine microstructure (Fig.14c) what can be imputed to the presence of fine ceramic particles at the grain boundaries that delays the motion of recrystallisation fronts during heating.

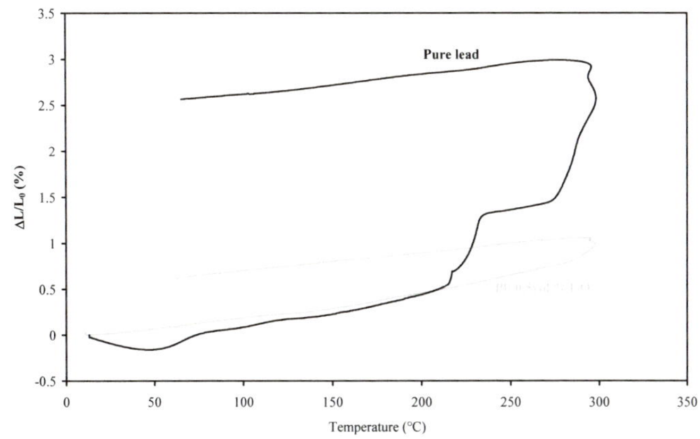

Fig. 12. Shrinkage curves of pure lead and Pb-0.5vol.%TiO_2 compacts

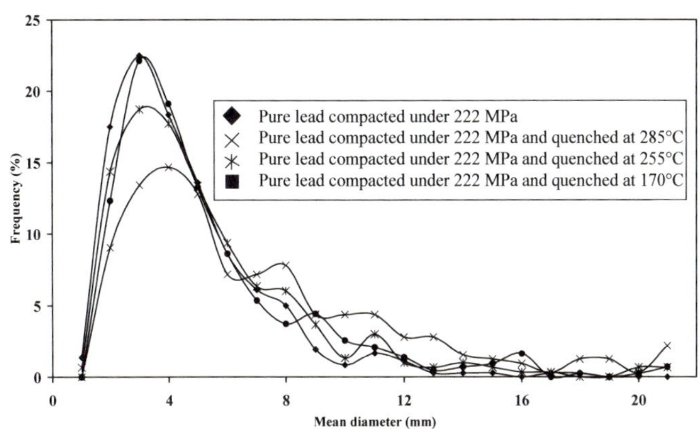

Fig. 13. Grain size distribution for pure lead compact as a function of thermal treatment

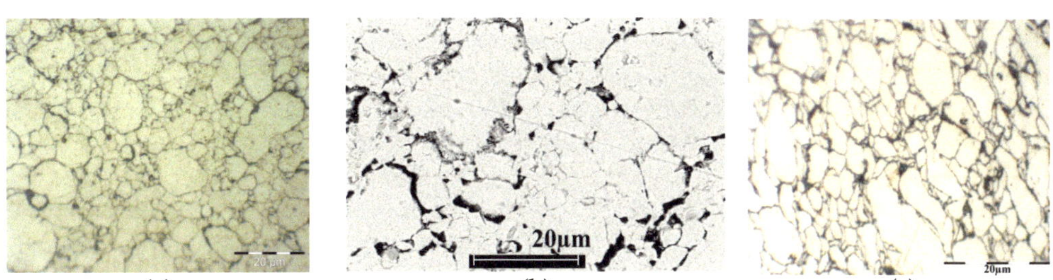

(a) (b) (c)

Fig. 14. Optical micrograph of non treated pure lead sample (a); SEM micrograph of pure lead sample treated until 260°C in the dilatometer (b); optical micrograph of the Pb/(0.5vol.%)TiO_2 composite treated until 260°C in the dilatometer (c)

To improve the densification process, sintering experiments under pressure were carried out for both pure lead and lead-TiO_2 composites. Several thermal and pressure cycling have been tested for pure lead as shown in Fig.15. The densification rate and the average grain size so-determined have been reported for pure lead in Table 2 as a function of experimental conditions. In densification terms, the better results (about 98% of the theoretical value) are obtained by coupling pressure and temperature as indicated in Fig.15b.

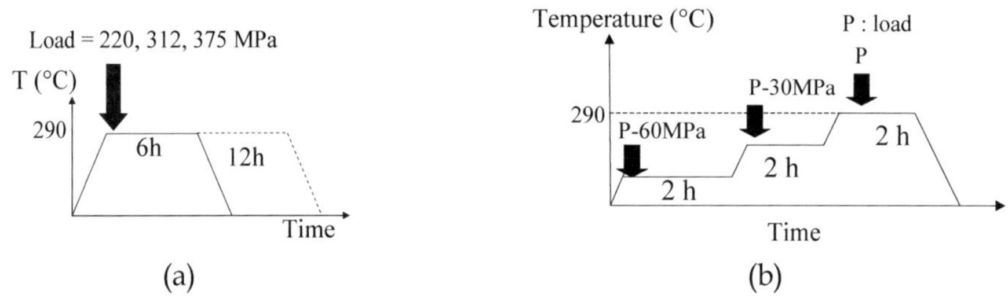

Fig. 15. Temperature and pressure cycling used for the sintering of pure lead compacts

	Sintering process as described in Fig.15a			Sintering process as described in Fig.15b	
	Load P (MPa) applied during 12 hours			Load P (MPa)	
	220	312	375	220	375
Average grain size (μm)	4.8	5.5	7.7	7.7	8.3
Relative density (ρ/ρ_0)	85.6	98.8	95.0	97.8	97.8
Initial hardness (HV_2)	15.7	10.7	10.4	8.9	9.4

Table 2. Main characteristics of lead samples sintered as mentioned in Fig.15 (with final temperature equal to 290°C)

In a second time, this optimized sintering process has been applied to the lead-ceramic composites. The main characteristics of the sintered samples (relative density, average grain size) have been reported in Table 3 as a function of ceramic used as strengthened agent (SiC, SiO_2, TiO_2 or Y_2O_3). The higher densification rate is reached when titanium dioxide powder is employed. This result can be associated to the good distribution of titanium dioxide particles at the grain boundaries (Fig.16). Moreover, the optimum TiO_2 volume fraction corresponds to 0.5 % because the relative density drops drastically for the higher ratio. This effect is probably due to both the residual porosity in TiO_2 aggregates and the removing of metal-metal junctions for the higher ceramic contents.

Finally, the sintering treatment reported in Fig.15b has been conducted to sinter the plasma arc and lead salt reduction powders. As an example, the relative density of pure lead plasma arc powder reaches up to 99.3% after hot pressing. No shrinkage or dedensification has been detected. In the same manner, the microstructure of hot pressed Pb-SiO_2 composites appears to be characteristic: dense zones (95% of the specimen surface) coexist with porous zones showing macro-cracks (5% of the specimen surface). (Fig.16a). Otherwise, the values of the relative density vary between 93 and 98.5%. From TEM investigations, these variations can be related to the ceramic content: the higher ceramic particle density, the higher residual porosity is. Moreover, silica particles appear to be amorphous, spherical and mainly located at the grain boundaries (Fig.16b).

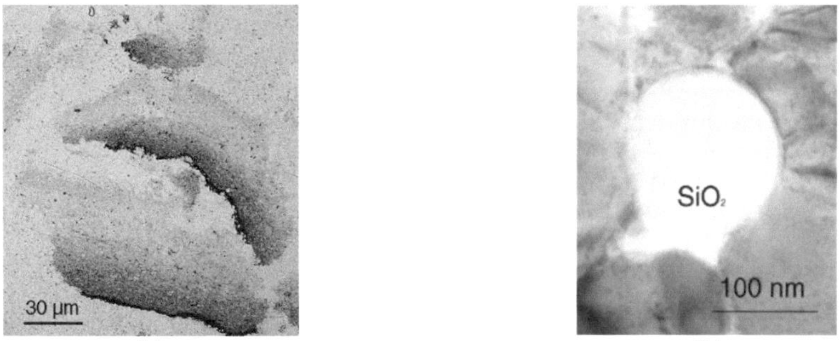

(a) (b)

Fig. 16. Hot compacted plasma arc powder (a) and silica particle at grain boundaries (b)

Nature of material	Volume fraction (%)	Average grain size (µm)	Relative density (ρ/ρ_0) (%)	Hardness (HV_2)	Stationary creep rate (s^{-1})
As-cast pure lead	-	> 300	-	-	$1.40\ 10^{-7}$
Pb-0.08%Ca-2.0%Sn alloy	-	150	-	-	$1.91\ 10^{-8}$
Sintered pure lead	-	7.2	97	-	$2.36\ 10^{-8}$
Pb/TiO_2 (Tioxyde)	0.5	5.6	97.8	13.3	-
	1	5.5	97.7	13.0	-
	3	5.0	96.6	15.1	-
	5	5.9	96.7	16.7	$6.47\ 10^{-10}$
Pb/Y_2O_3 (S.E.P.R.)	0.5	5.9	95.5	12.2	-
	1	5.8	96.7	13.9	-
	3	5.9	97.0	14.8	-
Pb/SiC (Cerac)	0.5	4.5	95.5	13.2	-
	1	4.9	96.3	13.8	-
	3	4.7	96.5	16.0	-
Pb/SiO_2 (Cerac)	0.5	6.2	96.4	10.9	-
	1	7.1	94.3	10.6	-
	3	7.7	93.9	10.4	-
Pure lead (arc plasma route)	-	0.7	99.3	6.0	$9.19\ 10^{-10}$
Pb/SiO_2 (arc plasma route)	2.5	0.25-10	98.5	15.0	$8.00\ 10^{-11}$
Pb/(6PbO-$PbBr_2$) (lead salt reduction route)	-	0.1-10	74	50.0	$4.67\ 10^{-11}$

(*) The creep tests have been performed under a 1.15 MPa pressure at 80°C.

Table 3. Average grain size, relative density, creep rate (*) values for the as-cast pure lead, the as cast and annealed Pb-0.08%Ca-2.0%Sn alloy, the sintered pure lead and the sintered lead-ceramic composites

Before sintering, the microstructure of the lead salt reduction powder compacts is characterized by dense bright and porous dark zones (Fig.17a). On the basis of EDXS analyses, the bright zones (Fig.17b) are only composed of lead whereas the dark zones (Fig.17c) are associated to high content of brome, oxygen and acicular particles. These observations are in a good agreement with those of the starting powder which revealed the presence of lead oxybromide phase giving to this powder a composite character.

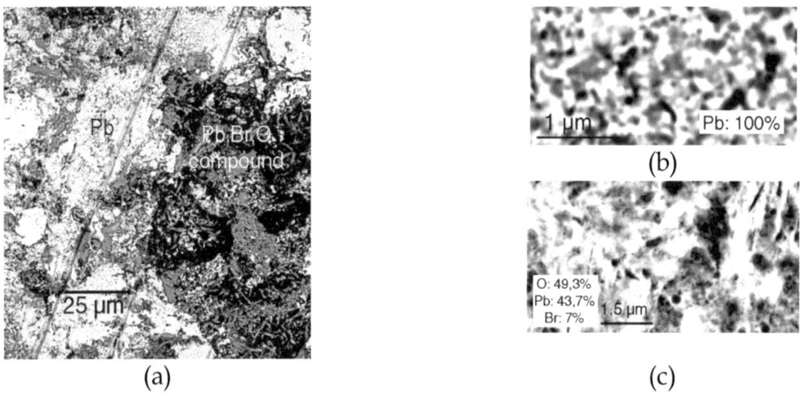

Fig. 17. SEM morphologies of green lead salt reduction compacted powder (a); SEM observations at higher magnification of the bright zone (b) and of the dark one (c)

Dilatometric tests have been carried out on lead salt reduction powders compacted under a 220MPa applied load. Shrinkage can be observed from 350K (Fig.18) and could be related to both the weigh loss detected on the TGA thermogram (Fig.18) and the endothermic peaks revealed on the corresponding DSC thermogram (Fig.11b). All these effects can result from the vaporization of organic compounds. The relative density value of compacts ranges between 64% and 74% before and after sintering, respectively. It can be noticed that the heat treatment remains similar to that described in details in Fig.15b.

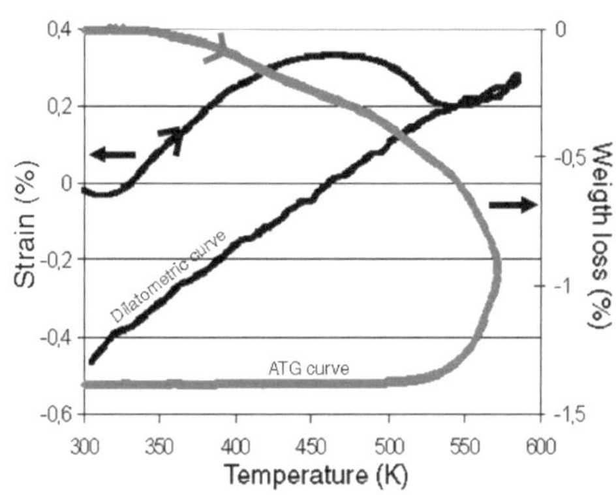

Fig. 18. Shrinkage and TGA curves during sintering of lead salt reduction powder

The microstructure of the sintered material is characterized by the coexistence of two zones which show different sintering behaviour (Fig.19): a full dense lead enriched zone and a porous lead oxybromide zone. In particular, SEM observation confirms the low densification of the $Pb_xO_yBr_z$ zone since this latter makes the propagation of microcraks easier (Fig.19).

Moreover, XRD investigations indicate that the sintering treatment lead to the crystallization of the oxybromide phase with the $6PbO\text{-}PbBr_2$ composition.

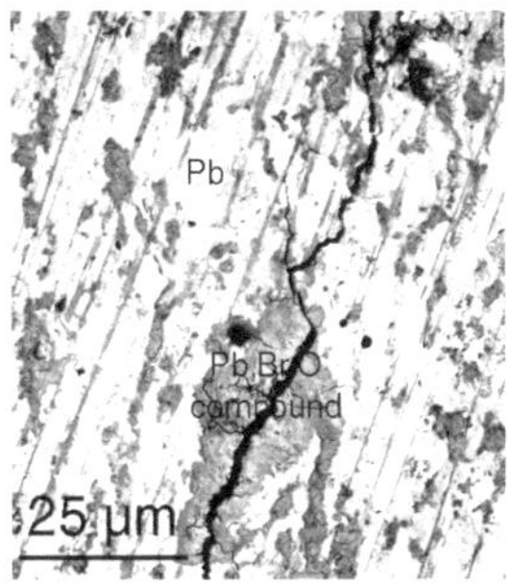

Fig. 19. SEM micrograph of sintered lead salt reduction powder

3. Working properties

3.1 Mechanical properties

3.1.1 Hardness tests

In a first step, the hardness evolution as a function of time at room temperature has been compared between the as-cast lead-based specimens and the lead-ceramic composites (Fig.20). Here, the composites specimens have been prepared from the commercial lead and ceramic powders according to the procedure detailed in section 2.2. In particular, the hardness curve of the as-cast Pb0.08wt.%Ca2.0wt.%Sn alloy then annealed for 2h at 90°C to prohibit the overageing appearance (Maître et al., 2003) has been reported in Fig.20a. This ternary composition is often considered as promising for the development of new acid battery grids (Bourguignon et al., 2003). From Fig.20a, it is well demonstrated that the ceramic addition to pure lead matrix leads to a significant hardness increase as well as its stabilization vs. time. The titanium dioxide addition allows reaching the higher hardness value, *i.e.* 13Hv, confirming the efficiency of the fine particle distribution. Conversely, the hardness of PbCaSn alloy widely fluctuates (Fig.20). These variations can be imputed to both the slow return to its thermodynamic equilibrium state.

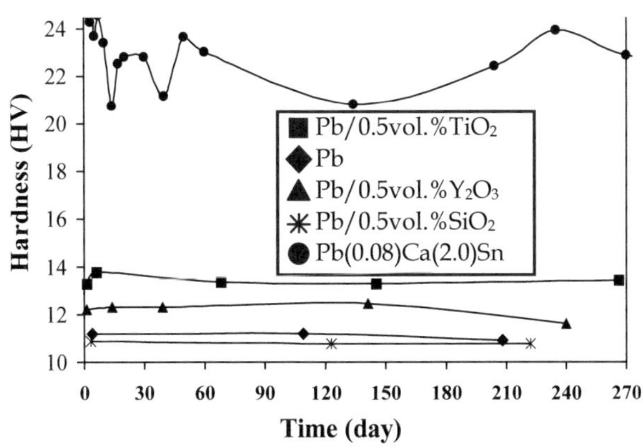

Fig. 20. Hardness evolution at room temperature of sintered pure lead and lead-ceramic composites as a function of time

It can be noticed that the stability of mechanical properties is also encountered for sintered specimens. So, the hardness of the Pb-SiO_2 specimens is about 15Hv_2 and remains constant for 20 months (Fig.21a). In the similar way, the hot pressed salt lead reduction powder shows hardness values near 50HV_2. Moreover, this latter is not significantly modified after a heat treatment of 15 days at 80°C or 110°C (Fig.21b).
In summary, although the microstructure of PbCaSn alloy is coarser ($\varnothing_m \approx 100$ µm (Bourguignon, 2003) in comparison with composite materials, its hardness value remains the higher. Consequently, it means that the $L1_2$ phase precipitation and the tin (sub)grain boundary segregation are only responsible of this hardening phenomenon.

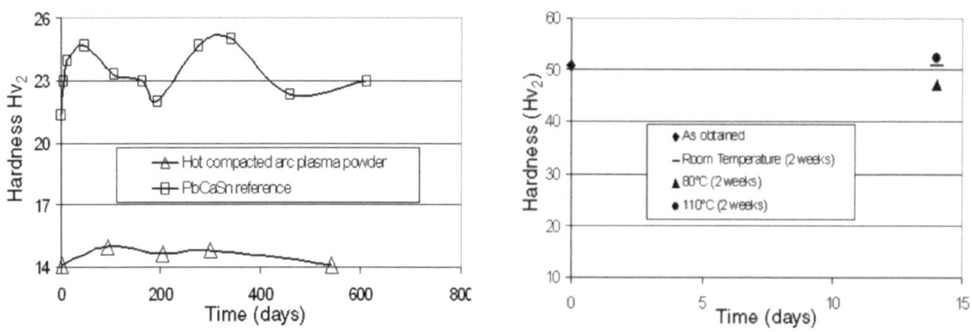

(a) (b)

Fig. 21. Evolution of the hardness value as a function of the ageing time for hot pressed arc plasma Pb-silica (a) powder and hot pressed lead salt reduction powder (b)

3.1.2 Creep resistance

In acid battery applications, the determination of working mechanical properties usually consists in evaluating the grid material creep resistance. Indeed, the positive electrode is chemically corroded during the charge process and the corresponding corrosion product (mainly lead dioxide) has a higher specific volume than the starting material (lead alloy). The tensile stresses on the grid so created during service lead to the growth of the positive plate. Therefore, the positive grid material should have a high creep resistance to prohibit the growth rate in service (Bagshaw, 1995).
The creep resistance of Pb-based materials has been measured thanks to compressive tests with a strain of 1.15MPa. A specific device has been developed in LCSM laboratory and has been previously described in details in literature (Cartigny et al., 2007).
In a first step, the evolution of the strain rate as a function of time under 1.15MPa at 80°C, for as-cast pure lead, annealed Pb0.08Ca2.0Sn alloy, sintered commercial pure lead and sintered Pb-(0.5%vol.)TiO_2 has been reported in Fig.22. For each positive electrode material, the creep rate is calculated in stationary conditions, *i.e.* for secondary creep step from the curves of Fig.22. The data so obtained are reported in Table 3 with the corresponding average grain size of sample. These experiments have been carried out to simulate the curing and the formation steps experienced by each alloy during the battery process (Cooper, 1998).
For as-cast pure lead, the creep rate values are compatible with those mentioned in previous works (Gifkins et al., 1967). Globally, the sintered lead creep rate is close to that of PbCaSn alloy which shows the higher mechanical properties of the as-cast grid alloys. Otherwise, the

creep rate of the Pb-TiO_2 composite (*i.e.* manufactured by mixing commercial powders) appears to be lower of two orders of magnitude than the annealed Pb-Ca-Sn alloys. In the same way, the composites produced from the arc plasma or the lead salt reduction powders show the better creep resistance: indeed, their creep rate varies between 5 and 8.10^{-11} s^{-1} (Table 3). Finally, the grain size of grid material as well as its composition (*i.e.* nature and volume fraction of ceramic) seems to play a major role on the improvement of its creep resistance. In particular, the creep rate of the sintered pure lead drastically decreases when the coarse commercial powder is preferred to the sub-microsized arc plasma powder (Table 3).

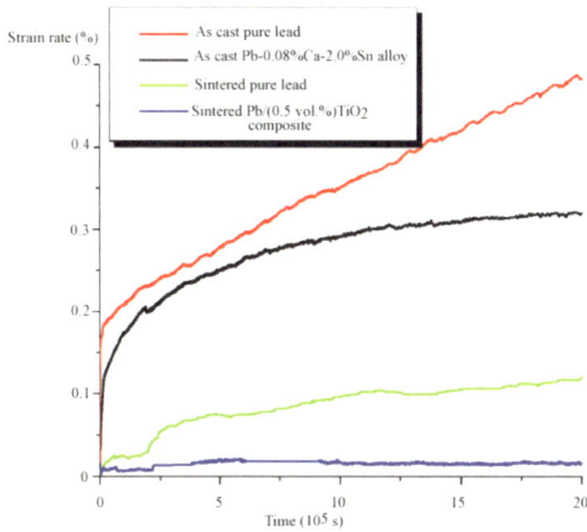

Fig. 22. Creep resistance at 80°C under a 1.15MPa load of as-cast pure lead, as-cast Pb0.08%Ca2.0%Sn alloy, sintered pure lead and sintered Pb-SiO_2 composite

3.2 Corrosion resistance

The overcharge conditions are simulated by a polarisation test at 1.5V for 5 days in 5M sulphuric acid maintained at 20°C. The corresponding curve i=f(t) for the annealed Pb0.08%Ca2.0%Sn alloy is reported in Fig.23a. It can be observed the presence of current variations probably due to the oxygen release. This latter leads to the lead oxide layer micro-cracking and the quasi-simultaneously electrode re-oxidation. Moreover, a low weigh loss is measured after the polarisation test achieved in overcharge conditions (Table 4).

Sample	Weight loss ($mg.cm^{-2}$)
As-cast lead	64.4
Annealed Pb0.08%Ca2.0%Sn	12.5
Sintered lead	166
Sintered Pb-0.5vol.%TiO_2	99

Table 4. Weight loss values measured after polarisation tests in overcharge conditions for as-cast lead, annealed PbCaSn alloy, sintered lead, sintered Pb-(0.5 vol.%)TiO_2

The i=f(t) curve of the sintered Pb-TiO_2 composite (Fig.23b) is characterized by a higher current density in overcharge conditions. This poor corrosion resistance of sintered samples relies on their residual open porosity. Indeed, the electrolyte, *i.e.* sulphuric acid, can preferentially seep in the pores (Figs 24a, 24b) leading to the grain loosening (Fig.24c). In these conditions, the weight loss measured can reach $130mg.cm^{-2}$. This interpretation can be applied to the corrosion behavior of composites manufactured from arc plasma or lead salt reduction powders since their respective i=f(t) curve (Fig.23c) and their microstructural features after the polarization test remain quite similar.

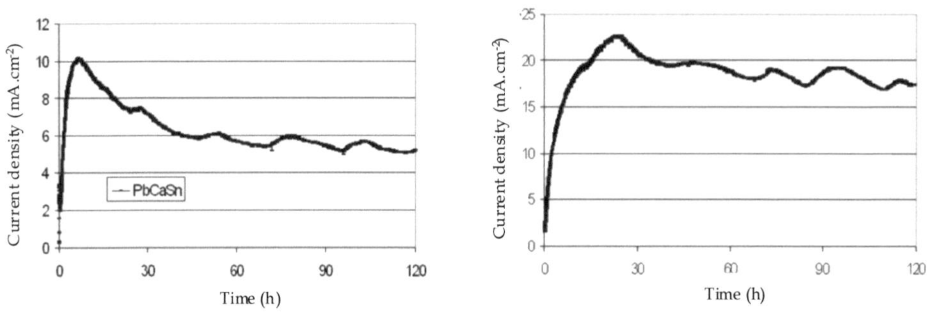

(a) (b)

Fig. 23. i=f(t) curves for annealed PbCaSn alloy (a), sintered Pb(0.5vol.%)TiO_2

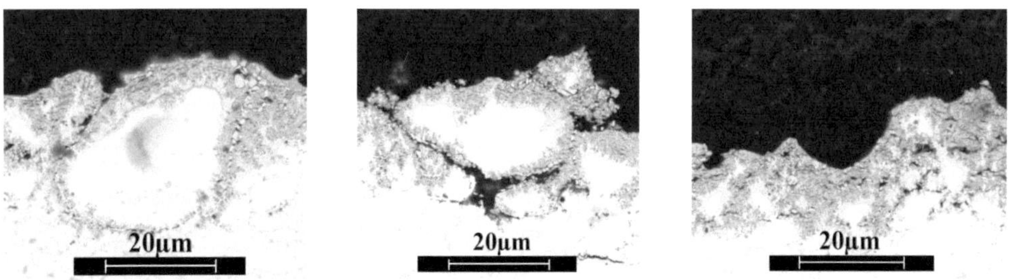

(a) (b) (c)

Fig. 24. SEM cross sections of sintered lead/corrosion products interface after the polarization test of the sintered lead-TiO_2 composite

3.3 Discussion and conclusion

The main properties of lead-based materials are reported in Tables 3 and 4. It clearly appears that the material elaborated from the lead salt reduction powder would present promising mechanical performances. Its hardness is higher than that reported for the other sintered or as-cast lead based materials. Although its densification is not complete, this material keeps the highest creep resistance in comparison with the other composites. Further comments and interpretations are detailed below.

The commercial powders usually lead to working properties depending on both the grain size and the oxide content of starting precursors. The presence of lead monoxide in all the batches allows blocking the grain boundary motion and, consequently, inhibiting the exaggerated grain growth. Otherwise, the lead oxide particles play the role of a strengthening agent within the matrix by displaying strong structural relationships with the lead particles. Consequently, the mechanical properties of composite would be enhanced in comparison with those of the reference (*i.e.* pure lead specimen): for example, the creep rate decreases by 40% and its hardness which remains stable for two years increases by 1.5 Hv_2. Nevertheless, the features of the commercial powders can deeply differ from one batch to another requiring the determination of the sintering parameters once again.

The compacts manufactured from pure lead arc plasma powder obviously show similar mechanical properties than those of the as-cast pure lead specimens. This result could be explained by considering the evolution of the grain boundary concentration throughout the elaboration process. Indeed, the pure lead powder appears to be partially sintered just after synthesis then the sintering treatment promotes the grain growth of the coarser lead particles. To explain their similar mechanical behaviour, one must consider that the grain boundary density is finally near between the two types of pure lead specimens.
The sintered bodies issued from Pb-SiO_2 and Pb-(6PbO-$PbBr_2$) are characterized by promising mechanical properties (Table 3). This observation results from the dual-acting of the homogeneous dispersion of hardening particles within the lead powder and the optimum content of non metallic grains. These conditions would be then fulfilled by using the arc plasma or the lead salt reduction process as synthesis method of composite powders. Nevertheless, some efforts should be done to obtain full dense materials by determining the adequate parameters of the sintering treatment. Actually, the presence of residual porosity for the current sintered Pb-(6PbO-$PbBr_2$) composite is probably the cause of the weakness of the creep resistance. Nevertheless, the weak corrosion resistance of this latter composite could limit its implementation as electrode materials in battery. Indeed, it is well known that the positive electrode in an acid battery reaches periodically voltage values ranging between 1.3 and 2.1V. This phenomenon is accompanied by the periodical conversion of dibrome (Br_2) to bromide (Br^-) according to the Pourbaix diagram. At this stage, it must be noticed that the dibrome specie could be liquid at the acid battery working temperature what could review the ability to use this composite as electrode.
Finally, the using of sintered composite materials are not affected by the appearance of softening structural transformations such as recrystallisation or overageing since these systems have reached their thermodynamical equilibrium. As a result, their microstructures as well as their mechanical properties remain stable as a function of time. Conversely, variations by 25% are usually registered for the hardness value of a Pb0.08%Ca2.0%Sn alloy. Otherwise, it appears that the weakness of the sintered materials comes from their electrolytic permeability which depends on their residual porosity. To avoid this phenomenon, the deposition on the composite system of thin coating could fill the residual open porosity without decreasing their mechanical properties.

4. Protective coatings against the corrosion

In a first step, the using of electrolytic coatings of pure lead or lead-tin alloy represents a promising method to promote the corrosion resistance of sintered composite in concentrated sulphuric acid. Indeed, these coatings allow both protecting the surface and keeping the high mechanical properties of the sintered substrate.
In a second step, the improvement of the corrosion resistance can be reached by implementing a coating process of active elements. Indeed, the lifetime of positive electrodes is related to the active matter decohesion which leads to the loss of electron conductivity and battery electrical capacity. So, the role of the active elements on the adhesion of oxide layers would claim to be explored. Previous studies (Bourguignon, 2003) have shown that the presence of a mixed oxide phase at the alloy-corrosion layer could improve the corrosion resistance of the lead-based alloy in the concentrated sulphuric acid.
This section is then focused on the implementation of lead-based coatings and of silica-based thin layers. To understand well the role of ceramic coatings on the electrochemical

corrosion of electrode materials, coatings have been deposited on reference-like substrates of PbCaSn alloys for which the corrosion process in H_2SO_4 is well known.

4.1 Lead-based coatings

4.1.1 Experimental procedure

Before coating, the substrates were polished, washed, dried and weighted. The cleaning resulted from the using of trichloroethylene as a solvent for 30s then the washing with water. The pickling was conducted with either the HBF_4 (8%)/H_2O_2 or acetic acid/H_2O_2 mixtures for 30s before a cleaning with HBF_4 to avoid the contamination of electrolytic bath. The electrolytic bath is a solution of HBF_4 containing 130g.L^{-1} of lead (Jobst *et al.*, 1997). The preparation of this bath is described in a previous study (Subramanian *et al.*, 1966). According to the literature, the current density was chosen ranging between 1 and 8A.dm^{-2}

4.1.2 Influence of the current density

Low current density values (*i.e.* between 2 and 3A.dm^{-2}) provide a homogeneous lead layer on a substrate manufactured by sintering of as-treated commercial lead powder (Fig.25a). The thickness of this layer so-obtained is about of 60µm. To improve the adhesion of this coating on lead-based substrate, an annealing treatment for 6h at 80°C has been performed. Nevertheless, the residual stresses probably due to the hydrogen insertion in the coating lead to its desquamation (Fig.25b). To solve this problem, it was decided to decrease the current density up to 1A.dm^{-2} during deposing process to inhibit the residual stresses.

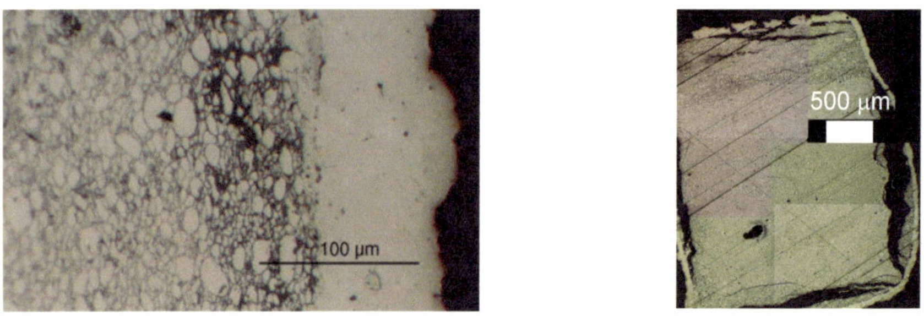

(a) (b)

Fig. 25. Cross sections of the lead electrolytic coatings obtained with a current density of 3A.dm^{-2} (a) and after an annealing of 6h at 80°C (b)

4.1.3 Influence of the annealing temperature

A current density of 1A.dm^{-2} was then employed to elaborate new lead-based coatings. A first annealing treatment which was performed at 80°C seems not to modify the microstructural features of the coating. Since the results obtained for this latter under simulating overcharge conditions were not convincing, further experiments were carried out for higher annealing temperatures. For example, an annealing treatment has been conducted at 110°C under vacuum. This treatment leads to both the decrease of the residual porosity in the substrate and prohibit the breakdown of the coating or substrate.

The cyclic corrosion test is the most representative because it tends to the working conditions of the lead acid battery. So, the potential varies between the battery overcharge 1.5V SHE^{-1} (Hg/$HgSO_4$ electrode) and the battery discharge 0.7V SHE^{-1} during 5 days at room temperature. In these conditions, the lead-based coating, annealed at 80°C, would be characterized by a drastic increase of the current density after the third cycle (Fig.26). Finally, the current density determined after the corrosion test is two times higher than that of the reference PbCaSn alloy. On the other hand, the corrosion behaviour of the covered specimen, annealed at 110°C, appears to be very different. Indeed, the current density is two times lower than that of the reference. These results are in a good agreement with the measurement of weight losses (*Table 5*) after corrosion.

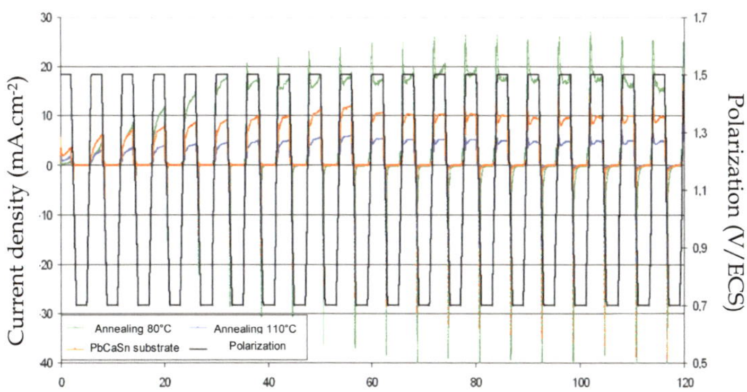

Fig. 26. Evolution of the current density in cyclic conditions for annealed lead coatings

Type of coating	Weight loss (mg.cm^{-2})
Reference (Pb0.08%Ca2.0%Sn)	20.8
Coated specimens treated at 80°C	147.2
Coated specimens treated at 110°C	16.8

Table 5. Weight loss values measured after corrosion tests for coated, annealed specimens

In particular, the weight loss remains significant for the coating annealed at 80°C which can be associated to the presence of a dense duplex layer. The outer layer would be composed of lead monoxide whereas the inner part corresponds to the lead sulphate compound. This sequence seems to be abnormal with regards to that obtained for the PbCaSn alloy (reference). Moreover, the lead oxide layer is characterized by the presence of lead particle. All these morphological characteristics show that the coating would not be adherent to the substrate. Indeed, the infiltration of electrolyte within the coating could promote the local fluctuation of pH and, more particular, the medium acidification. This latter would prohibit the lead monoxide formation but in the same time would favour the appearance the lead sulphate from the following reaction:

$$Pb + 2H^+ + SO_4^{2-} \quad = \quad PbSO_4^- + H_{2(g)} \tag{6}$$

This phenomenon could come from the residual porosity of the lead coating annealed at 80°C. Conversely, the lead coating treated at 110°C displays an uniform corrosion layer which can be related to the higher density and adhesion of the corrosion products (Fig.27a). The stacking sequence of the corrosion layers is in a good agreement with that observed in previous studies (Bourguignon et al., 2003): an internal thin and continuous layer of lead monoxide surmounted by coarse $PbO_2\alpha$ and $PbSO_4$ grains. Few residues of corrosion products can be detected under the coating nevertheless, these forays do not cross the lead coatings (Fig.27b).

(a) (b)

Fig. 27. Cross sections after cyclic corrosion test ($1A.dm^{-2}$) of the lead coating annealed at 80°C (a), at 110°C (b) (local infiltrations of corrosion products under the coating)

4.2 Silica-based coatings

A thin silica layer has been elaborated to improve the anchorage of the active matter during the corrosion of grids.

4.2.1 Experimental protocol

The substrates ($15\times15\times1mm^3$) were made of the reference PbCaSn alloy, were previously polished thanks to the using of SiC papers (grade 1200) and then were stored under argon. The Plasma Enhanced Chemical Vapor Deposition (PECVD) method was applied to elaborate thin silica layers. These experiments have been conducted in the SPCTS laboratory in Limoges (France). This way rests on the using of the silane (SiH_4) as silicon precursor. The first coatings have been synthesized by undertaking the pre-treatment of the substrate surface. This latter consists in the polarization at 150V under an argon/hydrogen flow for 15 minutes in order to clean the substrate surface. The microwave power is then fixed to 1000W. In a second step (*i.e.* the deposition step), the silane and the oxygen flows have been chosen as being equal 1sccm and 360sccm, respectively (P_T = 9Pa). The thickness of silica layer so-obtained can range between 80 and 200nm. Further coatings have been prepared without pre-treatment to form 100-1000nm thick layers (Régnier *et al.*, 1996).

4.2.2 Morphological characterization of the coatings

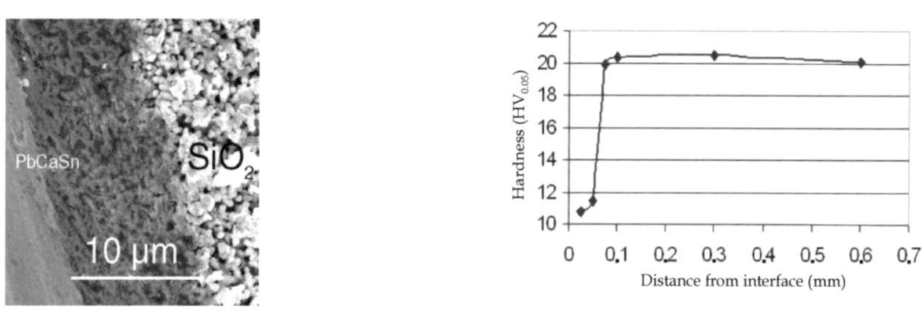

(a) (b)

Fig. 28. SEM micrograph of a breakage of a silica coated specimen (a) ; evolution of the hardness for a silica coated specimen after pre-treatment (b)

The coating of the pre-treated specimens appears to be porous and composed of not agglomerated spherical silica particles. The pre-treatment seems to modify the

microstructure of substrate. In particular, a melted zone can be detected at the silica coating-PbCaSn substrate interface (Fig. 28a). The disappearance of the cast and annealed microstructure of ternary alloys has been confirmed from hardness measurements: the zone submitted to the hydrogen ions beam can be associated to a hardness drastic decrease (Fig.28b).
Conversely, the untreated specimens are characterized by a covering silica layer. No melted zone has been detected from SEM observations or mechanical tests.

4.2.3 Electrochemical behaviour

In a first approach, the electrochemical behaviour of the silica coatings on the PbCaSn substrate has been evaluated from dynamical polarization tests between the free potential and a potential equal to +2V SHE^{-1} ($Hg/HgSO_4$ electrode). The corresponding voltamperogram (Fig.29) clearly evidences the significant effect of the ceramic coating. It cannot be observed for the silica coated specimens the different peaks corresponding to the lead oxidation at -0.9V ($Pb \rightarrow PbSO_4$,PbO) and at 1.45V ($Pb^{II} \rightarrow PbO_2$) (Bourguignon, 2003). Moreover, the peak of solvent oxidation leading to the oxygen release is shifted toward the higher potential value. From these results, it can be expected that the corrosion behaviour under overcharge conditions (1.5V SHE^{-1}) of the covered specimens should be improved by decreasing the released oxygen content and, consequently, by reducing the overall current and the damaging mechanical effect of the gaseous oxygen emission on the corrosion layers.

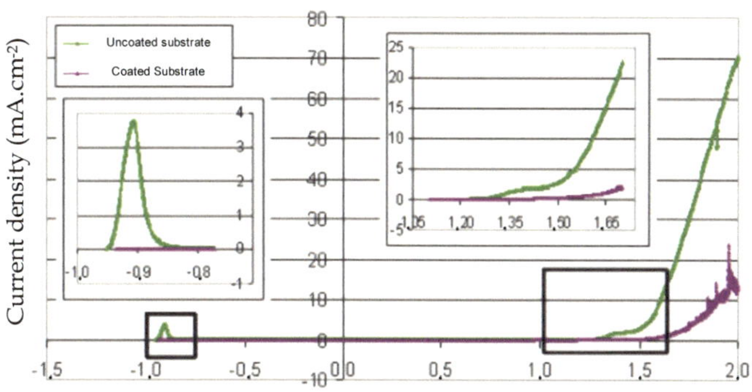

Fig. 29. Voltamperometry curves from the free potential up +2V SHE^{-1} for the uncoated substrate and the coated one

The cyclic voltamperometry curves obtained for potential ranging between -1.8 and 2V show that the silica coatings drastically inhibit the substrate corrosion during the first cycles. Then, several potential scans damage the silica layer and, finally, lead to the electrochemical response of the uncoated substrate. In particular, when the number of cycles exceeds three, the substrate oxidation becomes more sensitive and the $Pb^{+II} \rightarrow Pb^{+IV}$ transformation is clearly visible in the corresponding voltamperograms. In a same manner, the oxidation of the solvent is definitively located at anodic potential values lower than 1.4V.

4.2.4 Corrosion behaviour in the overcharge conditions

The pre-treated and coated specimens present a high current density after 24h at 1.5V in concentrated sulphuric acid (5M) (Fig.30a). The corresponding i=f(t) curve show variations which could be imputed to the oxygen release leading to the lead oxide layer micro-cracking and quasi-simultaneously electrode re-oxidation. This corrosion behaviour appears to be similar to that observed for sintered specimens even if in the present case the current density value remains lower. Therefore, in the early stages of the corrosion test, it can be noticed that the density current decreases as a function of the layer thickness. Finally, the weight losses are near whatever the pre-treated specimen.

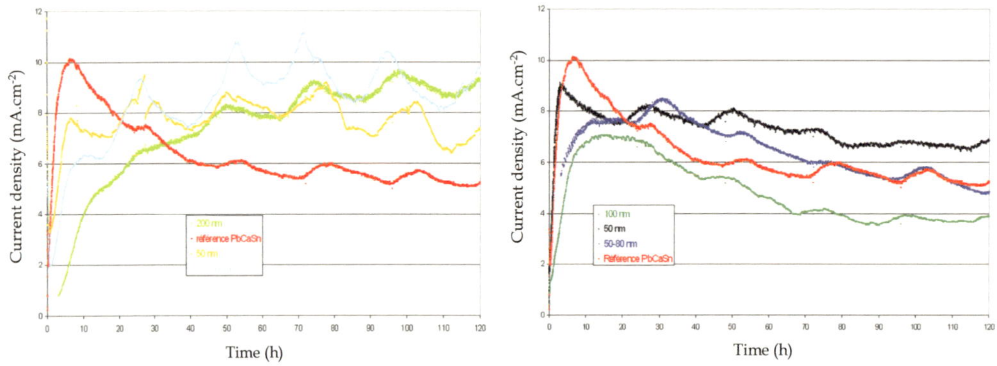

(a) (b)

Fig. 30. Current density in overcharge conditions for pre-treated and coated substrates (a); current density in overcharge conditions for untreated and coated substrates (b)

The benefit of silica coating on the electrochemical behaviour of untreated samples is demonstrated by the significant decrease of the current density after 24h of treatment (Fig.30b). The current corrosion would seem to be correlated to the silica layer thickness. Except for the thicker layer, all the coated substrates tend to keep the same corrosion resistance after one day as revealed by the comparison of the final weight losses. Otherwise, the sequences of corrosion layers remain similar between the coated and uncoated substrates: an inner dense corrosion layer and an outer porous part (Fig.31a). The thickness of the corrosion layer is significantly lower for coated specimens. The observation of SEM cross sections for corroded samples reveals that the corrosion phenomenon occurs along the grain boundaries of PbCaSn substrate (Fig. 31b). Chemical analyses by EPMA show that the silicon is only detected at the substrate/corrosion layer interface. Moreover, no silicate phase like $PbSiO_3$ is observed at this interface.

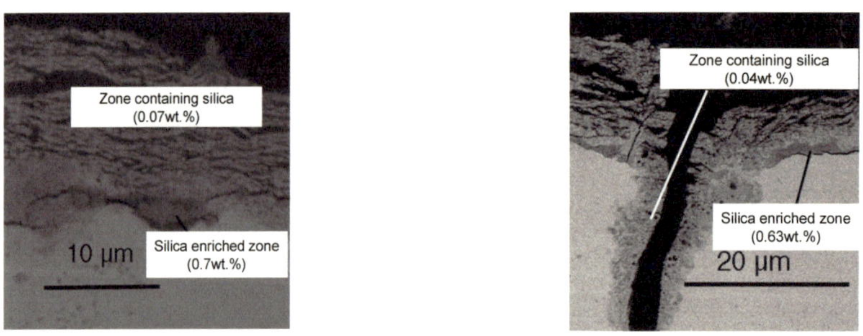

(a) (b)

Fig. 31. SEM observations of cross sections of silica coated specimens after corrosion test in overcharge conditions

4.2.5 Corrosion behaviour in cyclic conditions

The comparison of the current density evolution for different coating thicknesses shows up the slowing down effect of the silica layer on the corrosion kinetics. This positive influence is only significant at first (Fig.32). As soon as the third cycle is reached, the efficiency of the thicker coating (equal to 1000nm) is insufficient to soften the corrosion attack. Conversely, the silica coating for which the thickness is close to 300nm displays promising corrosion behaviour. The morphological features of corrosion layers are similar to those observed in overcharge conditions. The silicon element can be detected either at the corrosion layer/substrate interface or near the outer part of the corrosion layer. These results suggest that no specific corrosion mechanism is encountered for coated specimen but it occurs preferentially in zones where the silica layer is not covering or breakable.

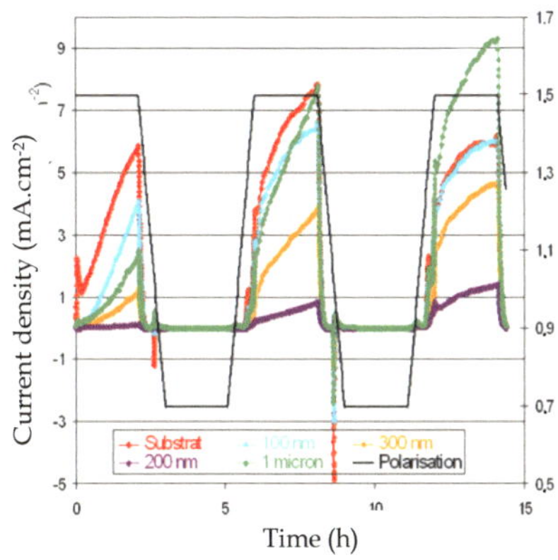

Fig. 32. Evolution of the current density as a function of time in cyclic conditions

4.3 Discussion and conclusion

The lead electrolytic coatings on sintered substrate have been elaborated by using a current lower than that reported in the literature. Indeed, the higher current leads to damaging effects because of the hydrogen trapping in the substrate pores. Hydrogen results from the reduction of proton of the acid electrolyte. An annealing treatment is required to guarantee the substrate protection against the corrosion. The annealing treatment seems to be more efficient at 110°C than at 80°C. Indeed, the increase of the annealing temperature leads to the decrease of both the current density and the weight loss after corrosion test. These promising results have been confirmed by the observations of the corrosion products: the morphology of corrosion layers remains similar to that obtained for polarization achieved at low potential (+0.7V). Concerning the silica coatings issued from PECVD experiments, it is necessary to analyse the specific effect of the "ionic cleaning". The pre-treated specimens usually show non covering silica layer on modified substrate. Indeed, it infers the softening of the PbCaSn substrate by surface melting. This effect is well characterized by the decreasing of mechanical properties (*e.g.* hardness) for pre-treated samples in comparison with those of cast and annealed PbCaSn substrates. These observations allow explaining the weak corrosion resistance of pre-treated samples and are in a good agreement with previous studies (Albert et al., 1997) which mentioned that the volume of released oxygen and the weight loss are more important for non aged alloy. Otherwise, the untreated specimens are characterized by the presence of covering silica layers whereas the substrate appears not to

be affected by the deposition process. The free potential in sulphuric acid is not modified but the current corrosion drastically decreases. In the same manner, the water oxidation moves toward the higher potential values and the lead oxidation ($Pb^{+II} \rightarrow Pb^{+IV}$) is not detected. This behaviour explains the decrease of the current corrosion density that can be observed for the silica coated samples (thickness layer < 300nm) in the early stages of the overcharge test. In this case, the corrosion layer is denser and is composed of silicon-doped phases at the corrosion layer-substrate interface. It can be noticed that the weight losses remain less important for silica coated samples. Thicker silica coatings (e>1µm) have undergone a reversal effect: PECVD technique does not provide adherent silica coatings.
From these results, the doping effect of silicon must be retained to explain the growth of a covering layer of PbO_{1+x} for which the predominant defect is the interstitial anion. This effect is identical to that mentioned in the literature (Bourguignon et al., 2003) concerning the formation of lead monoxide (in overcharge conditions) on the surface of cerium or lanthanum-doped PbCaSn alloys. The equilibrium of the defect formation can be written:

$$SiO_2 \quad = \quad Si^{\bullet\bullet} + O_O + O_i'' \tag{7}$$

It shows that the introduction in the lead oxide lattice of cation belonging an oxidation degree higher than that of lead such as Si^{4+}, La^{3+} or Ce^{4+} leads to enhance the content of predominant defects and promote the growth rate of oxide. Therefore, it can be noticed that a dense corrosion layer is detected just above silica-enriched zones (see Fig. 31a) and Fig. 31b). That means the silicon element could improve the anchorage of corrosion layer on lead-based substrate by forming mixed phases.

5. Conclusion

Three routes for the synthesis of lead-based powders have been investigated. The using of commercial powders which are previously treated in acetic acid bath provides after sintering compacts with a relative density higher than 97% and hardness near 12.5 Hv_2. These performances are linked to the presence of residual oxide and of a wide grain size distribution. Otherwise, the lead powder elaborated by arc plasma method seems to be characterized by agglomeration and pre-sintering effect at low temperature. Consequently, the mechanical properties of pure-lead sintered compacts remain similar to those encountered for as-cast pure lead specimens. Conversely, the arc plasma method provides nanosized, spherical and non agglomerated of Pb-SiO_2 powders. After sintering, the presence of silica intergranular inclusions leads to the mechanical reinforcement of lead-based matrix: in particular, the hardness value reaches 15 Hv_2 and the creep rate is decreased by two orders of magnitude in comparison with the behaviour of the PbCaSn alloy. The third synthesis route consists in reducing lead salts in organic medium. This route gives better mechanical properties for sintered specimens. The powder so-obtained is composed of not well crystallized phases containing brome, oxygen and lead elements. After hot compaction, these oxybromide phases confers promising properties to the lead matrix: a low relative density (74%), a high hardness (>50Hv_2) and a creep resistance that has multiplied fourfold in comparison with that of commercial lead powders. The decrease of creep rate is probably limited by the presence of cracks due to both the solvent removing and the expansion coefficient mismatch between the matrix and reinforcing agent.
Nevertheless, all these sintered materials show a residual porosity and a high grain boundary density. The corrosion tests in overcharge conditions have demonstrated that

these microstructural features are damaging for the electrochemical resistance of lead-based materials. Due to this assessment, a new route has been devised consisting in coating the sintered lead-base substrate with an electrolytic layer of lead. When higher current density values are targeted, it can be observed the mechanical deterioration of the substrate. This phenomenon could be attributed to the stresses correlated to the hydrogen release. So, the electrolytic coatings have required annealing treatments to reach homogeneous and free stress substrates. The electrochemical behaviour appears to be improved for lead-based substrate annealed at 110°C.
The corrosion resistance in concentrated sulphuric acid can be enhanced by putting a thin silica coating on the substrate. This coating could produce mixed oxide layers with the corrosion products making corrosion layers better anchored on the substrate. The silica coating has been elaborated by PECVD. When no pre-treatment is conducted, the substrate microstructure remains unchanged and keeps the aged state. The silica is then covering with a thickness ranging between 100nm and 1µm. The thicker silica layers seem to be subjected to a preferential detachment. Conversely, the silica layers which belongs a thickness varying from 100 to 300nm, a silica-enriched phase is located at the corrosion layer-substrate interface after overcharge conditions. This phenomenon leads to the decrease of the corrosion current during the electrochemical test. These positive trends could be coupled by manufacturing a new lead-based composite allowing to substitute the actual alloys. So, this composite could be composed of a substrate elaborated by hot compaction of lead-based powders synthesized from the reduction lead salts route. This sintered substrate should be covered by a silica layer obtained by PECVD. Such composite should have the mechanical properties of the substrate and electrochemical resistance of the coating.

6. References

Albert L., Chabrol A., Torcheux L., Steyer Ph., Hilger J.P., Improved lead alloys for lead/acid positive grids in electric-vehicle applications. *J. Power Sources*, 67, (1997) pp.257-265.

Bagshaw, N.E., Lead alloys: past, present and future. *J. Power Sources*, 53, (1995) pp.25-30.

Balan L., Schneider R., Billaud Denis, Fort Y., Ghanbaja J., A new synthesis of ultrafine nanometre-sized bismuth particles, *Nanotechnology*, 15 (2004) pp.940–944.

Balaz P., Godocikova E., Takacs, L., Skorvanek, I., Choi, W.S., Kim, B.S, Preparation of nanocrystalline metals by mechanochemical reaction of sulphides with reducing element. *Proc. of the Korea-Japan Int. Symposium on Powder Science and Technology*, pp.237-241, Busan, Korea, November 2005, 2005, Busan.

Ben David T., Lereah Y., Deutscher G., Kofman, R. Cheyssac P., Solid-liquid transition in ultra-fine lead particles. *Phil. Mag. A*, 71 (1995) pp.1135-1143.

Bouirden L., Hilger J.P., Hertz J., Discontinuous and continuous hardening processes in calcium and calcium—tin micro-alloyed lead: influence of 'secondary-lead' impurities. *J. Power Sources*, 33 (1991) pp.27-50.

Bourguignon G., Maître A., Rocca E., Steinmetz J., Torcheux L. Electrochemical study in sulphuric acid of the hardening $L1_2$ phases of Pb–Ca–Sn alloys. *J. Power Sources*, 113 (2003) pp.301-307.

Brochin F., Devaux X., Ghanbaja J., Scherrer H., Study of BiSb-SiO_2 nanocomposite powders produced by an arc plasma processing. *NanoStructured Materials*, 11(1999) pp. 1-8.

Bui, N., Mattesco P., Simon P., Steinmetz J., Rocca E., The tin effect in lead-calcium alloys. *J. Power Sources*, 67 (1997) pp.61-67.

Cartigny Y., Fiorani J.M., Maître A., Vilasi M., Pb-based composites materials for grids of acid battery. *Materials Chemistry and Physics*, 103 (2007) pp.270-277.

Cooper A., The Brite–EuRam lead-acid electric-vehicle battery project—progress report. *J. Power Sources*, 73 (1998) pp.127-145.

Dailly A., Schneider, R., Billaud D., Fort Y., Ghanbaja J. Nanometric antimony powder synthesis by activated alkaline hydride reduction of antimony pentachloride, *J. Nanoparticle research,* 5 (2003) pp.389-393.

Devaux X., Brochin F., Dauscher A., Lenoir B., Martin-Lopez R., Scherrer H., Scherrer S., Production of ultrafine powders of Bi-Sb solid solution. *NanoStructured Materials*, 8 (1997) pp.137-147.

Giess, H.. The influence of calcium, tin and grid thickness on corrosion induced grid growth. *J. Power Sources,* 53, (1995) pp.31-43.

Gifkins R.C., Snowden, K.U., Stress sensitivity of creep of lead at low stresses. *Trans. AIME*, 239 (1967) pp.910-915.

Hilger J.-P., Bouirden L., New representation of the hardening processes of lead alloys by transformation-time-temperature (TTT) diagrams. *J. Alloys and Compounds*, 236 (1996), pp.224-228.

Jiang W., Yatsui K., Pulsed wire discharge for nanosize powder synthesis. *IEEE Transactions on Plasma Science*, 26 (1998) pp.1498-1501.

Jobst, A. Grahl, W. Gruner, Warlimont H. Hardening and Smoothing effects of Lignin Sulfonates on the Electrodepostion of Lead. *J. Applied Electrochemistry*, 27 (1997) pp.455-461.

Knowles L.M., Thermal analysis of the system PbBr2-PbO, *J. Chem. Phys.*, 19 (1951) pp.1128-1130.

Maître, A., Bourguignon G., Fiorani J.-M., Ghanbaja J., Steinmetz J. Precipitation hardening in Pb–0.08wt.%Ca–x%Sn alloys—the role of the pre-ageing. *Materials Science and Engineering A*, 358 (2003) pp.233-242.

Miraglio L., Albert L., Ghachcham A. El., Steinmetz J., Hilger J.P., Passivation and corrosion phenomena on lead-calcium-tin alloys of lead/acid battery positive electrodes. *J. Power Sources*, 53 (1995) pp.53-61.

Reppich B., On the dispersion strenthening mechanisms in ODS materials, *Zeitschrift für Metallkunde*, 93 (2002) p.7.

Régnier C., Tristant P., Desmaison J., Remote microwave plasma-enhanced chemical vapour deposition of insulating coatings (SiO_2) on metallic substrates: film properties, *Surface and Coatings Technology*, 80 (1996) 18-22.

Roberts David H., Extruded Lead strengthened by dispersed oxide, *GB patent n°970259* (1964).

Subramanian R., Ramachandran S., Electrodeposition of Lead-Tin alloy for Use as Anodes, *Metal Finishing*, (1996) pp.53-56.

Takahashi K., Yasuda Y., Hagegawa H., Horie S., Karatsuki K., Eight years of experience with valve-regulated batteries for automotive use. *J. Power Sources*, 53 (1995) pp.137-141.

Tilman M.M., Crosby R.L., Desy D.H., Dispersion strengthening of lead by coprecipitation, *U.S., Bur. Mines, Rep. Invest* (1971).

12

Polymer Nanocomposites: Synthesis and Physical Properties

Gleb Yu. Yurkov[1,2], Alexandr S. Fionov[1,3], Oleg V. Popkov[1],
Igor D. Kosobudskii[4], Nikolay A. Taratanov[5] and Olga V. Potemkina[5]
[1]Baikov Institute of Metallurgy and Materials Science, Russian Academy of Sciences,
[2]Moscow Power Engineering Institute (Technical University),
[3]Institute of Radio Engineering and Electronics, Russian Academy of Sciences,
[4]Department of Chemistry, Saratov State Technical University,
[5]Ivanovo Institute of State Fire Service of EMERCOM of Russia
[1,2,3,4,5]Russia

1. Introduction

Development of the science of metal-containing nanoparticles, including those stabilized in polymer matrices, is stimulated by the steady growth of interest paid to this problem in many areas of chemistry, physics, and materials science (Pomogailo et al., 2004; Gubin et al., 2005, a; Gubin et al., 2005, b; Scomski, 2003; Hyeon, 2003; Lin & Samia, 2006; Gubin et al., 2002; Ushakov et al., 2005).

As is well known, materials based on metal-containing nanoparticles are thermodynamically unstable (Gubin, 2000). Isolation within a matrix is a prospective method for stabilization of nanoparticles. Various polymers can be used for this purpose, e.g. polyethylene (Gubin et al., 2005, b; Zanetti & Costa, 2004), polypropylene (Gubin et al., 2005, b; Zanetti et al., 2001), polytetrafluoroethylene (Korobov et al., 2004; Yurkov et al., 2006), etc. These polymers have relatively high thermal stability, unique rheologic properties, high dielectric strength, they are chemically inert and processable, which provides means for fabrication of objects of desired form and dimensions. Methods for syntheses of these polymers are well elaborated.

Most polymer materials are good dielectrics with stable physical and chemical properties (Xia et al., 2006). Chemical stability of polymers allows using them in severe conditions (Hong et al., 2003). Modification of polymers with various inorganic fillers is used for altering their mechanical, electric, and other operational properties (Brosseau et al., 2001). The intensity of these properties' change depends on the nature, composition, morphology, and concentration of filler, as well as parameters of the flow process of filler's introduction into the matrix.

Nanoscaled fillers are especially interesting for creation of composite materials based on polymer matrices, as the first have a number of properties which are not specific for bulk materials (Kawasumi et al., 1997; Wang & Herron, 1991; Siegel, 1993; Gleiter et al., 2001; Gubin et al., 2003). Supposedly, introduction of metal-containing nanoparticles into polymers and variation of their composition, concentration, and sizes would provide means for controlling mechanical, electrical, magnetic, and other properties of composite nanomaterials produced this way (Gangopadhyay & De, 2000).

Herein we present results of a few years of studies concerned with development, creation, and investigation of composite nanomaterials based on metal-containing nanoparticles embedded in the low density polyethylene (LDPE) matrix. The selection of LDPE as the stabilizing matrix was made due to the material's low cost and well-developed production technology. Besides that, LDPE can be readily mixed with both organic and inorganic fillers (Pomogailo et al., 2004; Gubin et al., 2005, b; Yurkov et al., 2006; Xia et al., 2006; Hong et al., 2003; Hong et al., 2005). LDPE is a thermoplastic polymer, which allows using it for making objects of desired form and dimensions in mild conditions. Due to the abovementioned properties, LDPE is widely used in electrical engineering and radio engineering industries. Nanoparticles containing iron, cobalt, copper, cerium oxide and semiconductor CdS were used as the fillers in the course of this work.

2. Experimental

The general protocol for synthesizing nanoparticles embedded in a polymer matrix is as follows. A metal-containing precursor solution is added to the rapidly stirred LDPE-oil solution-melt at 270-300°C (Gubin et al., 2005, b; Gubin et al., 2003). Argon is supplied to the reactor during all the synthesis for removing air and gaseous reaction products (fig. 1).

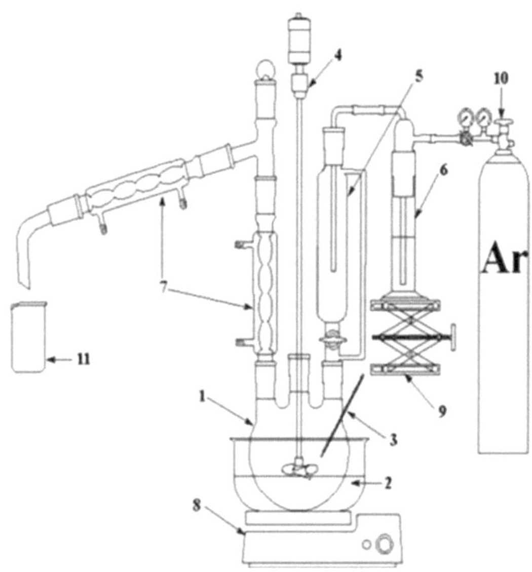

Fig. 1. Schematics of the laboratory setup for synthesis of composite materials based on metal nanoparticles and organic polymer matrices: 1 - glass reactor (3- or 4-necked flask); 2 - bath; 3 - thermometer; 4 - high-speed electric stirrer; 5 - dropping funnel with backpressure; 6 - bubbler with pyrogallol; 7 - condenser; 8 - electric heating mantle; 9 - lab jack; 10 - argon source; 11 - waste accumulation

Formates, acetates, oxalates, and carbonyls of iron and cobalt were used as the metal-containing precursors. Copper acetate, zinc and cadmium thioacetate complexes were used for syntheses of semiconductor nanoparticles; cerium nitrate and complex compounds of

cerium were used for syntheses of ceria nanoparticles. The synthesis of CdS nanoparticles was performed in two stages. First, $CdCl_2$ nanoparticles were prepared in LDPE-oil solution melt via dropwise addition of an acidified cadmium chloride solution. The second stage was chemical modification of the formed cadmium chloride nanoparticles via their interaction with hydrogen sulphide. The latter was bubbled through the LDPE-oil solution-melt with cadmium-containing nanoparticles ($CdCl_2$) using argon as the carrier gas.
Reactions of thermal decomposition of the precursors used are listed below:

$$Fe(CO)_5 \rightarrow Fe + 5CO \text{ (the main reaction)},$$

$$Fe(CO)_5 \rightarrow Fe_2O_3 + CO + C,$$

$$Fe(CO)_5 \rightarrow FeO + CO + C,$$

$$Fe(CO)_5 \rightarrow FeC + CO + CO_2.$$

$$Fe(HCOO)_3 \rightarrow Fe + CO_2\uparrow + 2CO\uparrow + H_2O.$$

$$Fe(CH_3COO)_3 \rightarrow Fe_2O_3 + CO_2\uparrow + 2CO\uparrow + H_2O;$$

$$Co(HCOO)_2 \rightarrow Co + CO_2\uparrow + 2CO\uparrow + H_2O.$$

$$Co(CH_3COO)_2 \rightarrow CoO + CO_2\uparrow + 2CO\uparrow + H_2O.$$

$$Cu(CH_3COO)_2 \rightarrow Cu + 2CO + H_2 + 2CH_2O$$

$$CdCl_2 + H_2S \rightarrow CdS + 2HCl\uparrow$$

$$Ce(NO_3)_3 \cdot 6H_2O \rightarrow CeO_2 + H_2O + NO_2\uparrow;$$

$$(Et_4N)_2Ce(NO_3)_6 \rightarrow CeO_2 + H_2O + N_2O\uparrow + CO_2\uparrow;$$

$$(NH_4)_2Ce(NO_3)_6 \rightarrow CeO_2 + H_2O + N_2O\uparrow + O_2\uparrow;$$

$$[Me(CH_3COO)_2 \cdot (NH_2)_2CS \cdot H_2O] \rightarrow MeS + \text{thiocarbamide decomposition products}$$

It is worthwhile to say that side reactions can take place during a synthesis, which will have effect on chemical composition of resulting nanoparticles. Besides that, nanoparticles of like composition can be prepared using different precursors.
The samples produced were separated from oil with benzene using a Soxhlet extractor and were dried in vacuum afterwards. All the samples were stored in air. The materials produced were coloured powders, with colours dependant on nanoparticles' sizes and composition.
Samples in the form of tablets were produced via hot pressing in a steel die using a hand screw press (6 kN). The pressing temperature was selected in such a way as to transform all the powder to a plastic mass avoiding its thermal destruction. Tablets 12…25 mm in diameter and 0.25…3 mm thick were prepared this way.
The protocol described was used for preparation of nanocomposites based on LDPE containing nanoparticles (Kosobudsky & Yurkov, 2000; Yurkov et al., 2001; Yurkov et al., 2002; Yurkov et al., 2007; Fionov et al., 2008, b; Taratanov et al., 2009, a; Popkov et al., 2009;

Elkin & Yurkov, 2009; Taratanov et al., 2009, b; Volkov et al., 2010; Yurkov et al., 2009; Taratanov et al., 2010).

Structure, composition, and physical properties of the synthesized nanocomposite materials were studied using a set of instrumental methods (transmission electronic microscopy (TEM), extended X-ray absorption fine structure (EXAFS) spectroscopy, electron paramagnetic resonance (EPR), X-ray powder diffraction (XRD), etc.). Electrophysical properties of the composites prepared are discussed below.

3. Electrophysical properties of nanocomposites

3.1 Electrophysical properties of composites containing magnetic nanoparticles

Five samples have been synthesized for conducting these studies. Samples 1 and 2 were prepared from $Fe(CO)_5$ and contain different amounts of iron-containing nanoparticles embedded in low density polyethylene. The nanoparticles are comprised of three phases: metallic iron, iron oxide and carbide. Sample 3 was prepared from iron (III) formate $Fe(HCOO)_3$ and contains nanoparticles which are mainly comprised of Fe_3O_4. Sample 4 was synthesized using iron (II) oxalate $FeC_2O_4{\cdot}2H_2O$; nanoparticles in it are comprised of iron oxide Fe_3O_4 and ε-Fe. Sample 5 was prepared via thermal destruction of cobalt (II) acetate and contains nanoparticles with complex structure; the nanoparticles are comprised of CoO, Co, and Co_3O_4 (Gudoshnikov et al., 2003; Yurkov et al., 2007, b).

Volume resistance, ρ_V, and permittivity, ε, values of the abovementioned samples are listed in tables 1 and 2. Figures 2, 3 depict dependencies of ρ_V and ε of the samples with Fe-containing nanoparticles as functions of concentration, *C*. The sample with $C = 0$ is low density polyethylene with no filler which was subject to all operations involved in preparation of the abovementioned samples.

Sample	C, wt. %	ρ_V, Ω·m at voltage		
		10 V	100 V	1000 V
Polyethylene	0	-	$4.9{\cdot}10^{14}$ ±25%	$3.9{\cdot}10^{14}$ ±5%
Sample 1	5	$1.4{\cdot}10^{14}$ ±50%	$1.2{\cdot}10^{14}$ ±25%	$0.9{\cdot}10^{14}$ ±5%
Sample 2	10	$5.5{\cdot}10^{13}$ ±25%	$5.5{\cdot}10^{13}$ ±10%	$4.25{\cdot}10^{13}$ ±5%
Sample 3	20	$4.5{\cdot}10^{12}$ ±25%	$3.0{\cdot}10^{12}$ ±10%	$1.0{\cdot}10^{11}$ ±5%
Sample 4	20	$9.5{\cdot}10^{13}$ ±25%	$5.1{\cdot}10^{13}$ ±10%	$1.5{\cdot}10^{13}$ ±5%
Sample 5	20	$2.8{\cdot}10^{13}$ ±25%	$1.6{\cdot}10^{13}$ ±10%	$5.5{\cdot}10^{12}$ ±5%

Table 1. Volume resistance of composite nanomaterials

Sample	C, wt. %	ε (1 kHz)	ε (1 MHz)	*d*, nm
Polyethylene	0	2.94	2.70	-
Sample 1	5	2.98	2.69	4.9
Sample 2	10	3.63	3.32	5.1
Sample 3	20	4.52	3.56	11.5
Sample 4	20	3.67	3.37	2.4
Sample 5	20	3.96	3.01	8.3

Table 2. Permittivity of composite nanomaterials

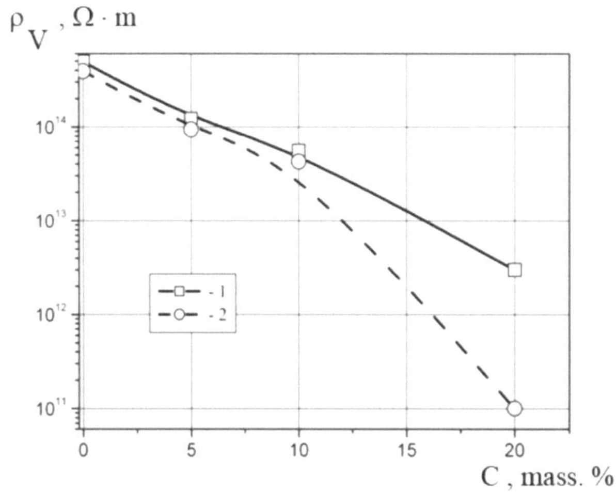

Fig. 2. Volume resistance of samples 1, 2, 3 as a function of nanoparticles weight concentration at different voltage: 1 – 100 V; 2 – 1000 V

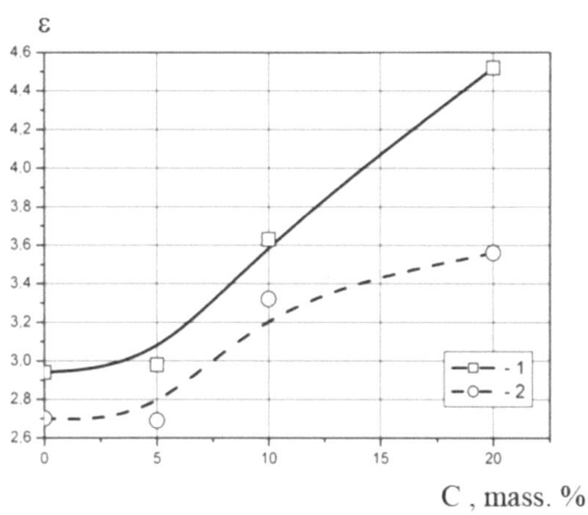

Fig. 3. Permittivity of samples 1, 2, 3 as a function of nanoparticles weight concentration at low frequencies: 1 – 1 kHz; 2 – 1 MHz

DC volume resistance of the samples (10^{11} - 10^{15} Ω·m) steadily lowers as the filler concentration *C* grows; the values are in the range typical for dielectrics. The lower volume resistance of the materials compared to that of the "empty" LDPE it is a result of polyethylene structure change towards higher amorphism and porosity which results in higher molecular mobility and lower ion dissociation energy in the polymer. This is also the reason for the steady growth of permittivity which occurs as the filler concentration increases.

The difference between volume resistance and permittivity values of the processed unfilled LDPE and the "normal" LDPE (10^{16} - 10^{18} Ω·m, 2.2-2.3) can be attributed to a higher amount of admixtures as a result of the treatment.

Volume resistance of samples with 5% and 10% wt. nanoparticles (samples 1 and 2), as well as that of the "empty" polyethylene, weakly depends on the field applied and linearly decreases as concentration of nanoparticles increases within all the operating field strength range.

As concentration of nanoparticles reaches 20% wt., dependence of volume resistance as a function of applied voltage becomes more pronounced. Samples 4 and 5 have 3 times less volume resistance at 1000 V than at 100 V, whereas the resistance drop of sample 3 is more than 10 times at the same voltages.

Additional contribution to admittance current can be made by polarization phenomena caused by nanoparticles' polarization: macro-shift polarization (within a sample) and micro-shift polarization (within some sites), as well as slowly developing dipole polarization.

Conductivity of filled LDPE depends on both composition and size of nanoparticles it contains. It is well illustrated by the lower volume resistance of sample 3 compared to that of samples 4 and 5 at 1000 V.

The steady growth of permittivity is caused by the contribution of metal-containing nanoparticles' polarization. The nanoparticles have more loose electron shells and hence higher polarizability compared to those of the polymer matrix, which results in the growth of permittivity ε as the filler concentration increases.

The difference between permittivity values of samples 3 and 4 which have equal nanoparticles concentrations is due to the higher susceptibility of larger particles.

Results of permittivity and electromagnetic radiation absorption measurements performed at microwave frequencies for samples with different filler concentrations are given in figs. 4 and 5.

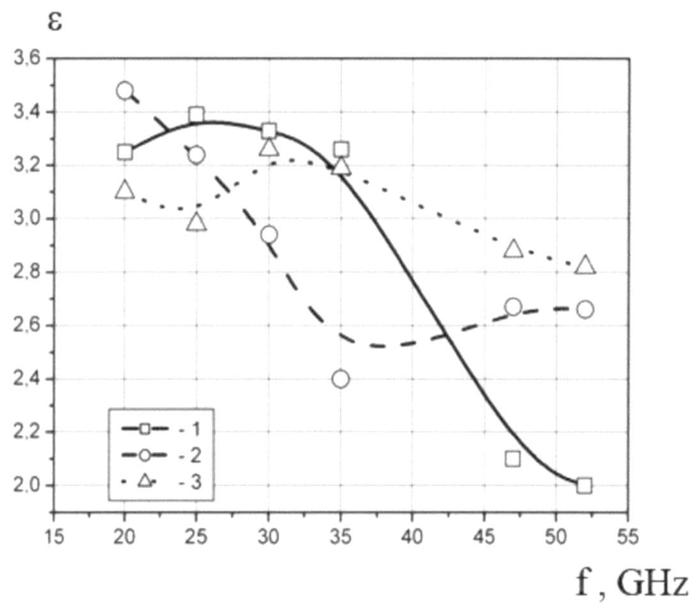

Fig. 4. Permittivity as a function of microwave radiation frequency: 1 –sample 1; 2 - sample 2; 3 - sample 3

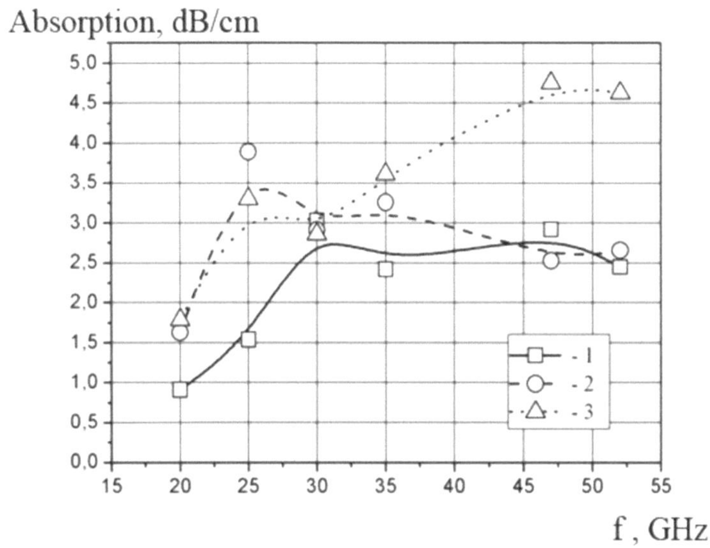

Fig. 5. Microwave absorption as a function of radiation frequency: 1 - sample 1; 2 - sample 2; 3 - sample 3

The data on permittivity and absorption in the microwave range can be used for estimating proportion of amplitude and time characteristics of relaxation processes. The decrease of permittivity and almost constant dielectric losses of samples 1 and 2 can be explained by decrease of relaxation processes' intensity with the relaxation rate unchanged. The increase of nanoparticles' concentration to 20% wt. (sample 3) results in decrease of relaxation time, but relaxation intensity doesn't change. There is a tendency towards increase of both dipole relaxation process intensity and rate as concentration of nanoparticles in the polymer grows. Just as in case of low frequencies, the dependencies mentioned above can be attributed to polarization of metal-containing nanoparticles whose susceptibility increases along with their size. The results of microwave measurements indicate that the samples produced have steady permittivity and absorption in a wide frequency range within the band studied.

Samples containing 30% wt. iron- or cobalt-containing nanoparticles have low volume resistance ($\approx 10^2\ \Omega\cdot$m). Nanoparticles in these samples are relatively large ($d_{av} \approx 8$ nm) and have well-formed metallic core. Electromagnetic radiation power reflection and extinction coefficients at 30 GHz are given in table 3 (Fionov et al., 2008, a).

Sample composition	C, % wt.	K_{ref}	K_{ext}, dB/cm
LDPE+Co_xO_y	30	0.10	77
LDPE+Fe_xO_y	30	0.53	45

Table 3.

Application of such composite materials in microwave equipment as nonlinear distribution elements (fillers of waveguide ducts and resonators, thin-film coats, electromagnetic screens, etc.) would lead to creation of a number of new devices for electromagnetic signal transformation and elements of active stealth technology.

3.2 Study of complex magnetic and dielectric permittivity at microwave frequencies

Samples prepared for measuring complex magnetic and dielectric permittivity at microwave frequencies using the cavity method were in the form of powders (granule size less than 300 μm) put in capillary vessels.

Among the materials with no more than 20% wt. nanoparticles, the one with 20% wt. Co-containing nanoparticles has the highest values of $\varepsilon' \approx 6.5$, $\varepsilon'' \approx 0.11$, $\mu' \approx 1.6$, and $\mu'' \approx 4.9 \cdot 10^{-3}$ at 6.5 GHz.

Results of measurements performed for powders containing Fe_3O_4 are given in fig. 6. Dielectric and magnetic permittivities, as well as dielectric and magnetic losses, grow as concentration of Fe_3O_4 increases. The highest values of ε' and μ' are observed for nanocomposites with 63% wt. Fe_3O_4 at the long-wave part of the studied frequency range (25.0 and 4.5, respectively); the highest dielectric and magnetic losses are detected at 5-12 GHz ($\varepsilon''_{max} \approx 10$) and 3-6 GHz ($\mu''_{max} \approx 2,7$), respectively.

As magnetic properties (μ' and μ'') of the nanocomposites are mainly determined by magnetic properties of nanoparticles, whereas dielectric properties are affected by concentrations of free and bound charges in the whole system, the unsteady behaviour of ε' and ε'' in case of different concentrations of metal-containing nanoparticles at different frequencies can be attributed to uncontrolled changes of composite structure.

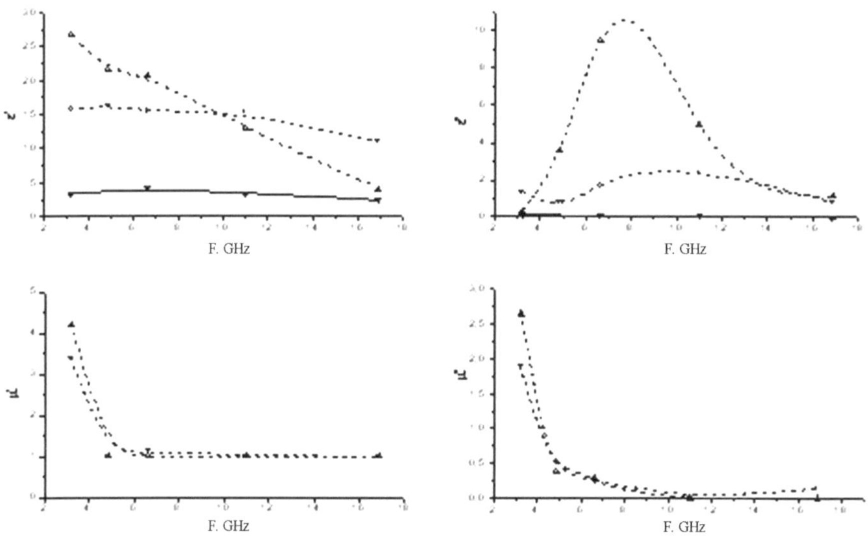

Fig. 6. Effect of Fe_3O_4 weight concentration of ε^* and μ^* of a composite as functions of frequency: —▼— 20% Fe_3O_4; - -○- -30% Fe_3O_4; - -△- -63% Fe_3O_4

For this reason, the influence of synthesis conditions on phase composition of metal-containing nanoparticles and electrophysical properties of composites has been investigated. Three types of materials have been studied: the initial one, annealed in inert atmosphere, and annealed in air.

Basing on Mössbauer spectra, nanoparticles in the initial sample with 63% wt. Fe_3O_4 were found to contain 11% wt. α-Fe with Fe_3O_4 being the rest. Concentration of α-Fe has dropped to 4% wt. after annealing in inert atmosphere. Nanoparticles in samples prepared by annealing in air were entirely comprised of γ-Fe_2O_3.

XRD data revealed some ordering of nanoparticles structure after annealing, perhaps as a result of their recrystallization.
The influence of the additional thermal treatment and oxidation of the initial composite which contained 63% wt. Fe_3O_4 can be seen in fig. 7. The treatment in inert atmosphere leads to increase of dielectric permittivity and both magnetic and dielectric losses in the composite.
The most prominent result of the additional oxidation of Fe_3O_4 nanoparticles to Fe_2O_3 is the decrease of dielectric losses and ε′. Magnetic losses also decreased a bit at 3 - 7 GHz.

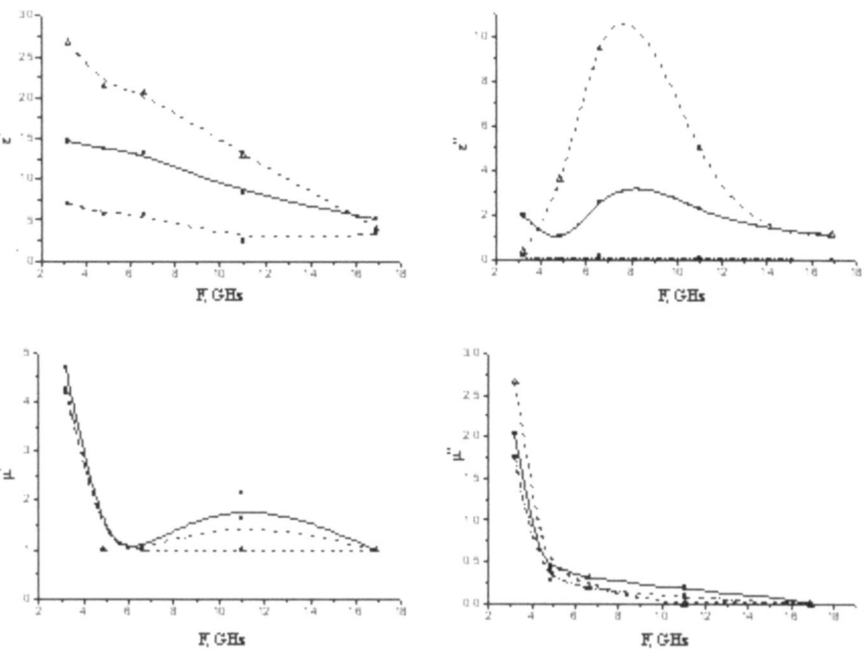

Fig. 7. Influence of the additional annealing in inert atmosphere and in air on properties of the composite with 63% wt. Fe_3O_4: —■— initial sample; - -△- - sample annealed in inert atmosphere; - · □ · - sample annealed in air

According to the percolation theory, dependencies of conductivity and dielectric permittivity of composites with disperse phase as functions of filler concentration can be described as (Efros & Shklovskii, 1976):

$$\sigma \sim \sigma_m(v_f^* - v_f)^{-q}$$

$$\varepsilon' \sim \varepsilon_m(v_f^* - v_f)^{-q}, \qquad v_f < v_f^*,$$

where σ_m is the specific conductivity of the polymer matrix;
ε_m is the dielectric permittivity of the polymer matrix;
v_f is the volume concentration of the filler;
v_f^* is the percolation threshold.
Therefore, conductivity and dielectric permittivity of a composite with a conducting filler are exponential functions, which provides means for explanation of ε' and ε'' dependencies as functions of nanoparticles concentration.

As it follows from the results of studies of the influence of thermal treatment on ε' and ε'', these values are mainly determined by the inner structure of nanoparticles or ratio of phases they are comprised of, which can have a significant effect on concentrations of free and bound charges in them. Contribution of charge states on both the surface of nanoparticles and metal-polymer interface can be another important factor.

Magnetic losses in these materials as a function of frequency can be attributed to the resonance of domain borders in single-domain particles and natural ferromagnetic resonance..

3.3 Electrophysical properties of composites containing CdS nanoparticles

Nanocomposites based on semiconductor nanoparticles stabilized in LDPE matrix are of special practical interest (Ushakov et al., 2006; Ponomareva et al., 2007, a; Kul'batskii et al., 2009; Radchenko et al., 2009). Electrophysical properties of a series of such materials containing 20% wt. CdS nanoparticles have been studied. The samples differed in average sizes of nanoparticles which were in the range from 4.9 to 19.5 nm.

Electric conductivity values (σ) of the samples measured at 300…420 K were in the range of 0.01÷0.54 μS. A dependence of σ as a function of reciprocal temperature during heating and cooling of a nanocomposite with CdS nanoparticles 4.9 nm in size is depicted in fig. 8 (the arrows indicate the process direction). A small hysteresis of σ is observed during cooling, which is obviously connected to reorganization of the dielectric matrix material. The temperature hysteresis of electric conductivity is a new property of the nanocomposite medium. Also, two dramatic changes of conductivity are observed during heating at *ca.* 75 and 95^{0}C. They can be probably attributed to a change of energy of charge carriers, which enables the latter to overcome potential barriers in the nanocomposite material.

Fig. 9 depicts a dependence of the real part of dielectric permittivity ε as a function of temperature during heating and cooling (curve 1). The permittivity value increases during heating. Interestingly, this process continues during cooling after the temperature maximum has been reached. As a result, the value of ε changes after the heating-cooling loop and doesn't change for some time (for at least 7 days). If the heating-cooling cycle is repeated, capacitance of the nanocomposite medium will increase, but this time the change will only be 10-20% of that after the first loop. The hysteresis becomes hardly observable starting from the 3rd or 4th loop. This behaviour of the composite material is obviously connected to the thermally stimulated process of the dielectric matrix structure rearrangement and, as a result, emerging of metastable states of the medium. The same dependence for the "empty" LDPE is depicted by curve 2 in fig. 9. In general, the behaviuor of the dependence is almost the same as in case of the metal-polymer nanocomposite; the only prominent difference is the much smaller (at least, 2 times smaller) temperature susceptibility of dielectric permittivity of LDPE. Therefore, impregnation of nanoparticles into a polymer matrix makes dielectric permittivity of the resulting material more susceptible to temperature changes, retaining the nature of its dependence as a function of temperature, in general. Permittivity ε of nanocomposites based on LDPE with 20% wt. CdS nanoparticles is 5.60±0.32 at 35°C.

Fig. 10 depicts temperature dependencies of the relative value of dielectric permittivity of nanocomposite samples with 20% wt. CdS which differ in average sizes of the nanoparticles. Curves 1, 2, and 3 in fig. 10 correspond to the average sizes of 4.9, 5.6, and 6.2 nm, respectively. As it follows from the dependencies acquired, size of nanoparticles has little effect on dependencies of relative permittivity as a function of temperature during cooling.

However, a prominent effect is observed during heating. These dependencies are congruent for all the studied materials up to 50°C. As temperature rises further, the growth rate of susceptibility to temperature becomes higher in case of smaller particles. It's worthwhile to note that size of nanoparticles has almost no effect on the relative change of dielectric permittivity during heating and cooling.

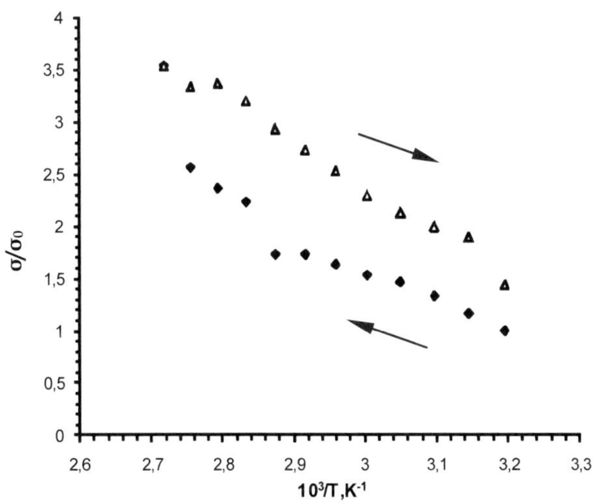

Fig. 8. Conductivity of a nanocomposite with 20% wt. CdS in LDPE as a function of reciprocal temperature

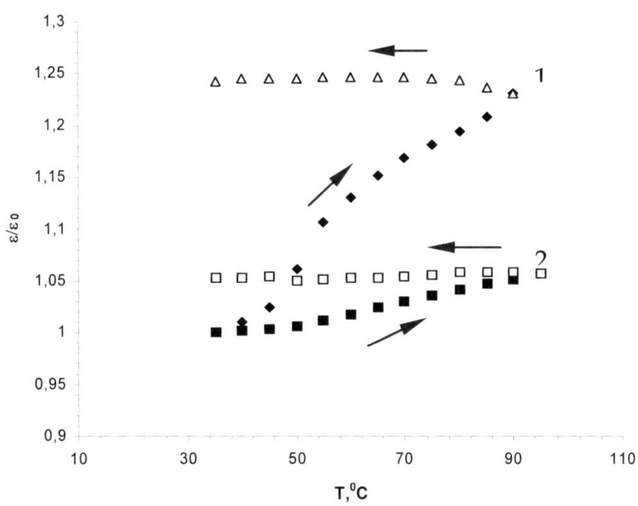

Fig. 9. The real part of dielectric permittivity as a function of temperature measured for: 1. nanocomposite comprised of LDPE and 20% wt. CdS nanoparticles; 2. - LDPE

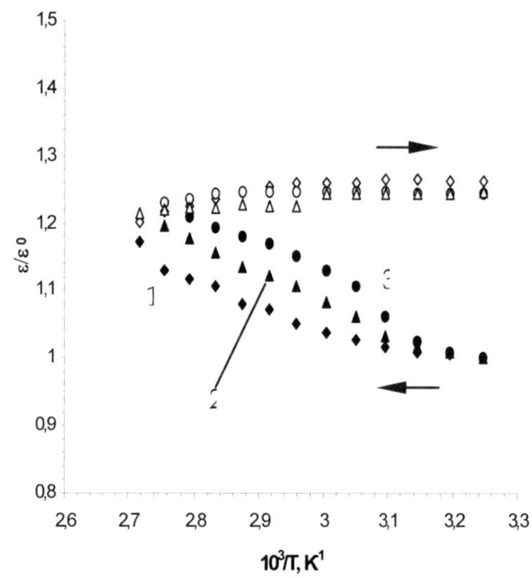

Fig. 10. Relative dielectric permittivity values as functions of temperature measured for nanocomposites with 20% wt. CdS nanoparticles in LDPE with the average size of nanoparticles equal to: 1) 4.9 nm; 2) 5.6 nm; 3) 6.2 nm

Changes of dielectric loss factor induced by heating and cooling have been estimated. The relative change of dielectric loss factor after the whole cycle was found to be 1.79. The value of dielectric loss factor of the samples studied was 0.0088 at 35°C (which was the initial temperature).

Fig. 11 depicts dependences of dielectric relaxation time as functions of temperature measured during the heating-cooling process. Heating leads to a monotonous growth of dielectric relaxation time. Also, a small temperature hysteresis of the relaxation time values is observed.

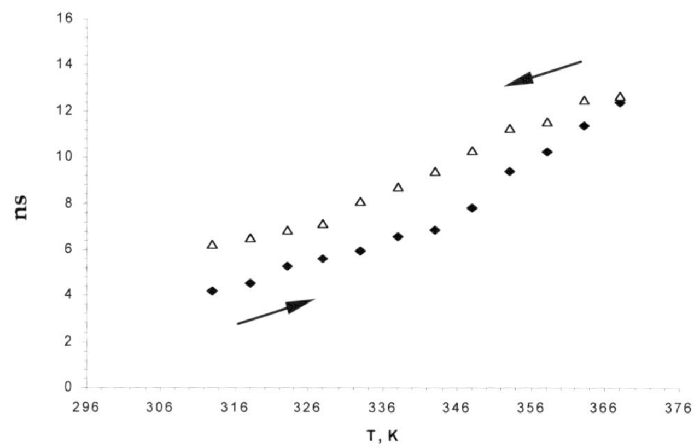

Fig. 11. Dielectric relaxation times of samples comprised of 20% wt. CdS in LDPE as functions of temperature

Dielectric relaxation times did not exceed 20 ns.
Fig. 12 depicts relative activation energy of relaxation processes as a function of temperature during heating and cooling of the sample containing 20% wt. Co. A common feature of the dependences measured is the growth of activation energy during heating and its decrease during cooling, which corresponds to a change of charge carrier energy with temperature changing and, therefore, a shift of Fermi level within the band gap of the material. It may result in release of trapped charges, therefore changing dielectric peoperties of a medium. It's worthwhile to note that this dependence is almost linear. The range of dielectric relaxation activation energy change was found to be 10 - 30 kJ/(K·mol).

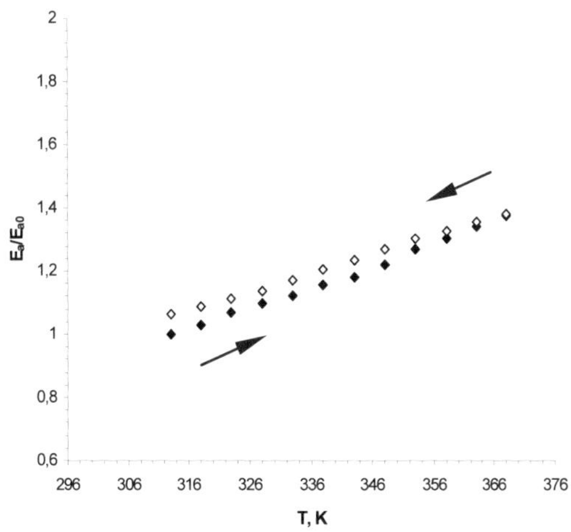

Fig. 12. Relative activation energy of dielectric relaxation processes in samples comprised of 20% wt. CdS nanoparticles in LDPE as a function of temperature

Therefore, the measured kinetic and energy parameters of temperature dependences in composite nanomaterials based on LDPE containing 20% wt. CdS nanoparticles allow qualitatively describe energy spectra of such materials. In our assumption, they are a set of mini-zones in the band gap near the top of the valent zone. In this case, the behaviour of dependencies of the parameters of the composites studied as functions of temperature can be attributed to charge carrier transitions between those mini-zones (Ul'zutuev et al., 2009). Further investigations revealed that such hysteresis behaviour of dielectric permittivity, dielectric losses, and time of dielectric relaxation as functions of temperature are also inherent to composite nanomaterials comprised of ceria nanoparticles in LDPE.

3.4 Electrophysical properties of composites containing Cu@Cu_2O nanoparticles

Fig. 13 depicts dependences of relative values of specific conductivity (G_x/G_o, curve 1), permittivity ($\varepsilon_x/\varepsilon_0$, curve 2), and dielectric loss coefficient (curve 3) as functions of nanoparticle concentration recorded at 1 MHz and T = 300 K for samples comprised of Cu@Cu_2O nanoparticles in LDPE which have different concentrations of nanoparticles. The initial values of conductivity, permittivity, and dielectric loss coefficient of the Cu@Cu_2O/LDPE sample with low nanoparticles concentration (x < 10% wt.) were G_b = 1.1 µS/m, ε_b = 2.5, and tand_b = 0.015, respectively. As concentration of nanoparticles grows, dielectric permittivity increases up to 6 times, dielectric losses at 1 MHz increase up to 3 times, and specific conductivity becomes almost up to 1.5 times higher at 30 mass.% nanoparticles. However, conductivity decreases as concentration of nanoparticles rises from 10% to 20% wt.

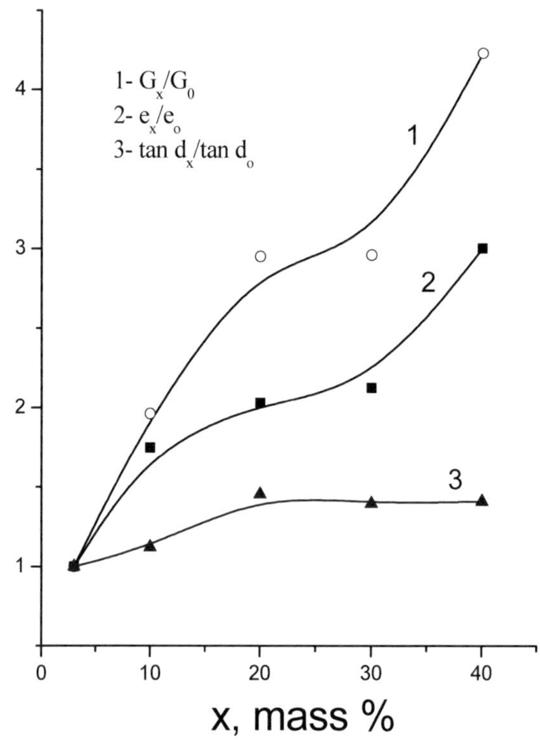

Fig. 13. Relative values of specific conductivity (G_x/G_o, curve 1), dielectric permittivity ($\varepsilon/\varepsilon_0$, curve 2), and dielectric loss coefficient (curve 3) as functions of nanoparticles concentration (x, wt.%) in composites comprised of $Cu@Cu_2O$ nanoparticles in LDPE

Studies of applications of Cu-containing nanocomposites in waveguide and resonator SHF devices as dielectric insertions revealed that such materials are prospective for controlling resonance frequencies of induced waves and field structure in such systems.

Logarithmic dependencies of current at 1 MHz as functions of reciprocal temperature measured for samples with 20, 25, and 30% wt. $Cu@Cu_2O$ in LDPE are given in fig. 14 (curves 1, 2, and 3, respectively).

Logarithmic dependences of current as functions of reciprocal temperature measured at 1 MHz for nanocomposites with different concentrations of copper-containing nanoparticles are depicted in fig. 14.

The current maximum (point A) shifts to higher temperatures as nanoparticles concentration grows. Different signs of derivatives of these functions are obviously due to different mechanisms of charge carrier dissipation in the nanocomposites.

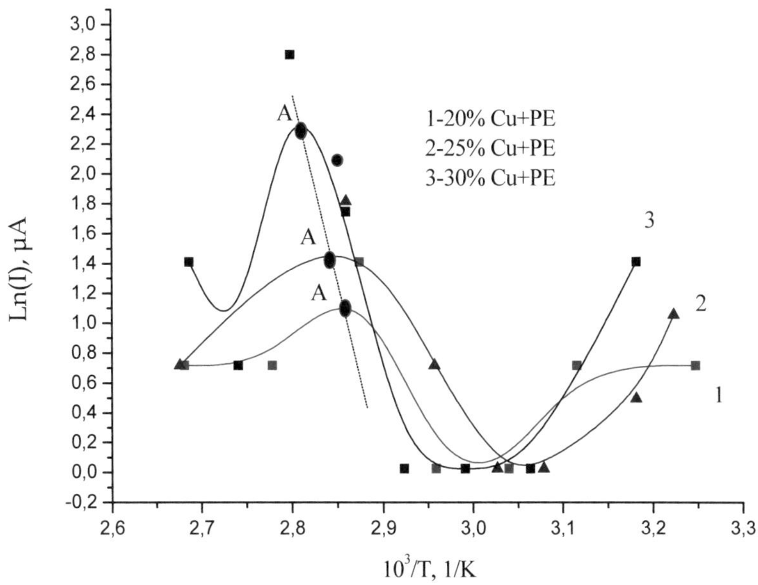

Fig. 14. Logarithmic dependences of current as functions of reciprocal temperature measured at 1 MHz for nanocomposites with different concentrations of copper-containing nanoparticles. Point A corresponds to a current maximum

3.5 Optical properties of the CeO_2 nanocomposites

The main measured values were the coefficient of reflection $R=I_r/I_o$ and coefficient of transmission $T = I_{out}/I_o$, where I_r, I_{out}, and I_o are the intensities of the reflected, transmitted and initial optical signals, respectively. Calculations of all the main spectral and optical properties of the samples were based on the measured dependencies R(λ) and T(λ). The "Kubelk-Munk" two-stream model was used as the experimental model; it has been successfully applied for optical investigations of biological environments.

The spectral dependencies of the optical absorption coefficient of the composite nanomaterials with the nanoparticle sizes ranging from 3.3 nm to 4.9 nm are shown in Figure 15. The measured dependences have absorption maxima shifted 15 nm far from each other. The maximum absorption corresponding to the larger nanoparticles is shifted towards the short wavelength region. It should be noted that the edge of the fundamental optical absorption of the low density polyethylene matrix is located in the short wavelength region (nm). It is known that the optical properties of composite materials mainly depend on the properties of excitons and polaritons. Excitons and polaritons are induced within the nanoparticles and at the nanoparticle-matrix boundaries (the Maxwell Wagner effect).

New aspects of exciton-radiation interaction are discussed from the viewpoint of microscopic nonlocal optical sponse developed for the study of nano-structures materials are reviewed in (Gorobinskii et al., 2007; Ushakov et al., 2008). Localization of excitons within the nanoparticles dramatically increases the binding energy (up to 20-30 meV) and the exciton oscillator strength. The increase of the binding energy is due to the decrease of the distance between electrons and holes which leads to intensification of the Coulomb interaction. Applying a matrix with small permittivity may increase the binding energy by an order of magnitude (up to 0.2-0.3 eV). The dielectric amplification caused by the

Coulomb energy concentration showed a large difference in the dielectric constants of nanoparticles in this matrix system. The measurement of the dielectric constant of the 10% wt. CeO_2-PE composite material showed its effective value to be 3.8±0.1 at a frequency of 1 MHz. The dielectric constant of LDPE is equal to 2.2±0.1. The collected optical absorption spectra (Figure 15) correspond to the exciton spectra of nanoparticles of different sizes. The coefficient of optical absorption K for the exciton transitions is determined by the following equitation [27]:

$$K = A\frac{8}{\pi^2}\left(\frac{\bar{a}}{a_{ex}}\right)^3\left(\frac{\hbar^2\pi^2}{M\bar{a}^2}\right)^{-1}\sum_n \frac{1}{n^2}\delta\left(\Delta + E_{ex} - \frac{\hbar^2\pi^2 n^2}{2M\bar{a}^2}\right) \quad (1)$$

where A is the variable proportional to the square of the modulus of the matrix dipole moment calculated based on the Blokhov functions; $\bar{a}$ is the average size of nanoparticles; a_{ex} is the exciton radius; $M = m_e + m_h$ is the total mass of an exciton (m_e is the mass of an electron, m_h *is* the mass of a hole); $\hbar$ is the reduced Planck constant; n is the number of electron levels ($n = 1$ is the lowest level); E_{ex} is the exciton energy; $\Delta = \hbar\omega - E_g$; $\hbar\omega$ is the photon energy; E_g is the band-gap energy.

According to Equation (1), the shift of the absorption maximum for the highest oscillation strength ($n = 1$) depends on the value of $M\bar{a}^2$; i.e., it depends on the mass and average radius of an exciton. In the case of ensembles of nanoparticles with the same average size, a decrease in their size without changing their mass shifts the absorption maximum towards the short-wave region of the spectrum (the so-called "blue shift"). In our case, the inverse situation occurred: the absorption maximum for the CeO_2 nanoparticles obtained from $(Et_4N)_2Ce(NO_3)_6$ with the size of 4.9 nm is shifted by 15 nm towards higher frequencies compared to the CeO_2 nanoparticles with the size of 3.3 nm obtained from $(NH_4)_2Ce(NO_3)_6$. Apparently, the shift can be explained by the fact that the mass of an exciton is noticeably lower for larger nanoparticles than for smaller ones. Hence, the multiplication product of the mass of an exciton by its radius gives a lower value for larger particles and results in a longer shift than for small nanoparticles. The widening of the optical absorption spectra at wavelengths longer than 550 nm observed for the nanoparticles with different average sizes (curves 1 and 2; Figure 15) is caused by the statistical size distribution of the particles.

The dispersion characteristics of the 10% wt. CeO_2-LDPE composite are shown in Figure 16. The index of refraction is noticeably higher for the smaller nanoparticles than for the larger ones. The exception is the point "**a**" (λ = 623 nm). The dispersion curve is very steep in the 400 - 500 nm region for the smaller nanoparticles, and this steepness is not observed for the larger nanoparticles.

The mean values of the index of refraction $<n_0>$ and the index of adsorption $<k>$ in the complex index of refraction ($n = n_0 - jk$) of the composite medium were calculated based on spectral data. For the pure polyethylene, these values are $<n_0> = 1.5$, and $<k> = 1.166 \cdot 10^{-5}$.

The dependencies of the index of refraction as subject to the nanoparticle size are shown in Figure 17 (a). As mentioned previously, composite nanomaterials are the media with the complex index of refraction $n = n_0 - jk$; where n_0 is the index of refraction, and k is the index of absorption.

The increase in the index of refraction with the decrease of the nanoparticle size can be explained by the growth of the polarization of the electric dipoles at the nanoparticle-matrix boundary (the Maxwell-Wagner effect) (see Figure 17).

The optical absorption, which is determined by the absorbing volume and the size of nanoparticles, increased along with an increase in concentration, in which case, the average value of the optical absorption coefficient should grow along with the size of the nanoparticles. Such behavior of both the optical refraction and absorption indexes is depicted in Figures 17 (b, c).

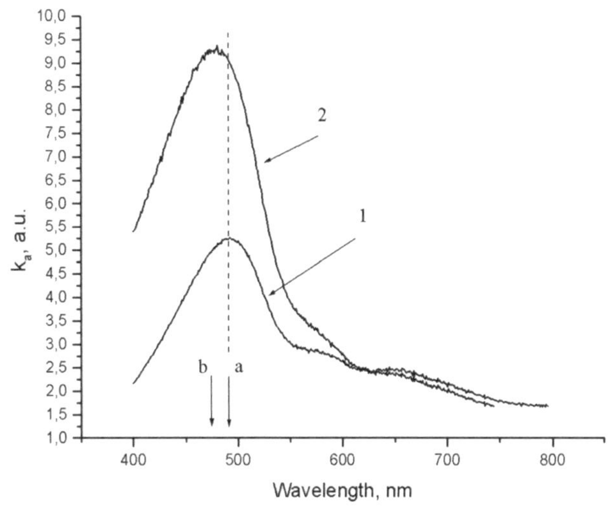

Fig. 15. Optical absorption in the 10% wt. CeO_2-LDPE composite nanomaterial: line 1 corresponds to the CeO_2 nanoparticles with the size of 3.3 nm; line 2 corresponds to the CeO_2 nanoparticles with the size of 4.9 nm. The points "a" and "b" are the maxima of the curves at 492 and 477 nm, respectively

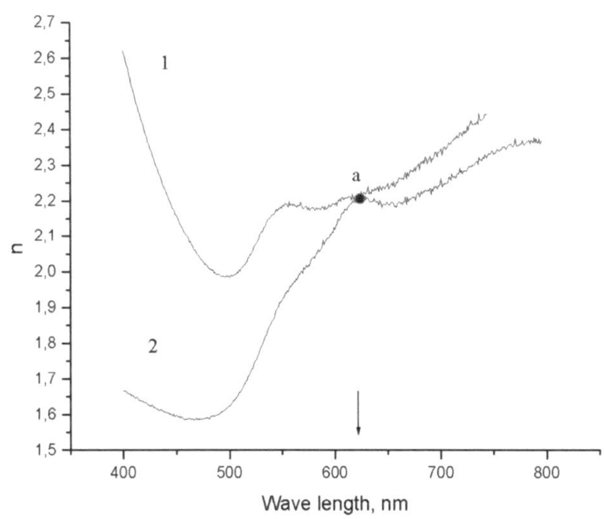

Fig. 16. The dispersion characteristics of the CeO_2-LDPE nanocomposite containing 10% wt. CeO_2 nanoparticles with different sizes. Curve 1: the average size of nanoparticles is 3.3 nm; curve 2: the average size of nanoparticles is 4.9 nm

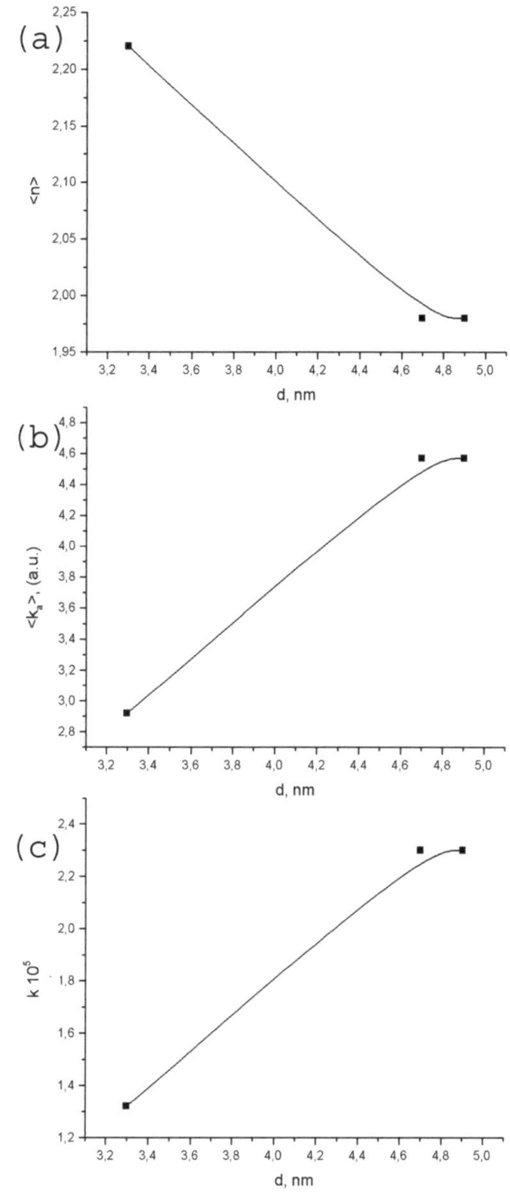

Fig. 17. Dependencies of the index of refraction (a), index of optical absorption (b), and absorption coefficient of the materials with nanoparticles having different average sizes (c)

Interesting optical properties of composites based on ZnO, CdS, and CuO nanoparticles were described elsewhere (Ushakov et al., 2006; Ponomareva et al., 2007, b; Yurkov et al., 2009; Kul'batskii et al., 2009).

4. Conclusion

1. Electrophysical properties of metal-polymer nanocomposites comprised of iron-, cobalt-, copper-containing or CdS nanoparticles embedded in low density polyethylene have been studied.
2. Measurements revealed that dielectric permittivity and electromagnetic radiation absorption of a nanostructured material grow as concentration of nanoparticles in the material increases.
3. Variation of size and concentration of nanoparticles has been found to change electrophysical properties of composite materials, namely: dielectric permittivity, volume resistance.
4. Microwave measurements characterize the samples studied as materials with steady absorption and permittivity in a wide band within the frequency range studied.
5. Dependences of dielectric permittivity, dielectric losses, and dielectric relaxation times of composite materials comprised of CeO_2 or CdS nanoparticles impregnated in LDPE have a hysteresis nature as functions of temperature.
6. Composites undergo additional polarization after cooling, and their dielectric permittivity increases by $\Delta\varepsilon$ (this value depends on nanoparticles size). $\Delta\varepsilon$ equals 1.3-1.4 for small nanoparticles (~5 nm) and lowers as nanoparticles size grows.

The composite medium changes its state stepwise during heating. The transition occurs at lower temperatures in case of smaller nanoparticles and longer times of dielectric relaxation, which is obvious because smaller nanoparticles are more chemically active. The higher the activation energy, the more rapid are the processes of dipole orientation in external magnetic field.

5. References

Brosseau, C.; Quéffélec, P. & Talbot, P. (2001). Microwave characterization of filled polymers. *J. Appl. Phys.* Vol. 89, No. 8, p. 4532-4540, ISSN 0021-8979.

Efros, A.L. & Shklovskii B.I. (1976). Critical behaviour of conductivity and dielectric constant near the metal-non-metal transition threshold. *Physica status solidi (b).* Vol. 76, Is. 2, p. 475-485, ISSN 0370-1972.

Elkin, P.K. & Yurkov, G.Yu. (2009). Ultrasound-provided synthesis of cobalt-containing nanoparticles. In: *Physics, chemistry and application of nanostructures* , Borisenko, V.E.; Gaponenko, S.V.; Gurin, V.S. (Ed.), p. 369-372, World Scientific, ISBN 978-981-4280-35-8, Singapore.

a) Fionov, A.S.; Yurkov, G.Yu.; Kolesov, V.V.; Taratanov, N.A. & Petrova, N.G. (2008). Poglotiteli elektromagnitnyh voln na osnove zhelezo- i kobal'tsoderjaschih nanochastic. Perspektivnye materialy. No. 6 (1), p. 192-196.

b) Fionov, A.S.; Yurkov, G.Yu.; Potapov, A.A.; Kolesov, V.V. & Taratanov, N.A. (2008). Perspektivnye nanostrukturirovannye polimernye kompozicionnye materialy dlya fizicheskih issledovanii ih metodami fraktal'nogo analiza. *Nelineinyi mir.* Vol. 6, No. 1, p. 37-41.

Gangopadhyay, R. & De, A. (2000). Conducting Polymer Nanocomposites: A Brief Overview. *Chem. Mater.* Vol. 12, No. 3, p. 608-622, ISSN 0897-4756.

Gleiter, H.; Weissmüller, J.; Wollersheim, O. & Würschum, R. (2001). Nanocrystalline materials: a way to solids with tunable electronic structures and properties? *Acta Materialia.* Vol. 49, No. 4, p. 737-745, ISSN 1359-6454.

Gorobinskii, L.V.; Yurkov, G.Yu. & Baranov, D.A. (2007). Production of high porosity nanoparticles of cerium oxide in clay. *Microporous and Mesoporous Materials.* Vol. 100, No. 1-3, p. 134-138, ISSN 1387-1811.

Gubin S.P. (2000). Chto takoe nanochastica? Tendencii razvitiya nanohimii i nanotehnologii. *Rossiiskii himicheskii jurnal.* Vol. 44. No. 6, p. 23-31, ISSN 0373-0247.

Gubin, S.P.; Spichkin, Yu.I.; Yurkov, G.Yu. & Tishin, A.M. (2002). Nanomaterials for high density magnetic data storage. *Russian Journal of Inorganic Chemistry.* Vol. 47, Suppl. 1, p. 32-67, ISSN 0036-0236.

Gubin, S.P.; Spichkin, Yu.I.; Koksharov, Yu.A.; Yurkov, G.Yu.; Kozinkin, A.V.; Nedoseikina, T.I.; Vlasenko, V.G.; Korobov, M.S. & Tishin, A.M. (2003). Magnetic and structural properties of Co nanoparticles in polymeric matrix. *Journal of Magnetism and Magnetic Materials.* Vol. 265, No. 2, p. 234-242, ISSN 0304-8853.

a) Gubin, S.P.; Koksharov, Yu.A.; Khomutov, G.B. & Yurkov G.Yu. (2005). Magnetic nanoparticles: preparation, structure and properties. *Russian Chemical Reviews.* Vol. 74, No. 6, p. 489-520, ISSN 0036-021X.

b) Gubin, S.P.; Yurkov, G.Yu. & Kosobudsky, I.D. (2005). Nanomaterials based on metal-containing nanoparticles in polyethylene and other carbon-chain polymers. *International Journal of Materials and Product Technology.* Vol. 23, No. 1-2, p. 2-25, ISSN 0268-1900.

Gudoshnikov, S.; Liubimov, B.; Matveets, L.; Ranchinski, M.; Usov, N.; Gubin, S.; Yurkov, G.; Snigirev, O. & Volkov. I. (2003) Magnetic properties of Fe-based nanoparticle assembly. *Journal of Magnetism and Magnetic Materials.* Vol. 258-259, p. 54-56, ISSN 0304-8853.

Hong, J.I.; Schadler, L.S. & Siegel, R.W. (2003). Rescaled electrical properties of ZnO'low density polyethylene nanocomposites. *Appl. Phys. Lett.* Vol. 82, No. 12, p. 1956-1958, ISSN 0003-6951.

Hong, J.I.; Winberg, P.; Schadler, L.S. & Siegel, R.W. (2005). Dielectric properties of zinc oxide/low density polyethylene nanocomposites. *Materials Letters.* Vol. 59, No. 4, p. 473-476, ISSN 0167-577X.

Hyeon T. (2003). Chemical synthesis of magnetic nanoparticles. *Chem. Commun.* p. 927-934, ISSN 1364-548X.

Kawasumi, M.; Hasegawa, N.; Kato, M.; Usuki, A. & Okada, A. (1997). Preparation and Mechanical Properties of Polypropylene-Clay Hybrids. *Macromolecules.* Vol. 30, No. 20, p. 6333-6338, ISSN 0024-9297.

Korobov, M.S.; Yurkov, G.Yu.; Kozinkin, A.V.; Koksharov, Yu.A.; Pirog, I.V.; Zubkov, S.V.; Kitaev, V.V.; Sarychev, D.A.; Buznik, V.M.; Tsvetnikov, A.K. & Gubin S.P. (2004). Metal-Containing Poly(tetrafluoroethylene): A Novel Material. *Inorganic Materials.* Vol. 40, No. 1, p. 26-35, ISSN 0020-1685.

Kosobudsky, I.D. & Yurkov G.Yu. (2000). Metall-containing nanoparticles in polymeric matrix.: II. Syntheses, physical and chemical properties. Application. *Izvestia vysshikh uchebnykh zavedeniy. Khimia I khimicheskay tekhnologia.* No. 5, p. 3-19.

Kul'batskii, D.M.; Ushakov, N.M.; Yurkov, G.Yu. & Podvigalkin V.Ya. (2009). Issledovanie opticheskikh kharakteristik kompozitnykh materialov na osnove nanochastits sul'fida kadmiya, stabilizirovannykh v matritse polietilena vysokogo davleniya. *Optika i spektroskopiya.* Vol. 106, No. 5, p. 780-784.

Lin, X.-M. & Samia, A.C.S. (2006). Synthesis, assembly and physical properties of magnetic nanoparticles. Journal of Magnetism and Magnetic Materials. Vol. 305, p. 100-109, ISSN 0304-8853.

Pomogailo, A.D.; Rozenberg, A.S. & Uflyand, I.E. (2004). *Nanochasticy metallov v polimerah.* Himiya, ISBN 5-7245-1107-X, Moskva.

a) Ponomareva, K.Yu.; Kosobudsky, I.D.; Tret'yachenko, E.V. & Yurkov, G.Yu. (2007). Synthesis and Properties of CdS Nanoparticles in a Polyethylene Matrix. *Inorganic Materials.* Vol. 43, No. 11, p. 1160-1166, ISSN 0020-1685.

b) Ponomareva, K.Yu. Kosobudskii, I.D.; Yurkov, G.Yu. & Kochubei, V.I. (2007). Issledovanie struktury nanochastic sul'fidov kadmiya i cinka, sintezirovannyh v matrice polietilena vysokogo davleniya. *Vestnik Saratovskogo gosudarstvennogo tehnicheskogo universiteta.* No. 2 (24), Is. 1, p. 60-65.

Popkov, O.V.; Yurkov, G.Yu. & Fionov A.S. (2009). Stabilization of nanoparticles on the surface of detonation nanodiamond. In: Physics, chemistry and application of nanostructures, Borisenko, V.E.; Gaponenko, S.V.; Gurin, V.S. (Ed.), p. 369-372, World Scientific, ISBN 978-981-270-599-0, Singapore.

Radchenko, M.V.; Lashkarev, G.V.; Sichkovskyi, V.I.; Arshakuni, A.A.; Gubin, S.P.; Yuhymchuk, V.O.; Domukhovski, V.; Story, T.; Piryatinskii, Yu.P. & Yurkov G.Yu. (2009). Nanomaterials Based on CdS Nanoparticles in Polyethylene Matrix. *Inorganic Materials.* Vol. 45, No. 5, p. 468-473, ISSN 0020-1685.

Scomski, R. (2003). Nanomagnetics. *Journal of Physics: Condensed Matter.* Vol. 15, p. R841-R896, ISSN 0953-8984.

Siegel, R. W. (1993). Nanostructured materials -mind over matter. *Nanostruct. Mater.* Vol. 3, No 1-6, p. 1-18, ISSN 0965-9773.

a) Taratanov, N.A.; Yurkov, G.Yu.; Fionov, A.S.; Koksharov, Yu.A.; Popkov, O.V. & Kolesov, V.V. (2009). Creation and Physical Properties of the Molybdenum-Containing Polyethylene-Based Nanomaterials. *Journal of Communications Technology and Electronics.* Vol. 54, No. 8, p. 937-946, ISSN 1064-2269.

b) Taratanov, N.A.; Yurkov, G.Yu.; Fionov, A.S.; Boimuratov, F.T.; Abdurahmanov, U. & Kosobudskii, I.D. (2009). Svinecsoderjaschie kompozicionnye nanomaterialy na osnove polietilena. *Izvestiya vysshih uchebnyh zavedenii. Himiya i himicheskaya tehnologiya.* Vol. 52, No. 7, p. 72-75.

Taratanov, N.A.; Yurkov, G.Yu. & Kosobudskii I.D (2010). Sintez reniisoderjaschih nanochastic na poverhnosti mikrogranul politetraftoretilena. *Vestnik Saratovskogo gosudarstvennogo tehnicheskogo universiteta.* No. 1 (44), p. 95-101.

Ul'zutuev, A.N.; Ushakov, N.M.; Yurkov, G.Yu. & Kosobudskii, I.D. (2009). Teplovoi gisterezis dielektricheskih svoistv v kompozitah na osnove nanochastic oksidov i sul'fidov perehodnyh metallov v matrice polietilena vysokogo davleniya. *Pis'ma v jurnal tehnicheskoi fiziki.* Vol. 35, No. 10, p. 80-86.

Ushakov, N.M.; Kosobudskii, I.D.; Yurkov, G.Yu.; Gubin, S.P.; Zapsis, K.V.; Kochubei, V.I. & Ul'zutuev, A.N. (2005). Novye kompozicionnye nanomaterialy s upravlyaemymi svoistvami dlya radiotehniki i elektroniki. *Radiotehnika.* No. 10, p. 105-108.

Ushakov, N.M.; Yurkov, G.Yu.; Baranov, D.A.; Zapsis, K.V.; Zhuravleva, M.N.; Kochubey, V.I. & Kosobudsky, I.D. (2006). Opticheskie i fotoluministsentnye svoistva kompozitnykh nanomaterialov na osnove nanochastits sul'fida kadmiya i polietilena vysokogo davleniya. *Optika i spektroskopiya.* Vol. 101, No. 2, p. 262-267.

Ushakov, N.M.; Yurkov, G.Yu.; Gorobinskii, L.V.; Popkov, O.V. & Kosobudskii, I.D. (2008). Nanocomposites based on the cerium oxide nanoparticles and polyethylene matrix: syntheses and properties. *Acta Materialia.* 2008. Vol. 56, No. 10, p. 2336-2343, ISSN 1359-6454.

Volkov, A.N.; Yurkov, G.Yu.; Ovchenkov, E.A.; Koksharov, Y.A.; Popkov, O.V.; Ivicheva, S.N.; Matveev, V.V. & Kargin Yu.F. (2010). Composite material based on SiO_2 micro-granules and cobalt nanoparticles. *Perspektivnye materialy.* No. 4, p. 56-59.

Wang, Y. & Herron, N. (1991). Nanometer-sized semiconductor clusters: materials synthesis, quantum size effects, and photophysical properties. *J. Phys. Chem.* Vol. 95, No, 2, p. 525- 532, ISSN 1089-5639.

Xia, X.; Cai, S. & Xie, C. (2006). Preparation, structure and thermal stability of Cu/LDPE nanocomposites. *Materials Chemistry and Physics.* Vol. 95, p. 122-129, ISSN 0254-0584.

Yurkov, G.Yu.; Kozinkin, A.V.; Nedoseikina, T.I.; Shuvaev, A.T.; Vlasenko, V.G.; Gubin, S.P. & Kosobudskii I.D. (2001). Copper Nanoparticles in a Polyethylene Matrix. *Inorganic Materials.* Vol. 37, No. 10, p. 997-1001, ISSN 0020-1685.

Yurkov, G.Yu.; Gubin, S.P.; Pankratov, D.A.; Koksharov, Yu.A.; Kozinkin, A.V.; Spichkin, Yu.I.; Nedoseikina, T.I.; Pirog, I.V. & Vlasenko V.G. (2002). Iron(III) Oxide Nanoparticles in a Polyethylene Matrix. *Inorganic Materials.* Vol. 38, No. 2, p. 137-145, ISSN 0020-1685.

Yurkov, G.Yu.; Baranov, D.A.; Dotsenko, I.P. & Gubin, S.P. (2006). New magnetic materials based on cobalt and iron-containing nanoparticles. *Composites part B: Engineering.* Vol. 37, No. 6, p. 413-417, ISSN 1359-8368.

a) Yurkov, G.Yu.; Baranov, D.A. & Gorobinskii, L.V. (2007). Synthesis of ceria oxide nanoparticles in polyethylene matrix, In: *Physics, chemistry and application of nanostructures* , Borisenko, V.E.; Gaponenko, S.V.; Gurin, V.S. (Ed.), p. 343-346, World Scientific, ISBN 978-981-270-599-0, Singapore.

b) Yurkov, G.Yu.; Fionov, A.S.; Koksharov, Yu.A.; Kolesov, V.V. & Gubin S.P. (2007). Electrical and Magnetic Properties of Nanomaterials Containing Iron or Cobalt Nanoparticles. *Inorganic Materials.* Vol. 43, No. 8, p. 834-844, ISSN 0020-1685.

Yurkov, G.Yu.; Taratanov, N.A.; Kosobudskii, I.D. & Naumenko, V.Yu. (2009). Med'soderjaschie kompozicionnye nanomaterialy. *Vestnik Saratovskogo gosudarstvennogo tehnicheskogo universiteta.* Vol. 39, No. 2. Is. 2, p. 37-43.

Zanetti, M.; Camino, G.; Reichert, P. & Mülhaupt, R. (2001). Thermal Behaviour of Poly(propylene) Layered Silicate Nanocomposites. *Macromol. Rapid Commun.* Vol. 22, No. 3, p. 176-180, ISSN 1022-1336.

Zanetti, M. & Costa, L. (2004). Preparation and combustion behaviour of polymer/layered silicate nanocomposites based upon PE and EVA. *Polymer.* Vol. 45, p. 4367-4373, ISSN 0032-3861.

13

Coupled Semiconductor Metal Oxide Nanocomposites: Types, Synthesis Conditions and Properties

Dr. Mohammad Reza Vaezi
Division of Nanotechnology and Advanced Materials,
Materials and Energy Research Center,
MERC, Karaj,
Iran

1. Introduction

A nanocomposite is a solid material containing two or more phases where one of the phases has the dimensions of less than 100 nanometers (nm), or structures having nano-scale repeat distances between the different phases that make up the material. One of the main types of nanocomposite materials is the coupled semiconductor metal oxides such as M_xO_y/ Me_zO_t (M and Me are referred to metal type and x, y, z, and t are the oxidation states in these metal oxides) which can be divided into four categories. The first is mixed oxide nanocomposite in which mixed oxides form solid solutions such as TiO_2/ SnO_2 nanocomposite. Into second category fall those mixed oxides that form distinct chemical compounds, such as $ZnCo_2O_4$ and $ZnSn_2O_4$. The third are systems that form neither compounds nor solid solutions such as TiO_2/WO_3 nanocomposite. Finally, there are those systems that form Core/ Shell nanocomposite structures.

These four categories of mixed oxide nanocomposites can be produced by several methods such as hydrothermal, chemical bath deposition, chemical vapor deposition, sol-gel and solid state techniques. Among the above- mentioned chemical methods that have been reported in literature, the most common ways for production of these nanosized coupled oxides are co-precipitation and hydrothermal methods. Recently, some of these nanosized coupled metal oxides are properly synthesized by the author and et- al and the studies have been focused on production, characterization and applications of these nanocomposites.

In this chapter, the principles, mechanism and experimental results of author's researches concerning to the synthesis of these nanosized coupled oxides (the above- mentioned four categories) such as ZnO/ TiO_2, ZnO/ SnO, ZnO/ Co_3O_4, Zn_2TiO_4 and $ZnTiO_3$ nanocomposites and ZnO/ Co_3O_4 and TiO_2/ SnO_2 core- shell nanocomposites by using the simple methods of co-precipitation and hydrothermall processes are reported. Due to the high specific surface area of the obtained powders, these nanosized coupled oxides have very good potential for applications in gas sensors, photo-catalysis and photo-electrochemical cells.

2. Mixed oxide nanocomposites which form distinct chemical compounds

2.1 Zn_2TiO_4 and $ZnTiO_3$ nanocomposites

Zinc titanates are promising candidates as dielectric materials [1-3]. It is reported that three compounds exist in ZnO-TiO_2 system, including Zn_2TiO_4 (cubic), $ZnTiO_3$ (hexagonal), and $Zn_2Ti_3O_8$ (cubic) [4-6]. Among these compounds, ilmenite-type hexagonal $ZnTiO_3$ compound has been reported to have superior electrical properties [3, 7, 8].

Li et al. [9] reported the formation of a new $ZnTiO_3$ (cubic) phase as precipitates inside the Zn_2TiO_4 matrix having the same structure and lattice parameter with Zn_2TiO_4 phase. Zinc orthotitanate, Zn_2TiO_4 can be prepared easily via the conventional solid-state reaction of 2ZnO.1TiO_2.

One of the main procedures for producing the mixed oxides is to prepare the conditions for diffusion of oxides into each other [10, 11]. Due to dependence of diffusion coefficient of oxides on structure, surface area, etc., the crystallography and physical properties of the obtained particles can be changed by changing their morphology and particles size.

Although a large volume of literature is available on the synthesis of oxide nano- composite powder, very little attention has been given to the effect of synthesis conditions such as synthesis temperature on the morphology and particle size distribution of these nano-composites. Furthermore, no previous studies have been conducted on the influence of synthesis temperature on the diffusion processing of oxide particles (ZnO and TiO_2) into each other. It is interesting to know how the morphology and size distribution of obtained particles change with the variation of synthesis temperature. Therefore, the purpose of this paper is to present results of such studies for chemical bath deposited Zn_2TiO_4 and $ZnTiO_3$ nano- composites.

The nano-composites of crystalline ZnO with amorphous TiO_2 before calcination processing can be synthesized via chemical bath deposition, CBD. After calcinations, the nano-crystalline Zn_2TiO_4 and $ZnTiO_3$ from ZnO–TiO_2 (1:1 mol %) were prepared.

XRD patterns of ZnO-TiO_2 nano- composite powder without calcinations treatment are shown in Fig. 1. ZnO peaks were identified while amorphous TiO_2 particles were detected by XRF (Table 1) [12].

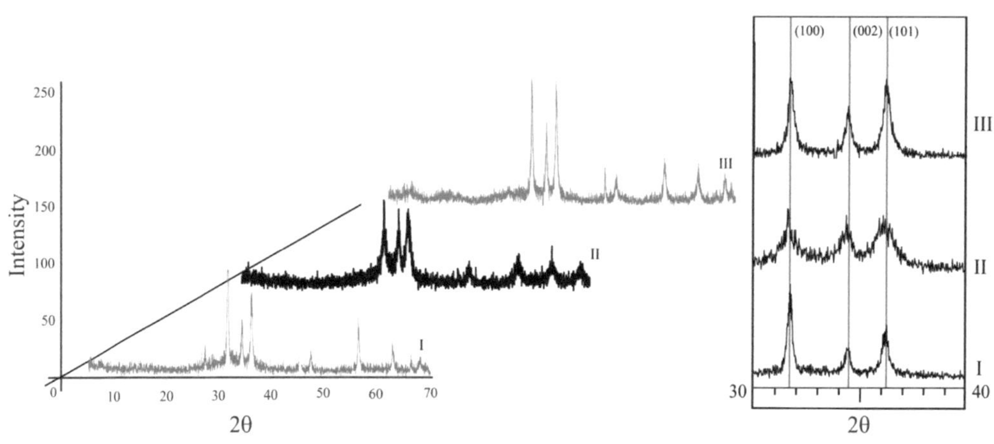

Fig. 1. XRD patterns of samples I (25 °C), II (45-50 °C) and III (70-75 °C) without calcinations [12].

Percentages of oxides(%wt)			Percentages of elements(%wt)				Samples
Other products	TiO_2	ZnO	Na	Cl	Ti	Zn	
18.1	39.7	42.2	5.3	14.1	30.2	48.6	Sample I
17.7	39.9	42.4	6.9	12.6	31.7	48.3	Sample II
20.4	34.8	44.8	6.2	10.2	31.3	51.2	Sample III

Table 1. XRF results of samples I, II and II [12].

With increasing the reaction temperature, the intensity of ZnO XRD peaks was increased. Also, with increasing the temperature, the sizes of obtained nano- particles was decreased and their morphologies tended to form semi spherical particles, which affected on the XRD peaks of obtained nano ZnO. This is well-agreed with reported observations of pure ZnO synthesis [12].

Based on the DTA of pure TiO_2 reported in the literatures [13, 14], the transformation from amorphous TiO_2 to crystalline TiO_2 is exothermic. Consequently, the temperature of 630 °C is selected for crystallization of amorphous TiO_2. All three samples were calcined for 4 hrs at cited temperature.

3.2 Formation of $ZnO/ZnTiO_3/Zn_2TiO_4$ nano- composite powders

Fig. 2 shows TGA/ DTA curves of the nano- composite powders produced by CBD method. As it can be seen, there is an undulating shape in the TGA curves (figures 2a, 2b, and 2c). The undulating shape of the TGA curves occurs in T > 350 °C. Also, a broad exothermic peak is clearly visible in the DTA curves close to temperature initialized the undulating behavior of TGA curves. The undulating shape of the TGA curves is caused by diffusion of the molecules of ZnO and TiO_2. Because of the diffusion of ZnO and TiO_2 phases, the compounds such as $ZnTiO_3$ and Zn_2TiO_4 may be formed. A broad exothermic peak observed at 350 °C <T < 600 °C in the DTA curves corresponds to atomic diffusion process occurred and therefore no considerable mass loss was detect in this region. A no sharp exothermic peak was also observed at ~ 630 °C in figures 2b and 2c. This peak corresponds to a phase transformation process of TiO_2 from amorphous to crystalline state.

After calcinations for 4hrs at 630°C, XRD patterns indicate that ZnO coexists with $ZnTiO_3$, Zn_2TiO_4 and Ti_3O_5. In the XRD pattern of sample I synthesized at 25°C, it can be detected $ZnTiO_3$ and Zn_2TiO_4 phases and the intensity of ZnO and Ti_3O_5 phases is negligible which can be ignored. It can be observed from these patterns that increasing the synthesis temperature decreases the amount of $ZnTiO_3$ and Zn_2TiO_4 phases and increases that of ZnO and Ti_3O_5 phases.

3. Mixed oxide nanocomposites which form neither compounds nor solid solutions

3.1 Synthesis of nanosized SnO/ZnO coupled oxides

Wang Cun et al reported the synthesis of ZnO/SnO_2 coupled oxides [15]. They compared the photocatalytic activity of ZnO/SnO_2 nano composites with their separate oxides. They observed that the photocatalytic activity of ZnO/SnO_2 nano composite was much more than SnO_2 and ZnO separately. They also proposed that its higher activity is because of lower recombination rate of electron-hole pairs in composite form. Similar results have been

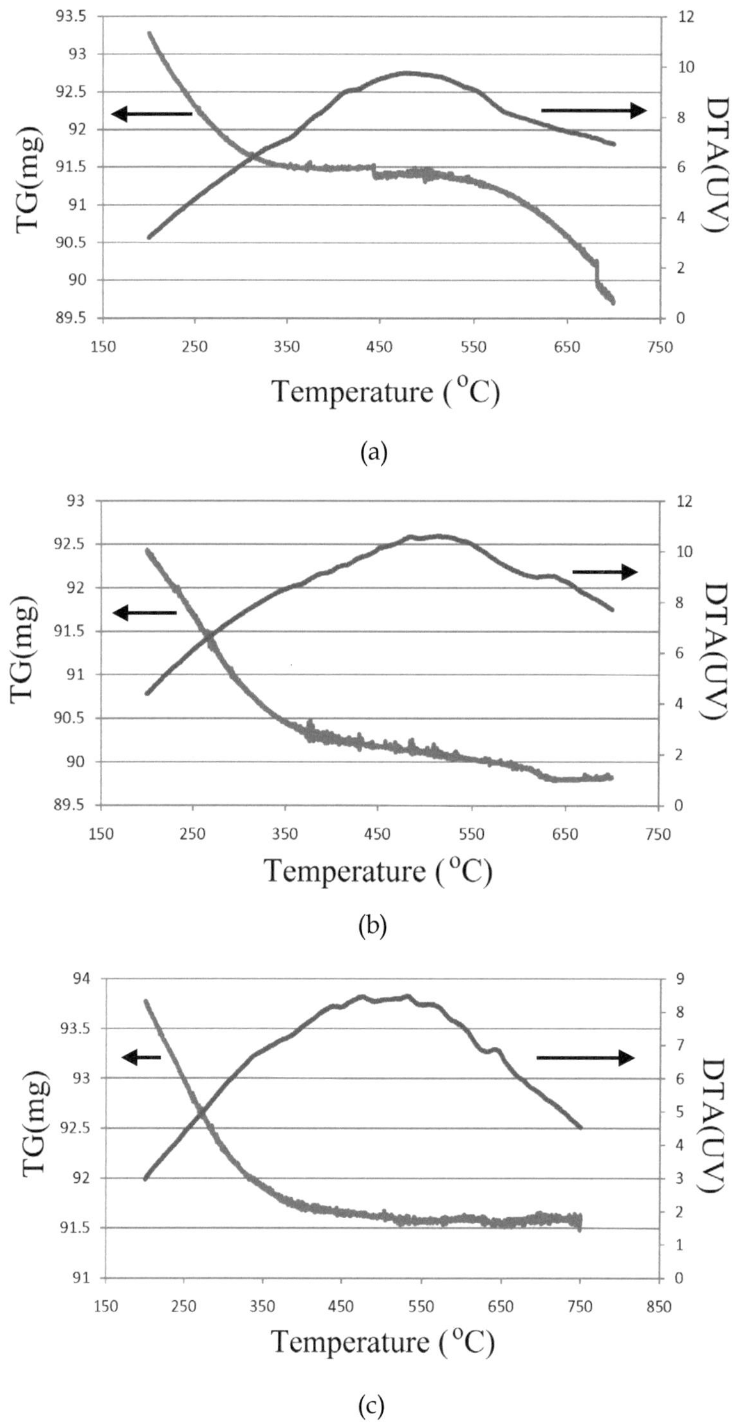

Fig. 2. TGA/ DTA curve of samples I (a), II (b), and III (c) [12].

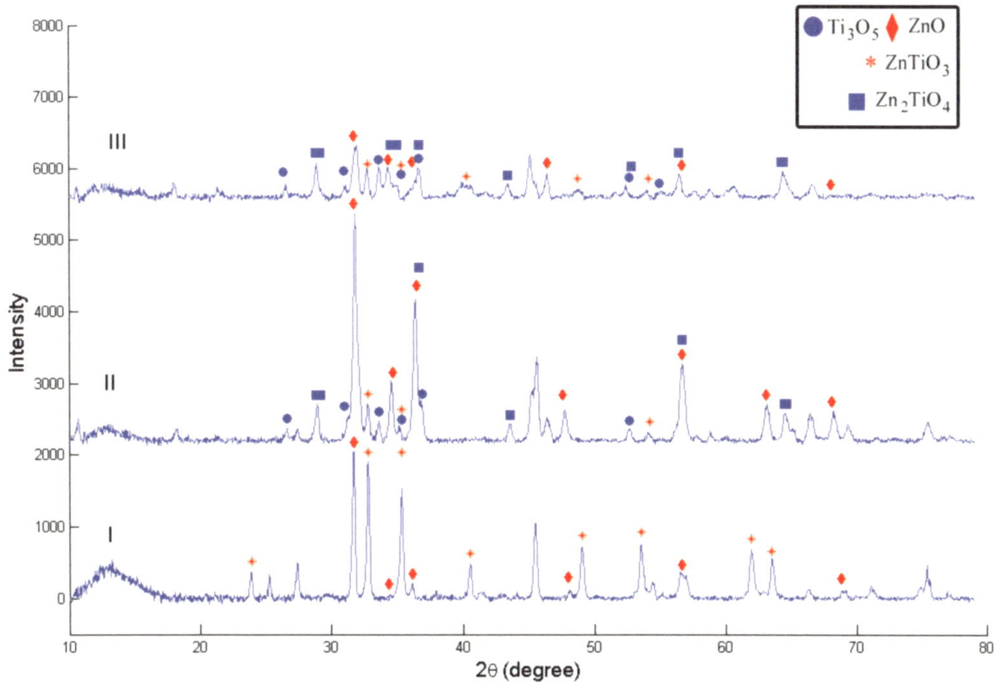

Fig. 3. XRD patterns of samples calcined at 630 ˚C (a) and the relative intensity of (101) peak for ZnO (b), (104) peak for $ZnTiO_3$ (c), and (211) peak for Zn_2TiO_4 (d) vs. the synthesis temperature [12].

reported in literature [16- 18]. Additionally Jinghong Li et al showed that by coupling nano oxides they usually show higher crystallinity and thermal stability, which can enhance their performance [18].

Recently, methods for obtaining nano- sized coupled SnO_2/ZnO are intensively developed and studies have been focused on production and characterization of these coupled oxides [16, 17, 19- 21].

A continual challenge for the researchers is to fabricate metal oxide materials, by controlling the oxidation state of multi valence metals ions in solutions. Although some studies have been conducted for production of nanosized SnO_2/ZnO coupled oxides, but the control of oxidation state of tin oxide in an aqueous solution in these coupled oxides has hardly been investigated.

Due to the fact that Sn(II) is easily oxidized to Sn(IV) , production of SnO/ZnO coupled nano oxide is relatively difficult. Thus, very few articles have been reported on the synthesizing of tin monoxide (SnO) [22-28] and to authors' best knowledge, no article on the synthesis of SnO/ZnO nanosized coupled oxide has been published till date.

In this section, fabrication of SnO/ZnO nanosized coupled oxide with oxidation state of II for tin in solution by using the simple method of co-precipitation is reported. These nanosized coupled oxides were obtained using $SnCl_2.H_2O$ and $ZnCl_2$ as precursors at three different temperatures. Due to the high specific surface area of the obtained powders, these nanosized coupled oxides have very good potential for applications in gas sensors, photo-catalysis and photo-electrochemical cells.

The procedure employed for preparing SnO/ZnO nano composites was as following: Two aqueous solutions of $SnCl_2$ and $ZnCl_2$ were added simultaneously dropwise to an aqueous solution of NaOH within about 30 min under vigorous stirring condition. This process was carried out at different temperatures of 25°C (sample I), 50°C (sample II) and 75°C (sample III) in beakers, each placed in a thermostatic cooling-heating water bath. Then the samples were kept for additional 2 hours under the same stirring condition and temperature. After synthesis, the powders were centrifuged and washed several times with distilled water and absolute ethanol and tested for removing of impurities specially NaCl with $AgNO_3$ and dried at 50°C in an oven.

The obtained phases are listed in Table 2. As shown in Table 2, final color of the synthesized sample at room temperature was pale yellow and the rest were dark green while becoming darker by increasing the temperature. During the synthesis of the sample I (at room temperature) the solution was always pale yellow, but for the other solutions after addition of certain amount of metal chloride solution that causes decrease in solutions pHs, the color of samples changed suddenly to dark green. It was observed that for sample III, this change in color occurred sooner than sample II. Also, by increasing the temperature from 25°C to 50°C the amount of the obtained powder increased drastically, as their initial solutions were the same.

sample	Synthesized temperature (°C)	Color	Obtained phases
I	25	Pale yellow	ZnO
II	50	Dark Green	SnO , ZnO
III	75	Dark Green	SnO , ZnO

Table 2. Condition of the synthesized powders

X-Ray diffraction patterns of the samples are shown in Fig.4. Brags peaks in sample I showed good agreement with ZnO (JCPDS no.36-1451), and no brag peaks of other compounds were found. But for the sample II and III the peaks are related to SnO (JCPDS no.06-0395) as well as ZnO (JCPDS no.36-1451).

Also, XRF analysis of sample I (Table 3) indicated existence of about 85.3 percent ZnO and 10.2 percent of SnO in the sample. Sn atoms can be doped in ZnO structure as two precursors of Zn and Sn are mixed together. As cited above, due to lack of SnO peaks in the XRD pattern of sample I, Sn should be formed as amorphous SnO or doped in ZnO during synthesis process. In fact this amount of doped Sn in comparison with the amount of Sn for Sample I is little enough to conclude that the major amount of Sn atoms have to be formed as individual amorphous phase instead of dopant in ZnO. J. Chouvin et al. reported synthesizing of SnO nano particles via CBD method. They used aquous solutions of $SnCl_2.2H_2O$ and NaOH as precursors of SnO and reported that by increasing temperature up to the boiling point, the mixtures' color turned from white to black and SnO crystals were formed during nucleation and growth of the previous amorphous phase [29]. This fact can be observed in this synthesis process too. The variation of obtained phase's ratio with increasing of synthesis temperature can be correlated to this phenomenon.

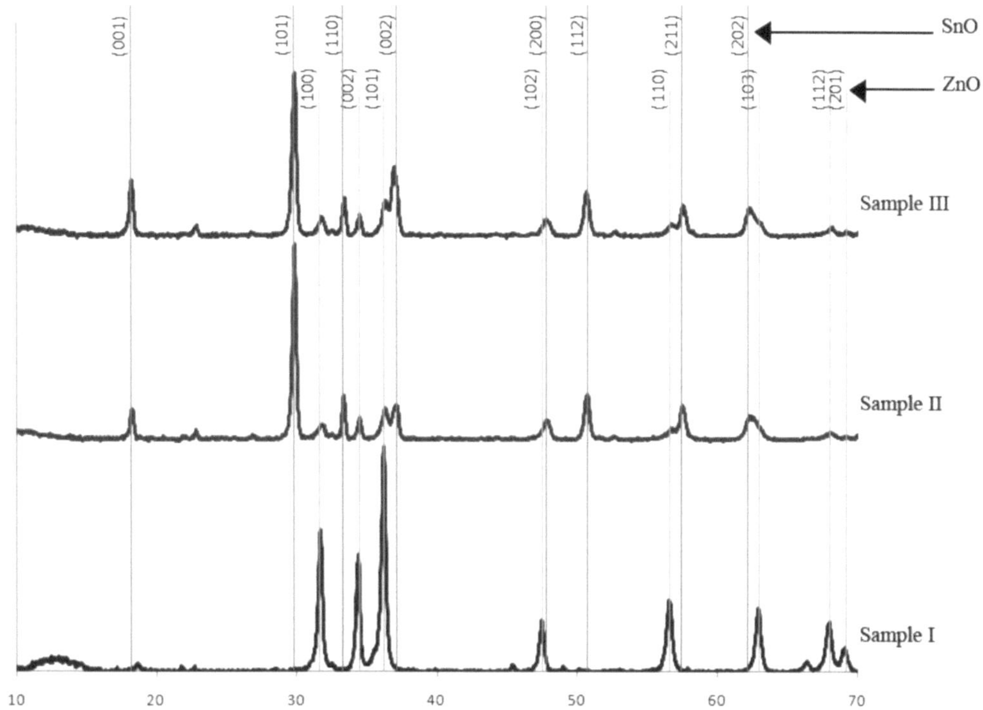

Fig. 4. X-Ray diffraction patterns of powders synthesized at 25 °C (sample I), 50 °C (sample II), and 75 °C (sample III).

Sample	Percentage of Products (%Wt)		
	ZnO	SnO	Other Products
I	85.3	10.2	4.5
II	48.2	48.6	3.2
III	51.7	45.6	2.7

Table 3. Results of XRF analysis of samples

The average crystallite size is calculated from the full width at half maximum (FWHM) of the diffraction peaks using the Debye-Scherer formula:

$$D = k\lambda / \beta \cos\theta \tag{1}$$

where D is the mean crystallite size; K is a grain shape dependent constant (here assumed to be 0.89); λ is the wavelength of the incident beam; θ is the Bragg reflection peak; and β is the full width at half maximum [30]. As can be seen in Table 4, the mean crystallite size of SnO particles from their (101) planes are 23.95 and 24.90 for 50 and 75°C, respectively. It can be seen from (002) plane of ZnO XRD patterns, at 25°C the mean crystal size of ZnO is about 23.56 nm but by increase in temperature to 50 and 75°C it decreased to 14.97 and 15.34 nm respectively. This decrease can be concluded from changing of obtained morphologies and growth conditions while increasing temperature [31, 32]. The results are listed in Table 4. Moreover the XRD results, at room temperature SnO wouldn't be formed, so the final pH of this solution would be higher because of higher hydroxyl groups that remained in the solution. As existence of higher hydroxyl groups favors the growth of the crystalline phases; this condition favors the faster growth of ZnO crystals. As a result their crystallite sizes would be higher. Also it has been reported in literature that in this method, co-existence of seconds phase hinders the growth of crystals [33].

	ZnO		SnO	
sample	crystallite size (nm)	c/a	crystallite size (nm)	c/a
I	23.56	1.600	---	---
II	14.97	1.599	23.95	1.280
III	15.34	1.598	24.90	1.284

Table 4. Mean crystallite sizes and lattice constants of prepared samples

SEM images of samples are shown in Fig. 5. For particles of sample I, at the first glimpse it may look like that they are disc like particles but more precise investigation reveals that they are agglomerations of many tiny particles with 456.2 nm mean particle size. Mean particles sizes of samples are measured and listed in Table 5.

Sample	BET surface area (m^2g^{-1})	Mean aggregates size (nm)	Mean particle size (nm)
I	16.28	58 in width 260 in diagonal	45.2
II	22.88	58.4	34.4
III	19.84	52.4	36.5

Table 5. Specific surface area from BET, Mean aggregates size from SEM and mean particle size from TEM.

It was found that at 25°C, ZnO particles produce flake like agglomerates with median size of about 58 nm in width and 260 nm in diagonal as shown in Fig. 6a, but in fact this agglomerates are formed by smaller particles. At 50°C and 75°C the median sizes of obtained semispherical agglomerates of ZnO/SnO coupled oxides were 58.4 nm and 52.4 nm, respectively. From TEM images, these agglomerates as shown in Fig. 7, are formed from individual nanoparticles. The median sizes of obtained nanoparticles from TEM images and the agglomerates median sizes are listed in Table 5. As it can be seen, by change in temperature the sizes of obtained powders haven't changed intensively from sample II to III. But as listed in Table 5, due to enhancement in agglomeration, the specific surface area of the obtained samples are decreased by increasing synthesis temperature from 50 to 75°C.

The specific surface area of the obtained samples was calculated from Brunauer-Emmett-Teller (BET) equation and listed in Table 5. As it can be seen, BET surface area of obtained powders are 16.28, 22.88, 19.84 m^2g^{-1} for synthesized powders at 25, 50 and 75 °C, respectively. The lower surface area of sample I should be attributed to intense agglomeration of the particles in the form of plates. As mentioned above higher surface area of sample II is because of its lower agglomeration and also lower particle size.

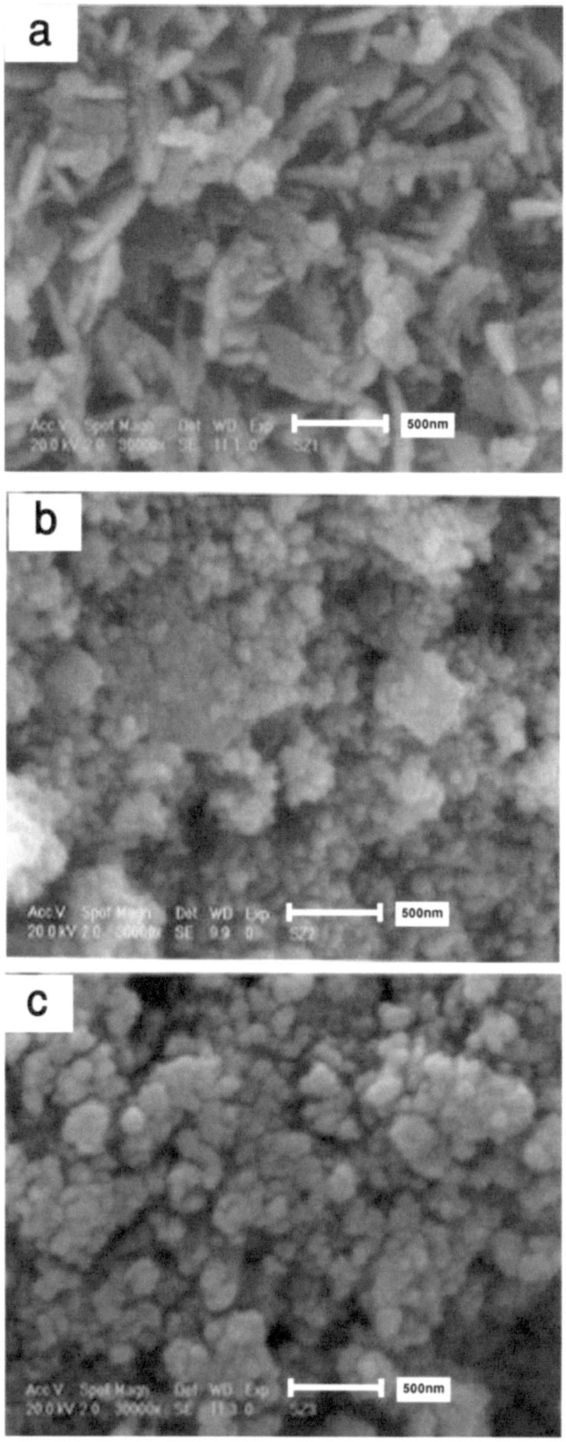

Fig. 5. SEM images of obtained powders at (a) 25, (b) 50 and (c) 75°C.

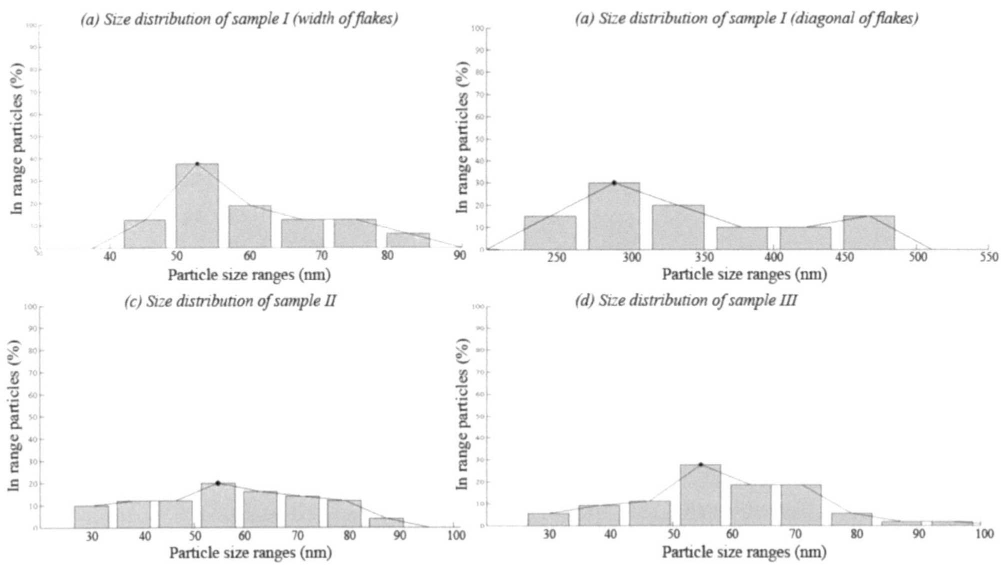

Fig. 6. Agglomerates size distribution of samples: (a) I (flakes width), (b) I (flakes diagonals), (c) II and (d) III.

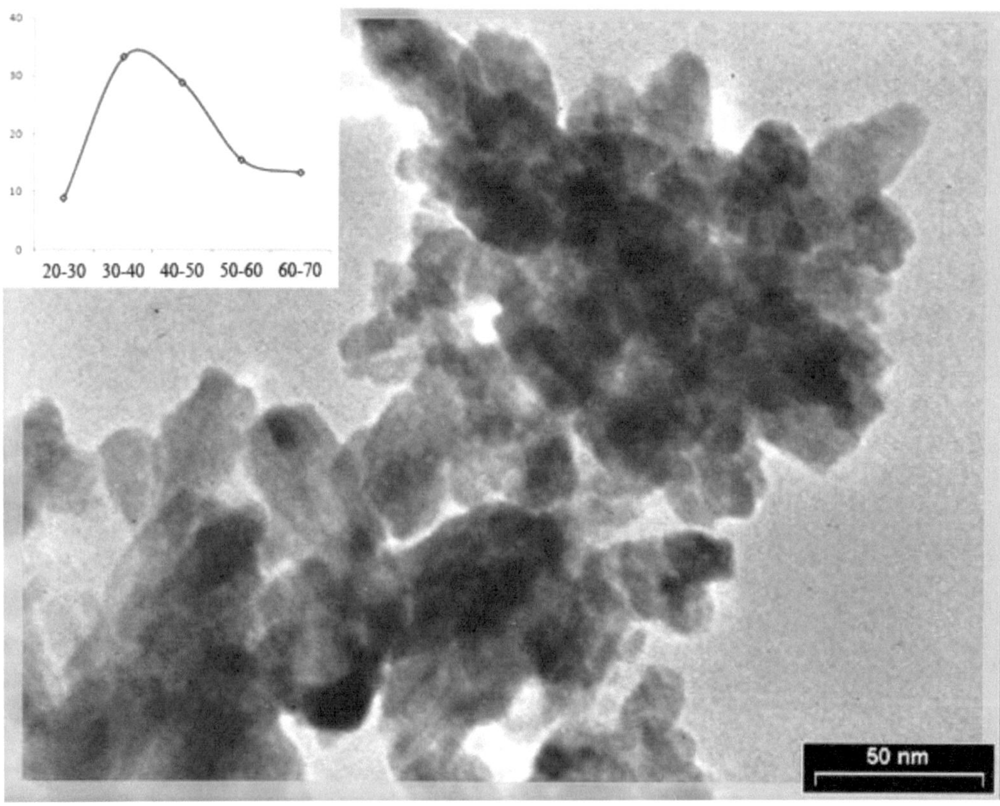

Fig. 7. TEM image of sample II with measured size distribution.

3.2 Synthesis of nanosized ZnO/Co_3O_4 coupled oxides via CBD method

Cobalt oxides are important materials that find applications in different fields such as catalysis, various type of sensor, electrochemical, electrical, and other opto-electronic devices [34]. Cobalt oxide is prepared by calcinations of $Co(OH)_2$ precursor synthesized by the chemical precipitation.

The main cobalt oxides are CoO and mixed Co(II) and Co(III) oxides ,Co_3O_4. When cobaltaus oxides are heated about 400-500 °C in oxygen, Co_3O_4 is readily obtained [35]. Co_3O_4 has typical spinel structure [36] and has various applications such as solid sensors, catalysts, electrode materials, magnetic materials, etc. In recent years, demands for monitoring of toxic and harmful gases have become more serious all over the world [37]. Various semiconductor gas sensors based on ZnO, SnO_2 and TiO_2 have been researched owing to their low costs and simple sensing method [38-40]. On the other hand, there still exist some disadvantages of them, for example, the poor sensitivity of SnO_2 [41] and high working temperature of ZnO [42]. In order to improve gas sensing properties of these sensors, many studies have been focused on novel metal catalysts such as Ag, Pt and Pd , materials doping, filming and oxides multiplicity (mixed oxides) [43] .

In this section, ZnO/ CoOH (before calcinations) and also ZnO/Co_3O_4 nanocomposites (after calcination at 300 °C) were synthesized by CBD method at different synthesis temperatures.

Two separate aqueous solutions of Zn $(NO_3)_2.4H_2O$, KOH and $Co(NO)_3.6H_2O$ were prepared. $Zn(NO_3)_2.4H_2O$ and $Co(NO)_3.6H_2O$ solutions were added dropwise to KOH aqueous solution, simultaneously, within about 30 min ($\left[Zn^{2+}\right]/\left[OH^-\right]=1/2$ and $\left[Co^{2+}\right]/\left[OH^-\right]=1/2$) in a beaker under high stirring condition. After 30 minutes, the samples were kept for 2 hrs in the same operating conditions. The conditions used for synthesis of different samples are listed in Table 6.

The final samples were centrifuged and then, the precipitated products were washed with distilled water and absolute methanol several times. Finally, the precipitates were collected and dried at about 50 - 60 °C for 24 hrs. Also, the temperature of calcination was selected about 300 °C for all of the samples.

Sample	*Temperature °C*	*Stirring time (hrs)*
I	25-35	2.5
II	65-75	2.5
III	85-95	2.5

Table 6. Experimental conditions used for synthesis of nano- composite powders.

3.2.1 Synthesis before calcination

The results of XRD analysis of samples are shown in Fig. 8. In all samples, ZnO and $Co(OH)_2$ were detected. Comparing the patterns obtained from the samples, it can be seen that the peak intensities grow with increasing temperature. When the temperature was increased the intensity of $Co(OH)_2$ peaks became higher than that of ZnO peaks. The major

peaks can be noticed corresponding to (110) plane of ZnO for sample I and II whereas in sample III major peak is related to (101) plane of $Co(OH)_2$. Also, the intensity of peaks related to ZnO is grown by increasing the temperature.
Esmaielzadeh and his co-workers found that the morphology of ZnO converted from flake state to nano particles. They reported that, via increasing the temperature, the particle size decreased and the intensity of ZnO peaks increased too. We can detect this fact in the pattern of ZnO- $Co(OH)_2$ nano- composites [44] .

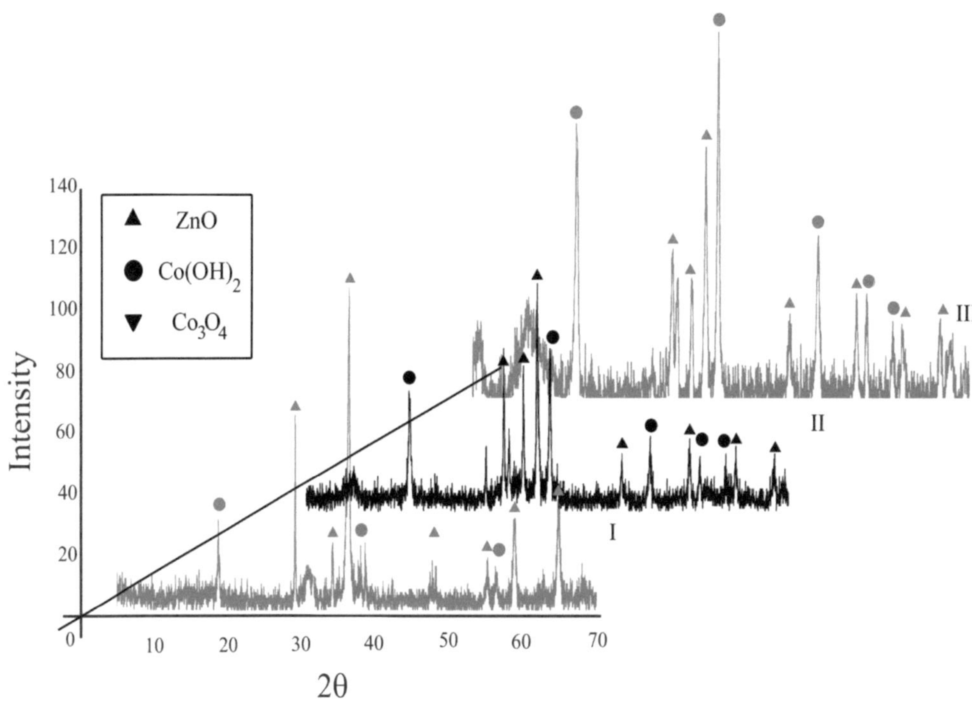

Fig. 8. XRD patterns of samples I, II, and III before calcinations.

The SEM images of the samples are shown in Fig. 9. The effect of synthesis temperature can be detected from these images. The SEM micrographs clearly indicate that the nano-composite samples are composed of spherical and semi- spherical nano- particles as well as flake shaped crystallites at ambient temperature. Also, from SEM images (Fig. 9) the maximum, minimum and median sizes with respect to samples size distributions can be obtained from the image processing program programmed with visual C++. The size distributions of samples I, II and III as well as the particle size ranges obtained from changing the synthesis temperatures are illustrated in Fig. 10.
The maximum, minimum and median sizes of obtained samples are shown in Table 7.

(a) Sample I

(b) SampleII

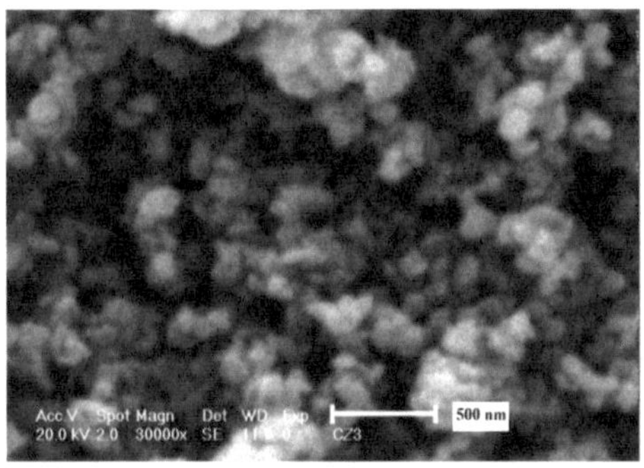

(c) Sample III

Fig. 9. SEM images of samples synthesized at (a) 25-35 °C (b) 65-75 °C (c) 85-95 °C .

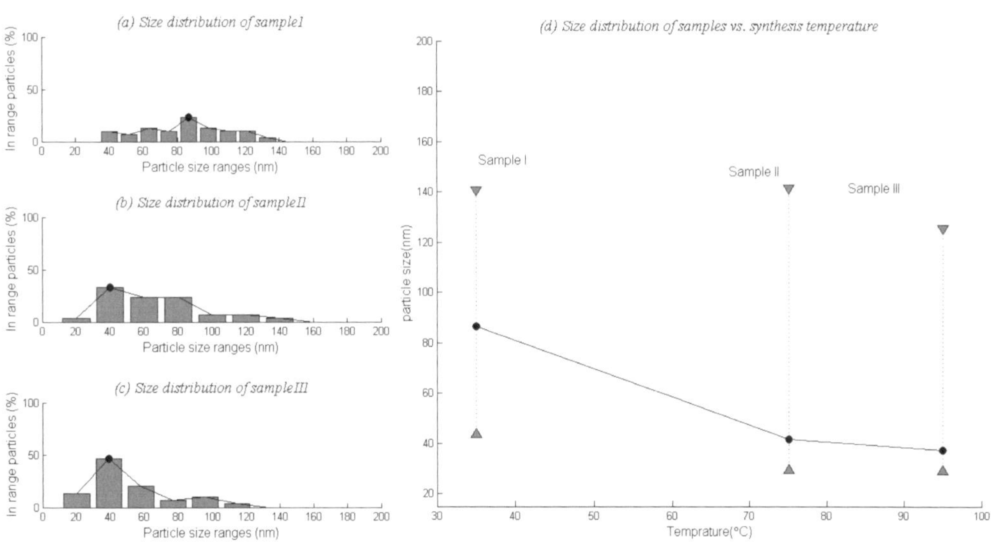

Fig. 10. Histogram of particle size distribution of (a) sample I, (b) sample II, (c) sample III and (d) The effect of synthesis temperature on the spherical particle's diameters for samples I, II and III. The particles size ranges are indicated as vertical bars on the figure.

Sample	*Synthesis temperature °C*	*Maximum particle size (nm)*	*Minimum particle size (nm)*	*Median particle size (nm)*
I	25-35	140	43	86.67
II	65-75	41	29	41.52
III	85-95	125	28	37.21

Table 7. Maximum, minimum and median sizes of samples I, II and III.

3.2.2 Synthesis after Calcination at 300 °C

From other investigations about converting $Co(OH)_2$ to Co_3O_4, DTA analysis indicated that $Co(OH)_2$ converts to Co_3O_4 at about 280 °C [45]. Moreover, no peaks of pure ZnO (exothermic or endothermic) were shown in DTA curve at 280 °C indicating no exothermic or endothermic transformations occurred at this temperature.

The thermal behaviors of obtained samples were carried out at argon atmosphere. Fig. 11 illustrates thermal behavior of sample I. In all of these analysis, because of coexistence of ZnO beside of $Co(OH)_2$, the distinct $Co(OH)_2$ peaks which had been detected in other papers would not detected here. But an endothermic tiny peak has been detected at about 284 °C (see Fig. 11).

Consequently, all of three samples calcinated for about 4 hrs at 300 °C in air atmosphere. The results of XRD analysis of samples are shown in Fig. 12.

The SEM images of nano- composite powders I, II and III calcinated at 300 °C are shown in Fig. 13. As can be seen, calcinations processing causes eliminating the flakes and the calcinated nano- composite samples are predominantly composed of spherical and semi-spherical nano- particles. Also, from SEM images presented in Fig. 13 the maximum, minimum, median sizes with respect to particle size distributions can be obtained from image analyzing process. The results can be seen in Fig.14 where the vertical lines are particle size ranges.

The maximum, minimum and median sizes of obtained samples are shown in Table 8.

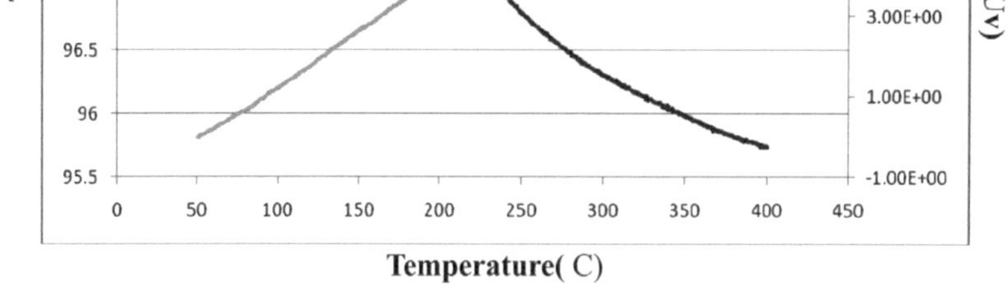

Fig. 11. DTA/ TGA analysis of sample I.

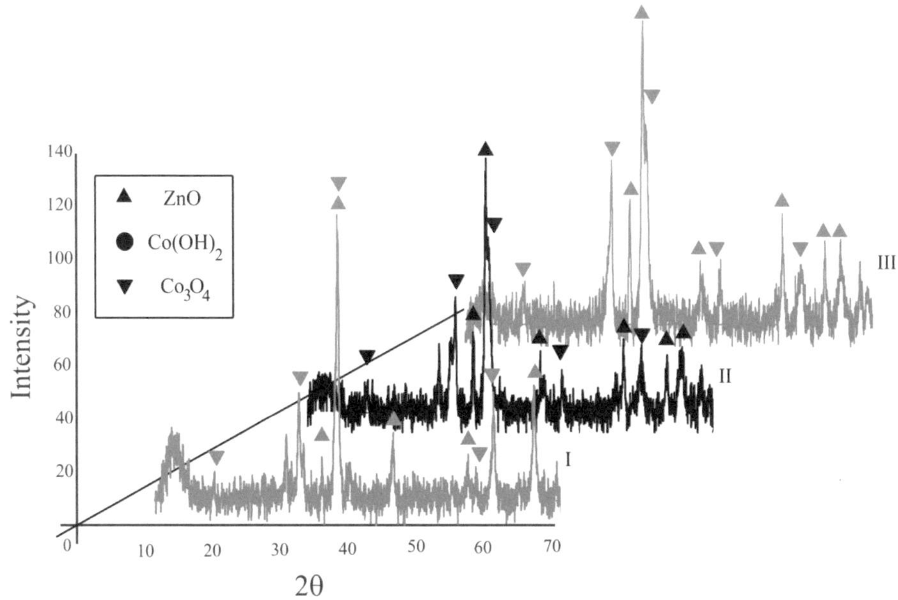

Fig. 12. XRD patterns of samples I, II and III after calcination in 300 °C for 4 hrs at air atmosphere.

(a) Sample I

(b) SampleII

(c) Sample III

Fig. 13. SEM images of samples (a) I, (b) II and (c) III after calcinations.

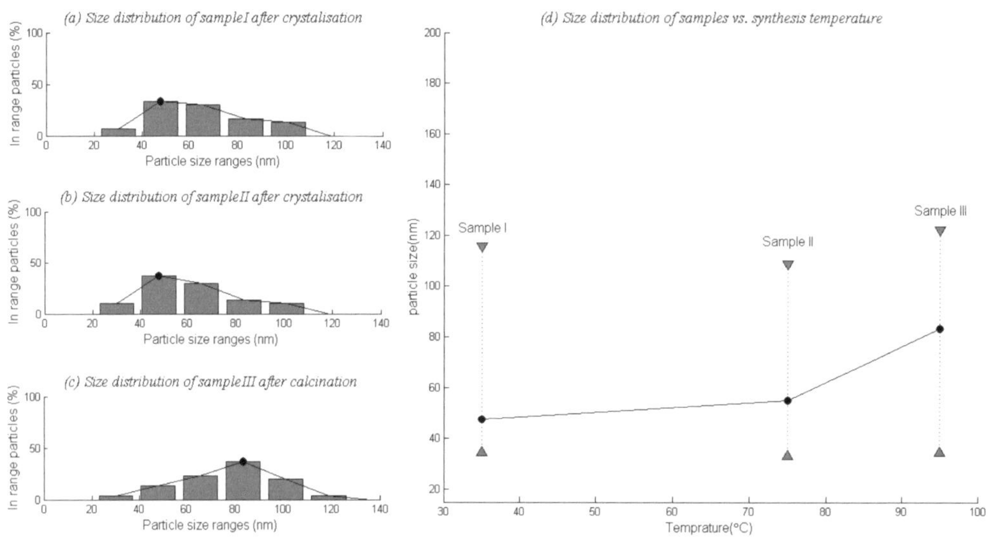

Fig. 14. Histogram of particle size distribution of (a) sample I, (b) sample II, (c) sample III after calcination and (d) The effect of synthesis temperature on the spherical particle's diameters for samples I, II and III after calcination. The particles size ranges are indicated as vertical bars on the figure.

Sample	*Synthesis temperature °C*	*Maximum particle size (nm)*	*Minimum particle size (nm)*	*Median particle size (nm)*
I	25-35	115	34	47.8
II	65-75	150	33	55.3
III	85-95	130	34	83.33

Table 8. Maximum, minimum and median sizes of samples I, II and III after calcinations.

3.3 Synthesis of nanosized ZnO/Co(OH)$_2$ coupled oxides via hydrothermal method

Recently, lamellar materials have been widely researched; layered hydroxide materials have attracted much interest in the production of catalyst, sorbent, ionic exchangers, ionic conductors, and electrochemical materials [46]. β-Cobalt hydroxide, has recently received increasing attention due to its application in the electric, magnetic and catalytic materials [47]. The atomic arrangement of $Co(OH)_2$ represents a member of the brucite type structure family (space group: $P\bar{3}ml$, point symmetry $\bar{3}2ml$). The cobalt hydroxide has hexagonal layered structure in which a divalent metal cation is located in an octahedral site generated by six hydroxyl oxygen atoms [48].

Three different theories were reported for studying the growth behaviour of materials. The BFDH law starting from planar mesh density, and considering the effects of screw axis and glide plane on crystal growth habit, provides a predicted theoretical growth habit of the crystal [49]. The PBC theory provides an ideal growth habit of crystals in terms of bond chain types between molecules and attachment energy [50]. But the above mentioned models have some shortages in explaining or predicting the crystal growth habit. The Growth Unit Model is based on the individual units which they contribute in the growth procedure. This theory hypothesizes that in the crystallization stage of oxides, cations exist in the form of complexes with OH^- ligands which their coordination numbers are equal to those of the crystal formed. These complexes are called growth units. W. Zhong was the first to put forward the Growth Unit Model [51] and nowadays many investigations were carried out by considering this theory for describing the growth behavior of the different materials [52].

The effects of surfactants on the morphologies produced in hydrothermal synthesis of different compounds were studied extremely [53-55]. The surfactants can be divided into two major categories including ionic and non-ionic surfactants. These surfactants affect the morphologies of the synthesized nanoparticles. In fact for the reaction system in the presence of surfactants, the surface tension of solution is reduced due to the existence of surfactant, which reduces the energy needed for the formation of a new solid phase [56].

CTAB is a well known cationic surfactant which can be ionized completely in water. The structure of this surfactant is shown in Fig. 15(a) [57].

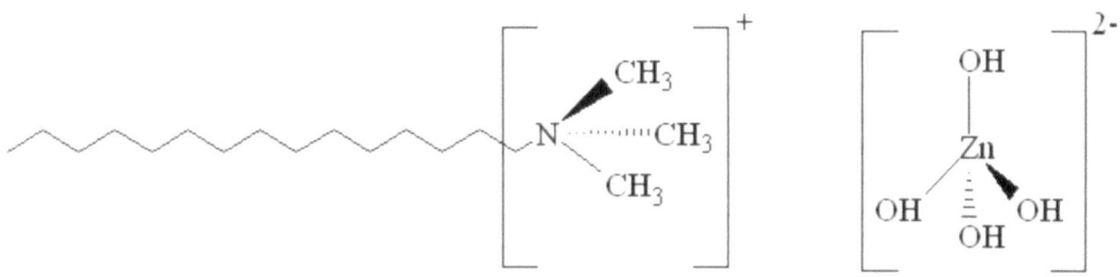

Fig. 15. The structure of (a) CTAB and (b) $Zn(OH)_4^{2-}$

In this section, ZnO/ $Co(OH)_2$ nano-composites were synthesized by hydrothermal method at 160 °C for different synthesis periods. CTAB was used as ionic surfactant.

Three separate aqueous solutions of $Zn(NO_3)_2.4H_2O$, KOH and $Co(NO_3)_2.6H_2O$ were prepared. $Zn(NO_3)_2.4H_2O$ and $Co(NO_3)_2.6H_2O$ solutions were added to KOH aqueous solution simultaneously under high stirring conditions. After mixing, the obtained solution was divided into 4 portions and poured to separate beakers. Then the surfactants were added to beakers while the solutions were being stirred with high intensity. After 5 minutes, the solutions were poured into a 35 mL Teflon-lined autoclave with a filling capacity of about 80%, then sealed and maintained at appropriate temperature for different periods. The conditions of synthesis are shown in Table 9. The prepared samples were filtered and washed with distilled water and absolute methanol several times. Then samples were dried at 60 °C for 24 hrs.

Sample	Surfactant		Time (hrs)
	Type	Amount (g)	
I	---	---	5
II	CTAB	0.1	5
III	---	---	24
IV	CTAB	0.1	24

Table 9. Synthesis conditions.

XRD patterns of samples I, II and III are shown in Fig. 16. From these patterns, $Co(OH)_2$ and ZnO were detected. From XRD patterns and SEM images (Fig. 17), the effect of surfactant on

XRD patterns and the obtained morphologies can be understood. In sample I, the absence of surfactant causes producing sharper XRD peaks, but the surfactant in other samples affects the growth conditions and thereby particle sizes and morphologies, so the sharpness of the peaks in sample II is different to sample I.

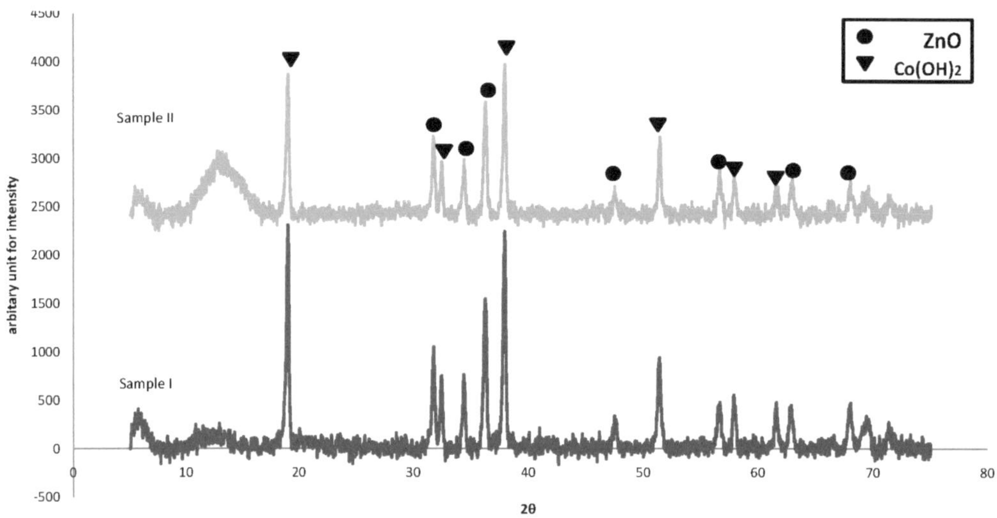

Fig. 16. (a) XRD patterns of samples I and II.

(a)Sample I

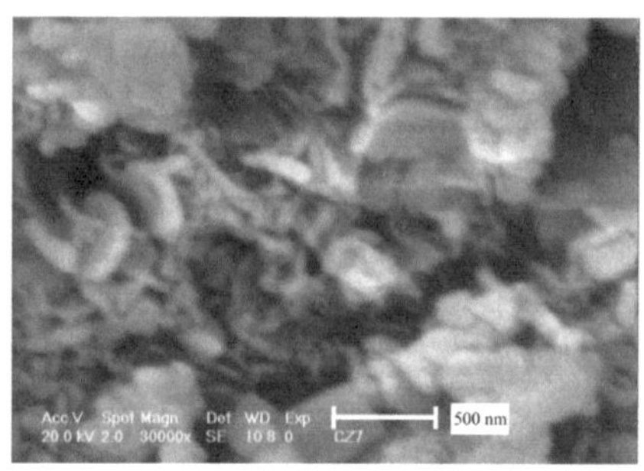

(b)Sample II

Fig. 17. SEM images of samples synthesized for 5 hrs at 160 °C

From SEM images and XRD patterns, the intensity of the peaks can be contributed to the type of morphologies obtained. For example, by comparing sample I to sample II, the morphologies were converted from semi spherical to needle like morphologies, due to presence of CTAB as a surfactant, and the intensities of XRD peaks were differed too.

For other samples synthesized for 24 hrs, the XRD patterns are shown in Fig. 18. Same as samples I and II the XRD patterns with and without CTAB are different. Also, the intensity of peaks is different like the previous samples and this can be related to size distribution and type of morphology. SEM images of samples III and IV are shown in Fig. 19.

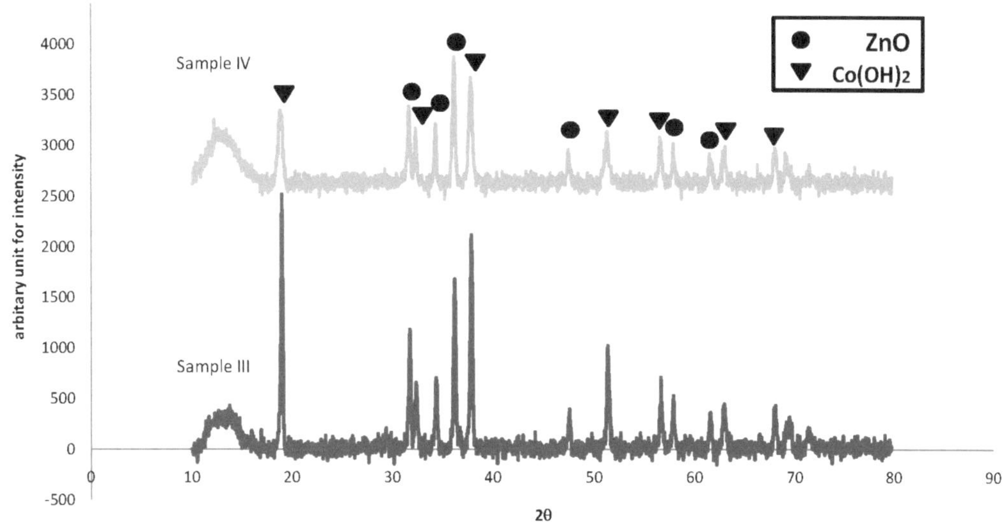

Fig. 18. XRD patterns of samples III and IV.

Fig. 19. SEM images of samples synthesized for 24 hrs at 160 °C

By comparing Fig. 18 to Fig. 19, the effect of periods of hydrothermal synthesis on the morphology and growth rates can be detected. As it can be seen, in presence of CTAB, the growth rate of samples II and IV is reduced in comparison with samples I and III.
The results of median sizes and crystallite size of samples are listed in Table 10.

Sample	ZnO plane	Mean Crystallite Size (nm)	$Co(OH)_2$ plane	Mean Crystallite Size (nm)
I	(101)	24	(101)	29
II	(101)	23	(101)	21
III	(101)	33	(101)	32
IV	(101)	27	(101)	23

Table 10. The mean crystallite sizes of the samples.

3.3.1 Synthesis of ZnO without surfactants

The main efforts have been carried out on hydrothermal synthesis of ZnO nanoparticles by Wen-Jun Li et al. [58]. They synthesized ZnO from $Zn(AC)_2$ and Na(OH) precursors at 200-300°C by hydrothermal method and conclude from the IR and Raman analyses that the growth units of ZnO are Zn complexes with OH^- ligands in the form of $Zn(OH)_4^{2-}$. The main reactions in forming ZnO can be divided into four stages, including:

Stage I) Ionization of zinc salts and sodium hydroxide in water:

$$Zn(M)_2 \xleftrightarrow{ionization} Zn^{2+} + 2(OH)^- \tag{2}$$

$$Na(OH) \xleftrightarrow{ionization} Na^+ + (OH)^- \tag{3}$$

Where M represents negatively charged groups of Zinc salts.

Stage II) Precipitation of Zinc hydroxide:

$$Zn^{2+} + 2(OH)^- \xrightarrow{solved\ in\ water} Zn(OH)_2 \downarrow \tag{4}$$

Stage III) Formation of $Zn(OH)_4^{2-}$ in hydrothermal conditions:

$$Zn(OH)_2 + 3(OH)^{2-} \xrightarrow{hydrothermal} Zn(OH)_4^{2-} \tag{5}$$

Stage IV) dehydration of $Zn(OH)_4^{2-}$ complexes:

$$x(Zn(OH)_4^{2-}) \xrightarrow[hydrothermal]{dehydration\ durring} (Zn_xO_y(OH)_z^{(x+2y-2x)-}) + (x-1)H_2O \tag{6}$$

Where, the x, y and z represent the numbers of Zn^{2+}, O^{2-} *and* OH^- respectively.
Therefore, the oxide particles can be produced by stacking of these complex polyhedrons by sharing their elements. In the interior parts, the oxygen ions are the connection centers where the exterior and the surface of the particles are $(OH)^-$ ligands, which have been proved by IR [59, 60] and Raman [61, 62] spectra in other investigations. So in the complex of $(Zn_xO_y(OH)_z^{(x+2y-2x)-})$, $(OH)^-$. ligands exist in the interface of the crystal while the interior parts consist of ZnO units as shown in Fig. 20. So stacking of these crystallites in hydrothermal condition can produce nanostructures [58]. Existence of Cobalt ions during this process can affect the morphologies and can be doped in the ZnO structures too [63].

3.3.2 Synthesis of $Co(OH)_2$ in hydrothermal processing

The formation mechanism of $Co(OH)_2$ is different to that of ZnO. When Cobalt salt is dissolved in water, it is ionized to Co^{2+} and M^- (where M represents negatively charged groups of Cobalt salts) and the first and second stages are same as ZnO, as is shown in equations (7) to (9).

Stage I) Ionization of Cobalt salts and Sodium hydroxide in water:

$$Co(M)_2 \xleftrightarrow{ionization} Co^{2+} + 2M^- \tag{7}$$

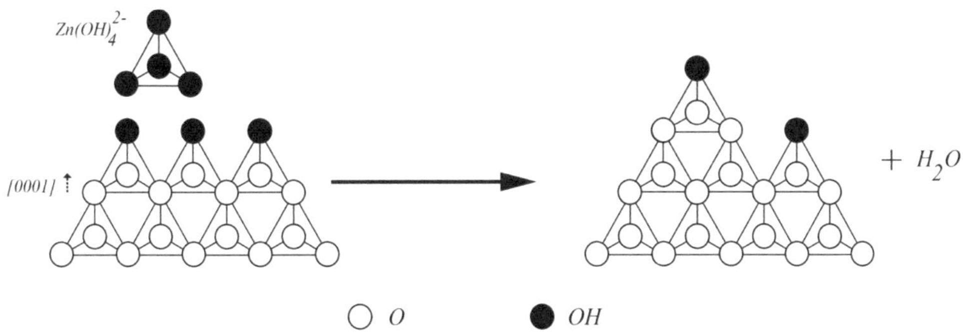

Fig. 20. Idealized interface structure image of ZnO crystal in [0001] direction and the Growth Unit Mechanism of crystallization of ZnO.

$$Na(OH) \xleftrightarrow{ionization} Na^{+} + (OH)^{-} \quad (8)$$

Stage II) Production of Cobaltous hydroxide:

$$Co^{2+} + 2(OH)^{-} \xrightarrow{solved\ in\ water} Co(OH)_2 \downarrow \quad (9)$$

B. Basavalingu et al. investigated the transformation of $Co(OH)_2$ to Cobalt oxides by hydrothermal synthesis. They calculated the minimum pressure and temperature for formation of CoO from $Co(OH)_2$. The calculated values are in good agreement with experimental data [64].
The saturated pressure of water in 160°C can be calculated from Antoine equation [65]:

$$\log P = A - \frac{B}{C+T} \quad (10)$$

Where A, B and C are 8.07131, 1730.63 and 233.426, respectively.
At 160°C, the pressure calculated from equation (10) is about 2.98144 bars, while for formation of CoO in this temperature higher pressure is necessary. The pressure for this transformation is higher than 500 bars. So by considering growth unit model and thermodynamics of formation of cobalt oxide from $Co(OH)_2$, the growth units of cobalt oxide do not form at 160°C and 3 bar pressure. So the nanoparticles of $Co(OH)_2$ grow by gathering the crystallites of $Co(OH)_2$.

3.3.3 Growth in the presence of CTAB

The effects of CTAB on the final morphology in the hydrothermal synthesis have been investigated extremely [53, 65]. After ionization in water, CTAB produces a cationic structure with a positively charged tetrahedron and a long hydrophobic tail. This tetrahedron is positively charged and can act as a cationic surfactant and affect the final structure of synthesized particles due to its electrostatic forces and stereochemical effects. The complimentary between CTA^+ and $Zn(OH)_4^{2-}$ (Fig. 15) endows the surfactant the capability to act as an ionic carrier [59]. As it is shown in Fig. 20, due to presence of OH^- ligands on $(Zn_xO_y(OH)_z^{(x+2y-2x)-})$ surface in the hydrothermal processing, a negatively charged film forms on its surface when CTA^+ which has a good complimentary with

$Zn(OH)_4^{2-}$ attract on its surface and a networks of CTA^+ forms on $(Zn_xO_y(OH)_z^{(x+2y-2x)-})$ surface. When the surfactant molecules leave, zincate will be carried away in the form of ion-pairs, so that the barrier layer becomes thinner (as shown in Fig. 21), and forms a rod-like morphology [53]. These phenomena can become more effective after CMC point of CTAB [57].

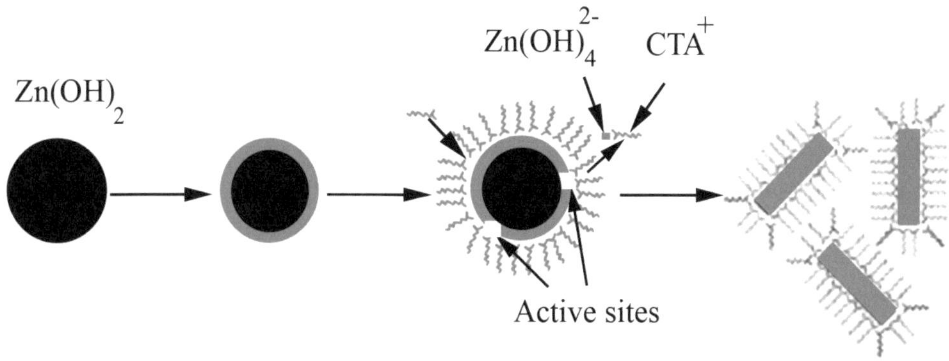

Fig. 21. Schematic illustration of the erosion process in the presence of CTAB. Black color, $Zn(OH)_2$, gray color, $Zn(OH)_4^{2-}$. The wave -like patterns indicates the CTA+ ions. A CTA+ ion carrying a zincate ion is shown.

In the crystallization process, surfactant molecules may serve as a growth controller, as well as an agglomeration inhibitor, by forming a covering film on the newly formed ZnO crystals. This can be resulted not only from absorption of CTA^+ on the surface of zinc complexes but also the stereochemical effect of its hydrophobic tail.

Due to thermodynamically limits for formation of growth units of cobalt oxide, as cited before, CTAB absorption on cobalt ions is restricted. The main effect of CTAB on the morphologies of $Co(OH)_2$ can be deduce from the stereochemical effects of surfactants which gathered on adjacent Zinc complex surfaces and also the repulsion of CTA^+ and Co^+ which restrict the in-situ ion concentrations of Co^+ and Zn^+.

Thus as shown in Fig. 17 and 19 the morphology of nano-composite consists of ZnO nano-rods and $Co(OH)_2$ nano-hexagonals.

4. Core/ Shell nanocomposite

4.1 TiO_2/ SnO_2 core/ shell nanocomposites

Recently, a lot of interest in core– shell nanocomposites has arisen due to their advantages such as stability, high catalytic activity, controllable compositions and structure, and so on [66,67]. In the coupled semiconductor systems, two different NPs are in contact and thus holes and electrons could be accessible on different particle surfaces, if needed for selective oxidation and reduction processes. In the core/shell geometry, a shell layer is coated onto the core particles and only one of the charge carriers is available at the surface of the shell for redox reaction and the other one is trapped inside (see Fig. 22) [68].

In this section, TiO_2/ SnO_2 core shell nanocomposites were produced via sol- gel method and the ethanol sensing behavior of the TiO_2/ SnO_2 nanocomposites was investigated and compared with the pure TiO_2 nanoparticles.

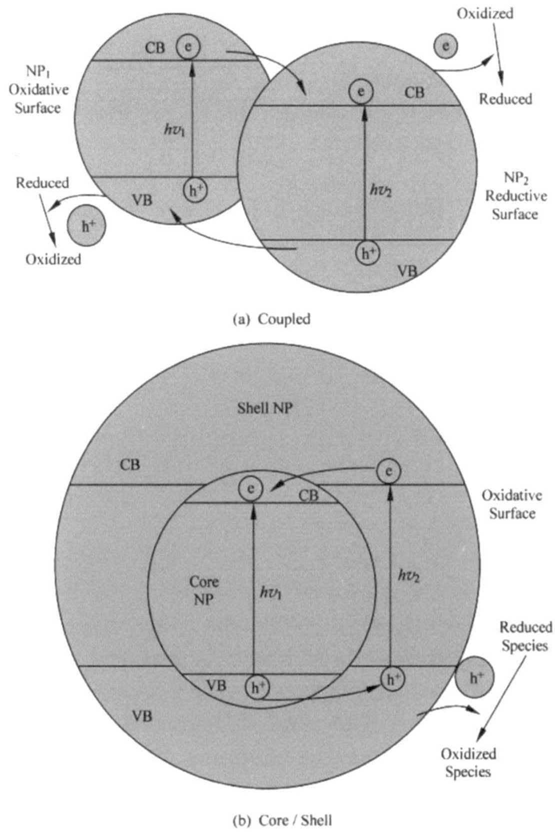

Fig. 22. Schematics of photoinduced charge separation and subsequent interfacial redox processes in (a) coupled and (b) core/shell geometry of semiconductor NPs [68].

All the chemical reagents used in this study are of analytical grade and used without any purification and treatment. TiO_2 nanoparticles were prepared from initial aqueous Ti^{4+} stock solution and hydrothermal treatment. The initial solutions were obtained by adding of 5.4 ml of titanium isopropoxide (TTIP) and same amount of isopropanol to 56.6 ml of distilled water at pH 1.5 achieved by an acid. TTIP solution was maintained under stirring conditions for a day at room temperature. Certain amount of triethylamine (TEA) was then added dropwise to solution containing Ti till pH value reached 7. It has been reported that the use of amines could act as shape controller template [69]. The obtained solution was then poured into a 35 ml Teflon lined stainless steel autoclave up to 70 vol% and maintained at 120 °C for 12 h. The autoclave was then cooled naturally to room temperature. TiO_2 powders were obtained by centrifugation and 3 times washing with distilled water. Thus, the obtained precipitates were then filtered and dried at 120 ˚C for 12 h.

The coating of SnO_2 nanoparticles on TiO_2 nanoparticles (core shell nanocomposite) was performed at room temperature by the following process:

0.2 gr. titania nanoparticles obtained were dispersed in a solution of NH_4OH and de-ionized water. Then, $SnCl_4.5H_2O$ diluted with de- ionized water was poured into the above-mentioned solution and hereby a white slurry was produced. After stirring 24 h., the resulting product was filtered and dried at 60 ˚C. The various molar proportions of Ti to Sn were examined. The symbol of TS1-1, TS1- 1.5 and TS1-2 are denoted as core shell nanoparticles based on the molar ratio of Ti to Sn 1:1, 1:1.5 and 1:2, respectively. All as-synthesized samples were calcined as 500 °C for 1 h.

Fig. 23 shows the SEM image and ED analysis of TiO_2/ SnO_2 nanocomposite with molar ratio 1:2. As it can be seen, the composite nanoparticles are spherical particles and have homogeneous morphology.

Also, according to Fig. 23 (b), the presence of Ti, Sn and Au atoms can be confirmed. Au atoms correspond to coating of Au on the particles surface.

The TEM micrographs of typically TiO_2/ SnO_2 nanoparticles are shown in Fig. 24. It can be clearly seen from the micrographs that a SnO_2 coating is on the TiO_2 surface forming core shell structure of TiO_2/ SnO_2 nanoparticles. The particle size in TiO_2/ SnO_2 nanocomposite sample TS1-2 is greater than that of TS1-1. ED pattern (Fig. 24c) shows diffuse rings, indicating the formed TiO_2/ SnO_2 is polycrystalline.

A certain amount of nanocomposite powder was pressed to form a disk with diameter of 1 cm. and thickness of 1mm. Then, the obtained sample sintered at 500 °C for 1 h. was considered as sensing element. The details of construction and schematic illustration of the fabricated gas sensor has been reported in the reference [70].

The responses of all sensors to ethanol have been measured at different concentrations ranging from 500 to 5000 ppm and at operating temperature in the range from 140 to 420 °C to investigate the gas sensing properties. Fig. 25 shows the sensor responses to ethanol for TiO_2 and nanocomposite samples at various operating temperatures.

Working temperature is one of the most important parameters for gas sensors. The conventional gas sensors based on SnO_2 and TiO_2 materials operate at the temperature region from 300 to 400 °C [71], but the response of sensors based on TiO_2/ SnO_2 nanocomposite to 1000 ppm ethanol was at around 200 °C. For the molar ratios of 1:1.5 (TS1-1.5) and 1:2 (TS1-2), the responses to 1000 ppm ethanol were 3.2 and 2.95, respectively. Meanwhile, at lower concentration of SnO_2 (TS1-1), there was no response. This observation can't be clearly explained yet and is under investigation by authors.

The dependence of the response on ethanol concentration at optimum operating temperature is shown in Fig. 26. It can be seen that nanocomposite sensors present more sensitivity than TiO_2 sensor at the concentration range from 500 to 5000 ppm ethanol. TS1-1.5 sample indicates the best sensitivity than the others. It was found that the response and recovery times of the sensors are less than 50 s.

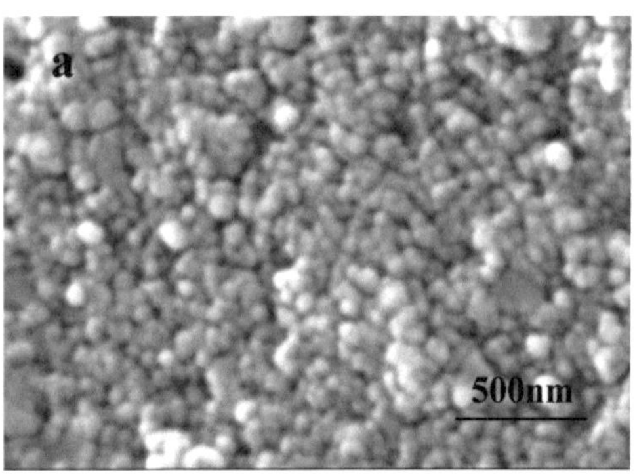

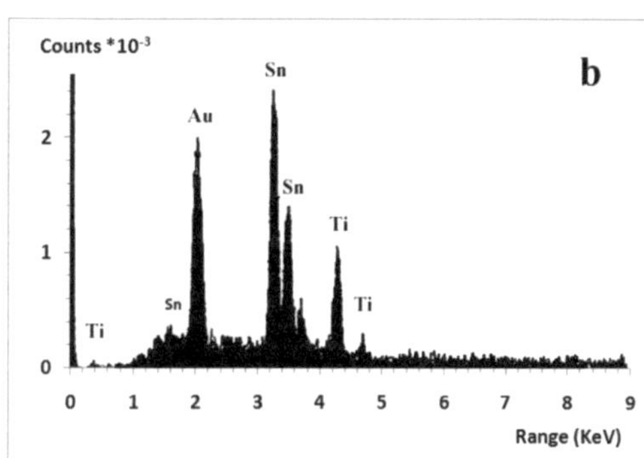

Fig. 23. a) SEM image and b) SEM- EDS element analysis of sample TS1-2.

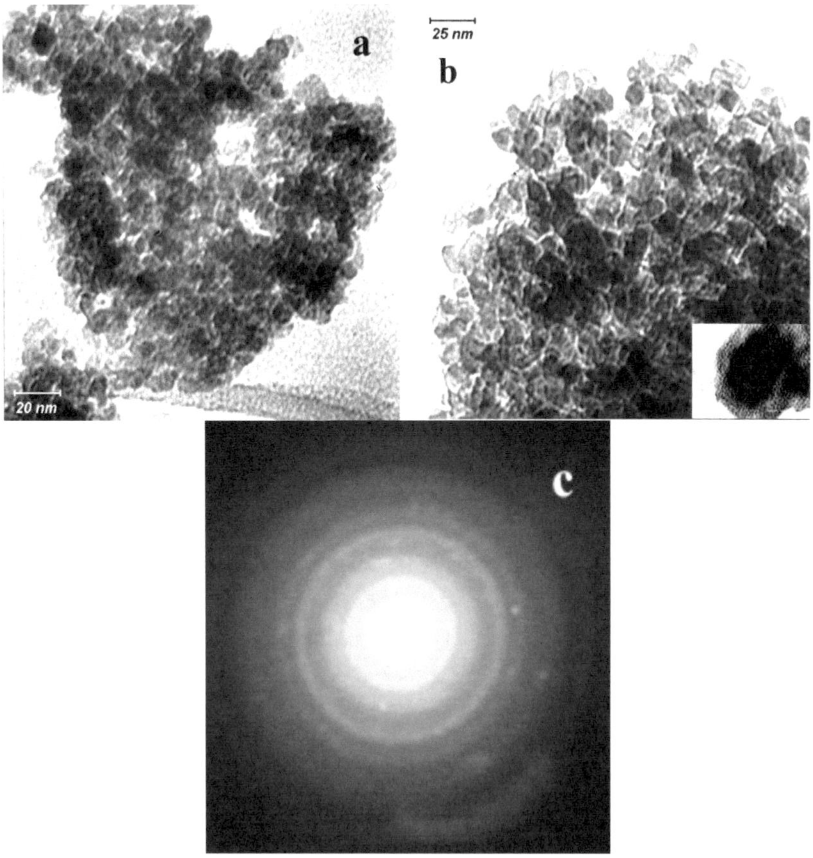

Fig. 24. TEM micrographs of TiO_2/ SnO_2 nanoparticles samples (a) TS1-1 and b) TS1-2 together with ED ring pattern of TiO_2/ SnO_2 nanoparticles.

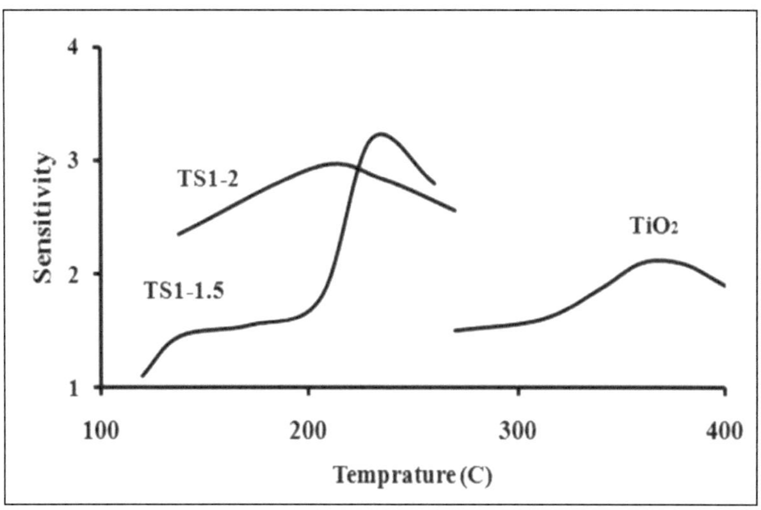

Fig. 25. The dependence of response on operating temperature for TiO_2, TS1-1.5 and TS1-2 samples.

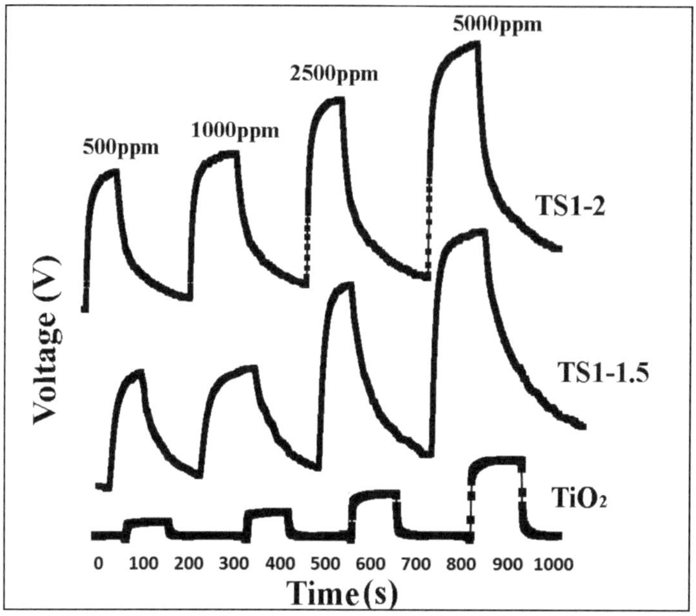

Fig. 26. The dependence of the response on ethanol concentration at optimum operating temperature.

4.2 Co_3O_4/ZnO Core/Shell nanoparticles

By emerging nano-materials a tremendous effort has been carrying out for investigating on their significant properties which make progressive improvements in many scientific fields. Among these nano-materials, nano-semiconductor materials have become one of the major fields in which attentions of different researchers, from bioscience to optoelectronics, have been attracted due to unique properties of these nanoparticles and dependency of their final properties on the size of particles. Nano-photocatalysts are one of the well known subroutine categories of semiconductors' applications. Their advanced and vast applications including cancer treatment, purification and sterilization of water and air, self cleaning coatings, etc make them one of the most important nano materials for daily uses.

Cancer is the top cause of the death in the world wide. Conventional methods for cancer treatments, surgical, radiological, immunological, thermo-therapeutic, and chemo-therapeutic treatments are well-known. The first attempts for cancer treatment by photocatalysts are related to A. Fujishima et al. efforts, in which TiO_2 nanoparticles were injected in the cancer tumors and exposed to high intensities of light. The results indicated on the fact that nano-photocatalysts can inhibit the tumor growth. Also, high ratio of surface to volume in nanoparticles causes the surface reactions occur better by contributing of major numbers of consisting molecules of a particle in surface reactions. Regardless the considerable benefits of nano-drugs, the side effects of nano material usage are considerable. The inability in controlling drug delivery endangers healthy cells on chemical toxicity. Thus, nowadays production of well controllable delivery systems has been increasing among researchers.

In environmental applications, un-purified water which contains considerable amounts of photocatalysts could become a problem. By exposing light to the remained photocatalysts, for each photon an electron-hole pair produces and subsequently it degrades the adjacent media. Due to unselective manner of aforementioned phenomenon, harmful and useful cells both are attracted with produced electron-hole pairs. The conventional methods for separating nano-particles from the media have low efficiency as well as high costs. Also, two major properties should be considered while using photocatalysts: specific surface area and band gap energy of prepared photocatalysts. High amount of specific surface area is one of the main functional variables for adsorbing cells on particles surface and increasing the active sites for photocatalytic reactions. On the other hand, most of the photocatalytic properties of semiconductors are related to their optical band gap. These properties show the workability of a photocatalyst.

Core/shell nanocomposites are one of the solutions for many difficulties in which a bifunctional nano-architecture or a modification in properties, which can't be achieved by using one type of nano-particles, is needed. Using a magnetic core enables directing particle using safe external magnetic field. If a core/shell composite contains magnetic core and semiconductor shell with high surface area and proper photocatalytic properties, the delivery and also purification of the particles become possible.

Co_3O_4 is a well known magnetic p-type semiconductor material. The performance of tri-cobalt tetra-oxide particles is highly dependent on the nanoparticles size and their specific surface areas. Thus, many attempts have been done to synthesize Co_3O_4 with high specific surface area and low dimension, such as sol-gel, chemical bath deposition (CBD), chemical vapor deposition (CVD) and hydrothermal methods. Among these, hydrothermal processing is widely used due to production of fine and uniform structures with high controllability in growth and also it can be used for preparation of fine and uniform nano Co_3O_4 particles.

Zinc oxide (ZnO) is one of the most important semiconductors. This material in nano scale shows specific properties which make it applicable as promising photocatalysts, light emitting diode, sterilization, etc. Many different chemical approaches have been reported for synthesizing ZnO nanoparticles such as hydrothermal, sonochemical, sol-gel, etc. Among these synthetic routes, sol-gel is known as one of the main synthesis routes for applying homogenous films on different surfaces and particles and thus, it is one of the main and important routes for making core/shell hetero-structures.

In this section, a promising route for synthesizing Co_3O_4/ZnO core/shell via soft chemical route has been reported. H_2O_2- assisted hydrothermal method was used for synthesizing Co_3O_4 core nanoparticles, while ZnO shell was obtained via sol-gel method.

4.2.1 Core nanoparticles

NH_4OH aqueous solution (pH=10) with appropriate amount of H_2O_2 were added drop by drop to $Co(Ac)_2$ aqueous solution in 30 minutes. The obtained solution was poured into 90 ml Teflon lined autoclaves and filled up to 80% of its volume. Then, the autoclaves were kept at 180 °C for 24 hours. After that, the autoclaves were cooled to room temperature naturally, and precipitates were filtered and washed with distilled water and ethanol for several times. Finally, the obtained powders were dried at 50 °C for 24 hours. The procedures are shown in Fig. 27.

4.2.2 Core/shell nanoparticles

A simple sol-gel procedure was used for producing ZnO shells on the Co_3O_4 nanoparticles. 0.3 g Co_3O_4 was dispersed in 50 ml solution of pure methanol and TEA with weight ratio of

$\left[\frac{TEA}{Zn(Ac)_2}=1\right]$. Another solution prepared by dissolving $Zn(Ac)_2$ in 50 ml deionised water. The concentration of $Zn(Ac)_2$ in final solution was 0.3 M. Both solutions were sonicated for 30 min. then, the second solution poured into the first solution under stirring condition and was kept for 2 hours at 60 °C. The obtained core/shell particles were filtered and dried at 50 °C for 24 hours. Then, the dried powder was calcined for 2 hours at 400 °C. The schematic diagram of procedure is shown in Fig. 28.

XRD patterns of synthesized nanoparticles are shown in Fig. 3. As it can be seen from Fig. 29, the peaks are attributed to Co_3O_4 (JCPDF 42-1467) for core nanoparticles and no other peak is detected while in core/shell just some undetectable traces can be detected from core nanoparticles. In core/shell nanoparticles, all major peaks are contributed to wurtzite ZnO (JCPDF 36-1451). The main reason for fading core particles in the XRD pattern of core/shell can be related to considerable amounts of ZnO nanoparticles synthesized adjacent to core nanoparticles. If thickness of produced shell films on the surface of the core particles were

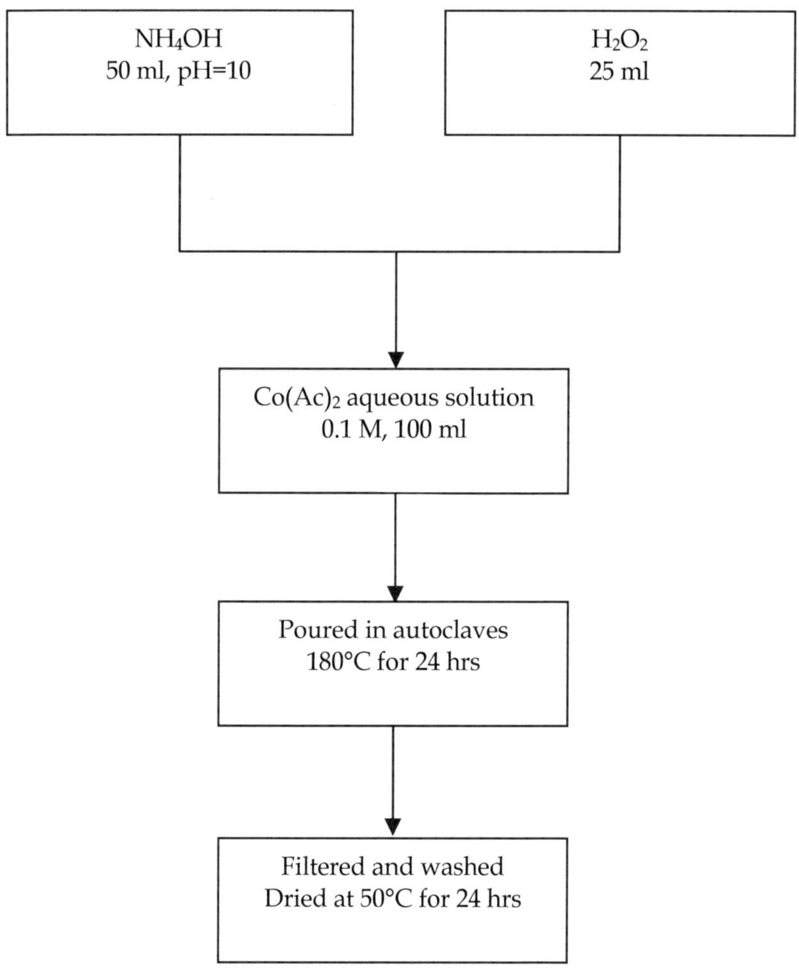

Fig. 27. Flowchart for synthesizing Co_3O_4 nanoparticles.

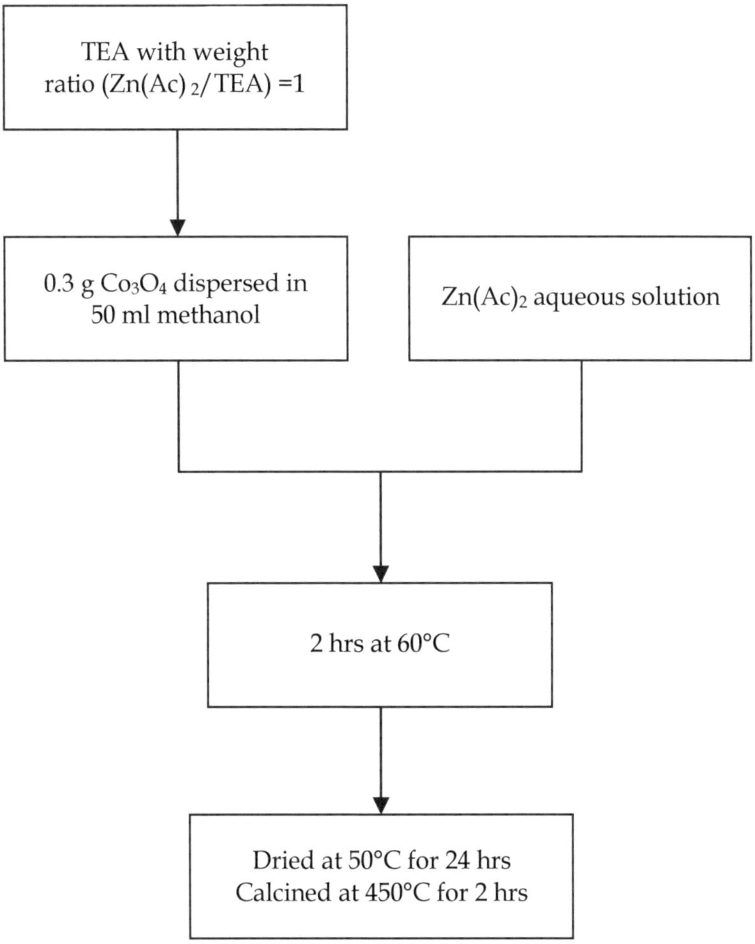

Fig. 28. Flowchart for synthesizing Co_3O_4/ZnO nanoparticles.

considerable in comparison with core nanoparticles, a reduction in the intensity of core peaks can be detected. Also, if the mass fraction of produced shell was much more than used core nanoparticles, it could result in reduction of core peaks in XRD patterns. The mentioned phenomena can intensify reduction in peaks intensities of core nanoparticles and also increase the noises in the core/shell patterns.

TEM images for the obtained core and core/shell nanoparticles are shown in Fig. 30. It can be seen that the median particle size of Co_3O_4 is about 22 nm where in core/shell nanoparticles, existence of the shell around the core particles makes the particles bigger and thus the median particle size is about 56 nm. Also, the darker parts of the image (Fig. 30b) in core/shell are related to core where brighter ones are related to ZnO shells. From BET analysis, the specific surface areas for core and core/shell were detected 98.6 and 54.6 m^2/g, respectively. BET adsorption isotherms for core and core/shell are shown in Fig. 31. The plotted isotherms represent highly isotherm of Type III. It has been shown, if a solid contains no porosity, its isotherm tends to appears as a Type III isotherm. Also, decrease in the specific surface area could be another reason for formation and growth of ZnO shells on Co_3O_4 surface and subsequence decrease in specific surface area.

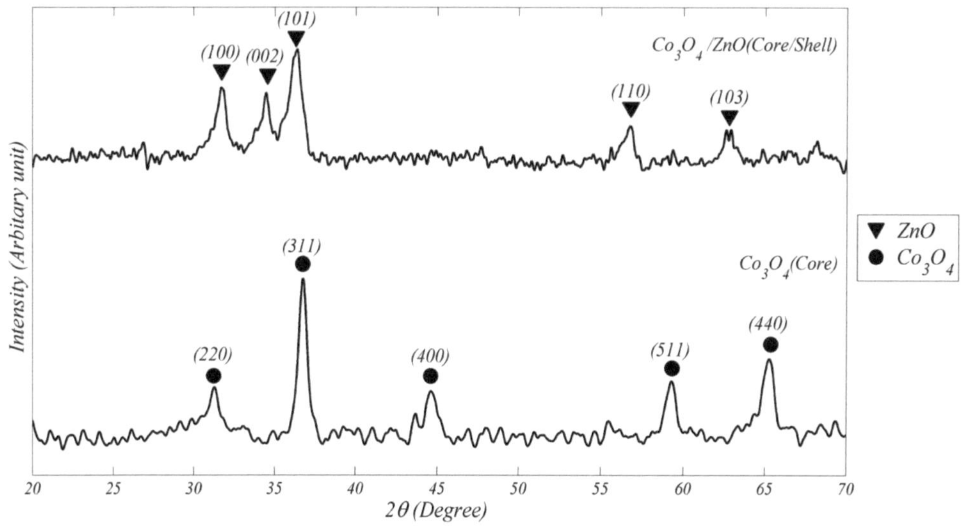

Fig. 29. XRD patterns of core and core/shell nanoparticles.

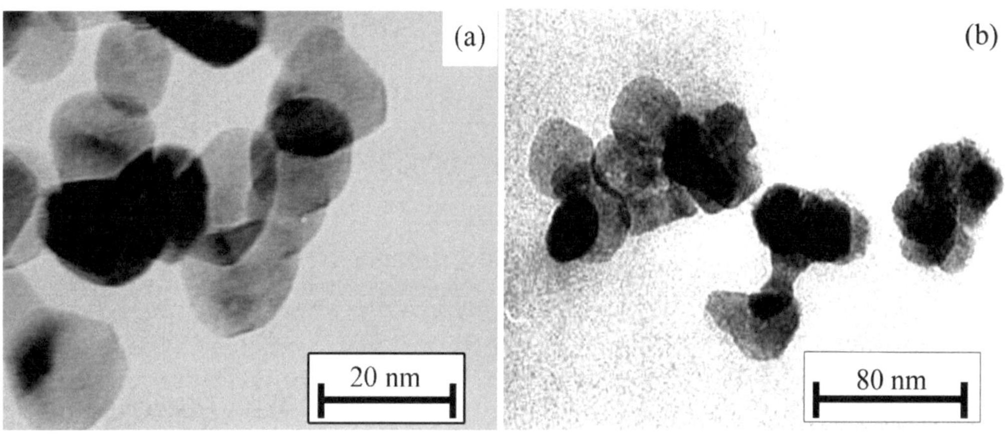

Fig. 30. TEM images of a) core and b) core/shell nanoparticles.

Optical absorptions of the obtained nanoparticles are shown in Fig. 32. Absorption coefficient, $a(\lambda)$, for allowed direct transition of semiconductors is given by the following expression:

$$\alpha = A\frac{\left(h\nu - E_g\right)^n}{h\nu} \tag{11}$$

where A is coefficient of the given electronic transition probability, E_g is band gap energy and n is equal to 0.5 and 2 for allowed direct and indirect transitions and 1.5 and 3 in case of forbidden direct and indirect transitions, respectively. In current study, the best fit of $(\alpha h\upsilon)^{1/n}$ versus photon energy was obtained 0.5 for both of core and core/shell.

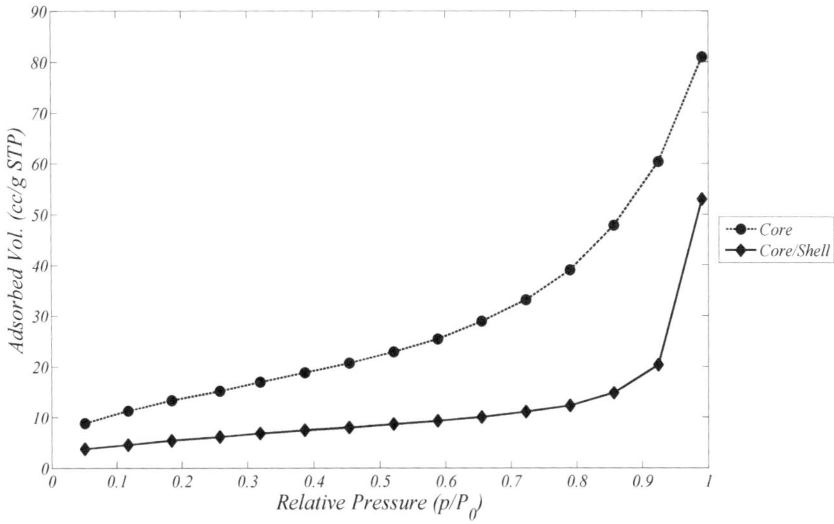

Fig. 31. BET adsorption isotherms.

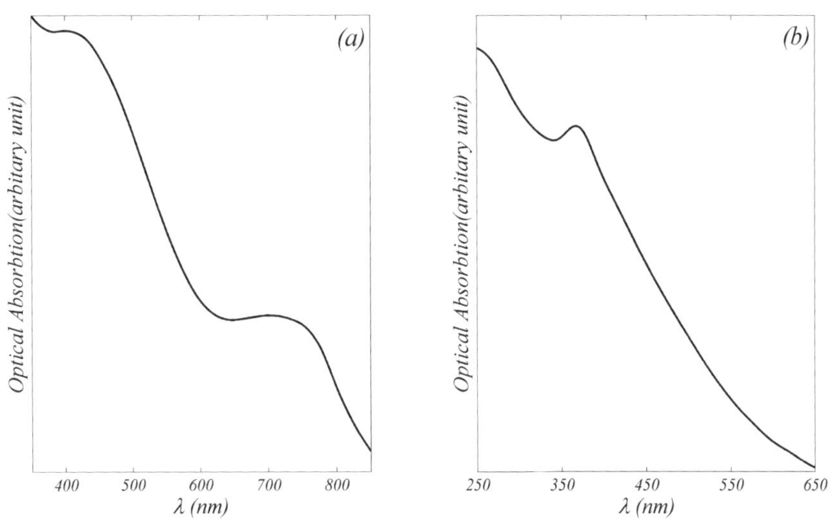

Fig. 32. UV-Vis spectra of a) core and b) core/shell nanoparticles.

Fig. 33 shows plots of $(\alpha h\upsilon)^{1/n}$ versus photon energy for both core and core/shell nanoparticles. As can be seen from these plots, in core (Fig. 33a) there are two regions in optical absorption. First one with 1.89 ev energy is related to nature of spinel structure of Co_3O_4. In spinel structure of Co_3O_4, Co(III) ions locate in center of structure. This fact affects the band structure and thus, gives rise to a sub- band in the bandgap energies. Thus two bands could be detected in absorptions, first related to excitation of O(II)→Co(III) which results in emerging 2nd band gap energy in 1.4 ev. This gap should be located inside of band gap and thus the bandgap should have larger amounts. 1.89 ev is the real optical band gap of Co_3O_4. The band gap shows a reduction in comparison with bulk Co_3O_4, there are many reports in which main reason for red shift in nano Co_3O_4 was contributed to quantum-confinement effects.

The optical properties of Core/shell are somehow different from core itself (Fig. 33b). As could be seen in Fig. 6, the obtained band gap is 3.42 ev which shows a blue shift in comparison with bulk ZnO ($E_g^{bulk}=3.3ev$). In our previous work, we reported red shift in ZnO band gap (E_g=3.27 *ev*) prepared by same way. This blue shift could be resulted from two phenomena: first, in large band gap semiconductors, mainly, a reduction in size results in an extension in band gaps. As shown in TEM images some parts of ZnO shells in nanocomposite are thinner than 25nm which could make a blue shift. Also, Existence of exciton shoulder in absorption spectra of nanocomposite is another claim for making ZnO with thicknesses low enough for emerging exciton shoulder at room temperature. Second, existence of p-type Co_3O_4 in core adjacent to n-type ZnO affect the final properties of nanocomposite and make a blue shift in final optical band gap of nanocomposite.

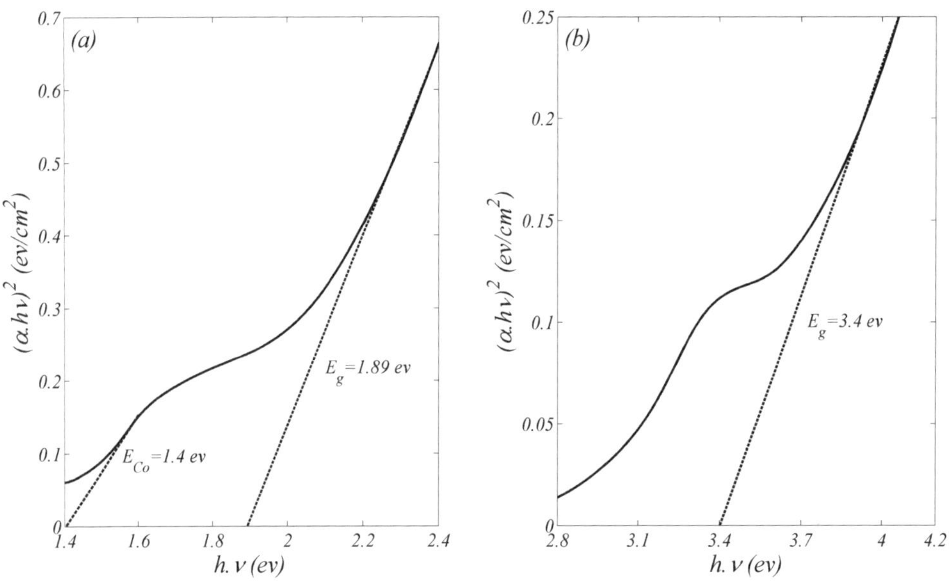

Fig. 33. $(\alpha h\upsilon)^{1/n}$ vs. photon energy for obtained a) Co_3O_4 and b) Co_3O_4/ZnO.

Fig. 34 shows magnetic behaviours of Co_3O_4 and Co_3O_4/ZnO core/shells. Also, in Table 11 the magnetic properties of prepared materials are listed. Co_3O_4 due to its spinel structure shows anti-ferromagnetic characteristics, however, the core shows dilute ferromagnetic characteristic. This might be due to existence of oxygen in spinel structure of Co_3O_4 which avoids direct interaction of magnetic momentums.

As could be seen from Table 11, the amount of saturation magnetism in core/shell nanocomposites shows a decrease in comparison with core. In core/shell Co_3O_4/ZnO nanocomposites there are two phenomena. ZnO is a diamagnetic material. These materials in VSM analysis show a negative magnetization and no hysteresis loop. ZnO can change its magnetic characteristic by doping of some impurities such as Co, Mn, Ni, etc. These dilute dopants can change the diamagnetic behaviour of ZnO into carrier induced ferromagnetic

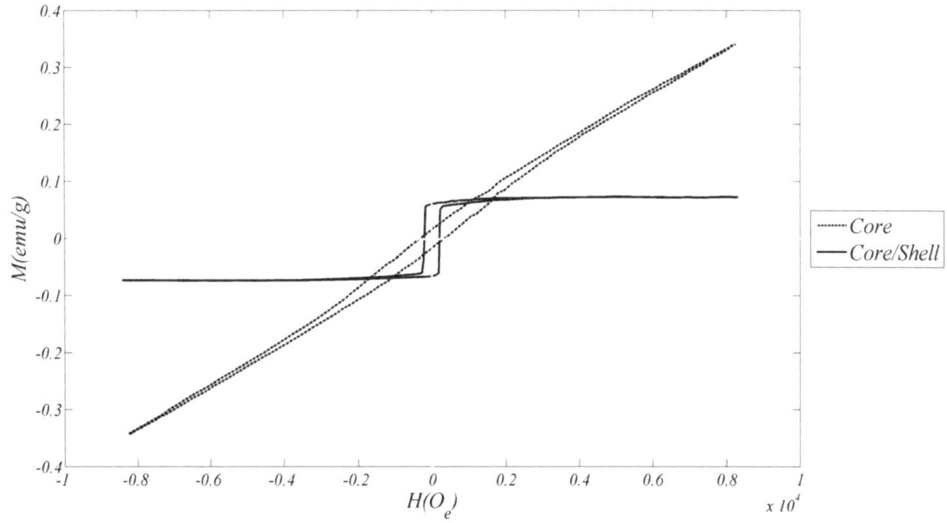

Fig. 34. VSM spectra of a) core and b) core/shell nanoparticles.

Sample	M_S(emu/g)	H_C(Oe)
Core	0.34	325
Core/shell	0.08	222

Table 11. Magnetic properties of core and core/shell.

materials. Existence of Co_3O_4 core adjacent to ZnO films results in a dilute ferromagnetic characteristic due to ability in ordering disorder spins arrays in surface of Co_3O_4 which in magnetic field could be ordered and compensation diamagnetic characteristic of shell component. On the other hand, by applying a diamagnetic shell on a magnetic material, saturation magnetism of core decreases. This decrease could be intensified by thickening of diamagnetic shell.

5. References

[1] Haga K., Ishii T., Mashiyama J., Ikeda T., Japanese Journal of Applied Physics 31 (1992) 3156.

[2] Kim H.T., Byun J.D., Kim Y., Materials Research Bulletin 33 (1998) 963.

[3] Kim H.T., Kim S.H., Nahm S., Byun J.D., Journal of the American Ceramic Society 82 (1999) 3043.

[4] Dulin F.H., Rase D.E., Journal of the American Ceramic Society 43 (1960) 125.

[5] Chang Y.S., Chang Y.H., Chen I.G., Chen G.J., Chai Y.L., Wu S., Fang T. H., Journal of Alloys and Compounds 354 (2003) 303.

[6] Kim H.T., Nahm S., Byun J.D., Kim Y., Journal of the American Ceramic Society 82 (1999) 3476.

[7] Golovchanski A., Kim H.T., Kim Y.H., Journal of the Korean Physical Society 32 (1998) S1167.

[8] Sohn J.H., Inaguma Y., Yoon S.O., Itoh M., Nakamura T., Yoon S. J., Kim H. J., Japanese Journal of Applied Physics 33 (1994) 5466.

[9] Li C.F., Bando Y., Nakamura M., Kimizuka N., Kito H., Mater. Res.Bull. 35 (2000) 351.

[10] Zakrzewska K., Thin Solid Films, 391 (2001) 229.

[11] Esmaielzadeh Kandjani A., Farzalipour Tabriz M., Pourabbas B., Mater. Res. Bull., 43 (2008), 645.

[12] M.R. Vaezi, S. K. Sadrnezhaad, A. Esmaielzadeh Kandjani, L. Nikzad, N.A. Arefian, S. Alibeigi, M. Farzalipour Tabriz, S.H. Mir Shah Ghassemi, J. Samei, Materials Science Poland, 25, No. 4, 2007, 1109- 1117.

[13] Haro-poniatowski E., Ostuski H., Sada E., J. Mater. Res., 9 (1994) 2102.

[14] Tang Z., Zhang J., cheng Z., Zhang Z., Journal of materials chemistry and physics 77 (2002) 314.

[15] Cuna W., Jincai Z., Xinming W., Bixian M., Guoying S., Pingan P., Jiamo F., Appl. Catal. B-Environ., 39 (2002), 269.

[16] Zhang M., An T., Hu X., Wang C., Sheng G., Fu J., Appl. Catal. A-Gen., 260 (2004) 215.

[17] Bandara J., Tennakone K., Jayatilaka P.P.B., Chemosphere, 49 (2002), 439.

[18] Wen Z., Wang G., Lu W., Wang Q., Zhang Q., Li J., Cryst. Growth. Des., 7 (2007), 1722.

[19] B.P.J. de Lacy Costello, Ewen R.J., Guernion N., Ratcliffe N.M., Sensor. Actuat. B-Chem., 87 (2002), 207.

[20] Sun S., Meng G., Zhang G., Zhang L., Cryst. Growth. Des., 7 (2007), 1988.

[21] Bagheri-Mohagheghi M. M., Shokooh-Saremi M., Thin Solid Films, 441 (2003), 238.

[22] Jiang J.C., Lian K., Meletis E. I., Thin Solid Films, 411 (2002), 203.

[23] Mukashev B. N., Tokmoldin S.Zh., Beisenkhanov N.B., Kikkarin S.M., Valitova I.V., Glazman V.B., Aimagambetov A.B., Dmitrieva E.A., Veremenithev B.M., Mat. Sci. Eng. B, 118 (2005), 164.

[24] Lorrain N., Chaffron L., Carry C., Delcroix P., Caër G. L., Mat. Sci. Eng. A, 367 (2004), 1.

[25] Pires F.I., Joanni E., Savu R., Zaghete M.A., Longo E., Varela J.A., Mater. Lett., 62 (2008), 239.

[26] Wu D. S., Han C. Y., Wang S. Y., Wu N. L., Rusakova I. A., Mater. Lett., 53(2002), 155.

[27] Jia Z. j., Zhu L. p., Liao G. h., Yu Y., Tang Y. W., Solid State Commun., 132 (2004), 79.

[28] Kobayashi T., Kimura Y., Suzuki H., Sato T., Tanigaki T., Saito Y., Kaito C., J. Cryst. Growth, 243 (2002), 143.

[29] Chouvin J., Branci C., Sarradin J., Fourcade J. O., Jumas J. C., Simon B., Biensan Ph., J. Power Sources, 81–82 (1999), 277.

[30] Culity B.D., *Elements of X-ray diffraction*, Addison-Wesley Company, USA, 1978.

[31] X. Zhong, R. Xie, Y. Zhang, T. Basche´ , W. Knoll, Chem. Mater., 17 (2005) 4038.

[32] Wu C., Qiao X., Chen J., Wang H., Tan F., Li S., Mater. Lett., 60 (2006), 1828.

[33] Zakrzewska K., Thin Solid Films, 391(2001), 229.

[34] M. Ando, T. Kobayashi, S. Iijima, et al. J. Mat. Chem. 7, 1779-1783, (1997)

[35] F. A. Cotton, G. Wilkinson, Advanced Inorganic Chemistry, Wiley, Newyork, 1972

[36] P. Cossee, Rec. Trav. Chim. Pays-bas 75 (1956) 1089

[37] L. Armelao, D. Barreca, S.Gross, A.Martucci, M.Tieto, E.Yondella, Journal of Non-Crystalline solids 293-295 (2001) 477-482
[38] C. H. kwon, H. K. Hong, D. H. Yun, K. Lee, S. T. Kim, Y. H. Roh, B. H. Lee, Sens. Actuators B, 24-25,610-613, (1995)
[39] G. Behr, W. Fhegel, Sens. Actuators B, 33-37, 2627-2632, (1995)
[40] C. Garzella, E. Comini, E. Tempesti, C. Frigeri, G. Sberveglieri, Sens. Actuators B, B 68 189–196, (2000)
[41] N. Barsan, M. Schweizer Berberich, W. Gopel, J. Anal. Chem., 365, 287-304, (1999)
[42] T.Seiyama, H. Futata, F. Era, N. Yamazoe, Mater. Sci. , 8 , 63-65, (1972)
[43] N. Xinshu, D. Weiping, D. Weimin, Sensors and actuators B 99(2004) 405-409
[44] X. Zhou, Y. Kobayashi, V. Romanyuk, N. Ochuchi, M. Takeda, S. Tsunekawa, A. Kasuya, Appl. Surf. Sci., 242 (2005) 281.
[45] Z. W. Zhao, Z. P. Guo, H. K. Liu, J. Power Sources, 147(2005) 264- 268.
[46] L. Wu, Y. Wu, X. pan, F. Kong, Opt. Mater. 28 (2006) 418-422.
[47] A. Shokuhfar, J. Samei, A. Esmaielzadeh Kandjani, M.R. Vaezi, Deffect Diffus. Forum, 273-276 (2008) 626-631.
[48] F. Pertlik, Monatsh. Chem. 130, (1999), 1083-1088.
[49] J.D.H. Donnay, D. Harker, Amer. Mineral. 22 (1937) 446.
[50] P. Hartman, W.G. Perdok, Acta Crystallogr. 8 (1955) 525.
[51] W.-Z. Zhong, G.-Z. Liu, Sci. China (B) 24 (4) (1994) 394.
[52] J. Du, Z. Liu, Y. Huang, Y. Gao, B. Han, W. Li, G. Yang, J. Cryst. Growth 280 (2005) 126–134.
[53] Y.H. Ni, X.W. Wei, X. Ma, J.M. Hong, J. Cryst. Growth 283 (2005) 48–56.
[54] G. Sun, M. Cao, Y. Wang, C. Hu, Y. Liu, L. Ren, Z. Pu, Mater.Lett. 60 (2006) 2777–2782.
[55] Z. Li, Y. Xiong, Y. Xie, Nanotechnology 16 (2005) 2303–2308.
[56] D. J. Shaw, Introduction to Colloid and Surface Chemistry, Butterworth-Heinemann, Oxford, 1992.
[57] R. Atkins, V.S.J. Craig, E.J. Wanless, S. Biggs, Adv. Colloid Interfac. 103 (2003) 219–304.
[58] W.J. Li, E.W. Shi, W.Z. Zhong, Z.W. Yin, J. Cryst. Growth 203 (1999) 186-196.
[59] W.Z. Zhong, D.Y. Tang, J. Cryst. Growth 166 (1996) 91.
[60] W.Z. Zhong, X.L. Yu, Sci. China Ser. E 41 (4) (1998).
[61] Y. Vronko, K. Gorbachev, V.V. Osiko, ICCG-10 Poster, B 34 (1992) 57.
[62] H. Kouta, Y. Kuwano, K. Ito, F. Marumo, J. Cryst. Growth 114 (1991) 676.
[63] Y. Wang, L. Sun, L.G. Kong, J.F. Kang, X. Zhang, R.Q. Han, J. Alloy Compd. 423 (2006) 256–259.
[64] B. Basavalingu, J. A. K. Tareen, G. T. Bhandage, J. Mater. Sci. Lett. 5 (1986) 1227-1229.
[65] I. N. Smallwood, Handbook of organic solvent properties, John-Wiley & Sons Inc., New york,1996.
[66] Y. Lu, Y.D. Yin, Z.Y. Li, Y.N. Xia, Nano Lett. 2, 785, (2002).
[67] N. Yu, X.P. Hao, X.G. Xu, M.H. Jiang, Mater. Lett. 61, 523, (2007).
[68] W.Z. Lin, L. Yi, Z. Ze, Handbook of nanophase and Nanostructured Materials, (Plenum, Kluwer Academic 2002).
[69] M.C. Hidalgo, M. Aguilar, M. Maicu, J.A. Navío, G. Colón, Catalysis Today. 129, 50, (2007

[70] M.R. Vaezi, Mater. Chem. Phys. 110, 89, (2008).
[71] N.V. Duy, N. V. Hieu, P. T. Huy, N. D. Chien, M. Thamilselvan, J. Yi, Physica E. 41, 258, (2008).

Part 3

Characterization and Fabrication

14

Strength Prediction of Composite Materials from Nano- to Macro-scale

Konstantinos I. Tserpes
Laboratory of Technology & Strength of Materials, University of Patras
Greece

1. Introduction

The primary scope behind composite materials development was to reduce structural weight and cost. Nowadays, after more than 30 years of development, this scope has been partially achieved through the extensive use of CFRP laminates in aeronautical, automotive and marine applications. During these years, the evolution of composite materials has being mainly accomplished experimentally; the contribution of theoretical modeling although necessary is limited due to the difficulties in stress and failure analyses originating from material complexity. In order to extend the use of composite materials, specific structural problems, such as the increase in out-of-plane strength of laminated composites and effective joining between composite parts, must be resolved. Towards the resolution of these problems, textile composites with 3D reinforcement have been proposed and adhesive bonding is being forwarded as an alternative to mechanical fastening. However, the understanding of the mechanical performance of textile composites is still evolving, while the integrity of adhesively composite bonded joints needs to be improved.

In the last decade, another area has come to the front in the science of composite materials; namely nanocomposites. The spunk for the offset of this area was the discovery of carbon nanotubes and the thereafter findings about their extraordinary mechanical properties. Since then, several works have been reported on the reinforcement of polymers by carbon nanotubes and other nanoparticles. The first findings are very promising; however, there are many issues, such as the strength of the nanotube/polymer interface, the role of nanotube alignment and the structural defects in nanotubes, that must be clarified before nanocomposites are set to production. In the case of nanocomposites, theoretical modeling is more important than in conventional composites since experimenting in the nano-scale is a very tedious task.

It is obvious that theoretical modeling plays a major role in the future developments in the science of composite materials. Among methods used to predict strength of composite materials and structures, progressive damage modeling (PDM) is probably the most adequate since it is capable of simulating the initiation and progression of all damage types in composites and predict strength of the specimen or structure. As the PDM simulates reality, it serves as an ideal complement to experiments because it can be directly compared to any type of experimental result, used for further evaluation of the experimental findings and shed light in damage details which cannot be explained experimentally. In the following, a short historical overview and a description of the method are given.

2. The PDM method

Damage is a multi-scale phenomenon that occurs from the atomic scale (nm) to structural scale (m). It is also of progressive nature as it evolves through the stages of initiation and propagation to catastrophic fracture. Prediction of damage is a diachronic engineering challenge. Most of the models developed for this purpose have been concentrated to one stage; either to damage initiation, thus being unable to predict strength, or to catastrophic fracture, thus neglecting damage history. A progressive damage model takes into account all three stages by predicting damage initiation and progression using failure criteria, simulating the influence of damage presence using material property degradation rules and finally, predicting strength using a final failure criterion. Furthermore, as it simulates the actual phenomenon, it can be implemented, by incorporating the appropriate tools, to engineering problems of any scale ranging from atomic to macroscopic.
The origin of the progressive damage modeling (PDM) method is traced back on 1958 to the damage model of (Kachanov, 1958). However, the onset for the systematic use of the method waited until 1987 when (Chang & Chang, 1987) demonstrated its capability in predicting strength of composite materials. Since then, the method has been evolved rapidly and has found several applications in materials and structures lying in different scales.
The PDM method comprises the modules of *stress analysis*, *failure analysis* and *material property degradation*. These modules integrate in an iterative procedure which is completed as soon as a *final failure criterion* is fulfilled. The iterative procedure in its general form is explained by means of the flowchart shown in Fig.1. In the various applications of the method, many alterations of this flowchart have been adopted. However, the general idea has been kept the same. As the procedure is repeated for each load-level, all data regarding the stress- and strain-field as well as the damage state can be stored and recalled thus, enabling the full description of the component's mechanical behavior.

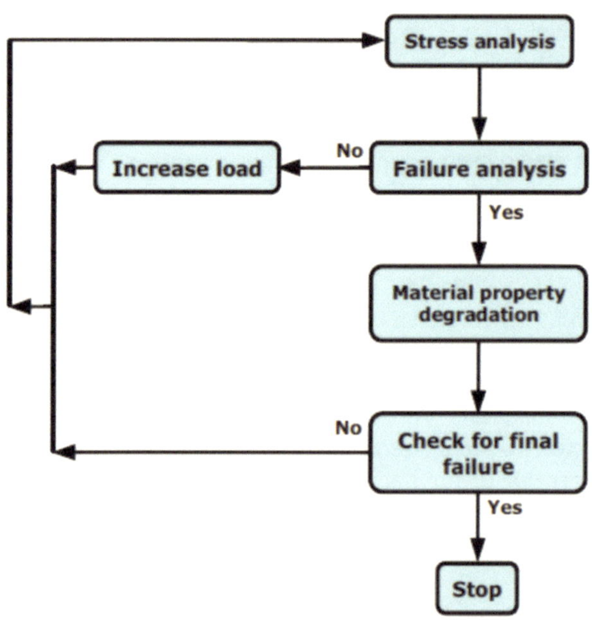

Fig. 1. Flowchart of the PDM method.

3. Summary of the chapter

In this chapter, the capability of the PDM method in predicting strength of composite materials lying in different scales is demonstrated through the description of method's application to 4 different structural problems lying from nano- (nm) to macro-scale (m). The cases considered are schematically described and scaled in Fig.2.

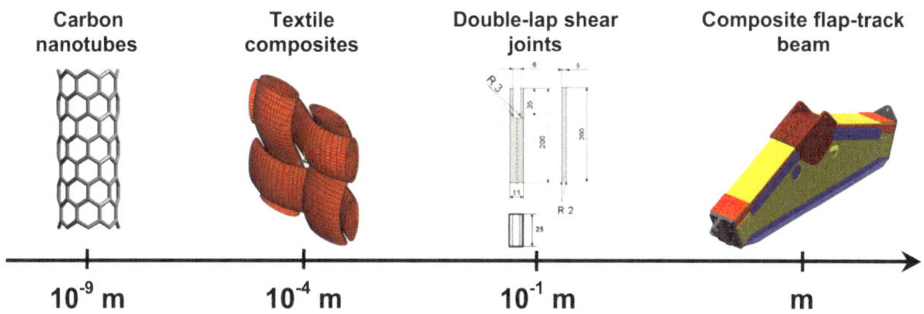

Fig. 2. Schematic description and scaling of the cases in which the PDM method has been applied.

4. Carbon nanotubes

Carbon nanotubes (CNTs) are fullerene-related structures that became public by (Iijima, 1991). They can be visualized as graphene sheets rolled into hollow cylinders composed of hexagonal carbon rings. The nanotube structure is described in terms of a chiral vector defined by the pair of indices (n,m). According to the values of the indices, three nanotube types arise; namely, the *armchair* (n,n), the *zigzag* $(n,0)$ and the *chiral* (n,m) CNTs. As the hexagonal pattern is repeated periodically, it leads to binding of each carbon atom to three neighboring atoms via covalent bonds. This covalent bond is one of the strongest chemical bonds in nature. Thus, it was expected to lend impressive mechanical properties to CNTs. Indeed, it has been experimentally and theoretically confirmed by several studies reported in the last decade that carbon nanotubes possess extraordinary mechanical properties (elastic modulus higher than 1 TPa, tensile strength in the range of 150 GPa, elastic strain up to 5% and failure strain up to 20%).

Combining extraordinary mechanical properties and fiber-like structure, CNTs offer unique potential for reinforcing polymers either as replacements of conventional fibers or as fillers to enhance the properties of the existed advanced composites. In (Qian et al., 2002) it has been demonstrated that with only 1% (by weight) of CNTs added in a matrix material, the stiffness of a resulting composite film can increase between 36% and 42% and the tensile strength by 25%. In (Yokozeki et al., 2007) it was observed a clear retardation in matrix cracking onset and accumulation in composite laminates filled with CNTs. Before establishing CNTs as mechanical reinforcements, their mechanical performance needs to be fully understood. At present, since experimentation at the nanoscale is still evolving, the only available tool for accomplishing this task is modeling. To date, two approaches have been mainly adopted: the atomistic and the continuum mechanics approaches. The atomistic simulations are limited in both time and length scales. For example, a 10 nm long single-walled carbon nanotube involves more than 2000 atoms and its atomistic study is computationally intensive. The lack of continuum mechanics models combined with the limited practical applications of atomistic models necessitates the development of continuum mechanics models able to simulate the mechanical performance of CNTs with large number of atoms under complex mechanical loading conditions, by considering also the topological and/or vacancy defects that may have introduced in the nanotube during its synthesis process. In (Tserpes & Papanikos, 2006), an atomistic-based progressive fracture model, which fulfills these requirements, was developed. The model will be described in the following sections.

4.1 A progressive fracture model for CNTs

The progressive fracture model is an atomistic-based continuum approach as it uses the FE method to analyze the structure of CNTs and an interatomic potential to describe the non-linear behavior of the C-C bonds. In the model, CNTs are treated as space-frame structures (Tserpes & Papanikos, 2005). As the FE model was developed using the ANSYS FE code, the model was entirely implemented using the ANSYS APDL macro-language.

4.1.1 Simulation of non-linear behavior of the C-C bonds

For simulating the non-linear behavior, the pairwise modified Morse potential (Belytschko et al., 2002) was used. This decision was strongly enforced by the simplicity of the specific potential over many-body potentials. To date, the modified Morse potential has been adequately applied in a number of cases where there are large deviations from equilibrium due to the presence of large strains.
According to the modified Morse potential, the potential energy of the nanotube system is expressed as

$$E = E_{stretch} + E_{angle} , \tag{1}$$

$$E_{stretch} = D_e\{1 - e^{-\beta(r-r_0)}]^2 - 1\} , \tag{2}$$

$$E_{angle} = \frac{1}{2}k_\theta(\theta-\theta_0)^2[1 + k_{sextic}(\theta-\theta_0)^4] , \tag{3}$$

where $E_{stretch}$ is the bond energy due to bond stretching and E_{angle} the bond energy due to bond angle-bending, r is the current bond length and θ is the current angle of the adjacent bond. The parameters of the potential are:

$$r_0 = 1.421\times10^{-10}\,m\,,\ D_e = 6.03105\times10^{-19}\,Nm\,,\ \beta = 2.625\times10^{10}\,m^{-1}\,,\ \theta_0 = 2.094\,rad\,,$$
$$k_\theta = 0.9\times10^{-18}\,Nm/rad^2\,,\ k_{sextic} = 0.754\,rad^{-4}\,.$$

As bond stretching dominates nanotube fracture and the effect of angle-bending potential is very small, only the bond stretching potential has been considered.
By differentiating Eq.(2), the stretching force of atomic bonds is obtained in the molecular force-field as

$$F = 2\beta D_e(1 - e^{\beta(r-r_0)})e^{-\beta(r-r_0)} \tag{4}$$

Fig.3 plots the relationship between force F and bond strain ε for the C-C bonds. The strain of the bond is defined by $\varepsilon = (r - r_0)/r_0$. As may be seen, the force-strain relation is highly non-linear at the attraction region especially at large strains. The inflection point (peak force) occurs at 19% strain. The repulsive force $(\varepsilon < 0)$ increases rapidly as the bond length shortens from the equilibrium length with less non-linearity than the attractive force.

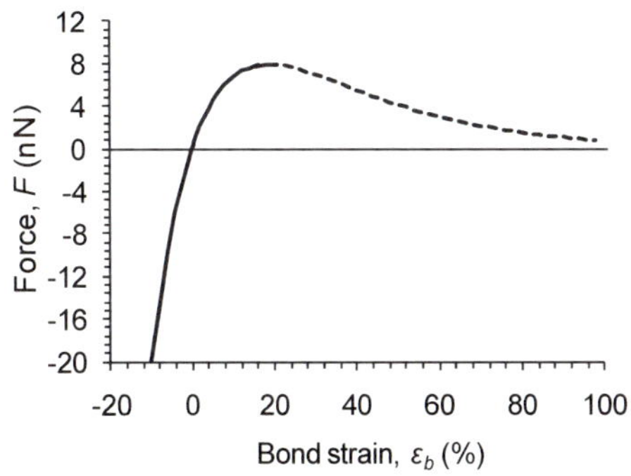

Fig. 3. Force-strain curve of the modified Morse potential.

4.1.2 Model algorithm

For modeling the C-C bonds, the 3D elastic ANSYS BEAM4 element was used. The non-linear behavior of the C-C bonds, as described by the interatomic potential, was assigned to the beam elements using the stepwise procedure of progressive fracture modeling, which is briefly described in the following lines. Initially, the stiffness of the beam elements is evaluated from the initial slope of the force-strain curve of the modified Morse potential (Fig.3) using the element's cross-sectional area A. The initial stiffness is 1.16 TPa. The nanotube is loaded by an incremental displacement at one of each ends with the other end being fully constrained. Zero transverse displacement was applied to the loading end in order to prevent buckling of nanotube at high loads. Fig.4 shows the FE mesh of the (20,0) nanotube along with the applied boundary conditions. At each load step, the stiffness of each element is set equal to $F/A\varepsilon$, where ε is the axial strain of the element as evaluated from the FE model and F is the interatomic force calculated using Eq.(4). This calculation of stiffness is applied until the inflection strain of the interatomic potential. When the axial strain of a bond reaches the inflection strain (19%) its stiffness is suddenly degraded (10^{-6} of the initial value) and the bond is disabled from carrying load. The next displacement increment is then applied to the nanotube and this iterative procedure goes on until catastrophic failure of the nanotube takes place.

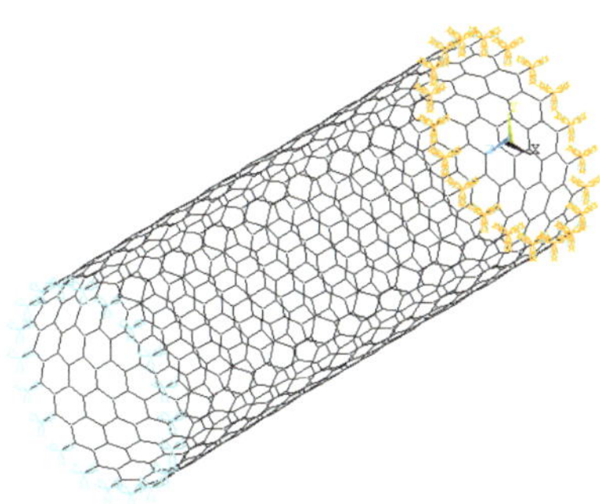

Fig. 4. FE mesh and boundary conditions of the (20,0) nanotube under axial tension.

4.2 Tensile behavior of pristine and defected CNTs

The progressive fracture model has been applied to simulate the tensile behavior of the zigzag (20,0) nanotube containing a defect of 10% weakening of one bond and a vacancy defect (1 missing atom). Numerical predictions are compared with the experimental measurements of (Yu et al., 2000) and the molecular mechanics simulations of (Belytschko et al., 2002). Fig.5 shows the comparison between the stress-strain curves. In (Yu et al., 2002) several experiments have been performed. The measured stress-strain curves show very large dispersion. As the comparison with the whole number of the curves would be valueless, three of those, which show the best correlation, have included in Fig.5. Curves that concern both types of initial defects are displayed. As may be seen, both theoretical models significantly overestimate the Young's modulus and strength of CNTs. The most possible cause for this discrepancy is the defects of unknown type and amount that appear in the nanotubes tested by (Yu et al., 2000) and not considered by the two models. In (Mielke et al., 2004) it was found that large approximately circular holes, which would be consistent with damage resulting from harsh oxidative purification processes, may substantially reduce the failure stresses and failure strains of CNTs, providing a likely explanation for the discrepancy. Specifically, in (Mielke et al., 2004) it was found that the one- and two-atom vacancy defects reduced the failure stresses by as much as 26% approaching thus the experimental failure stresses. Another possible cause, as stated in (Belytschko et al., 2002), is the slippage that possibly occurred at the attachments for the high strain cases reported in (Yu et al., 2000) resulting in a decrease of the measured values of nanotube Young's modulus. This kind of slippage would also imply that the failure strains are actually smaller than reported.

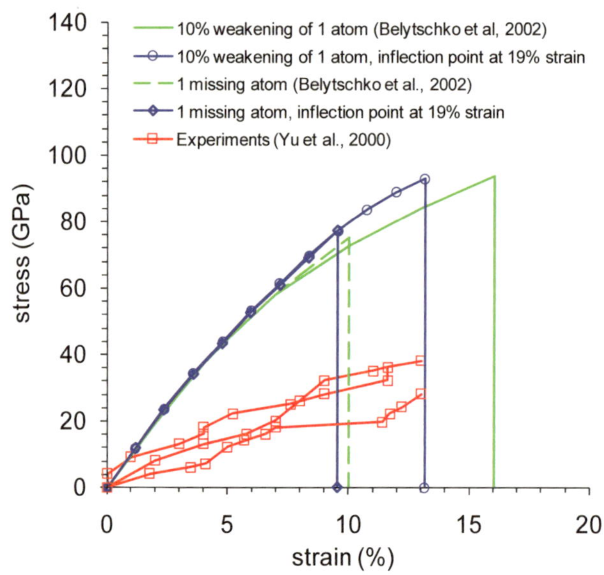

Fig. 5. Comparison between predicted tensile stress-strain curves with experimental and theoretical curves obtained from the literature.

Concerning the comparison between the progressive fracture model and the simulations of (Belytschko et al., 2002), a very good agreement is achieved regarding the Young's modulus and tensile strength of the nanotubes. The curves obtained by the two methods for both

types of initial defects coincide up to 8% strain giving the same Young's modulus for the nanotubes. For strains larger than 8%, the model predicts higher Young's modulus than the molecular mechanics simulations. Nevertheless, the fracture strain predicted by the model in this case (13.12%) was equal to the one obtained by two different experiments of (Yu et al., 2000).
Fig.6 shows the evolution of fracture as function of nanotube elongation in the (20,0) tube, in which the vacancy defect was modeled. As may be seen, in this case, fracture propagated circumferentially at the same row of bonds without any spreading until fracture of all bonds around circumference (separation of nanotube). The same evolution of fracture has been also obtained by (Belytschko et al., 2002) for the same nanotube and initial defect.

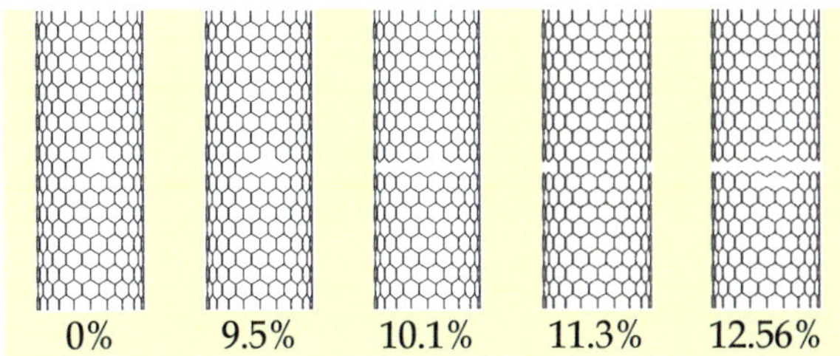

Fig. 6. Predicted evolution of fracture at the (20,0) nanotube as a function of applied strain.

The progressive fracture model described in this section is in advantage over classical atomistic approaches, such as molecular mechanics, in the sense that is simpler and performs in much smaller CPU times. All the analyses conducted within the frame of the current work took less than 1 minute in a Pentium 4 CPU 3.2 GHz, 1.0 GB RAM personal computer. Consequently, the model can be applied in CNT systems with very large number of atoms where the use of molecular mechanics simulations is unpractical. Moreover, the adoption of the FE method gives the model the potentiality to also simulate CNT-based nanostructures.

5. Textile composites

Textile composites, especially those with 3D reinforcement, offer enhanced through-thickness performance which is the weak point of laminated composites. Understanding of the mechanical behavior of textile composites is in progress. Both experimental and theoretical tools are being adopted. Modeling of textile composites is a very tedious task due to their complicated micro-geometry. Existing numerical tools and analytical relations, as have been developed for unidirectional laminates, are inapplicable. For instance, the widely used layered elements available in commercial FE codes cannot be used for the stress analysis of textile composites. Multi-scale approaches, that are capable to efficiently transfer the material's behavior from the micro- to the macro-scale, offer a base for facing this demanding modeling task. Several works have been reported on the modeling of textile composites. Most of them are based on a representative volume element (RVE) homogenized using finite elements to derive the mechanical properties of the textile composite material. Only few works have considered failure initiation and progression in order to evaluate the mechanical response of the material. Therefore, there is still a need for modeling the mechanical response of textile composites based on simulations of failure behavior. The most proper way to accomplish this is homogenized PDM incorporating the appropriate tools. In (Tserpes & Labeas, 2010) a multi-scale PDM was used to simulate the mechanical response of plain weave composites and cellular solids. In this section, the first part of this work concerning plain weave composites will be summarized.

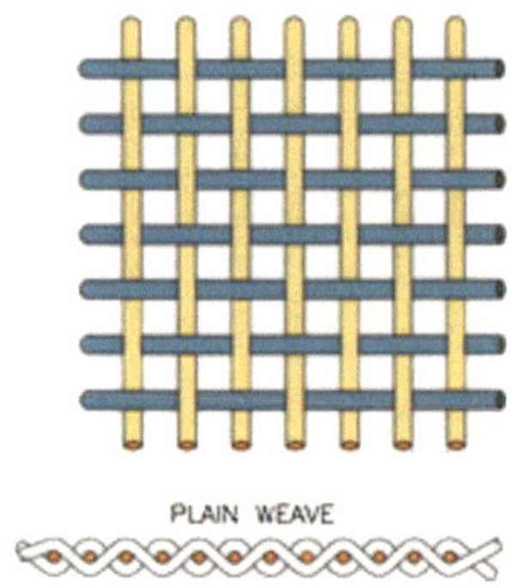

Fig. 7. Schematic of the plain weave composite material.

5.1 The multi-scale damage model

The flowchart of the multi-scale damage model is shown in Fig.8. The basic characteristic of the model is the proper definition of the RVE both in terms of geometry and size. Innovation of the model is the performance of homogenized PDM using up-to-date theoretical tools that ensures the accurate simulation of the mechanical response of the material system. In the following, the adjustment of each model component to the two material systems is described.

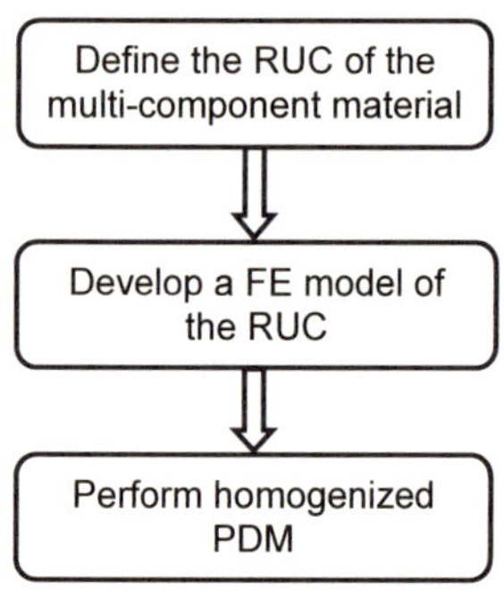

Fig. 8. Flowchart of the multi-scale damage model.

5.1.1 Development of the RVE

The RVE of the plain weave material system is defined as a rectangular volume containing two fill and two warp tows, with a length being representative of the crimp between them, and the surrounding matrix. The geometry of the RVE, taken from (Tang & Whitcomb, 2003) to enable comparison, was created using the software package WiseTex (Verpoest & Lomov, 2005) which possess specific capabilities in the creation of the micro-geometry of textile composite materials. The software requires as input the number of fill and warp tows, the crimp function and the geometry of the cross-sectional area of the tows. The RVE of the plain weave material is schematically described in Fig.9.

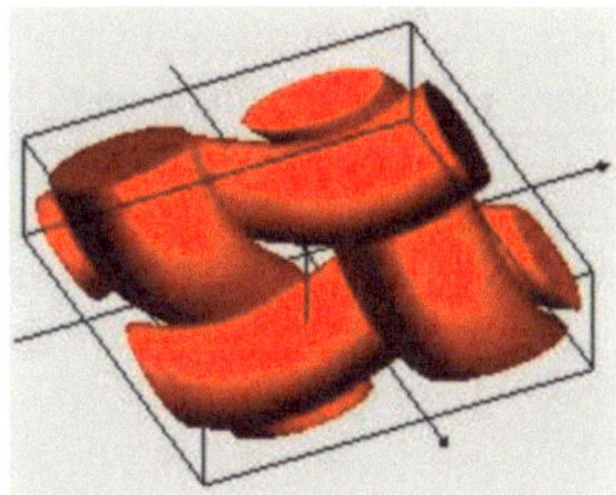

Fig. 9. Schematic representation of the plain weave's RVE.

5.1.2 FE modeling of the RVE

Creation of the FE model of the RVE was done by means of a semi-automated procedure: volumes, areas and lines were initially automatically generated in the ANSYS code by importing the RVE geometry of the RVE from WiseTex (Verpoest & Lomov, 2005) and the FE mesh of the RVE was generated manually in order to achieve compatibility between the different parts. Due to mesh compatibility, the coincident nodes of different parts were merged, thus implying a fully bonded structure. All parts have been modeled using the 3D 8-noded SOLID185 element which is appropriate for 3D modeling of solid structures. Typical FE meshes of the RVE and fiber tows are shown in Fig.10. Loading conditions and periodic boundary conditions have been also applied through the WiseTex software by means of a separate input file. This procedure requires as input the direction and magnitude of the applied strain and the software generates automatically the periodic boundary conditions for the specific load-case. In the present application, an incremental axial strain was applied to simulate axial tension of the plain weave.

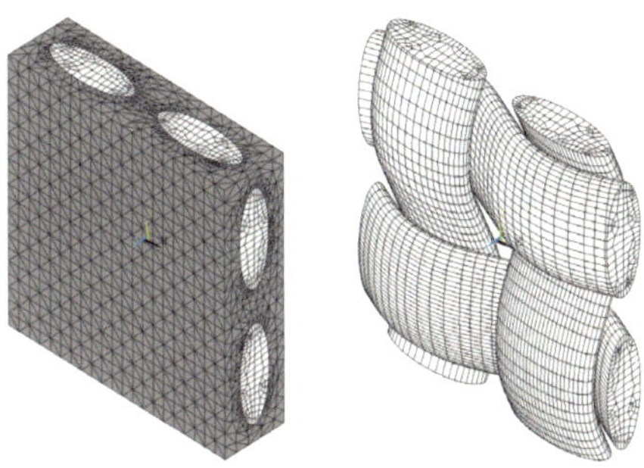

a b

Fig. 10. FE mesh of a. the full RVE of the plain weave, and b. the tows.

5.1.3 Damage analysis

Failure analysis of textile composites is a tedious task since the available failure theories have been developed for unidirectional laminates. However, it has been shown recently that the Hashin-type polynomial failure criteria give in some cases accurate predictions for textile composites also. In (Tserpes & Labeas, 2007) the stress-based Hashin-type failure criteria have been successfully applied to simulate the tensile behavior of non-crimp fabric composites, while in (Xiao et al., 2007) the strain-based Hashin-type failure criteria have been used to simulate progressive damage and delamination in PW S-2 glass/SC-15

composites under quasi-static punch-shear loading. In the present work, the strain-based criteria were used as they fit well with the damage mechanics approach used to imply material property degradation.
For the tows, four different failure modes have been considered; namely: fiber tensile/shear failure, matrix cracking/shear failure, fiber compressive failure, matrix compressive failure, fiber in-plane shear failure and delamination failure. The failure criteria for the detection the failure modes are described in the following
Fiber tensile/shear failure:

$$r_1^2 = \left(\frac{E_f \varepsilon_f}{S_{ft}}\right)^2 + \left(\frac{G_{fn}\varepsilon_{fn}}{S_{fn}}\right)^2 \quad (f\text{-direction}) \tag{5}$$

Matrix cracking/shear failure:

$$r_2^2 = \left(\frac{E_w \varepsilon_w}{S_{wt}}\right)^2 + \left(\frac{G_{wn}\varepsilon_{wn}}{S_{wn}}\right)^2 \quad (w\text{-direction}) \tag{6}$$

where *f*, *n* denotes the in-plane fill and out-of-plane directions, respectively. *t*, *s* denote the tensile and shear strengths, respectively. *E* and *G* are the axial and shear elastic moduli, respectively, while ε is the tensile strain. S_{ft} and S_{wt} are the tensile strengths in the fill and warp directions, respectively, while S_{fn} and S_{wn} the shear failure strengths in the *fn* and *wn* planes, respectively.
Fiber compressive failure:

$$r_3^2 = \left(\frac{E_f \cdot \varepsilon'_f}{S_{fc}}\right)^2, \varepsilon'_f = -\varepsilon_f - \langle \varepsilon_n \rangle \frac{E_n}{E_f} \quad (f\text{-direction}) \tag{7}$$

Matrix compressive failure:

$$r_4^2 = \left(\frac{E_w \cdot \varepsilon'_w}{S_{wc}}\right)^2, \varepsilon'_w = -\varepsilon_w - \langle \varepsilon_n \rangle \frac{E_n}{E_w}, \quad (w\text{-direction}) \tag{8}$$

where S_{fc} and S_{wc} are the compressive strengths in the fill and warp directions, respectively.
Fiber in-plane shear failure:

$$r_5^2 = \left(\frac{G_{fw} \cdot \varepsilon_{fw}}{S_{fw}}\right)^2 \tag{9}$$

where S_{fw} is the tow shear strength due to matrix shear failure.
Delamination:

$$r_6^2 = S^2 \left[\left(\frac{E_n \cdot \varepsilon_n}{S_{nt}}\right)^2 + \left(\frac{G_{wn} \cdot \varepsilon_{wn}}{S_{wn0}}\right)^2 + \left(\frac{G_{nf} \cdot \varepsilon_{nf}}{S_{fn0}}\right)^2\right] \tag{10}$$

where S_{nt} is the through-thickness tensile strength, S_{wn0} and S_{fn0} are the interlaminar shear strengths in *wn* and *fn* planes, respectively and *S* is a factor to take into account stress concentration on the growth of delamination equals to 1.2.

It is well known that textile composites and especially the PW possess a non-linear behavior due to matrix cracking and shear failures that appear at resin-rich pockets and tows. As these failure modes result from complex stress states, in order to predict them failure criteria that incorporate coupling between different stresses must be adopted. Such criteria are those described above. Along with failure criteria, a gradual material property degradation concept must be applied. Such a concept is the damage mechanics concept of (Matzenmiller et al., 1995) that incorporates also features for controlling strain softening. Note that the specific combination of the failure criteria and material property degradation has been used successfully in (Xiao et al., 2007). According to the damage mechanics concept, the presence of damage is simulated by the reduction of stiffness based on the functions

$$E_i = (1-\omega_i)E_{i0},\ G_i = (1-\omega_i)G_{i0} \tag{11}$$

where E_{i0} and G_{i0} are the initial values of the moduli and ω_i is the damage variable given by

$$\omega_i = \max\{q_{ij}\phi_j\},\ j = 1,\ldots,6 \tag{12}$$

$$\phi_i = 1 - e^{\frac{1}{m}(1-r_j^m)},\ r_j \ge 1 \tag{13}$$

where r_j is the failure index derived from Eq.(5), m is the softening parameter equals to 2 and q_{ij} is the vector-value function indicating coupling between damage variables and damage modes $(j = 1,\ldots,6)$ described in the following table

$$[q] = \begin{bmatrix} 1 & 1 & 1 & 0 & 0 \\ 0 & 0 & 1 & 0 & 0 \\ 0 & 0 & 1 & 0 & 1 \\ 1 & 1 & 1 & 1 & 1 \\ 0 & 0 & 1 & 0 & 0 \\ 1 & 1 & 1 & 0 & 1 \end{bmatrix} \tag{14}$$

Combining Eqs.(8) and (10) we get

$$\begin{aligned} \omega_1 = \max\{\phi_1,\phi_2,\phi_3\} \\ \omega_2 = \phi_3 \\ \omega_3 = \max\{\phi_3,\phi_5\} \\ \omega_4 = \max\{\phi_1,\phi_2,\phi_3,\phi_4,\phi_5\} \\ \omega_5 = \phi_3\ \omega_6 = \max\{\phi_1,\phi_2,\phi_3,\phi_5\} \end{aligned} \tag{15}$$

The damage variables apply with the above order to $E_f, E_w, E_n, G_{fn}, G_{wn}$ and G_{fn}, respectively.

The modules of stress analysis, failure analysis and material property degradation are integrated in an iterative algorithm. To demonstrate the algorithm, consider the fiber

tensile/shear failure in fill direction. From the stress analysis the strains are derived for each element. Then, parameter r_1 is derived from failure criterion of Eq.(7), ϕ_1, ϕ_2 and ϕ_3 from Eq.(12) and ω_1 from Eq.(15) to finally evaluate the degraded value of E_f from Eq.(11).
For the epoxy resin, the Maximum Stress failure criterion was used to check for possible failures in all material directions. Whenever a stress exceeded strength, the stiffness in the respective material direction was totally degraded.

5.2 Tensile response of plain weave composites

A study on the effect of degradation factor m on the predicted tensile response of the PW was conducted. Fig.11 compares the predicted tensile stress-strain curves for the cases of m =0.001, m =2 and m =8 with the curve predicted by (Tang and Whitcomb, 2003) using the maximum stress failure criterion and sudden material property degradation rules. The different degradation concepts are explained in Fig.12 by means of the resulted tensile stress-strain curves. The simple sudden degradation concept implies a constant Young's modulus until failure is predicted by the failure criterion. Then, the Young's modulus drops to a very small value, thus implying a continuously increasing strain with a constant stress. On the other hand, the damage mechanics concept of (Matzenmiller et al., 1995) implies a non-linear behavior from the first stages of loading as the Young's modulus decreases with increasing strain until a large drop to take place at a certain strain threshold. From that point and on the material behavior is governed by the degradation factor. For very small values of m (e.g. 0.001), the concept tends to coincide with the sudden degradation concept. With increasing m, a softening part appears. The softening rate increases with increasing m. The difference in the degradation concepts, in terms of the predicted tensile response of the plain weave material, is descriptively illustrated in Fig.8. The sudden degradation concepts give a bilinear behavior of the plain weave material. Transition between the two linear regimes is due to the accumulation of matrix cracking in the matrix and warp tows. The most significant finding is the large deviation in the predicted maximum load between the sudden degradation concepts and the damage mechanics concept. The deviation between the prediction of (Tang & Whitcomb, 2003) and the present prediction taken with m =0.001 is attributed to the different failure criteria used. In the case of the damage mechanics concept, the use of m =8 led to a more degraded behavior of the material compared to m =2 especially at large strains.
A physical interpretation can be given to the damage mechanics concept by relating the degradation in material's behavior with the damage mechanisms. Matrix cracking in warp tows as well as fiber in-plane shear failure initiate at the very early stages of loading and continue to accumulate until final failure to occur. Those failures are responsible for the non-linear behavior of the material. Final failure, considered to take place at maximum sustained load, is due to fiber tensile/shear failure in fill tows initiated at the areas of maximum curvature where longitudinal strain maximizes (see Fig.13). At this point, a small amount of delamination is also predicted to occur mainly at specimen edges attributed to stress redistribution caused by the accumulation of the aforementioned failure modes.
To conclude the present section, a non-linear tensile response, governed by matrix cracking at the resin-rich areas and warp tows, is predicted for the plain weave material. Final failure is due to fiber tensile failure in the fill tows. The comparison between the damage mechanics concept and the sudden degradation concept showed significant deviations mainly at the predicted maximum load. Due to the non-linear response of the material, the damage

mechanics concept is more appropriate as it incorporates features for controlling material softening. However, the proper use of the concept requires calibration of the degradation factor based on experiments.

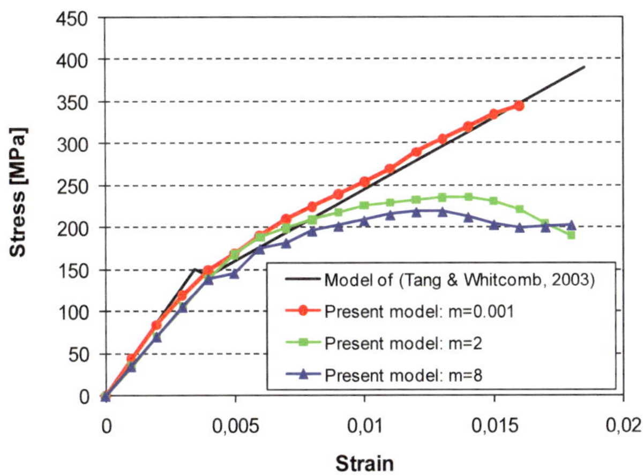

Fig. 11. Comparison of the predicted tensile stress-strain curves of the plain weave material obtained using m =0.001, m =2 and m =8 with the curve obtained in (Tang & Whitcomb, 2003) using the maximum stress failure criterion and the sudden degradation concept.

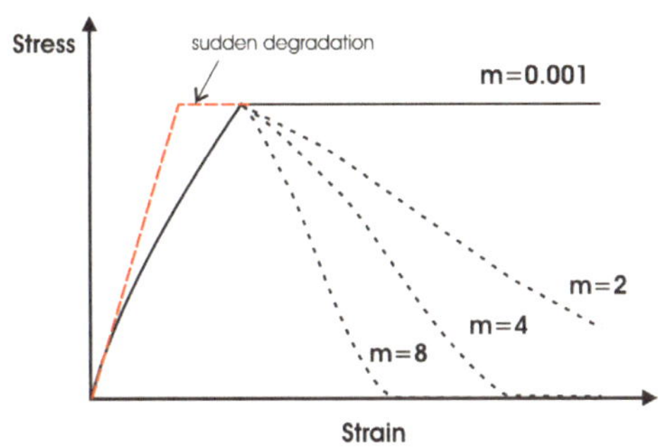

Fig. 12. Illustration of the different degradation concepts.

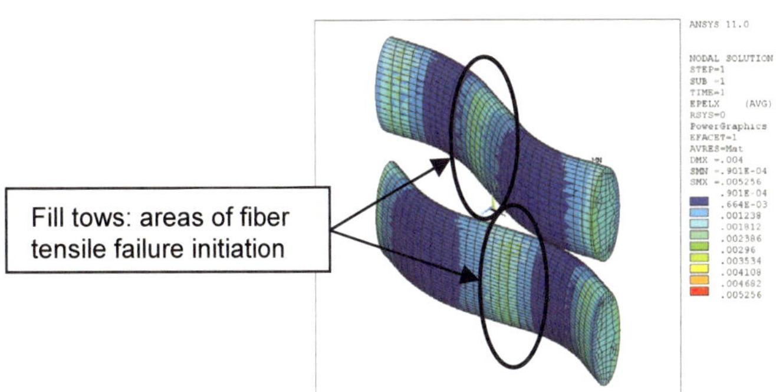

Fig. 13. Contour of longitudinal strain in fill tows and indication of fiber tensile failure initiation.

6. Double-lap shear joint

Currently in aerostructures carbon-fiber reinforced plastics are being mainly assembled by mechanical fasteners. This type of design implies weight penalties emanating from the need to deal with the stress concentrations developed around the bolts. Adhesive bonding serves as the most compatible alternative joining method for composite structural parts providing significant cost and weight savings. For establishing adhesive bonding as a reliable joining method, the ability of bonded joints to efficiently transfer load between assembled parts must be fully ensured. This pertains equally to the integrity of the joining element and bondline. As bonded joints are designed such that the load is transferred through shear, normal tensile loads arise in specific areas of the composite joining element. In bonded joints between traditional composite laminates such loads may lead to delamination in either the joining element or the assembled parts. Therefore, for this kind of applications, new composite materials with enhanced through thickness properties must be employed. Such materials are the 2D and 3D woven fabrics and the non-crimp fabric (NCF) composites.

In (Llopart et al., 2010) several aspects of adhesive bonding technology mentioned in this paragraph have been addressed by using the mesomechanical model developed in (Tserpes & Labeas, 2009) to study the influence of imperfect bonding, in terms of partial lack of adhesion, on the tensile strength of NCF Pi-shaped double-lap shear adhesively bonded joint schematically described in Fig.14. Manufacturing of the NCF joint elements and bonding between the assembled parts were carried out using novel techniques. Quality of adhesive bonding was tested using ultrasonic C-scan inspection.

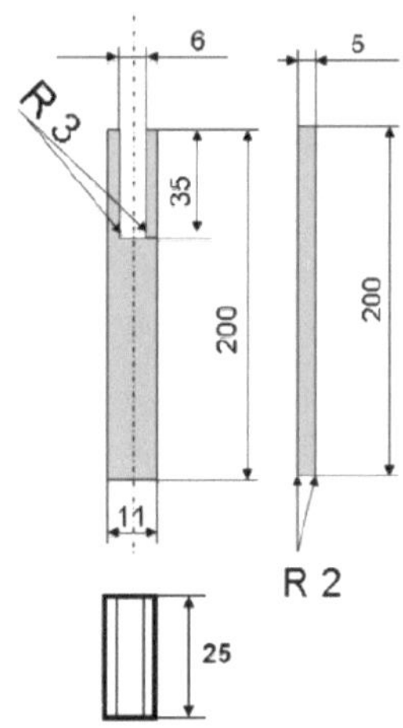

Fig. 14. Schematic representation of the NCF Pi-shaped DLS joint.

6.1 The mesomechanical model

A general description of the mesomechanical model is given by means of the flowchart shown in Fig.15. In brief, the model comprises the steps of:

- definition of the RVE of the woven fabric composite,
- characterization of the mechanical behavior of the RVE using local homogenized PDM,
- development of the FE model of the joint,
- assignment of the behavior of the RVE to the elements of the FE model and implementation of PDM to simulate the global mechanical performance of the joint.

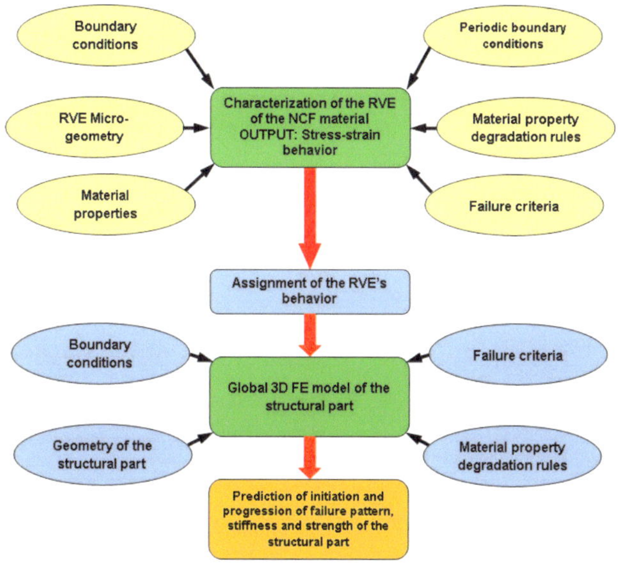

Fig. 15. The flowchart of the mesomechanical model.

6.1.1 Characterization of the mechanical behavior of the RVEs

The DSL joint comprises a combination of the dual layers $0^\circ/90^\circ$, $90^\circ/0^\circ$, $45^\circ/135^\circ$, $135^\circ/45^\circ$ and the unidirectional mono-layers 0°. A schematic description of the joint's lay-up is shown in Fig.16(a). The first two modules of the model concern the material characterization of these NCF HTS/RTM6 material system. For these dual layers, suitable RVEs have been defined in (Tserpes & Labeas, 2009). The result of this characterization is the stress-strain behavior of the two dual layers and summarized in Tables 1 and 2, respectively.

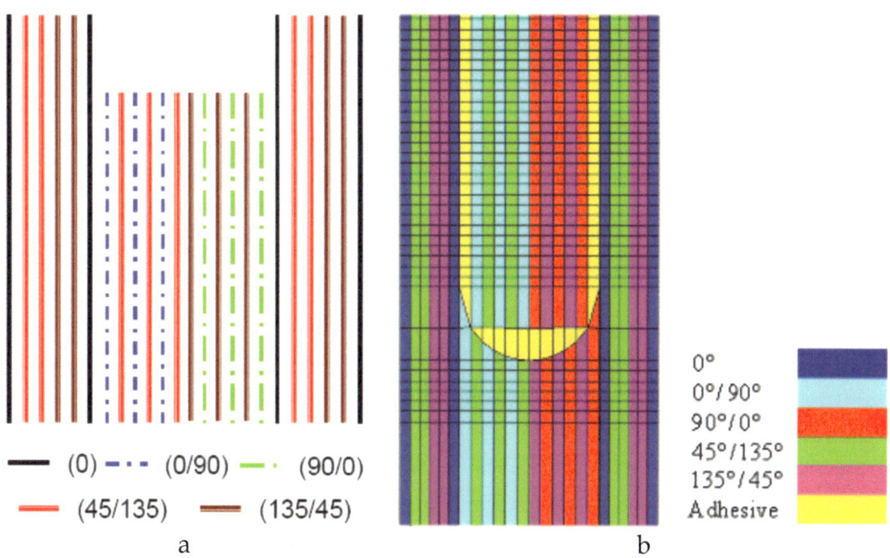

Fig. 16.a. Schematic description of the lay-up and b. Front-view of the FE mesh of the DLS joint.

	Simulated behavior of dual layer $0^\circ/90^\circ$	
Load- case	**Stiffness (MPa)**	**Strength (MPa)**
Longitudinal tension	$E_1^T = -6E+10\varepsilon_1^3 + 7E+08\varepsilon_1^2 - 3+06\varepsilon_1 + 75859$	$S_1^T = 629$
Longitudinal compression	$E_1^C = 76474$	$S_1^C = 764$
Transverse tension	$E_2^T = -595543\varepsilon_2 + 72379$	$S_2^T = 820$
Transverse compression	$E_2^C = 70413$	$S_2^C = 766$
Normal tension	$E_3^T = 11398$	$S_3^T = 74$
Normal compression	$E_3^C = 11398$	$S_3^C = 262$
In-plane shear	$G_{12} = 4250$	$S_{12} = 47$
Out-of-plane shear, *XZ*	$G_{13} = 4270$	$S_{13} = 47$
Out-of-plane shear, *YZ*	$G_{23} = 4290$	$S_{23} = 47$

Table 1. Numerically characterized behavior of the NCF HTS/RTM6 $0^\circ/90^\circ$ dual layer.

	Simulated behavior of dual layer $45^\circ/135^\circ$	
Load- case	**Stiffness (MPa)**	**Strength (MPa)**
Longitudinal tension	$E_1^T = 3E+10\varepsilon_1^3 - 5E+08\varepsilon_1^2 + 571209\varepsilon_1 + 22434$	$S_1^T = 111.4$
Longitudinal compression	$E_1^C = -1E+09\varepsilon_1^2 - 3E+06\varepsilon_1 + 19090$	$S_1^C = 48$
Transverse tension	$E_2^T = -9E+07\varepsilon_2^2 - 209462\varepsilon_2 + 21475$	$S_2^T = 117$
Transverse compression	$E_2^C = -2E+08\varepsilon_2^2 + 1E+06\varepsilon_2 + 20256$	$S_2^C = 145$
Normal tension	$E_3^T = 11398$	$S_3^T = 74$
Normal compression	$E_3^C = 11398$	$S_3^C = 278$
In-plane shear	$G_{12} = -7E+08\gamma_{12}^2 + 155230\gamma_{12} + 24864$	$S_{12} = 41$
Out-of-plane shear, *XZ*	$G_{13} = 4200$	$S_{13} = 46.5$
Out-of-plane shear, *YZ*	$G_{23} = 4261$	$S_{23} = 47$

Table 2. Numerically characterized behavior of the NCF HTS/RTM6 $45^\circ/135^\circ$ dual layer.

Index 1 refers to the direction of fibers, 2 to the transverse direction (matrix) and 3 to the normal direction. Index *T* refers to tensile and *C* to compressive.

6.1.2 FE modeling of the DLS joint

A 3D FE model of the DLS joint was constructed using the commercial FE code ANSYS. All parts were modeled using the structural solid SOLID45 element. Each of the composite biaxial layers was represented by a single row of elements. At each element, the behavior of the corresponding RVE described in Tables 2 and 3 was assigned. Fig.16 shows the correspondence between the dual layers and the element rows. At the elements with lay-up $135^\circ/45^\circ$ and $90^\circ/0^\circ$, the behaviors of the $45^\circ/135^\circ$ and $0^\circ/90^\circ$ RVEs were assigned, respectively. For the UD layer, the elastic properties and strengths of the unidirectional HTS/RTM6 lamina were used. A linear behavior up to failure was assumed for the UD layer.

In order to control and assure that the adhesive has spread out to the full surface on both sides of the insert/Pi-walls interface, non destructive tests (NDT) were carried out on the bonded specimens by means of ultrasonic C-scan. In order to quantify the lack of adhesive, the Imperfect Bonding Severity (IBS) ratio, equals to the total area without adhesive over the desired side bonded area (35 x 250 mm^2), is defined. The IBS ratio was automatically derived for each bonded plate using the statistic function of C-scan analysis.

Fig.17 shows the C-scan images of the DLS specimens for the three sets of tests. The two C-scan images correspond to the two sides of the Pi joint. The meaning of the different colors is explained in the figure. These sets were chosen because they are of representative imperfect bonding severity. As indicated by the values of the IBS ratio, sets 1, 2 and 3 comprise cases of medium, large and small severity, respectively. After NDT inspection, six specimens were cut out from each bonded plate.

Three different cases of imperfect bonding were modeled, namely C1, C2 and C3, being representative of the C-scan findings shown on Fig.17. Case C1 corresponds to the ideal situation (IBS=0% at both sides), case C2 corresponds to set 1 and case C3 to set 2. Lack of adhesive was simulated by subtracting from the FE model the elements of the adhesive included in the misbonded areas. Fig.18 shows the FE meshes of the adhesive for the three modeled cases. In C2 case, at one side of the joint 22% of the adhesive was subtracted. Likewise, in C3, 18% and 81% of the adhesive were subtracted from the two sides of the joint, respectively.

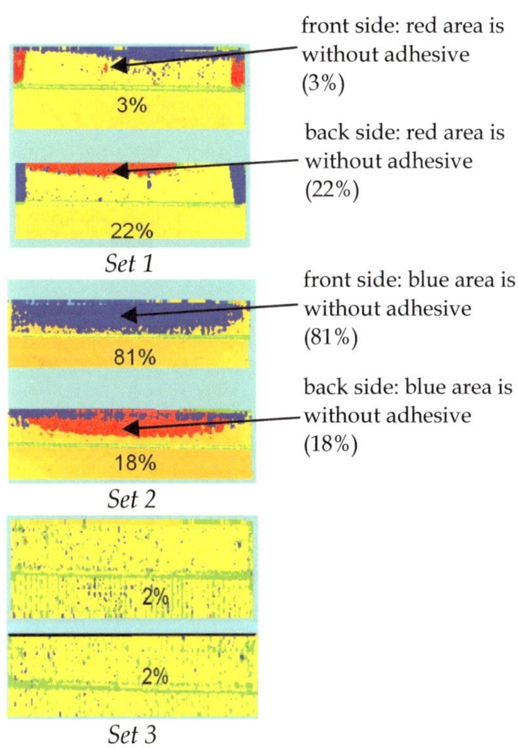

Fig. 17. C-scan images of the specimens indicating the amount of imperfect bonding.

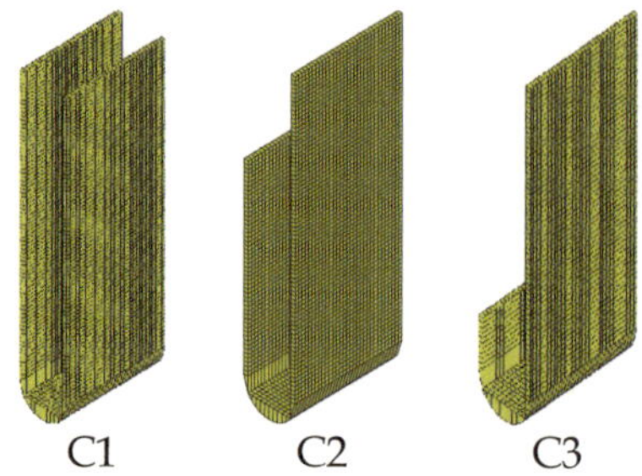

Fig. 18. FE meshes of the adhesives for the three cases modeled.

6.1.3 Damage analysis

At each load step, failure analysis and material property degradation are performed consecutively both at the NCF material (Pi-profile and insert) and the adhesive. Element failures are predicted by comparing stresses with material strengths at each direction (Maximum Stress failure criterion). As soon as failure is reached, the stiffnesses of the failed elements are degraded according to the severity of the failure. Failure in the fibers direction is assumed to be catastrophic and thus, all stiffnesses are degraded in order to totally disable the elements from carrying load. On the contrary, when failure is predicted in the directions transverse and normal to the fibers, the corresponding stiffnesses are degraded such as to disable the load-carrying capability only at the specific directions. Details about the material property degradation rules can be found in (Llopart et al., 2010) Prediction of debonding is fundamental since it is expected to be the primary failure mode of the DLS joint. Debonding is mainly due to shear failure of the adhesive between the Pi-flanges and insert caused by large shear stresses. Secondary debonding may occur due to tensile fracture owing to large normal tensile axial stresses developed between insert's head and Pi-slot. The aforementioned failures are respectively predicted using the following two criteria

$$\tau_{\max} = \frac{\sigma_1 - \sigma_3}{2} \geq \tau_a \tag{16}$$

$$\sigma_y \geq S^T \tag{17}$$

where $\tau_{\max}$ is the maximum shear stress at the adhesive, σ_1, σ_3 are the maximum and minimum principal stresses, respectively, and σ_y is the normal axial stress at the loading direction. As soon as failure is predicted in an element of the adhesive, its stiffness is totally degraded to simulate debonding. Plasticity of the adhesive has not been taken into account.

6.2 Numerical results

Fig.19(a) compares the predicted force-displacement curves for the three cases modeled. The curves have been plotted up to the maximum sustained force, which as in the experiments is set to be the strength of the specimen. Comparison reveals that the lack of adhesive causes a small decrease in the stiffness of the specimen and a dramatic decrease in strength. For the effect in strength to be clearer, predicted strengths along with the average experimental strengths of sets 1 and 2 are shown together in Fig.19(b). Set 3 has been excluded from the comparison due to the reasons discussed in section 4.1. Predicted strengths of imperfect bonded specimens C2 and C3 are degraded by 6.25 and 43.75%, respectively, with regard to the strength of the perfectly bonded C1 specimen. Similarly, the average experimental strength of set 2 specimens is degraded by 29.6% with regard to the strength of set 1 specimens.

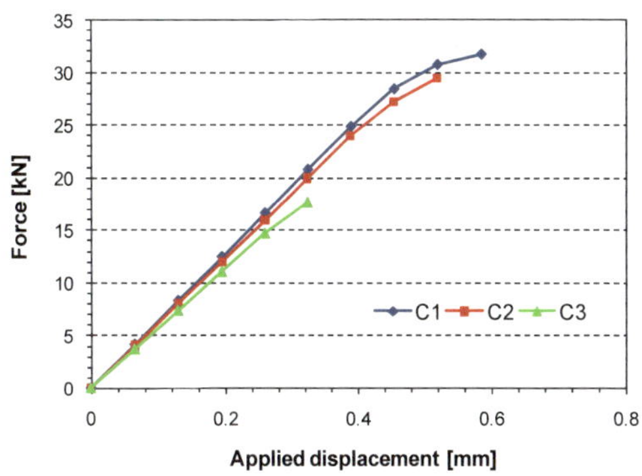

a

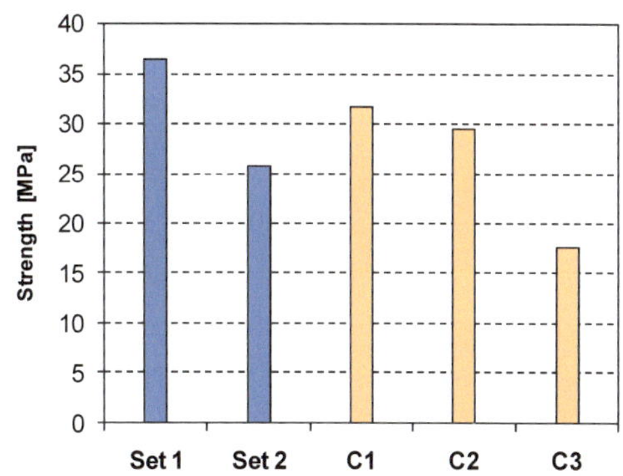

b

Fig. 19. a. The predicted force-displacement curves of the DLS joints, and b. Experimental and numerical failure loads of the DLS joints.

Due to its progressive nature, the model has the capability of simulating the failure initiation and progression at the NCF material and the adhesive. In the C1 specimen, shear failure of the adhesive (debonding) and transverse tensile failure (matrix cracking) of the UD composite layer are the primary failure modes predicted. Debonding initiated almost simultaneously at the area between insert's head and Pi-slot and at the upper edge of the adhesive where the respective stresses maximize. As the load increases, debonding propagated towards the center of the bonded area. Above a certain load level, propagation of debonding from the lower edge is due to shear stress. Initiation and progression of debonding as a function of pull-out force for the C1 case are illustrated in Fig.20a. Soon after debonding initiated, matrix cracking did so at the UD composite boundary layer as illustrated in Fig.20b. Specimen failed when more than half of the bonded area was debonded. At this point, matrix cracking had accumulated at almost half of the UD composite boundary layer. In C2 specimen, the same failure events with C1 took place with the difference that matrix cracking accumulated in much less degree at the UD composite layer. In C3 specimen, failure is due to adhesive shear failure and no other failures were predicted. Lack of adhesive at the imperfect bonded specimens causes the development of stress concentrations, since the same force is being transferred by less adhesive, which lead to early initiation and fast progression of adhesive shear failure. Thus, according to the amount of missing adhesive, failure at the NCF material either do not initiate or do not accumulate much due to the small loads being developed at the joint.

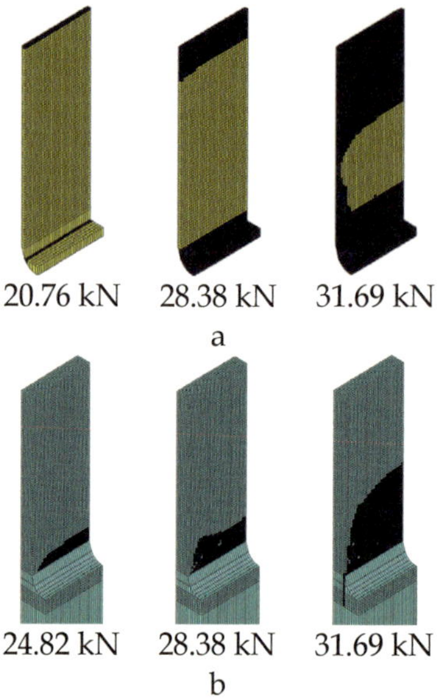

Fig. 20. a. Predicted initiation and progression of debonding (dark area) as a function of pull-out force at the C1 specimen and b. Predicted initiation and progression of matrix cracking as a function of pull-out force at the composite boundary layer of C1 specimen.

7. Adhesively bonded flap-track beam

In the frame of the European Project 'MOJO' (Modular Joints for Composite Aircraft Structures) (MOJO, 2006), a material-driven design for composite modular bonded joints was developed by treating both aspects of adhesive bonding and composites of enhanced through-thickness strength. The outcome of the research is the development of modular NCF Pi(π)-, H-, L- and T-shaped joining elements for adhesively bonded joints. The geometry and material of the elements were optimized with regard to strength of the element itself and the load-carrying capability of the bondline using experimental tests and numerical modeling. As a verification structure and demonstrator of the proposed design concept, the flap-track beam (FTB) of the Airbus A400M, originally made from state-of-the-art metallic materials and assembled exclusively by mechanical fasteners, was selected. For the verification and demonstration purposes, all parts, apart from the fittings, have been replaced by composite NCF parts and specific groups of fasteners have been replaced by adhesively bonded joints based on the developed joining elements. The criterion for the proposed design concept to be successful is for the modified FTB to have the capability of carrying the amount and type of load which the FTB has been originally designed to carry. In (Tserpes et al., 2010), the evaluation of the structural integrity of the FTB, performed by using a complicated experimental program and numerical modeling. In this chapter, the numerical part of the work will be briefly presented.

7.1 The FTB: structure, materials and loading conditions

A schematic representation of the FTB, drawn using the CATIA software, is shown in Fig.21a (D2.3.2, 2008). It consists of NCF panels assembled by mechanical fasteners through metallic connectors and by adhesive bonds through NCF joining elements. The disassembled FTB, in which the different parts are shown and the joining elements are indicated, is depicted in Fig.21b. The structure is symmetric in the three axes except from the differences at the edges where the boundary conditions are applied. The composite panels are the side and top yellow panels and the bottom gray panel. The remaining parts are metallic. Fasteners have been retained in the metal-composite connections, while adhesive bonds have been adopted in the composite-composite connections.

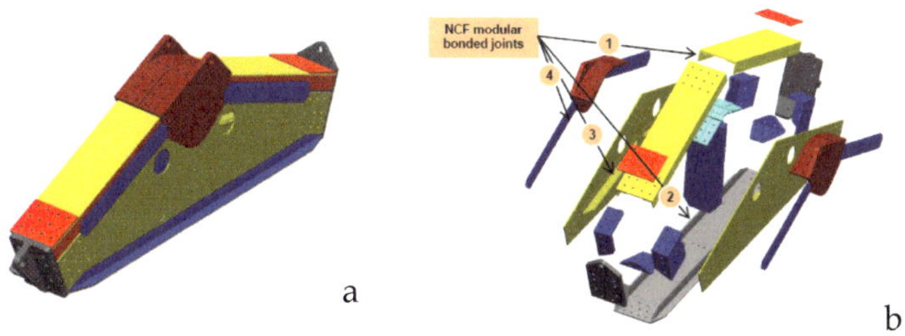

Fig. 21. a. Schematic representation of the FTB drawn with CATIA software (D2.3.2, 2008), and b. the disassembled structure.

The composite panels and joining elements are made of the NCF HTS/RTM6 composite material. The lay-up of the NCF modular joints and panels is quasi-isotropic comprising alterations of the layers except from the L-shapes that are made of layers. The number of layers varies according to panel thickness of the components. Metallic fittings are made either of aluminum or steel. Titanium fasteners have been used for metal-composite junctions and steel ones for metal-metal junctions. For the bonded junctions, the film adhesive Hysol EA9695 and the paste adhesive Hysol 9396 mixed with the 9395 were used. The loading conditions chosen represent realistic distressing situations occurring during service of the FTB. Specifically, the selected load-case corresponds to a landing case of the aircraft at maximum landing weight with applying of full throttle, and with fully deployed flaps but with spoilers retracted. At the left and top edges, the beam is constrained by inserting pins in the holes located there, while in the right corner, loading is applied. An alteration between the static and fatigue load-cases has been adopted in order to examine the fatigue damage tolerance capability of the FTB.

7.2 FE modeling

Complementary and prior to mechanical testing, a 3D detailed FE model of the FTB was built in order to assist design of the experiments, shed light in failure features that cannot be assessed by testing and propose a numerical approach that can be used for the design and analysis of large-scale complicated structural parts. Between the model and experiments, there is a two-way feedback, since the first numerical results aided the design of the experiments, and after the execution of the experiments, the model was improved and validated through the comparison with the experimental results.

The FE model of the FTB was created using the ANSYS FE code. Due to the large number of parts -most of them of complicated geometry- the large number of fasteners and the contact cases, creation of geometry using the ANSYS pre-processor was impossible. To this end, CATIA drawings of the FTB made by SABCA (D2.3.2, 2008) were imported in the ANSYS

ICEM CFD mesh generator as iges files (D2.3.2, 2008). After importing the files, only volumes of the parts, areas and border lines were available for building the mesh. So, in order to build a customized mesh, further modifications had to be made by defining sub-areas and sub-volumes. In order to reduce the total number of elements, specific actions were taken. For instance, a detailed mesh was adopted only in the stress concentration areas and bonded joints and one element was used through the thickness of the parts. The FE mesh of the FTB is shown in Fig.22. All parts of the structure were modelled using the SOLID185 element which is a 8-node structural solid suitable for modeling 3D solid structures. In total, 1,380,000 solid elements were used in the model. Despite that the SOLID185 element has a layered option, it was used as a solid orthotropic element in which the effective properties of the NCF layered material have been assigned. This was done because a layer-wise stress analysis would require an extremely huge computational effort due to the large number of layered elements that should be solved.
The parts of the FTB are assembled by mechanical fasteners and bonded joints. In the model, both joining types have been considered by simulating the exact joining conditions. Fasteners have been modeled as 3D solids consisting of the body and washers following their exact geometry, while bonded joints have been modeled by representing the adhesive with a different single layer of elements.

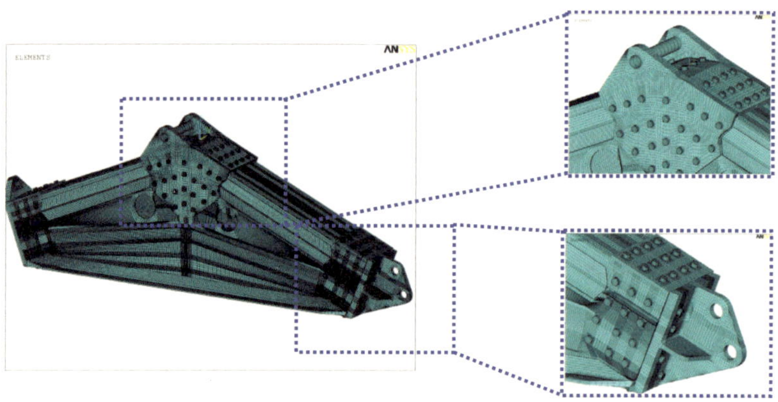

Fig. 22. FE mesh of the FTB.

In the model, all possible contacts between the assembled parts have been considered, namely the surface/surface, the fastener/hole-surface as well as the washer/surface contacts. Contact modeling was done using the ANSYS 3D 4-node surface-to-surface CONTAC173 element. The extremely complicated contact surfaces were generated using the ICEM's feature for automatic generation of contact elements between overlapped areas. The necessary corrections were performed manually afterwards. Almost 272,000 contact elements have been created. Presented in Fig.23 is a close view of the contact elements of the different contact cases. For the friction contacts, a friction coefficient of 0.3 was used. In the model, the exact boundary conditions applied for testing were simulated by creating fully constrained pins inside the holes at the left and top edges of the structure and applying the forces at the internal nodes of the hole at the right edge. As the forces are off-axis with respect to the global coordinate system, they were analyzed into on-axis components.

7.3 PDM

Due to the huge number of composite elements in the model, neither progressive damage modeling in its classical form, comprising the implementation of complicated failure analysis and material property degradation in element basis, nor the mesomechanical approach (Tserpes & Labeas, 2009), which simulates material's behavior based on stiffness-strain functions, could be applied. Therefore, a simplified scheme was adopted for simulating failure in the FTB. A linear material behavior is assumed and each stress component is compared with the corresponding strength (Maximum Stress criterion) and as soon as it exceeds it, the corresponding element stiffness is totally degraded. For the simulation of debonding, the stress intensity at the elements of the adhesive is compared to shear strength.

7.4 Numerical results

7.4.1 Predicted failure initiation

The model has the capability of simulating initiation and progression of the different types of damage in the composite material as well as debonding in the bonded joints. First analyses were conducted prior to the execution of the experiments in order for the predictions of damage initiation to be used for the evaluation of the selected load-cases in the experiments. This was done in order to avoid undesired catastrophic fracture of the specimen.

The predicted failure initiation patterns in the side-panels and bottom panel for the ultimate load are shown in Fig.23. As expected, failure predicted to initiate mainly in the fastener and hole areas due to the stress concentrations. Both tensile and compressive failure modes occurred due to the bending of the structure. The same failure modes have been also predicted in all composite parts connected by the mechanical fasteners. Predicted also is delamination in tension and compression at the edges of all NCF profiles at load levels starting at J=0.22 of the It is important to notice that the aforementioned failures are localized. Regarding bonded joints, it has been found to be whole since no extensive debonding has been predicted by the model. Contrary to the model, in the actual structure, some debonded areas have been detected using ultrasonic inspection after completion of loading (D7.1.1, 2009). These areas are mainly located in the junctions between the H profiles, the top and side panels and they have been resulted by the spreading of initial imperfect bonded areas created in the manufacturing phase. In any case, the bonded joints retain the minimum required integrity. These debonds were not captured by the model because perfect bonding has been assumed in the simulations.

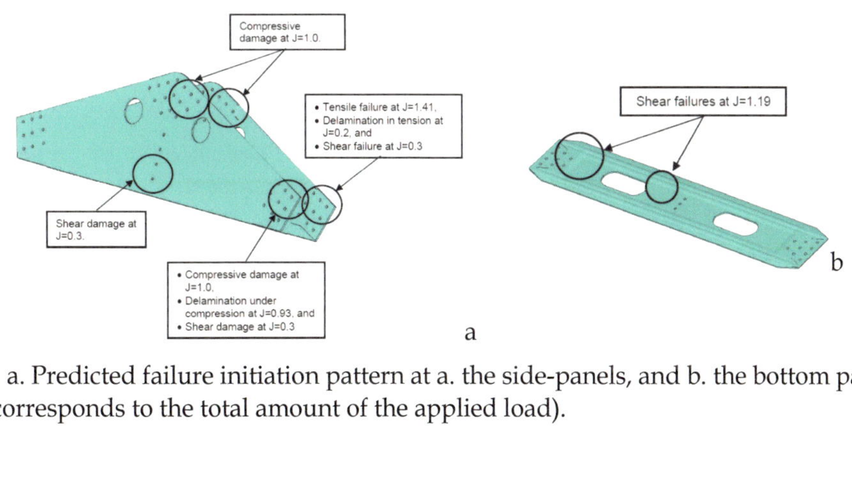

Fig. 23. a. Predicted failure initiation pattern at a. the side-panels, and b. the bottom panel (J=1.0 corresponds to the total amount of the applied load).

7.4.2 Comparison between measured and predicted strain field

In order to monitor the overall behavior of the FTB, a network of 50 strain gauges were mounted on its body. Some of the strain gauges were placed in critical areas with stress concentration (fastener areas and geometrically complex areas) and bonded areas. The measured and computed (nodal) strains for these locations are compared in Table 3. In general, a good agreement is obtained considering the uncertainties in the experimental procedure and the increased numerical errors. The average deviation between two methods is 29.15%. However, in most locations the deviation is lower than 15%. The average deviation is raised by few values exceeding 20%. These values refer to measurement points located inside fastener areas where the strain field is complicated and changes from tension to compression occurs in a very small distance.

The main conclusion from the study on the structural integrity of the novel FTB is that it is capable to effectively transfer the load which the original metallic fastened FTB has been designed to carry, thus validating the proposed design-concept. Localized failures having no significant effect on the integrity of the structure have been detected at fastener areas and bonded joints containing initial defects.

Strain gauge	**Experiment [μstrain]**	**Model [μstrain]**	**Deviation [%]**
1	2282.57	1978	13.3
2	1099.11	996	9.38
5	-262.51	-211	19.62
6	1366.24	1300	4.85
7	-308.23	-354	-14.85
8	-769.23	-701	8.87
9	-1016.07	-975	4.04
10	-575.09	-541	5.93
11	-411.51	-401	2.56
12	348.91	297	14.88
13	251.16	231	8.03
14	239.12	218	8.83
15	171.65	167	2.71
20	1098.15	821	25.24

Table 3. Comparison between measured and predicted strains of the FTB.

8. Conclusions

In this chapter, the scale-independent capability of the PDM method in predicting the strength of composite materials and structures is demonstrated by presenting the application of the method in 4 structural problems lying from nano- to macro-scale. Specifically, after a short historical overview and an introduction to the PDM method, the applications of the method for predicting tensile strength of pristine and defected CNTs lying in the nano-scale, the tensile response of plain weave composites lying in the micro-scale, the effect of imperfect bonding on the tensile strength of DLS joints lying in the macro-scale and the structural integrity of a novel FTB lying in the large-scale are described. In all cases, the method's predictions have been validated against numerical and experimental results. The following specific conclusions referring to progress made in the development of the method through these applications are drawn:

- Application to CNTs was the first attempt to use PDM method in the nano-scale. The attempt can be rated as successful since the simulated results have been validated successfully against established molecular mechanics methods. Furthermore, the PDM method gave the opportunity to combine the effectiveness of the FE method with the accuracy of the Morse interatomic potential, thus creating a method having significant advantages over classical molecular methods.
- Application of the PDM method to textile composites by means of multi-scaling and homogenization allows for the transfer of material's mechanical response from micro- to macro-scale. Note that without the consideration of PDM modeling, only the elastic properties of such complex materials can be evaluated.
- Application to the large-scale FTB confirms that such models can be proved very helpful to the design of composite structures through virtual experimentation.

9. References

Belytschko, T., Xiao, S.P., Schatz, G.C. & Ruoff, R.S. (2002). Atomistic simulations of nanotube fracture. *Physical Review B*, 65, 235430.

D2.3.2 MOJO Deliverable by SABCA (2008). Final drawing set of A and B level components.

D7.1.1 MOJO Deliverable by EADS-IWF (2009). NDT parameters and industrial NDT for MOJO.

Iijima, S. (1991). Helical microtubules of graphitics carbon. Nature, 354, 56-8.

Kachanov, M.L. (1958). On the creep fracture time. *Izv. Akad. Nauk. SSR.* 8, 26-31.

Llopart P. Ll., Tserpes, K.I., Labeas, G.N. (2010). Experimental and theoretical investigation of the influence of imperfect bonding on the strength of NCF double-lap shears joints. *Composite Structures*, 92, 1673-1682.

Matzenmiller, A. Lublinear, J. & Taylor, R.L. (1995). A constitutive model for anisotropic damage in fiber-composites. *Mechanics of Materials*, 20, 125–152.

Mielke, S.L., Troya, D., Zhang, S., Li, J-L., Xiao, S., Car, R., Ruoff, R.S., Schatz, G.C. & Belytschko, T. (2004). The role of vacancy defects and holes in the fracture of carbon nanotubes. *Chemical Physics Letters*, 390, 413-420.

Modular Joints for Aircraft Components (MOJO) (2006). Research Project funded by the European Commission within the frame of 6th Framework-Aeronautics and Space. Description of Work.

Qian, D., Dickey, E.C., Andrews, R. & Rantell, T. (2000). Load transfer and deformation mechanisms in carbon nanotube-polystyrene composites. *Applied Physics Letters*, 76(20), 2868-2870.

Tang, X. & Whitcomb, J.D. (2003). Progressive Failure Behaviors of 2D Woven Composites, *Journal of Composite Materials*, 37, 1239-1259.

Tserpes, K.I, & Labeas, G.N. (2009). Mesomechanical analysis of non-crimp fabric composite structural parts. *Composite Structures*, 87, 358-369.

Xiao, J.R., Gama, B.A. & Gillespie Jr., J.W. (2007). Progressive damage and delamination in plain weave S-2 glass/SC-15 composites under quasi-static punch-shear loading. *Composite Structures*, 78, 182-196.

Verpoest, I. Lomov, S.V. (2005). Virtual textile composites software WiseTex: Integration with micro-mechanical, permeability and structural analysis.
Composites Science and Technology, 65, 2563-2574.
Yu, M.F., Lourie, O., Dyer, M.J., Moloni, K., Kelly, T.F. & Ruoff, R.S. (2000). Strength and breaking mechanism of multiwalled carbon nanotubes under tensile load. *Science,* 287, 637.
Yokozeki, T., Iwahori, Y. & Ishiwata, S. (2007). Matrix cracking behaviors in carbon fiber/epoxy laminates filled with cup-stacked carbon nanotubes (CSCNTs), *Composites Part A*, 38, 917-924.

15

Effect of Spark Plasma Sintering in Fabricating Carbon Nanotube Reinforced Aluminum Matrix Composite Materials

Hansang Kwon[1] and Akira Kawasaki[2]
[1]*Advanced Materials Processing, Empa-Swiss Federal Laboratoriesfor Materials Science and Technology, Thun,*
[2]*Department of Materials Processing, Tohoku University, Sendai,*
[1]*Switzerland*
[2]*Japan*

1. Intorduction

Carbon nanotubes (CNT) are attractive next generation materials due to their unique properties, which lead to high mechanical, electrical, and thermal performance (Iijima, 1991; Endo et al., 1976; Niyogi et al., 2002; Komarov & Mirnov, 2004). This unique nano order material can not only be utilized on its own in precision industrial fields but can also provide high performance functionality in conjunction with conventional materials. For this reason, many researchers are investigating the fabrication of CNT reinforced metal, ceramic, and polymer matrix composite materials. Despite their research efforts, the fabrication of CNT-reinforced metal matrix composite materials, particularly with an aluminum (Al) matrix, is still facing several problems, such as difficulties in homogeneously dispersing the CNT in the Al matrix, (Salvetat-Delmotte and Rubio, 2002; Hilding et al., 2003; Xu et al., 1999) producing highly densified composite materials without any degradation of the CNT, (Kuzumaki et al., 1998; Sridhar & Narayana, 2009) and achieving enough interface strength between the Al matrix and reinforcement of CNT (Deng et al., 2007; Bakshi et al., 2009; Lahiri et al., 2009). To overcome these problems, Deng et al. fabricated the CNT-Al alloy composite materials by cold isostatic pressing and then subsequent hot extrusion techniques, with which they have achieved 45% incremental increase in tensile strength (Deng et al., 2007) Esawi et al. attempted to fabricate a CNT-Al matrix composite by mechanical milling and rolling or extrusion processes (Esawi & Morsi, 2007; Esawi & Borady, 2008; Esawi et al., 2009). Morsi et al. produced CNT-Al matrix composites by a unique powder metallurgy route using spark plasma extrusion (Morsi et al., 2009; 2010). Agawal et al. introduced several fabrication methods for CNT-Al and Al alloy composites based on thermal and cold spray forming technologies (Laha et al., 2004; Bakshi et al., 2008; Bakshi & Agarwal, 2010). Recently, we have also demonstrated the feasibility of making aluminum-carbon nanotube (Al-CNT) composite materials, producing not only a highly densified composite but also enhanced interface bonding between the Al matrix and CNT, by a combination of spark plasma sintering (SPS) followed by hot extrusion processes (Kwon et al., 2009; Kwon & Kawasaki, 2009; Kwon et al., 2010). However, the specific effect of SPS for the final product of Al-CNT composite is not yet clear. In the present study, we have fabricated CNT reinforced aluminum matrix composite materials with two different amounts of CNTs (1 and 5vol%), by a combination of SPS followed by hot extrusion processes. Pure Al bulks also have been fabricated using the same route for comparison. In particular, the SPS behavior of the Al-CNT mixture powder was discussed. Firstly, the morphologies of Al grains in spark plasma sintered (SPSed) Al-CNT and pure Al compacts were observed. The chemical stability between the Al and CNT and the real temperature between the each Al particles during the SPS process were discussed based on the two particles model. Finally, high resolution transmission electron microscopy was carried out to understand the materials at the boundary of the SPSed Al-CNT compact. Moreover, the tensile strength and elongation of the composites was measured and discussed in relation to the different amounts of CNT additions.

2. Experimental procedure

A precursor aimed at dispersing CNTs well was produced by the nanoscale dispersion method (Kwon et al., 2009; Noguchi et al., 2004). The precursor consisted of commercial gas atomized Al powder (ECKA Granules Co. Ltd., Japan, purity 99.85%, average particle size 14.82μm), multi-walled carbon nanotube (MWCNT, ILJIN Co. Ltd., Korea, purity 99.5%, average diameter and length 20 nm and 15㎛), and natural rubber (NR). The powder composition was adjusted to 1 and 5vol% of CNT. The CNT was mixed with NR in benzene and then Al powder was added. This Al-CNT-NR (precursor) was roll-milled to a thickness of 2 mm and a sheet shape was formed. A thermal gravimetric analysis (TGA6300, SEIKO Instruments, Inc., Japan) was carried out to check the specific decomposition temperature of the NR in argon atmosphere from the precursor. The heating and flow rate were fixed at 10 °C/min and 50 m*l*/min, respectively. The precursor was heat-treated in a quartz tube furnace (Shimadzu Co. Ltd., Japan) at 500 °C for 2h under an argon atmosphere with a 1*l*/min flow rate to evaporate the NR. The obtained Al-CNT mixture powder was sintered at 600 °C, with a holding time of 20 min, a heating rate of 40 °C/min, and a pressure of 50MPa in a ϕ15 mm carbon mold, using a SPS device (SPS-S515) manufactured by Sumitomo Coal Mining Co. Ltd, Japan. The sintered compact was extruded in a 60° conical die at 400 °C with a 500kN press (UH-500kN1, Shimadzu Corporation, Japan). The extrusion velocity and extrusion ratio were fixed at 2mm/min and 20, respectively. The microstructures of the samples after each step were observed by optical microscopy (PMG-3, OLYMPUS, Japan), field-emission scanning electron microscopy (FE-SEM) (FE-SEM6500, JEOL Ltd, Japan) and high-resolution transmission electron microscopy (HR-TEM) (HITACHI 200, Japan) with energy dispersive spectrometry (EDS, 5 nm spot size) and selected area diffraction pattern (SADP, under 10 nm nano-beam spot size and 1.2 m camera lens distance from specimen). Raman spectroscopy (SOLAR TII Nanofinder, Tokyo Instruments Co. Ltd, Japan) was employed to evaluate the disorder ratio of the CNTs after each step. The chemical stability between the CNT and Al powder was analyzed using a differential scanning calorimeter (DSC6300, SEIKO Instruments, Inc., Japan) in a purified argon atmosphere at 800 °C and a scanning rate of 10 °C/min. To evaluate the tensile strength, the extruded composites were machined into test pieces 3 mm in diameter, in accordance with ICS 59.100.01. The tensile strength was measured with a universal testing machine (AUTOGRAPH AG-I 50 kN, Shimadzu Co. Ltd, Japan)

3. Results and discussion

Figure 1 shows FE-SEM and TEM micrographs of the raw Al and CNT as-received powders. The Al particles show several size distributions and a spherical shape, as shown in Figure 1a. The morphology of the raw CNT has an extremely zigzagged shape with a high aspect ratio, as shown in Figure 1b. The out-wall of the CNT consisted of some disordered regions and amorphous carbon as shown by the black arrow in Figure 1c. Two different tip shapes of CNT were observed, opened and closed as shown by the white arrows in Figure 1c. It was found that the I_D/I_G (intensity ration of graphite and defect peaks in the CNT) ratio of the raw CNT observed from the Raman spectra (Keszler et al., 2004; Zhao & Wagner, 2004) was about 0.8, which means a relatively high number of defects and contaminants, as indicated in Table 1.

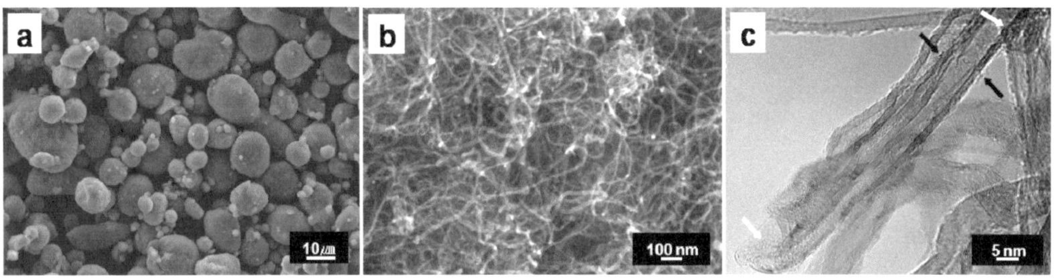

Fig. 1. FE-SEM micrographs of (a) as-received Al powders and (b) CNT. (c) HR-TEM micrograph of the pristine CNT. The white and black arrows indicate some disordered and amorphous impurity.

	Relative density (%)	I_D/I_G ratio (±10%)	Boundary thickness (nm)	Tensile strength (MPa ±10%)	Elongation (% ± 10%)
Heat treated Al-CNT powder	-	Raw CNT:0.877 Al-CNT:0.831	-	-	-
SPSed Al-1vol.% CNT compact	96.3	072	~100	-	-
SPSed Al-5vol.% CNT compact	96.1	0.85	~300		
Extruded Al-1vol.% CNT composite	98.6	0.79	~60	198.8	21.9
Extruded Al-5vol.% CNT composite	98.0	0.843	~200	187.3	10.5
Extruded pure Al bulk	100	-	15	52	19.5

Table 1. Various properties of SPSed Al-CNT compact and extruded Al bulk and Al-CNT composites.

TGA measurement was carried out to confirm the specific decomposition temperature for NR in the Al-CNT-NR mixture, as shown in Figure 2. A slight weight loss was observed for the mixture in the region of around 200 °C, which may be due to the removal of some organic contaminants and adsorbed moisture. A dramatic weight loss of about 15% was detected between 350-450 °C, which was due to the complete removal of NR. This demonstrated that the NR as a mixing medium could be completely removed from the Al-NR-CNT mixture at less than 450 °C. Based on the TGA result, we carried out a heat treatment for the precursor at 500 °C for 2 h.

Figure 3 shows FE-SEM micrographs of the Al-CNT mixture powders after removal of NR from the Al-CNT-NR precursor. The Al particles maintained their initial spherical morphology and particle size after heat treatment, as per those of the starting powder, as shown in Figure 2a and 3a (1vol%) and b (5vol%). Some regions with agglomerated CNTs were observed after the removal of NR from the Al-NR-CNT precursor, which may be due

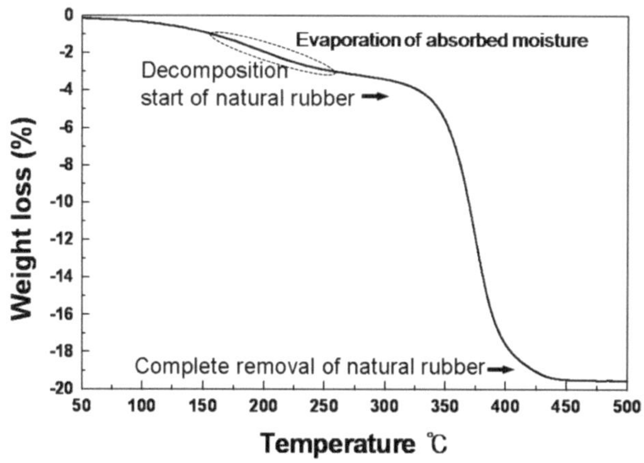

Fig. 2. TGA curve of Al-NR-CNT precursor.

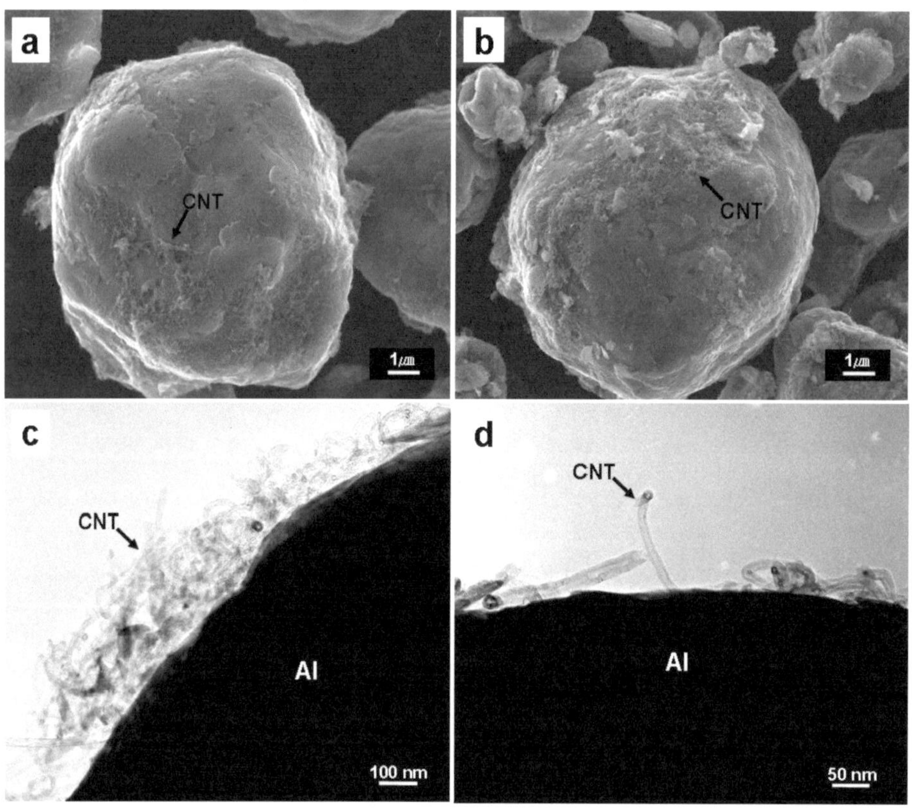

Fig. 3. FE-SEM micrographs of (a) Al-1vol% CNT and (b) Al-5vol% CNT mixture powders. HR-TEM micrographs of (c) condensation and (d) dispersed region of CNT on the Al particle in the Al-5vol% CNT mixture powder.

to the selective condensation of CNTs located where the NR was removed from between Al particles. Moreover, this CNT agglomerate ratio was relatively higher in the 5vol% CNT mixture powder than in the Al-1vol% CNT mixture powder, as shown in Figure 3a and b. We believe that this problem can be addressed by adjusting the Al particle size distribution. Overall, our findings indicated that most of the CNTs were well dispersed on the surface of Al particle, regardless of the amount of CNT added.

The relative density of pure Al bulk and Al-1 and 5vol% CNT composites subjected to the same processes were measured by an Archimedes method (Table 1). It was found that the extruded pure Al bulk was fully densified and the SPSed Al-CNT compact and extrudate were also highly densified, demonstrating that this combination of SPS and hot-extrusion processes is highly effective in densifying the Al-CNT mixture powder.

The tensile strength and elongations of the pure Al bulk and the composites are indicated in Table 1. The composites showed around 300% enhancement in tensile strength compared to that of the pure Al bulk. In particular, no degradation of elongation for the Al-1vol% CNT composite was observed, in spite of the highly enhanced tensile strength. In general, the nominal tensile strength of pure Al (http://en.wikipedia.org/wiki/Tensile_strength) was reported as about 40-50 MPa whereas the tensile strength of the extruded pure Al bulk obtained in this study was 52 MPa (Table 1). This small disagreement originated from differences in fabrication and testing conditions.

We have considered the effect of work hardening on tensile strength due to large plastic deformation in the processes employed. According to our previous report that used nano-indentation for work hardening, there was no change in the hardness of Al particles before and after the extrusion process (Kwon et al., 2009). It is suggested that the stress accumulation was lower than stress released during the extrusion process by dynamic recrystallization (Knocks, 2004; Kelly & Tyson, 1965; Cox, 1952; Serajzadeh, 2004). Therefore, the enhanced tensile strength of the composites with a small amount of CNT (1 and 5 vol%) is mainly due to the addition of CNT itself.

The tensile and elongation behavior should be considered with regard to the amount of CNT. It is assumed that the thickness of the dispersed CNT on the Al particle will be different depending on the amount of CNT added. i.e. the thickness of this CNT cluster could affect the mechanical properties of the composites, resulting in a large difference in elongation, as indicated in Table 1.

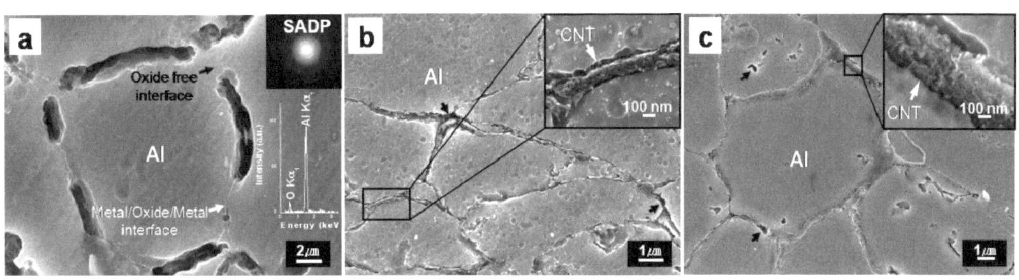

Fig. 4. FE-SEM micrographs of (a) the SPSed pure Al compact; the insets indicate the SADP and EDS of the boundary region (Kwon et al., 2010) (b) the SPSed Al-1vol% CNT compact, and (c) SPSed Al-5vol% CNT compact; the insets in (b) and (c) show a high magnification of the boundary region between Al and CNT.

Microstructures of the SPSed pure Al and Al-CNT compacts were observed in order to clarify the boundary structure. Firstly, the SPSed pure Al compact was investigated, as shown in Figure 4a. Several boundary grooves were observed after the etching process, which made use of liquid etchant (5 % NaOH). Thin layers (10-20 nm) of amorphous Al oxide and some oxide-free region (black arrow in Fig. 4a) in the boundary zone were observed by SADP and EDS, as shown in Figure 4a. Some of the Al oxide-free region was created by decomposition of Al oxide during the SPS process by micro-plasma (Zadra et al., 2007; Omori, 2000; Xie et al., 2001; Kwon et al., 2010). In the case of the SPSed Al-CNT compacts, the boundary structure was completely different to that of the pure Al compact. We observed boundary layers of around 100 and 300 nm thick for the 1 and 5 vol% CNT added compacts, respectively, as shown in Figure 4b and c. Several micro-pores were also observed regardless of the different amounts of CNT addition (black arrows in Fig. 4b and c). These pores may also include those produced during the etching process. Moreover, no significant grain growth was observed for both SPSed pure Al and Al-CNT compacts compared to the starting Al particles (see Fig. 1a and Fig. 4). We believe that the grain growth was restrained by the pinning effect of the CNTs, some of the existing Al oxide being present at the boundaries, and the quick processing cycle of SPS. Our previous results showed that the boundary of the SPSed Al-1 and 5vol% CNT compact contained mainly CNT, a small quantity of Al carbide (Al_4C_3), carbon black, amorphous carbon, Al oxide, and Al, which was confirmed by SADP and EDS as shown in Figure 5. (Kwon et al., 2009) The formation of Al_4C_3 in spite of being under the Al melting point will be discussed in depth later. However, it is understood that the thickness of the boundary (mainly CNT) can be changed by the amount of CNT addition.

Figure 6 shows TEM micrographs of the extruded Al-1 and 5vol% CNT composites. The thickness of the boundaries was slightly reduced after the extrusion process for Al-1 and 5vol% CNT composites to ~60 and ~200 nm, respectively, compared to that of the SPSed compacts, as indicated in Table 1. This slightly changed boundary thickness accompanied the large plastic deformation during the extrusion process. Furthermore, the CNT in the boundary were aligned with the extrusion direction and in tight contact with the Al matrix, as shown in Figure 6a and b. Hence, the applied pressure of extrusion affected not only the alignment of the CNTs but also the thickness of the boundary structure, regardless of the amount of CNT added. The implanted Al_4C_3 between the Al matrix and the CNT interface was observed in the 5vol% CNT composite, as shown in Figure 6c (Kwon et al., 2009). Such a microstructure also contributed to the enhancement of the mechanical properties of the composites. Figure 7 shows HR-TEM of the Al-1vol% CNT composite subjected to the extrusion process.

As can be seen in Fig. 7, two different types of aluminum carbide (Al_4C_3) were observed in the boundary zone, dumbbell and tube types (Kwon et al., 2010). We believe that dumbbell-shaped Al_4C_3 originated from the tip of a CNT, whereas tube-shaped Al_4C_3 originated from defective CNTs.

Here, we have discussed how the formation of Al_4C_3 was able to occur under the Al melting point. Figure 8 shows the DSC curves of the pure Al powder, Al-5vol% CNT powder, and SPSed (at 600 °C) Al-5vol% CNT compact. In the case of the powders, they showed endothermic peaks near the Al melting point (about 660 °C), as shown in Figure 8a and b. However, the SPSed Al-CNT compact showed an endothermic peak followed by an exothermic peak, as shown in Figure 8c. It indicated that a chemical reaction occurred,

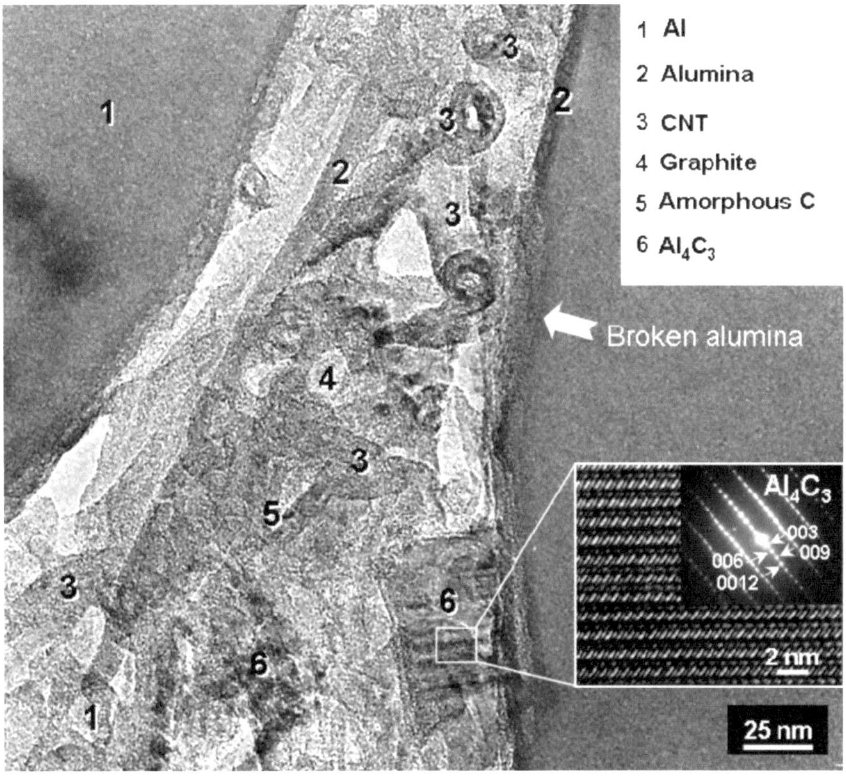

Fig. 5. TEM micrographs of the boundary layer of the SPSed Al-5vol% CNT compact. The boundary layer between the Al particles (1) consisting of alumina (2), CNT (3), amorphous carbon black (4), graphite (5), and Al_4C_3 (6) phases. The white arrow indicates broken alumina. The inset shows the SADP and HR-TEM micrographs of the Al_4C_3 phase (Kwon et al., 2009)

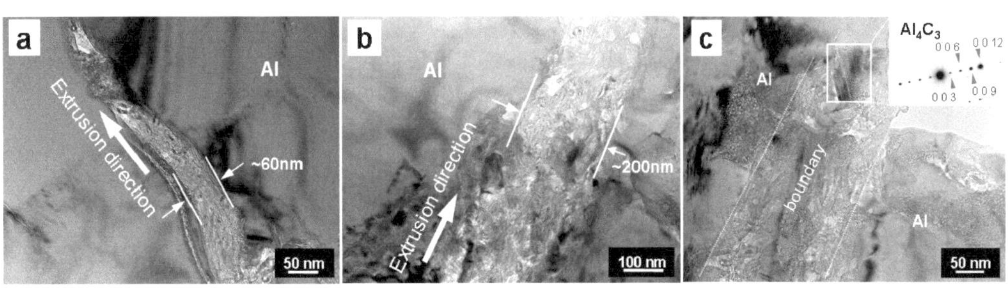

Fig. 6. TEM micrographs of (a) the extruded Al-1vol% CNT composite and (b) the extruded Al-5vol% CNT composite. (c) The implanted Al_4C_3 between the Al matrix and CNT in the Al-5vol% CNT composite (Kwon et al., 2009). The inset shows the SADP of the Al_4C_3.

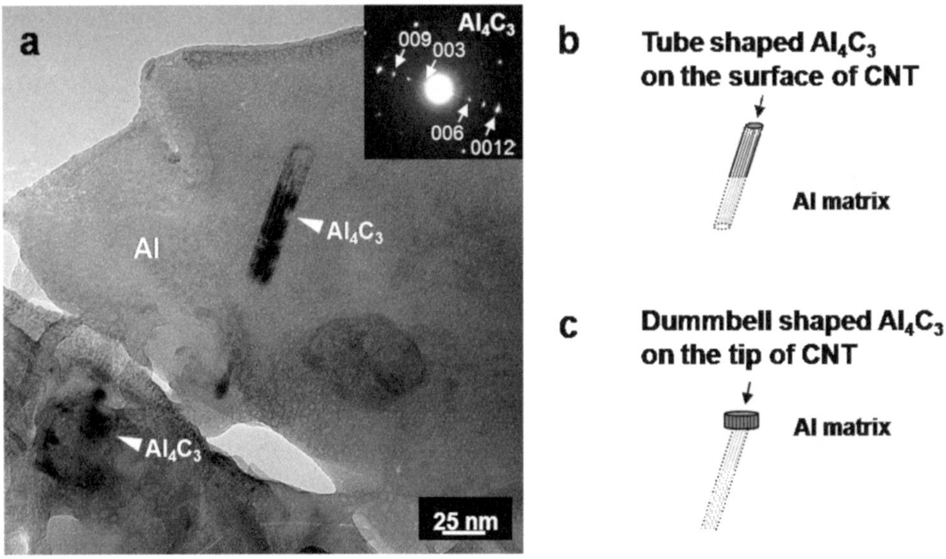

Fig. 7. HR-TEM images of the microstructure of Al–1vol% CNT subjected to the extrusion process. (a) The formation of Al_4C_3 was observed near the boundary zone. The inset confirms the crystal structure of Al_4C_3 by its selected-area diffraction (SAD) pattern. (b), (c) Tube-shaped Al_4C_3 was generated on the surface of defective CNTs, whereas particle-shaped Al_4C_3 was generated on the tips of CNTs. Both forms (b) and (c) were implanted into the Al matrix (Kwon et al., 2010).

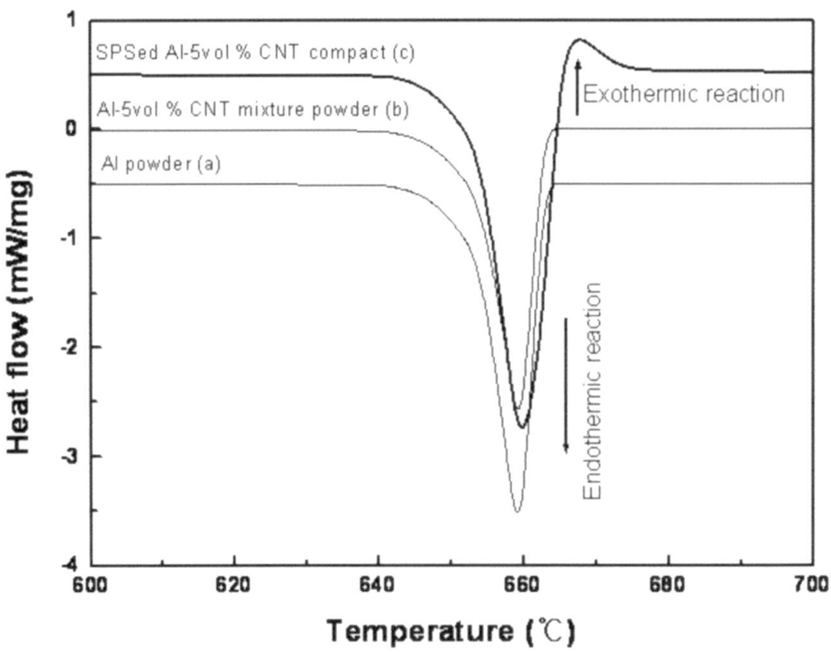

Fig. 8. DSC curves of the pure Al powder, Al-5vol% CNT powder, and 600 °C SPSed Al-5vol% CNT compact.

creating the exothermic peak right after the endothermic reaction. In general, it is known that the pure Al particle has a stable oxide layer on the surface of the particles (Zadra et al., 2007; Omori, 2000; Kwon et al., 2010). Therefore, the liquid Al generated is not able to flow out due to the presence of the Al oxide layer (melting point: <2000 °C) on the surface of the Al particle, even though the pure Al particle is completely melted. as illustrated by Figure 9a. For this reason, the Al-CNT mixture powder demonstrates the same phenomenon as in the case of the pure Al powder (Fig. 9b).

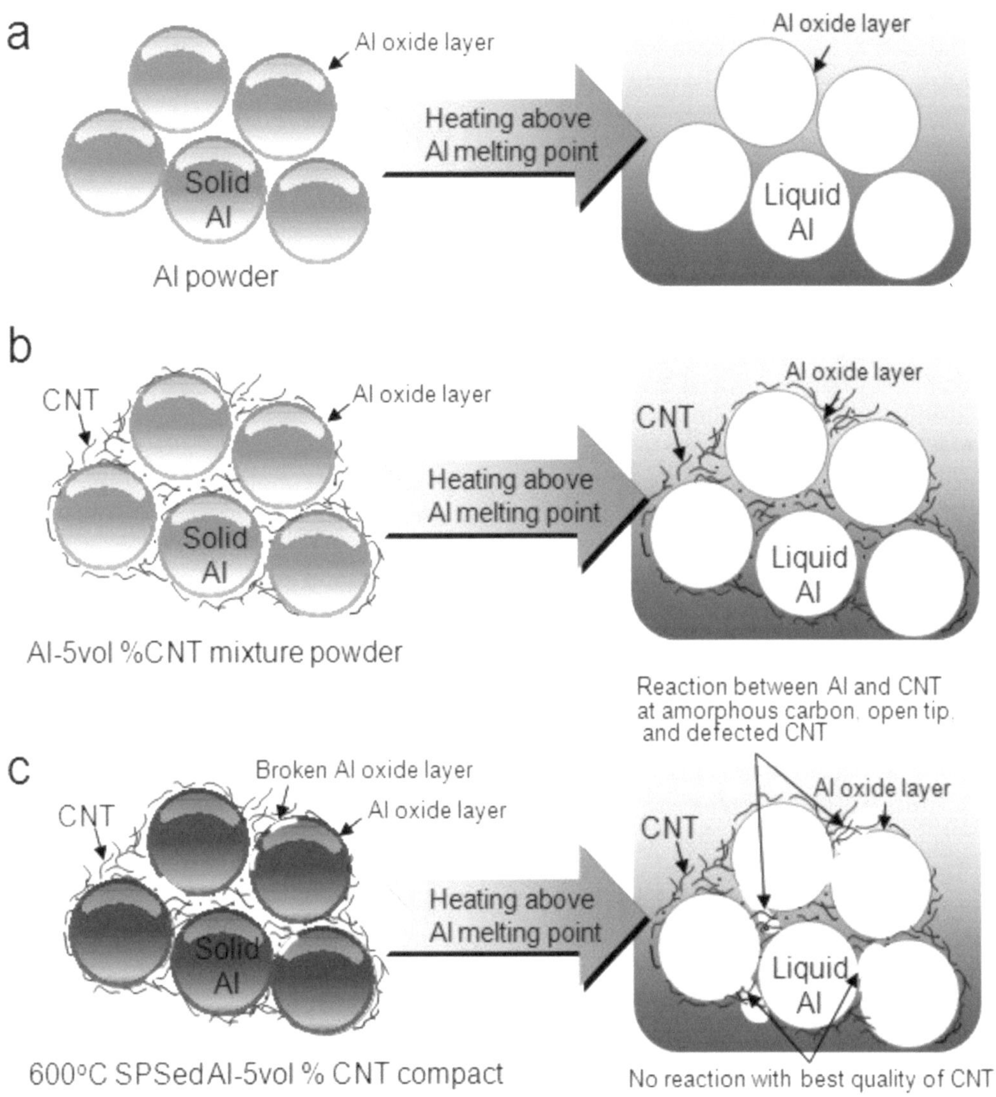

Fig. 9. Chemical stability of the CNT in the Al powder: (a) in the case of the pure Al powder, (b) Al-5vol% CNT mixture powder, and (c) 600 °C SPSed Al-5vol% CNT compact.

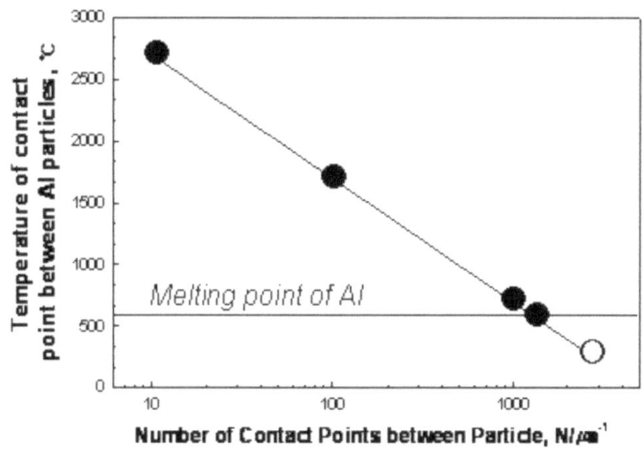

Fig. 10. Calculated temperature depends on the number of contact points between Al particles during the SPS process.

However, in the case of the SPSed Al-CNT compact, a different phenomenon is expected, because the SPSed Al-CNT compact possesses several broken parts of Al oxide layer induced during the SPS process by micro-plasma (Omori, 2000; Kwon et al., 2010). In this case, the liquid Al phase is able to flow out through several broken parts of the Al oxide layer, thereby making contact with the CNTs in the boundary layer when the SPSed Al-CNT compact is heated above the Al melting point, as shown in Figure 9c. In general, there is no chemical reaction between the Al liquid phase and CNTs with best-quality CNTs (which have no defects) (Ci et al., 2006). But disordered CNT, the amorphous carbon surface of the CNT, and opened-tips of the CNT lead to reaction with the Al liquid phase (Kwon et al., 2009; Ci et al., 2006; Laha & Argawal, 2008). According to the I_D/I_G ratio and TEM observation result of the raw CNT, it contained many structural defects, amorphous carbon, and opened-tip CNTs, as shown in Table 1 and Figure 1c, indicating that the CNTs used in the present work were highly likely to react with the Al liquid phase. Therefore, the exothermic peak that followed immediately after the endothermic peak can be attributed to the chemical reaction occurred between the Al and CNT, resulting in the formation of Al_4C_3 that was observed in the SPSed compact, as shown in Figure 5.

Nagae et al. reported the feasibility of generating an Al liquid phase under the Al melting point during the SPS process using a two-particle model based on the real temperature between the Al particles during the SPS process (Nagae et al., 2001). According to their calculations, the real temperature of the initial stage of sintering between the Al particles was about 1270 °C at the initial stage of the SPS process which is supposed to proceed at 570 °C. Figure 10 shows the calculated real temperatures between the Al particles using the two-particle model based on the SPS temperature of 600 °C used in the present study. The real temperature between Al particles decreased due to the increase in contact points as time passed. Thus, we believe that the real temperature between Al particles in the initial stage of the SPS process was higher than the sintering temperature employed. For this reason, the Al liquid phase can be generated at the SPS temperature employed in the present study, thereby creating the possible conditions for the formation of Al_4C_3.

The I_D/I_G ratio of the 5vol% CNT composite was higher than that for the 1vol% CNT composite (Table 1), resulting in the predominant formation of Al_4C_3 in the 5vol% CNT composite. The 5vol% CNT composite showed a slightly lower tensile strength and much lower elongation than the 1vol% CNT composite, which may be due to the excessive Al_4C_3 generated, as indicated in Table 1. However, a moderate amount of Al_4C_3 can contribute to better chemical bonding between the Al matrix and CNT but too large an excess of Al_4C_3 causes a deterioration in the mechanical properties. In other words, the Al_4C_3 generated during the SPS process greatly affected the degradation of elongation due to its brittleness. That is why the elongation of 1vol% CNT reinforced composite was enhanced significantly compared to that of the 5vol% CNT reinforced composite, as indicated in Table 1.

We have discussed the effect of the formation of Al_4C_3 on the mechanical properties in the composite materials obtained. It is understood that some regions of the interfaces between the Al matrix and CNT in the SPSed compact and the extruded composite were combined by stable chemical bonding through the Al_4C_3 formed, as shown in Figure 11a and b. These structures were observed in our actual SPSed compact and the extruded composite (Fig. 4 and 6). It is believed that this chemically bonded structure is useful for stress transfer (Knocks, 2004; Kelly & Tyson, 1965; Lourie & Wagner, 1999; Fukuda & Chou, 1982) from the Al matrix to the CNT reinforcement. Moreover, the shape and position of the Al_4C_3 formed supported the enhancement of the mechanical properties. This is because the dumbbell shape of the Al_4C_3 which was formed on the tip of CNT can be held firmly and makes for an efficient stress transfer between the Al matrix and CNT, as shown in Figure 11c. The convex Al_4C_3 which is formed on the amorphous surface of the CNT not only functioned as a barrier to crack propagation due to the restraint of sliding, but also contributed to efficient stress transfer, as shown in Figure 11d. However, the existence of Al_4C_3 and well dispersed CNTs could also have interrupted the dislocation movements in the composites, resulting in the improvement in the mechanical properties of the composites (George et al., 2005; Oh et al., 2009; Jeong et al., 2007).

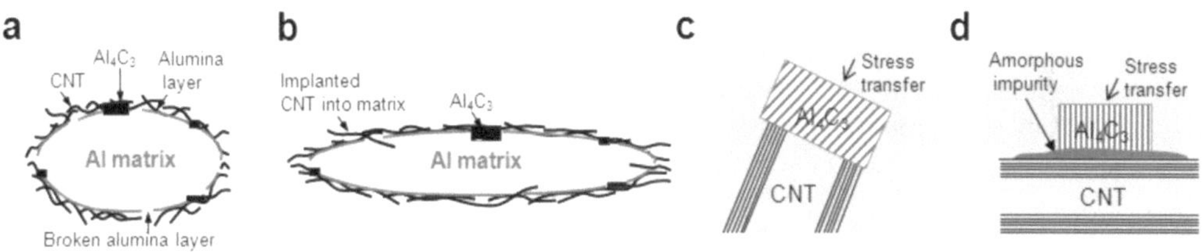

Fig. 11. Illustrations of structures of SPSed and extruded Al-CNT composites: (a) the generated Al_4C_3 and some parts of the broken alumina in the SPSed Al-CNT compact; (b) some of the CNT and Al_4C_3 implanted in the Al matrix in the extruded composites by mechanical pressure; (c) the dumbbell-shape and the convex-shape of Al_4C_3 which formed on the tip and surface of CNT and contributed to efficient stress transfer and a barrier to sliding.

Figure 12. shows the fracture surface of the Al-1and 5vol% CNT composites. Many dimples were observed in the Al matrix due to strong adhesion between particles. Several CNT bridges were observed near the cleavage regardless of the amount of CNT addition, as shown by the white arrow in Figure 12a and b. This CNT bridging also significantly contributed to the improvement in the mechanical properties. In particular, the 5vol% CNT composite showed a relatively large agglomeration of the CNTs compared to the 1vol% CNT composite (see black arrows in Fig. 12b and c). Moreover, several pointed-shape CNT tips were observed in the fracture surface of the 1vol% CNT composite, which means that these CNTs were broken while extended, due to efficient stress transfer with strong chemical bonding created by Al_4C_3 between the Al matrix and CNT, as shown by the white arrows in Figure 12c. We also observed no shape change of the CNT tips in the fracture surface of 1vol% CNT composite. This indicated that this CNT did not include formation of Al_4C_3, which resulted in the pulling out of CNT being relatively easy, as shown by the black arrows in Figure 12c. Therefore, the Al_4C_3 generated during the SPS process is helpful in enhancing the Al-CNT composite materials. However, it must be accompanied by careful control in the quantity and size of Al_4C_3, which is brittle and hydrophilic (Ruch et al., 2006; Khalid et al., 2004).

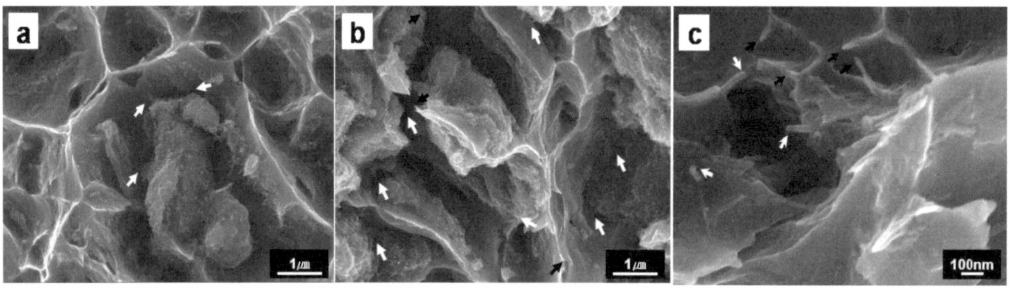

Fig. 12. FE-SEM micrographs of (a) the fracture surface of Al-1vol% CNT composite and (b) the Al-5vol% CNT composite after the tensile test; the white and black arrows indicate bridging and agglomeration of the CNTs. (c) High magnification of (a), where the white and black arrows indicate broken and pulled out CNT.

4. Conclusions

We have attempted to understand the effect of spark plasma sintering in fabricating Al-CNT composite materials which had highly enhanced tensile strength and no significant degradation of elongation compared to that of the pure Al bulk. Some parts of the Al oxide layer on the surface of the Al particles in the SPSed Al-CNT compact were broken during the SPS process, as confirmed by the DSC analysis and HR-TEM observation. Moreover, a small quantity of the Al_4C_3 was observed in the Al-CNT compact even though this was made under the solid state SPS temperature. The shape and formation position of Al_4C_3 also played an important role in the mechanical properties of the composites. The relative quantity of formed Al_4C_3 can be controlled by controlling the amount of CNT additions. We believe that the composites obtained were strengthened as a result of a complex phenomenon not only by the reinforcement of CNT but also through an efficient stress transfer effect by the chemical bonding between the Al matrix and CNT reinforcement through the Al_4C_3 formed during SPS process.

5. Acknowledgements

This research was partially supported by the New Energy and Industrial Technology Development Organization (NEDO) of Japan. We also would like to acknowledge Nissin Kogyo Co. Ltd for their technical support.

6. References

[1] Iijima, S. (1991). Helical microtubules of graphitic carbon. Nature, Vol. 354, pp. 56-58, ISSN 0028-0836.

[2] Endo, M.; Koyama, T. & Hishiyama, Y. (1976). Structural improvement of carbon fibers prepared from benzene. Jpn. J. Appl. Phys. Vol.15, pp. 2073-2076, ISSN 0021-4922.

[3] Niyogi, S.; Hamon, M. A.; Hu, H.; Zhao, B.; Bhowmik, P.; Sen, R.; Itkis, M. E. & Haddon, R. C. (2002). Chemistry of single-walled carbon nanotubes. Acc. Chem. Res. Vol. 35, pp. 1105-1113, ISSN 0001-4842.

[4] Komarov, F. F. & Mironov, A. M. (2004). Carbon nanotubes: presents and future. Phys. Chem. Solid State Vol. 5, pp. 411-429, ISSN 1729-4428.

[5] Salvetat-Delmotte, J. P. & Rubio, A. (2002). Mechanical properties of carbon nanotubes: a fiber digest for beginners. Carbon Vol. 40, pp. 1729-1734, ISSN 0008-6223.

[6] Hilding, J.; Grulke, E. A.; Zhang, Z. G. & Lockwood, F. (2003). Dispersion of Carbon Nanotubes in Liquids. J. Dispersion Sci. Tech. Vol. 24, pp. 1-41, ISSN 0193-2691.

[7] Xu, C. L.; Wie, B. Q.; Ma, R. Z.; Liang, J.; Ma, X. K. & Wu D. H. (1999) Fabrication of aluminum–carbon nanotube composites and their electrical properties. Carbon Vol. 37, pp. 855-858, ISSN 0008-6223.

[8] Kuzumaki, T.; Hayashi, T.; Miyazawa, K.; Ichinose, H.; Ito, K. & Ishida, Y. (1998). Processing of ductile carbon nanotube/C60 composite. Mater. Trans. Vol. 38, pp. 574-577, ISSN 1347-5320.

[9] Kuzumaki, T.; Miyazawa, K.; Ichinose, H. & Ito, K. (1998). Processing of carbon nanotube reinforced aluminum composite. J. Mater. Res. Vol. 9, pp. 2445-2449, ISSN 0884-2914.

[10] Sridhar, I. & Narayana, K. R. (2009). Processing and characterizations of MWCNT reinforced aluminum matrix composites. J. Mater. Sci. Vol. 44, pp. 1750-1756, ISSN 1573-4803.

[11] Deng, C. F.; Zhang, X. X.; Wang, D.; Lin, Q. & Li, A. (2007). Preparation and characterization of carvon nanotuves/aluminum matris composites. Mater. Lett. Vol. 61, pp. 1725-1728, ISSN 0167-577X.

[12] Bakshi, S. R.; Keshri, A. K.; Singh, V.; Seal, S. & Agarwal, A. (2009). Interface in carbon nanotube reinforced aluminum silicon composites: thermodynamic analysis and experimental verification. J. Alloys Compd. Vol. 481, pp. 207-213, ISSN 0925-8388.

[13] Lahiri, D.; Bakshi, S. R.; Keshri, A. K.; Liu, Y. & Agarwa, A. (2009). Dual strengthening mechanisms induced by carbon nanotubes in roll bonded aluminum composites. Mater. Sci. Eng. A Vol. 523, pp. 263-270, ISSN 0921-5093.

[14] Deng, C.; Zhang, X.; Ma, Y. & Wang, D. (2007). Fabrication of aluminum matrix composite reinforced with carbon nanotubes. Rare Metals Vol. 26, pp. 450-455, ISSN 1001-0521.

[15] Deng, C.; Wang, D.; Zhang, X. & Li, A. (2007). Processing and properties of carbon nanotubes reinforced aluminum composites. Mater. Sci. Eng. A Vol. 444, pp. 138-145, ISSN 0921-5093.

[16] Esawi, A. M. K.; & Morsi, K. (2007). Dispersion of carbon nanotubes (CNT) in aluminium powder. Composites Part A Vol. 3, pp. 646-650, ISSN 1359-835X.

[17] Esawi, A. M. K. & Borady, M. (2008). Carbon nanotube-reinforced aluminum strips. Compos. Sci. Technol. Vol. 68, pp. 486-492, ISSN 0266-3538.

[18] Esawi, A. M. K.; Morsi, K.; Sayed, A.; Gawad, A. A. & Borah, P. (2009). Fabrication and properties of dispersed carbon nanotube - aluminum composites. Mater. Sci. Eng. A Vol. 508, pp. 167-173, ISSN 0921-5093.

[19] Morsi, K; El-Desouky, A.; Johnson, B.; Mar, A. & Lanka, S. (2009). Spark plasma extrusion (SPE): Prospects and potential. Scr. Mater. Vol. 61, pp. 395-398, ISSN 1359-6462.

[20] Morsi, K.; Esawi, A. M. K.; Lanka, S.; Sayed, A. & Taher, M. (2010). Spark plasma extrusion (SPE) of ball-milled aluminumand carbon nanotube reinforced aluminum composite powders. Composites: Part A Vol. 41, pp. 322-326, ISSN ISSN 1359-835X.

[21] Laha, T.; Agarwal, A.; Mckechnie, T. & Seal, S. (2004). Synthesis and characterization of plasma spray formed carbon nanotube reinforced aluminum composite. Mater. Sci. Eng. A Vol. 381, pp. 249-258, ISSN ISSN 0921-5093.

[22] Bakshi, S. R.; Singh, V.; Balani, K.; Mccartney, D. G.; Seal, S. & Agarwal, A. (2008). Carbon nanotube reinforced aluminum composite coating via cold spraying. Surf. Coat. Technol. Vol. 202, pp. 5162-5169, ISSN 0257-8972.

[23] Bakshi, S. R.; Lahiri, D. & Agarwal, A. (2010). Carbon nanotube reinforced metal matrix composites -a review. Inter. Mater. Rev. Vol. 55, pp. 41-64, ISSN 1743-2804.

[24] Kwon, H.; Estili, M.; Takagi, K.; Miyazaki, T. & Kawasaki, A. (2009). Combination of hot extrusion and spark plasma sintering for producing carbon nanotube reinforced aluminum matrix composites. Carbon Vol. 47, pp. 570-577, ISSN 0008-6223.

[25] Kwon, H. & Kawasaki, A. (2009). Extrusion of spark plasma sintered aluminum-carbon nanotube composites at various sintering temperatures. J. Nanosci. Nanotechno Vol. l 9, pp. 6542-6548, ISSN 1550-7033.

[26] Kwon, H.; Park, D.; Silvain, J. F. & Kawasaki, A. (2010). Investigation of carbon nanotube reinforced aluminum matrix composite materials. Compos. Sci. Technol. Vol. 70, pp. 546-550, ISSN 0266-3538.

[27] Noguchi, T.; Magario, A.; Fukazawa, S.; Shimizu, S.; Beppu, J. & Seki, M. (2004). Carbon Nanotube/Aluminium Composites with Uniform Dispersion. Mater. Trans. Vol. 45, pp. 602-604, ISSN 1347-5320.

[28] Keszler, A. M.; Nemes, L.; Ahmad, S. R. & Fang, X. (2004). Characterisation of carbon nanotube materials by Raman spectroscopy and microscopy - A case study of multiwalled and singlewalled samples. J. Optoelectron. Adv. Mater. Vol. 6, pp. 1269-1274, ISSN 1454-4164.

[29] Zhao, Q. & Wagner, H. D. (2004). Raman spectroscopy of carbon-nanotube-based composites. Philos. Trans. R. Soc. Vol. 362, pp. 2407-2424, ISSN 1471-2962.

[30] http://en.wikipedia.org/wiki/Tensile_strength

[31] Kocks U. F. (2004). Solution hardening and strain hardening at elevated temperatures. American Society for Metal. Metals Park, OH, pp. 89-107.

[32] Kelly, A. & Tyson, W. R. (1965). Tensile properties of fibre-reinforced metals: Copper/tungsten and copper/molybdenum. J. Mech. Phys. Solids Vol. 13, pp. 329-338, ISSN 0022-5096.

[33] Cox, H. L. (1952). The elasticity and strength of paper and other fibrous materials. Br. J. Appl. Phys. Vol. 3, pp. 72-79, ISSN 0508-3443.

[34] Serajzadeh, S. (2004). Prediction of dynamic recrystallization kinetics during hot rolling. Modelling Simul. Mater. Sci. Eng. Vol. 12, pp. 1185-1200, ISSN 1361-651X.

[35] Zadra, M.; Casari, F.; Girardini, L. & Molinari, A. (2007). Spark plasma sintering of pure aluminium powder: mechanical properties and fracture analysis. Powder Metall. Vol. 50, pp. 40-45, ISSN 1743-2901.

[36] Omori, M. (2000). Sintering, consolidation, reaction and crystal growth by the spark plasma system (SPS). Mater. Sci. Eng. A Vol. 287, pp. 183-188, ISSN 0921-5093.

[37] Xie, G.; Ohashi, O.; Yoshioka, T.; Song, M.; Mitsuishi, K.; Yasuda, H.; Furuya, K. & Noda, T. (2001). Effect of Interface Behavior between Particles on Properties of Pure Al Powder Compacts by Spark Plasma Sintering. Mater. Trans. Vol. 42, pp. 1846-1849, ISSN 1347-5320.

[38] Kwon, H.; Park, D.; Park, Y.; Silvain, J. F.; Kawasaki, A. & Park, Y. (2010). Spark plasma sintering behavior of pure aluminum depending on various sintering temperatures. Met. Mater. Int. Vol. 16, pp. 71-75, ISSN 1598-9623.

[39] Ci, L.; Ryu, Z.; Jin-Phillipp, N. Y. & Rühle, M. (2006). Investigation of the interfacial reaction between multi-walled carbon nanotubes and aluminum. Acta Mater. Vol. 54, pp. 5367-5375, ISSN 1359-6454.

[40] Laha, T. & Agarwal, A. (2008). Effect of sintering on thermally sprayed carbon nanotube reinforced aluminum nanocomposite. Mater. Sci. Eng. A Vol. 480, pp. 323-332, ISSN 0921-5093.

[41] Laha, T.; Kuchibhatla, S.; Seal, S.; Li, W. & Agarwal, A. (2007) Interfacial phenomena in thermally sprayed multiwalled carbon nanotube reinforced aluminum nanocomposite. Acta Mater. Vol. 55, pp. 1059-1066, ISSN 1359-6454.

[42] Nagae, T.; Tomiya, S. & Yokota, N. (2001). Compaction and application of the aluminum powder by spark plasma sintering process. Toyama Industrial Technology Center *Vol.* 89, pp. 89-93 (In Japanese), ISSN 0914-2207

[43] Lourie, O. & Wagner, H. D. (1999). Evidence of stress transfer and formation of fracture clusters in carbon nanotube-based composites. Compos. Sci. Technol. Vol. 59, pp. 975-977, ISSN 0266-3538.

[44] Fukuda, H. & Chou, T. W. (1982). A probabilistic theory of the strength of short-fibre composites with variable fibre length and orientation. J. Mater. Sci. Vol. 17, pp. 1003-1011, ISSN 1573-4803.

[45] George, R.; Kashyap, K. T.; Rahul, R. & Yamadagni S. (2005). Strengthening in carbon nanotube/aluminium (CNT/Al) composites.Scr. Mater. Vol. 53, pp. 1159-1163, ISSN 1359-6462.

[46] Oh, S. H.; Legros, M.; Kiener, D. & Dehm, D. (2009). In situ observation of dislocation nucleation and escape in a submicrometre aluminium single crystal. Nature Mater. Vol. 8, pp. 95-100, ISSN 1476-1122.

[47] Jeong, Y. J.; Cha, S. I.; Kim, K. T.; Lee, K. H.; Mo, C. B. & Hong, S. H. (2007). Synergistic Strengthening Effect of Ultrafine-Grained Metals Reinforced with Carbon Nanotubes. Small Vol. 5, pp. 840-844, ISSN 1613-6829.

[48] Ruch, P. W.; Beffort, O.; Kleiner, S.; Weber, L. & Uggowitzer, P. J. (2006). Slective interfacial bonding in Al(Si)-diamond composites and its effect on thermal conduvtivity. Compos. Sci. Technol. Vol. 66, pp. 2677-2685, ISSN 0266-3538.
[49] Khalid, F. A.; Beffort, O.; Klotz, U. E.; Keller, B. A. & Gasser, P. (2004). Microstructure and interfacial characteristics of aluminum-diamond composite materials. Diamond Relat. Mater. Vol. 13, pp. 393-400, ISSN 0925-9635.

16

Formation of Various Nanocarbon and Composite Structures by Laser Ablation

Fumio Kokai
Mie University
Japan

1. Introduction

Nanocarbon and composite materials have attracted considerable research attention due to their unique properties and potential applications. Laser ablation of graphite or graphite containing a small amount of metals can produce various nanocarbon structures such as single-walled (SW) (Guo et al., 1995) and multi-walled carbon nanotubes (MWCNTs) (Hirahara et al., 2000), SW carbon nanohorn (CNH) aggregates (Iijima et al., 1999) and polyhedral graphite (PG) particles (Kokai et al., 2003). In this chapter, we describe a simple method to form various nanocarbon and composite structures using laser ablation in high-pressure Ar gas up to 0.9 MPa, with particular focus on the composite nanostructures. Graphite targets containing metals or compounds (Si, Fe, Co, Ni, Cu, Ag, B_4C, Y_2O_3, La_2O_3, and Gd_2O_3) were used as source materials. To effectively form nanocarbon and composite structures by laser irradiation onto the targets at room temperature, we used a continuous wave Nd:YAG laser to eject C and other species with relatively low kinetic energies into the high-pressure Ar gas. As a result, hot C and metal species are confined in a space surrounded by Ar gas. For example, the initial temperature of the vaporized C species was ~5000 °C estimated for laser irradiation from emission spectroscopy (Kokai et al., 2001). In addition, emission imaging and shadowgraphic studies implied that vaporized species had low expansion velocities of 10^2–10^3 cm/s (Kokai et al., 1999) due to the high-pressure Ar gas restricting their expansion. Unlike a laser ablation method combined with an electric furnace for SW and MWCNT growth, where vaporized C and metal species kept at 800–1300°C are essential for an efficient CNT growth (Gorbunov et al., 1999), the heat sources available for the growth of various nanocarbon and composite structures in this study are laser-ablated C and metal species themselves confined by high-pressure Ar gas. The control of the resident densities and the maintenance of high-temperature of laser-ablated C and metal species, based on adjusting the metal content in graphite and the Ar gas pressure, results in various composite nanostructures with high yields in the deposits. The nanostructures were characterized using scanning electron microscopy (SEM), transmission electron microscopy (TEM), selected area electron diffraction (SAED) and x-ray diffraction (XRD) patterns, and Raman spectroscopy. Depending on the type and content of metals and compounds, characteristic nanocarbon and composite structures such as hybridized SWCNH particles and Cu- or SiC-filled one-dimensional (1D) structures were formed. We discuss the metal-dependent growth of the various nanocarbon and composite structures with an emphasis on graphitization processes such as thermal graphitization, catalytic graphitization and vapor-liquid-solid (VLS) growth (Wagner & Ellis, 1964), and unconventional VLS growth without a catalyst.

2. Experimental

A set of elements, Si, Fe, Co, Ni, Cu, and Ag, and compounds, B_4C, Y_2O_3, La_2O_3, and Gd_2O_3, were mixed with graphite and with metal contents typically from 5 to 30 atomic (at.) %. For B_4C and Si, higher contents were used up to 60 (B content) and 70 at.%, respectively. The average particle sizes of the graphite, Si, Fe, Co, Ni, Cu, Ag, B_4C, Y_2O_3, La_2O_3, and Gd_2O_3 powders were 5, 45, 20, 2, 3, 2, 100, 60, 30, 40, and 30 μm, respectively. The mixed powders were ground and then pressed to form pellets (10 mm in diameter, 2 mm thick) under a pressure of 10 MPa in a stainless-steel die press. A continuous wave Nd:YAG laser (500 W peak power) was used for the laser ablation at room temperature. The laser beam was focused on the pellets through a quartz window installed on a cylindrical stainless-steel chamber (110 mm in diameter and 150 mm long) filled with Ar gas at pressures ranging from 0.05 to 0.9 MPa, as in previous studies (Kokai et al., 2003 and 2004). The size of the laser spot and the power density were adjusted to 2 mm and approximately 13–23 kW/cm^2, respectively, and the laser irradiation time was set to 2 s. After laser irradiation of several shots on the fresh surface of a rotating target, the depsits produced on the chamber wall were collected. The deposits were examined with a SEM and TEMs operated at 100 kV and 300 kV. XRD measurements of the deposits were carried out using Cu-Kα radiation operated at 40 kV and 150 mA. Raman spectra of the deposits were taken with excitation by means of the 488-nm line of an Ar^+ laser.

3. Characterization of nanocarbon and composite structures

3.1 Hybridized SWCNH and other particles from graphite containing Fe, Co, Ni, or Ag

SWCNH particles with diameters of 80–100 nm are almost spherical aggregates of many tubule-like structures made of graphitic sheets (Iijima et al., 1999). Room-temperature CO_2 laser ablation of graphite can produce SWCNH particles with a high yield of over 90%. Comparing the growth of SWCNH particles with those of other nanostructured graphitic carbons, platelet graphite (PLG) (Kokai et al., 2004) and PG particles (Kokai et al., 2003), grown under different Ar gas pressure conditions up to 0.8 MPa has suggested that the formation of tubule-like SWCNH structures and their aggregation in Ar gas was important for the formation of SWCNH particles (Kokai et al., 2004). The generation of a new hybrid structure, a SWCNH particle including a Ni-encapsulated carbon nanocapsule (CNC) in its center, was reported using a submerged arc method (Sano et al., 2004). The CNC is a hollow graphitic cage filled with iron-group or rare-earth metals and it has a typical diameter of 10–200 nm (Ruoff et al., 1993; Saito, 1995). However, the major product that was formed was Ni-encapsulated CNCs using the submerged arc method. The SWCNH particles hybridized with the CNCs had a very low (less than 1%) yield.

The laser ablation of graphite containing Fe, Co, Ni, or Ag produced SWCNH particles hybridized with CNCs with a high yield of ~70% and other nanostructures at Ar gas pressures of 0.1–0.3 MPa (Kobayashi et al., 2007). Figures 1a–d show typical TEM images of four products obtained by the laser ablation of a graphite-Fe target. The Fe content was 20 at.% in the target. The products shown here, which include Fe core particles, are a SWCNH particle, a PLG particle (Kokai et al., 2004), an amorphous carbon (AC) particle, and a CNC.

The Fe cores in the SWCNH, PLG, and AC particles were smaller than those of the CNCs. In addition to the core particles, small Fe particles often attached with the AC particles (Fig. 1c). This may be due to a higher concentration of defects in the AC particles.

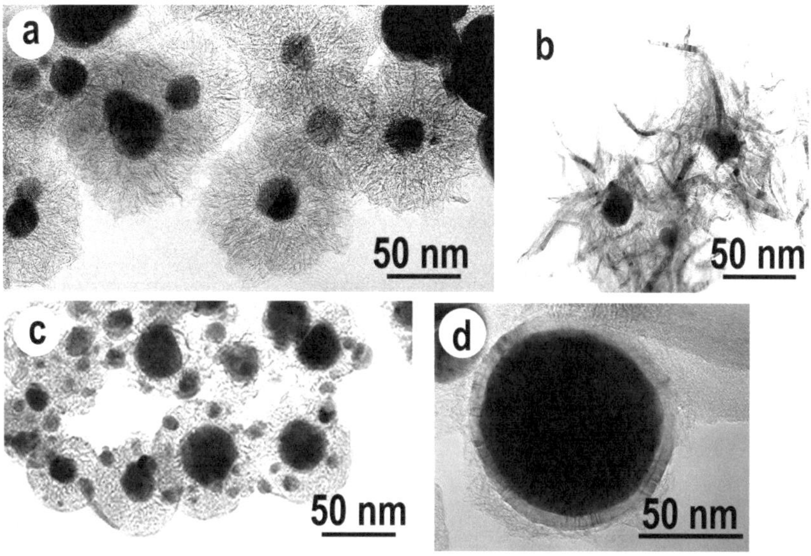

Fig. 1. TEM images of hybridized (a) SWCNH, (b) PLG, and (c) AC particles and (d) CNC obtained by laser ablation of graphite containing 20 at.% Fe and at Ar gas pressures of 0.1-0.3 MPa.

The outer region of one of the hybridized SWCNH particles is magnified in Fig. 2a, showing horn structures, similar to those of unhybridized SWCNH aggregates. The core region is also magnified in Fig. 2b. Graphitic layers cover the core just like in CNCs. These layers ranged in number from 10 to 25. The presence of these layers is similar to the SWCNH particles hybridized with Ni-containing CNCs. Similar graphitic layers were frequently observed as the cores of the PLG and AC particles. To identify the phase of the Fe core in the hybridized SWCNH particle, a SAED pattern and its corresponding lattice image were taken (Fig. 2c). The SAED pattern consists of two types of patterns: annular rings from graphitic layers and spots from the core. Several sets of spots were superimposed because the core was polycrystalline. Spots diffracted from a grain can easily be distinguished by comparing them in the lattice image. The direction of the spot must be normal to the lattice fringes. The spots, indicated by the arrows in the image, can be assigned to the (1-22) and (211) reflections of an orthorhombic Fe_3C crystal according to the data of Joint Committee on Powder Diffraction Standards (JCPDS) No. 6-0670 (*Pbnm, a=0.452, b=0.507, and c=0.674 nm*). The measured d-values calculated from the nominal camera length were 0.198 and 0.188 nm, and the interplanar angle was 76°. These values can not be explained by assuming α-Fe or γ-Fe phases.

The yields of the four products shown in Figs. 1a-d were strongly dependent on the Ar gas pressure and the Fe content in the graphite target. We roughly estimated the yields of the four products by comparing their areas in TEM images measured for several parts in the deposits after laser irradiation. Figure 3 shows the yields of the four products as a function of the Ar gas pressure. Graphite targets containing 20 at.% Fe were used for each laser

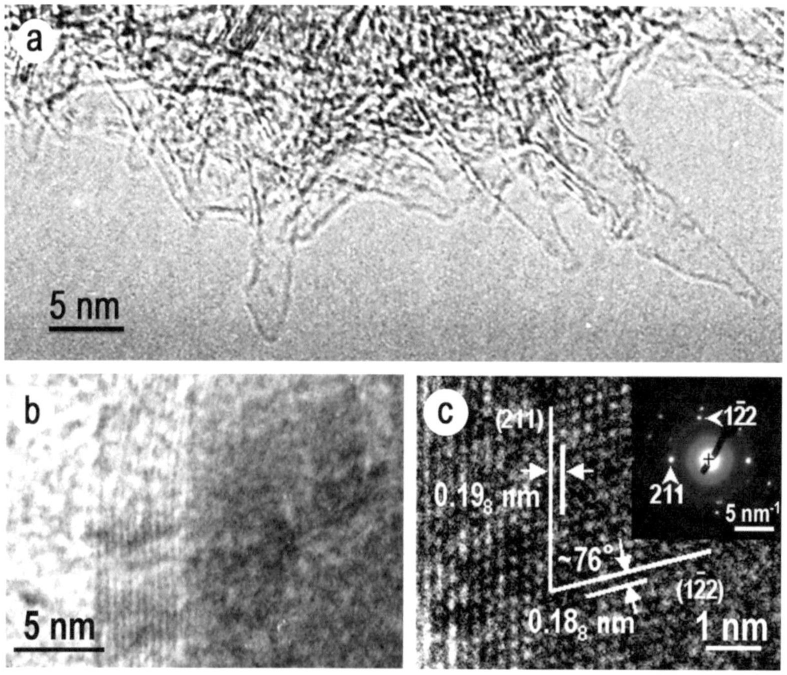

Fig. 2. TEM images observed in (a) the outer region and (b) near the core of a hybridized SWCNH particle and (c) the corresponding SAED pattern and lattice image.

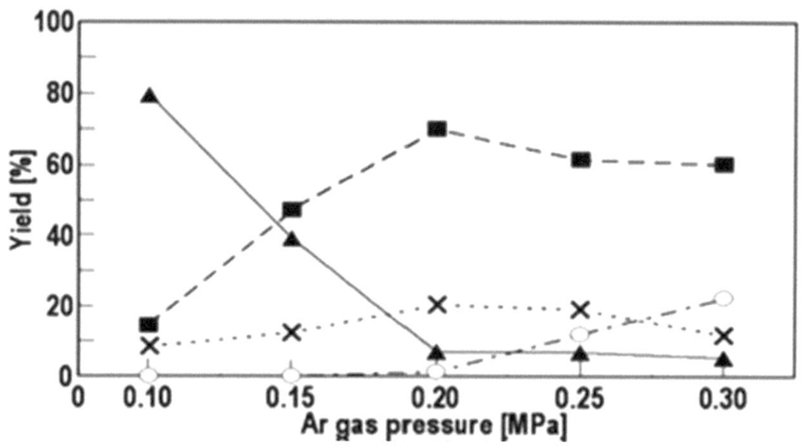

Fig. 3. Yields of hybridized SWCNH (■), AC (▲), and PLG (○) particles and CNCs (×) as a function of Ar gas pressure.

irradiation. At 0.1 MPa, the major product (~80%) was AC particles hybridized with Fe_3C particles or Fe_3C-filled CNCs. As the Ar gas pressure was increased, the yield of the hybridized AC particles decreased, while the yield of the SWCNH particles hybridized with Fe_3C-filled CNCs significantly increased. At 0.2 MPa, the yield of the hybridized SWCNH particles reached a maximum value of about 70%. Under this condition, almost all of the SWCNH particles contained Fe_3C-filled CNCs and the yield of unhybridized SWCNH particles, which did not contain CNCs, was less than 1%. When the Ar gas pressure was further increased, the yield of the hybridized SWCNH particles decreased and the yield of hybridized PG particles only slightly increased.

In an XRD pattern (Fig. 4) of the product obtained from a 20 at.% Fe target at 0.2 MPa, where the hybridized SWCNH particles attained the highest yield, all the peaks, except for the one at 50.5°, were assigned to the reflections of a Fe_3C crystal. This was consistent with the SAED analysis shown in Fig. 2c. The XRD pattern also indicated that the Fe in the other products (~30%) was also in the carbide phase. We note that when we used Fe-free graphite for laser ablation, the yield of the SWCNH particles was highest at the same Ar gas pressure of 0.2 MPa and the yield of the PLG particles increased with a further increase in the Ar gas pressure. The formation of the SWCNH particles hybridized with Fe_3C-filled CNCs thus seems to depend on the aggregation of SWCNHs.

Figure 5 shows the dependence of the Fe content in a laser-irradiation graphite target on the yields of products. The Ar gas pressure was kept at 0.2 MPa for each laser ablation. As the Fe content increased from 5 at.%, the yield of the hybridized SWCNH particles increased and reached a maximum (~70%) at 20 at.%. For 5 and 10 at.% Fe, unhybridized SWCNH particles were observed and caused a reduction in the yield of hybridized SWCNH particles. After the maximum yield was reached, the hybridized SWCNH particle yield decreased, while the CNC yield increased drastically. In addition to the Fe_3C-filled CNCs, the number of Fe-filled CNCs, confirmed by an SAED pattern having spots of metallic α-Fe, increased. This Fe content dependence indicates that the existence of a suitable amount of Fe relative to the C in an Ar atmosphere is important for forming high-yield SWCNH particles hybridized with Fe_3C-filled CNCs. The CNC formation in environments of excess Fe hinders the formation of SWCNH particles like the Fe content of 30 at.% of Fig. 5.

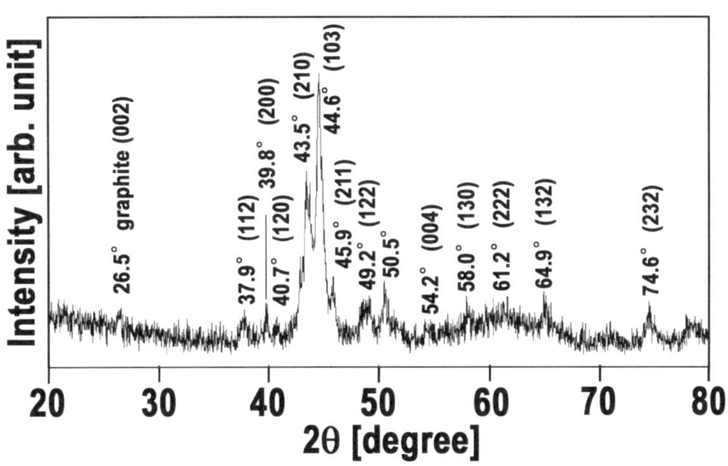

Fig. 4. XRD pattern of the product by laser ablation of graphite containing 20 at.% Fe at Ar gas pressure of 0.2 MPa. Most of the peaks are assigned to Fe_3C. The (110) and (200) reflections of α-Fe may overlap with the (103) and (132) reflections of Fe_3C, respectively, but the intensities of the (110) and (200) reflections are low, indicating a minor α-Fe product.

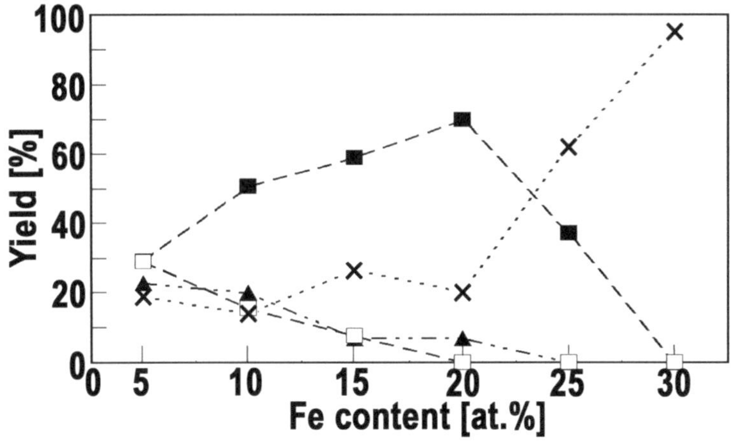

Fig. 5. Yields of unhybridized SWCNH particles (□), hybridized SWCNH particles (■), hybridized AC particles (▲), and CNCs (×) as a function of Fe content in the graphite target. The yield of the hybridized PLG particles (not shown) was at most 1%.

The formation of a metal- or carbide-filled CNCs is believed to occur through the precipitation of carbon from a supersaturated molten metal-C particle at a high temperature (Ruoff et al., 1993; Saito, 1995). On the basis of binary phase diagrams of metal and C (Massalski et al., 1990), the C solubility in a metal particle at a high temperature probably differs significantly depending on the type of metal used. Therefore, we attempted to fabricate hybrid structures of SWCNH particles and CNCs using graphite targets containing various metals. As a result, Co, Ni, and Ag were encapsulated to form hybridized SWCNH particles similar to that of Fe. Their yields were the highest (~70%) under the processing conditions, which were also similar to that used for Fe. For example, a TEM image of a hybridized SWCNH particle and the corresponding SAED pattern of a specimen formed using a graphite target containing 20 at.% Co and an Ar gas pressure of 0.2 MPa are shown in Fig. 6. According to JCPDS No. 15-0806, the spot indicated by the arrow in the SAED is assigned to β-Co (*fcc, a=0.354 nm*). Fast quenching may stabilize the meta-stable β-phase as was observed in the CNCs (Saito, 1995; Jiao et al., 1996). An XRD measurement (Fig. 7) also supported the quenching. All the peaks were successfully assigned to β-Co, except for the peak at 26.6° that reflected from graphite (002). It is important to note that the core was not a carbide but a metal, although the Co-C system contains carbides, such as Co_2C and Co_3C just like in the Fe-C system. The Ni and Ag cores were also metals rather than metal carbides. These metallic cores were also covered with graphitic layers, but they were thinner (1-15 layers) than those of the Fe_3C particles (10–25 layers).

In order to further analyze the detailed structures of the hybridized SWCNH particles, comparative TEM examination was performed for three types of hybridized SWCNH particles synthesized from graphite containing 20 at.% Fe, Co, or Ag (hereafter Fe, Co, and Ag hybrids) (Kokai et al., 2009). As shown in binary phase diagams (Massalski et al., 1990), Fe, Co, and Ag have significantly different C solubilities: 25, 4.2, and 0.036 at.%, respectively. From more than 200 samples for each hybridized structure (some samples can be seen in Fig. 8), the diameters of the hybrids and the encapsulated CNCs and the thicknesses of the SWCNH layers surrounding the CNCs were measured and compared.

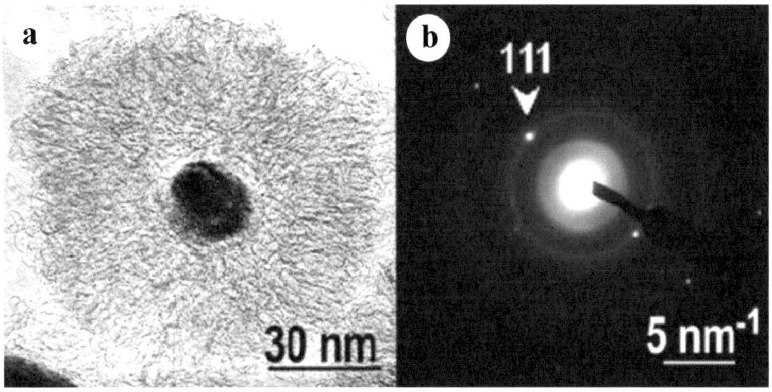

Fig. 6. (a) TEM image and (b) the corresponding SAED pattern of a product from graphite containing Co. The Co content in graphite was 20 at.%. The laser ablation experiments were carried out at Ar gas pressure of 0.2 MPa.

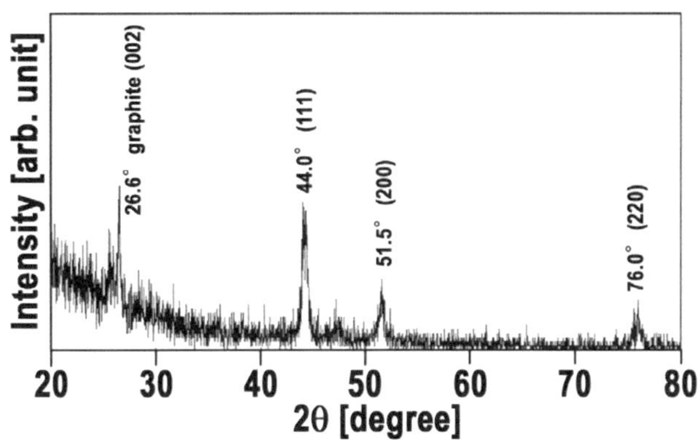

Fig. 7. XRD pattern of the product by laser ablation of graphite containing 20 at.% Co at Ar gas pressure of 0.2 MPa. The peaks are assigned to β-Co.

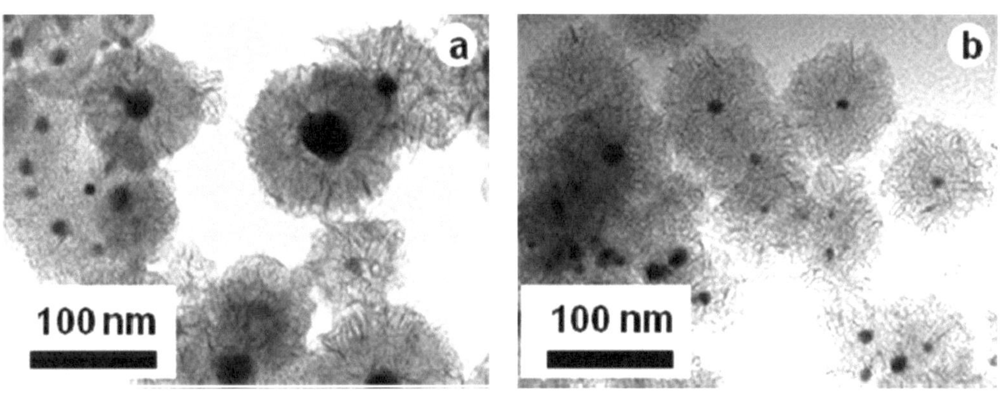

Fig. 8. TEM images of (a) Co and (b) Ag hybrids obtained by laser ablation of graphite containing 20 at.% Co or Ag at Ar gas pressure of 0.2 MPa.

Figure 9 shows diameter distributions of the hybrids (d_h) and CNCs (d_c) for the three hybrids. As shown in Fig. 9a, the d_h values for the three structures range from 40 to 150 nm; however the peak position and d_h distribution differ among the structures. The average d_h values are 96, 90, and 85 nm for the Fe, Co, and Ag hybrids, respectively. As shown in Fig. 9b, the d_c values range from 5 to 55 nm. The d_c distributions differ among the structures; the average diameters are 24, 19, and 16 nm for the Fe, Co, and Ag hybrids, respectively. Note that large CNCs with diameters up to ~140 nm, which were not surrounded by SWCNHs, were also observed, as in the previous study (Kobayashi et al., 2007). In addition to d_h and d_c, the distributions of the SWCNH layer thicknesses (t_n) were also measured for the three hybrids (not shown). The t_n values ranged from 15 to 65 nm. The average t_n values were

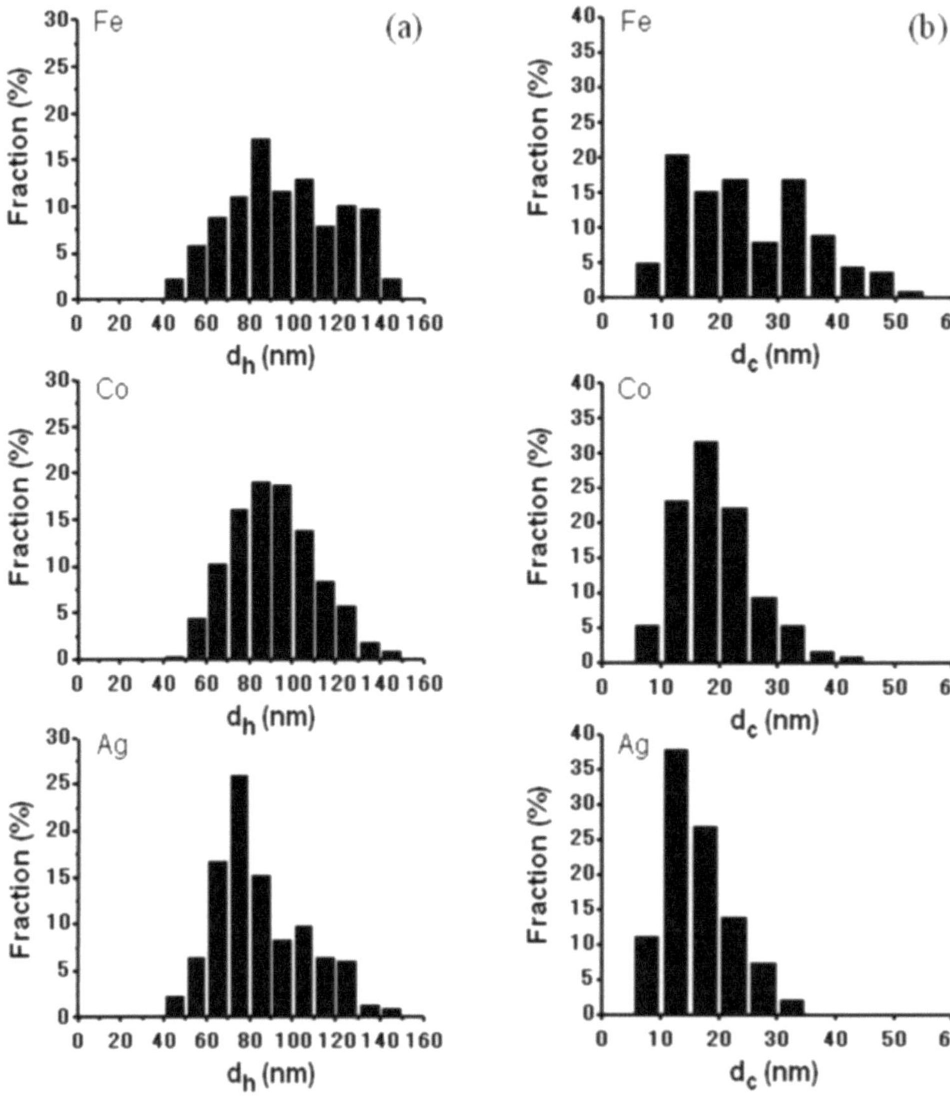

Fig. 9. Diameter distributions of the (a) hybrids (d_h) and (b) CNCs (d_c) for Fe, Co, and Ag hybrids.

18 nm for the Fe and Co hybrids and 17 nm for the Ag hybrid. Unlike the d_h and d_c values, no significant difference was observed in the average t_n values among the three hybrids.

The distributions of d_h, d_c, and t_n and their average values indicate that the difference in the diameters of the three hybrids is due to the formation of CNCs with different diameters in accordance with the type of metal. From the solublites of C for Fe, Co, and Ag, we believe that the diameters of the CNCs in the hybrids are governed by the differences in the C solubilities and precipitation processes. The stronger ability of Fe to form a carbide probably results in CNCs with larger sizes.

For the three hybrids, we investigated the correlation of d_h with d_c or $2t_n$. Figures 10 and 11 show plots of d_h versus d_c or $2t_n$ for the three hybrids. Although d_h increases with both d_c and $2t_h$, there is a stronger correlation between d_h and $2t_n$. For the Fe, Co, and Ag hybrids, the correlation coefficients were calculated to be 0.82, 0.74, and 0.70 between d_h and d_c and 0.94, 0.96, and 0.97 between d_h and $2t_n$. The correlation of d_h with $2t_n$ may be stronger because the lengths of the SWCNHs account for a considerable portion of the hybrid diameter.

In addition to the different d_h distributions for the three hybrids and stronger correlation between d_h and $2t_n$, some interesting features were observed regarding the formation mechanism of the hybridized SWCNH particles. Change in the metal contents in the graphite target resulted in the formation of unhybridized products (Fig. 5). In particular, when the content of the metal in graphite was lower than 20 at.%, we observed both hybrids and unhybridized SWCNH particles.

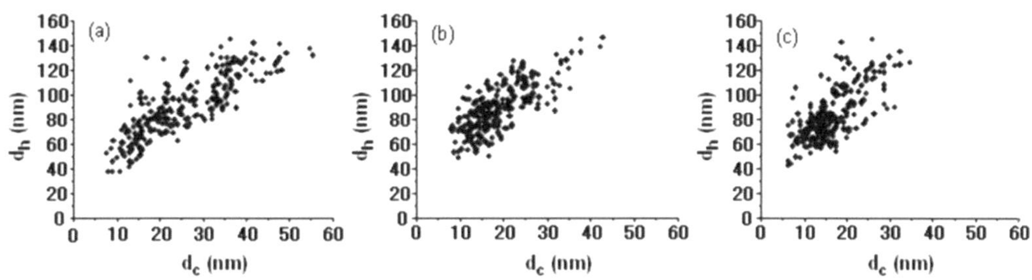

Fig. 10. Plots of hybrid diameter (d_h) versus CNC diameter (d_c) for (a) Fe, (b) Co, and (c) Ag hybrids.

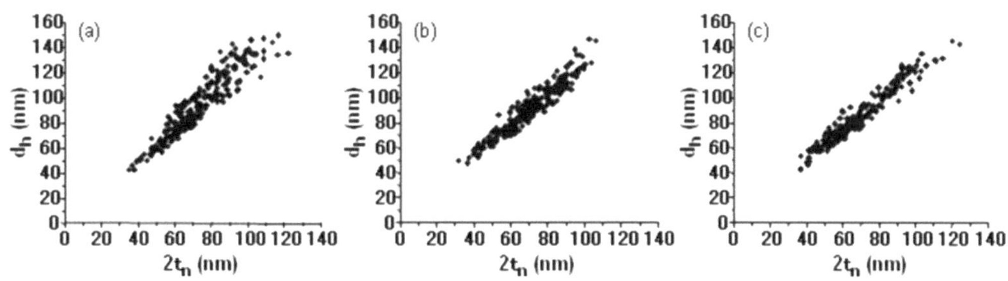

Fig. 11. Plots of hybrid diameter (d_h) versus twice the SWCNH layer thickness ($2t_n$) for (a) Fe, (b) Co, and (c) Ag hybrids.

Figure 12a shows a typical TEM image of the products produced at an Ar pressure of 0.2 MPa and with an Fe content of 10 at.%. There are Fe hybrids and unhybridized SWCNH aggregates; the diameters of the unhybridized SWCNH aggregates are less than those of the Fe hybrids. As the metal content in the graphite target was increased from 20 at.%, the numbers of hybrids and SWCNH aggregates decreased and the number of unhybridized CNCs rapidly increased (Fig. 5). Figure 12b shows a typical TEM image of the products produced at an Ar gas pressure of 0.2 MPa and with an Fe content of 30 at.%. There are many CNCs with diameters of 20–210 nm. The preferable formation of molten metal-C particles probably consumes the carbon source and hinders the formation of SWCNHs into hybrid structures (Fig. 5).

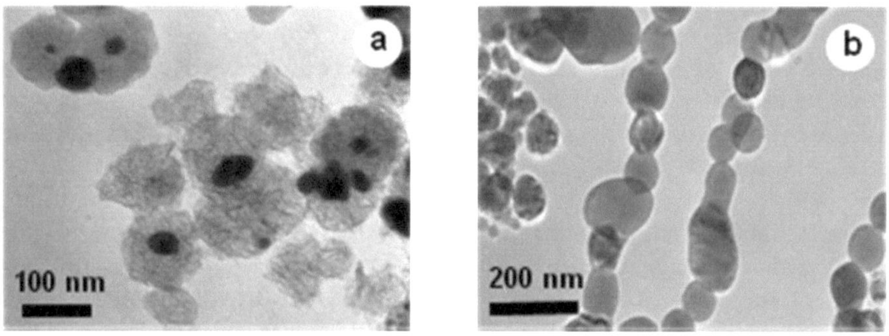

Fig. 12. TEM images of products obtained by laser ablation of graphite containing (a) 10 and (b) 30 at.% Fe at Ar gas pressure of 0. 2 MPa.

From some of the present findings, we propose a mechanism for hybrid formation, which is based on SWCNHs assembling around a molten metal-C particle as a nucleation core. After laser ablation of graphite containing metal, vaporized hot C and metal species form molten metal-C particles together with many SWCNHs and/or their precursors like graphene patches (Kawai et al., 2002) in a space confined by Ar gas. By using a differential mobility analyzer, the evolution in the size distribution of C particles was analyzed after laser ablation and C particles, 20–120 nm in size, were detected (Cheng at al., 2007), while the formation of SWCNHs and/or their precursors probably due to thermal graphitization is not clear The strong correlation between the hybrid diameter and the SWCNH layer thickness for Fe, Co, and Ag hybrids (Fig. 11) suggests that after the SWCNHs with a certain range of lengths are formed, they aggregate around molten metal-C particles with a certain range of diameters.

The contribution of only small CNCs (Figs. 9b and 10) to the hybrid formation suggests that the mutual interaction among SWCNHs near the center of the hybrid structure plays a crucial role in maintaining the resultant aggregate structure, while CNCs with diameters of 20–210 nm were formed. The increase in the numbers of covering graphitic layers due to the annealing of the CNCs at 600–1100°C (Jiao & Seraphin, 1998) and the graphitization of carbon (diamond and AC) by metal (catalytic graphitization) at 600–1000°C (Yudasaka et al., 2002) indicates that C precipitation, which leads to the formation of CNCs in SWCNH aggregates, occurs after SWCNH assembly. The growth of the hybridized SWCNH particle involves two graphitization processes, one thermal and the other catalytic.

3.2 Sea-urchin type SWCNTs and carbide-filled PG particles from graphite containing Y_2O_3, La_2O_3, and Gd_2O_3

The laser ablation of graphite containing Y_2O_3, La_2O_3, and Gd_2O_3 did not produce hybridized SWCNH particles and provided products composed of sea-urchin type SWCNTs and C polyhedra filled with carbide particles dependent on the Ar gas pressure (Kokai et al., 2008). For example, Fig. 13 shows a TEM image of the product made from graphite containing 20 at.% Y_2O_3 at an Ar gas pressure of 0.1 MPa. Sea-urchin type SWCNTs, which are radially arranged around a catalytic particle as formed in a previous study using an arc discharge method (Subramoney et al., 1993), are grown from a yttrium carbide particle with a diameter of ~20 nm. The SWCNTs were short, with lengths up to only ~50 nm.

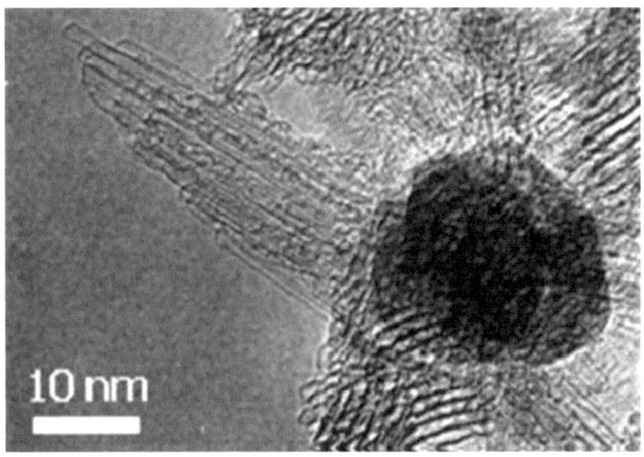

Fig. 13. TEM image of the product from graphite containing 20 at.% Y_2O_3 at Ar gas pressure of 0.1 MPa.

With increasing Ar gas pressure, the formation of C polyhedra filled with carbide particles was prominent. Figure 14 shows a TEM image of the product made from graphite containing 20 at.% Y_2O_3 at an Ar gas pressure of 0.2 MPa. Many carbide filled polyhedra are seen with sizes of 100–500 nm. Figure 15 shows a magnified TEM image and the corresponding SAED pattern of a polyhedron, indicating the formation of a PG particle having facets (Kokai et al., 2003) filled with a YC_2 crystallite particle according to JCPDS No. 11-0602. The diameters of the YC_2 particles were in the range of 30–150 nm. Compared to the diameters of the CNCs in hybridized SWCNH particles, larger YC_2 particles were included in the PG particle. The morphologies of the SWCNTs and PG particle are quite different from those of hybridized SWCNH particles.

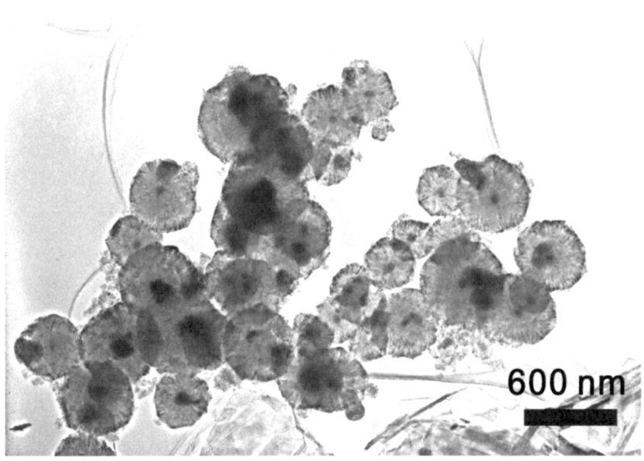

Fig. 14. TEM image of the product from graphite containing 20 at.% Y_2O_3 at Ar gas pressure of 0.2 MPa.

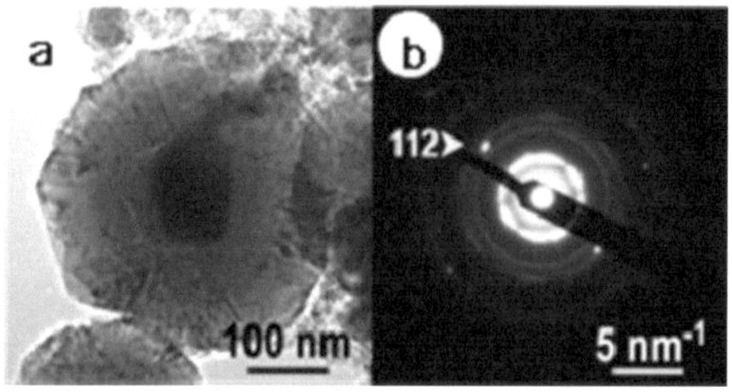

Fig. 15. (a) High-magnification TEM image and (b) the corresponding SAED pattern of the product from graphite containing 20 at.% Y_2O_3 at Ar gas pressure of 0.2 MPa.

The solubilities of C in liquid La and Y are very high, 42.9 and 70 at.%, respectively. The solubility of C in liquid Gd has not been reported, but Gd is known to form a carbide (GdC_2). Therefore, we believe that liquid-like molten metal-C particles are easily formed for these metals after laser ablation. For the growth of sea-urchin type SWCNTs under a relatively low Ar gas pressure condition, we believe that VLS growth, originally proposed to grow Si wires (Wagner & Ellis, 1964), play an important role through C dissolution, diffusion, and precipitation at the molten particles. The surface curvatures of the molten particles may change at the nucleation stage of SWCNTs, as reported from in situ observation of the dynamic deformation of metal particles resulting in a structure having a protrusion at the nucleation stage of a CNT (Hofmann et al., 2007; Yoshida et al., 2008). A SWCNT diameter may be determined by the size and shape of the highly curved structure at a molten particle that is saturated with C, resulting in a lift-off of a C cap to start the growth of a SWCNT. Unlike the SWCNTs grown efficiently from a graphite containing a small amount (~ 1 at.%) of Ni/Co as a catalyst (Guo et al., 1995; Yudasaka et al., 1997), the solubilities of C in liquid La, Gd, and Y, higher than those of Ni (2.7 at.%) and Co (4.2 at.%), are thought to govern the sea-urchin type SWCNT growth from a large (>20 nm) molten carbide particle.

For the formation of PG particles in laser ablation of pure graphite, high-temperature C species with a high resident density, established by confining C species with high-pressure Ar gas (0.5–0.9 MPa), are required (Kokai et al., 2004). In addition, PG particles were easily produced in the presence of 1 at.% Si or B in Ar gas, even at a low pressure of 0.1 MPa (Kokai et al., 2007). It was speculated that Si and B atoms could bind to a C atom and act as a seed to enhance the formation of C clusters. Similar to the behavior of Si and B atoms, carbide-filled PG particles in the presence of La, Gd, and Y may be grown by assembling C clusters around molten metal-C particles followed by graphitization at high-temperature in the range of 2000–3000 °C.

3.3 MWCNTs and BC-filled CNCs from graphite containing B_4C

Unlike laser ablation of graphite containing Fe, Co, and Ag, the laser ablation of graphite containing B_4C produced deposits containing MWCNTs with a high yield. The growth of MWCNTs was attempted by laser ablation combined with an electric furnace (Hirahara et al., 2000; Sabbaghzadeh et al., 2009); however, the dominant product was a CNC and the conditions for the nucleation and growth of MWCNTs by laser ablation are still not known well. Figure 16 shows typical SEM and TEM images of the deposit obtained by laser ablation at a B content of 20 at.% in the target and at an Ar gas pressure of 0.1 MPa. In the SEM image (Fig. 16a), relatively straight 1D structures are seen together with particle-like products. Some of the 1D structures are up to 30 μm long. In the TEM images (Figs. 16b and 16c), the 1D structures are hollow MWCNTs. In the higher resolution TEM micrograph (Fig. 16c), a MWCNT with a 29 nm in diameter is observed. The MWCNT is composed of 31 graphitic layers with interlayer spacings of ~0.35 nm, indicating fairly good graphitization.

As seen in Figs. 16b and 16d, particle-like products, small graphitic C particles and boron carbide (BC) particles encapsulated in CNCs, were also generated together with the MWCNTs. However, hybridized SWCNH particles observed for Fe-, Co-, and Ag-containing targets, were not present in the deposit.

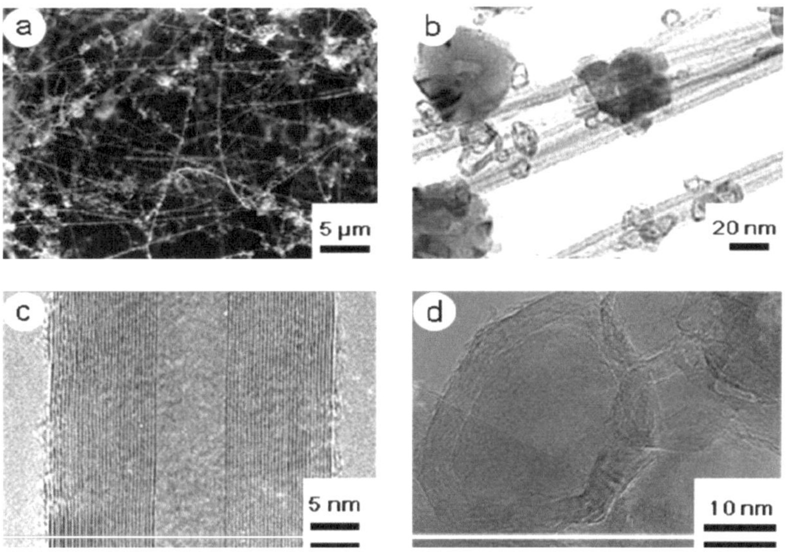

Fig. 16. (a) SEM and (b)TEM images of a deposit synthesized at B content of 20 at.% in the target and Ar gas pressure of 0.1 MPa. (c) Higher magnification TEM image of a MWCNT, 29 nm in diameter. (d) Higher magnification TEM image of CNCs filled with BC.

Further SEM and TEM studies on deposits produced by laser ablation under different conditions indicated that the fractions of MWCNTs in the deposits were highly sensitive to the B content in the graphite and the Ar gas pressure. The fractions of MWCNTs were roughly estimated by areas in TEM images for several parts in the deposit. Figure 17a shows the MWCNT fraction versus the B content. A constant Ar gas pressure of 0.1 MPa was used for the laser ablation. The MWCNT fraction is higher (50–60%) at B contents of 20–30 at.%. This fraction is much higher than previous studies using a nanosecond pulsed laser and an electric furnace (Hirahara et al., 2000; Sabbaghzadeh et al., 2009).

For the B contents lower and higher than 20–30 at.%, the MWCNT fraction decreased. In the lower B contents (1–10 at.%), small graphitic and PG particles were mainly produced. In the higher B contents (50–60 at.%), the major products were large BC-filled CNCs. Compared with the efficient growth of SWCNTs where the catalyst Ni/Co content in graphite is ~1 at.% (Guo et al., 1995; Yudasaka et al., 1997), the B content of 20–30 at.% in graphite for the optimum growth of MWCNTs is significantly different. Figure 17b shows the MWCNT fraction versus the Ar gas pressure. A constant B content of 20 at.% was used for the laser

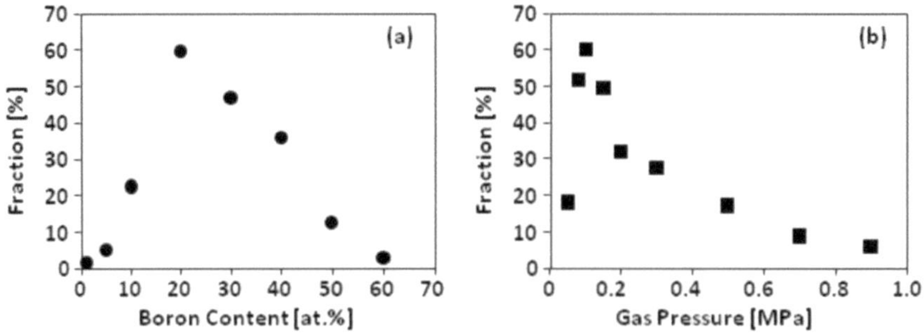

Fig. 17. Fractions of MWCNTs in the deposits versus (a) B content at Ar gas pressure of 0.1 MPa and (b) Ar gas pressure at B content of 20 at.%.

ablation. The MWCNT fraction shows a maximum (~60%) at 0.1 MPa. For Ar gas pressures lower and higher than 0.1 MPa, the MWCNT fraction decreases. At higher Ar gas pressures, large particles such as a PG particle and BC-filled CNCs, and thick rod-like whiskers were observed. At a low Ar gas pressure of 0.05 MPa, the major product was AC particles. The optimum Ar gas pressure of 0.1 MPa for the MWCNT growth is slightly higher than that of 0.07–0.08 MPa for the SWCNT growth (Guo et al., 1995; Yudasaka et al., 1997). We measured the diameters of more than 150 MWCNTs under each laser ablation condition. The diameter of the grown MWCNTs ranged from 7 to 70 nm.

Similar to previous observations (Hirahara et al., 2000), the presence of two types of BC particles, one encapsulated in CNCs (Fig. 16d) and another at a tips of MWCNTs, was confirmed in the present study. Figure 18a shows a TEM image showing a spherical BC particle about 39 nm in diameter at a tip of a MWCNT. In addition to the spherical BC particle, BC particles with various shapes such as ellipsoidal and elongated ones were observed as shown in Fig. 18b. Figure 18c shows a high-magnification TEM image of the portion of a BC particle surrounded by graphitic layers. Lattice fringes with spacings of 0.23 and 0.37 nm were observed and assigned to the distances between (021) and (012) planes in the B_4C crystal on the basis of the data of JCPDS No. 6-0555. The corresponding SAED pattern consists of two types: annular rings from surrounding graphene layers and spots from the BC core. Spots assigned to the (012) and (021) reflections of the B_4C crystal are seen. On the other hand, Figure 18d shows a high-magnification TEM image of a BC particle encapsulated in a CNC, where lattice fringes with spacings of 0.23 and 0.37 nm were also observed.

The diameter distributions of the two types of BC particles were compared (Hirahara et al., 2000). The diameters of the BC particles at MWCNT tips were 1–8 nm, while those of the BC particles in CNCs were 2–23 nm. As a result, they concluded that smaller BC particles seemed to be indispensable for the tubular growth and their critical diameter was found to be ~5 nm. In the present study, a similar tendency in diameters for the formation of MWCNTs and CNCs was observed in the two types of BC particles. Figure 19 shows diameter distributions of the two types of BC particles obtained for laser ablation at an Ar gas pressure of 0.1 MPa and a B content of 20 at.%. The diameters of BC particles at MWCNT tips and in CNCs are 10–100 nm and 20–280 nm, respectively. In this study, BC particles much larger than those of 1–8 nm seem to contribute to the growth of MWCNTs with large diameters up to 70 nm.

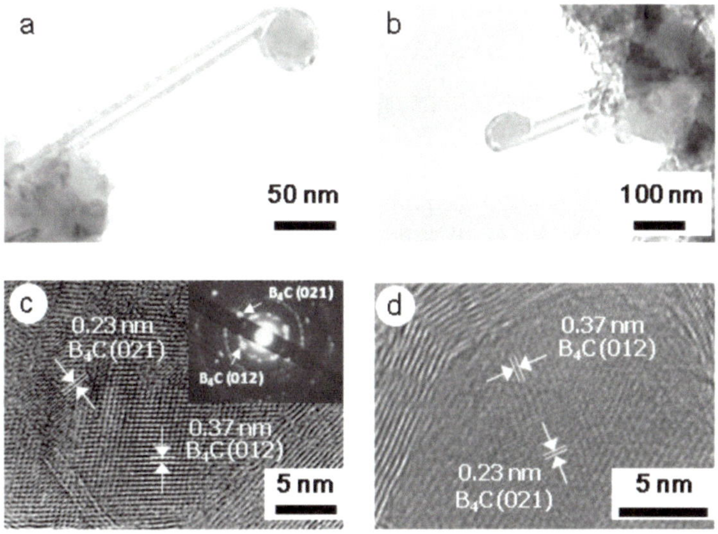

Fig. 18. (a, b) Low-magnification TEM images of the tips of MWCNTs with encapsulated BC particles. (c) High-magnification TEM image of portion of an encapsulated BC particle. The inset in (c) is the corresponding SAED pattern. (d) High-magnification TEM image of portion of an encapsulated BC particle in a CNC.

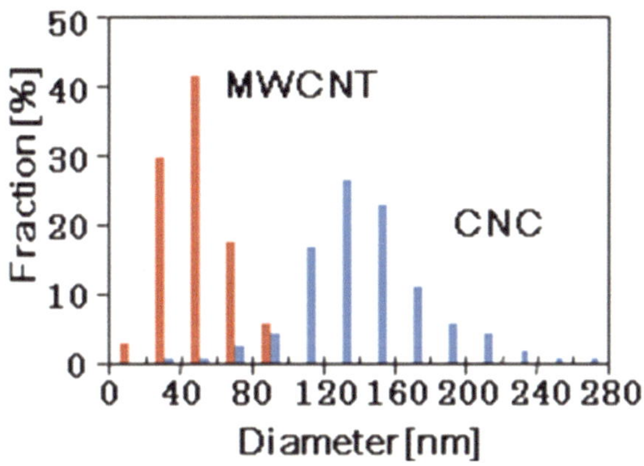

Fig. 19. Diameter distributions of two types of BC particles at tips of MWCNTs and in CNCs obtained from laser ablation at Ar gas pressure of 0.1 MPa and B content of 20 at.%.

As previously proposed (Hirahara et al., 2000), a B (in the form of BC) catalyzed growth mechanism of MWCNTs is considered because MWCNT tips encapsulating BC particles were also observed in the present study. However, much larger molten BC particles (10–100 nm) are thought to act as seeds at high temperature to grow MWCNTs that have up to 80 graphitic walls. The sizes of the catalyst particles were found to be crucial in terms of controlling the diameters and wall numbers to synthesize SW and MWCNTs in thermal chemical vapor deposition (Cheung et al., 2002). Other research groups have made similar observations (Li et al., 2001; Nerushev et al., 2003; Rümmeli et al., 2007). In the present laser ablation synthesis of MWCNTs, the B content in the graphite target and resident densities of laser-ablated B and C species confined by ambient Ar gas play an important role in

determining the size of molten BC particles. Unlike the thin SWCNT grown efficiently from a graphite target containing ~1 at.% Ni/Co, the observed optimum B amount of 20–30 at.% in the target (Fig. 17a) is consistent with the growth of MWCNTs with large diameters.
The VLS mechanism is thought to play a role in the growth of these MWCNTs, similar to the sea-urchin type SWCNTs in Section 3.2. The VLS growth of MWCNTs consists mainly of a three-part process: C is dissolved into a molten BC particle, diffused through the bulk and/or surface of the particle, and precipitated from the particle. According to the binary phase diagrams (Massalski et al., 1990), B has an eutectic temperature of 2375°C, which is much higher than those of Ni (1326.5°C) and Co (1320°C). B also has a C solubility (24.3 at.%) higher than those of Ni (2.7 at.%) and Co (4.2 at.%). Therefore, MWCNTs, which accompany the precipitation of large amounts of C are thought to be grown efficiently on large molten BC particles at temperatures higher than that (800–1300°C) for SWCNTs (Gorbunov et al., 1999). Although the actual temperature needed for C precipitation to form MWCNTs is not clear, temperatures up to ~2400°C may create favorable conditions for efficient MWCNT growth. Various shapes of BC particles observed in MWCNT tips such as spherical and elongated particles suggest that the surface curvatures of molten BC particles changes at the nucleation stage of MWCNTs as reported from in situ observation of the dynamic deformation of metal particles (Hofmann et al., 2007; Yoshida et al., 2008). A MWCNT diameter may be precisely determined by the size and shape of the highly curved structure resulting in a lift-off of a C cap.
Compared with the sizes of the two types of BC particles observed previously (Hirahara et al., 2000), much larger BC particles (10–280 nm) were observed at tips of MWCNTs (10–100 nm) and in CNCs (60–280 nm) in the present study (Fig. 19). In addition to the importance of the size of the molten BC particle for tubular growth (Hirahara et al., 2000), successive occurrence of the C dissolution-diffusion-precipitation process suitable for molten BC particles with specific sizes is thought to be indispensable for tubular growth. The growth of MWCNTs up to 70 nm in diameter may be governed by the action of molten BC particles with specific sizes. In addition to the B content in the target, the ambient Ar gas pressure also influence determining the sizes of BC particles. The C supply to the BC particles and its successive diffusion and precipitation are also influenced by Ar gas pressure, resulting in varying yields, diameters, and lengths of MWCNTs. Since larger molten BC particles could maintain C during the C supply, CNCs may form due to C not being precipitated at the high temperatures effective for the MWCNT growth. The formation of MWCNTs and CNCs is thought to relate to the balance among complex factors such as the resident densities of C and B species, C solubility in B particles, and the rates of C dissolution, diffusion, and precipitation. To form CNCs, C precipitation from core BC particles is believed to occur at the temperature lower than those to grow MWCNTs.

3.4 Cu-filled CNTs and CNCs from graphite containing Cu

Depending on the Ar gas pressure and Cu content, the deposits showing different morphologies were obtained (Kokai at al., 2009a). Figure 20 shows typical SEM images of deposits obtained at Ar gas pressures of 0.1, 0.5, and 0.9 MPa. Graphite targets with a constant Cu content of 20 at.% were used for laser ablation. As seen in Fig. 20a, sphere like particles with diameters of 30–130 nm were dominant at lower pressure regions of Ar gas (0.1–0.3 MPa). With increasing Ar gas pressure up to 0.9 MPa, the size of the particles was increased and the deposits became to contain straight 1D structures in addition to the sphere like particles (Figs. 20b and 20c).

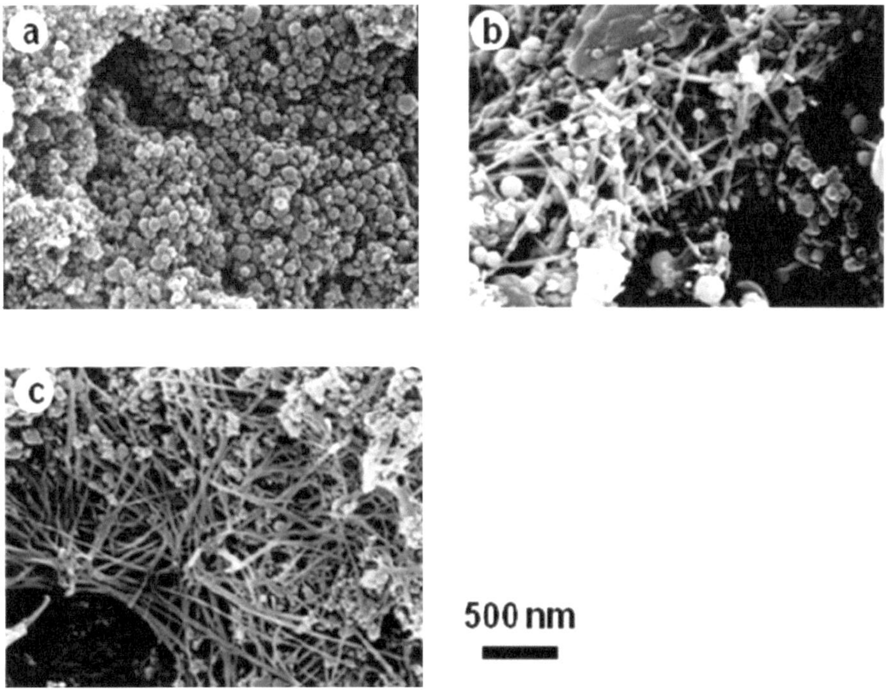

Fig. 20. SEM images of deposits produced at Ar gas pressures of (a) 0.1, (b) 0.5, and (c) 0.9 MPa.

TEM observation indicated that the sphere like particle obtained at lower Ar gas pressures were mainly composed of AC and SWCNH particles. Both particles include CNCs filled with Cu like the products shown in Section 3.1. Figure 21a is a typical TEM image of SWCNH particles, revealing diameters of 80−150 nm, including Cu-filled CNCs (dark areas). Figure 21b shows a TEM image observed near a core Cu particle of a SWCNH particle. Two graphitic layers with an interlayer spacing of about 0.34 nm are seen around the Cu particle as indicated by an arrow. The fraction of the hybridized SWCNH particles was ~70% at an Ar gas pressure of 0.2 MPa and with a Cu content of 20 at.% in the target.

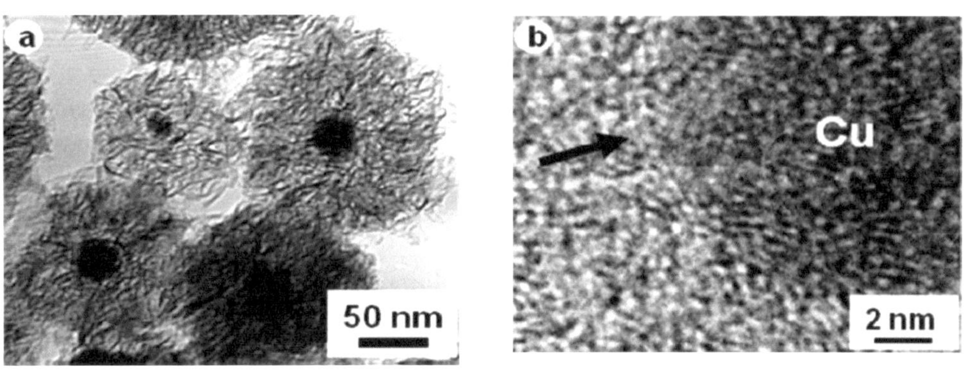

Fig. 21. (a) TEM image of SWCNH particles including Cu-filled CNCs. (b) High-magnification TEM image observed near the core Cu particle in a SWCNH particle.

Figure 22a is a typical TEM image of the deposit obtained at an Ar gas pressure of 0.9 MPa, which consists of 1D structures and sphere like particles. As seen in Fig. 22b, high-magnification TEM images of the 1D structures showed composite structures consisting of Cu nanowires with diameters of 18–20 nm surrounded by graphitic layers. The interlayer spacing in the graphitic layers is in the range 0.34–0.35 nm, which is slightly larger than that of graphite. Cu nanowires with up to seven graphitic layers were observed and Cu nanowires with 1-3 graphitic layers were dominant. Unlike previous studies using arc discharge, thermal chemical vapor deposition, and so on (Setlur et al., 1996; Zhang et al., 2003; Tao et al., 2006; Haase et al., 2007), in which both filled and unfilled CNTs were observed in deposited samples, in the TEM observation of more than 300 Cu-filled CNTs, unfilled or intermittently filled CNTs were not observed and all the grown CNTs contained Cu from root to tip.

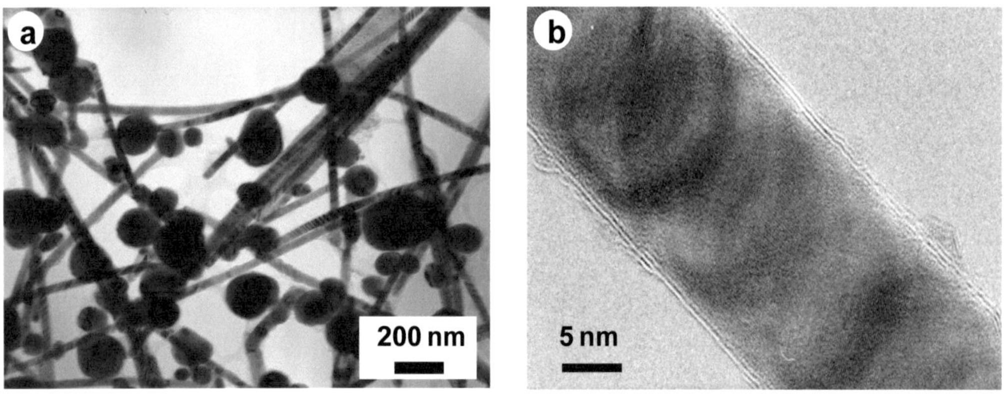

Fig. 22. (a) TEM image of a deposit obtained at Ar gas pressure of 0.9 MPa. (b) High-magnification TEM image of 1D structure with two graphitic layers.

Further characterization of the encapsulated Cu nanowire was carried out by the measurements of a high magnification TEM image and a SAED pattern (Fig. 23). According to JCPDS No. 4-0836, in the TEM image of a Cu wire, lattice fringes with spacing of 0.21 nm are seen corresponding to the distance of the Cu (111) planes. Also, in the SAED pattern, reflections assigned to graphite (002) and Cu (111), (200), and (220) planes are observed. These observed characteristics indicate the encapsulated Cu nanowire forms a polycrystal with a fcc structure. The fcc structure is predicted for a Cu nanowire encapsulated in a CNT by a theoretical study (Choi et al., 2003).

TEM examination of outer regions of the sphere like particles formed together with Cu-filled CNTs indicated that the particles were Cu-filled CNCs (Fig. 24). Cu is known as an element, which is not easily encapsulated in CNCs compared to other metals such as Co and Ni (Seraphin et al., 1996). A few studies (Lin et al., 1994; Jiao & Seraphin, 1998; Athanassiou, et al., 2006) showed that Cu nanoparticles are coated by only 3−4 graphitic layers. The Cu-filled CNCs in our study showed 3−9 graphitic layers. Thin graphitic layers were continuously curved to form spherical shell structures (Fig. 24a). Some thicker graphitic layers had flat sides joined at angles of 130°−150° (Fig. 24b).

In addition to the Ar gas pressure, we found the Cu content in the graphite target was also crucial for growing Cu-filled CNTs. Deposits obtained at different Cu contents at a constant Ar gas pressure of 0.9 MPa were investigated by TEM. As seen in Fig. 25a, when Cu content

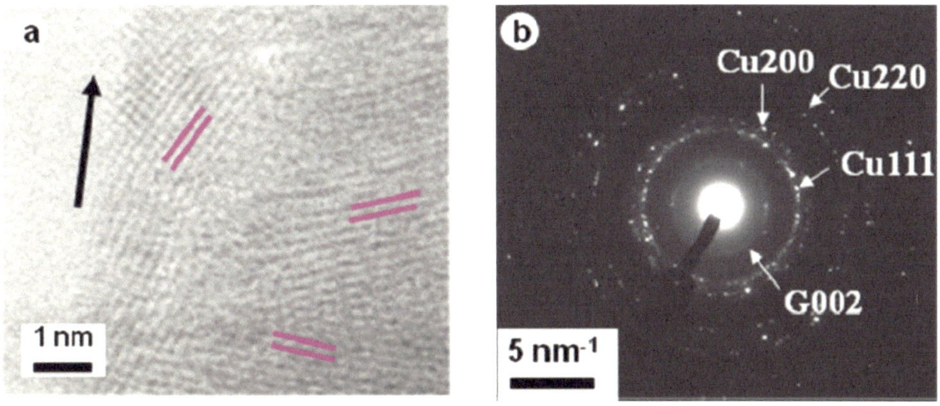

Fig. 23. (a) High magnification TEM image, focusing on the part of a Cu nanowire, and (b) its corresponding SAED pattern. In the TEM image, the arrow indicates the growth direction and the distance between the parallel fringes is close to the space of Cu (111) planes.

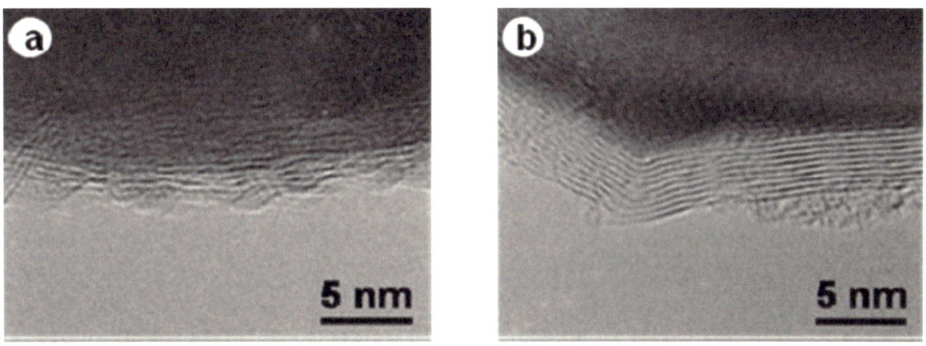

Fig. 24. TEM images of outer regions of CNCs with (a) thin and (b) thick graphitic layers.

was 1 at.%, Cu-filled CNTs were not formed and the major product was PG particles with diameters of 100–600 nm, similar to the product from the laser ablation of pure graphite (Kokai et al., 2003) and graphite containing 1 at.% Si or B (Kokai et al., 2007). With increasing the Cu content, Cu-filled CNTs were formed together with Cu-filled CNC with diameters of 10−120 nm (Fig. 25b). The yield of Cu-filled CNTs became a maximum around 20-30 at.% of the Cu content as shown in Fig. 22a. However, further increase in the Cu content led to a decrease in the yield of Cu-filled CNTs and the major products were Cu-filled CNCs. It is a unique feature that no unfilled CNT is seen even for low Cu content and all the grown CNTs contain Cu nanowires.

An interesting morphological feature of the Cu-filled CNTs in this work is that the diameter distribution is extremely narrow. The histogram shown in Fig. 26 is the diameter distribution of Cu-filled CNTs obtained from TEM images taken for more than 300 Cu-filled CNTs. The Cu-filled CNTs show uniform diameters ranging from 10 to 50 nm, and their mean diameter was calculated to be 19.8 nm. For comparison, the diameter distribution of Cu-filled CNCs is also shown in Fig. 26, and their diameters are 10–160 nm (59.7 nm mean diameter). The lengths of the Cu-filled CNTs ranged from 0.5 to 3 μm and their fraction was found to be approximately 60% in the total deposit from SEM and TEM examination of the deposit produced by using a graphite containing 20 at.% Cu at an Ar gas pressure of 0.9 MPa.

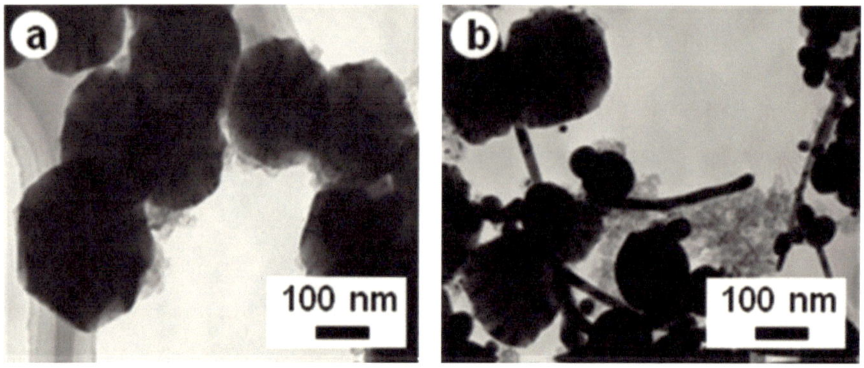

Fig. 25. TEM images of deposits obtained from graphite containing Cu of (a) 1 and (b) 5 at.%. A constant Ar gas pressure of 0.9 MPa was used.

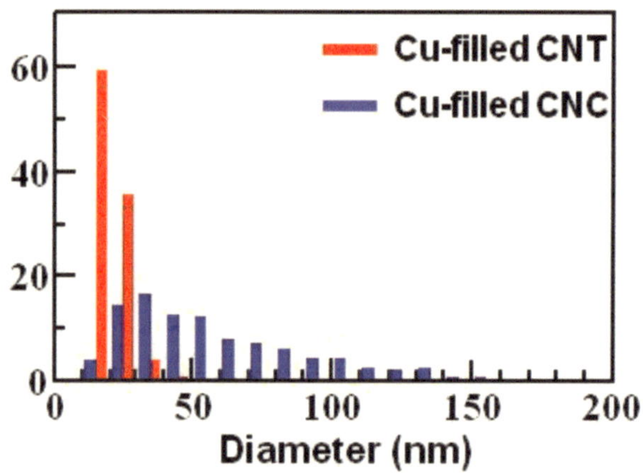

Fig. 26. Diameter distributions of Cu-filled CNTs and CNCs obtained for the deposit produced at Ar gas pressure of 0.9 MPa from graphite containing 20 at.% Cu.

To understand the possible growth mechanism for these unique Cu-filled CNTs, their edges were carefully examined by TEM. As shown in Fig. 27a, straight Cu-filled CNT structures, possessing sphere like particles at their edges, were observed. As seen in Fig. 27b, graphitic layers similar to those covering Cu nanowires were observed for the sphere like particles. The presence of the sphere like particles indicates the contribution of molten particles to form the Cu-filled CNTs. However, a well-known VLS mechanism (Wagner & Ellis, 1964) does not seem to act because of the absence of catalytic particles at the edges of the Cu-filled CNTs. In addition, formation mechanism of metal-filled CNT based on the growth of CNTs and simultaneous metal filling proposed for an arc discharge process (Demoncy et al., 1998) can not be applied because unfilled CNTs were not present.

CNTs completely filled with Cu nanowires are preferentially fabricated together with Cu-filled CNCs in a high pressure Ar gas environment of 0.9 MPa. In the space confined by Ar gas of high pressure such as 0.9 MPa, the collisions between C and Cu species are expected to be enhanced, although the exact size and feature of C and Cu species are not clear. We believe that strong interaction between C and Cu species results in the formation of molten Cu-C particles with unusually C-rich compositions at elevated temperature, while C interaction

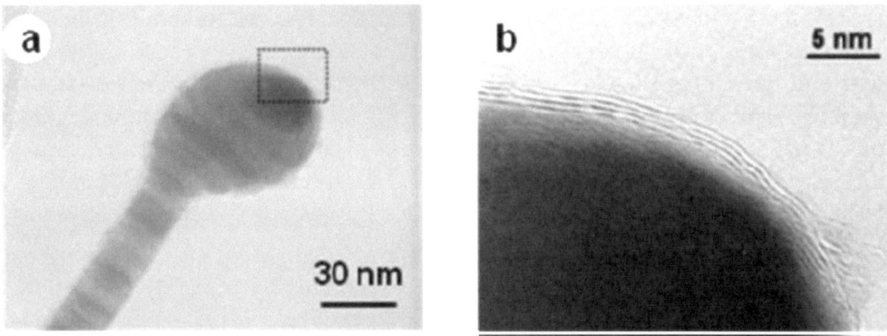

Fig. 27. (a) TEM image of a Cu-filled CNT with a spherical particle at its edge and (b) high magnification TEM image of outer region of the sphere like particle (rectangular area in Fig. 27a).

with Cu at normal atmospheric pressure is known to be weak (binding energy of 0.1–0.144 eV/atom (Seraphin et al., 1996; Tao et al., 2006)) and the C solubility is very low (0.04 at.%) (Massalski et al., 1990).

From the formation of Cu-filled CNTs together with CNCs including larger Cu particles and a narrow diameter distribution of Cu-filled CNTs (Fig. 26), we propose that molten Cu-C composite particles with certain small sizes play a crucial role in the formation of Cu-filled CNTs. The unusual composition of C and Cu, probably a C-rich one, in the molten particles and successive supply of C and Cu to the molten particles result in the precipitation of C and Cu. The growth of a Cu-C 1D composite structure followed by phase separation to form graphitic layers around the Cu nanowires finally leads to formation of Cu-filled CNTs. Escaping from unusually C-rich state for strain relaxation may be a driving force to precipitate C and Cu without the need of a catalyst.

3.5 SiC platelets and nanowires and SiC-filled MWCNTs from graphite containing Si

Deposits exhibiting particle and 1D structures with amorphous and crystalline features were observed for various Si content (5–70 at.%) and Ar gas pressure (0.05–0.9 MPa), instead of hybridized SWCNH particles. Figure 28 shows a typical TEM image of a deposit produced at an Si content of 20 at.% and an Ar gas pressure of 0.05 MPa. Sphere-like and irregular shaped amorphous particles with sizes of 5–150 nm are seen.

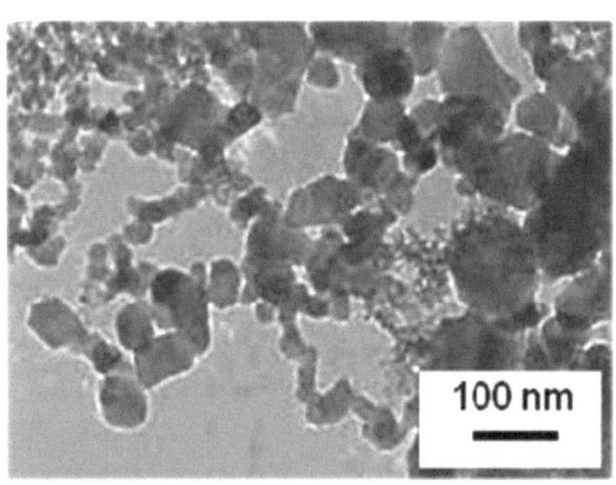

Fig. 28. TEM image of a deposit produced at Si content of 20 at.% and Ar gas pressure of 0.05 MPa.

An increased Si content resulted in platelet-like particles. Figure 29a shows a TEM image of a deposit produced at an Si content of 60 at.% and an Ar gas pressure of 0.05 MPa. Triangular and hexagonal platelet-like particles with sizes of 50–300 nm are seen together with sphere-like and irregular shaped particles. In the SEM image of the deposit as shown in Fig. 29b, cross sections of some platelets were observed and indicated the platelet thicknesses were in the range of 30–60 nm. The high-magnification TEM image of the platelet-like particle shown in Fig. 29c reveals lattice fringes of a spacing of 0.25 nm assigned to the (111) planes of cubic β-SiC based on the data of JCPDS No. 1-1119. The SAED also shows a typical pattern of the β-SiC phase consisting of spots forming a hexagon. These results suggest that the {111} crystal plane is the basal plane.

Fig. 29. (a) TEM and (b) SEM images of a deposit produced at Si content of 60 at.% and Ar gas pressure of 0.05 MPa and (c) higher magnification TEM image of a platelet particle. The inset in (c) is the corresponding SAED pattern.

Figure 30a shows a Raman spectrum of the deposit containing the platelet-like particles. Figure 30b shows the Raman spectrum in the 750–850 cm^{-1} region of Fig. 30a in the enlarged scale. Three peaks are seen at 520, 789, and 970 cm^{-1}. From previous Raman spectroscopic studies of SiC films (Chehaidar et al., 2001; Hu et al., 2004; Xu et al., 2005), the two weak bands at 789 and 970 cm^{-1} are assigned to those due to transverse and longitudinal optical phonon modes of SiC. However, the strong band observed at 520 cm^{-1} is attributed to an Si-Si transverse optical phonon mode. The weak band observed at 970 cm^{-1} is probably overlapped with a band due to the second order Si-Si longitudinal optical phonon mode. These results lead us to believe that nanoparticles of Si and/or Si-rich SiC exist in addition to crystalline SiC platelets in the deposit, while the scattering intensity of the Si-Si band at 520 cm^{-1} relative to that of the SiC band at 789 cm^{-1} is much higher. Similar hexagonal crystalline SiC platelets were synthesized using an autoclave heated at 350–500 °C for 20–40 h (Ju et al., 2009). The formation of hexagonal platelets was thought to be due to the preferential development of {111} and {110} planes of crystalline SiC. We believe that a similar development of {111} and {110} planes is involved in the growth of triangular and hexagonal SiC platelets by assembling a suitable amount of Si and C species after laser ablation.

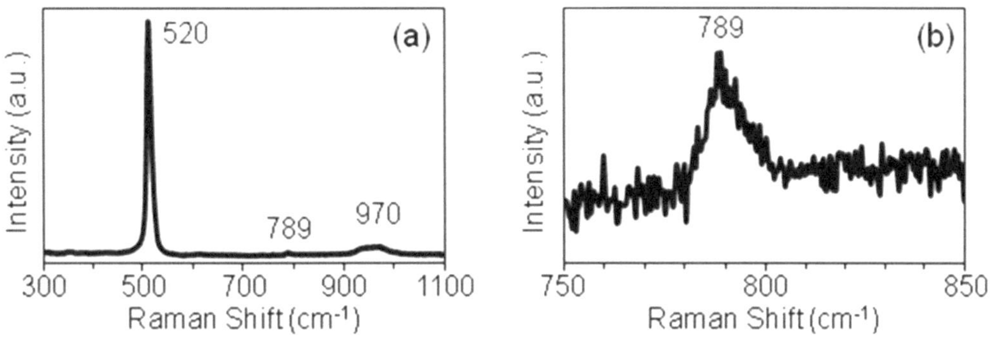

Fig. 30. (a) Raman spectrum of a deposit produced at Si content of 60 at.% and Ar gas pressure of 0.05 MPa. (b) Raman spectrum in the 750–850 cm^{-1} region of Fig. 30a in the enlarged scale.

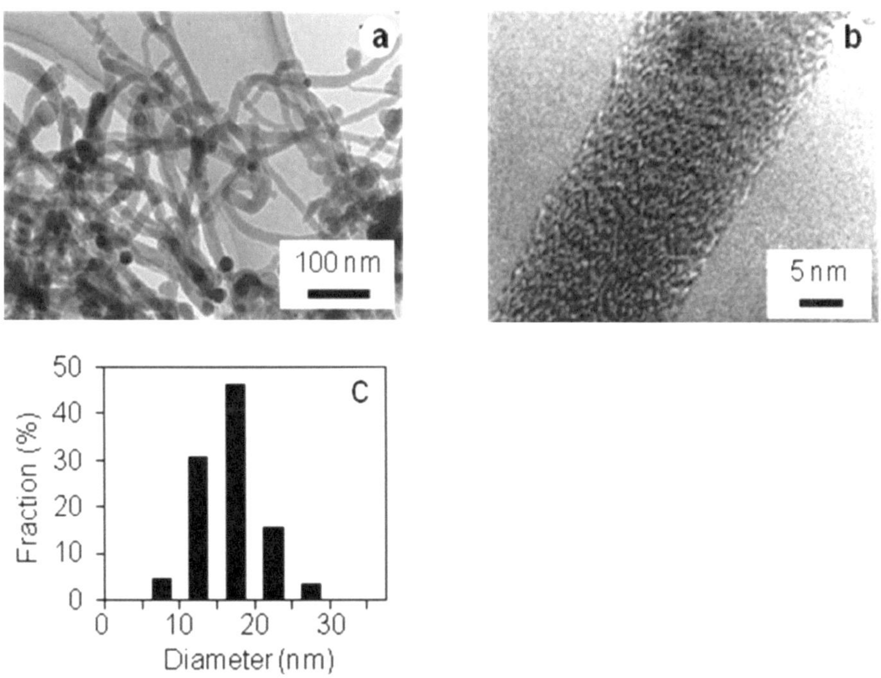

Fig. 31. TEM images of (a) thin 1D structures (amorphous nanowires) and (b) one amorphous nanowire produced at Si content of 70 at.% and Ar gas pressure of 0.1 MPa. (c) Distribution of diameters of amorphous nanowires.

Further increased Si content and Ar gas pressure resulted in 1D nanostructures with different thicknesses, lengths, and linearities (Kokai et al., in 2010). Using SEM, TEM, SAED, and Raman spectroscopy for the 1D nanostructures, we found that there were two types of

1D nanostructures: (1) an amorphous SiC nanowire and (2) a MWCNT filled with a crystalline SiC nanowire. Figure 31a shows a TEM image of the thin 1D nanostructures grown at an Si content of 70 at.% and an Ar gas pressure of 0.1 MPa. Curved nanowires can be observed. A further magnified TEM image (Fig. 31b) indicates an amorphous feature. Nanowires with diameters of 5–30 nm were observed. The diameter distribution obtained from the measurement of more than 200 samples is shown in Fig. 31c. Nearly 80% of the nanowires have diameters of 10–20 nm; the average diameter is 16 nm. The lengths of the nanowires were 0.5–2 μm from the observation of SEM images.
Figure 32 shows Raman spectra of the deposit including curved 1D nanostructures. Similar to the Raman spectra of Fig. 30, three peaks are seen at 520, 795, and 973 cm^{-1}. Compared to the intensity distribution in Fig. 30, the intensity of the band at 795 cm^{-1} relative to those of other two bands is decreased. Therefore we conclude the 1D nanostructure is an Si-rich amorphous SiC nanowire. The presence of the strong band at 520 cm^{-1} may indicate that nanowires composed of only Si are also included in the deposit.

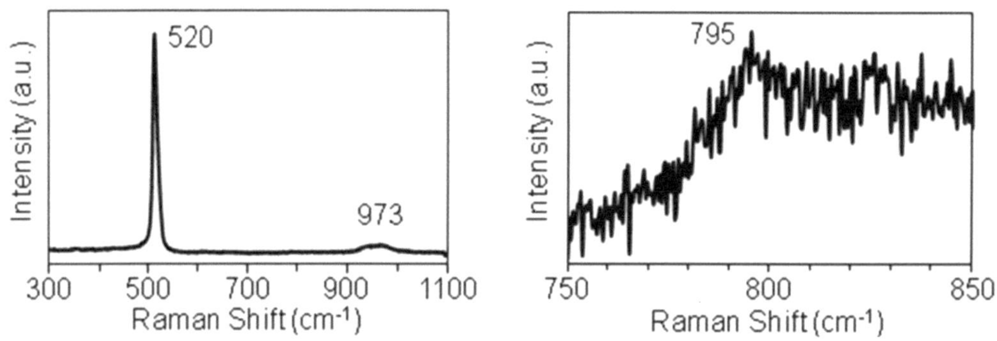

Fig. 32. (a) Raman spectra of a deposit including curved 1D nanostructures produced at Si content of 70 at.% and Ar gas pressure of 0.1 MPa. (b) Raman spectrum in the 750–850 cm^{-1} region of Fig. 32a in the enlarged scale.

With further increase of the Ar gas pressure, we found an increase in the amount of 1D nanostructures relative to the particles in the deposits. Figure 33a shows a typical TEM image of a deposit produced at an Si content of 70 at.% and an Ar gas pressure of 0.9 MPa. Straight and thick 1D nanostructures are seen. Figure 33b shows a higher-magnification TEM images of a 1D nanostructure. The different light and dark contrast observed in the 1D nanostructure indicates a coaxial structure of two different compositions. The TEM image of the outer portion seen in Fig. 33c indicates the presence of graphitic layers with an interlayer distance of ~0.35 nm. The TEM image focusing on the core seen in Fig. 33d exhibits lattice fringes of a spacing of 0.25 nm assigned to the (111) planes of cubic β-SiC as seen in Fig. 29c. The SAED also shows a typical pattern of the β-SiC phase. These results indicate the 1D nanostructure is a MWCNT filled with a crystalline SiC nanowire. The lengths of the SiC-filled MWCNTs were 2–20 μm from the observation of SEM images. Figure 34 shows two Raman spectra observed in two different Raman shift regions for a deposit including the straight and thick 1D nanostructures. Similar to the Raman spectra of Figs. 30 and 32, three bands are observed at 519, 789, and 969 cm^{-1} in the region of 300–1100 cm^{-1}. Like the typical Raman spectrum of MWCNTs synthesized using arc discharge and thermal chemical vapor deposition (Enomoto et al., 2006), two bands assigned to D and G bands are observed at 1356 and 1583 cm^{-1}, respectively.

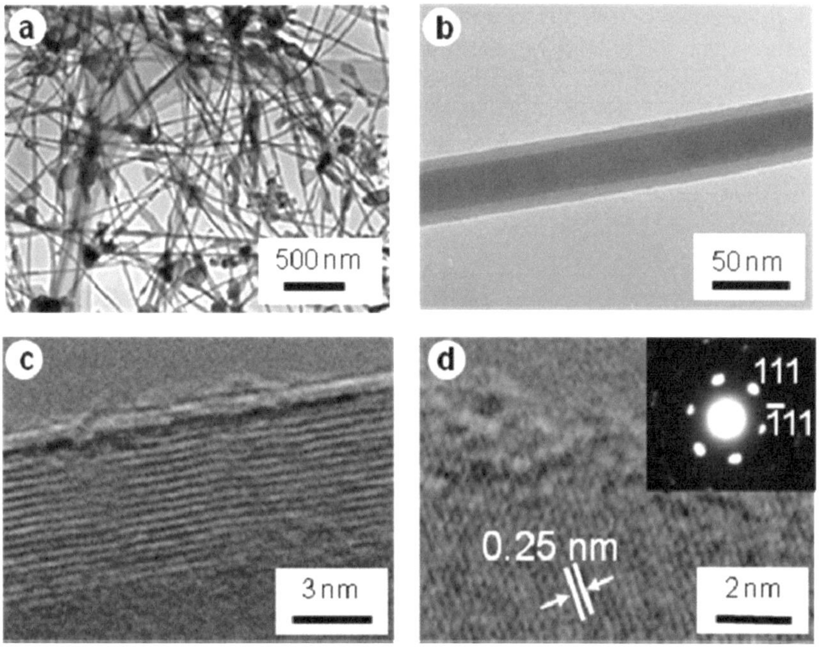

Fig. 33. (a) High-magnification TEM images of the 1D structure (crystalline SiC-filled MWCNT) produced at Si content of 70 at.% and Ar gas pressure of 0.9 MPa. (b) one 1D structure, (c) outer portion, and (d) lattice image of core SiC. The inset in (d) is the corresponding SAED pattern.

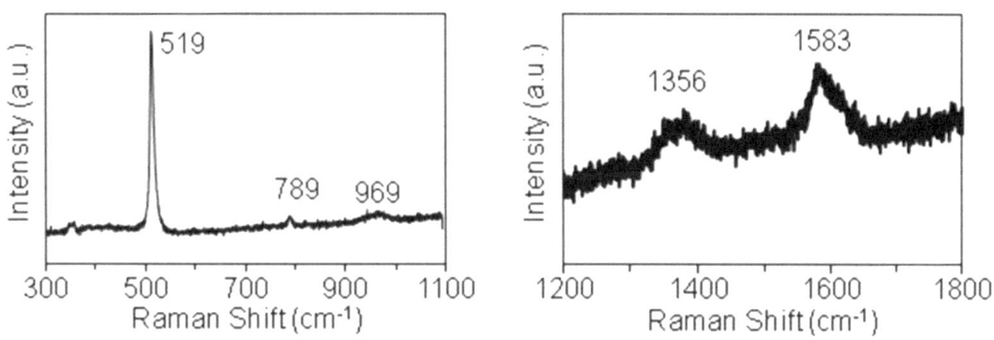

Fig. 34. Raman spectra of the deposit including the straight and thick 1D nanostructures observed in (a) 300–1100 and (b) 1200–1800 cm^{-1}.

A further detailed structure of the SiC-filled MWCNT was investigated by the measurement of the outer and SiC diameters of more than 200 SiC-filled MWCNTs. Figure 35a shows the distribution of the outer diameters of SiC-filled MWCNTs. The diameters range from 10 to 60 nm. More than 80% of the diameters are 20–40 nm; the average diameter is 32 nm. Figure 35b shows the distribution of the diameters of crystalline SiC nanowires encapsulated in MWCNTs. The diameters range from 15 to 50 nm, of which more than 80% of the diameters are 10–30 nm and the average diameter is 21 nm. These results indicate that the thicknesses of the graphitic layers of the MWCNTs are 2–9 nm (5–26 layers).

Unlike the thin amorphous SiC nanowires produced at an Ar gas pressure of 0.1 MPa, high Ar gas pressures of 0.5–0.9 MPa were essential for the formation of crystalline SiC-filled MWCNTs. The SiC-filled MWCNTs produced at low Si contents (30–50 at.%) were short and bending, probably due to the presence of defects. Increasing the Si content up to 70 at.%, enables the formation of straight and long SiC-filled MWCNTs. We checked more than 300 SiC-filled CNTs with TEM and found there were no MWCNTs without SiC filling from root to tip.

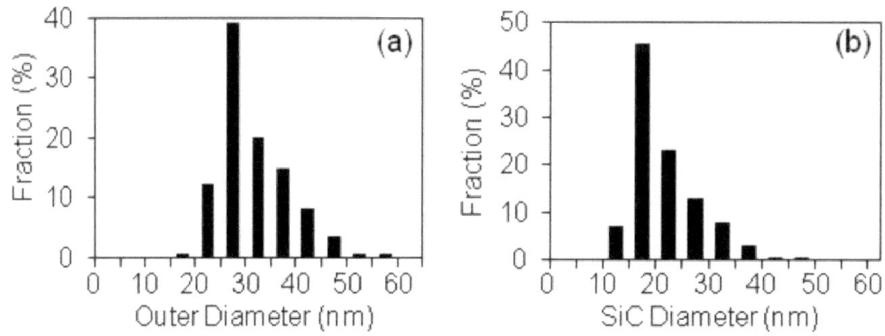

Fig. 35. Distributions of (a) outer diameters of SiC-filled MWCNTs and (b) diameters of the encapsulated SiC nanowires.

To understand the possible growth mechanisms for the two 1D nanostructures, their ends were carefully examined with TEM. For the amorphous SiC nanowires, sphere-like particles with uniform diameters of 25–30 nm were observed at their tips as shown in Fig. 36a. Some of the sphere-like particles are also seen in Fig. 31a as dark contrast portions. Figure 36b shows a high-magnification TEM image of a sphere-like particle, revealing that the sphere-like particle has an amorphous feature similar to the portion of the SiC nanowire as seen in Fig. 31b. On the other hand, for crystalline SiC-filled CNTs, angular polyhedron-like particles with sizes (100–400 nm) larger than the sphere-like particles were observed as shown in Figs. 36c and 36d. Figure 36e shows a high-magnification TEM image of the outer portion of a polyhedron-like particle, revealing that the particle is wrapped with graphitic layers as seen in the crystalline SiC-filled CNTs (Fig. 33c). The particle structure wrapped with graphitic layers at the tip is similar to that of the Cu-filled CNT in Section 3.4.

The presence of these sphere- and polyhedron-like particles indicates the contribution of liquid-like molten particles in the nucleation and continuous growth of the two 1D nanostructures. Similar to Cu-filled CNTs, catalyst materials are not present at the tips. When we used a Si powder of 99.999% purity instead of a 99% purity Si powder for laser-irradiation targets, same two types of 1D structures were formed. We measured x-ray photoelecton spectra of the deposits containing two 1D structures and detected Si, C, and O elements. However, other elements were not detected within the detection limit of 0.1 at.%. Therefore, the conventional VLS mechanism can not be applied for the growth of the two 1D nanostructures. Furthermore, an oxide-assisted growth mechanism (Zhang et al., 2003), proposed for Si and other nanowires together with the observation of oxide cover layers, and the conversion of CNTs to SiC nanowires (Dai et al., 1995: Ye et al., 2005) do not seem to contribute to the growth of these two 1D structures. We speculate that the somewhat different driving forces that nucleate the two 1D nanostructures are induced in the Si-containing molten particles without a catalyst. The yields and linearity of the two 1D nanostructures

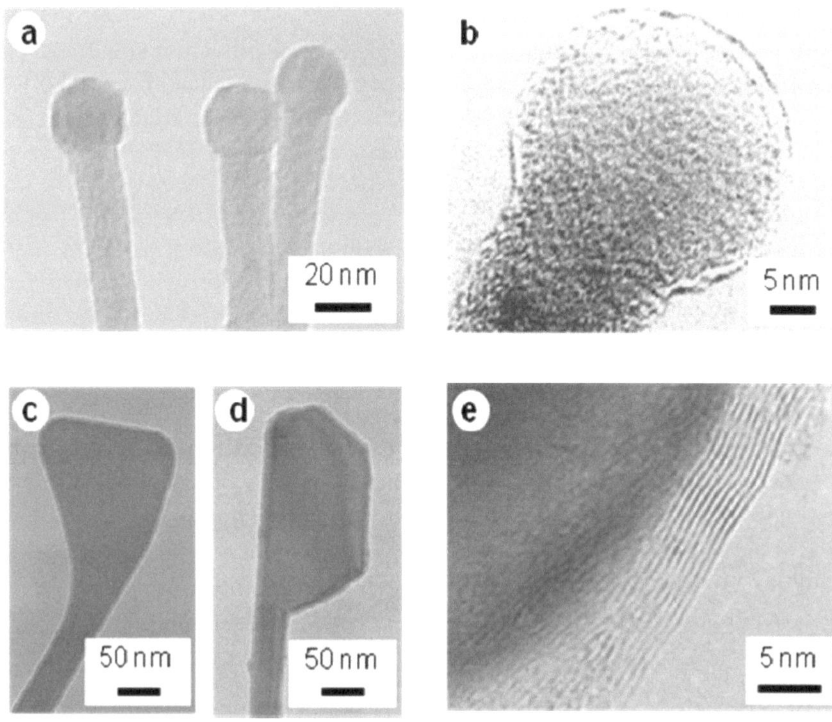

Fig. 36. Typical TEM images of the tips of (a, b) amorphous SiC nanowires and (c, d) crystalline SiC-filled CNTs. (e) TEM image of outer region of the tip of crystalline SiC-filled CNT

were strongly dependent on the compositions of Si and C in the laser-irradiation targets and Ar gas pressure. Therefore, Si-C molten particles with specific compositions, governed by the resident densities and temperatures of Si and C and their supply to the Si-C molten particles are thought to self-catalytically contribute the precipitation of Si and C to form the 1D nanostructures. We suggest that the instability of molten Si-C particles occurred together with the successive supplies of Si and C and resulted in Si and C precipitation. Silicon-rich molten particles may contribute to the formation of amorphous SiC nanowires as proposed in the formation of submicrometer SiC whiskers by the carbothermal reduction of silica at ~1500°C (Belmonte et al., 1996). Compared to the formation of amorphous SiC nanowires, ablated C and Si species with higher resident densities contribute to the growth of crystalline SiC-filled CNTs over a longer time at high temperature due to the stronger confinement of ablated Si and C, like the formation of PG particles and Cu-filled CNTs. This may result in the growth of a Si-C nanostructure followed by phase separation to form the outer graphitic layers. From the shapes of the tips and the grown SiC-filled CNTs in Figs. 36c and 36d, the deformation of the molten particles (formation of protrusion-like structures), as reported in-situ TEM images for the formation of CNTs (Hofmann et al., 2007; Yoshida et al., 2008), may be involved in the nucleation of Si-C nanostructures. The graphitic-layer-covered polyhedron-like shapes of the tips formed after growth of the SiC-filled CNTs probably results from the formation of crystalline SiC with facets of 50–300 nm.

4. Conclusion

We have presented a simple method to form various nanocarbon and composite structures using laser ablation in high-pressure Ar gas up to 0.9 MPa. Graphite containing Si, Fe, Co, Ni, Cu, Ag, B_4C, La_2O_3, Gd_2O_3, or Y_2O_3 was used as a source material. The laser ablation of graphite containing Fe, Co, Ni, and Ag enabled the high yield (~70%) fabrication of hybridized SWCNH particles including metal- or carbide-containing CNCs. Other by-products such as CNCs and hybridized PLG particles were also observed. For La_2O_3, Gd_2O_3, and Y_2O_3, hybridized SWCNH particles were not observed. Instead, products such as sea-urchin-type SWCNTs and PG particles including carbide particles were formed. Unlike the addition of these metals and compounds to graphite, different products were observed for B_4C, Cu, and Si by adjusting the Ar gas pressure and additive content. For B_4C, in addition to CNCs containing BC, MWCNTs were grown with a high yield of ~60%. For Cu, in addition to hybridized SWCNH particles including Cu-containing CNCs, CNTs filled with polycrystalline Cu nanowires were formed. For Si, three characteristic Si-C nanostructures, i.e., SiC platelets, amorphous SiC nanowires, and CNTs filled with crystalline SiC nanowires, were formed. Unlike previous work on the formation of CNTs filled with metals and compounds, the present method involves a high-pressure Ar gas of 0.9 MPa and has a unique feature in that all the grown CNTs contained Cu or SiC nanowires and no empty CNTs were present in the deposits. We have discussed the metal-dependent efficient growth of various nanocarbon and composite structures from high-temperature C and metal species confined by Ar gas. The co-existence of SWCNHs formed by thermal graphitization and smaller molten C-metal particles is probably essential to the formation of hybridized SWCNH particles; otherwise, the formation of CNCs and other structures are promoted. The strong correlation between the hybrid diameter and SWCNH layer thickness for Fe, Co, and Ag hybrids suggests that after SWCNHs with a certain range of lengths are formed, they aggregate around molten metal-C particles with a certain range of diameters. The C precipitation by catalytic graphitization, which leads to the formation of CNCs in SWCNH aggregates, occurs after SWCNH assembly. The VLS mechanism is thought to play a role in the growth of sea-urchin type SWCNTs and MWCNTs through C dissolution, diffusion, and precipitation. The nanostructures (diameters and wall numbers) of the SW and MWCNTs may be precisely determined by the sizes and shapes of the highly curved structures at molten C-catalyst particle, which results in lift-offs of C caps. The B-catalyzed MWCNT and CNC formation is thought to relate to the balance among complex factors such as resident densities of C and B species, C solubility in B particles, and rates of C dissolution, diffusion, and precipitation. The presence of sphere- and polyhedron-like particles at the tips of grown 1D nanostructures of Cu-filled CNTs, SiC nanowires, and SiC-filled CNTs indicates the contribution of liquid-like molten particles in the nucleation and continuous growth of these 1D nanostructures. A catalyst-free unconventional VLS mechanism, in which self-catalytic action in molten Cu-C and Si-C particles that is dependent on their compositions may be involved, plays a significant role in the growth of the 1D nanostructures. Although the details of the growth mechanisms proposed here are not clear, we believe that the present method using a high-pressure gas atmosphere can be extended to the growth of nanowires of other metal and semiconducting materials. Applications such as drug delivery, catalyst supports, spot welding, thermal management, and structural devices are expected in the future by using such nanocarbon and composite materials.

5. Acknowledgements

The author gratefully acknowledges the "Kakenhi (15310072 and 20510104)" Grants-in-Aid for Scientific Research provided by the Japan Society for the Promotion of Science in support of this work. This work was partly supported by the Cooperation of Innovative Technology and Advanced Research in Evolution Area (City Area) Project of the Ministry of Education, Culture, Sports, Science, and Technology.

6. References

Athanassiou, E.K.; Grass, R.N. & Stark, W.J. (2006). Large-scale production of carbon-coated copper nanoparticles for sensor applications, *Nanotechnology, 17*, 1668-1673, 0957-4484

Belmonte, T.; Bonnetain, L. & Ginoux, J.L. (1996). Synthesis of silicon carbide whiskers using the vapour-liquid-solid mechanism in a silicon-rich droplet, *J. Mater. Sci.*, 31, 2367-2371, 0022-2461

Chehaidar, A.; Zwick, A. & Carles, R. (2001). Investigation of structural and chemical ordering in Si-rich amorphous SiC alloys via Raman spectroscopy and numerical modelling, *J. Phys., Condens. Matter*, 13, 10743 -10755, 0953-8984

Cheng, M.-D., Lee, D.-W.; Zhao, B.; Hu, H.; Styers-Barnett, D.J.; Puretzky, A.A.; DePaoli, D.W.; Geohegan, D.B.; Ford, E.A. & Angelini, P. (2007). Formation studies and controlled production of carbon nanohorns using continuous in situ characterization techniques, *Nanotechnology*, 18, 185604 (8p), 0957-4484

Cheung, C.L.; Kurtz, A.; Park, H. & Lieber, C.M. (2002). Diameter-controlled synthesis of carbon nanotubes, *J. Phys. Chem. B*, 106, 2429-2433, 1520-6106

Choi, W.Y.; Kang, J.W. & Hwang, H.J. (2003). Structures of ultrathin copper nanowires encapsulated in carbon nanotubes, *Phys. Rev. B*, 68, 193405-1-193405-4, 0163-1829

Dai, H.; Wong, E.W.; Lu, Y.Z.; Fan, S. & Lieber, C.M. (1995). Synthesis and characterization of carbide nanorods, *Nature*, 375, 769-772, 0028-0836

Demoncy, N.; Stéphan, O.; Brun, N. ; Colliex, C. ; Loiseau, A. & Pascard, H. (1996). Filling carbon nanotubes with metals by the arc-discharge method: the key role of sulfur, *Eur. Phys. J. B*, 4, 147-157, 1434-6028

Enomoto, K.; Kitakata, S.; Yasuhara, T.; Ohtake, N.; Kuzumaki, T. & Mitsuda, Y. (2006). Measurement of Young's modulus of carbon nanotubes by nanoprobe manipulation in a transmission electron microscopy, *Appl. Phys. Lett.*, 88, 153115 (3p), 0003-6951

Gorbunov, A.A.; Friedlein, R.; Jost, O.; Golden, M.S.; Fink, J. & Pompe, W. (1999). Gas-dynamic consideration of the laser evaporation synthesis of single-wall carbon nanotubes, *Appl. Phys. A*, 69, S593-S596, 0947-8396

Guo, T.; Nikolaev, P.; Thess, A.; Colbert, D.T. & Smalley, R.E. (1995). Catalytic growth of single-walled nanotubes by laser vaporization, *Chem. Phys. Lett.*, 243, 49-54, 0009-2614

Haase, D.; Hampel, S.; Leonhardt, A.; Thomas, J.; Mattern, N. & Bücher, B. (2007). Facile one -step-synthesis of carbon wrapped copper nanowires by thermal decomposition of copper(II)-acetylacetonate, *Surf. Coat. Technol.*, 201, 9184-9188, 0257-8972.

Hirahara, K.; Suenaga, K.; Bandow, S. & Iijima, S. (2000). Boron-catalyzed multi-walled carbon nanotube growth with the reduced number of layers by laser ablation, *Chem. Phys. Lett.*, 324, 224–230, 0009-2614, 0009-2614

Hofmann, S.; Sharma, R.; Ducati, C.; Du, G.; Mattevi, C.; Cepek, C.; Cantoro, M.; Pisana, S.; Parvez, A.; Cervantes-Sodi, F.; Ferrari, A.C.; Dunin-Brorkwski, R.; Lizzit, S.; Petaccia, L.; Goldoni, A. & Robertson, J. (2007). In situ observations of catalyst dynamics during surface-bound carbon nanotube nucleation. *Nano Lett.*, 7, 602–608, 1530-6984

Hu, Z.; Liao, X.; Diao, H.; Kong, G.; Zeng, X. & Xu, Y. (2004). Amorphous silicon carbide films prepared by H_2 diluted silane-methane plasma, *J. Cryst. Growth*, 264, 7–12, 0022-0248

Iijima, S.; Yudasaka, M.; Yamada, R.; Bandow, S.; Suenaga, K.; Kokai, F. & Takahashi, K. Nano-aggregates of single-walled graphitic carbon nano-horns, *Chem. Phys. Lett.*, 309, 165–170, 0009-2614

Jiao, J. & Seraphin, S. (1998). Carbon encapsulated nanoparticles of Ni, Co, Cu, and Ti, *J. Appl. Phys.*, 83, 2442–2448, 0021-8979

Jiao, J.; Seraphin, S.; Wang, X. & Withers, J.C. (1996). Preparation and properties of ferromagnetic carbon-coated Fe, Co, and Ni nanoparticles, *J. Appl. Phys.*, 80, 103–108, 0021-8979

Ju, Z., Xu, L.; Pang, Q.; Xing, Z.; Ma, X. & Qian, Y. (2009). The synthesis of nanostructured SiC from waste plastics and silicon powder, *Nanotechnology*, 20, 355604 (6p), 0957-4484

Kawai, T.; Miyamoto, Y.; Sugino, O. & Koga, Y. (2002). Nanotube and nanohorn nucleation from graphitic patches: Tight-binding molecular-dynamics simulations, Phys. Rev. B, 66, 033404-1–033404-4, 1098-0121

Kobayashi, K.; Shimazu, T.; Yamada, Y.; Koshio, A.; Takahashi, Y. & Kokai, F. (2007). Formation of single-wall carbon nanohorn aggregates hybridized with carbon nanocapsules by laser vaporization, *Appl. Phys. A*, 89, 121–126, 0947-8396

Kokai, F.; Ishihara, M.; Koshio, A.; Nakayama, A. & Koga, Y. (2007). Fabrication of some graphitic polyhedra and balloon-like particles, *Diamond Rela. Mater.*, 16, 1264–1268, 0925-9635

Kokai, F.; Koshio, A.; Kasuya, D.; Hirahara, K.; Takahashi, K.; Nakayama, A.; Ishihara, M.; Koga, Y. & Iijima, S. (2004). Three nanostructured graphitic particles and their growth mechanisms from high-temperature carbon vapor confined by Ar gas, *Carbon*, 42, 2515–2520, 0008-6223

Kokai, F.; Koshio, A.; Kobayashi, K. & Deno, H. (2008). Formation of nanocarbon and composite materials by laser vaporization of graphite and eleven metals, *Proceedings of photon processing in microelectronics and photonics VII*, Vol. 6879, pp. 68791F-1–68791F-1-9, 0277-786X, San Jose, SPIE

Kokai, F.; Shimazu, T.; Adachi, K.; Kosho, A. & Takahashi, Y. (2009a). Fabrication of completely filled carbon nanotubes with copper nanowires in a confined space, *Appl. Phys. A*, 97, 55–62, 0947-8396

Kokai, F.; Tachi, N.; Kobayashi, K.; & Koshio, A. (2009b). Structural characterization of single-wall carbon nanohorn aggregates hybridized with carbon nanocapsules and their formation mechanism, *Appl. Surf. Sci.*, 255, 9622–9625, 0169-4332

Kokai, F.; Takahashi K.; Kasuya, D.; Nakayama, A.; Koga, Y.; Yudasaka, M. & Iijima, S. (2003). Laser vaporization synthesis of polyhedral graphite, *Appl. Phys. A, 77*, 69-71, 0947-8396

Kokai, F.; Takahashi, K.; Kasuya, D.; Yudasaka, M. & Iijima, S. (2001). Growth of single-wall carbon nanotubes dependent on laser power density and ambient gas pressure during room-temperature CO_2 laser vaporization, *Appl. Phys. A*, 73, 401-407, 0947-8396

Kokai, F.; Takahashi, K.; Yudasaka, M. & Iijima, S. (1999). Emission imaging spectroscopic and shadowgraphic studies on the growth dynamics of graphitic carbon particles synthesized by CO_2 laser vaporization, *J. Phys. Chem.* B, 41, 8686-8693, 1520-6106

Kokai, F.; Uchiyama, K.; Shimazu, T.; & Koshio, A. (2010). Fabrication of two types of one-dimensional Si-C nanostructures by laser ablation, *Appl. Phys. A*, 101, 497-502, 0947-8396

Lin, X.; Wang, X.K.; Dravid, V.P.; Chang, R.P.H. & Ketterson, J.B. (1994). Large scale synthesis of single-shell carbon nanotubes, *Appl. Phys. Lett.* 64, 181-183 , 0003-6951

Li, W.Z.; Wang, D.Z.; Yang, S.X.; Wen, J.G. & Ren, Z.F. (2001). Controlled growth of carbon nanotubes on graphite foil by chemical vapor deposition. *Chem. Phys. Lett.*, 335, 141-149, 0009-2614

Massalski, T.B.; Okamoto, H.; Subramanian, P.R. & Kacprzak, L. (1990). *Binary Alloy Phase Diagrams*, The Materials Information Society ASM International, 0-87170-404-8, Metal Park (USA)

Nerushev, O.A.; Dittmar, S.; Morjan, R.E.; Rohmund, F. & Campbell, E.E.B. (2003). Particle size dependence and model for iron-catalyzed growth of carbon nanotubes by thermal chemical vapor deposition, *J. Appl. Phys.*, 93, 4185-4190, 0008-6223

Rümmeli, M.H.; Kramberger, C. Schäffel, F.; Borowiak-Palen, E.; Gemming, T.; Rellinghaus, B, Jost, O.; Löffler, M.; Ayala, P.; Pichler, T. & Kalenczuk, R.J. (2007). Catalyst size dependencies for carbon nanotube synthesis, *Phys. Stat. Sol., B*, 244, 3911-3915, 0031-8965

Ruoff, R.S.; Lorents, D.C.; Chan, B.; Malhotra, R. & Subramoney, S. (1993). Single crystal metals encapsulated in carbon nanoparticles, *Science*, 259, 346-348, 0036-8075

Ryu,Y.-H.; Park, B.-T.; Song, Y.-H. & Yong, K. (2004). Carbon-coated SiC nanowires: direct synthesis from Si and field emission characteristics, *J. Cryst. Growth*, 271, 99-104, 0022-0248

Sabbaghzadeh, J.; Jafarkkani, P.; Dadras, S. & Torkamany, M.J. (2009). Synthesis of multi-wall carbon nanotubes by copper vapor laser. *Appl. Phys. A*, 94, 293-297, 0947-8396

Saito, Y. (1995). Nanoparticles and filled nanocapsules, *Carbon*, 33, 979-988, 0008-6223

Sano, N.; Kikuchi, T.; Wang, H.; Chhowalla, M. & Amaratunga, G.A.J. (2004). Carbon nanohorns hybridized with a metal-included nanocapsule, *Carbon*, 42, 95-99, 0008-6223

Seraphin, S.; Zhou, D. & Jiao, J. (1996). Filling the carbon nanocages, *J. Appl. Phys., 80*, 2097-2104, 0021-8979

Setlur, A.A.; Lauerhaas, J.M.; Dai, J.Y. & Chang, R.P.H. (1996). A method for synthesizing large quantities of carbon nanotubes and encapsulated copper nanowires, *Appl. Phys. Lett., 69*, 345-347, 0003-6951

Subramoney, S.; Ruoff, R.S.; Lorents, D.C. & Malhotra, R. (1993). Radial single-layer nanotubes, *Nature*, 366, 637, 0028-0836

Tao, X.Y.; Zhang, X.B.; Cheng, J.P.; Luo, Z.Q.; Zhou, S.M. & Liu, F. (2006). Thermal CVD synthesis of carbon nanotubes filled with single-crystalline Cu nanoneedles at tips. *Diamond Rela. Mater., 15,* 1271–1275, 0925-9635.

Wagner R.S. & Ellis, W.C. (1964). Vapor-liquid-solid mechanism of single crystal growth, *Appl. Phys. Lett.,* 4, 89–90, 0003-6951

Xu, J.; Yang, L.; Rui, Y.; Mei, J.; Zhang, X.; Li, W.; Ma, Z.; Xu, L.; Huang, X. & Chen, K. (2005). Photoluminescence characteristics from amorphous SiC thin films with various structures deposited at low temperature, *Solid State Commun.,* 133, 565–568, 0038-1098

Ye, H.; Titchenal, N.; Gogotsi, Y. & Ko, F. (2005). SiC nanowires synthesized from electronspun nanofiber templates, *Adv. Mater.,* 17, 1531–1535, 0935-9648

Yoshida, H.; Takeda, S.; Uchiyama, T.; Kohno, H. & Homma, Y. (2008). Atomic-scale in-situ observation of carbon nanotube growth from solid sate iron carbide nanoparticles, *Nano Lett.,* 8, 2082–2086, 1530-6984

Yudasaka, M.; Kasuya, Y.; Kokai, F.; Takahashi, K.; Takizawa, M.; Bandow, S. & Iijima, S. (2002). Causes of different catalytic activities of metals in formation of single-wall carbon nanotubes, *Appl. Phys. A,* 74, 377–385, 0947-8396

Yudasaka, M.; Komatsu, T.; Ichihashi, T. & Iijima, S. (1997). Single-wall carbon nanotube formation by laser ablation using double-targets of carbon and metal, *Chem. Phys. Lett.,* 278, 102−106, 0009-2614

Zhang, G.Y. & Wang, E.G. (2003). Cu-filled carbon nanotubes by simultaneous plasma-assisted copper incorporation, *Appl. Phys. Lett.,* 82, 1926–1928, 0003-6951

Zhang, H.-F.; Wang, C.-M. & Wang, L.-S. (2002). Helical crystalline SiC/SiO_2 core–shell nanowires, *Nano Lett.,* 2, 941–944, 1530-6984

Zhang, R.-Q.; Lifshitz, Y. & Lee, S.-T. (2003). Oxide-assisted growth of semiconducting nanowires, *Adv. Mater.,* 15, 635–640, 0935-9648

17

On the Transport Phenomena in Composite Materials using the Fractal Space-Time Theory

Vizureanu Petrică and Agop Maricel
The "Gheorghe Asachi" Technical University from Iasi
Romania

1. Introduction

A new way to analyze the dynamics of the physical systems is to consider that the particle movements take place on continuous but non-differentiable curves, i.e. on fractals. Then, the complexity of these dynamics is substituted by fractality. There are some fundamental arguments which can justify such hypothesis: i) by interaction, the trajectory is no longer everywhere differentiable. The "uncertainty" in tracking the particle is eliminated by means of the fractal approximation of motion; ii) the complex dynamical systems, which display chaotic behavior, are recognized to acquire self-similarity and manifest strong fluctuations at all possible scales. Every type of "elementary" process of motion induces both spatio-temporal scales and the associated fractals. Moreover, the movement complexity is directly related to the fractal dimension: the fractal dimension increases as the movement becomes more complex. Different definitions were given for the fractal dimension (Kolmogorov dimension, Hausdorff dimension, etc.), but once we choose the fractal-type dimension in the study of motion we must work with it until the end.

Therefore, considering that the complexity of the physical processes (from the system's interactions) is replaced by fractality (situation in which the particle movements take place on fractal curves), it is no longer necessary to use notions as collision time, mean free path, etc., i.e., the whole classical "arsenal" of quantities from the dynamics of physical systems. Then, the physical systems will behave as a special interaction-less "fluid" by means of geodesics in a fractal space-time. The theory which treats the interactions in the previously mentioned manner is the Scale Relativity (SR).

The SR is based on a generalization of Einstein's principle of relativity to scale transformations. Namely, "one redefines space-time resolutions as characterizing the state of reference systems scale, in the same way as speed characterizes their state of motion. Then one requires that the laws of physics apply whatever the state of the reference system, of motion (principle of motion-relativity) and of scale (principle of SR). The principle of SR is mathematically achieved by the principle of scale-covariance, requiring that the equations of physics keep their simplest form under transformations of resolution".

Another way of analyzing the system dynamics by means of the "fractals" is given by the Transfinite Physics (TP). The Transfinite Physics theory uses the Cantorian geometry as a working method. This geometry is a compromise between the discrete and the continuum. It is not simply discrete. It is transfinite discrete and has the cardinality of the continuum although it is not continuous. Seen from a far it looks as if it were continuous. The result was startling because it is possible to simulate four dimensionality using infinitely many weighted Cantor sets. This created a geometry and topology for space-time that is similar to that of radiation and obeys the same statistical distribution, namely a discrete gamma distribution which is known in physics as Planck distribution. El Naschie defined this space-time as $\varepsilon^{(\infty)}$ space-time.

Many applications of the fractal space-time and particularly of $\varepsilon^{(\infty)}$ space-timewere given in references. In the present paper, using the extended model of the Scale Relativity theory, the transport phenomena (speed and temperature fields) in composite materials are analyzed.

2. A short reminder of the Nottale's scale relativity theory in correspondence with Cresson's mathematical procedures

Let us suppose that the motion of particles take place on continuous but non-differentiable curves (fractal curves). The non-differentiability, according with Cresson's mathematical procedures and Nottale's physical principles, implies the followings:

i. a continuous and a non-differentiable curve (or almost nowhere differentiable) is explicitly scale dependent, and its length tends to infinity, when the scale interval tends to zero. In other words, a continuous and non-differentiable space is fractal, in the general meaning given by Mandelbrot to this concept;
ii. there is an infinity of fractals curves (geodesics) relating any couple of its points (or starting from any point), and this is valid for all scales;
iii. the breaking of local differential time reflection invariance. The time-derivative of a function F can be written two-fold:

$$\frac{dF}{dt} = \lim_{dt \to 0} \frac{F(t+dt) - F(t)}{dt} = \lim_{dt \to 0} \frac{F(t) - F(t-dt)}{dt} \tag{1}$$

Both definitions are equivalent in the differentiable case. In the non-differentiable situation these definitions fail, since the limits are no longer defined. "In the framework of scale relativity, the physics is related to the behavior of the function during the "zoom" operation on the time resolution δt, here identified with the differential element dt (substitution principle), which is considered as an independent variable. The standard function $F(t)$ is therefore replaced by a fractal function $F(t,dt)$, explicitly dependent on the time resolution interval, whose derivative is undefined only at the unobservable limit $dt \to 0$". As a consequence, this lead us to define the two derivatives of the fractal function as explicit functions of the two variables t and dt,

$$\begin{aligned} \frac{d_+F}{dt} &= \lim_{dt \to 0_+} \frac{F(t+dt,dt) - F(t,dt)}{dt} \\ \frac{d_-F}{dt} &= \lim_{dt \to 0_-} \frac{F(t,dt) - F(t-dt,dt)}{dt} \end{aligned} \tag{2a,b}$$

The sign, +, corresponds to the forward process and, -, to the backward process;

iv. the differential of a fractal function $F(t,dt)$ can be expressed as the sum of two differentials, one which is not scale-dependent, $dF'(t)$, and the other dependent on it, $dF''(t,dt)$, therefore

$$dF(t,dt) = dF'(t) + dF''(t,dt) \tag{3}$$

Particularly, the differential of the generalized coordinates, $d_{\pm}\boldsymbol{X}(t,dt)$, can be decomposed as follows

$$d_{\pm}\boldsymbol{X}(t,dt) = d_{\pm}\boldsymbol{x}(t) + d_{\pm}\boldsymbol{\xi}(t,dt) \tag{4a,b}$$

where $d_{\pm}\boldsymbol{x}(t)$ is the "classical part" and $d_{\pm}\boldsymbol{\xi}(t,dt)$ is the "fractal part". Starting from here, multiplying by dt^{-1} and using the substitutions

$$\boldsymbol{V}_{\pm} = \frac{d_{\pm}\boldsymbol{X}}{dt}, \quad \boldsymbol{v}_{\pm} = \frac{d_{\pm}\boldsymbol{x}}{dt}, \quad \boldsymbol{u}_{\pm} = \frac{d_{\pm}\boldsymbol{\xi}}{dt} \tag{5a-c}$$

we obtain the velocity field

$$\boldsymbol{V}_{\pm} = \boldsymbol{v}_{\pm} + \boldsymbol{u}_{\pm}; \tag{6a,b}$$

v. the fractal part of F, i.e. F'', satisfies the relation

$$|F''(t) - F''(t')| \approx |t - t'|^{\delta} \tag{7}$$

where δ depends on the fractal dimension D_F (for detail see references).
Particularly, the differential of the "fractal part" of $d_{\pm}\boldsymbol{X}$, becomes

$$d_{\pm}\xi_i \sim dt^{\frac{1}{D_F}} \tag{8a,b}$$

or more, as an equality relation (fractal equation):

$$\left(\frac{d_{\pm}\xi_i}{\lambda}\right) = \left(\frac{dt}{\tau}\right)^{\frac{1}{D_F}} \tag{9a,b}$$

Written as

$$d_{\pm}\xi_i = \frac{\lambda}{\tau}\left(\frac{dt}{\tau}\right)^{\left(\frac{1}{D_F}\right)-1} dt \tag{10a,b}$$

equations (9a,b) imply the temporal scales δt and τ, and the length scale λ, respectively. The significances of the time dt and τ result from the Random Walk (Brownian motion) or its generalization, Levy motion. The differential time dt is identified with the resolution time ("substitution principle"), $\delta t \equiv dt$, while τ corresponds to the fractal - non-fractal transition time. λ is a characteristic length, for example of Planck's or de Broglie's type (for details see references);

vi. by the relation (10 a, b) the velocity field $V_{\pm}^{i}$ becomes

$$V_{\pm}^{i} = v_{\pm}^{i} + u_{\pm}^{i} = v_{\pm}^{i} + \frac{\lambda}{\tau}\left(\frac{\tau}{dt}\right)^{1-\left(\frac{1}{D_F}\right)} \quad (11a,b)$$

The transition scale τ yields two distinct behaviors of the speed, depending on the resolution at which it is considered, since $V_{\pm}^{i} \to v_{\pm}^{i}$ when $dt >> \tau$, and $V_{\pm}^{i} \to u_{\pm}^{i}$, when $dt << \tau$;

vii. the local differential time reflection invariance is recovered by combining the two derivatives, d_+ / dt and d_- / dt, in the complex operator

$$\frac{\hat{\partial}}{\partial t} = \frac{1}{2}\left(\frac{d_+ + d_-}{dt}\right) - \frac{i}{2}\left(\frac{d_+ - d_-}{dt}\right) \quad (12)$$

We call this procedure "an extension by differentiability" (Cresson's extension). Applying this operator to the "position vector" yields a complex speed

$$\begin{aligned} \boldsymbol{V} = \frac{\hat{\partial}\boldsymbol{X}}{\partial t} &= \frac{1}{2}\left(\frac{d_+\boldsymbol{X} + d_-\boldsymbol{X}}{dt}\right) - \frac{i}{2}\left(\frac{d_+\boldsymbol{X} - d_-\boldsymbol{X}}{dt}\right) = \frac{\boldsymbol{V}_+ + \boldsymbol{V}_-}{2} - i\frac{\boldsymbol{V}_+ - \boldsymbol{V}_-}{2} = \\ &= \frac{1}{2}\left[(v_+ + v_-) + (u_+ + u_-)\right] - \frac{i}{2}\left[(v_+ - v_-) + (u_+ - u_-)\right] = v - iu \end{aligned} \quad (13)$$

with

$$\begin{aligned} v &= \frac{\boldsymbol{V}_+ + \boldsymbol{V}_-}{2} = \frac{1}{2}\left[(v_+ + v_-) + (u_+ + u_-)\right] \\ u &= \frac{\boldsymbol{V}_+ - \boldsymbol{V}_-}{2} = \frac{1}{2}\left[(v_+ - v_-) + (u_+ - u_-)\right] \end{aligned} \quad (14a,b)$$

The real part, v, of the complex speed represents the standard classical speed which is differentiable and independent of resolution, while the imaginary part, u, is a new quantity arising from fractality, which is non-differentiable and resolution-dependent. In the usual classical limit, $dt >> \tau$,

$$v_+ = v_- = \overline{v}, \; u_+ = u_- = 0 \quad (15a,b)$$

so that

$$\boldsymbol{V} = \overline{v}, \; u = 0 \quad (16)$$

In the limit, $dt << \tau$,

$$v_+ = v_- = 0, \; u_+ = u_- = \overline{u} \quad (17a,b)$$

and

$$\boldsymbol{V} = \overline{u}, \quad u = 0; \quad (18)$$

viii. "in order to account for the infinity of geodesics in the bundle, for their fractality and for the two valuedness of the derivative which all come from the non-differentiable geometry of the space-time continuum, one therefore adopts a generalized statistical fluid like description, where instead of a classical deterministic speed or of a classical fluid speed field, one uses a doublet of fractal functions of spaces coordinates and time which are also explicit functions of resolution time". Thus, the average values of the quantities must be considered in the previously mentioned sense. Particularly, the average of $d_{\pm}\boldsymbol{X}$ is

$$\langle d_{\pm}\boldsymbol{X}\rangle = d_{\pm}\boldsymbol{x} \tag{19}$$

with

$$\langle d_{\pm}\xi\rangle = 0 \tag{20a,b}$$

ix. in such an interpretation, the "particles", are identified with the geodesics themselves. As a consequence, any measurement is interpreted as a sorting out (or selection) of the geodesics by the measuring device.

3. Extended model of the scale relativity

Let us now assume that the curves describing the movement (continuous but non-differentiable) is immersed in a 3-dimensional space, and that X of components X^i (i $=\overline{1,3}$) is the position vector of a point on the curve. Let us also consider a function $f(\boldsymbol{X},t)$ and expand its total differential up to the third order:

$$d_{\pm}f = \frac{\partial f}{\partial t}dt + \nabla f\cdot d_{\pm}\boldsymbol{X} + \frac{1}{2}\frac{\partial^2 f}{\partial X^i\partial X^j}d_{\pm}X^i d_{\pm}X^j + \frac{1}{6}\frac{\partial^3 f}{\partial X^i\partial X^j\partial X^k}d_{\pm}X^i d_{\pm}X^j d_{\pm}X^k \tag{21a,b}$$

where only the first three terms were used in the Nottale's theory (i.e. second order terms in the equation of motion).

The relations (21a,b) are valid in any point of the space manifold and also for the points X on the fractal curve which we have selected in relations (21a,b).

From here, the forward and backward average values of this relation, using the notations $dX_{\pm}^{\ i} = d_{\pm}X^i$, take the form:

$$\langle d_{\pm}f\rangle = \left\langle\frac{\partial f}{\partial t}dt\right\rangle + \langle\nabla f\cdot d_{\pm}\boldsymbol{X}\rangle + \frac{1}{2}\left\langle\frac{\partial^2 f}{\partial X^i\partial X^j}d_{\pm}X^i d_{\pm}X^j\right\rangle + \frac{1}{6}\left\langle\frac{\partial^3 f}{\partial X^i\partial X^j\partial X^k}d_{\pm}X^i d_{\pm}X^j d_{\pm}X^k\right\rangle \tag{22a,b}$$

We make the following stipulations: the mean values of the function f and its derivates coincide with themselves, and the differentials $d_{\pm}X^i$ and dt are independent, therefore the averages of their products coincide with the product of average. Thus equations (22a,b) become:

$$d_{\pm}f = \frac{\partial f}{\partial t}dt + \nabla f\langle d_{\pm}\boldsymbol{X}\rangle + \frac{1}{2}\frac{\partial^2 f}{\partial X^i\partial X^j}\langle d_{\pm}X^i d_{\pm}X^j\rangle + \frac{1}{6}\frac{\partial^3 f}{\partial X^i\partial X^j\partial X^k}\langle d_{\pm}X^i d_{\pm}X^j d_{\pm}X^k\rangle \tag{23a,b}$$

or more, using equations (4a,b) with the property (20a,b),

$$d_{\pm}f = \frac{\partial f}{\partial t}dt + \nabla f d_{\pm}\boldsymbol{x} + \frac{1}{2}\frac{\partial^2 f}{\partial X^i \partial X^j}\left(d_{\pm}x^i d_{\pm}x^j + \left\langle d\xi_{\pm}^i d\xi_{\pm}^j \right\rangle\right) + \frac{1}{6}\frac{\partial^3 f}{\partial X^i \partial X^j \partial X^k}\left(d_{\pm}x^i d_{\pm}x^j d_{\pm}x^k + \left\langle d\xi_{\pm}^i d\xi_{\pm}^j d\xi_{\pm}^k \right\rangle\right) \quad (24a,b)$$

Even the average value of the fractal coordinate, $d\xi_{\pm}^i$, is null (see (20a,b)), for the higher order of the fractal coordinate average the situation can be different. First, let us focus on the mean $\left\langle d\xi_{\pm}^i d\xi_{\pm}^j \right\rangle$. If $i \neq j$ this average is zero due the independence of $d\xi^i$ and $d\xi^j$. So, using (10a,b) we can write:

$$\left\langle d\xi_{\pm}^i d\xi_{\pm}^j \right\rangle = \pm\delta^{ij}\frac{\lambda^2}{\tau}\left(\frac{dt}{\tau}\right)^{\left(2/D_F\right)-1} dt \quad (25a,b)$$

with

$$\delta^{ij} = \begin{cases} 1, & \text{if} \quad i = j \\ 0, & \text{if} \quad i \neq j \end{cases}$$

and we had considered that:

$$\begin{cases} \left\langle d\xi_{+}^i d\xi_{+}^j \right\rangle > 0 \text{ and } dt > 0 \\ \left\langle d\xi_{-}^i d\xi_{-}^j \right\rangle > 0 \text{ and } dt < 0 \end{cases}$$

Then, let us consider the mean $\left\langle d\xi_{\pm}^i d\xi_{\pm}^j d\xi_{\pm}^k \right\rangle$. If $i \neq j \neq k$ this average is zero due the independence of $d\xi^i$ on $d\xi^j$ and $d\xi^k$. Now, using equation (10a,b), we can write:

$$\left\langle d\xi_{\pm}^i d\xi_{\pm}^j d\xi_{\pm}^k \right\rangle = \delta^{ijk}\frac{\lambda^3}{\tau}\left(\frac{dt}{\tau}\right)^{\left(3/D_F\right)-1} dt \quad (26a,b)$$

with

$$\delta^{ijk} = \begin{cases} 1, & \text{if} \quad i = j = k \\ 0, & \text{if} \quad i \neq j \neq k \end{cases}$$

and we considered that:

$$\begin{cases} \left\langle d\xi_{+}^i d\xi_{+}^j d\xi_{+}^k \right\rangle > 0 \textit{ and } dt > 0 \\ \left\langle d\xi_{-}^i d\xi_{-}^j d\xi_{-}^k \right\rangle > 0 \textit{ and } dt < 0 \end{cases}$$

Then equations (24a,b) may be written under the form:

$$d_{\pm}f = \frac{\partial f}{\partial t}dt + \nabla f d_{\pm}\boldsymbol{x} + \frac{1}{2}\frac{\partial^2 f}{\partial X^i \partial X^j}d_{\pm}x^i d_{\pm}x^j \pm \frac{\partial^2 f}{\partial X^i \partial X^j}\delta^{ij}\frac{\lambda^2}{2\tau}\left(\frac{dt}{\tau}\right)^{\left(2/D_F\right)-1} dt + \frac{1}{6}\frac{\partial^3 f}{\partial X^i \partial X^j \partial X^k}d_{\pm}x^i d_{\pm}x^j d_{\pm}x^k + \frac{\partial^3 f}{\partial X^i \partial X^j \partial X^k}\delta^{ijk}\frac{\lambda^3}{6\tau}\left(\frac{dt}{\tau}\right)^{\left(3/D_F\right)-1} dt \quad (27a,b)$$

If we divide by dt and neglect the terms which contain differential factors, equations (27a,b) are reduced to:

$$\frac{d_{\pm}f}{dt} = \frac{\partial f}{\partial t} + v_{\pm}\nabla f \pm \frac{\lambda^2}{2\tau}\left(\frac{dt}{\tau}\right)^{(2/D_F)-1}\Delta f + \frac{\lambda^3}{6\tau}\left(\frac{dt}{\tau}\right)^{(3/D_F)-1}\nabla^3 f \tag{28a,b}$$

with $\Delta = \sum_i \partial^2 / \partial X_i^2$ and $\nabla^3 = \sum_i \partial^3 / \partial X_i^3$. These relations also allows us to define of the operator,

$$\frac{d_{\pm}}{dt} = \frac{\partial}{\partial t} + v_{\pm}\nabla \pm \frac{\lambda^2}{2\tau}\left(\frac{dt}{\tau}\right)^{(2/D_F)-1}\Delta + \frac{\lambda^3}{6\tau}\left(\frac{dt}{\tau}\right)^{(3/D_F)-1}\nabla^3 \tag{29a,b}$$

Under the circumstances, let us calculate $(\hat{\partial} f/\partial t)$. Taking into account equations (29a,b), (12) and (13), we obtain:

$$\frac{\hat{\partial} f}{\partial t} = \frac{1}{2}\left[\frac{d_+ f}{dt} + \frac{d_- f}{dt} - i\left(\frac{d_+ f}{dt} - \frac{d_- f}{dt}\right)\right] = \frac{\partial f}{\partial t} + \boldsymbol{V}\cdot\nabla f - i\frac{\lambda^2}{2\tau}\left(\frac{dt}{\tau}\right)^{(2/D_F)-1}\Delta f + \frac{\lambda^3}{6\tau}\left(\frac{dt}{\tau}\right)^{(3/D_F)-1}\nabla^3 f \tag{30}$$

This relation also allows us to define the fractal operator:

$$\frac{\hat{\partial}}{\partial t} = \frac{\partial}{\partial t} + \boldsymbol{V}\cdot\nabla - i\frac{\lambda^2}{2\tau}\left(\frac{dt}{\tau}\right)^{(2/D_F)-1}\Delta + \frac{\lambda^3}{6\tau}\left(\frac{dt}{\tau}\right)^{(3/D_F)-1}\nabla^3 \tag{31}$$

We now apply the principle of scale covariance, and postulate that the passage from classical (differentiable) mechanics to the "fractal" mechanics which is considered here can be implemented by replacing the standard time derivative d/dt by the complex operator $\hat{\partial}/\partial t$ (this results is a generalization of the principle of scale covariance given by Nottale). As a consequence, we are now able to write the equation of geodesics (a generalization of the first Newton's principle) in a fractal space-time under its covariant form:

$$\frac{\hat{\partial}\boldsymbol{V}}{\partial t} = \frac{\partial \boldsymbol{V}}{\partial t} + \boldsymbol{V}\cdot\nabla\boldsymbol{V} - i\frac{\lambda^2}{2\tau}\left(\frac{dt}{\tau}\right)^{(2/D_F)-1}\Delta\boldsymbol{V} + \frac{\lambda^3}{6\tau}\left(\frac{dt}{\tau}\right)^{(3/D_F)-1}\nabla^3\boldsymbol{V} = 0 \tag{32}$$

This means that the global complex acceleration field, $\hat{\partial}\boldsymbol{V}/\partial t$, depends on the local complex acceleration field, $\partial_t \boldsymbol{V}$, on the non-linearity (convective) term, $\boldsymbol{V}\cdot\nabla\boldsymbol{V}$, on the dissipative term, $\Delta\boldsymbol{V}$, and on the dispersive one, $\nabla^3\boldsymbol{V}$.

If the motions of the fractal fluid are irrotational, i.e. $\boldsymbol{\Omega} = \nabla\times\boldsymbol{V} = 0$ we can choose $\boldsymbol{V}$ of the form:

$$\boldsymbol{V} = \nabla\phi \tag{33}$$

with ϕ a complex speed potential. Then, equation (32) becomes:

$$\frac{\partial \boldsymbol{V}}{\partial t}+\nabla\left(\frac{\boldsymbol{V}^2}{2}\right)-i\frac{\lambda^2}{2\tau}\left(\frac{dt}{\tau}\right)^{\left(2/D_F\right)-1}\Delta \boldsymbol{V}+\frac{\lambda^3}{6\tau}\left(\frac{dt}{\tau}\right)^{\left(3/D_F\right)-1}\nabla^3 \boldsymbol{V}=0 \tag{34}$$

and more, by substituting equation (33) in equation (34), we have by integration,

$$\frac{\partial \varphi}{\partial t}+\frac{1}{2}(\nabla\phi)^2-i\frac{\lambda^2}{2\tau}\left(\frac{dt}{\tau}\right)^{\left(2/D_F\right)-1}\Delta\phi+\frac{\lambda^3}{6\tau}\left(\frac{dt}{\tau}\right)^{\left(3/D_F\right)-1}\nabla^3\phi=F(t) \tag{35}$$

with $F(t)$ a function of time only. We note that equation (34) has been reduced to a single scalar relation (35), i.e. a generalized Bernoulli (GB) type equation.
Let us choose the complex speed potential in the form:

$$\phi=-i\frac{\lambda^2}{\tau}\left(\frac{dt}{\tau}\right)^{\left(2/D_F\right)-1} ln\psi \tag{36}$$

where ψ behaves both as speed potential and wave function. Then, ψ by means of equation (35) satisfies a generalized Schrödinger (GS) type equation:

$$\frac{\lambda^4}{4\tau^2}\left(\frac{dt}{\tau}\right)^{\left(4/D_F\right)-2}\Delta\psi+i\frac{\lambda^2}{2\tau}\left(\frac{dt}{\tau}\right)^{\left(2/D_F\right)-1}\partial_t\psi+\left(i\frac{\lambda^5}{12\tau^2}\left(\frac{dt}{\tau}\right)^{\left(5/D_F\right)-2}(\nabla^3 ln\psi)+\frac{F(t)}{2}\right)\psi=0 \tag{37}$$

When the transport phenomenon in a fractal space-time implies the temperature fields, T, the heat transfer equation has the form

$$\frac{\hat{\partial} T}{\partial t}=\frac{\partial T}{\partial t}+\boldsymbol{V}\cdot\nabla T-i\frac{\lambda^2}{2\tau}\left(\frac{dt}{\tau}\right)^{\left(2/D_F\right)-1}\Delta T+\frac{\lambda^3}{6\tau}\left(\frac{dt}{\tau}\right)^{\left(3/D_F\right)-1}\nabla^3 T=0 \tag{37a}$$

4. The dissipative approximation of transport phenomenon in fractal structures and some applications. Extended fractal hydrodynamic model

Let us consider that the dissipative and convective effects are dominant in comparison with the dispersive ones. Consequently, the covariant form of the first Newton's principle in the fractal space-time is reduced to equation

$$\frac{\hat{\partial} \boldsymbol{V}}{\partial t}=\frac{\partial \boldsymbol{V}}{\partial t}+\boldsymbol{V}\cdot\nabla\boldsymbol{V}-i\frac{\lambda^2}{2\tau}\left(\frac{dt}{\tau}\right)^{\left(\frac{2}{D_F}\right)-1}\Delta\boldsymbol{V}=0\ . \tag{38}$$

For $\boldsymbol{V}$ of the form (see relation (33) with (36))

$$\boldsymbol{V}=-i\frac{\lambda^2}{\tau}\left(\frac{dt}{\tau}\right)^{\left(\frac{2}{D_F}\right)-1}\nabla ln\psi \tag{39}$$

equation (38) becomes a Navier-Stokes type equation,

$$\frac{\hat{\partial} \mathbf{V}}{\partial t} = \frac{\partial \mathbf{V}}{\partial t} + \nabla\left(\frac{\mathbf{V}^2}{2}\right) - i\frac{\lambda^2}{2\tau}\left(\frac{dt}{\tau}\right)^{\left(\frac{2}{D_F}\right)-1} \Delta \mathbf{V} = 0 \tag{40}$$

with an imaginary viscosity coefficient, ν

$$\nu = i\frac{\lambda^2}{2\tau}\left(\frac{dt}{\tau}\right)^{\left(\frac{2}{D_F}\right)-1} \tag{41}$$

while in terms of the ψ function, up to an arbitrary phase factor which may be set to zero by a suitable choice of the phase, a "Schrödinger" type equation results

$$\frac{\lambda^4}{4\tau^2}\left(\frac{dt}{\tau}\right)^{\left(\frac{4}{D_F}\right)-2} \Delta\psi + i\frac{\lambda^2}{2\tau}\left(\frac{dt}{\tau}\right)^{\left(\frac{2}{D_F}\right)-1} \frac{\partial \psi}{\partial t} = 0\,. \tag{42}$$

The presence of an imaginary viscosity coefficient specifies the followings: i) at macroscopic scale, the behavior of the fractal fluids is of viscoelastic type or hysteretic type. Such a result is in agreement with the opinions given in references: the fractal fluid can be described by Kelvin-Voight or Maxwell rheological model with complex structure coefficients (particularly, the imaginary viscosity coefficient (41)). Thus, such "materials" are endowed with "memory"; ii) at microscopic scale, the scalar field of the complex velocity has a stochastic behavior. Particularly, at Compton scale ($D = \lambda^2/2\tau = \hbar/2m_0$, with $\hbar$ the reduced Planck constant and m_0 the rest mass of the microparticle) and in the fractal dimension D_F =2 (the microparticle motion take place on Peano`s curves), the scalar field of the complex velocity is also a wave function such that, equation (42) takes the form of the "standard" Schrödinger equation:

$$\frac{\hbar^2}{2m_0}\Delta\psi + i\hbar\frac{\partial \psi}{\partial t} = 0 \tag{43}$$

This means that, the Schrödinger equation results from a Navier-Stokes type equation with the imaginary viscosity coefficient, $\nu = i\hbar/2m_0$, for irrotational movements of a fractal fluid at Compton scale.

Let us choose the scalar function ψ in the form $\psi = \sqrt{\rho}e^{iS}$, with $\sqrt{\rho}$ being amplitude and S phase. Thus, the complex speed field (13), has the components

$$v = \frac{\lambda^2}{\tau}\left(\frac{dt}{\tau}\right)^{\left(\frac{2}{D_F}\right)-1} \nabla S, \qquad u = \frac{\lambda^2}{2\tau}\left(\frac{dt}{\tau}\right)^{\left(\frac{2}{D_F}\right)-1} \nabla \ln \rho\,. \tag{44a,b}$$

where v is the real (differentiable) part and u is the imaginary (non-differentiable or fractal) part.

The equations (44a, b) which define the components of the complex speed field of the fractal fluid are more general than those from Nottale's SR model. For the fractal dimension $D_F = 2$, the Nottale's results are obtained:

$$\mathrm{v}=\frac{\lambda^2}{\tau}\nabla S, \qquad \mathrm{u}=\frac{\lambda^2}{2\tau}\nabla \ln\rho \tag{45a,b}$$

Introducing (13) with (44a,b) in (40) and separating the real and imaginary parts, i.e. through the separation of the movements at differentiable scale from those at non-differentiable scale, we obtain

$$\begin{gathered}\frac{\partial \boldsymbol{v}}{\partial t}+\nabla\left(\frac{\boldsymbol{v}^2}{2}-\frac{\boldsymbol{u}^2}{2}-\frac{\lambda^2}{2\tau}\left(\frac{dt}{\tau}\right)^{\left(\frac{2}{D_F}\right)-1}\nabla\cdot\boldsymbol{u}\right)=0\\ \frac{\partial \boldsymbol{u}}{\partial t}+\nabla\left(\boldsymbol{v}\cdot\boldsymbol{u}+\frac{\lambda^2}{2\tau}\left(\frac{dt}{\tau}\right)^{\left(\frac{2}{D_F}\right)-1}\nabla\cdot\boldsymbol{v}\right)=0\end{gathered} \tag{46a,b}$$

With equation (45a,b), equation (46b) takes the form

$$\nabla\left(\frac{\partial \ln\rho}{\partial t}+\mathrm{v}\cdot\nabla \ln\rho+\nabla\cdot\mathrm{v}\right)=0 \tag{47}$$

or, by integration with $\rho\neq 0$

$$\frac{\partial\rho}{\partial t}+\nabla.(\rho\boldsymbol{v})=T(t) \tag{48}$$

with T(t) a function which depends only an time. In these condition, the equations (46a) and (48) with T(t)=0 become:

$$\begin{gathered}m_0\left(\frac{\partial \boldsymbol{v}}{\partial t}+\boldsymbol{v}\cdot\nabla\boldsymbol{v}\right)=-\nabla(Q)\\ \frac{\partial\rho}{\partial t}+\nabla\cdot(\rho\boldsymbol{v})=0\end{gathered}, \tag{49a,b}$$

with Q the fractal potential,

$$Q=-m_0\frac{\lambda^4}{2\tau^2}\left(\frac{dt}{\tau}\right)^{\left(\frac{4}{D_F}\right)-2}\frac{\Delta\sqrt{\rho}}{\sqrt{\rho}}=-\frac{m_0\boldsymbol{u}^2}{2}-m_0\frac{\lambda^2}{2\tau}\left(\frac{dt}{\tau}\right)^{\left(\frac{2}{D_F}\right)-1}\nabla\cdot\boldsymbol{u}\,. \tag{50}$$

The fractal potential depends only on the imaginary part $\boldsymbol{u}$ of the complex speed field, $\boldsymbol{V}$, and it comes from the non-differentiability of the fractal space-time. Equation (49a), i.e. the momentum conservation law, and equation (49b), i.e. the probability density conservation law, form the fractal hydrodynamic model.

The wave function of $\psi(\boldsymbol{r},t)$ is invariant when its phase changes by an integer multiple of 2π. Indeed, equation (44a) gives:

$$\oint m_0 v d\boldsymbol{r}=m_0\frac{\lambda^2}{\tau}\left(\frac{dt}{\tau}\right)^{\left(\frac{2}{D_F}\right)-1}\oint dS=2\pi n m_0\frac{\lambda^2}{\tau}\left(\frac{dt}{\tau}\right)^{\left(\frac{2}{D_F}\right)-1},\quad n=0,\ \pm1,\ \pm2,\ \ldots \tag{51}$$

a condition of compatibility between the extended SR hydrodynamic model and the wave mechanics. Particularly, at Compton scalar $\lambda^2/\tau = \hbar/m_0$ and for D_F=2 equation (51) takes the standard form

$$\oint \boldsymbol{p} \cdot d\boldsymbol{r} = nh \tag{52}$$

i.e. the Ehrenfest relation.

The set of equations (49a,b) represents a complete system of differential equations for the fields $\rho(\boldsymbol{r},t)$ and $v(\boldsymbol{r},t)$; relation (51) relates each fractal hydrodynamic solution $(\rho,v)_n$ with the wave solution ψ in a unique way.

The field $\rho(\boldsymbol{r},t)$ is a probability distribution, namely the probability of finding the particle in the vicinity $d\boldsymbol{r}$ of the point $\boldsymbol{r}$ at time t,

$$dP = \rho d\boldsymbol{r}, \quad \iiint \rho d\boldsymbol{r} = 1, \tag{53a,b}$$

the space integral being extended over the entire area of the system. Any time variation of the probability density $\rho(\boldsymbol{r},t)$ is accompanied by a probability current ρv pointing towards or outwards, the corresponding field point $\boldsymbol{r}$ (equation (49b)).

The real velocity field $v(\boldsymbol{r},t)$ (equation (49a)), varies with space and time similar to a hydrodynamic fluid placed in a fractal potential (50). The fractal fluid (in the sense of a statistical particles ensemble) exhibits, however, an essential difference compared to an ordinary fluid: in a rotation motion $v(\boldsymbol{r},t)$ increases (decreases) with the distance from the center $\boldsymbol{r}$ decreasing (increasing) (equation (51)).

The expectation values for the real velocity field and the velocity operator $\hat{v} = -i\left(\lambda^2/\tau\right)\left(dt/\tau\right)^{\left(\frac{2}{D_F}\right)-1}\nabla$ of wave mechanics are equal,

$$\langle v \rangle = \iiint \rho v d\boldsymbol{r} = \iiint \Psi^* \hat{v} \Psi d\boldsymbol{r} = \langle \hat{v} \rangle_{WM} \tag{54}$$

but in the higher-order, |n|>2, similar identities are invalid, namely $\langle v^n \rangle \neq \langle v^n \rangle_{WM}$. The expectation value for the 'fractal force' vanishes at all times (theorem of Ehrenfest), i.e.

$$\langle -\nabla Q \rangle = \iiint \rho(-\nabla Q) d\boldsymbol{r} = 0 \tag{55}$$

since

$$m_0 \frac{\lambda^4}{2\tau^2}\left(\frac{dt}{\tau}\right)^{\left(\frac{4}{D_F}\right)-2} \iiint \rho \nabla \left(\frac{\nabla^2 \sqrt{\rho}}{\sqrt{\rho}} \right) d\boldsymbol{r} = m_0 \frac{\lambda^4}{4\tau^2}\left(\frac{dt}{\tau}\right)^{\left(\frac{4}{D_F}\right)-2} \oint (\rho \nabla \nabla \ln \rho) \cdot d\boldsymbol{\sigma} = 0 \tag{56}$$

Two types of stationary states are distinguished:

i. Dynamic states. For $\partial / \partial t = 0$ and $v \neq 0$, i.e. at the differentiable scale, equations (49a,b) give

$$\nabla\left(\frac{m_0 v^2}{2}-\frac{m_0 \boldsymbol{u}^2}{2}-m_0\frac{\lambda^2}{2\tau}\left(\frac{dt}{\tau}\right)^{\left(\frac{2}{D_F}\right)-1}\nabla\cdot\boldsymbol{u}\right)=0 \tag{57a,b}$$

$$\nabla\cdot(\rho v)=0$$

namely,

$$\frac{m_0 v^2}{2}-\frac{m_0 \boldsymbol{u}^2}{2}-m_0\frac{\lambda^2}{2\tau}\left(\frac{dt}{\tau}\right)^{\left(\frac{2}{D_F}\right)^{-1}}\cdot\nabla\cdot\boldsymbol{u}=E \tag{58a,b}$$

$$\rho v=\nabla\times\boldsymbol{F}$$

Consequently, the non-fractal inertia, $m_0 v\cdot\nabla v$, and the fractal force, $-\nabla Q$, are in balance at every field point - equation (57a). The sum of the non-fractal kinetic energy, $mv^2/2$, and fractal potential, Q, is invariant, i.e., equal to the integration constant $E\neq E(\boldsymbol{r})$ - equation (58a). $E\equiv\langle E\rangle$ represents the total energy of the dynamic system. The probability flow density ρv has no sources - equation (57b), i.e., its streamlines are closed - equation (58b). In an external potential U the equation (58a) becomes:

$$\frac{m_0 \boldsymbol{v}^2}{2}-\frac{m_0 \boldsymbol{u}^2}{2}-m_0\frac{\lambda^2}{2\tau}\left(\frac{dt}{\tau}\right)^{\left(\frac{2}{D_F}\right)-1}\nabla\cdot\boldsymbol{u}+U=E \tag{59}$$

ii. Static states. For $\partial/\partial t=0$ and $v=0$, i.e. at the non-differentiable scale, equations (49a,b) give

$$\nabla\left(-\frac{m_0 \boldsymbol{u}^2}{2}-m_0\frac{\lambda^2}{2\tau}\left(\frac{dt}{\tau}\right)^{\left(\frac{2}{D_F}\right)-1}\nabla\cdot\boldsymbol{u}\right)=0, \tag{60}$$

i.e.

$$-\frac{m_0 \boldsymbol{u}^2}{2}-m_0\frac{\lambda^2}{2\tau}\left(\frac{dt}{\tau}\right)^{\left(\frac{2}{D_F}\right)-1}\nabla\cdot\boldsymbol{u}=E\ . \tag{61}$$

Thus, the fractal force, $-\nabla Q$ has the zero value - equation (60). The fractal potential, Q, is invariant, i.e. equal to the integration constant $E\neq E(\boldsymbol{r})$ - equation (61). $E\equiv\langle E\rangle$ represents the total energy of the static system. Equation (49b) is identically satisfied.
In an external potential U the equation (61) becomes:

$$-\frac{m_0 \boldsymbol{u}^2}{2}-m_0\frac{\lambda^2}{2\tau}\left(\frac{dt}{\tau}\right)^{\left(\frac{2}{D_F}\right)-1}\nabla\cdot\boldsymbol{u}+U=E \tag{62}$$

As an illustration of the fractal hydrodynamic formalism, stationary and time-dependent fractal systems are further analyzed.

If the transport phenomenon implies temperature fields then, according with the fractal operator (31) in which we neglected the dispersive effects, the heat transfer equation becomes:

$$\frac{\hat{\partial} T}{\partial t}=\frac{\partial T}{\partial t}+\boldsymbol{V}\cdot\nabla T-i\frac{\lambda^2}{2\tau}\left(\frac{dt}{\tau}\right)^{\left(2/D_F\right)-1}\Delta T=0 \tag{63}$$

From here, by separating the real and imaginary parts and then by their summing, the usual heat transfer equations results:

$$\frac{\partial T}{\partial t}+(\mathrm{v}-\mathrm{u})\cdot\nabla T=\frac{\lambda^2}{2\tau}\left(\frac{dt}{\tau}\right)^{\left(2/D_F\right)-1}\Delta T \tag{64}$$

5. The dispersive approximation of transport phenomenon in fractal structures. Some properties of matter

Let us consider that the dissipate effects can be neglected in comparison with the convective and dispersive ones. Then, through the equation (34), the microparticle movements are described by a generalized Korteweg-de Vries (GKdV) type equation

$$\frac{\partial \boldsymbol{V}}{\partial t}+\boldsymbol{V}\cdot\nabla\boldsymbol{V}+\frac{\lambda^3}{6\tau}\left(\frac{dt}{2}\right)^{\left(\frac{3}{D_F}\right)-1}\nabla^3\boldsymbol{V}=0 \tag{65}$$

By substituting (44a,b) in equation (65), and separating the real and the imaginary parts, we obtain the following system:

$$\begin{aligned}&\frac{\partial v}{\partial t}+\nabla\left(\frac{\boldsymbol{v}^2}{2}-\frac{\boldsymbol{u}^2}{2}\right)+\frac{\lambda^3}{6\tau}\left(\frac{dt}{\tau}\right)^{\left(\frac{3}{D_F}\right)-1}\nabla^3 v=0\\&\frac{\partial \boldsymbol{u}}{\partial t}+\nabla(v\cdot\boldsymbol{u})+\frac{\lambda^3}{6\tau}\left(\frac{dt}{\tau}\right)^{\left(\frac{3}{D_F}\right)-1}\nabla^3\boldsymbol{u}=0\end{aligned} \tag{66a,b}$$

In the one-dimensional differentiable case, u=0 or $\rho = const.$, using the dimensionless parameters,

$$\bar{\phi}=(v/v_0),\ \bar{\tau}=\omega_0 t,\ \bar{\xi}=k_0 X \tag{67a-c}$$

and the normalizing condition

$$\frac{v_0 k_0}{6\omega_0}=\frac{k_0^3}{\omega_0}\frac{\lambda^3}{6\tau}\left(\frac{dt}{\tau}\right)^{\left(\frac{3}{D_F}\right)-1}=1 \tag{68}$$

the equations (66a,b), takes the standard form of the KdV equation,

$$\partial_{\bar{\tau}}\bar{\phi}+6\bar{\phi}\partial_{\bar{\xi}}\bar{\phi}+\partial_{\bar{\xi}\bar{\xi}\bar{\xi}}\bar{\phi}=0 \tag{69}$$

Through the substitutions,

$$w(\theta)=\bar{\phi}(\bar{\xi},\bar{\tau}),\ \theta=\bar{\xi}-v_f\bar{\tau} \tag{70}$$

equation (69), by double integration, becomes

$$\frac{1}{2}w^2=F(w)=-(w^3-\frac{v_f}{2}w^2-gw-h) \tag{71}$$

with g, h two integration constants. If F(w) has real roots, they are of the form

$$e_1=\bar{w}+2a\left[\frac{E(s)}{K(s)}-\frac{1}{s^2}\right],\ e_2=\bar{w}+2a\left[\frac{E(s)}{K(s)}-1\right],\ e_3=\bar{w}+2a\frac{E(s)}{K(s)} \tag{72a-c}$$

with

$$a=\frac{e_3-e_2}{2},\ s^2=\frac{e_3-e_2}{e_3-e_1},\ K(s)=\int_0^{\pi/2}(1-s^2\sin^2\varphi)^{-1/2}d\varphi,$$

$$E(s)=\int_0^{\pi/2}(1-s^2\sin^2\varphi)^{1/2}d\varphi \tag{73a-d}$$

$\bar{w}$ a reference value, and K(s), E(s) the complete elliptic integrals of s modulus. The stationary solution of equation (69) has the expression,

$$\bar{\phi}(\bar{\xi},\bar{\tau})=\bar{w}+2a\left(\frac{E(s)}{K(s)}-1\right)-2a\cdot cn^2\left\{\frac{\sqrt{a}}{s}\left[\bar{\xi}-\left(6\bar{w}+4a\left(\frac{3E(s)}{K(s)}-\frac{1+s^2}{s^2}\right)\right)\bar{\tau}+\bar{\xi}_0\right];s\right\} \tag{74}$$

where cn is the Jacobi's elliptic function of s modulus and $\bar{\xi}_0$ an integration constant. As a result, the one-dimensional oscillation modes of the speed field are of cnoidal type – and have the normalized wave length,

$$\lambda=2sK(s)/\sqrt{a} \tag{75}$$

- see figure 1a, the normalized phase speed,

$$v_f=6\bar{w}+4a\left[3\left(E(s)/K(s)\right)-\left((1+s^2)/s^2\right)\right] \tag{76}$$

- see figure 1b, and the normalized group speed,

$$v_g=6\bar{w}+4a\left[3\frac{E(s)}{K(s)}-1-\frac{1}{s^2}+3\frac{E^2(s)+2(s-1)E(s)K(s)-(s-1)K^2(s)}{E(s)K(s)+K^2(s)-sK^2(s)}+\right.$$
$$\left.+\frac{2(s-1)K(s)}{s^2[E(s)+K(s)-sK(s)]}\right] \tag{77}$$

- see figure 1c.

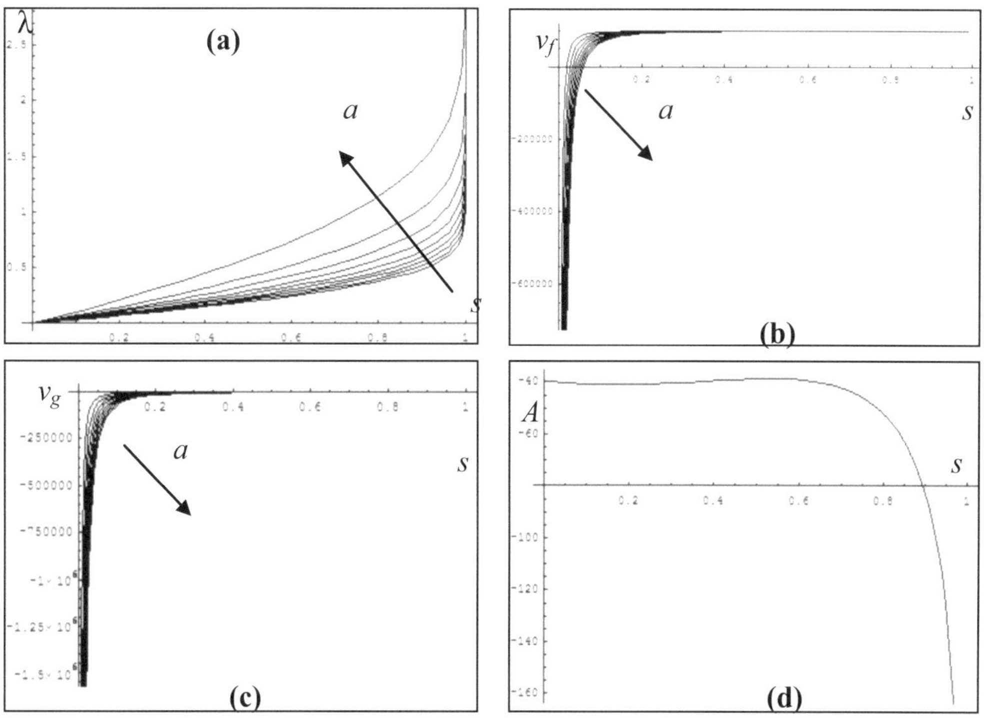

Fig. 1a-d. The dependences on s of the (a) normalized wave length λ, (b) normalized phase speed υ_f, (c) group velocity υ_g (various values of the parameter a), and (d) of the quantity A

Then, the followings result: (i) Through the λ,τ coefficients, the parameter s becomes a 'measure' of 'charge' transport type in the considered matter. Thus, the solution (74), for $s=0$, is reduced to one-dimensional harmonic waves, and for $s\to 0$ to one-dimensional waves packet. These two subsequences describe the 'charge' transport in a non-quasi-autonomous regime. For $s=1$, the solution (74) becomes an one-dimensional soliton while for $s\to 1$ one dimensional solitons packet results. These last two subsequences describe 'charge' transport in a quasi-autonomous regime; (ii) By eliminating the parameter a from relations (75) and (76), one obtains,

$$(v_f-6\overline{w})\lambda^2=A(s)\,,\ A(s)=16\left[3s^2E(s)K(s)-(1+s^2)K^2(s)\right] \qquad (78a,b)$$

where the quantity A(s) is plotted in figure 1d. We observe that for s = 0 ÷ 0.7, A(s) ≈ const. and consequently equation (78a) takes the form, $(v_f-6\overline{w})\lambda^2=const.$. Therefore, in the differentiable case, the 'charge' transport is controlled through the flowing regimes of the fractal fluid, and the separation between them is given by the 0.7 value of the parameter s; (iii) The previous results show through the normalized group speed (77) an increase of the 'charge' transport by means of quasi-autonomous structures. This theoretical result explain some "anomalies" that were experimentally observed in composite materials, e.g. the increase of the thermal conductance etc.

Let us study now the previous phenomenon in the non-differential case. This can be achieved by the substitutions, $\phi = (v_f/4) f^2$ and $i\eta = (v_f/4)^{1/2}\theta$ in equation (71). Moreover, this equation with $h = 0$, becomes, $\partial_{\eta\eta} f = f^3 - f$,i.e. a Ginzburg-Landau (GL) type equation. The followings result: (i) The η coordinate has dynamic significations and the variable f has probabilistic significance. The space-time becomes fractal. (ii) Since the general solution of GL equation can be expressed, with an adequate normalization and choice of the integration constants, by means of the elliptic function $f(\eta) = sn(\eta;s)$, then the 'charge' transport is controlled by the fractal potential (50)

$$Q = -(1/f)(d^2 f/d\eta^2) = \left(1 - f^2\right) = cn^2(\eta, s), \tag{79}$$

also through cnoidal oscillation modes. Thus, as in the previous differentiable case, $(s = 0, s \to 0)$ implies the non-quasi-autonoumous regime, while (s=1, $s \to 1$) implies the quasi-autonoumous regime; (iii) For $s = 1$ the general solution of GL equation is the fractal kink, $f_k(\eta) = tanh(\eta)$. In this case we can build a field theory with spontaneous symmetry breaking. The fractal kink spontaneously breaks the "vacuum" symmetry by tunneling, and generates coherent structures. This mechanism is similar with the one of superconductivity and can explain the properties of composites through the Cooper type pairs; (iv) The normalized fractal potential takes a very simple expression which is directly proportional with the density of states of the fractal fluid - see equation (79). When the density of states, f^2, becomes zero, i.e. in the absence of the vacuum symmetry spontaneous breaking, the fractal potential takes a finite value, $Q \to 1$. The fractal fluid is normal (it works in a non-quasi-autonomous regime) and there are no coherent structures in it. When f^2 becomes 1, i.e. in the presence of the vacuum symmetry spontaneous breaking, the fractal potential is zero,

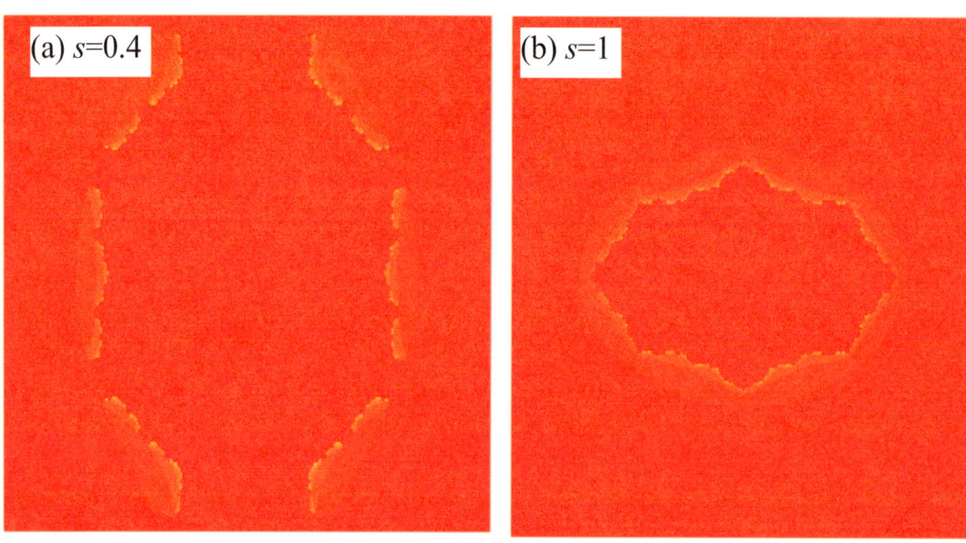

Fig. 2 a-b. The iterative map induced by the elliptic function cn^2 for various values of the s parameter

i.e. the entire quantity of energy of the fractal fluid is transferred to its coherent structures. Then the fractal fluid becomes coherent (it works in a quasi-autonomous regime). Therefore, one can assume that the energy from the fractal fluid can be stocked by transforming all the environment's entities into coherent structures and then 'freezing' them. The fractal fluid acts as an energy accumulator through the fractal potential (79); (v) the correlation between the differentiable and the non-differentiable scales implies the equivalence theorem in periods of two cn^2 elliptic functions. Then, the fractal space-time is of Cantor type.

Moreover, the 'charge' transport implies at any scale a fractal. Such result is obtained through the iterative map induced by the elliptic function cn^2 for various values of the s parameter - Figures 2a-b. For any object given in these figures the Hausdorff-Besicovitch theorem is respected. Evidently, all presented conclusions can be extended to the temperature field.

6. References

Agop, M., G. V. Munceleanu, O. Niculescu, T. Dandu-Bibire, *Physica Scripta* (in press).

Agop, M., C. Radu, T. Bontas, *Chaos Solitons and Fractals* 38, 5, 1243 (2008).

Agop, M., O. Niculescu, A. Timofte, L. Bibire, A. S. Ghenadi, A. Nicuta, C. Nejneru, G. V. Munceleanu, Non-differentiable mechanical model and its implications, *International Journal of Theoretical Physics*, DOI: 10.1007/s10773-010-0330-5, (2008).

Agop, M., P. E. Nica, P. D. Ioannou, A. Antici, V. P. Paun, *Euro Physics Journal* D 49, 239 (2008).

Agop, M., P. E. Nica, S. Gurlui, C. Focsa, V. P. Paun, M. Colotin, *Euro Physics Journal* D 10.11.40/epjd/e2009_00304-5, (2008)

Agop, M., P. Nica, M. Gartu, *Gen. Relativ. Gravit.* 401, 35 (2008).

Argyris, J., C. Ciubotariu, G. Mattutis, *Chaos, Solitons and Fractals* 12, 1 (2001).

Bowman F; *Introduction to Elliptic Function with Applications;* London: English University Press, 1955.

Buzea, C. Gh., I. Rusu, V. Bulancea, Gh. Badarau, V. P. Paun, M. Agop, *Physics Letters A* (in press).

Buzea, C. Gh., C. Bejinariu, C. Boris, P. V. Vizureanu, M. Agop, *International Journal of Nonlinear Sciences and Numerical Simulations* 10 (11-12), 1399 (2009).

Célérier, M.N., L. Nottale, *J. Phys. A: Math. Gen.* 37, 931 (2004).

Cresson, J., F. Ben Adda, *C. R. Acad. Sci. Paris* 330, 261 (2000).

Cresson, J., F. Ben Adda, *J. Math. Anal. Appl.* 262, 721 (2001).

Cresson, J., *Chaos, Solitons and Fractals*, 14, 553 (2002).

Cresson, J., *J. Math. Phys.* 44, 4907 (2003).

Cresson, J., F. Ben Adda, *Chaos, Solitons and Fractals*, 19, 1323 (2004).

Cresson, J., *J. Math. Anal. Appl.* 307, 48 (2005).

Cresson, J., *Int. J. of Geometric Meth. in Mod. Phys.*, 3, 1395 (2006).

Cresson, J., J-N. Denarie, *Lect. Notes in Physics*, Planat, Paris, 2000.

Cristescu, C.P., *Nonlinear Dynamics and Chaos in Science and Engineering*, Academy Publishing House, Bucharest, 2008.

Cristescu, C.P., B. Mereu, C. Stan, M. Agop, *Chaos Solitons and Fractals* 40, 975 (2009).

Da Rocha, D., L. Nottale, *Chaos, Solitons and Fractals* 16, 565 (2003).

El Naschie, M.S., O. E. Roessler, I. Prigogine, *Quantum Mechanics, Diffusion and Chaotic Fractal*, Elsevier, Oxford, 1995.

Feder, J., A. Aharony, *Fractals in Physics*, North Holland, Amsterdam, 1990.
Ferry, D. K.; Goodnick S. M.; *Transport in Nanostructures*; Cambridge: Cambridge University Press, 1997.
Gouyet, J.F., *Physique et Structures Fractals*, Masson, Paris, 1992.
Gurlui, S., M. Agop, M. Strat, G. Strat, S. Bacaita, A. Cerepaniuc, *Phys. Plasmas* 13, 063503 (2006).
Gurlui, S., M. Agop, P. Nica, M. Ziskind, C. Focsa, *Phys. Rev.* E 78, 026405 (2008).
Imry, Y.; *Introduction to Mesoscopic Physics*; Oxford: Oxford University Press,2002.
Ioannou, P.D., P. Nica, V. Paun, P. Vizureanu, M. Agop, *Physica Scripta* 78, 6, 065101.
Jackson, E. A.; *Perspectives in Nonlinear Dynamics*; Cambridge: Cambridge University Press, vol. I+II, 1991.
Madelbrot, B., *The Fractal Geometry of Nature*, Freeman, San Francisco, 1982.
Nelson, E., *Quantum Fluctuations*, Princeton Univ. Press, NY, 1985.
Nica, P. Vizureanu, M. Agop, S. Gurlui, C. Focsa, N. Forna, P. D. Ioannou, Z. Borsos, *Japanese Journal of Applied Physics 48, 066001 (2009).*
Niculescu, O., D. G, Dimitriu, V. P. Paun, P. D. Matasaru, D. Scurtu, M. Agop, *Phys. Plasmas* 17, 042305 (2010).
Nottale, L., *Chaos, Solitons and Fractals* 9, 1051 (1980).
Nottale, L., *Fractal Space-Time and Microphysics: Towards a Theory of Scale Relativity*, World Scientific, Singapore, 1993.
Nottale, L., L'univers et la lumière, *Cosmologie classique et mirages gravitationnels*, Flammarion, Paris, 1993.
Nottale, L., *La Relativité dans tous ses états : du mouvements aux changements d'échelle*, Hachette, Paris, 1998.
Nottale, L., *Chaos, Solitons and Fractals* 10, 459 (1999).
Nottale, L., J. Chaline, P. Grou, *Les arbres de l'évolution*, Hachette, Paris, 2000.
Nottale, L., *Chaos, Solitons and Fractals* 16, 539 (2003).
Nottale, L., *Chaos, Solitons and Fractals* 25, 797 (2005).
Nottale, L., M. N. Célérier, T. Lehner, *J. Math. Phys.* 47, 032303 (2006).
Nottale, L., Ch. Auffray, *Progress in Biophysics and Molecular Biology* 97, 115 (2008).
Nottale, L., Scale relativity and fractal space-time: theory and applications, in *Proceedings of First International Conference on the Evolution and Development of the Universe*, Paris, France, 2008.
Nottale, L., J. Phys. A: *Mathematical and Theoretical* 42, 275306 (2009).
Poole, C.P.; Farach, H.A.; Creswick, R. J. *Superconductivity*; San Diego-New York-Boston-London-Sydney-Tokyo-Toronto: Academic Press, 1995.
Stan, C. P. Cristescu, D. Alexandroaie, M. Agop, *Chaos Solitons and Fractals* 41, 727 (2009).
Weibel, P., G. Ord, G. Rössler (Editors), *Space-time Physics and Fractality*, Springer, Vienna, New York, 2005.

18

Dependence of Thermal and Electrical Conductivities of Actinide-Zirconium-Hydride Composite Materials on Hydrogen Concentration

B. Tsuchiya[1], K. Konashi[2] and M. Yamawaki[3]
[1]*Department of General Education, Faculty of Science and Technology, Meijo University, 1-501, Shiogamaguchi, Tempaku-ku, Nagoya 468-8502,*
[2]*International Research Center for Nuclear Materials Science, Institute for Materials Research, Tohoku University, Oarai-machi, Ibaraki-ken 311-1313,*
[3]*Department of Applied Physics, Graduate School of Engineering, Tokai University, 1117, Kitakaname, Hiratsuka, Kanagawa 259-1292,*
Japan

1. Introduction

High-level radioactive waste generated by the reprocessing of spent nuclear fuel from nuclear reactors includes long-lived radioactive nuclides. The current method for the disposal of such waste involves the vitrifying the waste under rigid control, storing it in monitored areas until the radiation decays to permissible levels, and then disposing it underground. Many types of transmutation methods have been studied to reduce the need for geological disposal. Recently, a transmutation method has been proposed for actinide radioactive waste; this method involves the use of hydride irradiation targets, which are loaded in the form of pellets into the core of fast breeder reactors containing mixed-oxide fuel (Yamamoto et al., 1997), (Yamawaki et al., 1998), and (Konashi et al., 2001). The irradiation hydride targets are composite materials, composed of titanium, zirconium, and hafnium hydrides, which contain hydrogen storage metals, and actinide elements such as ^{237}Np, ^{241}Am, and ^{243}Am. The hydrides in the irradiation targets act as neutron moderators to provide a high flux of the thermal neutron. During irradiation, a temperature gradient occurs between the center and edge of the targets, and the distribution of the hydrogen concentration changes with the hydrogen diffusion (Huang et al., 2000). In the design of irradiation hydride targets, it is extremely important to investigate the changes in the mechanical, thermal, and electrical properties of the hydrides including their various hydrogen concentrations and to understand the basic heat transfer processes. Thermal conductivity is the most important physical property.

In the present study, the effects of the hydrogen content on the electrical and thermal properties of metal-hydride composite materials such as uranium-zirconium hydrides (45 wt% U-ZrH_x; x = 1.60 and 1.90) and uranium-thorium-zirconium hydrides ($UTh_4Zr_{10}H_x$; x =

18–27) are investigated, and the heat conduction due to free electrons and that due to free phonons are discussed. From the results we can estimate the absolute values of the thermal conductivities for actinide-hydride composite materials including Np and Am. We also consider the correlation between the thermal transportation and the hydrogen content for the irradiation hydride targets since the hydrogen chemical potential of Th-hydride is close to those of Np- and Am-hydrides.

2. Experiments

Alloys with 45 wt% U-Zr and UTh_4Zr_{10} were prepared by melting the constituent elements—U and Zr with a purity of 99.9 wt% and Th with a purity of 99.99 wt%—in vacuum in a high-frequency induction furnace, which was manufactured by Mitsubishi Materials Corporation. The composition of the UTh_4Zr_{10} alloy was selected so as to ensure solid solution formation at high temperatures and more amount of absorbed hydrogen than any other U-Th-Zr alloy (Yamamoto et al., 1997) and (Yamawaki et al., 1998). The microstructure of the UTh_4Zr_{10} alloy mixing α-Th with β-UZr_{2+x} phases, as shown in Fig. 2(a), was observed using scanning electron microscopy (SEM) and X-ray diffraction (XRD).
The hydrogenation of 45 wt% U-Zr and UTh_4Zr_{10} alloys was carried out using Sieverts' apparatus, supplied by Mitsubishi Materials Corporation (Tsuchiya et al., 2000). The 45 wt% U-Zr and UTh_4Zr_{10} alloys, mounted into a quartz tube evacuated under a pressure of 1.3×10^{-4} Pa, were heated at 1173 K for 3.5 h to remove a small amount of residual hydrogen, absorbed into the alloys during the melting and quenching of the constituent elements. The alloys were then heated at temperatures ranging from 673 to 1173 K and exposed to pure protium gas in the quartz tube at pressures ranging from 1.0×10^3 to 1.0×10^5 Pa, where the temperature and pressure values were measured using a thermocouple and a baratoron, respectively. When equilibrium for the hydrogenation was achieved, the protium gas was removed from the quartz tube and the annealing temperature was then gradually reduced to room temperature, with the sample still in the tube.

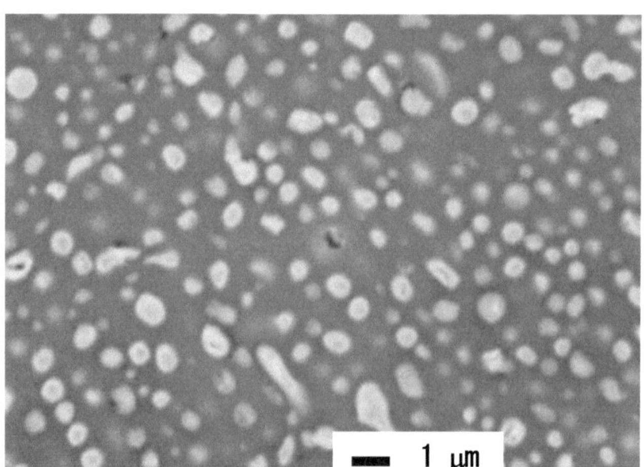

Fig. 1. SEM (BEI: backscattered electron image) micrograph of 45 wt% U-$ZrH_{1.60}$, composed of α-U (white area) and δ-$ZrH_{1.60}$ (dark area) phases

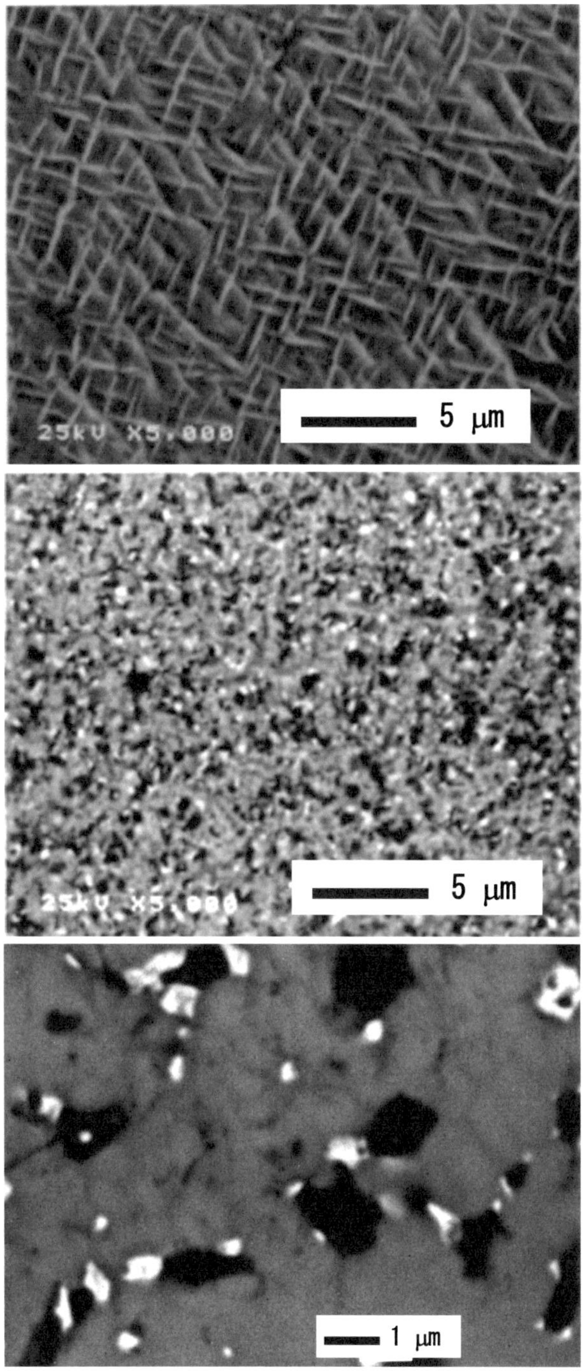

(a)
(b)
(c)

Fig. 2. SEM (BEI: backscattered electron image) micrographs of (a) UTh_4Zr_{10} alloy, which is composed of α-Th (white area) and β-UZr_{2+x} (black area) phases, (b) and (c) $UTh_4Zr_{10}H_{24}$ which is composed of α-U (white area), $ThZr_2H_{5.59}$ (gray area), and $ZrH_{1.86}$ (black area) phases

The compositions of the 45 wt% U-Zr and UTh_4Zr_{10} hydrides were measured based on the hydrogen pressure changes at equilibrium and the mass gains before and after hydrogenation. The fabricated products were 45 wt% U-ZrH_x (x = 1.60 and 1.90) and $UTh_4Zr_{10}H_x$ (x = 18–27). The surface morphology and crystalline structure of these two hydride composite materials were extensively examined using SEM and XRD. Figure 1 shows an SEM (BEI: backscattered electron image) micrograph of 45 wt% U-$ZrH_{1.60}$. It reveals that the U phase of approximately 1.0 μm in diameter is dispersed in the bulk of the $ZrH_{1.60}$ phase; this is indicative of a composite material mixing an actinide element with a hydride. The morphology of 45 wt% U-$ZrH_{1.90}$ is almost the same as that of 45 wt% U-$ZrH_{1.60}$. The structures of U, $ZrH_{1.60}$, and $ZrH_{1.90}$ are orthorhombic (α-phase), face-centered cubic (δ-phase), and face-centered tetragonal (ε-phase), respectively (Tsuchiya et al., 2001).

BEI-SEM micrographs with magnifying powers of 5000 and 10000 for $UTh_4Zr_{10}H_{24}$ are shown in Figs. 2(b) and (c), respectively. The α-U (white area) and $ZrH_{1.86}$ (black area) phases of approximately 1.0 μm in diameter are dispersed in the bulk of the $ThZr_2H_{5.59}$ (gray area) phase; this is indicative of a composite material mixing an actinide element with two kinds of hydrides. The values of x in $ThZr_2H_x$ and ε-ZrH_x were estimated based on the relationship between the measured XRD data and published data (Nakata et al., 1966), (Yamamoto et al., 1997), and (Tsuchiya et al., 2002). The other hydrides of $UTh_4Zr_{10}H_x$ were determined to be $ThZr_2H_{3.30}$ and δ-$ZrH_{1.69}$ ($UTh_4Zr_{10}H_{18}$), $ThZr_2H_{4.17}$ and ε-$ZrH_{1.74}$ ($UTh_4Zr_{10}H_{20}$), $ThZr_2H_{5.59}$ and ε-$ZrH_{1.83}$ ($UTh_4Zr_{10}H_{23}$), $ThZr_2H_{5.79}$ and ε-$ZrH_{1.87}$ ($UTh_4Zr_{10}H_{26}$), and $ThZr_2H_{5.94}$ and ε-$ZrH_{1.90}$ ($UTh_4Zr_{10}H_{27}$), respectively. Therefore, the hydrides of $UTh_4Zr_{10}H_x$ (x = 18–27) are mainly three-phase composite materials, composed of α-U, δ-phase ZrH_x ($1.5 \leq x < 1.7$) or ε-phase ZrH_x ($1.7 \leq x \leq 2.0$), and $ThZr_2H_x$ ($4.0 < x < 6.3$), although there are small quantities of residual Th, Th hydrides, and oxides such as ThO_2 and ZrO_2.

Figures 3(a) and (b) show the hydrogen release from 45 wt% U-$ZrH_{1.60}$, U-$ZrH_{1.90}$, $ZrH_{1.60}$, $ZrH_{1.90}$, $UTh_4Zr_{10}H_{20}$, $UTh_4Zr_{10}H_{24}$, and $UTh_4Zr_{10}H_{27}$ by isochronal annealing for 10 min at temperatures of 298–973 K. In Fig. 3(a) it is important to note that the decomposition temperature of 823 K for U-$ZrH_{1.60}$ is the same as that for $ZrH_{1.60}$, while the temperature of 773 K for U-$ZrH_{1.90}$ is higher than that for $ZrH_{1.90}$. The decomposition temperatures of $UTh_4Zr_{10}H_{20}$, $UTh_4Zr_{10}H_{24}$, and $UTh_4Zr_{10}H_{27}$ are approximately 823, 723, and 673 K, respectively, as shown in Fig. 3(b). There are two types of stages in the hydrogen molecular re-emission process. The first and second stages of the re-emission are due to hydrogen release from the $ThZr_2H_x$ and ZrH_x phases, respectively.

To avoid the reduction of the hydrogen concentration by hydrogen molecular re-emission, special specimen containers, made of sapphire, were used for the thermal diffusivity measurements with a laser flash method (Tsuchiya et al., 2002). The heating temperatures were successful in elevating to 900 K (U-$ZrH_{1.60}$, $UTh_4Zr_{10}H_{18}$, and $UTh_4Zr_{10}H_{20}$), 840 K (U-$ZrH_{1.90}$), 800 K ($UTh_4Zr_{10}H_{23}$ and $UTh_4Zr_{10}H_{24}$), 750 K ($UTh_4Zr_{10}H_{24}$ and $UTh_4Zr_{10}H_{26}$), and 700 K ($UTh_4Zr_{10}H_{27}$) during the thermal diffusivity measurements. In addition, an electrical resistivity measurement was carried out from room temperature to 700–800 K for each hydride using a four-contact method with a direct current (DC) of 350 mA, to clarify the heat conduction due to free electrons.

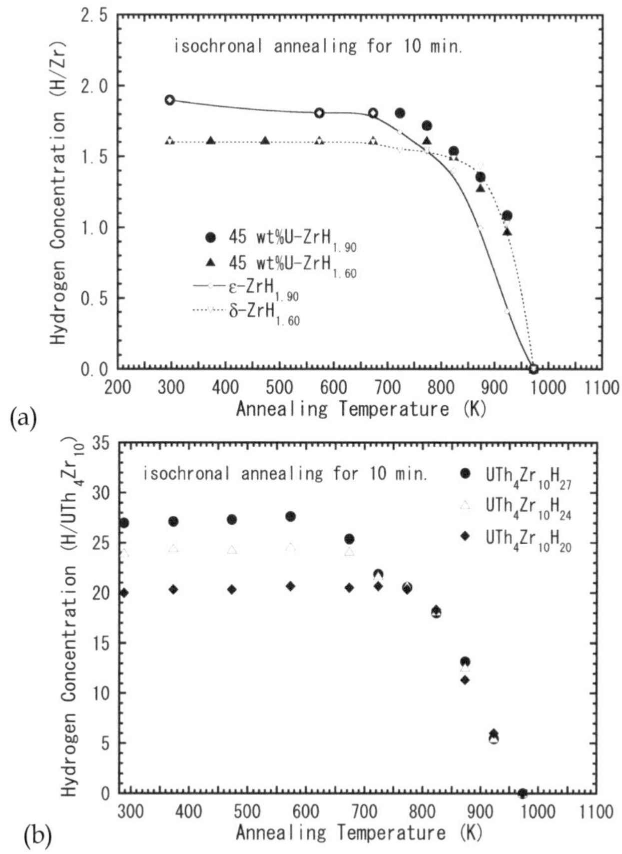

Fig. 3. Changes in hydrogen concentrations in (a) U-$ZrH_{1.60}$, U-$ZrH_{1.90}$, $ZrH_{1.60}$, and $ZrH_{1.90}$ and (b) $UTh_4Zr_{10}H_{20}$, $UTh_4Zr_{10}H_{24}$, and $UTh_4Zr_{10}H_{27}$ after isochronal annealing at different temperatures for 10 min

3. Results and discussion

Figures 4(a) and (b) show the thermal diffusivities of 45 wt% U-$ZrH_{1.60}$, 45 wt% U-$ZrH_{1.90}$, $UTh_4Zr_{10}H_{18}$, $UTh_4Zr_{10}H_{20}$, $UTh_4Zr_{10}H_{23}$, $UTh_4Zr_{10}H_{24}$, $UTh_4Zr_{10}H_{26}$, and $UTh_4Zr_{10}H_{27}$, measured while increasing (solid marks) and decreasing (open marks) the heating temperature. The agreement between the values on heating and cooling indicates that there is no hydrogen release from the hydrides during the thermal diffusivity measurements. The thermal diffusivities for U-$ZrH_{1.60}$ and U-$ZrH_{1.90}$ increased with increasing hydrogen concentration and with decreasing temperature. The temperature dependence of the thermal diffusivities for U-$ZrH_{1.60}$ and U-$ZrH_{1.90}$ is similar to that for $ZrH_{1.60}$ and $ZrH_{1.90}$ (Tsuchiya et al., 2002), although the absolute values are different, as shown in Fig. 4(a). This is because the thermal diffusivity of U is nearly constant in the temperature range up to 900 K (Takahashi et al., 1988). The present experimental values are approximately 1.5 times those of 10 wt% U-ZrH_x (Nakata et al., 1966), depending on the amount of doped-U.
On the other hand, the thermal diffusivities of $UTh_4Zr_{10}H_{23}$, $UTh_4Zr_{10}H_{24}$, $UTh_4Zr_{10}H_{26}$, and $UTh_4Zr_{10}H_{27}$ decreased with increasing temperature, while those of $UTh_4Zr_{10}H_{18}$ and

$UTh_4Zr_{10}H_{20}$ increased gradually up to a temperature of 940 K, as shown in Fig. 4(b). It is interesting to note that the thermal diffusivities at 640 K for $UTh_4Zr_{10}H_{18-27}$ are equal.

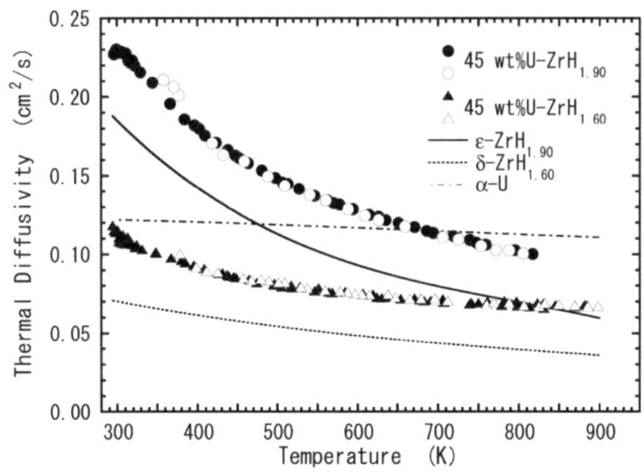

(a)

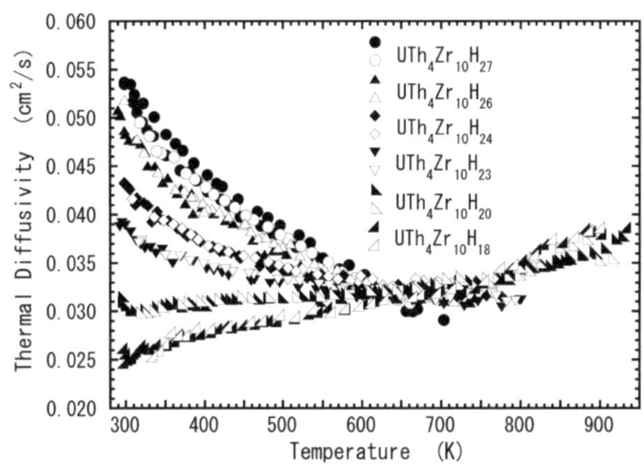

(b)

Fig. 4. Temperature dependence of the thermal diffusivities for (a) U-ZrH_x (x = 1.60 and 1.90), ZrH_x (x = 1.60 and 1.90), U, and (b) $UTh_4Zr_{10}H_x$ (x = 18, 20, 23, 24, 26, and 27)

To clarify the electronic heat conduction for U-ZrH_x and $UTh_4Zr_{10}H_x$, the electrical resistivity was measured when heating to 700–800 K and cooling to room temperature. Figures 5(a) and (b) show the electrical resistivities of 45 wt% U-$ZrH_{1.60}$, 45 wt% U-$ZrH_{1.90}$, $UTh_4Zr_{10}H_{18}$, $UTh_4Zr_{10}H_{20}$, $UTh_4Zr_{10}H_{23}$, and $UTh_4Zr_{10}H_{26}$ when heating (solid marks) and cooling (open marks). The electrical resistivities of U-ZrH_x increased as the temperature increased and the hydrogen concentration decreased, as shown in Fig. 5(a). In the case of U-$ZrH_{1.90}$, a slight distortion between the values obtained on heating and those obtained on cooling indicates a reduction in the hydrogen concentration, following hydrogen release from $ZrH_{1.90}$. The hydrogen concentration dependence on the resistivity of U-ZrH_x strongly dominates that of ZrH_x (Bickel & Berlincourt, 1970) and (Tsuchiya et al., 2002) and slightly dominates that of U (Bell, 1954). The electron scattering due to hydrogen vacancy in the hydrides significantly affects the resistivity. The scattering of electrons due to optical phonons as well as acoustic phonons affects the resistivity behavior at higher temperature.

The results for $UTh_4Zr_{10}H_x$ showed gradual decreases in the electrical resistivity with an increase in the hydrogen composition to approximately 690 K, which was higher than the temperature (640 K) for the thermal diffusivity, and an increase above 690 K, as shown in Fig. 5(b). The electron scattering due to hydrogen vacancies in both the ZrH_x and $ThZr_2H_x$ phases significantly reflects the temperature dependence of the electronic conduction for $UTh_4Zr_{10}H_x$. In particular, at higher temperatures the scattering of electrons due to optical phonons in the $ThZr_2H_x$ phase may essentially dominate the resistivity, although the resistivity of $ThZr_2H_x$ has not been measured yet.

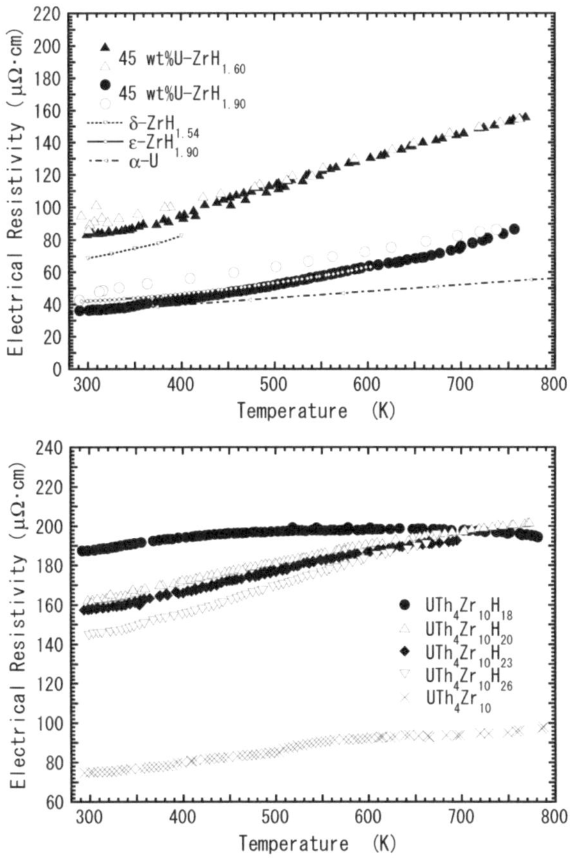

Fig. 5. Temperature dependence of electrical resistivities for (a) U-ZrH_x (x = 1.60 and 1.90), ZrH_x (x = 1.54 and 1.90), U, and (b) $UTh_4Zr_{10}H_x$ (x = 0, 18, 20, 23, and 26)

Figures 6 (a) and (b) show the thermal conductivities, λ, of 45 wt% U-$ZrH_{1.60}$ and 45 wt% U-$ZrH_{1.90}$, respectively, at temperatures up to 773 K, which were calculated via $\lambda = \alpha C_p d$, where α, C_p, and d represent the measured thermal diffusivity, the specific heat, and the density, respectively. The experimental values of d for U-$ZrH_{1.60}$ and U-$ZrH_{1.90}$ were d = 8.256 and 8.209 g/cm^3, respectively. The values of C_p for U-$ZrH_{1.60}$ and U-$ZrH_{1.90}$ were expressed by the following equations as functions of temperature, T: $C_p = 0.120 + 4.72 \times 10^{-4} T$ and $C_p = 0.146 + 4.71 \times 10^{-4} T$ [J/(g · K)], respectively, taking into account the weight fractions of U and ZrH_x and reported data on the specific heat of U; C_p = 0.120 J/(g · K) and

ZrH_x (x = 1.60–2.00); $C_p = [6.98 \times 10^{-2}\,T + \{34.4 + 14.8 \times (x - 1.65)\}]/M_{ZrHx}$ [J/(g • K)], where x and M_{ZrHx} are the composition of H in ZrH_x and the molecular weight of ZrH_x (Simnad, 1981). The temperature dependence of the thermal conductivity for 45 wt% U-$ZrH_{1.60}$ was almost constant with increasing temperature, while that for U-$ZrH_{1.90}$ showed a gradual decrease. The conductivities of approximately 0.27 W/(cm·K) (45 wt% U-$ZrH_{1.60}$) and 0.43 W/(cm·K) (45 wt% U-$ZrH_{1.90}$) at a temperature of 700 K were higher than the reported value (0.18 W/(cm·K)) for 10 wt% U-$ZrH_{1.6}$ without temperature dependence; this value has been used for TRIGA reactors. The conductivity strongly depends on the quantities of absorbed hydrogen and doped-U.

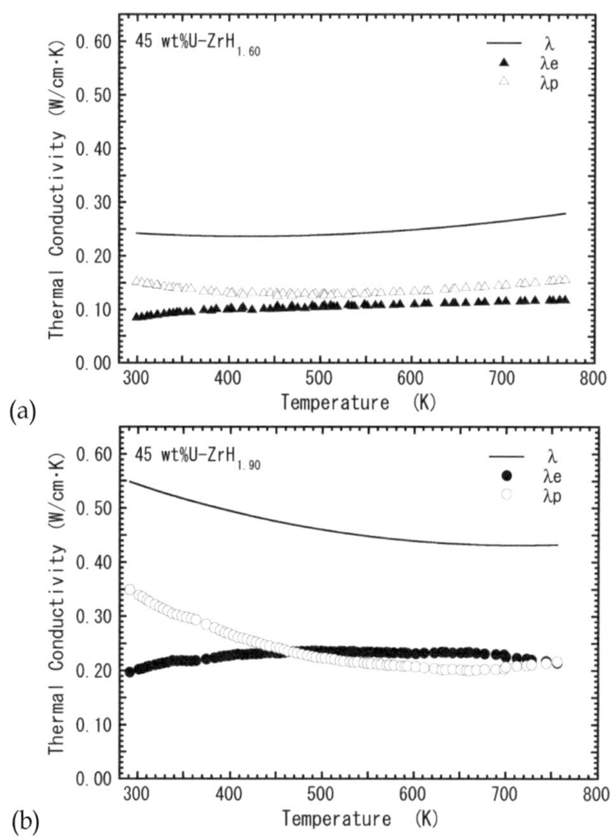

Fig. 6. Thermal conductivities, λ, of (a) U-$ZrH_{1.60}$ and (b) U-$ZrH_{1.90}$. λ_e and λ_p represent the thermal conductivity due to free electrons and that due to phonons, respectively

Figures 7 (a), (b), (c), and (d) show the thermal conductivities, λ, of $UTh_4Zr_{10}H_{18}$, $UTh_4Zr_{10}H_{20}$, $UTh_4Zr_{10}H_{23}$, and $UTh_4Zr_{10}H_{26}$, respectively, calculated using the equation given above. The density and specific heat values of $UTh_4Zr_{10}H_x$ (x = 18–27) are expressed by considering the experimental results and the reported values of specific heat as follows: $d = 8.40 - 2.99 \times 10^{-2}\,x$ [g/cm³] and $C_p = -0.110 + 6.87 \times 10^{-4}\,T + 6.36 \times 10^{-3}\,x$ [J/(g·K)], where x and T are the composition of H in $UTh_4Zr_{10}H_x$ and the temperature (Tsuchiya et al., 2000). The high thermal conductivity (approximately 0.13 W/(cm·K)) for $UTh_4Zr_{10}H_x$ (18–27) at a temperature of 700 K, which is close to the reported value (0.18 W/(cm·K)) for 10 wt% U-$ZrH_{1.6}$, increases the safety level in reactors when the linear power is high. For all the $UTh_4Zr_{10}H_x$, the temperature dependence of the thermal conductivities showed a gradual increase with increasing temperature. On the other hand, the hydrogen dependence of the thermal conductivity dramatically changed at the boundary temperature of 690 K, corresponding to the temperature and composition dependences of the electrical property and the thermal diffusion.

The thermal conductivity was evaluated using the values of the electronic conduction, λ_e, obtained from the relation $\lambda_e = L_e\sigma T$, which is based on the Wiedemann-Franz rule. In this equation, σ denotes the electrical conductivity ($\sigma = 1/\rho$) and is obtained from the measured electrical resistivity. Further, L_e is the Lorenz number corresponding to the electronic conduction, expressed as $L_e = (\pi^2/3)(k_B/e)^2 \approx 2.45 \times 10^{-8}$ WΩ/K², where k_B and e are the Boltzmann constant and the elementary electric charge, respectively. Consequently, the thermal conductivity associated with phonon conduction, λ_p, was determined by subtracting λ_e from λ (i.e., $\lambda = \lambda_e + \lambda_p$), as shown in Figs. 6 and 7.

For the U-ZrH_x, heat is conducted for each phase in the temperature range 300–800 K because of the migration of both free electrons and phonons, although the absolute values depend on the hydrogen content and, at temperatures below 450 K, the contribution from the phonons was greater than that from the electrons.

For $UTh_4Zr_{10}H_x$, in the temperature range 300–800 K, the contribution of the migration of electrons to heat conduction is greater than that of the migration of phonons. The thermal property of the hydride composite materials seems to be similar to that of metals rather than insulators. It may be related to the quantity of the free electrons, derived from the heavy metals of the actinides. The reduction of the conductivity with a decrease in the hydrogen content, indicative of an increase in hydrogen vacancy, shows that phonon scattering by electrons and phonons has a significant effect on the nonstoichiometric structures of the hydrides. In particular, it is interesting that the phonon conduction increases with increasing temperature. The migration of phonons in the $ThZr_2H_x$ phase may be enhanced at high temperatures, since the phonon conduction in the Zr_2H_x phase is usually proportional to $1/T$, where T is the temperature. The composite materials of $ThZr_2H_x$ are now being further studied to clarify the effect of hydrogen on the thermal transportation process of the actinide-hydride materials in more detail.

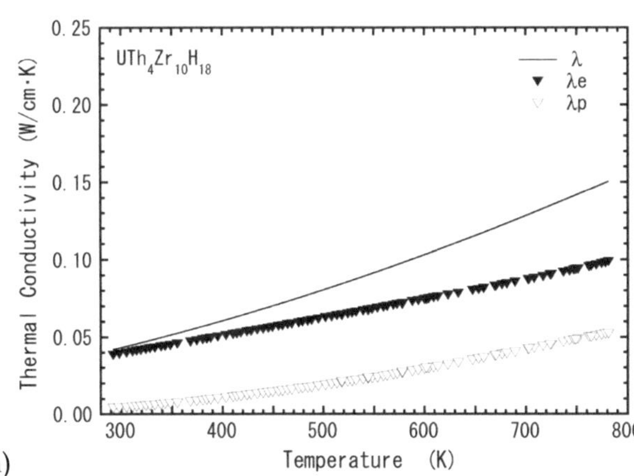

(a)

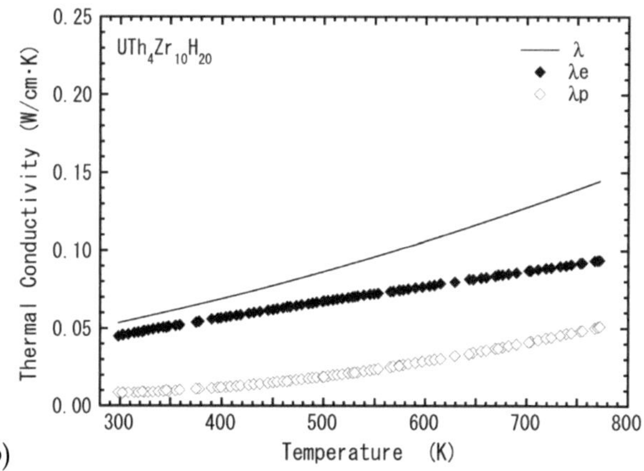

(b)

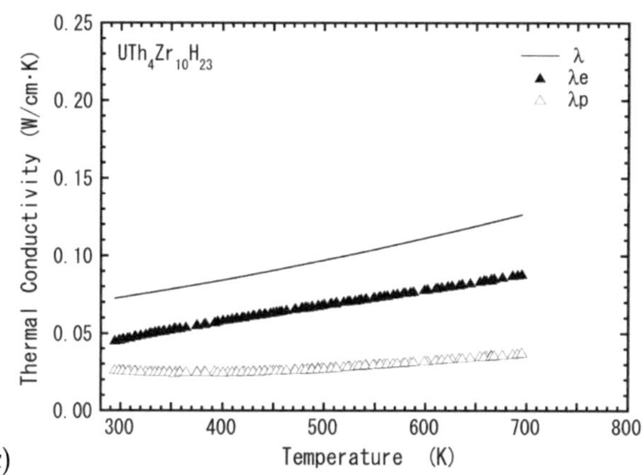

(c)

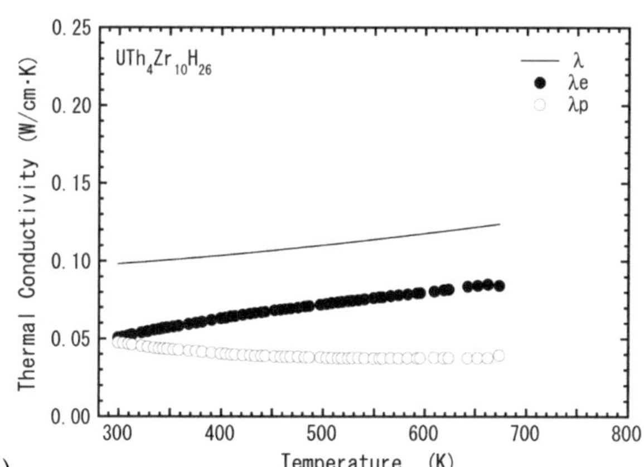

(d)

Fig. 7. Thermal conductivities, λ, of (a) $UTh_4Zr_{10}H_{18}$, (b) $UTh_4Zr_{10}H_{20}$, (c) $UTh_4Zr_{10}H_{23}$, and (d) $UTh_4Zr_{10}H_{26}$. λ_e and λ_p represent the thermal conductivity due to free electrons and that due phonons, respectively

4. Conclusions

The thermal diffusivities of the actinide-hydride composite materials such as 45 wt% U-ZrH_x (x = 1.60 and 1.90) and $UTh_4Zr_{10}H_x$ (x = 18–27) were measured by the laser flash method, and their thermal conductivities were calculated by taking into account the density and the specific heat. The thermal conductivity was significantly dependent on the quantities of doped actinide metals and formed hydrides in the composite materials and, in particular, on the hydrogen concentration. The heat conduction due to electrons and that due to phonons were determined using the Wiedemann-Franz rule on the basis of the electrical conductivity calculated using the measured electrical resistivity. In the case of thermal conduction by U-ZrHx at room temperature, phonon-phonon scattering is dominant, and the number of electrons and photons are approximately equal in the case of heat conduction at high temperatures of above 450 K. In addition, the thermal conductivity decreases with a decrease in the hydrogen content; this is indicative of an increase in the hydrogen vacancy, because phonon scattering by electrons and phonons has a significant effect on the nonstoichiometric structures of the hydrides.

The actinide-hydride composite materials of $UTh_4Zr_{10}H_x$ showed a gradual increase in the thermal conductivities with an increase in the temperature. The dependence of the thermal conductivities on the composition showed an increase with an increase in the hydrogen content at temperatures below approximately 690 K. In contrast, a decrease in the composition dependence was observed at temperatures above 690 K. For heat conduction at temperatures below 690 K, the contribution of the migration of the electrons is considerably larger than that of the migration of the phonons. The increase in the conductivity at higher temperatures may be attributed to the enhancement of phonon conduction in the $ThZr_2H_x$ phase.

It is concluded that the actinide-hydride composite materials with high thermal conductivities of approximately 0.27 W/(cm·K) (45 wt% U-$ZrH_{1.60}$), 0.43 W/(cm·K) (45 wt% U-$ZrH_{1.90}$), and 0.13 W/(cm·K) ($UTh_4Zr_{10}H_{18-27}$) at 700 K, which have thermal properties similar to those of metals, are potential candidates for fabricating the irradiation target used in the transmutation of actinide radioactive wastes.

5. References

Yamamoto, T., Suwaruno, H., Kayano, H. & Yamawaki, M. (1997). *J. Nucl. Mater.*, 247, 339-344.

Yamawaki, M., Suwarno, H., Yamamoto, T., Sanda, T., Fujimura, K., Kawashima, K. & Konashi, K. (1998). *J. Alloys Comp.*, 271-273, 530-533.

Konashi, K., Tsuchiya, B., Yamawaki, M., Fujimura, K. & Sanda, T. (2001). *Proceedings of International Conference on Back-End of the Fuel Cycle from Research to Solutions*, GLOBAL2001, Paris, France.

Tsuchiya, B., Huang, J., Konashi, K., Teshigawara, M. & Yamawaki, M. (2001). *J. Nucl. Mater.*, 289, 329-333.

Huang, J., Tsuchiya, B., Konashi, K. & Yamawaki, M. (2000). *J. Nucl. Sci. Technol.*, 37(10), 887-892.

Tsuchiya, B., Huang, J., Konashi, K., Saiki, W., Onoue, T. & Yamawaki, M. (2000). *J. Alloys Comp.*, 312, 104-110.

Tsuchiya, B., Konashi, K., Yamawaki, M. & Nakajima, Y. (2002). *J. Nucl. Sci. Technol.*, Suppl.3, 855-857.

Nakata, M. M., Ambrose, C. J. & Finch, R. A. (1966). *Proceedings of 6th Conf. On Thermal Conductivity*, pp. 479-507.

Tsuchiya, B., Teshigawara, M., Konashi, K., Nagata, S., Shikama, T. & Yamawaki, M. (2002). *J. Nucl. Sci. Technol.*, 39(4), 402-406.

Takahashi, Y., Yamawaki, M. & Yamamoto, K. (1988). *J. Nucl. Mater.*, 154, 141-144.

Bickel, P. W. & Berlincourt T. G. (1970). *Phys. Rev. B*, 2(12), 4807-4813.

Bell, I. P. (1954). *UKAEC*, RDB(c) TN-101, 1-16.

Simnad, M. T. (1981). *Nucl. Eng. Des.*, 64, 403-422.

19

Dye Doped Polymer-Filled Nanoporous Glass – a New Class of Materials for Laser Optics

Modest Koldunov and Alexander Manenkov
A.M.Prokhrov General Physics Institute, Russian Academy of Sciences, Moscow Russia

1. Introduction

A subject of this chapter is a dye-activated polymer-filled nanoporous glass as a new composite material for laser optics. It has been realized for the first time in 1988 [1] as a Q-switching optical element for lasers. Motivation for that work was to improve properties of Q-switchers based on dye-activated balk polymers possessing good optical properties, but having rather poor thermo-optical and mechanical properties, which lead to thermo-optical distortions of laser beams and to a mechanical deformation of an element surface at variations of temperature and humidity of an environment. To reduce these negative effects of the optical polymer an idea was proposed to insert the polymer element into a rigid cage. This idea has been realized at first in a triplex element: a polymer film clamped between glass plates. Another approach to solving the problem was synthesizing a polymer in pores of nonporous glass (NPG), i.e. making a polymer-filled nanoporous glass (PFNPG) composite

Note that a technology of NPG has been well developed by that time [2-4]. Mention also that attempts to make NPG-based laser elements were performed: laser operation, Q-switching, and mode-locking have been demonstrated in porous glass elements impregnated with liquid dye solutions [5-7].

Difficulty of making the composite NPG-polymer is connected, mainly, with peculiarity of a polymerization process in pores. A significant shrinkage effect takes place at polymerization that can create porous leading to light scattering. This problem has been solved in [1] and a technology of the PFNPG composite of high optical quality has been developed.-

Comprehensive studies [8-25] of properties of PFNPG composite elements made by the developed technology confirmed that the composite has a significant advantage over the bulk polymer. The composite preserving positive properties of modified optical polymers (high solubility of organic dyes, high laser-induced damage resistance) exclude its negative properties (poor thermo-optical quality, sensitivity to temperature and humidity variations).

A goal of this chapter is analyzing the PFNPG composite as the laser optical material. The chapter consists of four sections.

In the first section general requirements to laser optics elements are briefly discussed.

In the second section principles of the PFNPG composite technology are described.

In the third section the results of comprehensive studies of the composite properties are reported and analyzed.

In the forth section experimental results of studies of PFNPG laser elements doped with different type functional organic compounds are presented and discussed.

2. Requirements to laser optical elements: a general analysis

Before a detailed analysis of properties of the dye-impregnate polymer-filled nanoporous glass it is reasonable to considerer qualitatively, in a general form, the most important characteristics of laser optics materials. General properties, to which the laser optics materials for laser optics applications have to be satisfied, are as follows.

- Transparency in an operation wavelength range, due to both light absorption and scattering loses, has to be high.
- Mechanical strength has to be high enough for a treatment of the optical elements using a standard glass fabrication technology.
- Laser-induced damage resistance of the materials in both the single-shot (1×1) and multi-shot (1×N) irradiation regimes has to be high.
- Thermo-optical figure of merit of the material, characterized by the parameter $\mu = \chi(dn/dT)^{-1}$, where χ is thermal conductivity, n is refractive index, T is temperature, has to be high enough to avoid optical distortions of a propagating radiation in high-power laser systems operating, especially, at high repetition rates.
- Functional organic compounds impregnated to a host material to get a desirable optical effect, active-laser emission or passive-radiation control, i.e. Q-switching and mode-locking, have to be distributed inside the material homogeneously to avoid, for example, luminescence quenching in lasing elements.

Speaking about the material characteristics, note that measurements of absolute values of some of them are significant problems. On this reason properties of the PFNPG composite, reported in this chapter, were investigated in comparison with the same properties of the bulk modified polymethylmathacrilate (MPMMA) fabricated from the same monomer composition as the polymer component of the composite. For comparison, in Table 1 some characteristics of PFNPG and MPMMA [19,26] are presented.

	Bulk Polymer	PFNPG-composite
Solubility of dyes, mol/l	~ 10^{-2}	~ 10^{-2}
Optical transparency range, nm	300÷1600	300÷1600
Light scattering	Very weak	Weak
Thermo-optical Figure of Merit*, W/m	~ 10^{3}	~ 10^{5}
Laser-induced Damage Resistance**, 1×1, J/cm^2	~ 60	~ 70
Laser-induced Damage Resistance**, 1×200, J/cm^2	~ 5	~ 30
Micro-hardness, N/mm^2	~ 100	~ 2000
Climatic resistance	Low	High

* Thermo-optical Figure of Merit $\mu = \chi(dn/dT)^{-1}$, χ : thermal conductivity, n: refractive index, T : temperature.

** At pulsewidth 10 ns, wavelength 1064 nm. 1×1: single-shot regime, 1×200: multi-shot regime

Table 1.

The data, presented in Table 1, confirm the statement in the introduction above that the PFNPG composite has significantly better properties than those of the bulk polymer.

3. Technology of PFNPG-composite

3.1 Structure of NPG

Studies of porous glasses have been started in 1930th [2-4]. Technology of their fabrication is based on a liquation phenomenon consisting of the following.

Some of the multi-component glasses (for example, $SiO_2 \times B_2O_3 \times Na_2O \times R_2O_3$) at a thermal treatment (typically, at 550-600° C) are segregated into chemically different phases: a resistant silicate SiO_2 - enriched phase and a non-resistant sodium-borate phase containing, mostly, Na_2O and B_2O_3. It is important that the two phase segregation may be different depending on a relative content of the components: one phase can be distributed in the other phase (in a droplet form) constituting a "colloidal solution" or both phases can form a two interpenetrating set ("two-cage" structure).

The nanoporous glass is fabricated from compositions which form, in the segregation process, two interpenetrating nets. The example of such the composition is 59.0 $SiO_2 \times 33.2B_2O_3 \times 4.0Na_2O \times 3.8\ K_2O$. For obtaining the nanoporous glass the chemically unstable sodium-borate phase is removed by acid pickling. Porous glasses obtained by this technology content, typically, 93÷99% SiO_2, 2÷5% B_2O_3, 0.05÷0.5% NO_2 and K_2O.

This technology includes the thermal treatment, acid treatment, water washing, and drying. Porosity of the NPG defined as $P = V_p/V_g$, the ratio of total pores volume V_p to a glass volume V_g, and the porous size distribution are important characteristics for the PFNPG composite material intended for optical applications. The value of P depends on the initial content of the glass and the technology of its treatment. Typically, P=45%. Measurement of the pore size-distribution is based on investigation of an isothermal water vapor absorption in NPG (water porometery method [27]). Results of such the measurements show that the pore size distribution in NPG samples fabricated by the technology, described above, is narrow, single-modal. The typical example is shown in Fig 1.

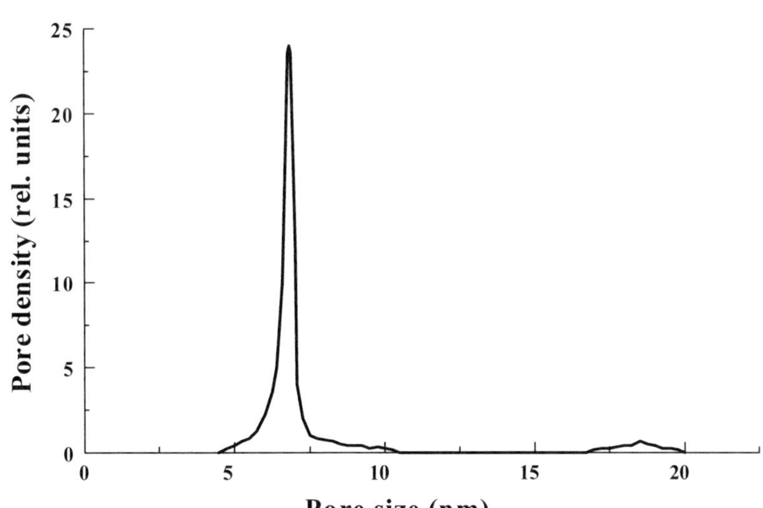

Fig. 1. Pore size distribution in NPG obtained with the water porometery method. The initial glass composition is $59{,}0SiO_2 \times 33{,}2B_2O_3 \times 4{,}0Na_2O \times 3{,}8K_2O$. (After Ref. [12])

Varying a content of the initial glass and a thermotreatment regime one can get the nanoporous glasses with pore sizes in the range of 1÷2000 nm. However, for optical applications the pore sizes not exceeding 20 nm are of the most interest.
Note that the water porometery method can not give information on a structure of the NPG. The more detailed information is obtained with electron microscopy. Application of this technique has shown that pores in NPG contain some amount of small silicate particles [28]. This type of particles is seen in Fig. 1 (weak peak at 18 nm).

3.2 Content and making technology of PFNPG

The polymer component of the PFNPG composite used in studies described in this chapter, consists of the polymethylmethacrelate (PMMA) modified with low molecular additives and activated with deferent type functional organic compounds [26]. A nature of the latter is chosen depending on a concrete function of the optical element (lasing, Q-switching, etc).
A procedure of making the composite consists of the following stages.
First, a monomer composition is prepared. This stage includes distillation, adding a low molecular modificator, a polymerization initiator, and the functional organic compounds.
Second, a NPG element is inserted into the monomer composition. At this stage NPG pores are quickly filled with the monomer composition due to the capillary effect.
Third, a free-radical polymerization is conducted at 30÷100°C.
The PFNPG samples, obtained in these three stages, are annealed and polished to get an optical element.
In a practical realization of PFNPG procedure, described above, some requirements have to be satisfied. Among these requirements the most important ones are the following.
A preparatory annealing of the NPG samples has to be done to remove molecules adsorbed on the pore surface. The annealing process has to be carried out in well-controlled conditions to avoid collapsing the pore. At the polymerization stage temperature regime has to be also well-controlled to avoid thermally induced stresses leading to peeling the polymer from the pore surfaces. Analyzing the polymerization process in nanoscale pores we have to point out that a structure of the polymer synthesized in the pores can be rather different from that of a bulk polymer synthesized at "open" conditions. In particular, crystallites, globules and other structural elements forming in such "the open" conditions can not be formed at the polymerization in nanopores.

4. Properties of PFNPG

4.1 Samples for investigation

All the data of the investigation results presented below have been obtained on the PFNPG samples made by the technology described briefly in Section 3 above. The NPG, the cage of PFNPG composite, has been made from the glass of the initial content $59.0SiO_2{\times}33.2B_2O_3{\times}4.0Na_2O{\times}3.8K_2O$. Porosity of the NPG samples was $P \cong 43\%$, the pore size distribution was single modal with the average pore size $d \cong 7$ nm (see Fig.1). The polymer component of the PFNPG was the modified polymethylmethacrelate (MPMMA). The optical transparency range of the PFNPG samples, limited by the transparency range of the polymer component, was 300÷1600 nm [29] (transparency range of the silica cage component is 200÷2700 nm [30]).
A microscopic testing showed high optical quality of the PFNPG samples: no visual defects were observed.

4.2 Mechanical properties and surface structure of optical elements

Microhardness of the PFNPG (and MPMMA for comparison) was measured using the method described in [31]. The samples were loaded at 1 N during 50 sec. Results of these measurements: the microhardness of the PFNPG composite was 1500÷2000 N/mm^2, whereas the microhardness of the MPMMA was in factor of 15÷20 less. For comparison: the microhardness of the sodium-borate glass and the quartz glass are 5000 N/mm^2 and 7000 N/mm^2 respectively [31]. It has been found that the microhardness of the PFNPG does not depend, in contrast to MPMMA, on contents of the polymer component of the composite, and, hence, is determined by the microhardness of the nanoporous glass.

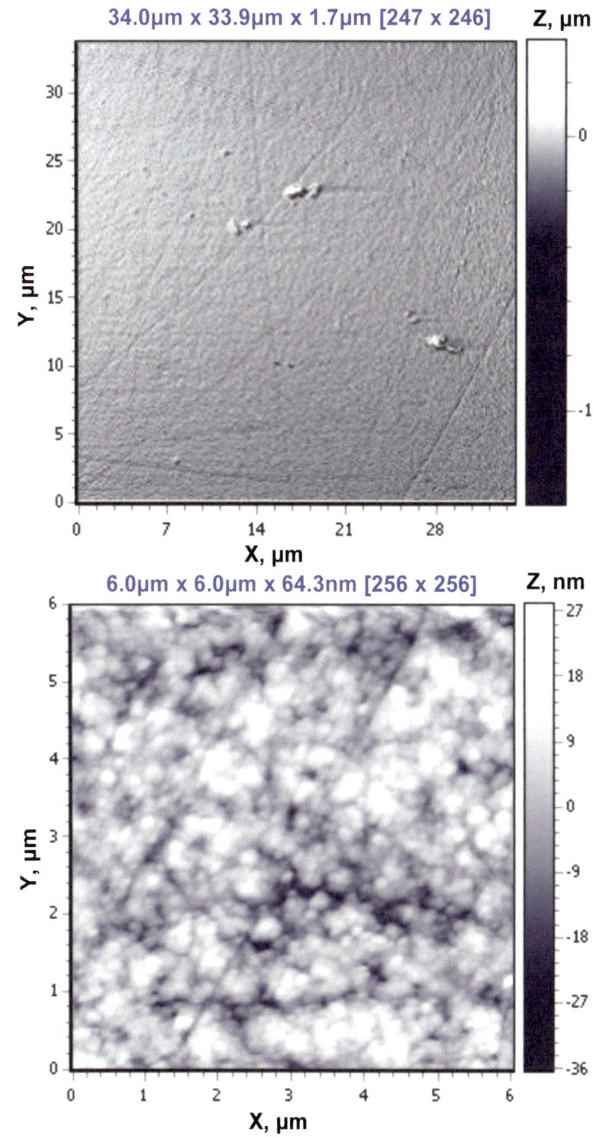

Fig. 2. Surface morphology of the PFNPG sample recorded on AF-microscope with the scan apertures 35×35 µm (top) and 6.5×6.5 µm (bottom). (After Ref. [20].)

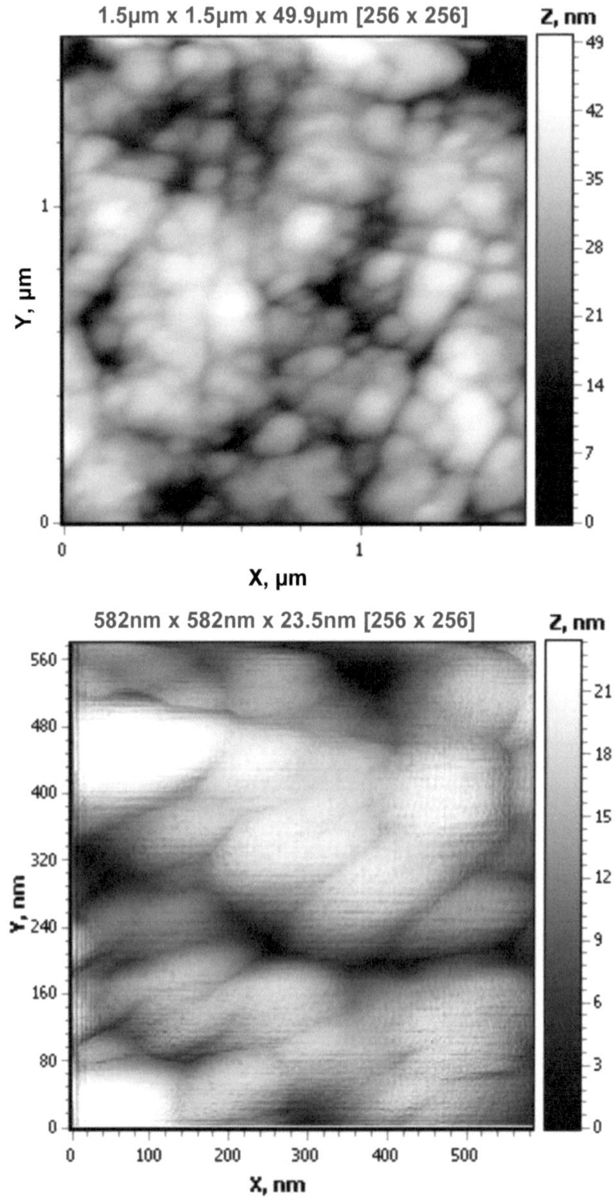

Fig. 3. Surface morphology of the PFNPG sample recorded on AF-microscope with the scan aperture 1.3×1.3 μm (top) and 0.3×0.3 μm (bottom). (After Ref. [20].)

Results of data processing of such the patterns characterized by parameters *Ra* - mean deviation from "ideal" surface, and *Sm* - mean step of roughness, for two PFNPG samples are presented in Table 2.

Three types of the structure defects are seen Fig.2 and Fig.3 on the PFNPG surfaces when the scan aperture varies: at 35×35 μm aperture - chain roughness and scratches on the smooth surface, and at 6.5×6.5 μm aperture - nanostructure of the surface.

Scan aperture, μm	Sample 1		Sample 2	
	Ra, nm	Sm, nm	Ra, nm	Sm, nm
35 × 35	10.5	1642	8.8	1004
6,5 × 6,5	5.9	467	5.7	480
1,3 × 1,3	4.0	345	4.9	411
0,3 × 0,3	1.9	63	2.5	139

Table 2.

A nature of these surface defects and their connection with polishing technology is not clear yet, and further studies are required to clarify this problem.

4.3 Light scattering

The PFNPG composite is a heterogeneous medium. The refractive index difference of the quartz cage and the polymer component may be as high as $\Delta n \cong (2 \div 3) \times 10^{-2}$ (this estimate is based on the assumption that the refractive index of the polymer component in the NPG pores is the same as in the bulk polymer). Such the difference may lead to the significant light scattering losses. For estimation of these losses let us consider the following simple model of the PFNPG composite.

We will consider the two-component composite as "colloidal solution" consisting of spherical polymer inclusions, corresponding to pore sizes, homogeneously distributed in the quartz matrix. A single inclusion light scattering cross-section at $r << \lambda$ (r is pore radius, λ is radiation wavelength) may be obtained from the Raleigh approximation [32]

$$\sigma = \frac{8}{3}\pi k^4 \left| \frac{n_1^2 - n_2^2}{n_1^2 + 2n_2^2} \right| V \tag{1}$$

where k is the wave number, V is the pore volume, n_1 and n_2 are refractive indexes of the quartz matrix and the polymer, respectively. At $n_1 \sim n_2 >> \Delta n$ the relationship (1) is simplified:

$$\sigma = b \times 10^5 \frac{r^6}{\lambda^4} \left(\frac{\Delta n}{n} \right)^2 \tag{2}$$

where n is the refractive index of the quartz, b is a numerical coefficient of the order of 1. Light scattering-produced losses are characterized by the extinction coefficient $h = N\sigma$, where N is a concentration of scattering particles. For example, at porosity $P = 40\%$, $r = 3.5$ nm, $\Delta n = 3\times10^{-2}$, $n = 1.5$ and $\lambda = 500$ nm we get $h \cong 10^{-2} \div 10^{-3}$ cm-1. This rather high extinction doesn't agree with the experimental data for PFNPG samples we studied (much lower losses have been observed). This disagreement indicates that an assumption of independent scattering by each particle is not fulfilled in the case of the real PFNPG composite samples. Indeed, that assumption is correct only for low concentration of particles separated by a distance, l, lager than radiation wavelength, λ, whereas in the PFNPG the concentration of pores is very high: their size, r, and interpores distance, l, are

$\cong 7$ nm (i.e. $<< \lambda \cong$ 500 nm). At the conditions $l << \lambda$ particles (pores) scatter the light coherently (due to interference). The number of these "coherent particle" in λ^3 volume is

$$\frac{4}{3}\pi r^3 N \cong 0.4 \Rightarrow N_{coher} \sim \lambda^3 N \sim 0.1\left(\frac{\lambda}{r}\right)^3 \sim 10^5 \qquad (3)$$

Such the high-concentration particle medium can be characterized by an effective refractive index $\bar{n}$ and, according to Ewald-Oseen theorem [33], will produce a low scattering only at small angles in the propagation direction.
Investigation of the small-angle light scattering in the PFNPG composite and the MPMMA bulk polymer samples have been carried out using a setup shown in Fig. 4.

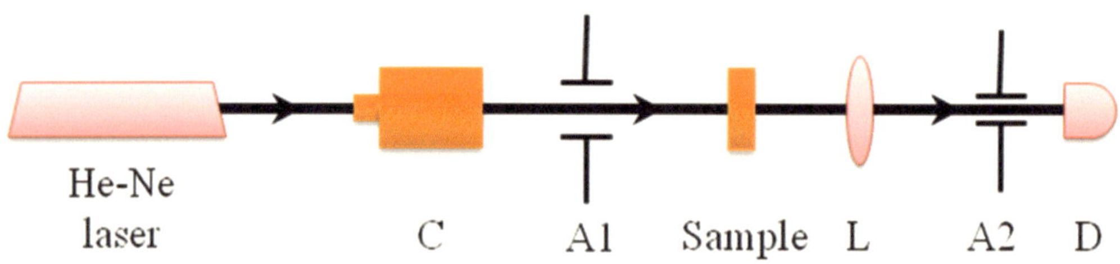

Fig. 4. Schematic of the experiment for investigating small-angle light scattering. **C:** collimator, **A1:** input diaphragm, **L:** collecting lens, **A2:** limiting diaphragm, **D:** photodetector. (After Ref. [12].)

The 633-nm He-Ne laser was used as a radiation source. The laser beam was expanded with a collimator (C) and was limited by an output diaphragm (A1) of the diameter 1 cm. The collimated laser beam was transmitted through a sample, focused on a diaphragm (A2) of a small diameter ϕ, and detected with the photodetector (D). The dependence of the transmitted radiation intensity on ϕ obtained for a number of PFNPG samples and a reference MPMMA sample are shown in Fig. 5.
As is seen in Fig.5 the transmittance of MPMMA does not depend on the diaphragm aperture. This means that scattering losses in this material is negligible (a deviation of the transmittance from unity is due to Fresnel refraction from the sample surfaces). The aperture size dependence of the transmittance observed for PFNPG samples indicates that small-scale inhomogeneity losses are also very low (except for a sample fabricated by "non-perfect" technology) , but large -size inhomogeneities responsible for a transmittance decrease at ϕ <150 µm rather significantly reduce the transparence of the samples. A size, $\bar{D}$, of these inhomogeneities can be estimated from the elementary diffraction theory:

$$\bar{D} \sim f\frac{\lambda}{d} \sim 4\times 10^{-3} f \qquad (4)$$

where f is a focal length of the lens, d is the diaphragm aperture size at which the transmittance decrease is observed. Taking concrete parameters of the setup and experimental data, the values of $\bar{D}$ in PFNPG samples studied have been estimated as $\bar{D} = 1 \div 3$ mm.

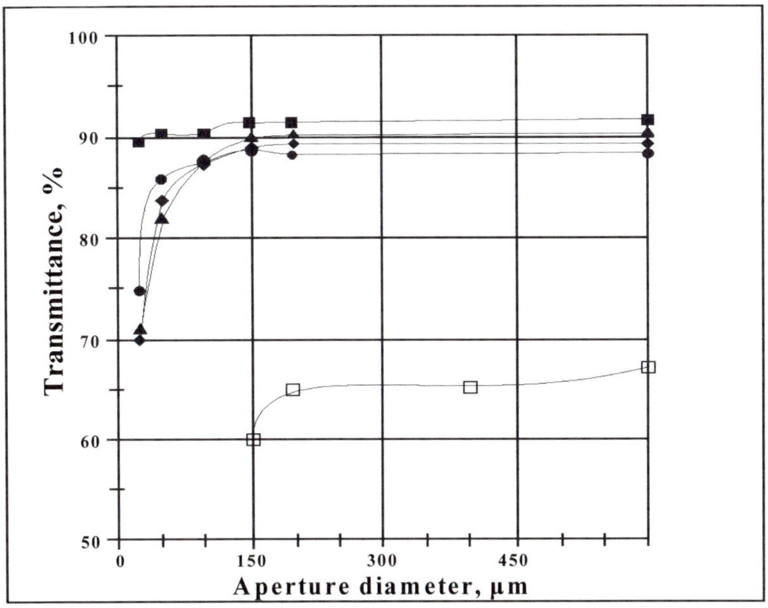

Fig. 5. Transmittance of the samples versus limiting diaphragm aperture
■ is MPMMA doped with PM-597 dye; ◆ is PFNPG doped with PM-580 dye; ● is PFNPG doped with PM-597 dye; ▲ is PFNPG doped with 11B dye; □ is PFNPG sample fabricated by with "non-perfect" technology. (After Ref. [12].)

An analysis of these data has shown that scattering in the PFNPG samples studied is due to large-scale inhomogeneities associated with the sample making technology. This is especially seen for the PFNPG sample fabricated by "non-perfect" technology.

Detail studies of optical inhomogeneity of PFNPG samples were performed using interferometric holography [34]. Schematic of experimental setup is shown in Fig. 6

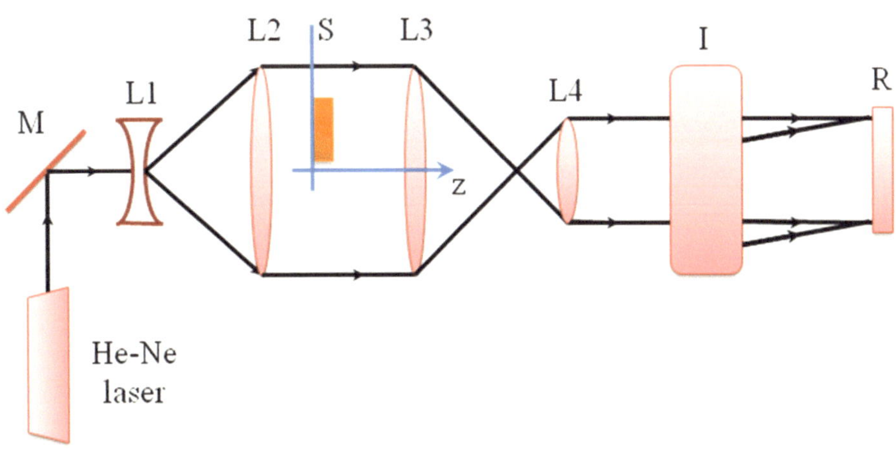

Fig. 6. Schematic of experimental setup for holography recording: **He-Ne** laser, **M:** mirror; **L1** and **L2:** telescopic system for beam forming, **S:** sample under study, **L3** and **L4:** input beam system, **I:** interferometer, **R:** hologram recording plate

Schematic of a hologram restoration is shown in Fig. 7.

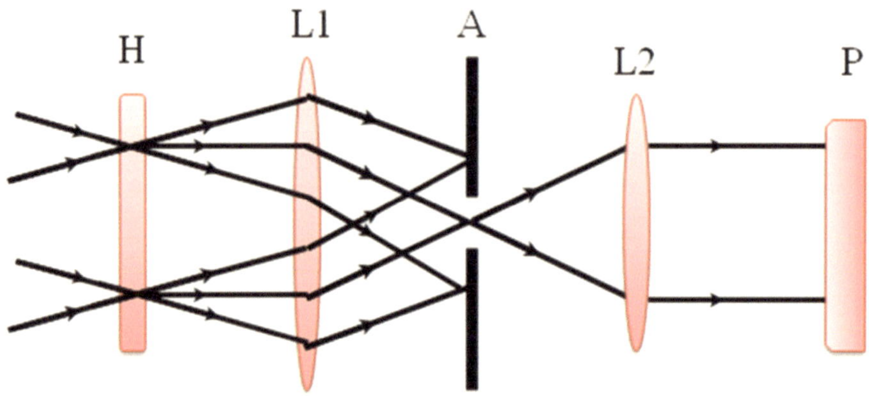

Fig. 7. Schematic of hologram restoration: **H:** hologram, **L1** and **L2:** focusing lenses, **A:** aperture, **P:** registration plate

Typical interference patterns of PFMPG sample are shown in Fig. 8.

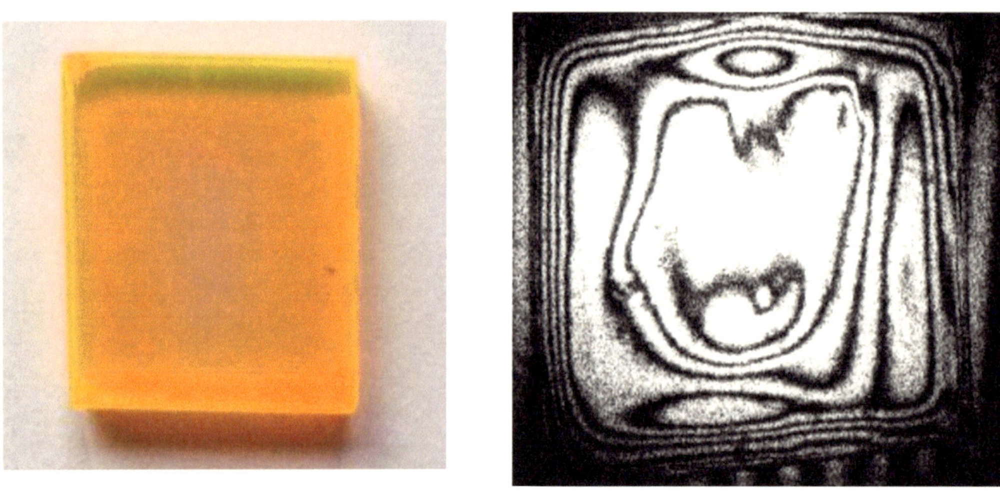

Fig. 8. PFNPG sample (left) and its interference pattern (right). (After Ref. [20].)

Result of these interferometic studies has shown that PFNG samples have high optical quality: the optical length difference in a central part of the sample doesn't exceed $0,02\lambda$ that corresponds to the refractive index variation $\delta n \leq 10^{-6}$. A rather large variation of the refractive index observed only in the periphery of the sample is due to structural changes associated with diffusion processes in silica.

4.4 Laser-induced damage resistance of PFNPG

Laser-induced damage of optical materials is important phenomenon limiting the intensity of the laser radiation propagating in optical components of high-power laser systems. Elucidation of the fundamental mechanisms of laser-induced damage in optical materials of different classes was a subject of many studies during more than 40 years since its first observation [35]. These studies resulted in establishing major features of the phenomenon (dependence of damage thresholds on radiation frequency, pulsewidth, and temperature, statistical behaviour at single-shot and multi-shot irradiation regimes, etc), a development of theoretical models of damage mechanisms and their experimental verification. This allowed working out the effective methods of a significant increase of the damage resistance of many type optical materials including glasses, crystals, and polymers (for a review of these achievements see [36-38] and reference therein).

In particular, it has been reliably established that in a majority of optical materials the laser-induced damage is initiated by absorbing inclusions (it's so called *extrinsic* mechanism), whereas *intrinsic* damage mechanisms (associated with the impact and multi-photon ionization) are realized only in very pure (inclusion-free) materials.

Many experimental facts confirm a dominating role of absorbing inclusions in laser-induced damage (LID) of optical materials. The most pronounced facts are:

- a variation of the damage threshold in samples of the same material obtained by identical technology,
- an increase of the damage threshold with purification of the materials,
- a variation of the damage threshold in different parts of the sample,
- a dependence of the damage threshold on the irradiation spot size.

The theory of the extrinsic laser-induced damage has been well developed and confirmed experimentally [36-40]. In particular, an accumulation effect, reduction of the laser-induced damage threshold at the multi-shot irradiation regime, the most pronounced in polymer material [26], is rather well understood [41-43].

The laser-induced damage in PFNPG has been investigated in both the single-shot and the multi-shot regimes at two laser wavelengths, 1063 nm and 532 nm (Nd:YAG laser and its second harmonic) and pulsewidth 15 ns and 26 ns, respectively. A spatial variation of the damage threshold has been observed indicating an extrinsic nature of damage. The average values of the damage threshold were 70 J/cm^2 and 35 J/cm^2 at 1036 nm and 532 nm respectively [9,12,17,19,24,25]. An influence of technological factors, the purification of the monomer and the impregnation of low molecular additives, on the damage resistance of the PFNPG and MPMMA samples have been also investigated. Results of these studies are shown in Table 3 [9,12,19].

Samples	Monomer purification	Low molecular additive	Single-shot (1×1) threshold, J/cm^2	Multi-shot (1×200) threshold, J/cm^2
MPMMA	-	-	17	1.6-1.9
MPMMA	+	-	63	2.2-2.6
MPMMA	-	+	15	5.4-6.3
PFNPG	-	-	58	31-38
PFNPG	+	-	76	35-43
PFNPG	-	+	56	25-28

Table 3.

The data presented in this table indicate the important features of the laser-induced damage. The damage thresholds of PFNPG in both the single-shot and the multi-shot regimes are higher of those for MPMMA. The most significant difference is observed at the multi-shot irradiation regime that has a principle advantage for practical applications. It's important also that both the purification of the monomer and the impregnation of the low molecular additives don't affect significantly the damage threshold of PFNPG, in contrast to the bulk PMMA.

These results may be simply explained by the assumption that the laser-induced damage in both materials (PFNPG and bulk MPMMA) has the extrinsic nature, i.e. associated with a presence of absorbing inclusions (this assumption is confirmed, as it had been pointed out above, by the fact of a spatial variation of the damage thresholds). Small-size pores (~ 7nm) in the NPG act as an effective filter for the monomer impregnated to the PFNPG at fabrication procedures. This self-filtration effect reduces the influence of any absorbing inclusions and increases the laser-induced damage resistance.

4.5 Activation of PFNPG with functional organic compounds

Impregnation of functional organic compounds (FOC) into PFNPG to get desirable laser optics elements (lasing, Q-switching, mode-locking, etc) requires special technological studies. The most important requirements to the FOC impregnation technology are as follows. Impregnation procedures have to be consistent with the PFNPG technology. FOC has to be homogeneously distributed in the composite, and does not interact with the composite matrix (NPG cage) to avoid undesirable changes of the properties.

In accordance with the PFNPG technology FOC is dissolved in a monomer composition. However, since NPG cage is an effective adsorbent, due to a very developed pore surface structure, the FOC can strongly interacts with the pore surface and changes the properties. An efficiency of this possible adsorption processes can be, naturally, very different for different types of FOC. Some results of the studies of this problem are presented below. These studies have been carried out with FOC of Pyromethene (PM 567, PM 580, PM 597, PM 650), Phenolemine (Ph 510, Ph 512, Ph 640), and some other series [9,10,13,15,16]. To investigate the homogeneity of FOC distribution in the PFNPG volume, transmittance of He-Ne laser radiation at 633 nm has been studied [12]. The 2 mm diameter laser beam was scanned over a sample surface. Experimental data for the PFNPG sample activated with PM 580 are presented in Fig 9. Similar results have been obtained for other FOCs: PM 597, PM 567, PM 650, Ph 512.

As is seen in Fig.9 the FOC is distributed in the sample rather homogeneously. Location of the FOC in the PFNPG was found to be strongly dependent of its chemical nature: experimental studies of the FOC impregnation process showed that Rodamine 11B strongly adsorbed on pore surface, whereas the FOC of Pyrromethene series do not show the adsorption effect [15].

The efficiency of FOC impregnation to the composite may be characterized by the saturation coefficient $\xi = c_v / c_m$, where c_m and c_v are the FOC concentration in the initial monomer composition and in the PFNPG composite, respectively. A concentration dependence of this coefficient has been investigated by measuring an absorption spectrum intensity of FOC in the monomer solution. It has been found that for all FOC investigated Buger law $D(\lambda) = \ln(I_0/I) = \varepsilon(\lambda)cL$ is satisfied in a solubility range (here $D(\lambda)$- optical density, $\varepsilon(\lambda)$ and c – extinction and concentration of FOC, respectively, L is sample thickness, I_0 and I are intensities of incident and transmitted radiation, respectively). This means that in the whole solubility range of FOC associates in monomer are not formed. It's important that the absorption spectra of FOC in the PFNPG composite and in the monomer composition were identical. This indicates that the Buger low is satisfied also for FOC impregnated to the composite, and, hence the saturation coefficient can be determined from the relationship $S = D_c L_m / D_m L_c$ where D_c and D_m are optical densities of the composite sample and FOC solution cell, respectively.

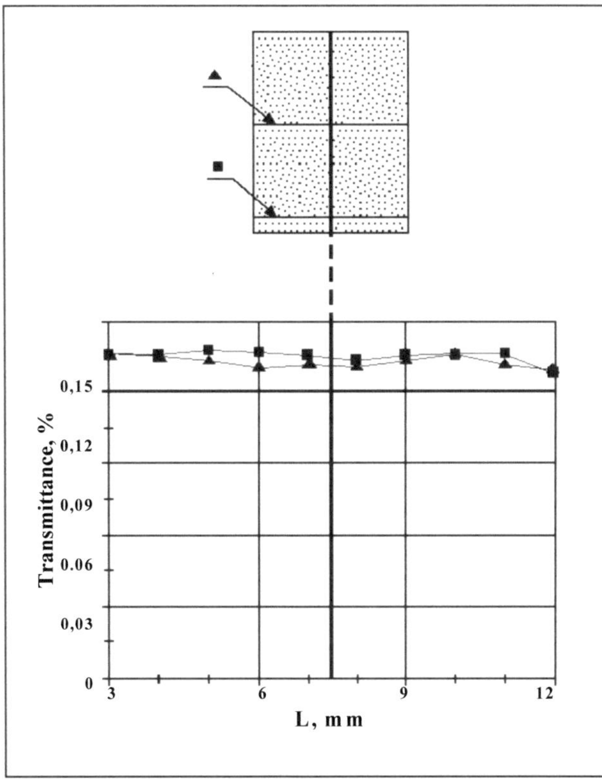

Fig. 9. Transmittance at 543 nm of PM 580 dye-activated PFNPG composite at various sites of the sample. Concentration of the dye in the monomer composition is 2.4 mmol/l. (After Ref. [12].)

Detailed investigations of the saturation process have been carried out for two classes of FOC: Pyromethenes (PM 567, PM 580, PM 597 and 560) ant Phenolemines (Ph 10, Ph 512 and Ph 640). It has been found that for all Pyrromemthene series FOC studied the values of ε are almost the same: $\xi = 0.6 \div 0.7$, and do not depend on a stand time of NPG in the monomer composition. In the contrast, for FOC of Phenolemine series the values of ξ exceed unity (ξ = 1÷1.2 for Ph 512, 1.7÷3 for Ph 510, and 7÷15 for Ph 640) and increase with the stand time of NPG in the monomer composition. Such the behaviour of the saturation coefficient is due to, perhaps, the difference of the dye polarity. In particular, Ph 640 has the ionic form and, taking into account physical-chemical properties of NPG pores [28], it is well adsorbed on the pores surfaces. Therefore, one may conclude that low polarity FOC (PM 567, PM 580, PM 597, PM 650) do not incline to adsorption on the pore surface, that correlates with rather low saturation coefficients, whereas the high polarity FOC, especially having the ionic form, are well adsorbed on the pore surface.

4.6 Optical distortions

A dependence of the refractive index of optical materials on different external factors is the important characteristic for many laser applications. Two dependences, on temperature and intensity, are most significant in high-power laser optics. They can induce significant undesirable distortions: divergence, self-focusing or self-defocusing of optical beams.

Thermo-optical effects associated with the temperature dependence of the refractive index (dn/dT) can be characterized by the thermo-optical figure of merit $\mu = \chi(dn/dT)^{-1}$. A thermo-optical lensing induced by the inhomogeneous laser-produced heating is determined by this parameter: a higher value of μ corresponds to lower optical distortions.

Thermo-optical distortions in the PFNPG composite have been investigated by measuring the focal length of the thermal lens induced by a propagating laser beam [12]. For comparison similar measurements have been done also for the bulk MPMMA samples. Schematic of the experiment is shown in Fig 10.

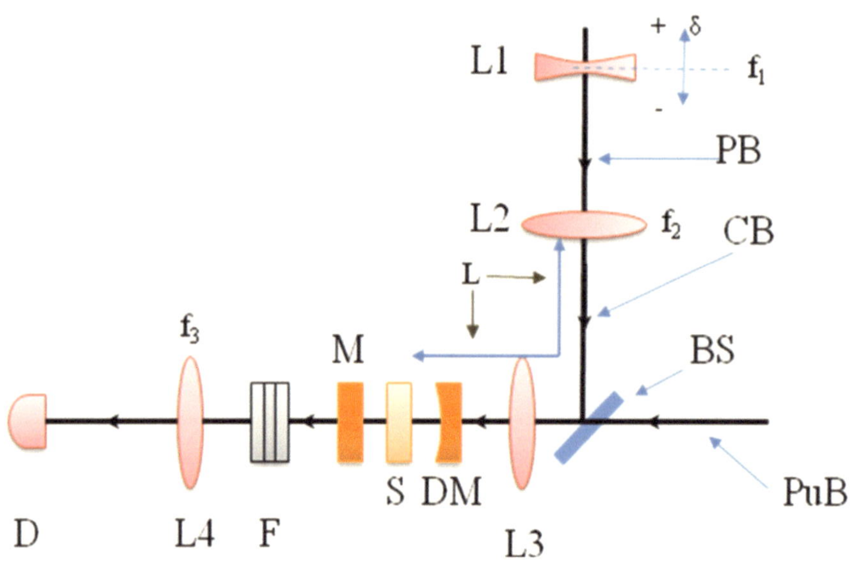

Fig. 10. Schematic of experimental setup for laser-induced thermal lens studies: **D:** photodetector, **L4:** lens, f_3=1000 mm, **F:** light filter, **M:** output cavity mirror, **S:** sample under study, **DM:** dichronic mirror, **L3:** large-aperture lens, **PuB:** pump beam (λ=532 nm), **BS:** beam splitter, **CB:** collimated beam, **L2:** lens f_2=300 mm, **PB:** He-Ne laser probe beam, **L1:** lens f_1=-25 mm. (After Ref. [12].)

The probe beam from a He-Ne laser (PB) at 633 nm was expanded with a collimator consisting of a diverging lens with the focal length $f_1 = -25$ mm (L1) and a converging lens with the focal length $f_2 = 300$ mm (L2). The beam cross section was limited by the 5 mm diaphragm. The pump beam (PuB) at 532 nm passed through the same diaphragm. Both beams were incident on a laser element (S) under study. The pump beam transmitted

through the laser element was rejected with a filter (F), and the probe beam was focused with a lens (L4) of the focal length $f_3 = 1000$ mm on a photo-detector (D). The lens (L1) with the focal length f_1 was displaced to obtain the maximum signal from the detector in the absence of the pump.
Upon pumping the output signal decreased due to defocusing of radiation by a thermally induced lens. Displacing the lens L1 by some distance δ one can get the signal increased to maximum. The value f_T is related to δ by the simple expression from geometrical optics:

$$f_T = (L - f_2) - f_2^2/\delta \tag{5}$$

where L is a distance from the L2 lens to the laser element.
Thermo-optical effects were studied in the laser elements doped with the PM 597 dyes, which absorbed the radiation at 532 nm. The dependence of f_1 on the average absorbed pump power was measured for the PFNPG laser element and the reference MPMMA element. The study showed that thermo-optical effects in the PFNPG composite are much weaker than those in the MPMMA (see Fig. 11).

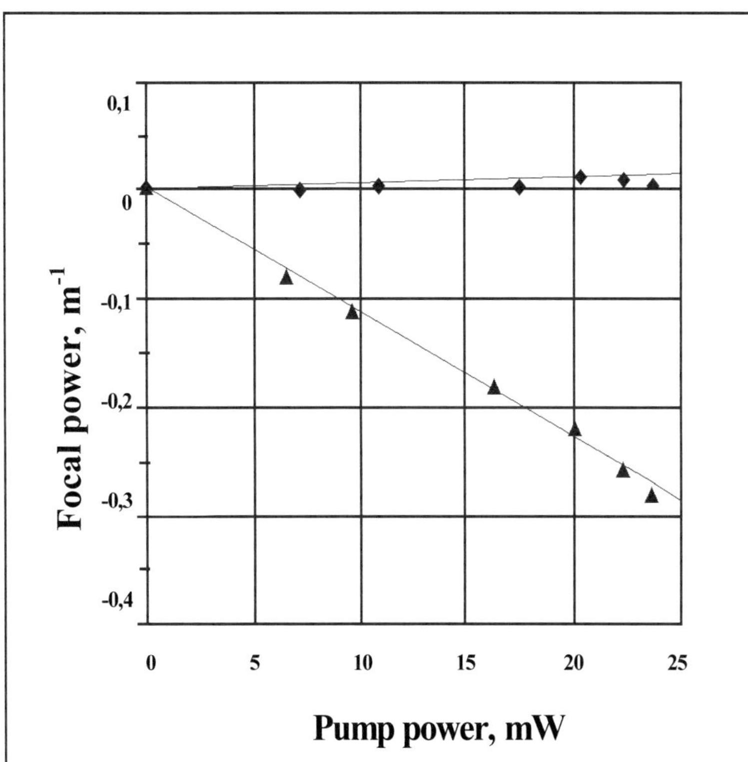

Fig. 11. Focal power dependence of the thermo-optical lens versus the average pump power for the PFNPG element (◆) and the reference MPMMA element. (▲). (After Ref. [12].)

The advantages of the composite over the bulk PMMA are caused by a number of factors. The main one is a considerably lower magnitude of dn/dT. Indeed, according to the Lorentz - Lorentz relation [33], the molar refraction index, R, is

$$R = \frac{M}{\rho}\frac{n^2 - 1}{n^2 + 2} \tag{6}$$

where M is the molar mass and ρ is the material density.
By differentiating (6) with respect to temperature and taking into account that M and R are independent of the temperature we obtain [12]:

$$\frac{dn}{dT} = -\alpha\frac{(n^2 - 1)(n^2 + 2)}{2n} \tag{7}$$

where α is the linear expansion coefficient of the material.
For PMMA $\alpha \cong 7 \times 10^{-5}$ K^{-1} [29]. Being impregnated into glass pores, PMMA cannot expand freely because of quartz cage limitations. This means that PMMA in the NPG pores expands with the temperature increasing in the same way as cage pores do. In this case, to estimate the value of dn/dT of PMMA impregnated into the NPG, one should use in (7) the linear expansion coefficient of fused silica, i.e. take the value $\alpha \cong 0.51 \times 10^{-6}$ K^{-1} [44]. Therefore the value of the dn/dT for the PFNPG composite is by factor of 100÷130 times lower than that for the bulk PMMA.
The second important factor, which reduces thermo-optical effects in the PFNPG composite as compared to PMMA, is the high thermal conductivity of the quartz cage of the composite. The thermal conductivity coefficient for quartz and PMMA are 1.3 and 0.16 kcal h^{-1} m^{-1} K^{-1}, respectively [30], so that their ration is approximately equals to eight. Taking into account that the glass porosity is 40% this ratio means that thermo-optical effects should be reduced by a factor of 4÷5.
Thus, the thermo-optical figure of merit of the PFNPG is a factor of 500÷600 times larger than that for the bulk PMMA.
A nonlinear refraction of PFNPG samples has been also investigated. A traditional Z-scan techniques [45] was used in the experiments. A setup and experimental details will be presented in the Section 5 below. Here we summarize only the results of these studies in comparison with data obtained for MPMMA. It has been found that PFNPG doesn't reveal an appreciable nonlinear refraction in the contrast to the bulk MPMMA that has a significant nonlinear susceptibility. Further studies are required to understand a reason for these results.

5. Laser applications of PFNPG elements

The PFNPG composite as a host material allows making laser optics elements of different types by varying a content of functional organic compounds impregnated. The Q-switchers [1, 8-10], lasing elements [11, 13-21], and laser radiation power limiters [22-25] have been made and successfully demonstrated to date. Other elements, such as the mode-lockers, can be also made by choosing a proper FOC. Below in this section some experimental results on lasing and power limiting properties of PFNPG are presented.

5.1 Absorption and luminescence spectra features of dye-impregnated PFNPG

PFNPG samples activated with Pyrromethene (PM-567, PM-597 and PM-560) and Phonolemine (Ph-510, Ph-512 and Ph-640) series dyes have been investigated. These dyes were chosen since they possess a high lasing efficiency in liquid solutions and widely used in experimental practice.

Dependence of absorption spectra of all these dyes has been investigated at different concentrations in the monomer and in the PFNPG composite. The concentration was varied in the wide range, from 10^{-5} mol/l to a solubility limit. Data on the solubility limit, C_s, characteristic luminescence quenching concentration, C_a, and "spectral shoulder" of a luminescence lines, C_{ex}, of some dyes in the monomer solutions are presented in Table 4.

Dye	PM567	PM580	PM597	PM650	Ph512	Ph510
C_s, mol/l	2×10^{-2}	2.2×10^{-2}	2×10^{-2}	1×10^{-2}	1×10^{-2}	0.5×10^{-2}
C_e, mol/l	1×10^{-2}	1×10^{-2}	1×10^{-2}	0.5×10^{-2}	2.5×10^{-3}	1×10^{-3}
C_{ex}, mol/l	1×10^{-3}	1×10^{-3}	1×10^{-3}	1×10^{-5}	1×10^{-3}	1×10^{-3}

Table 4.

Main results of the studies may be formulated as follows.

1. Absorption spectra of all dyes investigated in the monomer solutions remain unchanged in a whole concentration range, up to the solubility limit and Buger law is fulfilled. It means that dye associates are not formed. Typical absorption spectrum for PM-650 is shown in Fig. 12.

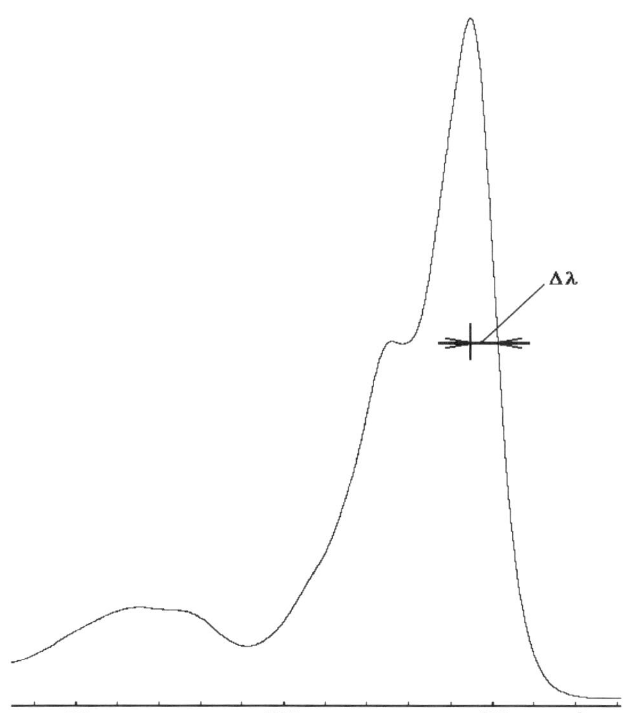

Fig. 12. Absorption spectrum of PM-650 in the monomer composition

Some spectral characteristics of dyes important for understanding their lasing properties are presented in Table 5.

Dye	λ_a, nm	λ_f, nm	$\Delta\lambda$, nm	$\varepsilon(\lambda_a)\times10^{-3}$	$\varepsilon(\lambda_p)\times10^{-3}$	$\varepsilon(\lambda_f)\times10^{-3}$
PM 567	518	536	10	96	28	14
PM 580	520	537	10	89	32	16
PM 597	524	561	16	80	64	3
PM 650	590	607	14	45	14	14
Ph510	524	593	46	22	21	1.5
PH512	533	578	24	25	25	2
Ph640	590	616	16	30	8.5	5.5

λ_a, λ_f, и λ_p are wavelengths of main pick of absorption and luminescence lines and pump respectively, $\varepsilon(\lambda)$ is an extinction, $\Delta\lambda=(\lambda_{1/2}-\lambda_a)$, $\lambda_{1/2}>\lambda_a$,, $\varepsilon(\lambda_{1/2})=(1/2)\,\varepsilon(\lambda_a)$.

Table 5.

2. Absorption spectra of the dyes impregnated to PFNPG composite are identical with those in the monomers mixture. This result is natural for Pyromethene dyes since the saturation coefficient (characterizing impregnation efficiency) $\xi<1$, so that the dyes located (as it has been discussed above in Sec. 4) in a volume of the polymer component. However, the identity of absorption spectra in PFNPG and in monomer mixture for dyes possessing high ($\xi>1$) saturation coefficients is not clear since this fact ($\xi>1$) indicates a strong interaction of dye molecules with the pore surface and has to lead to absorption spectrum change.
3. In the contrast to the absorption spectra features, just discussed above, luminescence spectra of dyes both in monomer solutions and impregnated to PFNPG, reveal a concentration dependence: at high concentrations, exceeding C_{ex}, a "spectral shoulder" appears at the long-wavelength side of the luminescence line. Such a behaviour of the luminescence spectra evidences, perhaps, of an eximer formation at high concentrations of the dyes.
4. At high dye concentrations (> 10^{-3} mol/l) in the monomer solution a luminescence quenching is observed (see, Fig 13).

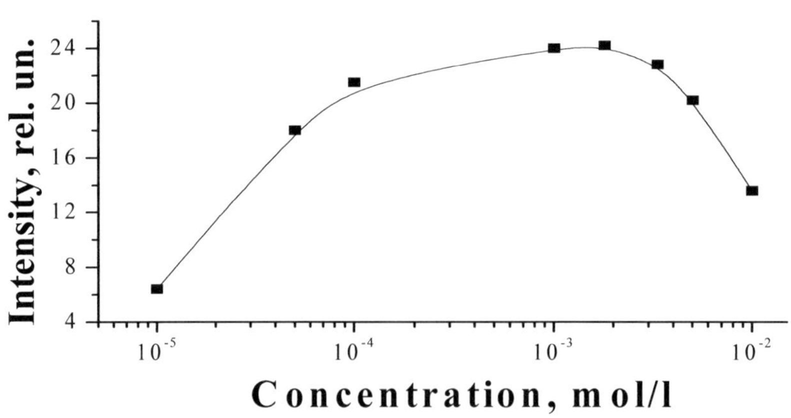

Fig. 13. Luminescence intensity versus dye concentration of Ph-512 in the monomer solution

5. An increase of the luminescence intensity is observed for the dyes impregnated to PFNPG as compared with the monomer solutions. This effect is due, perhaps, to a suppression of non-radioactive relaxation.

5.2 Lasing properties

Lasing characteristics of the dye-impregnated PFNPG composite have been investigated using experimental setup shown in Fig. 14.

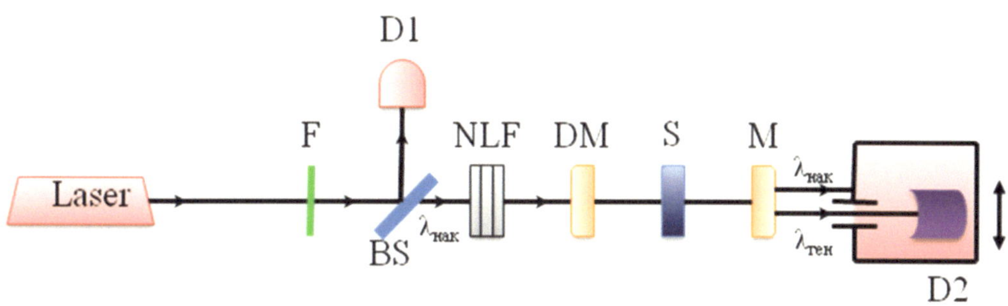

Fig. 14. Schematic of setup for investigation of lasing characteristics of dye-impregnated PFNPG elements. **F**: light filters **BS**: beam splitter, **NLF**: neutral light filters, **DM**: dichroic mirror, **S**: sample, **M**: mirror, **D1**: photodiode detector and **D2**: photodiode detectors or wavelength-meter

The pump source was a frequency-doubled Q-switched Nd: YAG laser. The longitudinal pumping scheme was used. A laser element (S) was placed into a cavity formed by a dichroic mirror (DM), transparent at the pump wavelength, λ_p =532 nm, and totally reflecting at the lasing wavelength, λ_g. The pump beam diameter at the laser element was 1.3 mm. The repetition rate, f, of the pump pulses was 3 Hz or 33 Hz., and the duration, τ_p, energy, E_p, and intensity, I_p, were 5 ns, 2mJ, and 30mW/cm^2 at $f = 3$ Hz, and 7ns, 0.7 mJ, and 7.5 MW/cm^2 at $f = 33$ Hz, respectively.

The pump-to-laser conversion efficiency, η, and the service life, $N_{0,7}$, defined as a number of pulses at which η drops to the 0.7 level of its maximum value at the first pump pulse, were measured in a broad range of the pump intensity variation. Also, the dependences of η and $N_{0,7}$ on the pump pulse reprate were investigated.

The conversion efficiency was determined as the ratio $\eta = \left(E_g / E_P\right)$ of the lasing pulse energy, E_g, to the pump pulse energy, E_P. The investigation of the dependence of the conversion efficiency on the optical density, D, of the laser elements has reviled its non-monotonous character. The typical shape of $\eta(D)$ observed for the PM597 dye-doped PFNPG element is shown in Fig.15. An existence of the optimal optical density, D_m, corresponding to the maximum value of the conversion efficiency, is explained as associated with the inhomogeneous pumping of the laser element. Indeed, at the longitudinal pumping condition, which took place in the experiments described above, the pump intensity gradually decreases from a front surface of the laser element to a back one if the optical density of the element is high enough. Therefore, at high densities, $D > D_m$, a region adjacent to the back surface remains "unpumped" and reabsorption processes at a laser transition reduce the lasing efficiency.

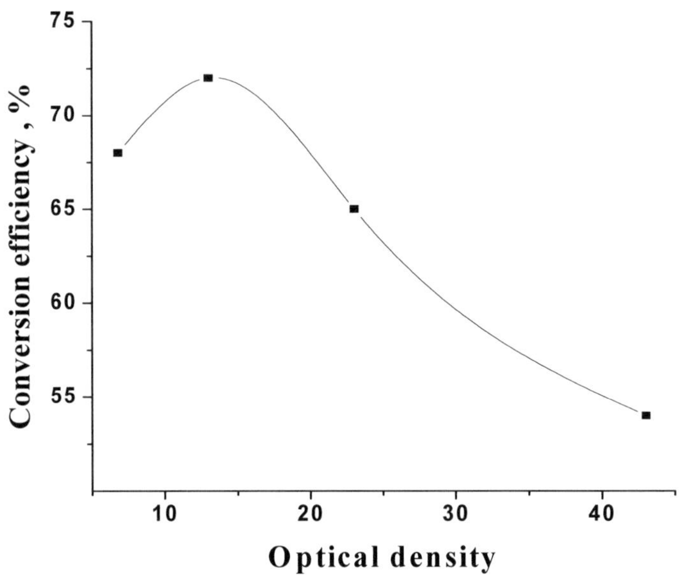

Fig. 15. Lasing conversion efficiency versus optical density of PM597 dye-doped PFNPG element pumped at λ_p =532 nm, I_p =30 MW/cm^2. (After Ref. [16].)

Experimental data for the conversion efficiency and the service life obtained for PFNPG elements doped with some Pyrromethene and Phenolemine families dyes are presented in Table 6.

Dyes	PM567	PM580	PM597	PM650	PH512	Ph510	Ph640
λ_g, nm	562	561	570	625	595	614	642
D_m	8.8	12	13	1.6	5.8	10	4.3
η_m (%)	55	55	72	33	42	30	35
$N_{0,7}\times10^3$	40	51	340	8	30	14	16

Table 6.

This data refer to elements with optimal optical densities ($D = D_m$) and have been obtained at the pump pulse reprate 33 Hz.

It has been observed that at the lower reprate, 3 Hz, the conversion efficiency for PFNPG elements remained unchanged and the service life increased in a factor of 2-3. The different behaviour has been observed for dyes dissolved in liquid solutions (monomer mixtures and ethanol): the conversion efficiency measured at the same pumping conditions, decreased at the higher (33 Hz) reprate. This result definitely demonstrates the advantage of the solid-state PFNPG laser elements over the liquid solution dye lasers.

5.3 Nonlinear absorption

Nonlinear absorption in the functional organic compounds was investigated since 80-th of the last century. These studies were motivated by needs to create the laser radiation power limiters for protection of receives (including eyes).

A majority of studies were carried out on liquid solutions of the FOC. However, for practical applications solid-state limiters are, naturally, more preferable. In this respect the PFNPG is a very suitable candidate for such power limiters. Taking into account this, the nonlinear absorption properties of some organic compounds impregnated to PFNPG have been recently studied [22-25].

Compounds of Phthalocyanine and Porphyrin families containing "heavy" atoms have been chosen for these studies since they showed the most suitable nonlinear properties in liquid solutions [46-49]. In particular, Zinc Porphyrin (PrZn), Zinc Phthalocyanine (PcZn) and Lead Phthalocyanine (PcPb) dyes have been chosen. The solubility limits of these compounds in the monomer solution are 5 mmol/l for PrZn and 1 mmol/l for PcZn and PrZn.

The absorption spectra of these compounds in monomer solutions and in PFNPG have been found identical in the whole concentration range up to solubility limits.

The samples of the bulk MPMMA impregnated with these dyes have been also studied for comparison.

Optical densities at 532 nm of the samples studied are presented in Table 7.

Sample	PFNPG PrZn	PFNPG PcZn	PFNPG PcPb	MPMMA PcZn	MPMMA PcPb
Optical density at 532 nm	0.58	0.24	0.38	0.5	0.8

Table 7.

Before presenting experimental data let us discuss briefly a possible mechanism of nonlinear absorption in the chosen compounds.

It is supposed that the nonlinear absorption in these compounds is due to radiation-induced triplet-triplet absorption [46-49]. An energy level diagram shown in Fig. 16 explains this mechanism.

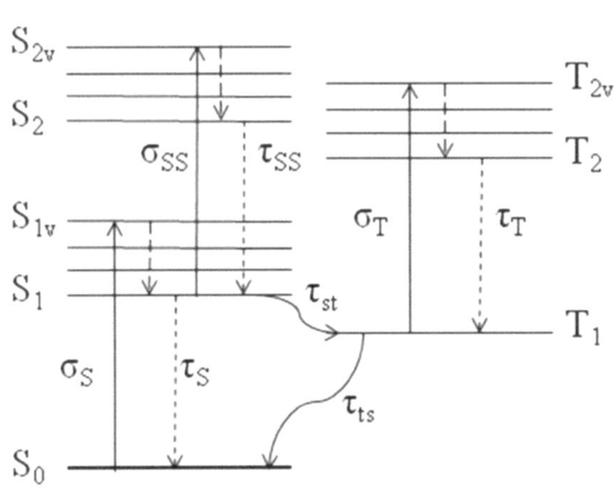

Fig. 16. Energy level diagram of organic dyes

A high power laser radiation at a resonant frequency corresponding to $S_0 \rightarrow S_{1v}$ transition, increases a population of the triplet state T_1 and induces the electromagnetic transitions between two triplet states $T_1 \rightarrow T_{2v}$. If the cross-section, σ_T, of the $T_1 \rightarrow T_{2v}$ transition exceeds the cross-section, σ_S, of the $S_0 \rightarrow S_{1v}$ transition, a transmittance of the sample at the laser frequency (transition $S_0 \rightarrow S_{1v}$) will be decreased, i.e. will lead to laser radiation power limiting. Besides this darkening mechanism, due to the triplet-triplet absorption, other mechanisms are possible (for example, due to $S_1 \rightarrow S_{2v}$ transition). However, in the organic dyes containing heavy atoms, the mechanism, just described above, is supposed to be dominant one.

Experimental studies of nonlinear absorption were carried out using Z-scan technique [24,25]. Schematic of the setup is shown in Fig. 17.

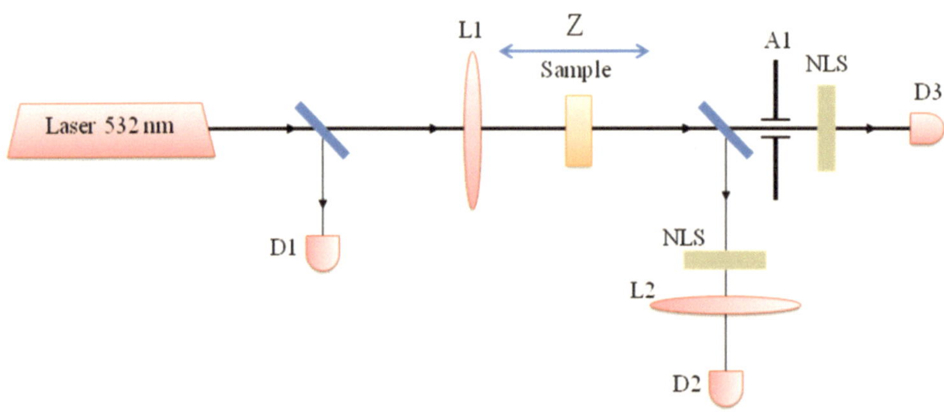

Fig. 17. Schematic of setup for investigation of nonlinear absorption: **A1:** diaphragm ∅3,5 mm, **L1:** lens (focal length 135 mm), **D1, D2** and **D3:** photodiode detectors, **NLF:** neutral light filters. (After Ref. [25].)

The second harmonic of the single mode, 40 ns - pulse duration Nd:YAG - laser has been used as a light radiation source. The radiation beam focused by 135 mm focal length lens, L1, passed through a sample under study, registered by the detector D2.

The nonlinear absorption in the sample was registered in "wide aperture configuration" (when whole beam was detected) when it scanned along the propagation direction. The nonlinear refraction was registered in the "narrow aperture configuration" (when 3.5. mm aperture diaphragm limit the transmitted beam diameter) by the detector D3.

The nonlinear absorption has been revealed in all samples studied. Typical transmission curves observed at the "wide" and "narrow" configurations in the PcPb sample are shown in Fig. 18.

As is seen, both the left and the right curves are symmetric relative to the focal position of the sample. This definitely indicates an absence of the nonlinear refraction in The PFNPG samples.

Different Z-scan plots have been detected in the bulk PMMA samples doped with the same dye: the symmetric shape plot in the "wide aperture" configuration and the asymmetric one - in the "narrow aperture" configuration (Fig. 19). This result indicates a presence of the nonlinear refraction in the PcZn-doped PMMA polymer.

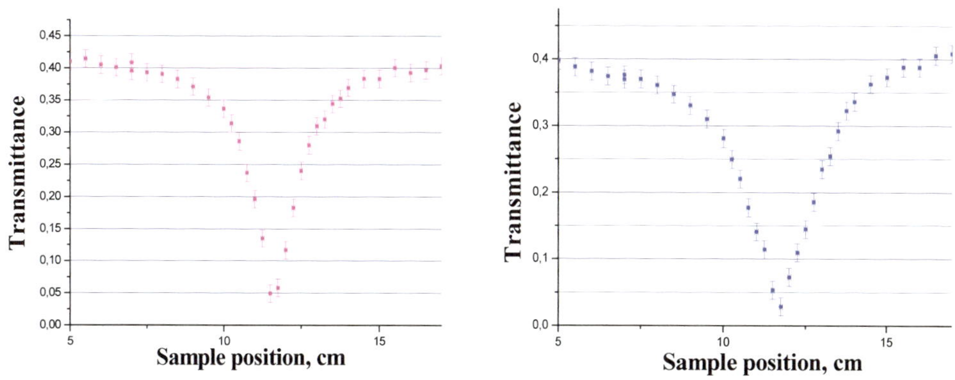

Fig. 18. Transmission versus Z-position of the sample made of PFNPG doped with PcPb, observed in wide aperture (left) and narrow aperture (right) configuration (After Ref. [25].)

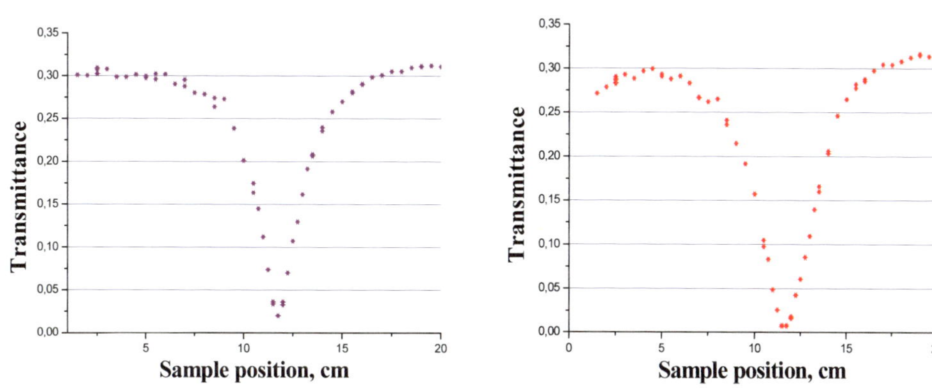

Fig. 19. Transmission versus Z-position of PcZn-doped PMMA sample, observed in "wide aperture" (left) and "narrow aperture" configurations. (After Ref. [25].)

To understand the difference of the nonlinear refraction properties of the PFNPG and the bulk PMMA polymer, the dye-free samples made from these materials, have been studied in "narrow aperture" Z-scan configuration. Z-scan curves observed in this case are shown in Fig. 20.

The important feature of these results is that the nonlinear refraction is definitely detected in the bulk PMMA polymer, where as it does not revealed in PFNPG composite at the same experimental conditions.

Note the other interesting results obtained in Z-scan experimental studies of dye-doped PFNPG: no laser-induced damage has been revealed at the laser radiation energy densities up to 4 J/cm^2, but destruction of the dyes has been observed at lower energy densities, beginning at 2 J/cm^2.

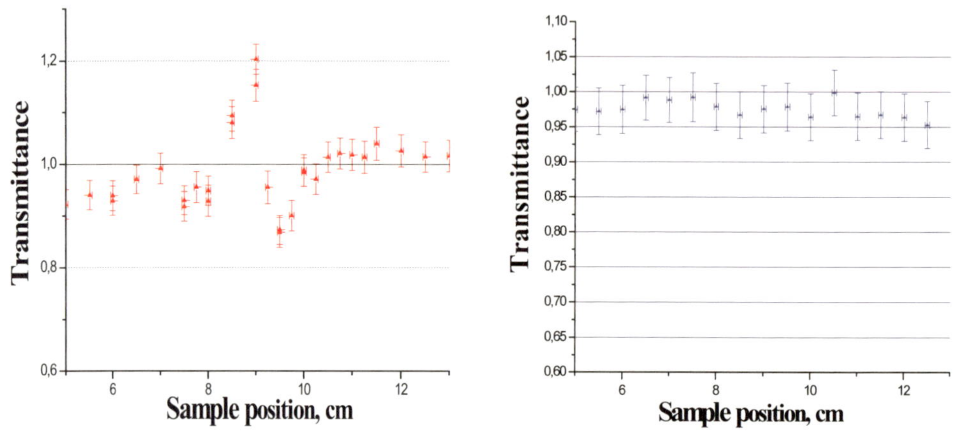

Fig. 20. Transmission versus Z-position of dye-free bulk PMMA polymer (left plot) and PFNPG composite (right plot) samples. "Narrow aperture" configuration. (After Ref. [25])

Characteristics of the nonlinear absorption in the dye-doped PFNPG have been studied using other experimental setup shown in Fig.21. The transmittance of the samples was measured at a wide range of the laser radiation intensity [22,23].

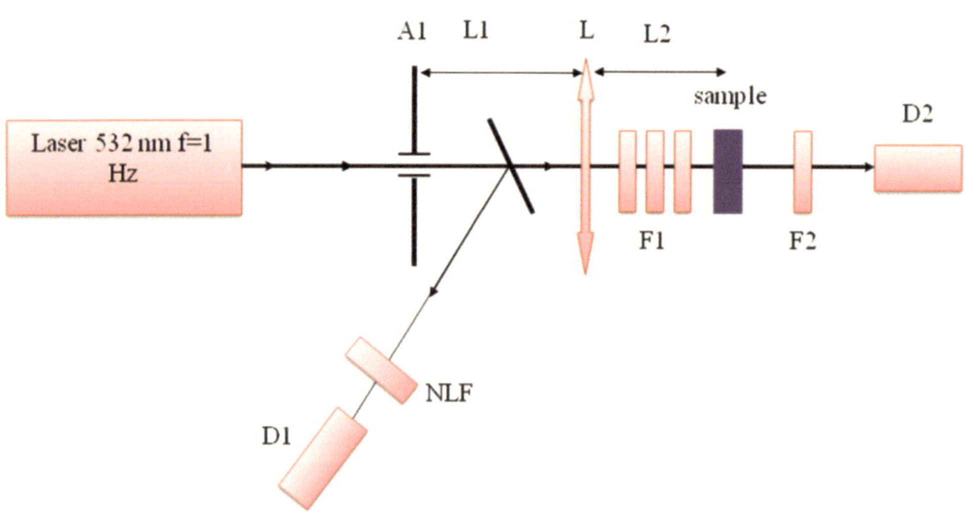

Fig. 21. Schematic of setup for investigating nonlinear absorption: **D1** and **D2:** photodetectors, **F1** and **F2:** neutral filters, **A1:** 2 mm aperture diaphragm, **L:** 8 mm focal length lens. (After Ref. [23].)

Variation of the energy density in the sample was made by a reciprocal replacing neutral filters F_1 and F_2. Experimental results for some dye-doped PFNPG samples are shown in Fig 22 and 23.

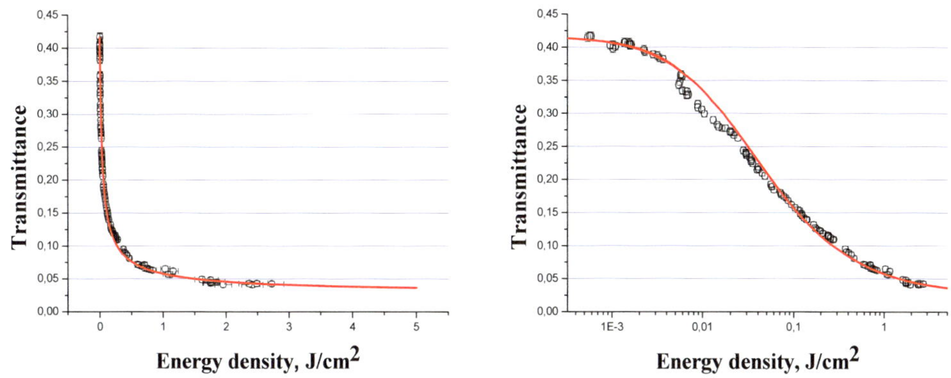

Fig. 22. Transmittance versus laser pulse energy density for PcPb - doped PFNPG composite: linear (left) and log (right) scales. (After Ref. [23].)

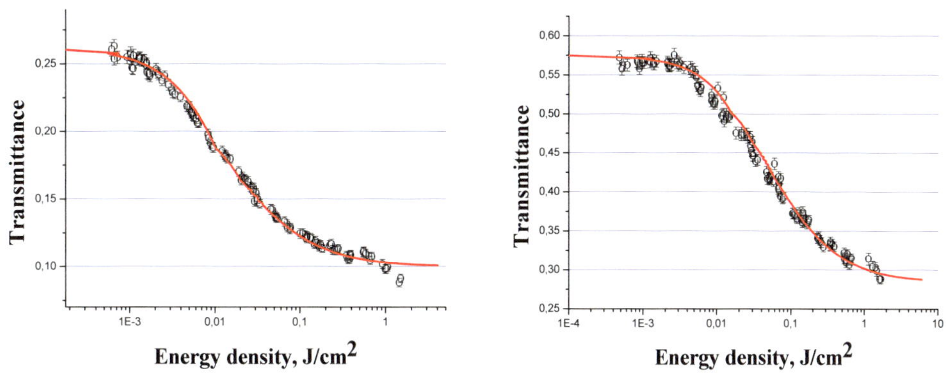

Fig. 23. Transmittance versus laser pulse energy density for PrZn-doped PFNPG (left) and PcZn-doped PFNPG (right) samples. Log scales. (After Ref. [23].)

The following parameters can characterize the transmittance $T(E)$ dependences: a saturation energy density, E_s, a contrast, C_T which is the ratio of the initial transmittance at low intensities to the transmittance at E_s, and a nonlinear absorption initiation energy $E_{th} = (T(0) + T(E_s))/2$ (parameter E_{th} is called somewhere, by mistake, "threshold energy density"). These parameters for some dye-doped PFNPG composite materials are presented in Table 8.

	PcPb	PcZn	PrZn
C_T	10	2.1	2.4
E_{th}, J/m^2	580	460	100
E_s, J/m^2	19520	7910	1680

Table 8.

5. Conclusions

The data presented in this chapter on the properties of the polymer-filled nanoporous glass showed that this composite is the perspective material of laser optics for making solid-state dye-lasers in the visible and near-IR spectral ranges and a different type of "passive" optical elements: saturable absorbers, radiation power limiters, etc.

To the present, the technology of this material including its impregnation with functional organic compounds (dyes), both "the active" (lasing) and "passive" (controlling radiation amplitude - temporal characteristics) has been developed. The comprehensive studies of structure and optical properties of the material have been carried out and the operation of the optical elements in lasing and controlling modes has been successfully demonstrated.

At the same time, it ought to be pointed out that these studies have been carried out in the laboratory experiments, and further efforts have to be made for improving the technology of the material (to bring it to an industrial level) to make possible the wide practical applications of this new material of laser optics. A relalization of these efforts will lead, as may be expected, to further development of works on dye lasers and laser radiation control elements, and their applications.

6. References

[1] A.M.Dolotov, M.F.Koldunov, N.M.Sitnikov, et al. USSR Patent AC №1814475 (1990).

[2] O.S.Molchanova, Sodium-borate-silicate and porous glasses. Moscow, *Oborongiz*, p. 162, (1961).

[3] N.S.Andreev, O.V.Mazurin, U.A.Poray-Koshits, G.P.Roskova, V.N.Filipovich, Liquation phenomena in glasses. Leningrad, *Nauka*, p. 220, (1974)

[4] O.V.Mazurin, G.P.Roskova, V.I.Averyanov, T.V.Antropova, Two-phase glasses: structure, properties, application. Leningrad, *Nauka*, p. 276 (1991).

[5] G.B.Al'tshiler, E.G.Dul'neva. A.V.Erofeev, et al., Phototropic shutters utilizing micro-porous glass activated by dye molecules. *Quantum Electron*, Vol. 12, p. 1094-1096, (1985)

[6] G.B.Al'tshiler, E.G.Dul'neva, A.V.Erofeev, I.A.Mokienko. Solid-state-liquid passive laser switch - *Pisma in J. Tech. Phys*. Vol. 14, p. 2290 (1988).

[7] G.B.Al'tshiler, E.G.Dulneva, I.K.Meshkovsky, K.I.Krylov. Solid-state active media based on dyes. *J. App. Spectroscopy*, Vol. 36, p. 592, (1982).

[8] S.M.Dolotov, M.F.Koldunov, A.A.Manenkov, et al., Composite material for laser elements based on polymer and microporous glass, *Quantum Electron*, Vol. 22, p. 1060-1062, (1992)

[9] M.F.Koldunov, A.A.Manenkov, N.M.Sitnikov, et al, *Proc. of Laser optics '93*. Russia, St. Peterburg, 1993, Vol.1, p. 246.

[10] M. F. Koldunov, A. A. Manenkov, N. M. Sitnikov, et al, Dye-impregnated polymer-filled porous glass: a new composite material for solid state dye lasers and laser beam control optical elements. *Proc. SPIE* Vol. 2114, p. 101 (1994)

[11] H. R. Aldag, S. M. Dolotov, M. F. Koldunov, et al., Efficient solid state dye lasers based on polymer-filled microporous glass, *Proc. SPIE* Vol. 3929, pp. 133-145 (2000).

[12] H.R.Aldeg, S.M.Dolotov, M.F.Koldunov, et al., Microporous glass - polymer composite as a new material for solid-state dye lasers. I. Material properties. *Quantum Electron*, Vol. 30 (11), pp. 954-958, (2000)

[13] H.R.Aldeg, S.M.Dolotov, M.F.Koldunov, et al., Microporous glass - polymer composite as a new material for solid-state dye lasers. II. Lasing properties, Quantum *Electron,* Vol. 30 (12), pp. 1055-1059, (2000)

[14] J. A. Russell, D. P. Pacheco, H. R. Aldag, S. M. Dolotov, M. F. Koldunov, et al., Beam-quality measurements on solid state dye lasers using nonconfocal unstable resonators. *Proc. SPIE* Vol. 4267, pp. 36-46 (2001)

[15] S.M.Dolotov, M.F.Koldunov, Ya.V.Kravchenko, et al., An efficient solid-state laser based on a nanoporous glass – polymer composite doped with phenalemine dyes emitting in the 600 - 660-nm region. *Quantum Electron,* Vol. 32 (8), pp. 669-674, (2002)

[16] M.F.Koldunov, Ya.V.Kravchenko, A.A.Manenkov, I.L.Pokotilo, Relation between spectral and lasing properties for dyes of different classes. *Quantum Electron,* Vol. 34, pp. 115-119, (2004).

[17] M. F. Koldunov and A. A. Manenkov, Polymer-filled nanoporous glass: a new material for solid-state dye lasers and nonlinear optical elements. *Proc. SPIE* Vol. 6054, pp. 605401-07 (2005)

[18] S. S. Anufrick, M. F. Koldunov, Yu. M. Kuznetsov, A. A. Manenkov, et al., Lasing characteristics of phenolemine 512 and pyrromethene 580 dyes, impregnated to polymer-filled nanoporous glass, at various excitation wavelengths, *Proc. SPIE* Vol. 6054, pp. 60540V-60545V (2005)

[19] A.A.Manenkov, M.F.Koldunov, S.S.Anufrik, New material for laser optics: nanoporous glass polymer composite activated by functional organic compounds. *Proc. of VI International Conf. «Laser physics and optical technology»* Belarus, Grodno, 2006, Vol. 1, pp. 13-17.

[20] S. S. Anufrik, M. I. Ihnatouski, M. F. Koldunov, et al., Optical, structural, and lasing properties of a composite material nanoporous glass filled with an organic dye activated polymer, *Proc. SPIE* Vol. 6735, pp. 67351X-673516X, (2007)

[21] V. Tarkovsky, S. Anufrik, M. Koldunov, and A. Manenkov, The space-angular characteristics of a microsecond solid laser on the basis of a nanoporous glass polymer composite activated with dyes. *Proc. SPIE* Vol. 6731, pp. 673137-673143. (2007)

[22] L.M.Koldunov, M.F.Koldunov, A.V.Petuhov, A.V.Sizuhin, Reversible nonlinear absorption in nanoporous glass polymer composite doped with functional dyes: experiment and background model. *Technical digest,* ICONO/LAT 2010, Russia, Kazan, August 23-26, 2010. ITuQ46

[23] L.M.Koldunov, M.F.Koldunov, A.V.Petuhov, A.V.Sizuhin, Reversible nonlinear absorption in nanoporous glass polymer composite doped with functional dyes: experiment and background model. *Proc. SPIE.* (in press)

[24] L.M.Koldunov, M.F.Koldunov, I.L.Pokotilo, et al., Nonlinear refraction and nonlinear absorption in nanoporous glass composite activated by functional dyes. *Technical digest,* ICONO/LAT 2010, Russia, Kazan, August 23-26, 2010, ITuQ11

[25] L.M.Koldunov, M.F.Koldunov, I.L.Pokotilo. et al, Nonlinear refraction and nonlinear absorption in nanoporous glass composite activated by functional dyes. *Proc. SPIE.* (in press)

[26] K.M.Dumaev, A.A.Manenkov, A.P.Maslukov, et al., Laser radiation interaction with optical polymers - Trudy IOF RAN the USSR, Vol. 33, p. 144, (1991)

[27] S.P.Zhdanov, Physics and chemistry of silicates. Leningrad, *Chemistry,* p. 195 (1987).

[28] W.Vogel, Glaschemie, Leipzig, p. 434, (1979)

[29] T.A.Speranskaj, L.I.Tarutina, Optical properties of polymers. Leningrad, *Chemistry*, p. 136 (1976).

[30] A.P.Babichev, N.A.Babushkin, A.M.Bratkovckii, et al., Fiz. velichiny. Moscow, *Energomashizdat*, p. 1232, (1991).

[31] D.S.Sanditov, G.M.Bartenev, Physical properties of disordered structures. Novosibirsk, *Nauka*, p. 259, (1982)

[32] H.C. van de Hulst. Light scattering by small particles. New York. John Wiley&Sons Inc. London. Chapman&Hall Ltd. p. 536, (1957).

[33] M.Born, E.Wolf, Principle of optics. Pergamon press, Oxford, p. 855, (1964).

[34] I.S.Zeilikovich, A.M. Lyalikov, Holographic methods for regulating the sensitivity of interference measurements for transparent media diagnostics. *Sov. Phys. Usp.* Vol. 34 (1) pp. 74–85 (1991)

[35] C.R.Giuliano, Laser-induced damage to transparent dielectric materials. - *Applied Physics Letters*, 1964, Vol. 5, p. 137-139

[36] A.A.Manenkov, A.M.Prokhorov, Laser-induced damage in solids. *Sov. Phys. Usp*. Vol. 29 pp. 104–122 (1986)

[37] A.A.Manenkov, Fundamental mechanisms of laser-induced damage in optical materials: understanding after a 40-years research. *Proc. SPIE*, Vol. 27132, p. 2-11, (2008)

[38] M.F.Koldunov, A.A.Manenkov, Recent progress in theoretical studies of laser-induced damage (LID) in optical materials: fundamental properties of LID threshold in the wide-pulse-width range from microseconds to femtoseconds. *Proc. SPIE* Vol. 3578, pp. 212-226 (1999)

[39] M.F.Koldunov, A.A.Manenkov, I.L.Pokotilo. Pulse-width dependenct of the laser damage threshold of the transparent dielectric containing the absorbing inclusions. *Bulletin of the Russian Academy of Sciences.* (Physics) Vol. 59, pp. 72-83, (1995).

[40] M.F.Koldunov, A.A.Manenkov, I.L.Pokotilo, Pulse-width and pulse-shape dependencies of laser-induced damage threshold to transparent optical materials. *Proc. SPIE* Vol. 2714, pp. 718-730, (1996)

[41] M.F.Koldunov, A.A.Manenkov, I.L.Pokotilo, Multishot laser-induced damage in optical materials: an analysis of main regularities. *Proc. SPIE* Vol. 3244, pp. 641-649, (1998)

[42] M.F.Koldunov, A.A.Manenkov, I.L.Pokotilo. Multishot laser damage in transparent solids: theory of accumulation effect. *Proc. SPIE* Vol. 2428, pp. 653-667, (1995)

[43] M.F.Koldunov, A.A.Manenkov, I.L.Pokotilo. Theoretical analysis of the accumulation effect in laser damage in transparent dielectrics under repeatad irradiation conditions. *Quantum Electron*, Vol. 22, pp. 674-678. (1995).

[44] H. Ebert, Physikalisches taschenbuch, Frieder, Vieweg&Sohn p. 552, (1957).

[45] M. Sheik-bahae, A.A. Said, T. Wei, et al, Sensitive measurement of optical nonlinearities using a single beam. *IEEE J. of Quantum Electronics*, Vol. 25 (4) pp. 760-769, (1990).

[46] Kamjou Mansour, Daniel Alvarez, Kelly J. Perry, et al., Dynamics of optical limiting in heavy-atom substituted phthalocyanines. *SPIE 1853*, pp, 132-140, (1993).

[47] J. S. Shirk, R.G.S. Pong, et al., Optical limiter using a lead phthalocyanine. *Appl. Phys. Lett* 63, pp. 1880-1882, (1993).

[48] Weijie Su and Thomas M. Cooper , Mark C. Brant. Investigation of Reverse-Saturable Absorption in Brominated Porphyrins. *Chem. Mater*. 10, pp. 1212-1213 (1998).

[49] Kai Dou, Xiaodong Sun, Xiaojun Wang et al. Optical limiting and nonlinear absorption of exited states in metalloporphyrin-doped sol gels. *IEEE Journal of Quantum Electronics*, 35 (7) pp.1004-1013, (1999)

20

Broadband Spectral Analysis of Non-Debye Dielectric Relaxation in Percolating Heterostructures

E. Tuncer[1], J. Belattar[2], M. E. Achour[2] and C. Brosseau[3]
[1]*Applied Superconductivity Group, Fusion Energy Division, Oak Ridge National Laboratory, Oak Ridge, TN 37831-6122,*
[2]*Laboratoire des Systèmes de Télécommunications et d'Ingénierie de la Décision, Département de Physique, Faculté des Sciences, Université Ibn Tofail, B. P. 133, 14 000 Kénitra,*
[3]*Université Européenne de Bretagne, Université de Brest, Lab-STICC and Département de Physique, CS 93837, 6 avenue Le Gorgeu, 29238 Brest Cedex 3,*
[1]*USA*
[2]*Morocco*
[3]*France*

1. Introduction

Dielectric relaxation in complex materials has been fascinating physicist for more than half a century[1-5]. Much of the present understanding of dynamics in a very wide range of complex dielectric systems (e.g. filled polymers, glasses, biopolymers, liquid crystals, and even concrete) currently revolves around the ideas of spatial and dynamic heterogeneity. These complex materials in general exhibit nontrivial relaxation dynamics on both microscopic and macroscopic scales. Fundamental questions remain not only about dynamics, but also about how randomness and connectedness give rise to the dielectric attributes. Addressing such questions is quite challenging because the structural aspects of random composites are still largely unknown.[1-7] Within this perspective, a number of experiments, e.g. dielectric spectroscopy experiments[8-9] and computer simulations, e.g. Monte Carlo and finite element calculations[6,7,10-13] have progressed in parallel. Models for dielectric relaxation in heterostructures can be broadly placed in two groups.

The first picture has been heavily investigated in various phenomenological models of dielectric relaxation. Previous results suggest that pure Debye response is virtually non-existent in complex solids. As an alternative to the Debye model, investigators have used the purely empirical analytical expressions, one of them is called Cole-Cole[14] and it explores the spectral dependence of (relative) complex permittivity ε as a function of angular frequency ω, $\varepsilon = \varepsilon_\infty + (\varepsilon_s - \varepsilon_\infty)\dfrac{1}{1+(i\omega\tau)^{1-\alpha}}$, where ε_s and $\varepsilon_\infty^{'}$ are the static $(\omega << \tau^{-1})$ and the high frequency $(\omega >> \tau^{-1})$ permittivity values for a given relaxation, which can be considered atomic, electronic, ionic or interfacial polarization. In addition, it is assumed that the non-negative empirical exponent a gauges the broadening of the dielectric loss spectrum. Jonscher[15] suggested an alternate expression for the imaginary part of the permittivity ε'', $\varepsilon'' = \frac{\varepsilon''_m}{(\omega\tau)^{-m} + (\omega\tau)^{1-n}}$, where the exponents m and n lie between zero and unity. They determine the shape of the response at frequencies lower and higher than the maximum frequency $\omega_{\max}$ at which the maximum value of ε'' is observed $[\varepsilon''_m = \max(\varepsilon'')]$, i.e. the loss peak. The physical process that gives rise to this peak is a dipolar re-orientation. Several studies of these spectral models have been carried out to analyze the fractional exponents m and n of the power-law decays. This behavior has been observed in carbon black (CB)-epoxy composites frequency over the range from 200 Hz to 15 MHz,[16] and other types of binary (conductor-insulator) composites containing nanoparticles.[17] We point out that permanent and induced dipoles, electron conduction in CB aggregates, Maxwell-Wagner-Sillars (MWS) interfacial polarization all contribute to the effective (collective) relaxation behavior of the composite material. Based upon Jonscher's framework, we assume that the medium consists of a cluster of fluctuating reorientable dipoles and/or small size polar entities. Jonscher's model relies on the notion that in a typical solid the microscopic dipolar, electronic or ionic transitions are very rapid (on the scale of ps or less) and they take place in a system in which electrostatic or strain interactions are invariably present but the adjustment of local equilibrium takes a longer time to be realized.[15] In other words, the cluster structure of a dipolar system may be considered as a natural consequence of the fact that when the electric field is on, only some of the dipoles have enough energy and time to reach a configuration state aligned along the electric field. Thus, the dielectric response originates with specific, spatially limited regions containing dipoles with positions altered by the applied electric field and their random environment. During the relaxation process the strongly coupled local (intracluster) motions are expected to be generated first, and then followed by the weakly coupled (intercluster) motions, which produce the long-range structure.[15]

An alternative approach to the above empirical expressions is the so-called continuous distribution of relaxation times [18-19] The distribution of relaxation times reveals the true relaxation spectrum of a system. The complex effective permittivity $\varepsilon(\omega)$ is defined for an ensemble of Debye processes with a continuous relaxation time distribution, $\varepsilon(\omega) = \varepsilon_\infty + (\varepsilon_s - \varepsilon'_\infty)\int_{-\infty}^{\infty} \frac{g[\ln(\tau)]}{1+i\omega\tau} d[\ln(\tau)]$, where $g(\tau)$ is the distribution of relaxation times. Each relaxation has its own relaxation strength $(\varepsilon_s - \varepsilon'_\infty)$, and it is therefore natural to keep relaxation time and strength of each Debye process together in a numerical approach, then $\int_{-\infty}^{\infty} g[\ln(\tau)] d[\ln(\tau)] = \varepsilon_s - \varepsilon'_\infty$. Although, the concept of distribution of relaxation times can be notoriously misleading by inducing numerical instabilities, and in certain cases lead to ambiguities in interpretation of dielectric data[18], linearization of the problem and using the Monte Carlo integration hypothesis puzzle out the instability and ill-conditioned nature of the problem.[19] Numerous examples of distribution of relaxation times are reported in different systems by one of the present authors (ET).[20] The distribution $g(\tau)$ constructed captures correctly the relaxation properties. One should note immediately that, except for some special cases, e.g. Cole-Cole, inverting these integrals is nontrivial since they are nonlinear.[8]

It should be emphasized that a direct comparison between these different approaches has been rarely attempted except in Tuncer and collaborators.[19-20] The major focus of this work is to provide such a comparison based on experimental data on CB filled epoxy composites, thereby establishing the bridge between these two widely used complementary theoretical approaches to broadband relaxation in dielectric materials.

Experiments were conducted with CB particles (Raven 2000, Raven 5000, Raven 7000 obtained from the Columbian Chemicals Company USA and Monarch 1100 purchased from Cabot Company USA) and diglycidylic ether of bisphenol F (DGEBF) epoxy (Araldite XPY 306, epoxy equivalent weight=172) obtained from Ciba Geigy Ltd. The mixture of DGEBF and CB particles was processed with an amine curing agent (4,9-dioxadodecan-1,12-diamine, equivalent weight=81, supplied by BASF, and used without further purification). Resin was cured at room temperature for 24 h. The neat DGEBF has a dc conductivity of 1.4 10^{-14} $\Omega^{-1}m^{-1}$, a density of 1.19 gcm^{-3}, and a glass transition temperature at. T_g=83 °C. The different series of samples were fabricated by mechanical mixing.[16] The CB volume fraction within the composite is denoted as ϕ_2. The values of the percolation threshold, ϕ_c, for DGEBF/Raven 2000, DGEBF/Raven 5000, DGEBF/Raven 7000, and DGEBF/Monarch 1100 were determined by dc conductivity measurements (linear current-voltage relationship) and are equal to 3.6, 10, 2.75 and 8 vol%, respectively. The effective (relative) complex permittivity of the composite samples was measured in the frequency ($F=\omega/2\pi$) range from 200 Hz to 15 MHz. using a HP 4194A impedance analyzer. The experimental setup and procedures for measuring $\varepsilon(F,T)$, where the temperature T ranges from ambient temperature up to 70 °C, are similar to those previously described.[16] Note that the contribution of the Ohmic conductivity, i.e. $\sigma_{dc}(\phi_2,T)/\varepsilon_0\omega$ was removed to the raw ε'' data.

Room temperature isotherms of complex permittivity in Cole-Cole representation are shown in Fig. 1 for several CB concentrations of the DGEBF/Raven 2000 series. Superimposed on these experimental data is a nonlinear least-squares fit to the Cole-Cole form with fitting parameters ε_s, ε'_∞, α, and τ. Evolution of the best-fit parameters with CB concentration is presented in Fig. 2. As concentration ϕ_2 increases from ≈ 0 to close to the percolation threshold, ε_s (resp. ε'_∞) increases by 300 % (resp. 75 %). The estimated exponent α deviates significantly from 1. For this range of CB volume fraction ϕ_2 extends over nearly 2 orders of magnitude.

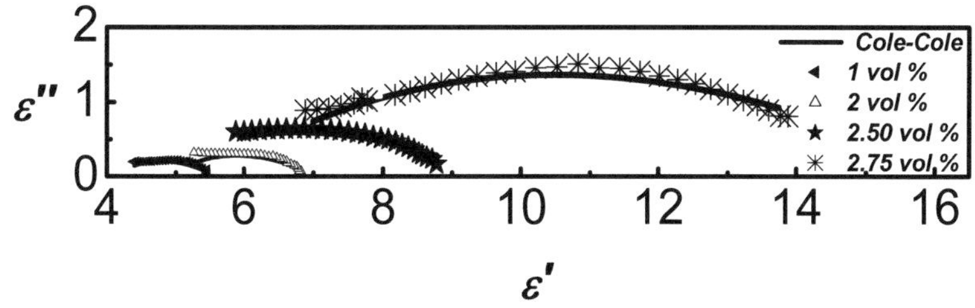

Fig. 1. Cole-Cole plot ε'' as a function of ε' for concentration ϕ_2 =0.1, 0.5, 1, and 2 vol % (Raven 2000-DGEBF samples). Room temperature. Symbols are experimental data and solid curves are 4-parameter fit to the Cole-Cole equation. The numbers in the graph indicate the value of the CB loading

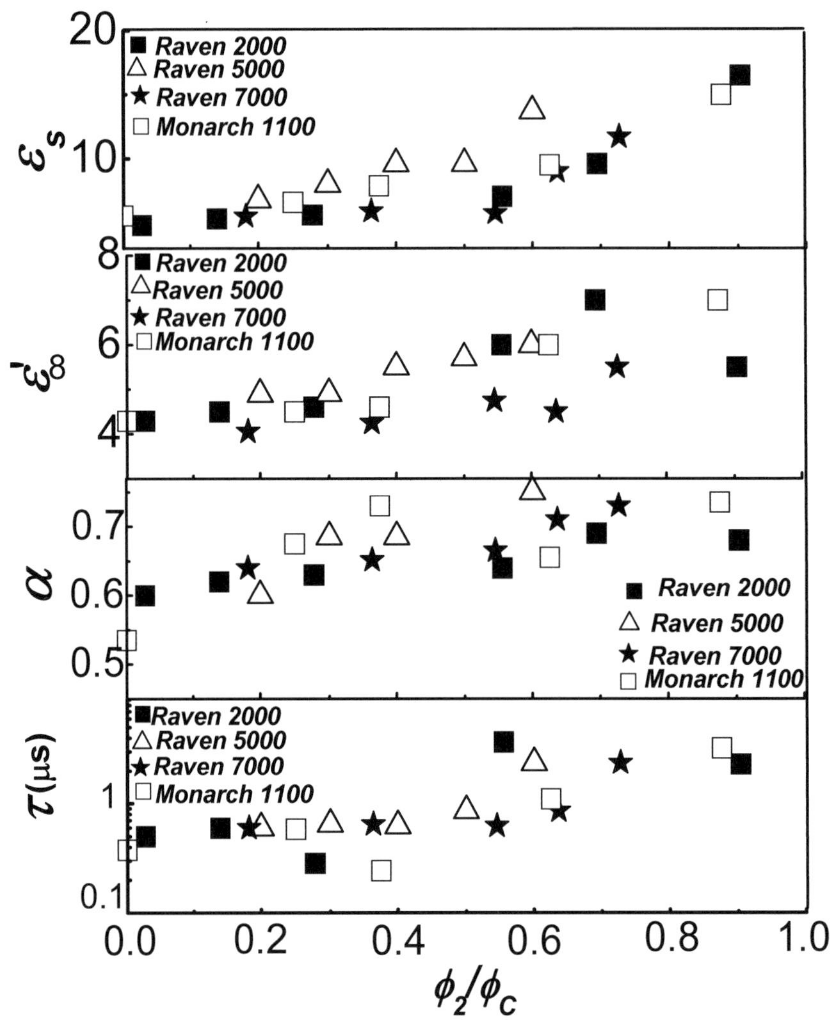

Fig. 2. (a) Plot of the static (relative) permittivity ε_s vs ϕ_2 / ϕ_c for the different series of samples. The symbols in the graph indicate the type of CB particles. (b) Same as in (a) for the limiting permittivity at high frequencies ε'_∞. (c) Same as in (a) for the Cole-Cole exponent α. (d) Similar to (a) for the relaxation time τ at room temperature

For $T \leq T_g$ the temperature dependence of τ typically follows the Vogel-Tammam-Fulcher (VTF) thermal activation law for the slowing down of relaxation processes found in many amorphous materials.[16,19] As shown in Fig. 3, with increasing temperature *T*, the τ isotherms exhibit decreasing evolution. Assuming that the temperature dependence of τ between 20°C and 70°C is of the form $\tau = \overline{\tau}_0 \exp[W/k(T-T_0)]$, the best fit of our data using maximum likelihood method yields $\overline{\tau}_0 \approx 10^{-8}$ s, $267 \leq T_0 \leq 292$ °C which is an ordering temperature lower than T_g whose significance has remained unclear (there is no microscopic prescription to calculate T_0), and apparent free energy of thermal activation $W=0.63 \pm 0.19$ meV, which we interpret as the average potential barrier between different sites inside the heterostructure. To put numbers in context, we note that the values for W reported in the literature for similar materials range are consistent with ours.[12-13,18] The value of $\overline{\tau}_0$ is almost 3 orders of magnitude larger than $\overline{\tau}_0 \approx 10^{-11}$ s, commonly used for the pre-exponential factor. A possible interpretation of this pre-exponential factor is to assume that a dipole on average visits $N = \overline{\tau}_0 / \tau_0 \approx 10^3$ sites inside the cluster before being relaxed. Similar behavior, i.e. long trapping time, has been reported for bulk silicate filled carboxylated nitrile rubber.[21-22] Meanwhile, the values of α remain $\approx 0.70 \pm 0.05$ in this range of T.

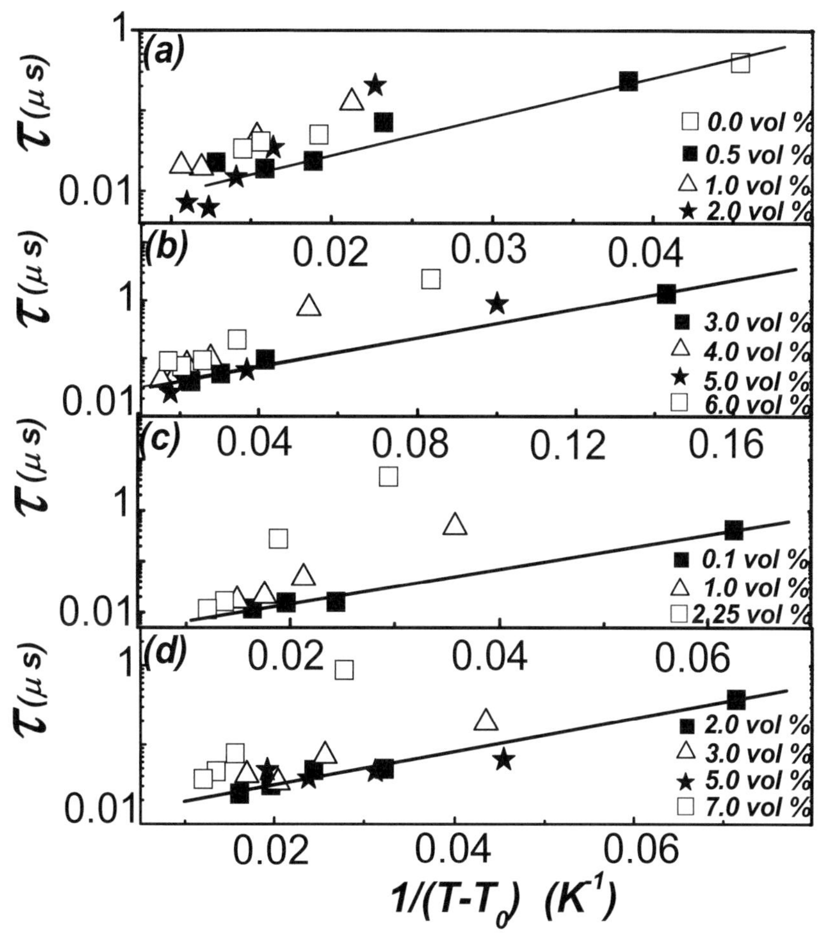

Fig. 3. (a) Plot of τ as a function of temperature for Raven 2000-DGEBF samples and different values of CB volume fractions. The solid curve is a fit to the Vogel-Tammam-Fulcher law to the data represented by solid squares (see text). The numbers in the graph indicate the value of ϕ_2. (b) Same as in (a) for Raven 5000-DGEBF samples. (c) Same as in (a) for Raven 7000-DGEBF samples. (d) Same as in (a) for Monarch 1100-DGEBF samples

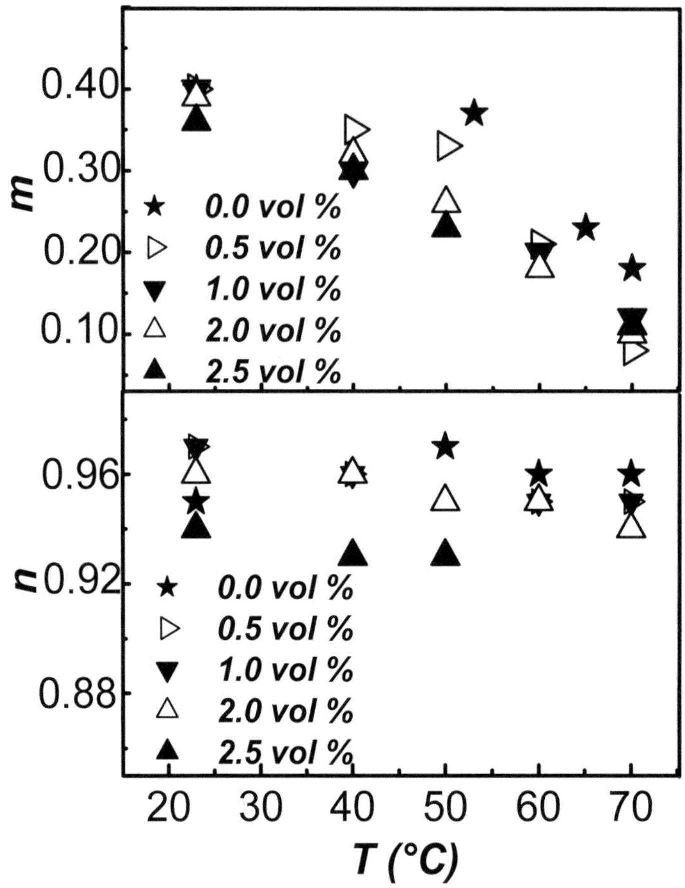

Fig. 4. (a) Plot of m vs T for the different concentrations of Raven 2000 in the Raven 2000-DGEBF samples. The symbols in the graph indicate the value of the CB loading. (b) Same as in (a) for n

So far, we have only discussed the Cole-Cole phenomenology. In order to correlate the temperature dependence of the relaxation mechanisms for the samples investigated here and better understand how temperature affects the values of n and m obtained from our variable-temperature measurements are graphed in Fig. 12 for the Raven 2000-DGEBF as an illustrative example. A similar behavior is observed for the other series of samples. Two typical behaviors corresponding to the temperature range explored are: m decreases strongly with T and n is close but smaller than 1. Such comparison can provide a comprehensive picture of the similarities and differences between the CB types in the overall relaxation behavior of these filled polymers. By contrast, our results are less clear for the m dependence as a function of CB content because the fits yield larger uncertainty for m. Jonscher suggested to associate the values of m and n with the degree of correlation between the flip-flop transitions (fast intra-cluster interactions) for m and between the flip transitions (slow inter-cluster interactions) for n, respectively.[15] According to this modeling, flip transitions represent tunnelling modes between different configurations giving a net change of the total dipole moment and flip-flop transitions represent local fluctuations of the dipole moment. The latter retains the average value of the total polarization and may be regarded as synchronous transitions in opposite sense at different points in the material, giving zero net change of dipole moment. The value m=0 corresponds to completely uncorrelated flip-flop transitions of the system. The limit n=0, corresponds to a complete screening, as in a free charge system, while n=1 corresponds to the absence of screening, as would be the case with immobile charges (Debye system) which are unable to follow local changes of potential.[13] In contrast to the classical Debye response, corresponding to a system of non-interacting permanent dipoles,[13] the dielectric response arises as being due to a highly dispersive and lossy system yielding a frequency independent $\varepsilon'' / \varepsilon'$ ratio.

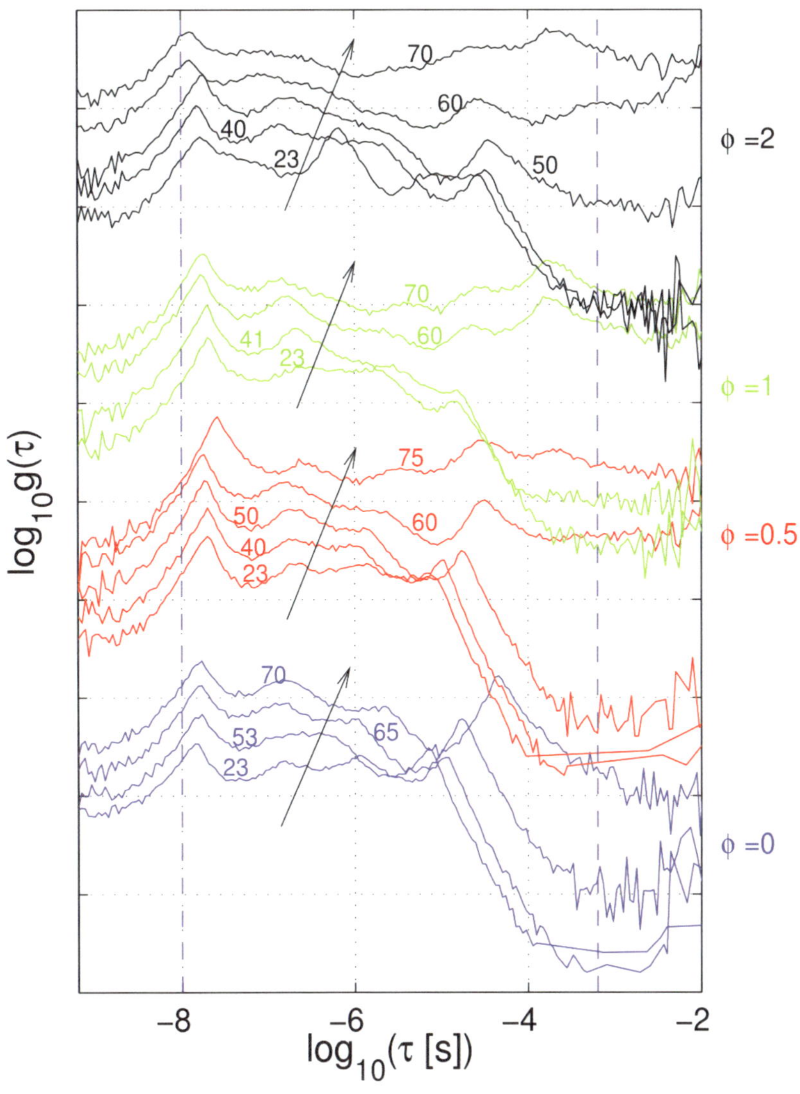

Fig. 5. (Color online) The relaxation time distributions g(τ) for all samples of Raven 2000-DGEBF. The numbers on each curve indicate the temperature. The volume fractions are shown on the right side of the graph. The distributions are shifted to unravel the relaxations

The principle of the numerical analysis of a set of experimental frequency-response data that involves a continuous distribution of relaxation times is identical to that described elsewhere.[19-20] Fig. 5 shows the results of this analysis and shows a comparison of $g(\tau)$ at different temperatures and CB content for Raven 2000-DGEBF samples. The data are shifted to guide the eyes, and each relaxation spectrum had similar baseline, which is shown in Fig 6. The vertical dashed lines in Fig. 5 are drawn to indicate the measurement window for the reciprocal frequency. The relaxations resolved at fast times had clear peaks due to coupling of the high frequency permittivity with high frequency relaxations—the relaxations were not fully observed within the current experimental frequency window. Even Kramers-Kronig transformation[23] would resolve an accurate relative permittivity when a relaxation is not observed fully. The relaxations resolved were broad and distributed over the relaxation time axis. The low temperature data for all the samples (23°C and 40°C) had the long relaxation times observable within the experimental frequency window—this is shown in Fig. 5 with a sharp drop in the distribution after 10^{-5}-10^{-4} s. This characteristic is most visible in the unfilled sample ($\phi_2 = 0$) over all the temperatures, where in other words the polarizations were finalized. Over these temperatures the distributions of the filled samples exhibited continuation of the polarization in the materials. The reason for the change in the relaxation spectra with temperature in filled samples was due to the influence of the CB particles on the energy landscape—the local field and the dynamics of polymer motion were altered. One can also speculate about the ionic species and their dynamics at long times with addition of fillers. Currently there is none systematic investigations on the change in relaxation times spectra in filled systems.

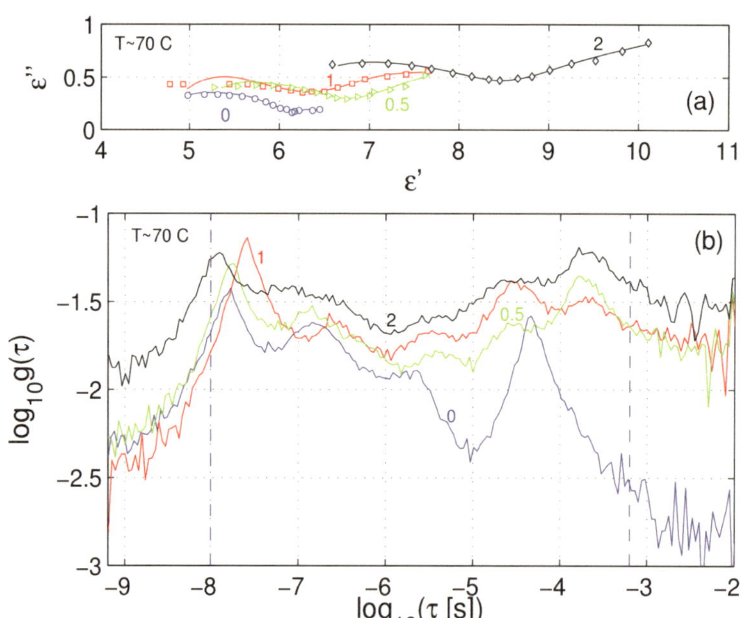

Fig. 6. (Color online) (a) Cole-Cole plot, and (b) resolved distributions for samples with CB volume percentage 0, 0.5, 1 and 2. The complex permittivities that are close to the glass transition are shown. The dashed lines in (b) indicate the reciprocal experimental frequency range that probes to the relaxation times

One striking observation of the analysis method is presented in Fig. 6 for different volume fraction of CB at high temperatures (>65°C). The relaxations resolved around medium times (10^{-7} s and 10^{-5} s) could be followed through all samples from unfilled to 2 vol%. Similar kind of response was previously observed and reported by Tuncer *et al.*[24] This sort of behaviour should be expected since the signature of the matrix material should be detectable with the impedance measurements and the applied analysis, unless the matrix physically changed with the addition of fillers. To illustrate how the resolved distributions were actually describing the data, the dielectric response generated from the distribution data are plotted on the experimental values with solid lines in Fig. 6. There is very good agreement with measurements and the distribution of relaxation times analysis. Finally, we show the relaxation spectra and dielectric data estimated for composite samples using room temperature data in Fig. 7. The evolution of the relaxation spectra with filler content is noteworthy. As shown with examples, the distribution of relaxation times approach is a compelling tool to better understand composite materials.

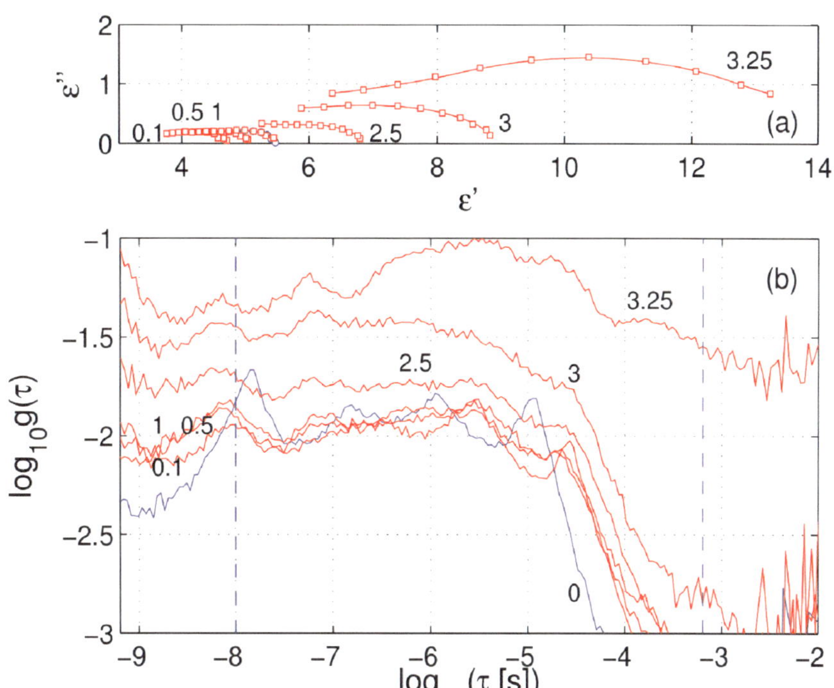

Fig. 7. (Color online) (a) Cole-Cole plots, and (b) resolved distributions for samples with different CB volume percentages. The data are obtained at room temperature, and concentrations of the samples are below the percolation threshold. The dashed lines in (b) indicate the reciprocal experimental frequency range that probes to the relaxation times

In summary, a combination of several complimentary analytical techniques has been used to investigate the dielectric relaxation in carbon black-epoxy composites. The results presented here confirm the non-Debye relaxation behavior in these percolative heterostructures that was identified by the analysis of the permittivity spectra of these heterostructures which can be accounted for by the Cole-Cole phenomenological equation. The relaxation time follows a

VTF temperature dependence with activation energy of the order of the meV. Comparison can be also made with an alternative explanation based on the concept of a set of dipoles making sudden jumps between fixed orientations in space and causing a slow screening response in the surrounding medium.[13] This deviation from Debye's behavior is assumed to be due to many-body effects on relaxation processes described by two fractional exponents m and n characterizing the power-law decays of this spectral model. The exponents m and n are both positive and smaller than unity. Interestingly, they have different temperature variations: while m is strongly decreasing with increasing temperature ($<T_g$), n takes a value close to 1. Our results suggest that many body interactions have a dominant role to explain the collective coupling between the dipole ensemble and the environment. However accurate they may be, each modeling method considered in this work has one important approximation which is difficult to overcome: effective medium approximation, i.e. when spatial inhomogeneities manifest on scales much smaller than the relevant length scale in the system, an average homogeneous treatment is justified. A continuous distribution of relaxation times has been obtained from experimental data following the method of one of the authors (ET). Our results raise the question of how the parameters of empirical expressions α, m and n can be interpreted. Unlike these empirical model parameters the $g(\tau)$ obtained from distribution of relaxation times is not based on a model, the relaxation spectrum is a true representation of the material—the spectrum does not change with choice of initial parameters or does not require a-priori assumption. A detailed physical (microscopic) interpretation of these quantities depends on both of the cluster structure and the collective nature of the interactions. Yet the problem is so important that numerical answers that include multi-fractal geometry are highly desirable. It would be very attractive to perform spectral density analysis on the composite response[25] to separate structural information from the intrinsic properties of the constituents. In that case the structure whether it is fractal or not would be resolved with the analysis, however, in that case one needs to have the full information on the electrical properties of the constituents.

We end by making a number of comments about some standing questions and future directions. Various questions can still be addressed. Interaction with interfaces is one of the main relaxation mechanisms for dipoles. In order to deal with interfaces it will be interesting to consider relaxation spectra based on *ab initio* Monte Carlo and molecular dynamics simulations. Since most properties of random composites are determined by topology, the beginning of any such study is the creation of experimentally credible structural models. Studies in this direction would benefit from basic information on the mesostructure obtained via atomic force microscopy.[26] A fully microscopic theory of relaxation in percolative heterostructures is outstanding and remains an exciting prospect. One should employ the spectral density representation to completely realize the electrical properties of composite materials.

2. Acknowledgments

M. E. A. thanks the Université de Brest for its financial support and hospitality where some of this work was completed. This work was supported by the Lab-STICC, Unité Mixte de Recherche CNRS 3192. E.T. was sponsored by the U.S. DOE Office of Electricity Delivery and Energy Reliability--Advanced Cables and Conductors, under contract DE-AC05-00OR22725 with Oak Ridge National Laboratory, managed and operated by UT-Battelle, LLC.

3. References

[1] R. Landauer, in *Electrical Transport and Optical Properties of Inhomogeneous Media*, edited by J. C. Garland and D. B. Tanner (American Institute of Physics, New York, 1978), vol. 40, pp. 2–43.

[2] A. H. Sihvola, *Electromagnetic Mixing Formulas and Applications* (IEE Publishing, London, 1999).

[3] D. J. Bergman and D. Stroud, in Solid State Physics, *Advances in Research and Applications*, edited by H. Ehrenreich and D. Turnbull (Academic, New York, 1992), Vol. 46.

[4] S. Torquato, *Random Heterogeneous Materials: Microstructure and Macroscopic Properties* (Springer, New York, 2002).

[5] M. Sahimi, *Heterogeneous Materials I : Linear Transport and Optical Properties* (Springer, New York, 2003).

[6] C. Brosseau, J. Phys. D 39, 1277 (2006); C. Brosseau and A. Beroual, Prog. *Mater. Sci.* 48, 373 (2003).

[7] E. Tuncer and S. M. Gubanski, *J. Appl. Phys.* 8, 310 (2001).

[8] H. Schäfer, E. Sternin, R. Stannarius, M. Arndt, and F. Kremer, Phys. Rev. Lett. 76, 2117 (1996); F. Kremers and A. Schönhals, eds., *Broadband Dielectric Spectroscopy* (Springer, Berlin, 2003), and J. R. Macdonald, ed., *Impedance Spectroscopy* (Wiley, New York, 1987); A. Bello, E. Laredo, and M. Grimau, *Phys. Rev.* B 60, 12764 (1999).

[9] Y. Feldman, A. Andrianov, E. Polygalov, G. Romanychev, I. Ermolina, Y. Zuev, and B. Milgotin, *Rev. Sci. Instrum.* 67, 3208 (1996); Y. Feldman, A. Puzenko, and Y. Ryabov, *Chem. Phys.* 284, 139 (2002).

[10] E. Tuncer and S. M. Gubanski, in Dielectric and Related Phenomena DRP'98: *Polymers and Liquid Crystals*, edited by A. Wlochowicz and E. Targosz-Wrona, Technical University of Lodz, Branch in Bielsko-Biala, Poland (SPIE, Bellingham, WA, 1998), pp. 136–142; E. Tuncer, *Licenciate thesis-Tech. Rep.* 338 L, Chalmers University of Technology, Gothenburg, Sweden, 2000; E. Tuncer, Dielectric relaxation in dielectric mixtures, PhD thesis, Chalmers University of Technology, Gothenburg, Sweden, 2001; E. Tuncer, S. M. Gubanski, and B. Nettelblad, *J. Appl. Phys.* 89, 8092 (2001); E. Tuncer, B. Nettelblad and S. M. Gubanski, *J. Appl. Phys.* 92, 4612 (2002); E. Tuncer, S. M. Gubanski, and B. Nettelblad, J. *Electrostatics* 56, 449 (2002); E. Tuncer, Turk. J. Phys. 27, 121 (2003); E. Tuncer and E. Tuncer, Turk. *J. Phys.* 23, 101 (2003); E. Tuncer. *Phys. D: Appl. Phys.* 37, 334 (2004); R. F. Hamou, J. R. Macdonald and E. Tuncer, *J. Phys.*: Condens. *Matter* 21, 025904 (2009).

[11] V. Myroshnychenko and C. Brosseau, *Phys. Rev.* E 71, 016701 (2005); V. Myroshnychenko and C. Brosseau, *J. Appl. Phys.* 97, 044101 (2005); V. Myroshnychenko and C. Brosseau, *J. Phys.* D 41, 095401 (2008); V. Myroshnychenko and C. Brosseau, *J. Appl. Phys.* 103, 084112 (2008); V. Myroshnychenko and C. Brosseau, *IEEE Trans. Dielectr. EI* 16, 1209 (2009). See also A. Mejdoubi and C. Brosseau, *Phys. Rev.* B 74, 165424 (2006); A. Mejdoubi and C. Brosseau, *Phys. Rev.* E 74, 031405 (2006); A. Mejdoubi and C. Brosseau, *Phys. Rev.* B 74, 165424 (2006); A. Mejdoubi and C. Brosseau, *J. Appl. Phys.* 100, 094103 (2006); A. Mejdoubi and C. Brosseau, *J. Appl. Phys.* 101, 084109 (2007).

[12] J. P. Calame, A. Birman, Y. Carmel, D. Gershon, B. Levush, A. A. Sorokin, V. E. Semenov, D. Dadon, L. P. Martin, and M. Rosen, *J. Appl. Phys.* 80, 3992 (1996); J. P. Calame, *J. Appl. Phys.* 94, 5945 (2003).

[13] L. C. Shen, C. Liu, J. Korringa, and K. J. Dunn, *J. Appl. Phys.* 67, 7071 (1990); C. Liu and L. C. Shen, Modelling and Simulation in *Materials Science and Engineering*, 1, 723 (1993); P. K. Ghosh and M. E. Azimi, *IEEE Trans. Dielec. Insul.* 1, 975 (1994).
[14] B. K. P. Scaife, Principles of Dielectrics (Clarendon Press, Oxford, 1998); K. S. Cole and R. H. Cole, *J. Chem Phys.* 9 (1941) 341.
[15] A. K. Jonscher, Dielectric Relaxation in Solids, (Chelsea Dielectric Press, London, 1987). See also A. K. Jonscher, Universal Relaxation Law, (Chelsea Dielectric Press, London, 1996) and more recently A. K. Jonscher, *J. Phys.* D 32, R57, (1999). See also L. A. Dissado and R. M. Hill, *J. Chem. Soc. Farady Trans.* 2, 291 (1984).
[16] M. E. Achour, C. Brosseau, and F. Carmona, *J. Appl. Phys.* 103, 094103 (2008); C. Brosseau and M. E. Achour, *J. Appl. Phys.* 105, 124102 (2009).
[17] J. Wu and D. S. McLachlan, *Phys. Rev.* B 58, 14880 (1998); D. S. McLachlan, W. D. Heiss, C. Chiteme, and J. Wu, *Phys. Rev.* B 58, 13558 (1998); J. Wu and D. S. McLachlan, *Phys. Rev.* B 56, 1236 (1997); ibidem 58, 14880 (1998).
[18] J. R. Mcdonald, *J. Appl. Phys.* 62, R51 (1987).
[19] E. Tuncer and S. M. Gubanski *IEEE Trans. Dielectr. EI*, 8, 310 (2001).
[20] E. Tuncer, M. Furlani, B.-E. Mellander, J. Appl. Phys. 95, 3131 (2004); E. Tuncer, M. Wegener, R. Gerhard-Multhaupt, *J. Non-Cryst. Solids* 351, 2917 (2005); E. Tuncer, M. Wegener, P. Frubing, and R. Gerhard-Multhaupt, *J. Chem. Phys.* 122, 084901 (2005); E. Tuncer, J.R. Macdonald, J. *Appl. Phys.* 99, 074106 (2006); J. R. Macdonald and E. Tuncer, *J. Electroanalytical Chem.* 602, 255 (2007); E. Tuncer, G. Polizos, I. Sauers, D. R. James, A. R. Ellis, J. M. Messman, T. Aytug *Cryogenics* 49, 463 (2009).
[21] N. G. McCrum, B. E. Read, and G. Williams, Anelastic and Dielectric Effects in Polymeric Solids (Dover, New York, 1991); G. Strobl, The Physics of Polymers (Springer, Berlin, 1996), and C. A. Angell, in Relaxations in Complex Systems 1, edited by K. Ngai and G. B. Wright, (*Natl. Technol. Inform. Ser.* U.S. Dept. of Commerce, Springfield, VA, 1985).
[22] J. Fritzsche, A. Das, R. Jurk, K. W. Stöckelhuber, G. Heinrich, and M. Klüppel, eXPRESS *Polym. Lett.* 2, 373 (2008).
[23] H. A. Kramers, *Nature* London 117, 775 (1926); R. d. L. Kronig, *J. Opt. Soc. Am.* 12, 547 (1926).
[24] E. Tuncer, N. Bowler, I. J. Youngs and K. P. Lymer *Phil. Mag.* 86, 2359 (2006).
[25] E. Tuncer Phys Rev B 71, 012101 (2005); E. Tuncer, *J. Phys. D: Appl. Phys.* 38, 223 (2005); E. Tuncer, Materials 3, 585 (2010); E. Tuncer, N. Bowler and I. J. Youngs, Physica B 373, 306 (2006); E. Tuncer and G. A. Niklasson, *Optics Comm.* 281, 4374 (2008); E. Tuncer, *J. Phys.: Condens. Matter* 17, L125 (2005); E. Tuncer, *Phil. Mag. Lett.*, 85, 269 (2005).
[26] J. Ravier, F. Houze, F. Carmona, O. Schneegans, and H. Saadaoui, Carbon 39, 314; (2001); F. Houze, R. Meyer, O. Schneegans, and L. Boyer L, *Appl. Phys. Lett.* 69, 1975 (1996); F. Carmona, and J. Ravier, Carbon 40, 151 (2002).

Advanced Cellulosic Nanocomposite Materials

Qingzheng Cheng[1], David DeVallance[1], Jingxin Wang[1] and Siqun Wang[2]
[1]*West Virginia University, Morgantown, WV,*
[2]*University of Tennessee, Knoxville, TN*
United States

1. Introduction

Biomass, especially woody biomass, represents the most important sustainable resource which can be used as feedstock for producing biofuels and bioproducts. In the past decade, overdependence on petroleum products has consistently increased. Biodegradable products from renewable materials are becoming increasingly more attractive due to escalating prices and shortages of crude oil. Cellulose, synthesized mainly in biomass by photosynthesis, is the most abundant natural biopolymer in the world. Natural cellulosic fibers, particles, fibrils (micro and nano scale), and crystals/whiskers are attractive replacements for man-made fibers (e.g., glass and aramid fibers) used as reinforcement and fillers when making environmentally friendly products. These cellulosic materials have many advantages including: renewable, low cost, low density, low energy consumption, high specific strength and modulus, high sound attenuation, nonabrasive, and a relatively reactive surface.
Recent research has focused on determining how to use biomass effectively as low cost, environmentaly friendly raw materials in many products while reducing the dependence on petroleum based resources. In the past two decades, research and manufacturing efforts have focused on using natural fibers to replace artificial fibers such as glass fibers for reinforcement material and fillers and to make environmentally safe products (George et al., 2001). Fibrils (in micro and nano scale) isolated from natural fibers have much higher mechanical properties as compared to original fibers. Given this, important research is being conducted on generating fibrils and combining them with polymers to form nanocomposites with the expectation of improving strength and stiffness as compared to composites reinforced with artificial fibers (Berglund, 2005; Cheng et al., 2007a; Cheng et al., 2007b; Cheng et al., 2009b; Herrick et al., 1983; Turbak et al., 1983;). Microfibres are defined as cellulose fibers 0.1-1 μm in diameter, with a corresponding minimum length of 5-50 μm (Chakraborty et al., 2006). Franks (1987) defined nanotechnology as "the technology where dimensions or tolerances in the range 0.1 to 100 nm (from the size of an atom to the wavelength of light) play a critical role". The definition of nano-size material has broadened significantly to include a large variety of systems that have at least one-dimension at the nanometer scale (1-100 nm) (Jordan et al., 2005). As one of the next generation of environmentally friendly materials, bio-based nanocomposites have grown rapidly and include composites that are completely biodegradable (both the fibril and polymer matrix are biodegradable) and partly biodegradable (at least the fibril is biodegradable).

It is still a challenge to isolate fibrils (in both micro and nano scales) at a reasonable cost with low degradation, and to effectively disperse them evenly in polymer matrixes. However, many methods have been developed to generate fibrils (e.g., chemical and mechanical) and to produce bio-nanocomposites (e.g., film casting and freeze dry followed by compressive molding and/or extrusion). Problems and challenges in nanocomposites reinforced with cellulose fibrils (micro and nano scales) include: 1) disintegration of cellulose to fibrils without severe degradation; 2) producing low cost, environmentally friendly nanocomposites; 3) characterizing the mechanical properties of single fibrils (to evaluate the effects of the isolation process) used in producing nanocomposites; 4) dispersing cellulose fibrils in polymer matrixes; 5) adhesion between the fibrils and polymer matrixes; 6) applications of bio-based nanocomposites in the ecological system; 7) biodegradability of nanocomposites partly or completely in the ecological system.

This chapter will review cellulosic nanocomposites development at the Division of Forestry and Natural Resources, West Virginia University and the Center for Renewable Carbon, University of Tennessee, as well as several other institutes. Cellulosic nanofibers/nanocrystals were isolated directly from hardwood residues and hybrid poplars, and used to reinforce poly(vinyl alchole) (PVA) when making nanocomposites at West Virginia University. At the Center for Renewable Carbon, cellulose fibrils were isolated from Lyocell fiber, a regenerated cellulose fiber, pure cellulose fiber, and pulp fiber through a high intensity ultrasonic treatment. The resulting fibrils were characterized for physical and mechanical properties and used as reinforcements for several polymers, such as poly(lactic acid) (PLA), PVA, and polypropylene when making cellulosic nanocomposites. These research efforts provided alternative methods to isolate cellulosic fibrils and nanofibers and to reinforce polymers when fabricating cellulosic composites in a manner that produces higher yields in an more environmentally friendly and energy efficient way, as compared to current techniques..

2. Cellulose

Cellulose is the world's most abundant natural, renewable, biodegradable polymer, and a classical extracellular high-performance skeletal biocomposite consisting of a matrix reinforced by fibrous biopolymer. The whisker like microfibrils, which are near defect free crystals with axial physical properties approaching those of perfect crystals, are biosynthesized and deposited in a continuous fashion by plants or animals. Wood, one of the most important plants for society , is not only a good building material, but also an abundant cellulose source.

Although the first investigations on cellulose were performed by Braconnot in 1819 and Payen in 1839, the structure of cellulose was clearly established by Staudinger in 1926, (Lima and Borsali, 2004). Cellulose is a high-molecular-weight linear homopolymer, consisting of repeating β-D-glucopyranosyl units joined by (1-4) glycosidic linkages in a variety of arrangements. Cellulose is a (1-4)-link β-D-glucoglycan, β-D-glupynanose unit, which is joined to form a linear-molecular chain (Fig. 1) (Chaplin, 2010). The hydrogen bonds formed between the ring oxygen atom of one glycosyl unit and the hydrogen atom of the C-3 hydroxyl group of the preceding ring hinder the free rotation of the rings on their linking glycosidic bonds resulting in chain stiffening. The adjacent cellulose chains fit closely together in an ordered crystalline region, so that high strength and insolubility (in common solvents) can be observed in plants, and in some cellulose-constituted animals. In nature, cellulose chains have a DP of approximately 10,000 glucopyranose units in wood cellulose and 15,000 in native cellulose cotton (Chaplin, 2010; Sjostrom, 1993).

Microfibrils are constituted by amorphous and crystalline domains, are biosynthesized, and are self-assembled from cellulose chains. The degree of crystallinity (ratio of the mass of crystalline domains to the total mass of the cellulose) and typical dimensions are dependent on their origin (Sarko and Muggli, 1974; Woodcock and Sarko, 1980). Cellulose exists basically in four different forms, cellulose I (I_α and I_β), II, III, IV, which can be interconverted by chemical and thermal processes (Lima and Borsali, 2004; Woodcock and Sarko, 1980). Cellulose was described displaying six different polymorphs, namely I, II, IIII, IIII_I, IVI , and IVI_I with the possibility of conversion from one form to another (O'Sullivan, 1997; Samir et al., 2005).

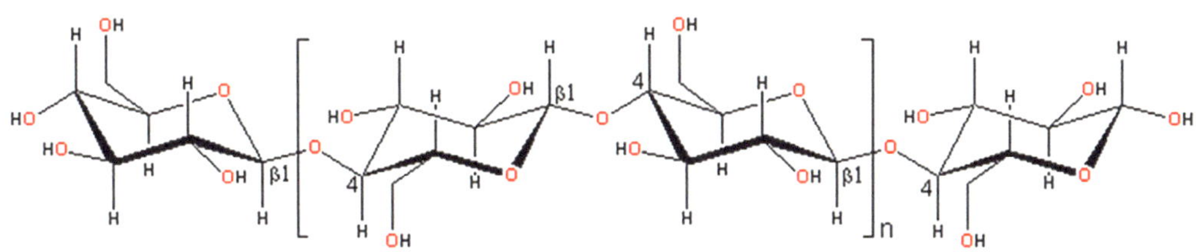

Fig. 1. Bassic chemical structure of cellulose (Reprinted from Chaplin, 2010 with permission)

3. Isolation and characterization of cellulosic nanocrystals/microfibrils

3.1 Cellulosic nanofibers/nanocrystals isolation

Two common methods, chemical and mechanical, have been developed to obtain fibrils. Stron acid hydrolysis, which removes the amorphous regions of cellulose fiber, is the main chemical method. This method can be used to generate cellulose crystals or nanofibers, commonly called cellulose whiskers or nanocrystals. The most common mechanical method utilizes strong shear forces, such as those obtained using a homogenizer, to separate cellulose fiber to fibrils. This method can be used to generate cellulose microfibril or bundles, commonly called microfibrillated cellulose (MFC).

Cellulose nanocrystals can be generated by chemical treatment, especially acid hydrolysis, from various biomass resources. A typical procedure to prepare cellulose nanocrystals has five steps: hydrolysis by acid, centrifugation and neutralization, rinsing with deionized water by centrifuge, dispersion by ultrasonic irradiation, and ultrafiltration to remove any remaining ions (Choi and Simonsen, 2006). Cellulose whiskers or nanocrystals have been obtained from many cellulosic materials, such as wood fibers, cotton, ramie, tunicate mantles, sugar beet pulp, and bacterial cellulose (Beck-Candanedo et al., 2005; Chakraborty et al., 2005; Dufresne et al., 1997b; Edgar and Gray, 2002; Kvien et al., 2005; Lu et al., 2005; Pu et al., 2007; Zimmermann et al., 2004). In terms of mechanical methods, a process using high-intensity ultrasonication (HIUS) was developed at the Center for Renewable Carbon to isolate fibrils from several cellulose resources, which is discussed in Section 3.3.

3.2 Cellulosic nanocrystals/microfibers characterization

Many methods have been used for fibril characterization, including morphological characterizations, degree of fibrillation, and crystalline characterizations. Several tools can be used for morphological observation, such as a scanning electron microscope (SEM), an atomic force microscope (AFM), and a transmission electron microscopy (TEM) (Bhatnagar and Sain, 2005; Bondeson et al., 2006; Chakraborty et al., 2005; Choi and Simonsen, 2006; Wang and Sain, 2007; Zimmermann et al., 2004). The degree of fibrillation of the fibers used to isolate MFC/microfibrils/aggregates by mechanical methods may be evaluated by water retention value (WRV) (Cheng et al., 2007a; Cheng et al., 2010a, b; Herrick et al., 1983; Turbak et al., 1983; Wang and Cheng, 2009; Yano and Nakahara, 2004). Wide angle X-ray diffraction (WAXD) is used to study the crystallinity of the fibrils (Bhatnagar and Sain, 2005; Chakraborty et al., 2006; Cheng et al., 2007a; Gindl and Keckes, 2005).

Mechanical properties of a single cellulose fibril are very important, especially when the ultimate use of the fibrils is for reinforcing composite materials. The elastic modulus of the crystalline region of cellulose has been determined either experimentally or theoretically, such as experimental deformation micromechanics of natural cellulose fibers using Raman spectroscopy and X-ray diffraction (Matsuo, 1990; Sakurada et al., 1962; Wiley and Atalla, 1987). Mechanical properties of a single polymeric nanofiber and bacterial cellulose nanofibers have been measured by an atomic force microscope (AFM) (Cheng and Wang, 2008; Cheng et al., 2009a; Tan and Lim, 2004; Wan et al., 2006).

3.3 Isolation of cellulosic fibril aggregates by ultrasonication

A novel process using high-intensity ultrasonication (HIUS) to isolate fibrils from several cellulose resources (Fig. 2), such as regenerated cellulose fiber (Lyocell fiber), pure cellulose fiber (TC40, 180, and 2500), microcrystalline cellulose (MCC), and pulp fiber, was developed (Wang and Cheng, 2009; Cheng et al, 2010a). HIUS produces very strong mechanical oscillating power (Abramov, 1998), which forms hydrodynamic forces of ultrasound to separate cellulosic fibrils in micro/nano scales.

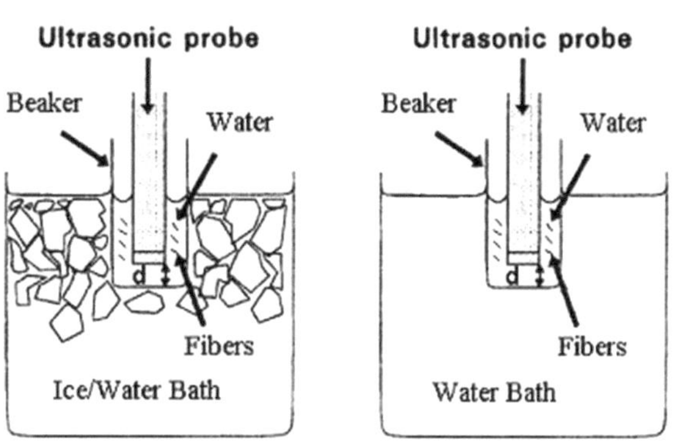

Fig. 2. A scheme for HIUS treatment of fibers using ice/water bath or water-only bath (Reprinted from Wang and Cheng, 2009 with permission from Wiley)

During the HIUS process, cellulose materials were soaked in distilled water until saturated, then a HIUS was directly applied to cellulose fibers suspended in water. Six factors, including ultrasonic power (P, %), treatment time (t, min), treatment temperature (T), fiber concentration (C, %), fiber size (FS, μm), and ultrasonic probe to container bottom distance (d, mm), were considered and three or four levels for each factor were used to check the effects of cellulose fibrillation. Water retention value (WRV) measured by a centrifuge system (Cheng et al., 2007a; Cheng et al., 2010b) was used to evaluate and compare the effects of different factors. The treated and separated fibrils in micro and nano scales by HIUS were investigated using polarized light microscopy, SEM, and AFM. The crystallinities of treated and untreated cellulose fibers were studied by Wide-angle X-ray diffraction (WAXD) and Fourier transform infrared spectroscopy (FTIR) (Cheng et al., 2010a).

The temperature of water suspension during HIUS treatment changes with different powers and cooling methods. Generally a higher temperature can be reached, and a higher speed of temperature increase can be obtained, by using more power. Fig. 3 shows the WRV of the untreated and treated Lyocell fibers for different treatment time and temperature. The degree of microfibrillation of the treated fibers increased as the treatment time increased. It indicates that the fibers became smaller and more surface area on the fibrils as treatment time and temperature increased because WRV is related to fibril and microfibril surface and volumetric phenomena. The longer the treatment time, the smaller the fibrils can be obtained in the suspension mixture. Higher temperature of the suspension was very helpful for cellulose microfibrillation. Measured by Kajaani FiberLab 3, the continuous fiber length distributions of Lyocell fiber and TC180 including before and after treatment, and separated fibrils indicate that there were more short fibers in the treated materials and separated fibrils (Wang and Cheng, 2009).

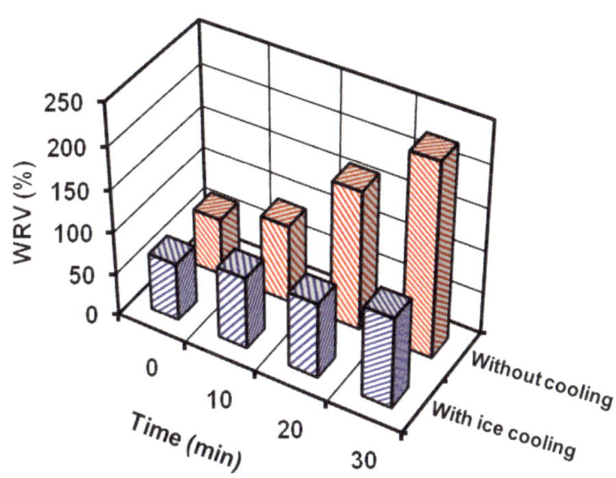

Fig. 3. WRVs of Lyocell fibers treated by HIUS for different durations and at different temperatures (Reprinted from Wang and Cheng, 2009 with permission from Wiley)

After HIUS treatment, a mixture suspension of fiber and fibril with a diameter ranging from microns to tens of nanometers was obtained. Many small fibrils with a diameter less than 1μm were peeled from Lyocell fibers. Some fibrils were still on the surfaces of the big ones that dominate the treated fibers, while some were already isolated from the big fibers. The structure and appearance of the Lyocell fibrils on silicon wafers observed by SEM and AFM are shown in Figs. 4 and 5. The diameters or widths of the fibrils isolated from Lyocell fibers were in a wide range of tens to hundreds nm, and have a wide range of aspect ratio (length/diameter).

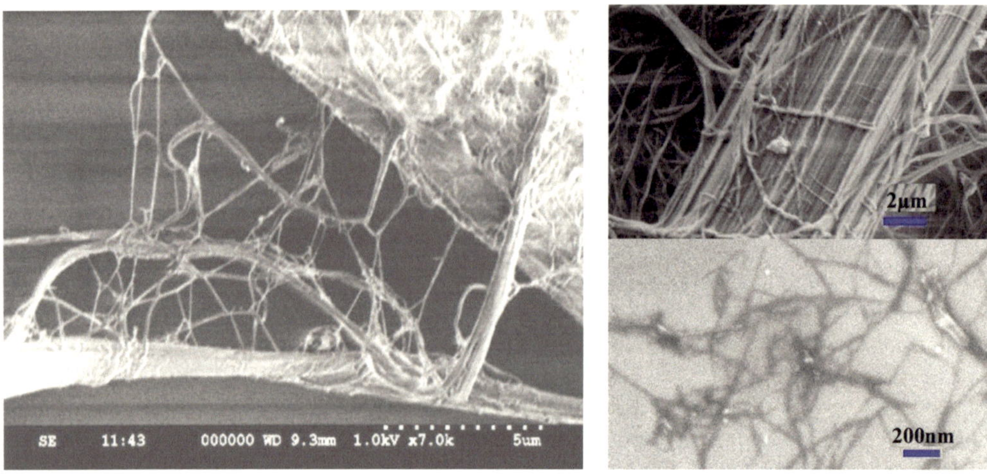

Fig. 4. SEM images of Lyocell fibers and fibrils treated 30 min by HIUS (Reprinted from Cheng et al., 2010a with permission from Wiley)

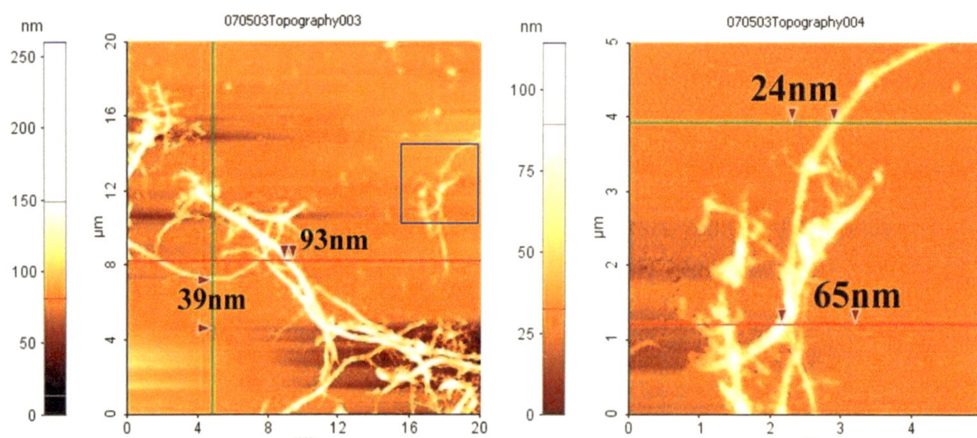

Fig. 5. AFM images of Lyocell fibrils treated 30 min by HIUS (Reprinted from Cheng et al., 2010a with permission from Wiley)

4. Cellulosic nanocomposites

4.1 Fabrication and characterization of cellulosic nanocomposites

The fabrication of nanocomposites with cellulose fibrils of a high aspect ratio for load-bearing applications is relatively new. A wide range of nanocomposites has been studied to investigate the reinforcements of the fibrils/crystals, but there are still some problems, such as the dispersion of cellulosic materials in polymer matrices, especially when they are hydrophobic. The most frequently used method is solvent evaporation casting, including water soluble and other solvent soluble. Nanocomposite materisls can be made by solvent casting method in Teflon or propylene dishes (Favier et al., 1995; Taniguchi and Okamura, 1998). Hot-pressing method after freeze-drying of fibrils can also be used

(Dufresne et al., 1997a). A mat formation followed by resin imersion and curing by UV light was used to make optically transparent composites reinforced with plant fiber-based cellulose microfibrils (Iwamoto et al., 2005). Another mat method was used to make PF resin composite using hot press with high pressures (Nakagaito and Yano, 2005). A method of filtration mats followed by compression molding was also used (Cheng et al., 2007b).

Various analysis instruments can be used to investigate the characterization of the obtained composites, including mechanical properties, morphological characteristics, and thermal properties. Tensile and bending tests are normally used to measure the mechanical properties (Bhatnagar and Sain, 2005; Cheng et al., 2007a, 2009b; Choi and Simonsen, 2006; Taniguchi and Okamura, 1998; Zimmermann et al., 2005; Nakagaito and Yano, 2005). Additionally, dynamic mechnical analysis (DMA) tests can also be used used to evaluate composite performance under various conditions of temperature and relative humidity (Choi and Simonsen, 2006; Favier et al., 1995). For morphological characterizations of nanocomposite, AFM, SEM, and TEM are common tools (Cheng et al. 2009a; Kvien et al., 2005; Pu et al., 2007; Taniguchi and Okamura, 1998; Zimmermann et al., 2004; Zimmermann et al., 2005). These techniques are helpful to investigate the homogeneity of the composite, presence of voids, dispersion level of the fibrils within the continuous matrix, presence of aggregates, sedimentation, the interface between the fibril and the polymer matrix, and possible orientation of the fibrils. Thermal properties are very important for nanocomposites. Differential scanning calorimetry (DSC), Thermogravimetric analyzer (TGA), and DMA have been used to evaluate thermal properties (Ljungberg et al., 2006; Orts et al., 2005; Samir et al., 2006; Wang et al., 2010).

4.2 PVA Cellulosic nanocomposites reinforced with cellulosic nanofibers

Pure cellulose and pulp fiber are the common raw materials for the isolation of nanocrystals and nanofibers in current research. Wood particles, such as hardwood residues and fast plantation poplar, were directly used to isolate cellulose nanifibers/nanocrystals by acid hydrolysis following lignin removal (Cheng et al., 2009c). In this section, we introduce cellulosic nanofibers isolated from juvenile poplar wood samples, and the mechanical and thermal properties of polyvinyl alcohol (PVA) nanocomposites reinforced by these nanofibers (Wang et al., 2010).

Two poplar samples, black cottonwood (*Populus trichocarpa*) (PT) and a hybrid poplar (HP), were used as raw materials. Microcrystalline cellulose (MCC) was used as reference material. Wood stems were debarked, dried, and milled to particles passing a 60-mesh screen. An acidified sodium chlorite solution was used to remove the lignin in wood particles (*Wise, 1946*). After lignin removal, wood particles were then placed in a 60% H2SO4 (v/v) solution for 2 h at 45°C with medium stirring. The obtained mixture was centrifuged and decanted after hydrolysis. Then the solid cellulosic particles were washed and then subjected to ultrasonic irradiation in deionized water to disperse the nanofibers. All the obtained samples were turbid suspension, however, after ultrasonic treatment, the suspensions were transparent. The diameters of most cellulose nanofibers were about 20-30 nm and lengths were in the hundreds nm range as illustrated in AFM height image (Fig. 6), indicating that the nanofibers were still bundles of single nanocrystals since a single wood cellulose crystal is about 3-5 nm in diameter.

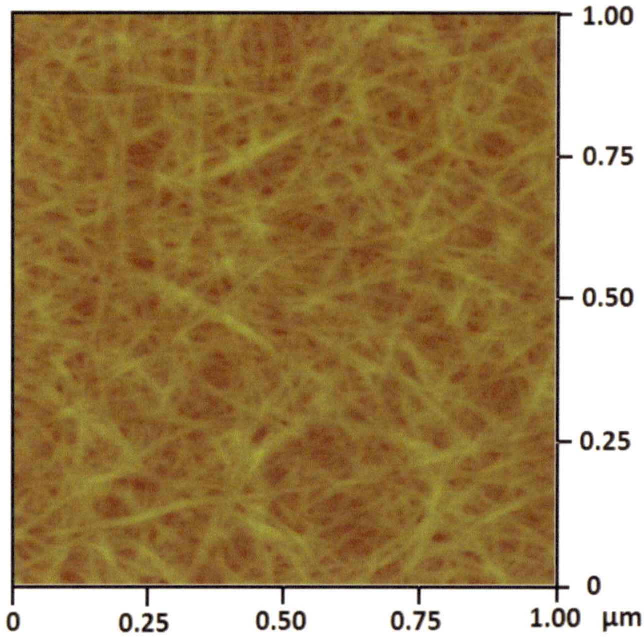

Fig. 6. AFM image of nanofibers from hybrid poplar

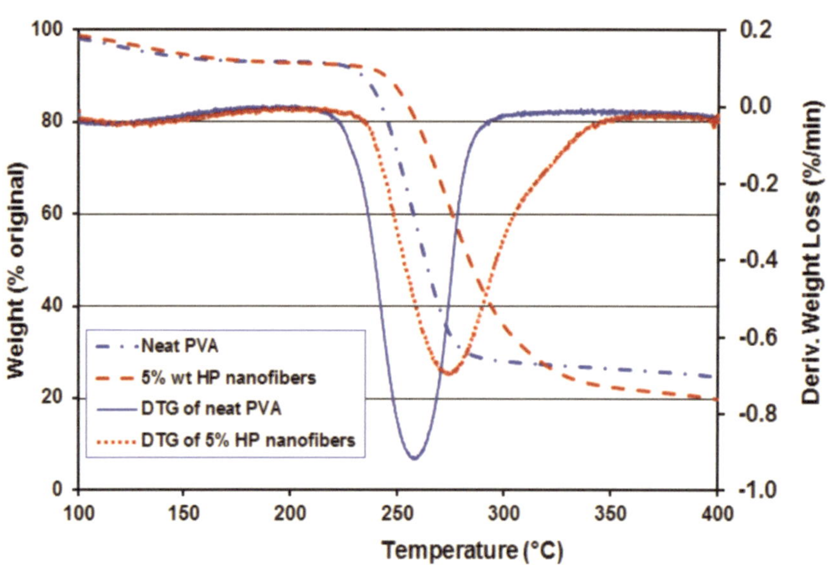

Fig. 7. TGA and DTG thermograms of neat PVA and its nanocomposites with 5% HP nanofibers

Adding cellulosic nanofibers and MCC particles decreased PVA's elongation to break, which indicated that the toughness of PVA increased by these cellulosic materials because the elongation is related to the toughness of a material. The nanofiber reinforced PVA nanocomposites had much higher mechanical properties (tensile modulus and strength) than those of neat PVA and MCC reinforced composites. PVA strength was not decreased by cellulosic nanofibers, but PVA tensile strength was significantly decreased by MCC particles. Presumably, the small nanofibers had higher mechanical properties and higher aspect ratios (length/diameter) than those of MCC. The modulus of PVA was significantly increased by only 2% cellulosic nanofibers (Wang et al., 2010).

Thermogravimetric analyzer (TGA) results indicated that the thermal degradation behavior of PVA was influenced by adding cellulosic nanofibers (Fig. 7). The weight loss of PVA started at its melting point with a very narrow temperature range of degradation, while pure nanofibers started to degrade at a lower temperature and showed a very broad degradation temperature range. Cellulosic nanofibers increased the degradation temperature of PVA. The decomposition peak temperature of the nanocomposites was clearly increased by cellulosic nanofibers. The derivative thermogravimetric (DTG) curves of PVA and 5% HP nanofiber reinforced composites are shown in Fig. 7. The nanocomposite showed thermal events with a shift toward higher temperature compared to the neat PVA, which indicated that the reinforced nanocomposites should be more thermal stable.

4.3 PVA cellulosic nanocomposites reinforced with cellulosic fibrils

Cellulose fibrils, in micro and nano scales, were generated from several cellulose sources, including regenerated cellulose fiber (RCF), pure cellulose fiber (PCF, TC180), and microcrystalline cellulose (MCC), by ultrasonic treatment to reinforce PVA to make biodegradable nanocomposites by film casting. A mixture of fiber and fibril in micro and nano scales was obtained from RCF, PCF, and MCC after HIUS treatment. The tensile properties of PVA were improved significantly by most cellulosic small fibrils, and seperated small fibrils provided better reinforcement than that of untreated big fibers (Cheng et al., 2009b).

Fig. 8 shows the tensile modulus and strength of neat PVA and its composites reinforced by untreated and treated MCC, and small MCC fibril with 2%, 6%, and 10% by weight. The PVA tensile modulus was decreased significantly by 2% of untreated MCC particles, but was increased by treated MCC and seperated small MCC fibrils. The PVA tensile strength was decreased by untreated and treated MCC particles, however, it was increased by small MCC fibrils. All the results indicated that untreated MCC was not suitable for PVA reinforcement because it has a low aspect ratio, while the treated MCC and seperated small fibrils had a high aspect ratio that could reinforce PVA mechanical properties.

The tensile modulus and strength of PVA and its composites reinforced with 2% untreated and treated PCF, and small and big fibrils from PCF are shown in Fig. 9. The composites with small fibrils had much higher modulus and strength properties than those of neat PVA, untreated and treated PCF reinforced composites. However, the mechnical properties of the composites with big fibrils did not significantly change. Both PVA modulus and strength were decreased significantly by untreated PCF. Seperated small PCF fibril reinforced composites had higher improvement for both modulus and strength than those of big fibrils, presumerly because small PCF fibrils had higher mechanical properties than those of big fibrils (Cheng and Wang, 2008; Cheng et al., 2009a).

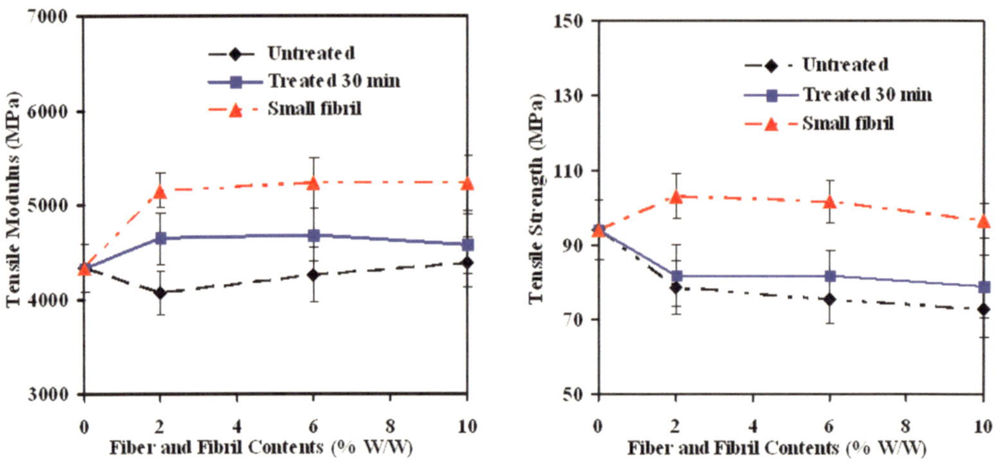

Fig. 8. Tensile modulus and strength of PVA and its composites reinforced by untreated and treated MCC, and small MCC fibrils (Reprinted from Cheng et al., 2009b with permission from Elsevier)

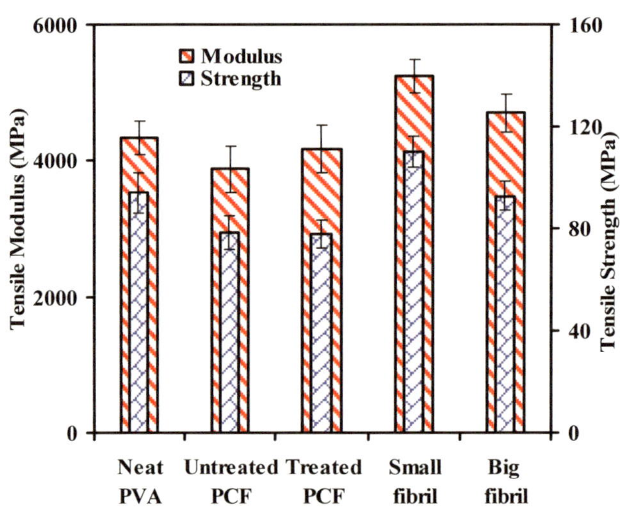

Fig. 9. Tensile modulus and strength of PVA and its composites reinforced by untreated and treated PCF, and small and big PCF fibrils (Reprinted from Cheng et al., 2009b with permission from Elsevier)

Table 1 shows the tensile modulus and strength of neat PVA and its composites reinforced by untreated and treated RCF fibers, small RCF fibrils, and commercial microfibrillated cellulose (MFC, used as a reference material) of 2%, 6%, and 10% by weight. PVA tensile modulus was increased significantly by all the fibers, fibrils, and MFC except 2% and 6% of untreated RCF, but the modulus changes were not significant among the filler loadings of 2%, 6%, and 10%, except between the 2% and 6% of MFC. PVA tensile strength was not significantly changed by untreated RCF and MFC except 10% of MFC. However, PVA tensile strenth was increased significantly by only 2% of treated RCF fibers and all the small fibrils. Adding more fiber and fibrils did not increase strength due to the dispersion problem for higher filler loadings as mentioned above. Again, small RCF fibrils had better reinforcement for PVA strength than the treated fiber; and the treated fiber had better reinforcement for both the modulus and strength of PVA than untreated fibers. The RCF fibrils had much better reinforcement for PVA than that of MCC and PCF fibrils, and again, possibly because the aspect ratio of RCF fibrils was higher and the elastic modulus of RCF fibrils was higher than that of natural fibers.

	Modulus (GPa) (PVA: 5.7±0.2)			Strength (MPa) (PVA: 112±6)		
Materials	2%	6%	10%	2%	6%	10%
Untreated	6.0±0.4	6.1±0.3	6.3±0.4	108±6	107±7	105±8
Treated	6.6±0.2	6.7±0.2	6.5±0.4	125±6	118±7	115±7
Small	6.5±0.4	6.8±0.4	6.8±0.3	127±6	124±7	130±7
MFC	6.3±0.3	7.1±0.3	7.2±0.4	112±10	113±8	101±7

Table 1. Tensile modulus and strength with standard deviations of PVA and its composites reinforced with untreated and treated RCF fibers, small RCF fibrils, and MFC (Reprinted from Cheng et al., 2009b with permission from Elsevier)

SEM images of fracture surface after tensile testing and AFM images of cross-sections after cutting by microtome from the PVA nanocomposites reinforced by small RCF fibrils are shown in Figs. 10 and 11, respectively. The dispersion of small fibrils was not perfectly uniform observed from the cross-sections. Some pulled out holes after tensile test were detected. These results indicated that the increments of tensile strength and modulus of the composites were not as high as expected, especially for higher fibril loadings because the fibril dispersion of 10% fibril loading composite was much worse than 2% fibrils loading samples, which possiblly was the main reason why increased amounts of filler did not further reinforce the PVA composites.

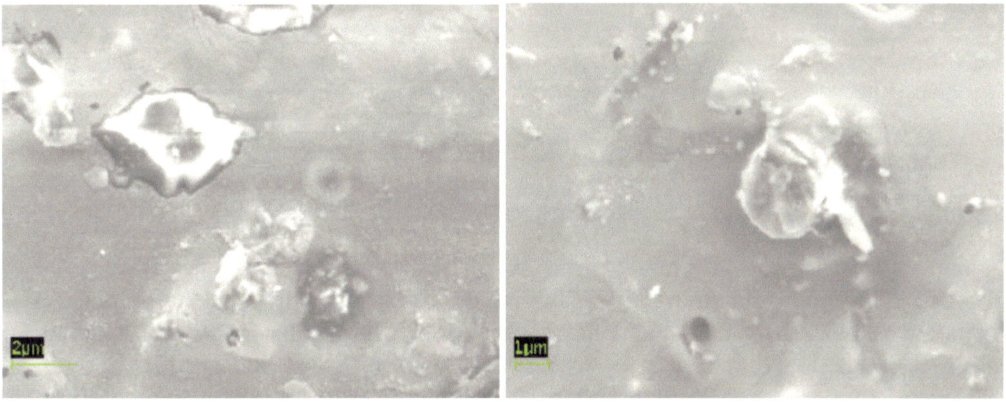

Fig. 10. SEM images of the fractured cross-sections of PVA composites reinforced with small RCF fibrils (Reprinted from Cheng et al., 2009b with permission from Elsevier)

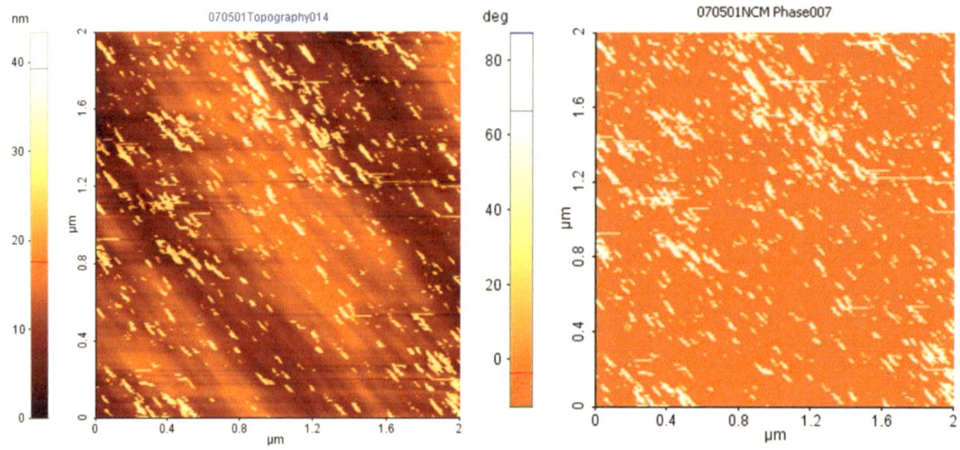

Fig. 11. AFM topography and phase images of the cross-sections of PVA composites reinforced with small RCF fibrils (Reprinted from Cheng et al., 2009b with permission from Elsevier)

4.4 PLA and PP cellulosic composites reinforced with fibril aggregates

Fibril aggregates were generated from a regenerated cellulose fiber (RCF, Lyocell fiber) by ultrasonication treatment. Composites of poly(lactic acid) (PLA, biodegradable) and polypropylene (PP, non-biodegradable) reinforced with fibril aggregates were prepared by compression molding. For PLA composites, the fibril aggregates and the PLA polymer were combined together using multiple layers followed by compressive molding (Fig. 12) (Cheng et al., 2007b). The PP composites were made from filtered mats of cellulose fibril aggregates and PP fibers by compression molding. The mats were made from the well-stirred mixture of cellulose fibril aggregates and PP fibers by a filtration system using the same setup above for fibrils mats. The formed mixture mats were oven dried, put in a round metal mold with a nominal thickness of 254 µm, and hot pressed at a temperature of 170 °C and a pressure of 5 MPa for approximately 5 minutes. The physical and mechanical properties of these composite materials were evaluated by SEM and tensile tests, respectively (Cheng et al., 2007a).

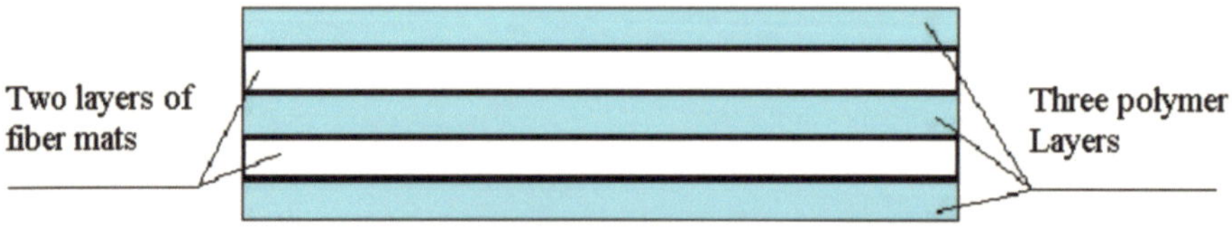

Fig. 12. Sandwich structure of PLA and fibril composites for compressive molding

The tensile modulus and strength of PLA and its composites reinforced by untreated and treated Lyocell fibers by loading of 10% and 20% indicated that all composites have a higher tensile modulus than that of pure PLA. The untreated and treated fibers of 10% loading did not change PLA tensile strength. The tensile strength of the 20% fiber loading composites with untreated fibers decreased 18%. However, the treated fibers increased the tensile strength about 14% compared to pure PLA, which may be attributed to the contribution of the fibrils isolated from the fibers and microfibrillation on the fiber surface. Some gaps around big fibers were observed by SEM images, indicating the adhesion between the PLA and the mixture of fibers and fibrils was poor (Fig. 13).

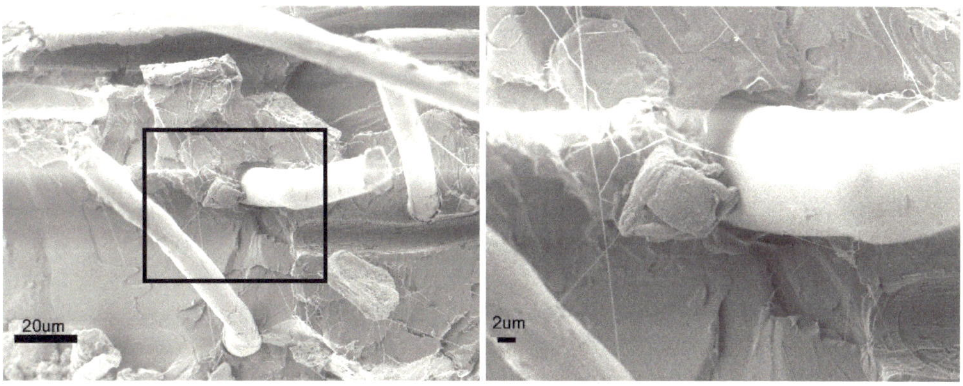

Fig. 13. The fractured cross-sections of PLA composites by SEM: (left) overview, and (right) detailed view

Some typical tensile curves of neat PP and its composites with 10% and 20% of untreated and treated Lyocell fibers are shown in Fig. 14. Higher fiber loading composites had higher tensile strength and lower elongation to break, indicating that the composites were becoming more brittle.

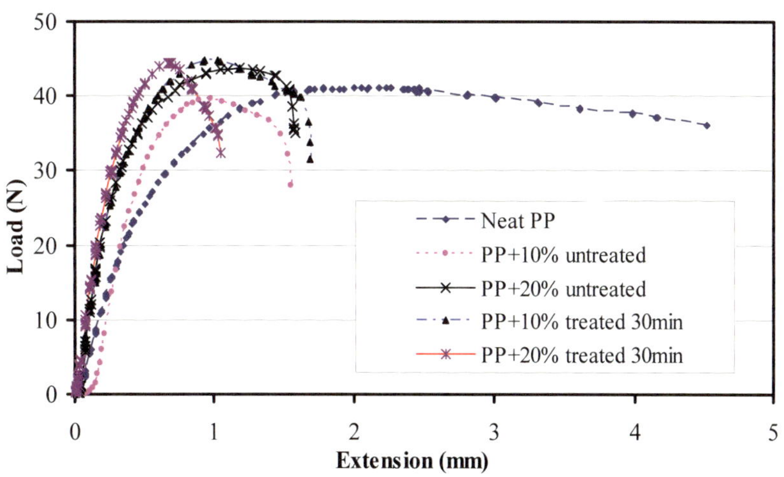

Fig. 14. Typical stress-strain curve of PP and its composites

The results of tensile modulus and strength of PP and its composites reinforced with untreated and treated Lyocell fibers indicated that all composites have higher tensile modulus and strength than those of neat PP. The treated fibers had higher mechanical reinforcement for PP composites. The PP tensile modulus was increased 81% and 96% by loadings of 10% and 20% untreated fibers, respectively. The PP tensile modulus was increased 93% and 130% by adding 10% and 20% of treated fibers, respectively. The fibers and fibril aggregates had a good distribution observed from the thickness of the composites, but some holes in the PP matrix and some gaps between fibers and PP matrix were observed. The presence of the discontinuities between the PP matrix and fibers from the selected broken cross-sections indicated that there was not good adhesion between the PP matrix and the fibrils and fibril aggregates (Cheng et al., 2007a).

5. Conclusions

Natural cellulosic fibers, particles, fibrils (in micro and nano scale), and crystals/whiskers are alternative reinforcement materials that are able to replace man-made fibers, such as glass and aramid fibers to make environmentally friendly products. Cellulose fibrils in micro and nano scales can be isolated from many resources, including wood particles, regenerated cellulose fiber, pure cellulose fiber, and pulp fiber by chemical and/or mechanical methods. Most current methods are low yield, not environmental friendly and/or not energy efficient. A method using high intensity ultrasonic (HIUS) was developed to isolate fibrils from several cellulose resources: regenerated cellulose fiber, pure cellulose fiber, microcrystalline cellulose, and pulp fiber. The obtained material is a mixture of fibrils in micro and nano scales. Small fibrils can be separated by centrifuge. Water retention values revealed that the cellulose fibrillation was significantly increased by HIUS treatment for all cellulose resources. The mixture, and separated small fibril aggregates, can be used to increase the tensile modulus and strength of PVA, PLA, and PP by fabrication methods of film casting and compression molding, respectively. Because there were not perfect adhesions between the polymers and the fibers and fibril aggregates without further modification of cellulose and/or polymers, the polymer reinforcements with these cellulosic fibrils were not as high as expected strength and stiffness. The small fibrils on the surfaces of big fibers and those isolated from the fibers reinforced the polymer composites with higher tensile modulus and strength compared with the composites with untreated fibers. Cellulose nanofibers/nanocrystals can be isolated from wood particles, such as poplar wood samples, using acid hydrolysis after lignin removel. Most nanofibers were bundles of single nanocrystals if the acid hydrolysis conditions are mild, e.g. lower temperature and/or lower acid concentration. These nanofibers can be used to reinforce PVA when fabricating bio-nanocomposites and had much better performance than commercial MCC for PVA reinforcement.

6. Acknowledgments

The authors thank the US DOE/NETL MARCEE Project (DE-FC26-04NT42136) and USDA Wood Utilization Research Special Program, and Tennessee Agricultural Experiment Station project # 96 for funding.

7. References

Abramov, O. (1998). *High-Intensity Ultrasonic Theory and Industrial Applications.* Gordon and Breach Science Publishers. ISBN 90-5699-041-1, Amsterdan

Beck-Candanedo, S.; Roman, M. & Gray, D.G. (2005). Effect of reaction conditions on the properties and behavior of wood cellulose nanocrystal suspensions. *Biomacromolecules*, 6, 2, 1048-1054, ISSN 1525-7797

Berglund, L. (2005). Cellulose-Based Nanocomposites. In: *Natural fibers, biopolymers, and biocomposites,* Mohanty, A.; Misra, M. & Drzal, L. (Ed.), 807-832, Taylor & Francis, ISBN 0-8493-1741-X, Boca Raton, FL

Bhatnagar, A. & Sain, M. (2005). Processing of cellulose nanofiber-reinforced composites. *Journal of Reinforced Plastics and Composites,* 24, 12, 1259-1268, ISSN 0731-6844.

Bondeson, D.; Mathew, A. & Oksman, K. (2006). Optimization of the isolation of nanocrystals from microcrystalline cellulose by acid hydrolysis. *Cellulose,* 13, 2, 171-180, ISSN 0969-0239

Chakraborty, A.; Sain, M. & Kortschot, M. (2005). Cellulose microfibrils: A novel method of preparation using high shear refining and cryocrushing. *Holzforschung,* 59, 1, 102-107, ISSN 0018-3830

Chakraborty, A.; Sain, M. & Kortschot, M. (2006). Reinforcing potential of wood pulp-derived microfibres in a PVA matrix. *Holzforschung,* 60, 1, 53-58, ISSN 0018-3830

Chaplin M. (2010). Water Structure and Science. http://www.lsbu.ac.uk/water/hycel.html, (September, 2010)

Cheng, Q. & Wang, S. (2008). A method for testing the elastic modulus of single cellulose fibrils via atomic force microscopy. *Composites Part A: Applied Science and Manufacturing,* 39, 1838-1843, ISSN 1359-835X

Cheng, Q.; Wang S. & Han, Q. (2010a). Novel process for isolating fibrils from cellulose fibers by high intensity ultrasonication. II: Fibril characterization. *Journal of Applied Polymer Science,* 115, 5, 2756–2762, ISSN 0021-8995

Cheng, Q.; Wang S. & Harper D. (2009a). Effects of process and source on elastic modulus of single cellulose fibrils evaluated by atomic force microscopy. *Composites Part A: Applied Science and Manufacturing,* 40, 583-588, ISSN 1359-835X

Cheng, Q.; Wang S. & Rials, T. (2009b). Poly(vinyl alcohol) nanocomposites reinforced with cellulose fibrils isolated by high intensity ultrasonication. *Composites Part A: Applied Science and Manufacturing,* 40, 218-224, ISSN 1359-835X

Cheng, Q.; Wang, J. & McNeel, J. (2009c). Biodegradable Nanocomposites from Wood Cellulose Residuals. *Proceedings of 10th International Conference on Wood & Biofiber Plastic Composites & Cellulose Nanocomposites Symposium,* pp. 388, ISBN 978-1-892529-55-8, May 2009, Madison, WI.

Cheng, Q.; Wang, J.; McNeel J. & Jacobson P. (2010b). Water retention value measurements of cellulosic materials using a centrifuge technique, *BioResources,* 5, 3, 1945-1954, ISSN 1930-2126

Cheng, Q.; Wang, S.; Rials, T. & Lee, S.H. (2007a). Physical and mechanical properties of polyvinyl alcohol and polypropylene composite materials reinforced with fibril aggregates isolated from regenerated cellulose fibers. *Cellulose,* 14, 6, 593-602, ISSN 0969-0239

Cheng, Q.; Wang, S.; Zhou, D.; Zhang, Y. & Rials, T. (2007b). Lyocell-derived cellulose microfibril/nanofibril and its biodegradable nanocompsites. *Journal of Nanjing Forestry University,* 31, 4, 21-26, ISSN 1000-2006

Choi, Y.J. & Simonsen, J. (2006). Cellulose nanocrystal-filled carboxymethyl cellulose nanocomposites. *Journal of Nanoscience and Nanotechnology,* 6, 3, 633-639, ISSN 1550-7033

Dufresne, A.; Cavaille, J. & Helbert, W. (1997a). Thermoplastic nanocomposites filled with wheat straw cellulose whiskers. 2. Effect of processing and modeling. *Polymer Composites*, 18, 198-210, ISSN 0272-8397

Dufresne, A.; Cavaille, J.Y. & Vignon, M.R. (1997b). Mechanical behavior of sheets prepared from sugar beet cellulose microfibrils. *Journal of Applied Polymer Science*, 64, 6, 1185-1194, ISSN 0021-8995

Edgar, C.D. & Gray, D.G. (2002). Influence of dextran on the phase Behavior of suspensions of cellulose nanocrystals. *Macromolecules*, 35, 19, 7400-7406, ISSN 0024-9297

Favier, V.; Canova, G.R.; Cavaillé, J.Y.; Chanzy, H.; Dufresne, A. & Gauthier, C. (1995). Nanocomposite Materials from Latex and Cellulose Whiskers. *Polymers for Advanced Technologies*, 6, 5, 351-355, ISSN 1042-7147

Franks, A. (1987). Nanotechnology. *Journal of Physics E: Scientific Instruments*, 20, 1442-1451, ISSN 0022-3735

George, J.; Sreekala, M.S. & Thomas, S. (2001). A review on interface modification and characterization of natural fiber reinforced plastic composites. *Polymer Engineering and Science*, 41, 9, 1471-1485, ISSN 1548-2634

Gindl, W. & Keckes, J. (2005). All-cellulose nanocomposite. *Polymer*, 46, 23, 10221-10225, ISSN 0032-3861

Herrick, F.W.; Casebier, R.L.; Hamilton, J.K. & Sandberg, K.R. (1983). Microfibrillated cellulose: morphology and accessibility. *Journal of Applied Polymer Science: Applied Polymer Symposium*, 37: 797-813, ISSN 0271-9460

Iwamoto, S.; Nakagaito, A.N.; Yano, H. & Nogi, M. (2005). Optically transparent composites reinforced with plant fiber-based nanofibers. *Applied Physics A-Materials Science & Processing*, 81, 6, 1109-1112, ISSN 0947-8396

Jordan, J.; Jacob, K.I.; Tannenbaum, R.; Sharaf, M.A. & Jasiuk, I. (2005). Experimental trends in polymer nanocomposites - a review. *Materials Science and Engineering A-Structural Materials Properties Microstructure and Processing*, 393, 1-2, 1-11, ISSN 0921-5093

Kvien, I.; Tanem, B.S. & Oksman, K. (2005). Characterization of cellulose whiskers and their nanocomposites by atomic force and electron microscopy. *Biomacromolecules*, 6, 6, 3160-3165, ISSN 1525-7797

Lima, M.M.D. & Borsali, R. (2004). Rodlike cellulose microcrystals: Structure, properties, and applications. *Macromolecular Rapid Communications*, 25, 7, 771-787, ISSN 1022-1336

Ljungberg, N.; Cavaille, J.Y. & Heux, L. (2006). Nanocomposites of isotactic polypropylene reinforced with rod-like cellulose whiskers. *Polymer*, 47, 18, 6285-6292, ISSN 0032-3861

Lu, Y.S.; Weng, L.H. & Cao, X.D. (2005). Biocomposites of plasticized starch reinforced with cellulose crystallites from cottonseed linter. *Macromolecular Bioscience*, 5, 11, 1101-1107, ISSN: 1616-5187

Matsuo, M. (1990). General-Analysis of the Measurement of the Crystal-Lattice Modulus of Semicrystalline Polymers by X-Ray-Diffraction. *Macromolecules*, 23, 13, 3261-3266, ISSN 0024-9297

Nakagaito, A.N. & Yano, H. (2005). Novel high-strength biocomposites based on microfibrillated cellulose having nano-order-unit web-like network structure. *Applied Physics A-Materials Science & Processing*, 80, 1, 155-159, ISSN 0947-8396

O'Sullivan, A.C. (1997). Cellulose: the structure slowly unravels. *Cellulose,* 4, 173-207, ISSN 0969-0239

Orts, W.J.; Shey, J.; Imam, S.H.; Glenn, G.M.; Guttman, M.E. & Revo, J.F. (2005). Application of cellulose microfibrils in polymer nanocomposites. *Journal of Polymers and the Environment,* 13, 4, 301-306, ISSN 1566-2543

Pu, Y.; Zhang, J.; Elder, T; Deng, Y.; Gatenholm P. & Ragauskas A. (2007). Investigation into nanocellulosics versus acacia reinforced acrylic films. *Composites Part B-Engineering,* 38, 3, 360-366, ISSN 1359-8368

Sakurada, I.; Nukushina, Y. & Ito, T. (1962). Experimental Determination of Elastic Modulus of Crystalline Regions in Oriented Polymers. *Journal of Polymer Science,* 57, 165, 651-660, ISSN 1099-0518

Samir, M.; Alloin, F. & Dufresne, A. (2005). Review of recent research into cellulosic whiskers, their properties and their application in nanocomposite field. *Biomacromolecules,* 6, 2, 612-626, ISSN 1525-7797

Samir, M.; Alloin, F. & Dufresne, A. (2006). High performance nanocomposite polymer electrolytes. *Composite Interfaces,* 13, 4-6, 545-559, ISSN 0927-6440

Sarko, A. & Muggli, R. (1974). Packing Analysis of Carbohydrates and Polysaccharides .3. Valonia Cellulose and Cellulose-Ii. *Macromolecules,* 7, 4, 486-494, ISSN 0024-9297

Sjostrom, E. (1993). *Wood chemistry fundamentals and applications,* Academic Press, ISBN 10: 0126474818, New York

Tan, E.P.S. & Lim, C.T. (2004). Physical properties of a single polymeric nanofiber. *Applied Physics Letters,* 84, 9, 1603-1605, ISSN 0003-6951

Taniguchi, T. & Okamura, K. (1998). New films produced from microfibrillated natural fibres. *Polymer International,* 47, 3, 291-294, ISSN 0959-8103

Turbak, A.F., Snyder, F.W. and Sandberg, K.R., 1983. Microfibrillated cellulose, a new cellulose product: properties, uses, and commercial potential. *Journal of Applied Polymer Science: Applied Polymer Symposium,* 37, 815-827, ISSN 0271-9460

Wan, W.K.; Hutter, J.L.; Millon, L. & Guhados, G. (2006). Bacterial cellulose and its nanocomposites for biomedical applications, In: *Cellulose Nanocomposites: Processing, Characterization, and Properties,* Oksman, K. & Sain, M. (Ed.), 221-241, ISBN13: 9780841239807, American Chemical Society, Washington DC

Wang, J.; Cheng, Q.; Adebayo, A. & Difazio, S. (2010). Polyvinyl alcohol nanocomposites reinforced with cellulosic nanofibers isolated from juvenile poplar. In: *Fiber-reinforced Composites,* Cheng, Q. (Ed.), (In print), Nova Science Publishers, Inc., ISBN: 978-953-307-468-9, Hauppauge, NY

Wang, B. & Sain, M. (2007). Isolation of nanofibers from soybean source and their reinforcing capability on synthetic polymers. *Composites Science and Technology,* 67, 11-12, 2521-2527, ISSN 0266-3538

Wang, S. & Cheng, Q. (2009). A novel process to isolate fibrils from cellulose fibers by high intensity ultrasonication. Part I: Process optimization. *Journal of Applied Polymer Science,* 113, 1270–1275, ISSN 0021-8995

Wiley, J.H. & Atalla, R.H. (1987). Band Assignments in the Raman-Spectra of Celluloses. *Carbohydrate Research,* 160, 113-129, ISSN 0008-6215

Wise L.E.; Murphy M. & Addieco A.A. (1946) Isolation of holocellulose from wood. *Paper Trade Journal,* 122, 35-43, ISSN 0031-1197

Woodcock, C. & Sarko, A. (1980). Packing Analysis of Carbohydrates and Polysaccharides. 11. Molecular and Crystal-Structure of Native Ramie Cellulose. *Macromolecules,* 13, 5, 1183-1187, ISSN 0024-9297

Yano, H. & Nakahara, S. (2004). Bio-composites produced from plant microfiber bundles with a nanometer unit web-like network. *Journal of Materials Science,* 39, 5, 1635-1638, ISSN 0022-2461

Zimmermann, T.; Pohler, E. & Schwaller, P. (2005). Mechanical and morphological properties of cellulose fibril reinforced nanocomposites. *Advanced Engineering Materials,* 7, 12, 1156-1161, ISSN 1438-1656

Zimmermann, T.; Pohler, E. & Geiger, T. (2004). Cellulose fibrils for polymer reinforcement. *Advanced Engineering Materials,* 6, 9, 754-761, ISSN 1438-1656

22

Composite Material for Shielding Mixed Radiation

Hu Huasi
School of Nuclear Science and Technology, Xi'an Jiaotong University (XJTU), Xi'an, 710049 Shannxi, China

1. Introduction

1.1 Composite material design idea for shielding mixed radiation

What we have done is to design a shielding material of high performance secures us or nuclear facilities from radiation. However, most of you are unfamiliar with nuclear physics, so here you may consider the nuclear radiation shielding, the transportation of neutrons and other particles in material, as small balls with velocity rolling on the surface of Mars. Admittedly, this hypothesis is questionable, but it may prepare you for understanding. Now, we take neutron transportation as an example. In left of Fig.1, Twenty small bolls of high speed (fast neutrons) roll towards right on Mars. Some of them, three or more, trapped by hollows disappear, the others have collided against the rocks scatter with lower energy, only a few of them may go across holding tiny energy (thermal neutrons). Nuclear reaction cross sections determine probability of trapping and collision.

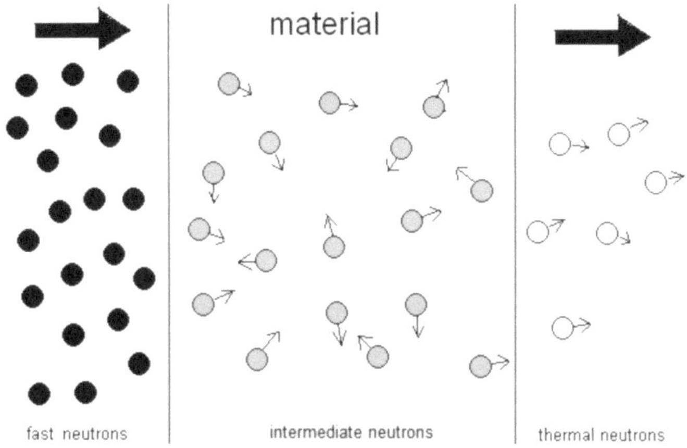

Fig. 1. The transportation of neutrons in material

This work was supported by the National Natural Science Foundation of China (Grant No.10576022, 10975113), the National High Technology Research and Development Program of China (Grant No.2009AA050705), and NINT Contract 200509006.
E-mail: huasi_hu@mail.xjtu.edu.cn

The following picture Fig.2 is the brief interaction of neutrons with other materials. The heavy metal elements, such as tungsten and iron, slow down the fast neutron to intermediate neutron by the inelastic scattering and resonance scattering; then, the low-Z elements by elastic scattering cool the intermediate neutron down into the thermal neutron group. Finally some elements such as boron capture the thermal neutrons. The high-Z elements absorb both original and secondary γ-rays.

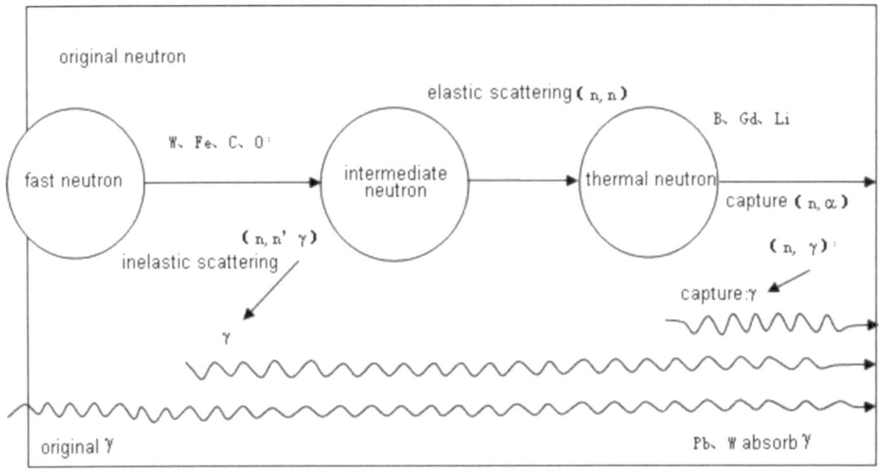

Fig. 2. The brief interaction mechanism

Nowadays, many kinds of materials were employed as nuclear radiation shielding material, while, the requirements of radiation shielding become more and more strictly. Especially in mobile nuclear devices and manned spacecrafts [1][2], the radiation shielding materials must have high effective shielding properties for the sake of the restriction the materials' weight and volume. Also, shielding materials have to service for a long time in quite inclement environment with high temperature, corrosion, and so on. Furthermore, shielding materials could be damaged by constant radiation due to interaction with each other [3][4].

1.2 Genetic algorithm

Genetic algorithm is an natural selection on a computer, it makes biological evolution in a computational model. John H.holland was the first one who establish the biological evolution as a model and developed this method. In the long period, the creature has gradually adapted to the environment during its continued existence of biological process, and it was improved. The base of evolution is the improvement of individual. Each individual has different ability for different living environment. According to Darwin's theory of evolution, those with a strong ability to adapt to environmental change are more viable organisms, have bigger chance to survive, and that creature may have more offsprings. Those with a weak ability to adapt to environmental change are doomed to disappear. This is called "natural selection, survival of the fittest." Through natural selection, species will gradually improving to survive in the living environment, and finally some developed species appear. Following the biological evolution, a GA allows a population which is composed of many individuals to evolve in specified selection rules to a state with best fitness

The basic theories of Genetic algorithms can be divided into four parts

1.2.1 Original population

GA is based on populational optimization algorithms. The process of searching optimized solution starts from one original population. The original population was often generated by a random method. The original population determines the time cost in searching. And experiential data can help to cut searching time, but, it may lead to a factually incorrect result. So this section requires elaborate work.

1.2.2 Construction fitness function

The key point in Genetic algorithms is the construction of a fitness function. Employing this function, we could evaluate the individual "healthy" which determines the chance of this individual to live longer or has descendants. In another words, it determines the individual or solution whether fits our requirement or not.

1.2.3 Selection and evolution

Natural selection in GA is associated with the fitness value. With a higher fitness value, one gets a bigger chance of survival and multiplication, and vice versa. The average fitness value of population may rise, it is the symbol of evolution.

1.2.4 Optimized solution obtained

If the average fitness value rises to a maximum, we may get a preferable individual with the highest fitness value in the population. Admittedly, this is, maybe, not the real best, but compared with the endless time of searching a real what we got is often acceptable.

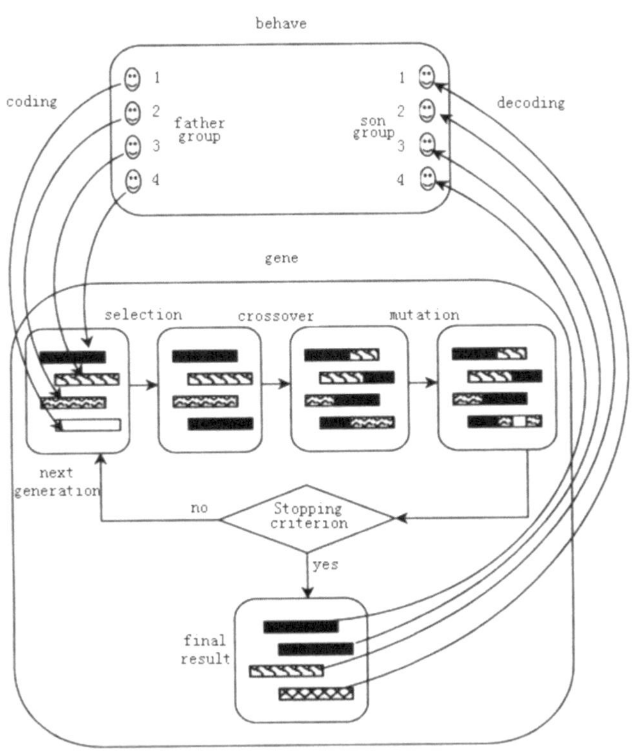

Fig. 3. Encoding Scheme

The following are the key points of GA. The first is population. This population represents one of potential solutions group. Generally speaking, the larger the population size is, the better result we will get. But, it will take more time for calculating. The second is Encoding Scheme (gene representation), Fig.3. Each individual in population is actually the entity with characteristic chromosome. Chromosome is a combination of multiple genes. The combination of genes determines the performance of individual. It requires help from genes to set up the relationship between chromosome and the solution to the optimization problem. That is the encoding of chromosome because of the complexity of encoding. We often employ a easy method, binary encoding, to simulate the gene encoding. The third is Genetic operator, it includes crossover and mutation. It simulates the natural process of reproduction, and it shows the relationship of chromosomes. The fourth is selection tactic. It reveals a natural law of "natural selection, survival of the fittest." in biological evolution.

1.2 GENOCOPII

In our work, GENOCOPII [5], which can prevent us from the trouble of details in Genetic algorithm code, is employed. It is a higher version of GENOCOP which is the acronym of Genetic Algorithm for Numerical Optimization Problem for Constrained Problems. It is a genetic algorithm-based program written in C for constrained and unconstrained optimization.

1.3 MCNP code

The MCNP code [6], an internationally recognized code for analyzing the transport of neutrons and gamma rays by the Monte Carlo method, is developed and maintained by Los Alamos National Laboratory. This code treats transport of neutrons, gamma rays, and secondary gamma rays stemming from neutron interactions. The MCNP code of a higher version can also deal with the transport of electrons, both primary source electrons and secondary electrons created in gamma-ray interactions. The MCNP code has an input file. In our case, there two named inpn and inpp to feed the code with the information calculating

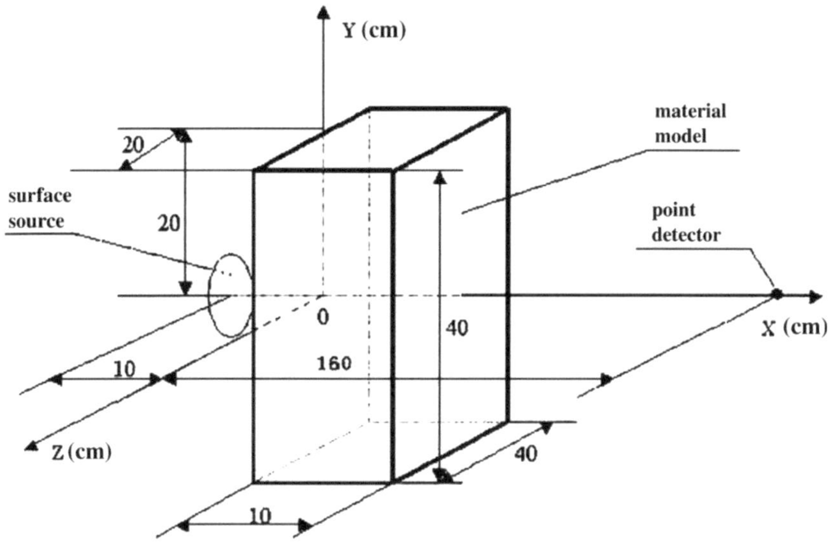

Fig. 4. Schematic diagram

neutrons and gamma rays. Then the calculation finished, the output file with the result is obtained. The figure, Fig 4, is the schematic map of a virtual experiment ran in MCNP code to evaluate the shielding capability of materials.

2. Mathematical modeling in optimal design of shielding material

The optimal design of shielding material focus on the dose equivalent left in the case that neutrons and gamma rays penetrate a specific thickness of the shielding material.
We employed a round surface with fission energy spectra as a radiation source, the dose equivalent corresponding to 2.407 neutrons and 7.77 photons (these are the average number released per fission event [7]) was selected applying the optimized objective function [8][9], this function subjects to certain constraints: equalities, inequalities and domain constraints. They depend on the uniformity of the composite material components, the component range of every element in composite from minimum to maximum, and the density of composite materials:

The Object and Fitness Function:

$$\min f(X) = [f_n(X), f_g(X)]^T \tag{1-1}$$

where:

$f_n(X)$ – total dose equivalent of neutrons and γ-rays with fission energy spectra in one time fission process penetrating a specific thickness of the shielding material, Sv

$f_n(\mathrm{x})$ – contribution of neutrons in dose equivalent, Sv;

$f_g(\mathrm{x})$ – contribution of gamma rays in dose equivalent, Sv;

X – the variable vector, mass components matrix in composite material, $X = \left[x_1, x_2, \ldots, x_p\right]^T$, where x_i (i =1,2,...,p) is the components material ratio.

The constrains:

$$\sum_{i=1}^{p} x_i = 1 \tag{1-2}$$

$$10/\rho(X) = \sum_{i=1}^{p} \frac{10}{\rho_i} x_i \tag{1-3}$$

$$\rho(X) \in D \tag{1-4}$$

$$L \le X \le U \tag{1-5}$$

where:

x_i – subjects to constraint of normalization

$\rho(X)$ – density function of single component and composite material,

D – vector of the density ranges.;

L , U – components vectors of minimum & maximum respectively

3. Process of optimal design of shielding material

3.1 Optimal design

The Fig.4 shows four steps in the GA optimization design program. These steps can be divided into two parts, the first and last steps belong to GENOCOPII part, and the others are MCNP part. The MCNP code here plays a role like one fitness function evaluating shielding capability of materials. There also lies constrains of materials density, components ratio we focus on in GENOCOPII. Firstly, the raw materials are selected from the present materials (simple elements or compound materials) in the world. The MCNP code is recognized for analyzing the transport of neutrons and gamma rays, but its materials part is not so good. It is next to impossible to simulate the real materials but the virtual one of homogeneous mixture of certain elements. In the process of selecting, the cost and the radiation shielding capability are mainly concerned. Their physical and chemical capability, mechanical and thermal properties are also considered by experience. Secondly, according to a component material ratio assumed originally, each virtual materials density, and all other parameters were edited in the proper MCNP input files: "inpn" and "inpp". After that, MCNP calculates one fission neutron dose equivalent Hn (in output file outpn), one

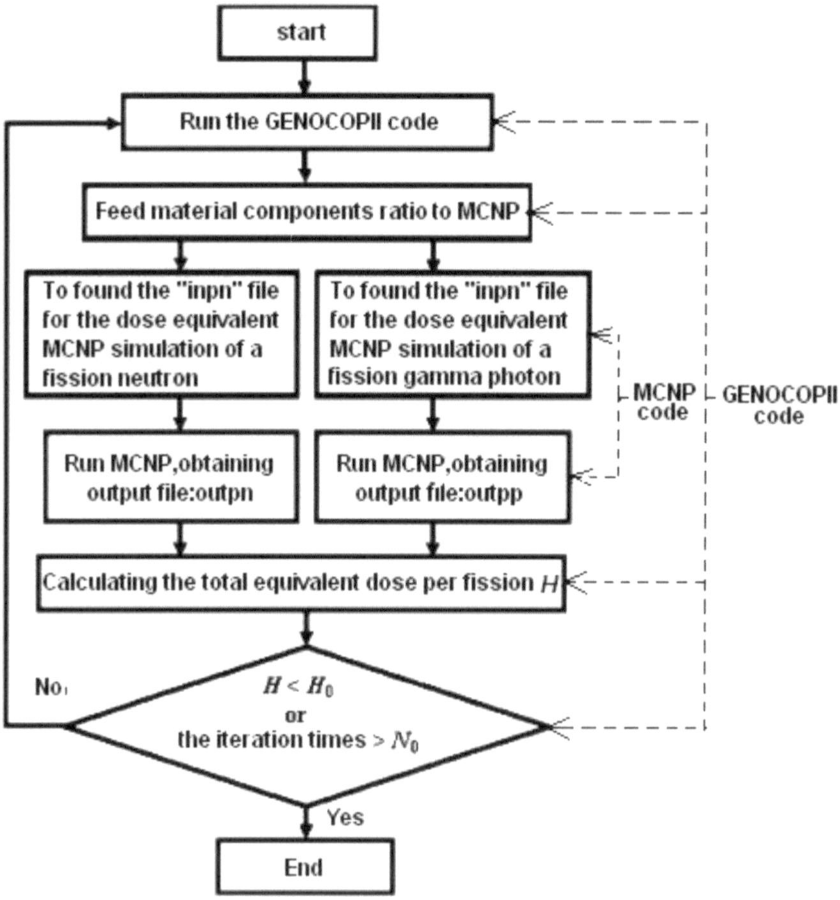

Fig. 5. Optimize procedures

fission photon dose equivalent Hp (in outpp), and the total dose equivalent of the neutrons and -rays per fission. Thirdly, the GA program can transfer the MCNP results at every generation in every count, if so needed. Finally, the object function for optimizing design is a certain function, based on constraint conditions confirmed by some calculations.
The achievement of the final goal, a qualified shielding material, requires an exploratory stage. This indispensable stage sloves some problems, such us, the selection of raw materials, evaluation of components vectors of minimum & maximum respectively, experimental research on manufacturing process, et al.

3.2 Simple cases of optimization design on shielding material

By running the optimal design program, a series of materials optimized ratios have been obtained. They are listed in Table1. Jxa1 and Jxa2 are polyamide composites, Pb6 is anhydride composite, Djy is another composite suited to powder forming with heat isostatic pressing (HIP). Interlayer is also polyamide composite without Fe, W and Pb.

element	H	^{10}B	^{11}B	C	O	Ti	Fe	Gd	W	P	density $/g.cm^3$
Jxa1	2.226	0.316	0.581	17.88	5.307	0.06	0	1.736	5.013	67.29	3.38
Jax2	2.45	0.078	0.31	18.5	5.31	0.06	0	1.74	12.67	57.73	3.53
Pb6	1.523	0.366	1.466	14.567	5.056	0.379	0	0	14.436	62.186	3.72
Djy	0	0	0	26.28	1.95	0	50.17	12.80	8.80	0	5.7
Interlayer	8.19	0.16	0.64	65.87	18.43	0.93	0	5.78	0	0	1.48

Table 1. Shilding material element contents optimal designed /W%

Components	Name	Formula	Density/ $g{\cdot}cm^{-3}$
1	Graphite	C	2.25
2	Carbonization boron	B_4C	2.52
3	Nano TiO_2	TiO_2	0.86
4	Oxidation gadolinium	Gd_2O_3	7.4
5	Iron	Fe	7.86
6	Tungsten carbide	WC	15.5
7	Lead	Pb	11.342

Table 2. The optional shielding materials components under Polyester composite process

The Fig.6 shows the relationship between total dose equivalent and generations of Jxa1.Its obviously that with the population evolution, the value of objective function, that is total dose equivalent, decreases. When the number of generation grows up to 40, the total dose equivalent seems steady. And the individual with best performance having components information is obtained.
Our program works well. Furthermore, according to the interaction mechanism of neutrons and other materials, we develop a multilayer material. The first layer iron is quite effective to slow down the fast neutrons to intermediate neutrons by inelastic scattering. The middle

is the Interlayer, cake, which not only slows down intermediate neutrons to thermal neutrons by elastic scattering on hydrogen (8.19w%) effectively, but also captures thermal and intermediate neutrons by boron-10 (0.16w%) and gadolinium (5.16w%). The last layer, lead, can attenuate the γ -rays and the secondary γ -rays of neutrons. Iron and lead are the common shielding materials for theirs low price and easy machining. But they cannot be used extensively due to the weight and volume limit. Therefore their thickness component ratio is optimized. The Cakes shielding capabilities were simulated.Fig.6.

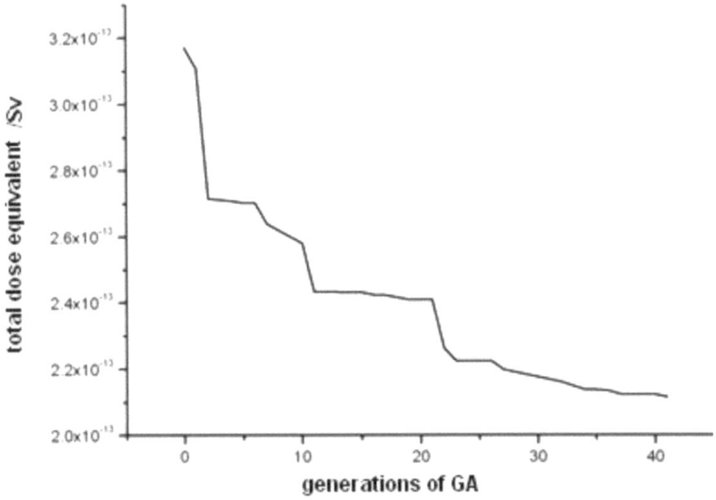

Fig. 6. Total dose equivalent

However, the MCNP code is of low efficiency due to Monte Carlo method. It may take a long time to combine the MCNP code with genetic algorithm for calculating, so we hope MCNP code may give way to some deterministic method. The method of fission neutrons removal cross sections to calculate radiation shielding can be found in some references [10]-[12]. But there lie two problems that have not been solved thoroughly: attenuation of low energy neutrons (thermal and epithermal) and inelastic scattering of fast neutrons.
Further studies may focus on developing a comprehensive optimization design which not only considering the shielding performance but also stability of material in irradiation and other requirements. One promising method is using the multi-scale modeling to predict life span of certain material [13].

4. Trial-producing the composite materials

4.1 The development of the polymer-based composites materials technology manufacture craft

The solid of the resin composites refers to the curing reaction of polymer materials, namely resin cross linking by linear molecular structure change into three-dimensional reticular molecular structure. With the difference of the resin curing temperature and the length time, its solid degree is different. So, having researched the process of the resin curing, we can obtain the relationship between the solid time of the resin curing and temperature. Then employing the variable time and temperature we could change the mechanics and thermal properties. In this paper we used the DSC to analysis solid degree [14].

4.1.1 Resin material

Lots of thermosetting and thermoplastics can be used as composite resin. From the present situation, the commonly used of thermosetting resins are: unsaturated polyester, ethylene polyester, epoxy, phenol, BMI and polyimide matrix, table 2, The common thermoplastic resin: polypropylene (PP), polycarbonate (PC), Nylon (Nylon), PEEK (PEEK), polyether (PES), etc. In the aspect of bearing temperature, heat-resistant performance is superior to thermoplastic resin.

Because the raw materials of the bisphenol A (2) phenol is easily obtain, and it is the cheapest, so the output is largest, about 85% of the total output of epoxy resin[15], people also referred to it as general epoxy resin. The temperature in this research is 273-573K, the bisphenol-A type of epoxy resin can meet the requirement. In addition, it has the advantage of curing, price, molding, etc, so we used the bisphenol A type of epoxy resin type which can meet the requirement, adding to the advantage of curing, price, molding, etc, so we used the bisphenol A type of epoxy resin composite material.

1. the structure of the bisphenol A type of epoxyresin[16]

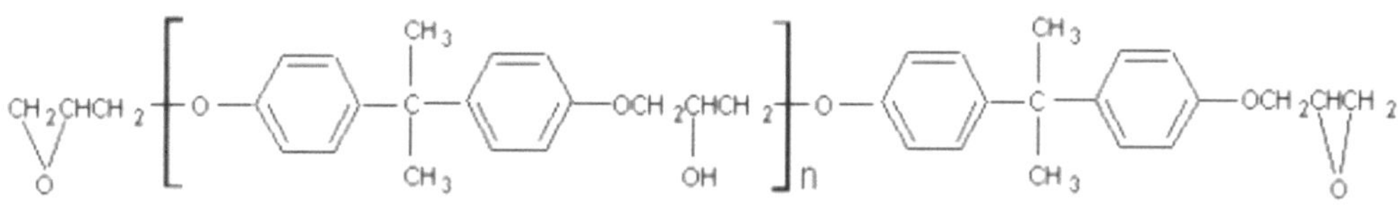

2. epoxy resin curing reaction

Through the analysis of the essential resin curing materials, this paper has selected the materials as belows: E - 51 bisphenol A, PMDA, MA, epoxy propane butyl ether and TiO_2 as the part of resin, the specific materials reaction process as follow [17]:

1. first the the epoxy resin reacts with anhydride then producing carboxylic acid which containing ester chain:

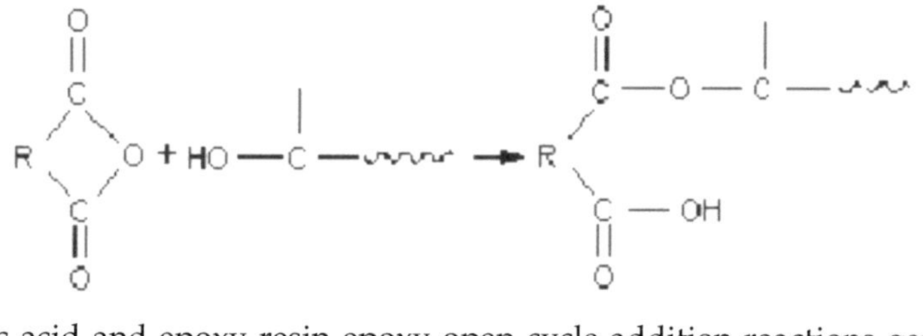

2. Carboxylic acid and epoxy resin epoxy open cycle-addition reactions generating para-ammonium hydroxyl:

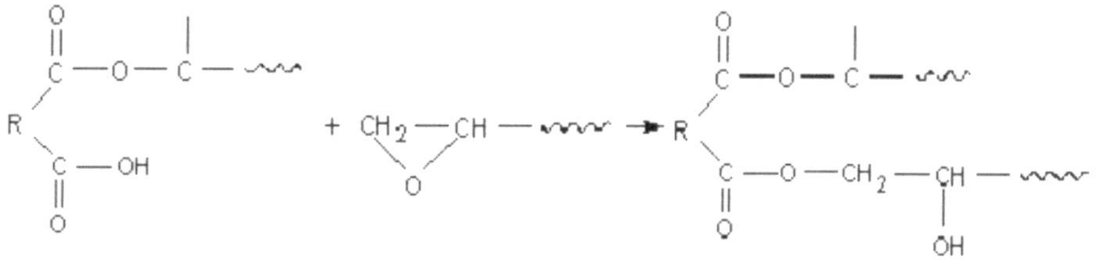

3. the produced hydroxyl react with another acid anhydride (the response equation is the same as (1), simultaneously with the reaction (4))

4. the produced para-hydroxyl again react with another epoxy

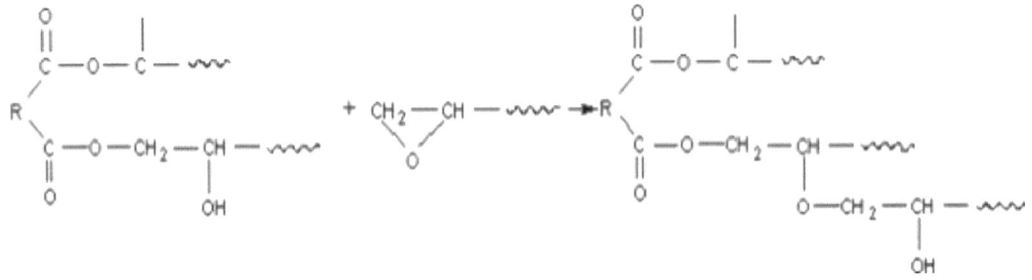

4.1.2 The experiment of differential scanning calorimeter (DSC)

The DSC (Q1000 V9.0 Build 275) used in this experiment is produced in the company TA of American, as shown in figure.7. The temperature range: 180-998K, the experiment using nitrogen gas conduct the heat, the flow speed is 50.0 ml/min, accuracy: ± 0.5K, the sensitivity: 0.2 0.2μW, error range: 1%.

Fig. 7. Q1000 Build 275 DSC instrument

1. Experimental material.

The raw materials used in this experiment as shown in table 3.

component
E-51 bisphenol A
PMDA
MA
Epoxypropane butyl ether
Pb
Fe
Rutile nanometer TiO_2
Total mass

Table 3. Raw materials for the experiment

2. the experiment scheme[18-20].

(1) put the mixed samples into DSC pool, choose the warming speed :2.5K/min, 5 K/min, 10 K/min and 20 K/min, and then scanning from 283K to 572K, obtain the trends warming curves as ,as shown in figure. 8.

(2) According to the dynamic scanning exothermic curve, The beginning, peak and the end temperature of solidification with constant surrounding temperature can be extrapolated.

(3) According to the initial curing temperature we selected four samples temperatures, we solidified the sample1, sample2, sample 3, samples4 and sample 5 separately at these constant temperatures.

(4) After the constant temperature solidification, cool the samples at the rate of 10 K/min to surrounding temperature. And then operating the DSC at the same rate. Finally, we obtain the reaction surplus heat and the vitrification temperature.

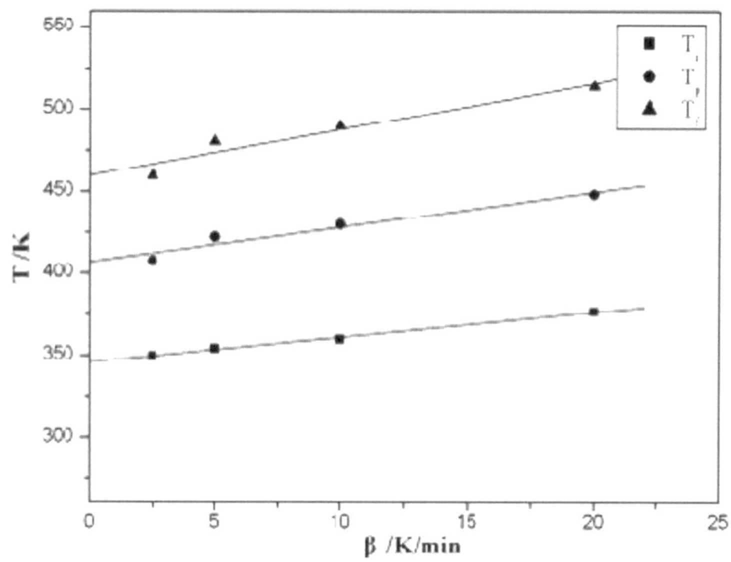

Fig. 8. The characteristicof sample 3 at difference sweepspeed

4.1.3 The thermal analysis results and discussion

1. The vitrification temperature

The vitrification temperature is in agreement with curing temperature, but high curing temperature will reduce curing degree, see figure.9. While the vitrification temperature is related with nanometer TiO2 Accounts for 3% of bisphenol A, the vitrification temperature reaches extreme, as shown in figure.10.

Through the nonisothermal curing DSC scans, it found that the TiO_2 can reduce the resin's curing temperature and promote the release heat of solidify effectively, and then the curing product's vitrification temperature also have certain effect, at the same temperature, when TiO_2 take up 3% of bisphenol A , the vitrification temperature is the highest one.

In a certain temperature range, the increase of isothermal curing temperature can increase the vitrification temperature of the curing product.

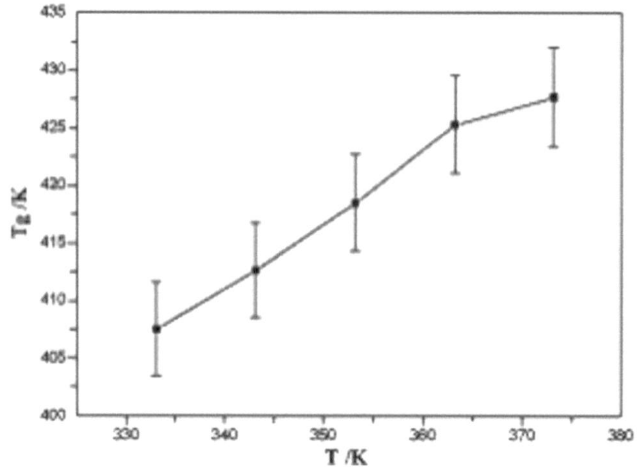

Fig. 9. The isothermal curing temperature determines the vitrification temperature

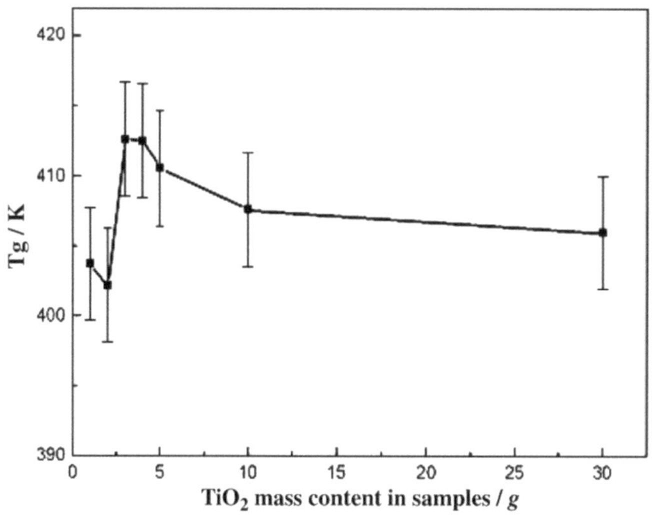

Fig. 10. The relationship between vitrification temperature and TiO2 content at 343K

4.2 The trial-producing process

According to trial-manufactures of composite materials and the characteristic of the shielding composites, the preparation process can be divided into four stages:

1. Preparation

(1) dube the silicon oil inside the surface of material molding, placed it around 1 day, standby, as shown in Fig.11.

(2) According to the proportion weighting PMDA and MA, make it into powder and blend equality, put it into the bottle, then placed the bottle into the heater heating until it completely melt.

(3) According to the proportion weighing iron powder (Fe), lead oxide (Pb), carbonation boron (B_4C), tungsten (WC), graphite (C), titanium dioxide (TiO_2 and trioxide gadolinium (Gd_2O_3) and propylene oxide butyl ether, etc.

Fig. 11. Materials moulds

(4) Put bisphenol A on electric stove heating, melting weighing.
(5) Blending the bisphenol A PMDA with MA and stir well, then pour epoxy propane butyl ephedrine and TiO_2, stir well.
(6) Put the enhanced mutually particle into the mixture solution.
2. The former dispose of material solidified
(1) Put the mixed well materials into the blender, stir well with different rotate speed
(2) Put the mixed well materials into the vacuum air exhaust containers, exhausting until the vacuum degree no longer change.
(3) Repeat a and b two or three times then pouring mold, put the forming materials into containers, again exhausting several times.

Fig. 12. Vacuum blender

Fig. 13. The polymer poured into the mold

3. Materials solidify

(1) Put the forming container into the designated temperature preheating constant temperature box, curing at constant temperature.

(2) The first stage of curing is over, making the constant temperature box rapid-heating until the curing temperature, and then curing.

Fig. 14. The constant temperature box

4. Mold unloading

Get out the cured materials, the mold unloading sample is shown in Fig.15.

Fig. 15. The mold unloading sample

4.3 The samples of materials

According to the result of the optimization design material composition and the established industrial art technological process, it design a series of new polymer radiation shielding materials, polyamide as the curing agent of samples Jxa1 (Fig.16) and Jxa2 (Fig.17), mixed anhydride as curing agent of sample Pb6 (Fig.18), and polyamide as curing agent and has no metal particles of inorganic polyester with filling samples Interlayer (Fig.19) without Gd_2O_3.

Fig. 16. xa1 samples (cylindrical)

Fig. 17. Jxa2 samples

Before curing material the air exhaust is very important, because it dose good to low material's internal porosity rate, thus improving the uniformity of the internal structure and the mechanical properties of materials.

Fig. 18. Pb6 samples

Fig. 19. Interlayer

Fig. 20. Djy samples without the hydro form formed

5. The test of shielding material performance

5.1 The neutron shielding assessment experiment device

5.1.1 The general situation of serial electrostatic accelerator neutron source system

1. the mainly configuration
   5SDH-2 serial electrostatic accelerator neutron source system (shown in Fig.21).

(1) Ion source
   SNICSII type sputtering ion source.

(2) Injection
   A double-slit lens, x-y roll cone, vacuum pump, and Michael Faraday deflection magnets, single lens.

(3) Accelerator form
   Metal tube, two ceramic acceleration transmission chains, peel, the corona pin and high voltage undammed controller, voltmeter, rotation system, vacuum pump and control console.

(4) The focus on magnetic analysis system
   Double the lens, switch magnets, Y guide.

(5) Beam pipeline
   Two 1.5 m long beam pipeline.

(6) Recycling systems
   SF_6 gas recovery system.

Fig. 21. The serial electrostatic accelerator system

2. Other supporting system

(1) The neutron monitoring system
   BF_3 long used for neutron probe, more, The beam from a different direction deviation from pipeline at different distances from the target to hit the neutron beams.

(2) Positioning system
   Contains the sideway, optical centre aligning, and TV monitoring system.

(3) Data acquisition
   NIM standard system.

(4) Main technical indices

Neutron emission strength: 10^9n/pulse,

d - D reaction: can produced in the 3-7MeV points neutrons,

d -T reaction: can produce under 3MeV point's neutrons.

5.2 Shielding material evaluation experiment

5.2.1 The layout of evaluation experiment

In the use of ^{252}Cf spontaneous fission neutron source shielding materials of samples of neutron shielding effect evaluation experiments, source, detector and neutron probe general layout as shown in Fig.22. The dotted line in the red circle diagram in front of detector have a homemade container, above it is a polyethylene cylinder. So, it can ensure that the backward play source efficient and accurate. Ensure radiation safety, what the picture have done also through the neutron shadow cone and alignment holes to keep out. Likewise, when shielding material sample absorbing neutrons, unload the neutron shadow cone, and put the thickness of shielding materials which to be examined on the neutron beams on channel, the subsequent transmission neutron probe is needed before receiving throughout the sample.

Fig. 22. The measurement of Sample neutrons shielding materials absorption

5.2.2 The experiment principle

Set the neutron various through the thickness measurement's flux data is $\phi_i i=1,2,3$, A sample tested before and once measured this bottom flux measurement and neutron flux material , take two measurement flux average ϕ_b and ϕ_0 , The through rate of the samples of neutron is $i=1,2,3$ By the type,

$$T_i = \frac{\phi_i - \phi_b}{\phi_0 - \phi_b} \tag{1-7}$$

5.2.3 The experimental results

To develop the jxa1, jxa2, pb6, Cake1, Cake2, Cake3 and multilayer compound shield Fe - CH_2 - Pb, All of which had experiment examined.

For example ,we only use polyester sandwich materials Cake1's (with iron and lead, equivalent density, 4.554g·cm^{-3}) experimental results of assessment of the data processing to analyzed and discussed and compare with the numerical simulation results and multilayer composite shielding experimental results numerical simulation results.

Composite multi-layer Fe shield - CH_2- Pb`s Data can be gain acceding to (1-7), every weaken ratio's indeterminacy lie on $\phi_i$ , $\phi_b$ , $\phi_0$ `s indeterminacy and themselves .

$$\frac{\Delta T}{T} = \pm\left[\frac{(\Delta\phi_i)^2 + (\Delta\phi_i)^2}{(\phi_i - \phi_b)^2} + \frac{(\Delta\phi_0)^2 + (\Delta\phi_b)^2}{(\phi_0 - \phi_b)^2}\right]^{\frac{1}{2}} \tag{1-8}$$

Due to ϕ_b and ϕ_0 is difference, in the experiment, the difference is 10 times, so the type can be simplified as follow

$$\frac{\Delta T_i}{T_i} \approx \pm\frac{[(\Delta\phi_i)^2 + (\phi_b)^2]^{\frac{1}{2}}}{\phi_i - \phi_b} \tag{1-9}$$

So, we can press get the uncertain result from (1-8) which gives the weaken ratio.

By the type of (1-8), we could see that with the more thickness of the shielding materials, ϕ_i is gradually approach ϕ_b , Transmittance T_i 's relative uncertain fierce raise, the behind data points can be even discarded.

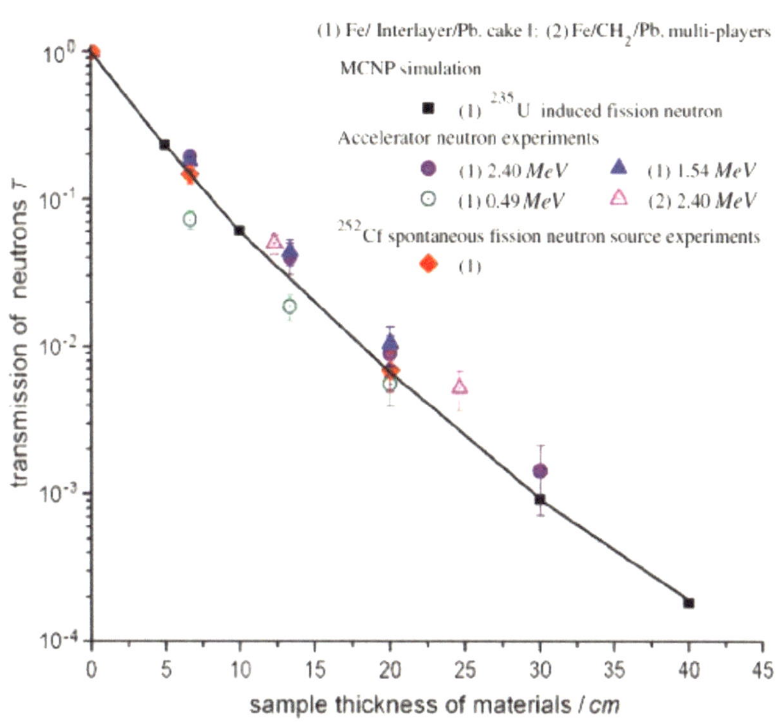

Fig. 23. Experimental test on neutron shielding performance of Cake1 and comparison to Fe/CH_2/Pb multi-layers

Fig.21 Cake1 gives the situation of, 2.4 MeV (purple, circular) 1.54 MeV (blue, triangles), 0.49 MeV (turquoise, circle), three accelerator neutrons` cross rate. The picture also gives experiment date (red, lozenge)of the ^{252}Cf fission spectrum source's through rate, and also gives the neutron penetration curve (black, square)of the MCNP's complete theoretical simulation of neutron spectrum (and the difference with ^{252}Cf fission spectrum are tiny). We found that the experimental through rate data of the neutron spectra neutron source was fall on the theory simulation curve, this show that within the thickness the neutron penetrate shielding materials simulate theory is correct, the simulation of materials from the design and manufacturing process through examination results are success.

As show in Fig.23, multilayer composite shielding body sample gives the date of two thickness at the energy points of 2.40 MeV (purple, upward triangle) ,The effect of neutron shielding is worse than the polyester sandwich material at the same conditions (with iron and lead).

As shielding materials is more and more thickness, it is obvious that the phenomenon of the through rate `s uncertain is fierce raise.

5.3 The experiment principle

As shown in Fig.22, it is a device to check the effect of the shielding materials, NaI (Tl) energy spectrum detector. The left two peaks in Fig.23 namely are the γ-rays of ^{60}Co two all-powerful peaks.

Set the various thickness measurement of rays all-around peak counts word address range is, $I_i, i = 1,2,3$, Before and behind the test, we test the sample without the material and counting , take an average of two counts as I_0, Then the various thickness of the sample`s weakened ratio is $T_{,i}\, i = 1,2,3$, the calculate formula as follow.

$$T_i = \frac{I_i}{I_0} \tag{1-10}$$

The biggest advantage is that the counting time can be control by people, increasing the count time can reduce the uncertainty of experimental data. But when the source of is low activity, the work efficiency is low. And when we choice higher activity source, proper handle radiation safety issues, it will make the efficiency greatly improved.

Fig. 24. The experiment of the device of radiation absorb

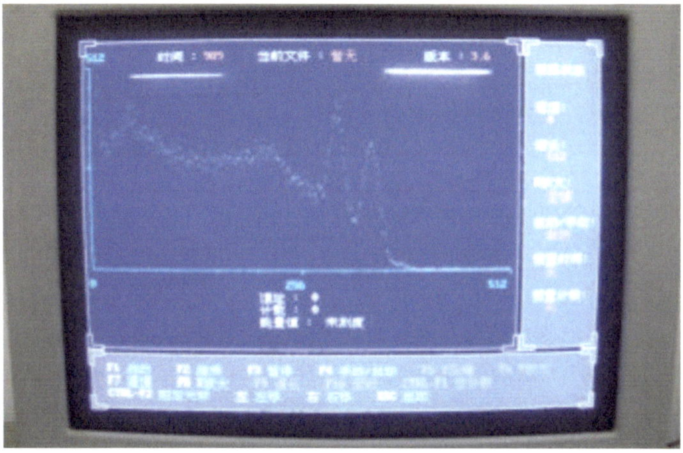

Fig. 25. The γ-rays energy spectrum of $^{60}C_O$ tested by NaI(Tl) device

The Measurement data isn't more, the focuses is to choice the function of the sample shielding material, then it could tell us the optimum design method and the development process whether meeting the requirements or not. The Material sample chooses Cake1 and multilayer composite shield Fe - CH2 - Pb (6/3/1).

density / $g \cdot cm^{-3}$	thickness / cm	totalize count	completely area	Weaken ratio	The simulate of Co	The simulate of fission spectrum
Cake 1 (4.554)	0	234550	128205	1	1	1
	6.67	54689	23255	0.1814	0.1841	0.2418
	10	27430	9616	0.0750	0.0790	0.1250
	16.67	8150	1250	0.0098	0.0148	0.0357
	20	5335	284	0.0022	0.0064	0.0194
multilayer (6.126)	0	234550	128205	1		
	6.67	43518	12828	0.1001		
	13.34	10143	832	0.0065		
	0	72839	31151	1		
Jxa1	9.47	16662	5822	0.1869		
Pb6	9.35	14356	5126	0.1646		
Jxa2	0	29718	16435	1		
	4.6	15661	7795	2.1000		

Table 3. The test result of two kind of materials $^{60}C_O$ radiation's weaken ratio

The measuring results see Table 3. The weaken ratio in the table counting according to the formula (1-10).

The uncertain of the weakened of the ratio`s depends on I_i, I_0 and itself,

$$\frac{\Delta T_i}{T_i} = \pm \left[\left(\frac{\Delta I_i}{I_i} \right)^2 + \left(\frac{\Delta I_0}{I_0} \right)^2 \right]^{\frac{1}{2}} \tag{1-11}$$

So it can estimated weaken ratio acceding to (1-11). In the experiments it observed the fundus, we discover that the rang of fundus' peak is so tiny, then it could be negligibled, this is the advantage to use the devise to test the weaken ratio.

Fig.20 presents the icon of table 1-4. From the picture we could see that the sample of Cake1 shielding the^{60}Co (red and black) is better than the fission spectrum rays (blue), and with the thickness increase it will be better and better. Then we only see ^{60}Co γ radiation, the theory simulate date (red) is higher than the experimental date (black), its reason is that the accumulation effect is action.

Theory simulation need materials large enough ,the cross-section ought to 40*cm**40cm, but experiments Cake1 diameter only 15cm, the Fig.25 is better effect that only part of the multiple scattering rays in the experiments Cake1 sample get point detector (the sensitive area is lesser). So Fig.26 gives a dotted line. The dotted line reflects the situation that the accumulation effect is corrected similar to date imitate. Then, we could draw the conclusion that the numerical simulation and experimental results are basically the same.

Get such conclusion no double illustrates that the design of the sample to shielding the □ray as we expect , and the materials design craft reach the standard , therefore the performance of the sample shielding ray is reliable.

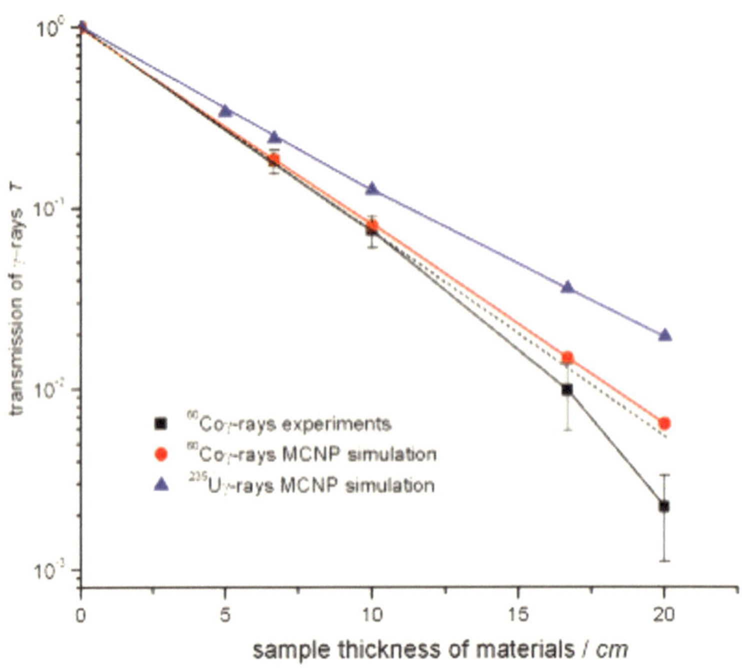

Fig. 26. The compare of the result of the Cake1sample neutron weaken ratio experiment and the result of date simulate

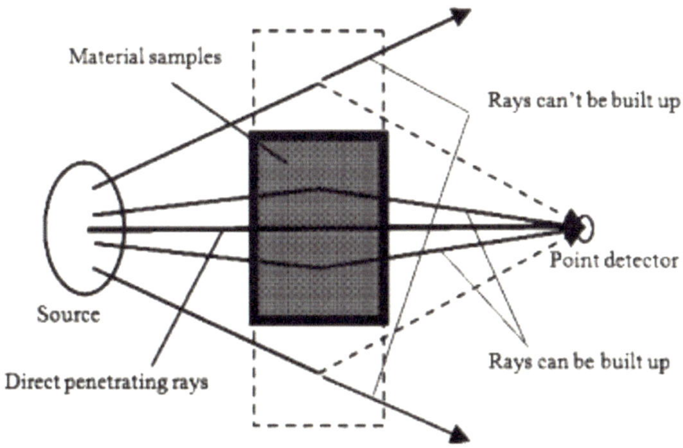

Fig. 27. The part accumulate effect of the γ-ray pierce the limited scale materials in experiment

5.4 The check experiment of the sample's machinery performance

The experiment mainly determined the draw intensity of the sample Jxa1 (including tensile rate) as shown in figure.25, the experiment test the compress stress of Jxa1, Interlayer and Pb6 as shown in figure.26, it list the comparison of the result of the shielding materials machinery performance and PB202 products.

a) b)

Fig. 28. The determine experiment of the samples Jxa1(including tensile rate)
a) The draw experiment of Jxa1 b) the breaked experiment sample of Jxa1

From the table1-6, we may could see:
(1) the draw strength of the sample Jxa1 is twice more than PB202 , therefore can used at the more acuity environment.
(2) From the result show when the compression stress is 50%, the two material, Jxa 1, Interlayer, could keep relative integrality after it endure great pressure.
This is great advantage when it mentioned to the situation to remain the integrality of the shielding materials. The solidify agent of Pb6 is mixed acid anhydride, the trial-producing

craft is complex, the combine uniformity is relatively poor, add to the precipitation phenomenon the heavy metal material s itself, the layering phenomenon still more obvious, therefore it exists scatter in the anti-compress experiment result. The behind two dates listed in table 3 is anti-compress strength.

(3) because the materials is polymer, the softening point temperature of it is large more than PB202 which containing large Numbers of polyethylene much, then the obtained materials in study will have more extensive application. The date listed in the table is the solidify temperature at which the materials has solidified, the softening point naturally higher than after solidify temperature.

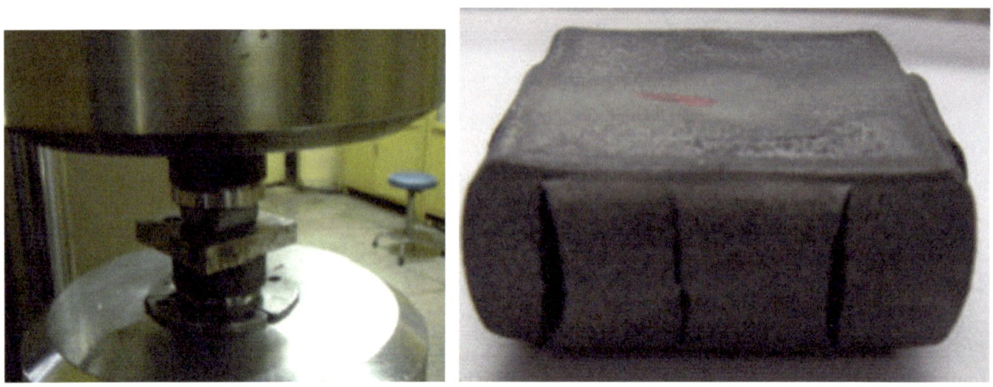

a) b)

Fig. 29. The experiment to determined the compressing stress of the shielding materials' sample. a) The compressing stress test of the sample, b) The compressing stress sample of Jxa1 when it was compressed to 50%

The code name of materials	density $/10^{-3}$ kg□m^{-3}	the soften point /℃	the draw srength /MPa	The draw rate* %	Stress at the compress of 50% /MPa
PB202[9]	3.42	99	12.63 +	– –	– –
Interlayer	1.48	>140	– –	– –	152.47
Jxa1	3.43	>140	25.82	8.6	148.07
Pb6	3.72	>160	– –	– –	133.09 , 148.96 , 193.05

*:refers to Chinese national standard GB/T6344-1996; #:refers to Chinese national standard GB6669-2001; +: literature[9] isn't say the experimental standard

Table 4. The compare of the shielding materials` machinery performance and the PB202 products

5.5 The experiment of the shielding materials sample's metallographic

5.5.1 The process of prepare the sample

(1) Sampling

Get a 10×10×20 cuboid from trial material, and then make a cylinder which diameter is 10mm.

(2) Embedded

Make the prepared cylinder stick to the prepared rob tool, use it after solidify.

(3) Grind

Firstly, Use the 280 # sand paper polished the burnishing surface according the requirement again and again, then repeat the process with 4000 #, 800#, 1000# unlit the surface of the sample is smooth.

(4) Polishing

Put the rubbed sample on the polish machine to polish, until the surface smooth and have mirror face effect.

The metallographic experiment samples prepare for the shielding materials sample as shown in figure 27, there have Pb6, Interlayer and Jxa1 three kind of sample, each parallelism the picture (a) , (b) and (c).

a) b) c)

Fig. 30. The metallographic experiment samples prepare for the shilding materials sample a) Pb6, b) Interlayer, c) Jxa1

5.5.2 The experimental observation and analysis

Take a picture and observe the prepared of three kinds of metallographic experiment samples, and every corresponding metallographic picture as shown in Fig.31.

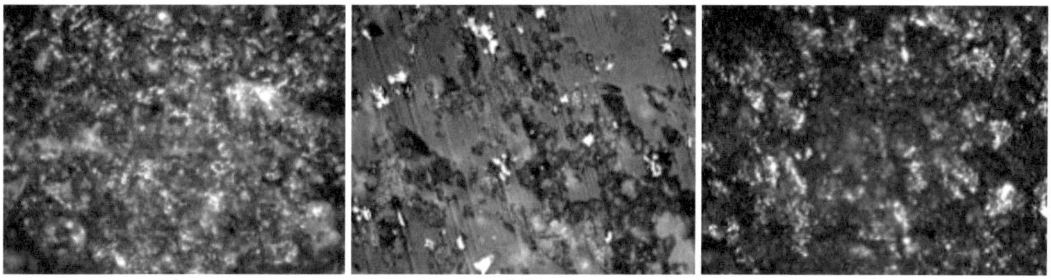

a) b) c)

Fig. 31. The picture of the shilding materials sample in metallographic experiment (magnify multiple: 400), a) Pb6; b) Interlayer; c)Jxa1

From the picture (a), the materials of Pb6 is a mixed acid anhydride, the metal particle combine with the polymer relative asymmetry, In the picture (b), the sample of the polyester of filling use the poly-amine as the solidify agent ,and only have the inorganic metal particle Gd_2O_3 , the B_4C, Gd_2O_3 powered disperse equality in the polymer, from the picture (c), Jxa1 use poly-amine as the solidify agent and add the inorganic power of W, Pb, because it emphasis the problem of precipitation, so the metal particles in polymer dispersion is well-distributed. Generally speaking, except Pb6, when the inorganic metal particle blend with the poly-amine as the solidify agent in material sample (below 200). This particle of inorganic metal powder is enough to develop nuclear shielding materials, the result of the radiation shielding experiments have proved this.

6. Discussion

In GA optimal design and shielding experiments, the selected material geometry model and experimental material samples were just slabs. If cylindrical and spherical geometry model and relative material samples were selected, actual shielding performance could be obtained. In other words, the shielding performance of the materials is better in applications than that in experiments because actual shields mostly have cylindrical or spherical geometry structure.

Due to lower intensities of radiation sources, the material samples are all too thin to test shielding effects perfectly in deep penetration situations in our experiments. The heat and radiation resistance of materials should be studied quantitatively in experiments. Therefore the related experiments should be carried out in the near future.

For the heavy metal powders, such as Fe, Pb and W, mixing with polymer, several precipitation phenomena appeared in the forming process when the mixture consists of high component of polymer, for example, more than 30%. In our study, the efforts to manufacture a new kind of special gradient material (SGM) by virtue of the precipitation phenomena are still under way. A permanent magnet is used to attract Fe powders from the polymer mixture, thus Pb and W powders remain precipitated in the mixture. Then the SGM with "middle heavy-light-heavy" density distribution could be produced. This kind of material is very suitable for the nuclear radiation shielding material.

Because Fe, Pb and W are separated completely to be independent parts of shield, the Cakes are the extreme product of the SGM. Furthermore, because Fe, Pb and W are ordinary materials, the production of Cakes could focus on just the Interlayer.

Fusion neutrons and -rays mixed radiation shielding material could be designed applying our GA optimal design method. It could design not only the shielding material component but also the thickness ratio of shield based on the interlayer products.

The tanks for nuclear fission waste and isotope transportation could be made adopting our interlayer forming technique partially.

7. Applications

The collimators in nuclear radiation measurement system were designed using the cakes to obtain favorable collimated beams requested by experimental physical scientists.

The trial production of the thick pinhole for fusion neutron penumbra imaging has been carried out according to the manufacturing process of shielding material[21][22].

8. Conclusions

The method to optimize in lightweight, compactness and high temperature sustaining for the neutrons and γ-rays mixed field shielding materials was established by genetic algorithms combined with the MCNP code. Several trial-manufacture samples were produced using the related manufacture technology.

Shielding tests of samples verified that the correctness of optimal design method and reliability of the manufacturing process. The results of research can be wide applied to the practices of nuclear science and technology.

9. Acknowledgments

The authors would like to thank Zhiqiang Wang at the Metrology Department of the China Institute of Atomic Energy (CIAE), who provided accelerator and neutron dose monitoring. The authors also acknowledge valuable discussions with Professor Mingguang Zheng at the Shanghai Institute of Nuclear Power Engineering, and Professor Yuangang Duan at the Chengdu Nuclear Power Institute of China.

10. References

[1] T. S. Gates and J. A. Hinkley, Computational Materials: Modeling and Simulation of Nanostructure Materials and Systems NASA/TM- 2003-212163, Mar. 2003, pp. 1-17.

[2] J. W. Wilson, F. A. Cucinotta, M. H. Kim, and W. Schimmerling, Optimized shielding for space radiation protection, *Phys. Med.*, vol. 17 (Suppl. 1), pp. 67–71, 2001

[3] National Council on Radiation Protection, Guidance on Radiation Received in Space Activities, *NCRP Rep.* No.98, Jul. 31, 1989, pp. 1–21.

[4] P. A. Lessing, Development of 'DUCRETE' INEL-94/0029, March 1995, pp. 1–16.

[5] Michalewicz Z, Janikow C. GENOCOP: A Genetic Algorithm for Numerical Optimization Problem with Linear Constrains[J]. *Communications of the ACM*, 1992:175-201.

[6]Briesmeister JF. MCNP : A General Monte Carlo N-Particle Transport Code , Version 4A: LANL Report LA2126252M . New Mexico : Los Alamos National Laboratory , 1993.

[7] N. M. Schaeffer, Ed., Reactor Shielding for Nuclear Engineers, TID-25952, *U.S. Atomic Energy Commission Office of Information Services*, Jan. 1973.

[8] H. Hu et al. Optimized design of shielding materials for nuclear radiation, (in Chinese) *Atom. Ener. Sci. Technol.*, vol. 39, no. 4, pp.363–366, Jul. 2005.

[9] Huasi Hu, et al. Study on Composite Material for Shielding Mixed Neutron and γ-Rays. *IEEE Transactions on Nuclear Science*, ISSN: 0018-9499, 2008, 55(4):2376-2384.

[10] R. D. Albert and T. A.Welton, A Simplified Theory of Neutron Attenuation and Its Application to Reactor Shield Design, Westinghouse Electric Corp., Atomic Power Division, Pittsburgh, PA, WAPD-15, Nov.1950

[11] G. T. Chapman and C. L. Storrs, Effective Neutron Removal Cross-Sections for Shielding, Oak Ridge National Laboratory, Oak Ridge, TN, ORNL-1843 (AECD-3978), Sep. 1955.

[12] A. W. Casper, Modified Fast Neutron Attenuation Functions, General Electric Corp., Atomic Products Div., Cincinnati, OH, XDC-60-2-76,Feb. 1960.

[13] G. R. Odette, B. D. Wirth. Radiation Effects in Fission and Fusion Reactors[R]. *Handbook of materials Modeling*, 999-1037.

[14] Wu Shizhen, Kang Zhuang, Wu Xiaoqing.Using differential scanning calorimetry (DSC) to research the curing system dynamics of TDE - 85/7 # anhydride. *Tianjin polytechnic university journals.* 2005, 24(6): 30-33.

[15] Mi Ae Choi, Mi Hye Lee, Jaeeon Chang, Seung Jong Lee. Three-Dimensional Simulations of the Curing Step in the Resin Transfer Molding Process, *Polymer Composites*, 1999, 4:543-552.

[16] Liu Xiangxuan. The effect of the Composite nanometer TiO_2 to the performance of epoxy resin anhydride(in Chinese). *Insulation communication.*

[17] Wang Dezhong. The production and application epoxy resin (in Chinese). Beijing: Chemical industry press.

[18]Wang Xiaojie, Xie QunHui, zhang wei. The curing study of epoxy resin matrix(in Chinese). *FRP/composite materials.* 2001, 2:1-3.

[19] Chen Xiang, YuanJian, YuLin. Bisphenol A linear phenolic resin curing epoxy resin(in Chinese). *Thermosetting resin.* 2001, 2:1-4.

[20] Shi Zixing. The Synthesis and characterization as well as curing behavior research of Polysulfone polysulfone resin/*Montmorillonite hybrid materials*(in Chinese).

[21] Wu Yuelei, Huasi Hu, et al. The Nature of Single Round Hole Neutron Penumbral Imaging. *Fusion Science and Technology* (FS&T), ISSN: 1536-1055, 2010, 57(3):292-297, American Nuclear Society.

[22] Wu Yuelei, Huasi Hu, et al. obtaining point spread function of penumbral encoding aperture with "expectation maximization" algorithm based on matched source-image pair experiment. *Review of Scientific Instruments* (RSI), 2010, 81(5): 053502-053502-7, 0034-6748.

Isotope Effects Induced by Exterior Actions on the Solid Surface

Nikolay N. Nikitenkov, Yurii I. Tyurin and Vitalii V. Larionov
Tomck Polytechnical University
Russia

1. Introduction

Many scientific and technological problems for which solution the examination of the isotope composition or its modification as a result of certain processes are required. So, for example, in geology it is the problems related to geochronology, isotope geochemistry (including shaping of the earth crust, examination of a glaciers etc.). In the space researches: a problems of the isotope space geochemistry, the space chemistry and the space geochronology (processes of shaping of a planets, comets, asteroids; the history of shaping of a chemical elements in the Universe) (Shukolyukov Yu.A., 1996). In the nuclear power: the fuel and other functional materials of the nuclear and thermonuclear reactors. And many other things.

Now are known many processes in which a diversions from the natural abundances of isotopes are observed. In the physics and chemistry the following processes are most known:

a. The kinetics processes;
b. The reactions of the isotopic interchanging;

These processes are responsible for a isotope modifications in the following physical and chemical phenomena: evaporation, condensation, melting, crystallization, diffusion and dissociation.

Thus it is known, that chemical bonds with a heavy isotopes more strongly than the same bonds with a light isotopes.

In general terms, all isotopic effects can be parted on two types:

1. Kinetic - based on different velocities of processes and caused by the fact that in the aggregate of interacting molecules the particles with smaller mass possess greater velocity;
2. Thermodynamic - which are reflection of a different energy states of the system. A composition with heavy isotopes possesses a smaller reserve of a free energy, than the same compositions with a light isotope (Galimov E.M., 1981).

It is now know also that deriving of the large isotopic effect in certain process are necessary: a high relative odds of masses, a high degree of chemical bond covalence, modifications of a oxides states, a modification of a phase or modular state.

Last researches of isotopic effects in the chemistry have led to discovery of one more view of the isotopic effect - fractionation of nuclear-isotopes on a nuclear magnetic moment (Buhanchenko A.L., 2007). The possibility of detection of nuclear polarization is a new research technique of mechanisms of a chemical reactions and it is a method of detection of a radicals and radical stages in a chemical process. The magnetic isotopic effect significantly differs by the nature from the usual kinetic isotopic effect related to difference in the isotope masses.

In condensed matter the isotope effects are manifested more brightly than in other states of matter.

Now the solid-state physics studies influence of isotope effects on elastic, thermal both vibrational and other properties of crystals. It is installed, for example, that at replacement of hydrogen on a deuterium in the LiH crystal the energies of a electron transitions increase by 2 orders (Plekhanov V.G., 2003). Now actual problems of isotope engineering are new medium for a data recording, a doping in a semiconductor by a neutron nuclear transmutation, control sensors betterment by ultrapure materials and many other things. Actual problem and major reaching in materials technology is the growing of crystals with the controlled isotope composition.

The purpose of the present chapter is a review of the latest work and new experimental data on changes in the isotope composition in the near-surface layers of solids as a result of the secondary ion emission, ion scattering from the surface, ion implantation, electrolytic saturation by hydrogen isotopes, thermo-diffusion from external sources and chemical solution action on the solid surface is presented. Such investigation is needed in materials technology, where atom migration mechanisms and isotope composition changes during various surface interactions occur. The change of isotope composition of the ion beams and near-surface layers of solids has been studied in many works on ion sputtering and secondary ion emission, scattering of ions from surfaces, hydrogen saturation and thermo diffusion from external sources and ion implantation. In all these phenomena, it is very important to identify the basic processes, which determine the change of isotope composition in the near-surface layers of solids.

The good understanding of physical mechanisms, which are reviewed here (and of others still unknown mechanisms) will become a basis for the obtaining of the monoisotopic or enriched surfaces with a certain isotopes. It, undoubtedly, will be claimed by future technologies. These investigations are also of the great academic interest as example of non-equilibrium systems.

2. Experimental results and discussion

2.1 Isotope effects during ion sputtering and the secondary ion emission

Slodzian et al were probably the first to study the isotope effect (IE) in secondary ion emission (SIE) (Slodzian et al, 1980), although preferential sputtering was noted earlier in 1977 (Lorin et al, 1982). It was established (Slodzian et al, 1980) that in SIE of some minerals, metals, copper oxide, and GaAl alloy produced by oxygen ions, in the energy range 0–20 eV the yield of the light isotope in comparison to the heavy one is greater than the natural isotope ratio. Thus, the enrichment of the sputtered particles by the light ions was found. The degree of enrichment depends on the isotope atomic number, on the matrix containing the isotope in question, and on the ion velocity. Almost at the same time, the space and energy distribution of the SIE isotope rate was studied (Shapiro et al, 1985). The data (Shapiro et al, 1985) confirmed the results and conclusions (Slodzian et al, 1980). The ratio $f_{LH}(E)=N^{+}_{L}(E)/N^{+}_{H}(E)$ for metals, (where $N^{+}_{L}(E)$ and $N^{+}_{H}(E)$ are the energy spectra of the secondary light and heavy ions) has been thoroughly studied. It was found that, in the ratio

$f_{LH}(E)$ in the energy range (0–80 eV) there is either peak or plateau at different energies for different metals. As a rule, this peak is located at low energies for the light elements and displaced to greater energies for the heavy ones.

The function $f_{LH}(E)$ in the energy range E=4–30 эВ was studied carefully (Gnaser & Hutcheon, 1987) to verify the proposed effect model. Lately IE in sputtering and SIE were studied too (Gnaser & Hutcheon, 1988), (Gnaser & Oechsner, 1990), (Shimizu & Hart, 1982), (Shapiro et al, 1985), (Shwarz, 1987). In order to explain isotope effect in SIE (IESIE), the standard ionization probability equations (Shapiro et al, 1985) and the experimental dependences of ionization probability on the atomic velocity (v) in the form of exp($-v_0/v$) (v_0 is constant) are used. The change of the isotope composition follows from the dependence of ionization probability on atomic mass. However, it does not explain the mechanism of the phenomenon.

The experimental investigation of IESIE and its mechanism is the main object of our research (Nikitenkov et al, 1987, 1988) – the functions $f_{LH}(E)$ for metals somewhat similar to those described were studied. In general, our data confirm the results of previous work, although our experimental values for some metals are greater (Fig.1).

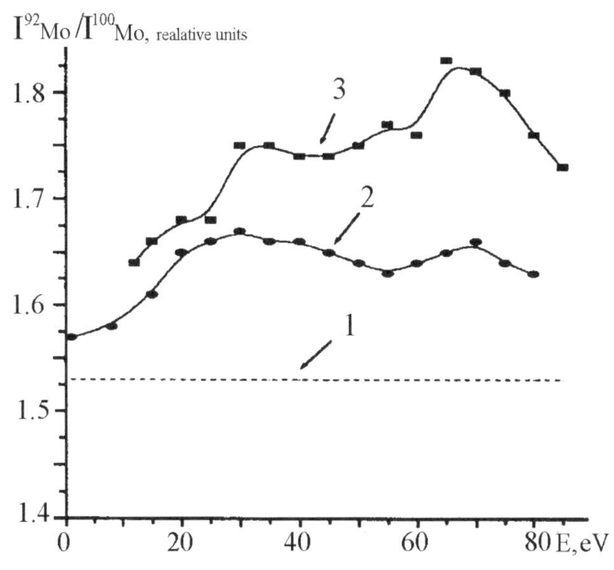

Fig. 1. The mass line intensity ratio of molybdenum ions 92 Mo and 100Mo as a function of energy: 1 – standard isotope ratio, 2 – (Shimizu & Hart, 1982) data, 3 – our data.

The experimental data confirm the existence of a peak or plateau in the $f_{LH}(E)$ function, although the correlation between the peak energy position and the isotope mass was not confirmed. At the same time, a decrease in the ratio of the maximum value $f_{LH}(E)$ to the natural abundance with decreasing isotope mass has been observed. It was also shown that the larger the nuclei neutron difference the greater is the IESIE.

From all established regularities, the most reliable are follows: there is the greater ionization probability of the light isotope in comparison with the heavy one, and there is an inverse dependence of the effect on atomic mass. The last dependence obviously correlates with the relative nuclear mass and volume change for the same deficit or excess of neutrons while moving from the light isotopes to the heavy ones. This fact shows that IESIE is due to the difference in composition of atomic electron shells. This difference is due to the isotope displacement (ID) of the atomic electron energy levels. ID depends mainly on the nuclear mass and dimension.

The isotopic displacement of electron levels due to the nuclear dimensions is caused by overlapping of the s-electron and nucleus wave functions. The probability of electron localization in the nucleus volume becomes rather great (Gangrskiy & Markov, 1984). The nucleus size is described by the mean square (proton) radius $<r^2>$. Its value depends on the number of neutrons in the nucleus (Fig. 2). Thus, the nucleus size affects the atom ionization potential and consequently the atom ionization probability during the sputtering process. This influence can manifest itself through the ionization potential value and the interaction length (λ) between atoms and the surface. These quantities depend on $<r^2>$ (indeed $\lambda \sim a \sim <r^2>$, where a is the atom radius).

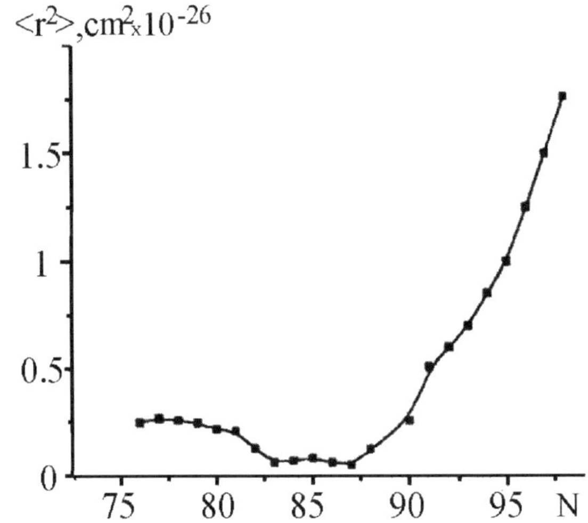

Fig. 2. The mean square nucleus radius as a function of the number of neutrons N for rubidium isotopes (Gangrskiy & Markov, 1984).

The isotope displacements of spectral line frequencies of some optical transitions are about 10^{-5}–10^{-4} eV. The ionization energy (I) of isotopes should differ with the ID values observed by optical experiments. The ionization probability at SIE depends on I exponentially. Thus, we can evaluate the relative difference in ionization probability for different isotopes. For example (Nikitenkov et al, 1987), the ionization probability R^+ may be:

$$R^+(E) \sim E^n \cdot \exp\left[-\frac{(I-\varphi)\cdot c}{E^{1/2}}\right], \tag{1}$$

E is the ion kinetic energy, I is the ionization energy, φ is work function, $c = \pi/2\gamma\hbar$, $\hbar = h/2\pi$ where h is Planck's constant; $n = \gamma_a\delta/4\gamma k T_s$, $\gamma_a = a_p/v$ is a quantity describing the interaction length between the surface and the escaping atom; a_p is a quantity of the lattice constant order; v is the velocity of the recoil atom; δ is a constant; k is the Boltzmann's constant; T_s is the temperature of an electron subsystem in the cascade developing region. With the help of the formula (1) we get the relative difference of ionization probabilities for two atoms with the potential increment ΔI and an interacting length increment $\Delta\lambda = \Delta(1/\gamma_a)$ as follows:

$$R(E)=\frac{R_1^+-R_2^+}{R_2^+}=\frac{R_1^+}{R_2^+}-1=E^{(\Delta\lambda)\cdot\kappa_1}\cdot\exp\left[\frac{\Delta I\cdot\kappa_2}{E^{1/2}}\right]-1 \qquad (2)$$

here $\lambda=1/\gamma_a$, κ_1 and κ_2 are dimensional constants (previously we assumed $\kappa_1=\kappa_2=1$).
Fig. 3 shows experimental and theoretical curves designed with using equation (2) and the following relation:

$$R(E)=\frac{\Delta R^+}{R^+}=E^{\Delta\lambda}-1 \qquad (3)$$

The formula (3) is obtained in an analogous way to formula (2) using the relation ($R^+\sim E^n$) instead of the formula (1). We assume that $\kappa_1\cdot\Delta\gamma\approx\kappa_2\cdot\Delta I\approx10^{-4}$. In addition, the curve 1 (Fig. 3) for ^{107}Ag and ^{109}Ag isotopes represents the experimental quantity:

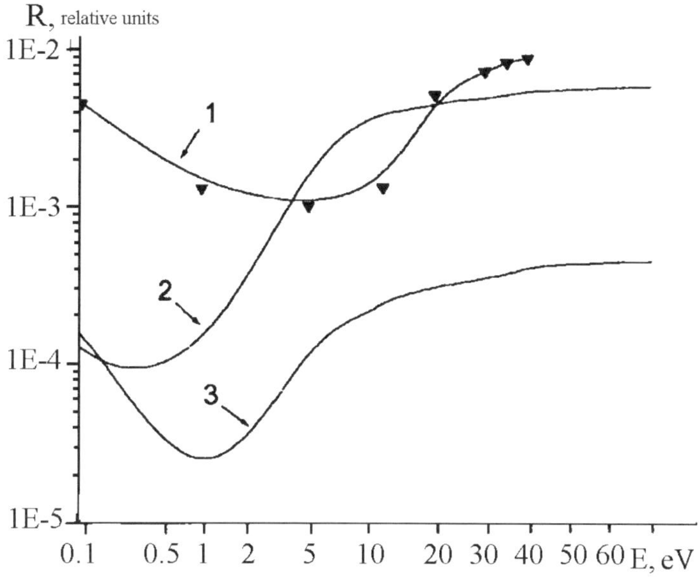

Fig. 3. Relative probabilities of ionization of the secondary ions as a function of energy:
1—experimental data in accordance with formula (4);
2 and 3—theoretical values in accordance with formulas (2) and (3).

$$R'(E)=\frac{\Delta M}{M}\cdot\left|\frac{R_{exp.}}{R_{st}}-1\right|, \qquad (4)$$

here $R_{exp.}=I^+_L(E)\,/\,I^+_H(E)$, I_i^+ ($i=L,H$) are the isotope mass line intensities in corresponding energy spectra, R_{st} is the standard isotope ratio and $\Delta M/M$ is the relative mass difference.
Fig. 4 shows the quantities $R'(E)$ (4) for some isotopes obtained in our experiments (Fig. 4, a) and additional result of Shimizu et al (Fig. 4, b). A comparison of Fig. 3 and Fig. 4 leads to the conclusion that the form of every experimental curve corresponds (in general) to the theoretical curve. The experimental and theoretical (R and R') quantities are often close at the minima. The different positions of the sharp minima of the theoretical and experimental

curves can be explained by the experimental errors caused, in particular, by wide energy windows (about 1 eV) of the energy analyzers. Thus, the given results suggest a possible relation between IESIE and isotope shift of the ionization level of an atom.

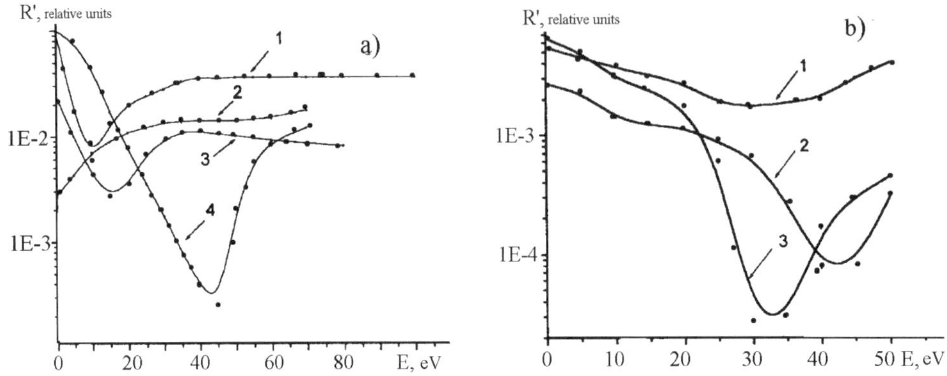

Fig. 4. The experimental quantities (R_{exp}/R_{st}) as a function of the secondary ion energy:
a - our data for: 1 — ^{24}Mg and ^{26}Mg, 2 — ^{92}Mo and ^{94}Mo, 3 — ^{50}Cr and ^{52}Cr, 4 — ^{114}Sn and ^{116}Sn;
b - data from (Shimizu & Hart, 1982). 1 — ^{28}Si and ^{30}Si, 2 — ^{50}Cr and ^{52}Cr, 3 — ^{24}Mg and ^{26}Mg.

2.1.2 Isotope effects during ion scattering on surface

The isotope effects in the ion fluxes backscattered from metallic surfaces have been mainly studied for $^3He^+$-$^4He^+$ and $^{20}Ne^+$-$^{22}Ne^+$ ion pairs (Helbig & Orvek, 1980), (Helbig & Adelman, 1977). The method of ion scattering double spectroscopy (ISDS) was applied. When using this method, the components of the ion pairs are present in the primary beam in equal concentrations. A method of alternating injection of these isotopes in the discharge chamber was used. The energy of the primary ions was in the range 200–3000 eV, the electric current from 0.1 up 500 nA, the incident and reflect ion angles 0–180 degrees.

Fig. 5 shows the IE data for the $^3He^+$-$^4He^+$ pairs (Helbig & Adelman, 1977). The scattering yield (Y) is represented in arbitrary units, although the relative intensity difference of $^3He^+$-$^4He^+$ pairs for every metal corresponds to the experiment. The abscissa axis origin for every metal (Zn, Pb, In, Sn) is designated by the dashed line below the chemical symbol. The values of the reciprocal velocity ($1/v=m/2Ve$, where m and e are the ion mass and charge, V accelerating potential) are shown on the abscissa axis. A very appreciable maximum structure of the curves for lead, indium, and tin has been explained as the result of competition between the scattering cross section increase and changes in neutralization probability when decreasing the beam energy (Smith, 1971).

At small reverse velocity, the scattered $^3He^+$ ion current is greater than that of $^4He^+$ ions for all ion-target combinations. This is quite natural, because the scattering cross section of light particles is greater than that for heavy particles under identical conditions (the force field, neutralization probability). For all studied systems at high values of inverse velocity, the scattered $^4He^+$ ion current tends to be greater than that of $^3He^+$ ions. This is obviously due to the fact that, at low energy, the $^4He^+$ ions neutralize with appreciably less probability then that for the $^3He^+$ ions. The difference between these isotopes of the same velocity is caused by their different momenta, which, in its turn, leads to a small deviation (less then 1 Bohr unit) in their scattering trajectories. The basic neutralization processes (the Auger

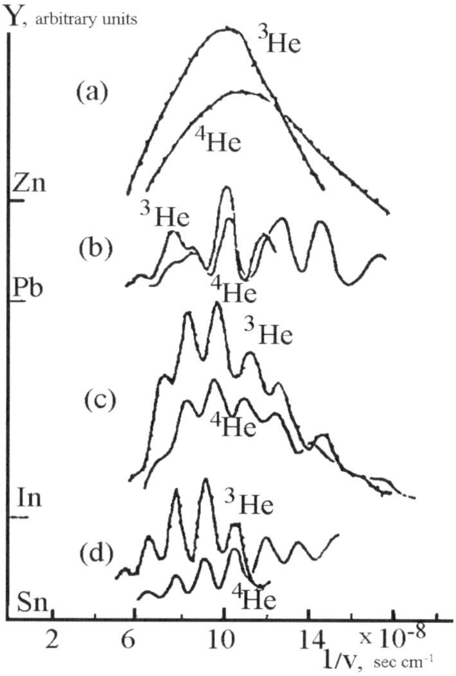

Fig. 5. The elastic scattered ion $^3He+$ and $^4He+$ yield (Y) as a function of inverse velocity for zinc, lead, indium, and tin surfaces; the scattering angle is 90°(Helbig & Adelman, 1977).

neutralization and resonance tunnel neutralization) are believed to occur rather far from the surface (about 4 Bohr units). It is difficult to understand why such small differences in their trajectories can affect the relative neutralization probability of these ionized isotopes.

Heavy ions produce greater modification of a surface than light ions in the processes of implantation, surface etching, and cleaning. However, these processes cannot be responsible for curves peculiarities at low energies.

The curve oscillations (Fig. 5) for helium ions scattered from lead, indium, and tin surfaces have been explained by quasi-resonance electron transportation (Erikson & Smith, 1975). A glance at these curves shows that the positions of maxima and minima for scattering of $^3He^+$ and $^4He^+$ ions from the same surface correspond to the identical inverse velocities (Erikson & Smith, 1975). However, a more detailed analysis leads to the conclusion that, in the low energy range, there are appreciable displacements of the maximum and minimum positions. While bombarding the lead surface (Fig. 5 c), the isotope shift (at the inverse velocity about $12{\cdot}10^{-8}$ sec·cm^{-1}) coincides with data for angular distributions (Erikson & Smith, 1975): i.e. increasing the scattering angle, at constant energy and mass of bombardment particles, leads to the same effects as increasing their mass at constant velocity and scattering angle. In both cases, the distance of maximum approach decreases. It is very interesting to note that the displacements of a low energy peak for indium and tin (Fig. 5 c, d) are of the opposite sign than those for the lead.

From Fig. 5 it follows that the heights of the $^3He^+$ and $^4He^+$ peaks are not identical and are not proportional each other. The differences of peak heights in the tin low energy spectra are probably due to the differences of the curve general form. The zinc non oscillatory data (Fig. 5, a) confirm this. In general, the curves have a parabolic form with a sinusoidal modulation.

About the possible reasons of such oscillations look (Tolk et al, 1976). The scattering $^3He^+$ from Pb is an exclusion from this rule. This curve has a small peak at the inverse velocity about $8{,}5 \cdot 10^{-8}$ sec·cm^{-1}, which is apparently due to the quasi-resonance electron transportation. The reflected light ion currents for $^3He^+$ and $^4He^+$ ions are about two times greater than those for heavy isotopes. For $^{20}Ne^+$ and $^{22}Ne^+$ ions, this increment is about 20 %, i.e. the heavier the isotope, the smaller the effect. It correlates with the isotope relative mass difference (just the same as for IE SIE). For a detailed explanation of ISDS energy distributions it is necessary (as well for SIE) to consider the ID of the ionization level.

2.1.3 Isotope effects during ion implanting

Our results on changing the isotope composition of copper, implanted in nickel, are represented in (Puchkareva et al, 2002). The ion implantation of copper atoms in the nickel polycrystals with average grain size about 20 micron has been produced by an implanter "Diana-2" (Aksenov, 1987) provided with a vacuum arc ion source in an impulse periodical regime with the impulse length 150–200 μs and frequency 50 Hz. The implantation was performed at the pressure $6 \cdot 10^{-3}$ and $4 \cdot 10^{-2}$ Pa at an ion impulse current density of 125 μA·cm^{-2} and average target current density of 1 μA·cm^{-2}. The implantation doses were $2 \cdot 10^{16}$ and $2 \cdot 10^{17}$ ion·cm^{-2}, the time interval for accumulating these doses was about 6 hours. The ions were accelerated to 50 kV. Previously electrochemically polished nickel samples were mounted upon a metallic support in such a way as to provide the tight contact with the metallic substrate surface. While implanting the sample the temperature was not more than 50 °C. Fig. 6 shows the distribution profiles for ^{65}Cu and ^{63}Cu through the nickel sample depth and the summed ^{65}Cu+^{63}Cu profile after implanting to a dose of $2 \cdot 10^{17}$ ion·cm^{-2} at a pressure of $6 \cdot 10^{-3}$ Pa. It is seen, that the isotope distribution patterns are different: a sloping curve for ^{65}Cu and a rising curve for ^{63}Cu. This means that, in the surface layer at a distance about 100 nm, the ^{65}Cu atoms have accumulated. After reaching the ^{63}Cu intensity line maximum, equalization of isotope composition begins, approaching a natural ratio at the depth 250 nm.
Fig. 6 and Fig. 7 show the depth concentration discrepancies for copper isotopes $C_i = I_i / \Sigma I_i$ (i – isotope, I – mass line intensity) to their natural abundances (R_i). At the depth of about 330 nm the data on isotope composition for implanted copper atoms coincide with the natural. A maximum deviation from the isotope natural abundance (about 54%) is observed at the sample surface. According to Figs. 6 and 7 there are three specific depth regions: 1) from zero to 80–100 nm, 2) from 100 to 250 nm, 3) deeper than 250 nm. Changes in isotope composition are more appreciable in the surface layers, where the rising implanted ion concentration (maximum is reached at the depth about 100 nm) has been observed (Fig. 6). The region of sharply decreasing implanted copper concentration in layers 120–250 nm corresponds to the isotope natural abundance, approximately.
A diffused "tail" of the implanted copper profile corresponds to the region, where the isotope composition coincides with the natural one. Thus, the maximum violation of isotope composition occurs in the region of maximum energy loss of implanted ions and maximum accumulation of implanted copper ion concentration. Fig. 8 shows analogous curves for the dose $2 \cdot 10^{16}$ ion·cm^{-2}. Comparing Figs. 7 and 8 we can see that the layer-by-layer mode of isotope composition does not change with the dose increasing although, for greater doses, the isotope composition change occurs at greater depths. Probably, it is due to increase in the total implanted copper concentration and its increase with depth is a result of radiation stimulated diffusion. This experiment was repeated at an ion implanted dose of $2 \cdot 10^{17}$

ion·cm^{-2} and pressure of 4·10^{-2} Pa. The copper isotope composition transformation effect did not change in general, but deviations in the copper isotopes implanted profiles have been found (Fig. 8). They become narrower, gathering towards the sample surface.

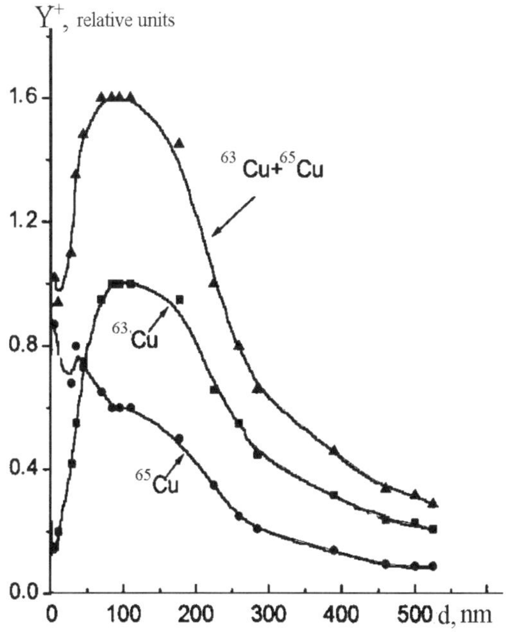

Fig. 6. The ^{65}Cu and ^{63}Cu depth distribution for a nickel sample, and summed ^{65}Cu+ ^{63}Cu profile.

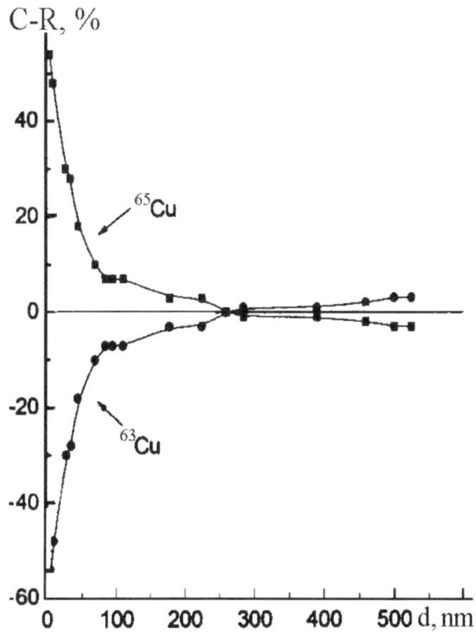

Fig. 7. Layer-by-layer deviations of the copper isotope concentration (C_i) from the natural abundance (R_i). Implantation dose is 2x10^{17} ions cm^{-2}.

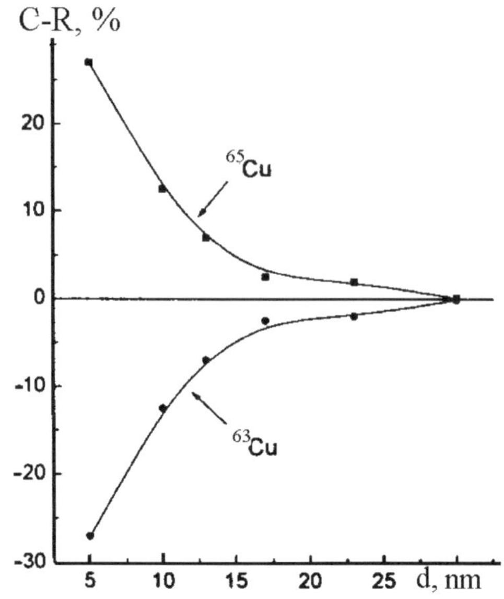

Fig. 8. Layer-by-layer deviations of the copper isotope concentration (C_i) from the natural abundance (R_i). Implantation dose is $2 \cdot 10^{16}$ ions·cm^{-2}.

Traditional diffusion mechanisms cannot explain the implanted copper isotope composition transformation. It is known that the ratio of isotope diffusion coefficients is inversely proportional to the square-root of their mass ratio: $D(^{63}Cu)/D(^{65}Cu) = (65/63)^{1/2} = 1.016$. However, the effect observed is several times greater.

2.2 The isotope effects during hydrogen saturation of materials

2.2.1 Massive metal samples of palladium and titanium

In table 1. are shown the data on a modification of the isotope composition of a massive (not thin-film) palladium samples after its saturation by the electrolytic expedient with use of electrolyte LiOD+D_2O (Chernov, Nikitenkov et. al., 2000).

The samples	Concentration of *i*-th isotope in an intermixture of isotopes, %				
A.m.u. of isotope	**104**	**105**	**106**	**108**	**110**
Natural intermixture	10.0	22.2	27.3	26.7	11.8
The initial sample	10.8	23.5	28.6	25.2	11.8
After saturation	16.4	25.0	21.4	22.9	14.3
Diversion from the natural	6.4	2.8	-5.9	-3.8	2.5

Table 1. Isotope composition of a palladium after saturation by a deuterium on observed datas on 10 samples (after surface refining by a sounding beam).

A procedure of the isotope compositions research was used the same what in part 2.2.2. From the Table 1 follows, that the isotope composition noticeably varies in surface layers of the palladium samples at electrolytic saturation by deuterium.

Fig. 9 shows a deviation of the content for the palladium isotopes in the course of the electrolytic saturation of a palladium and with an additional neutron exposure (Lipson, Maili et. al., 2003). In the Fig. 9: $\delta N = C - C_{ini}$, where C - the isotope concentration after and C_{ini} - before saturation.The purpose of the work was the determination of influence of mechanical stress on allocation of palladium isotopes. Mechanical stress in samples formed by both deuterium charging and an exposure by thermal neutrons.

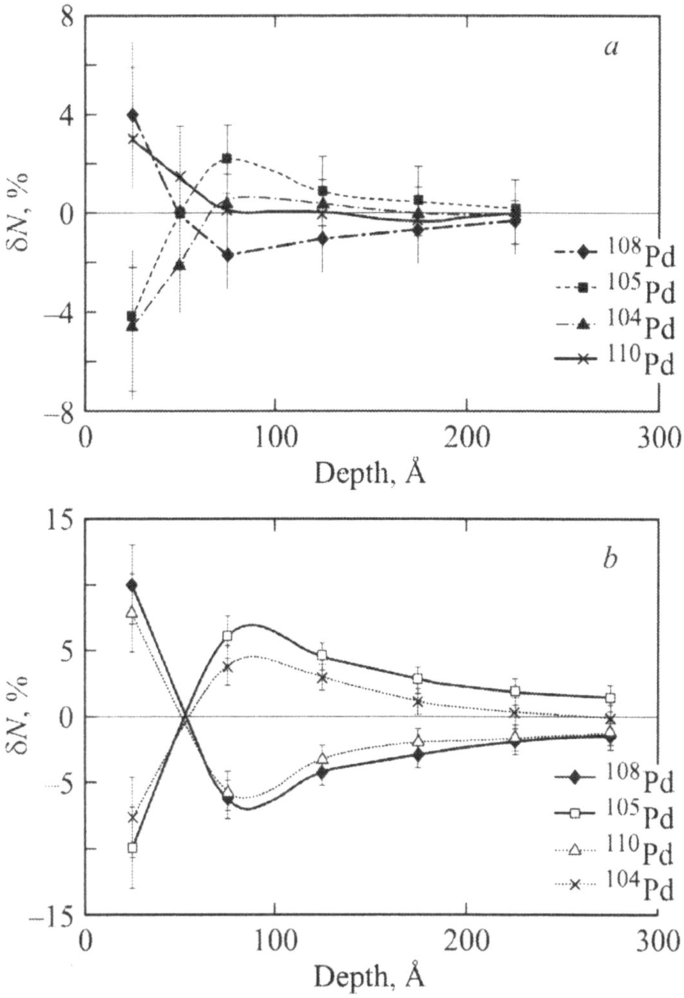

Fig. 9. Level-by-level diversions δN of the isotope concentration ^{104}Pd, ^{105}Pd, ^{108}Pd and ^{110}Pd as a result of a palladium samples saturation by deuterium: a - the sample subjected to an electrolysis within three days in the conditions of a natural neutron hum, b - the sample subjected to an electrolysis within three days at a simultaneous exposure by thermal neutrons (Lipson et al, 2003).

The deuterating of palladium samples made by electrolytic method in the 1 M NaOD solution at a cell with the parted anode and cathode spaces. Current density of electrolysis I_e=30 mA/cm^2; electrolysis time 1–3 days at room temperature. Composition of PdD_x

deuteride after the termination of an electrolysis: $0.72 < x < 0.80$. An exposure of samples be thermal neutrons with energy E_n=60 meV energy and Φ_n=200 n/s·cm^2 fluence made directly in the course of an electrolysis. After an electrolysis were measured a macroscopic buckling of the samples l and the residual plastic strain ε_p. It has been obtained, that at electrolysis in the conditions of a natural neutron hum average values of these parameters are equal $\langle l \rangle$=1.5 cm и $\langle \varepsilon_p \rangle = 3.0 \cdot 10^{-3}$. At the electrolysis in the presence of neutrons the straining parameters is a lot of above: $\langle l_n \rangle$=5.0 cm, $\langle \varepsilon_{pn} \rangle = 1.3 \cdot 10^{-2}$. The studying of the isotopic composition of the deuteride palladium samples was made by SIMS method on the CAMECA IMS 5f devise.

Analogously to a palladium there is a modification in the isotope composition in near-surface layers of the titanium at saturation by deuterium. In Table 2 and Fig.10 are shown the deviation from natural abundances for titanium isotope concentrations at saturation of titanium samples by deuterium (Chernov et al, 2000).

From Table 1 and Fig. 9 taking into account differences in experiment techniques reveals that in both independent experiments made by different authors, on different procedures of a sample preparation and effects examination follows that results are gained the similar.

Isotopies a.m.u.	Saturation dose, mA·min·cm^{-2}			
	4500	6000	12000	30000
	Maximum diversions, %			
	Large crystalline (LC)			
46	0.6	0.2	0.7	1.2
47	1.4	2.8	3.0	8.0
48	-2.0	-3.0	-3.7	-9.2
Submicrocrystalline (SMS)				
46	1.1	1.2	1.3	1.5
47	1.2	1.3	1.5	3.0
48	-2.3	-2.5	-2.8	-4.5

Table 2. The maximum deviations (from initial) in contents of the titanium isotopes in near surface area on two modifications of titanium samples depending on a dose of their saturation.

In the end of this partition we will descript some possible mechanisms of isotope separation (Lipson, Maili et. al., 2003) which can be responsible for given result.

Authors (Lipson et al, 2003) have viewed possible models of isotope separation for samples which similar viewed here (Fig. 9, 10; Tab. 1, 2). Thus for a palladium the following experimental quantities of effects are accepted (on the average): on depth h>100 Å concentration of ^{108}Pd isotope decreases and a ^{105}Pd isotope is increases by 5 % (for a pair ^{110}Pd−^{104}Pd on the average by 4 %). On this basis a factors of the full enriching are calculated according to (Kaplan, 1955).

$$A = \frac{N_F/(1-N_F)}{N_I/(1-N_I)}, \tag{5}$$

where $\langle N_F \rangle$=0.265 and $\langle N_I \rangle$=0.225 —average concentration, accordingly, ^{105}Pd and ^{108}Pd. From here follows -A_1=1.24. Analogously for a pair ^{110}Pd−^{104}Pd -A_2=2.55.

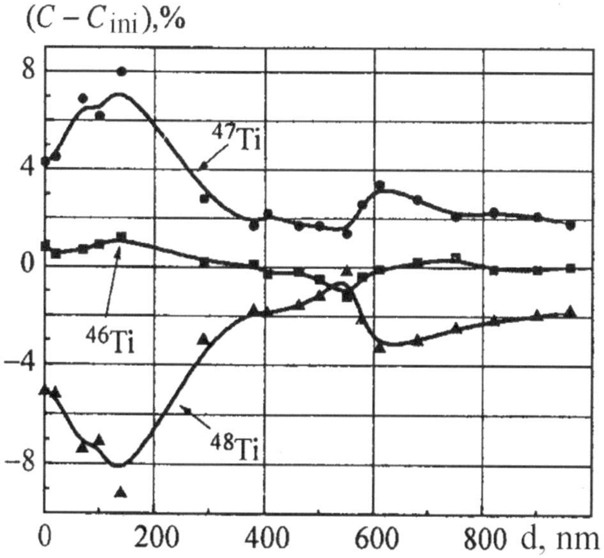

Fig. 10. Level-by-level distributions of the titanium isotope concentration in near-surface field of the large crystalline titanium after the electrolytic saturation by a dose of 30 000 mA·min·cm^{-2}.

On the basis of the gained results and calculations by (5) authors conclude that the diffusive model of isotope separation in this case does not work. Therefore it is supposed, that in a near-surface layers of the sample there are others, more effective reasons of isotope separation. As objects of such reasons can be the screw dislocations, dislocation loops or the spiral steps organized at cross of screw dislocations with a sample surface (Poluhin et al, 1982). If the rotation of such sites of the linear flaws happens with a velocity close to a sound velocity in Pd, than such objects can serve effective „nano-centrifuges" in which there is a hydrogen isotope separation. Really, a simple factor of partitioning α for such centrifugal machine is spotted as (Kaplan, 1955):

$$\alpha = \exp\left[\frac{(M_2 - M_1)(v_s)^2}{2RT}\right], \qquad (6)$$

where v_s - the effective velocity of rotation accepted to an equal half of a sound velocity v_0; R – the gas constant; T – temperature. According to (6), for a pair ^{108}Pd–^{105}Pd – α =2.60 and for a pair ^{110}Pd–^{104}Pd – α =6.50. As A = α_s an isotope separation rather efficiently in case of a nano-centrifuging also it can be realized in one stage (s≤1) in a layer h <50Å. Notice that at isotope separation by a centrifuging method heavy isotopes move to rim of rotation (opposed to diffusive method) whereas a light isotopes will remain on the spot, i.e. its captured by a dislocations. Therefore the surface will be enriched by heavy (mobile) isotopes and depleted by light isotopes (related to a kernel of dislocations). It is necessary notice that the discovered effects of isotope separation can be expressed much more clearly on more major depth for light impurities in metals (Wilson et al, 1989). As impurities tend to a segregation exclusively in areas of concentration of internal stresses, referred an isotope separation models are most applicable to diffusing impurity atoms.

2.2.2 Thin-films systems

While studying (Chernov et al, 1998, 1999, 2000), (Nikitenkov et al, 2004, 2006), (Larionov et al, 2010) the isotope effects by hydrogen saturation we used the secondary ion mass spectrometry (SIMS) method to analyze the samples of bulk titanium (having various crystal structure), palladium, and thin film Ti/Al and Ti/ceramics structures. In all cases, at sufficiently large saturation doses appreciate deviations from the natural isotope abundance were seen in the surface layer both for the matrix elements and for the introduced admixtures.

Ti/Al and Ti/ceramics structures have been gained by a method vacuum magnetron depositions on the cleansed aluminum and ceramics surfaces. The thickness of titanium films made 0.4 microns, a thickness of aluminium substrates - 50 microns and ceramics substrates - 1 mm. Saturation of structures by the hydrogen isotopes was made by the electrolytic method at use it as the cathode. Electrolyte $LiOD+D_2O$ and $LiOD + (1/2) D_2O + (1/2) H_2O$ were used current density of electrolysis from 15 to 100 мA/cm^2; a time of saturation from 3 till 10 h.

Fig. 11, 12 shows the experimental data on a titanium-aluminum system before and after hydrogen saturation in $D+(1/2)D_2O+(1/2)H_2O$ electrolyte. The Y coordinate represents quantities proportional to the absolute yield of secondary ions. The film surface and the "film-substrate" interface change appreciably during the saturation. The regularities found are as follows.

a. D, H, Li, and Ti compounds are produced at the film surface. While sputtering through the depth of 200–300 nm (Fig. 12), the molecular LiD^+ and LiH^+ ions are found in the mass spectra of a saturated sample and a decreasing Ti^+ ion yield is observed (Fig. 11);

b. An isotope effect on the molecular ions yield is observed (Fig. 12). The Li_xD_y and Al_xD_y compounds are located nearer to the sample surface than the Li_xH_y and Al_xH_y compounds. This holds for the sample surface as well as for the transitional film-substrate region.

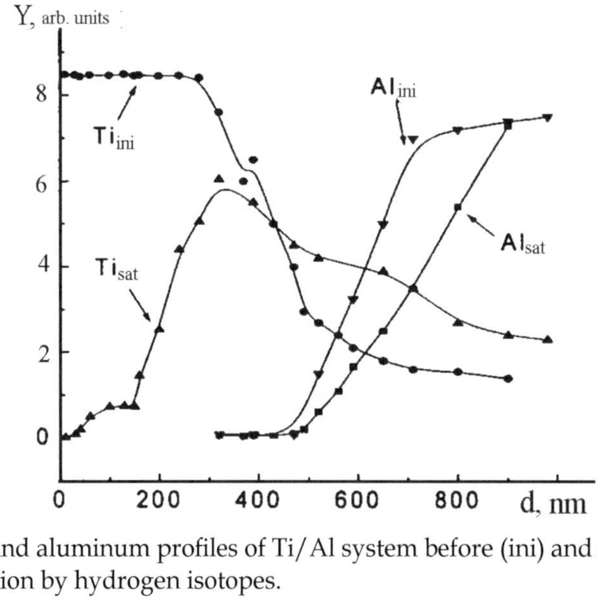

Fig. 11. Titanium and aluminum profiles of Ti/Al system before (ini) and after (sat) its electrolytic saturation by hydrogen isotopes.

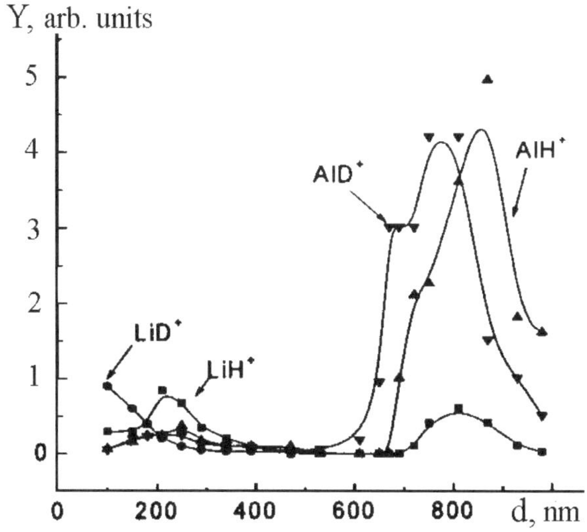

Fig. 12. Profiles of hydrogen and deuterium compounds after electrolytic saturation of the Ti/Al system by hydrogen isotopes.

Figs. 13 and 14 illustrate the appreciable deviations of lithium isotope concentration from its natural abundance in a Ti/Al system after saturation by the hydrogen isotopes. Fig. 11 shows the Li isotope profile. The maximum quantity of lithium is observed in the film formed on the titanium surface. In the film-substrate interface there is a plateau, and there is a recession at the point where the probe beam enters the aluminum substrate. At the depth 600–700 nm (the region previous to the region of AlH, AlD, and LiH ions maximum yield), a sharp peak in the ^{6}Li ions yield sharp peak is observed. This is a violation of the natural isotope abundance.

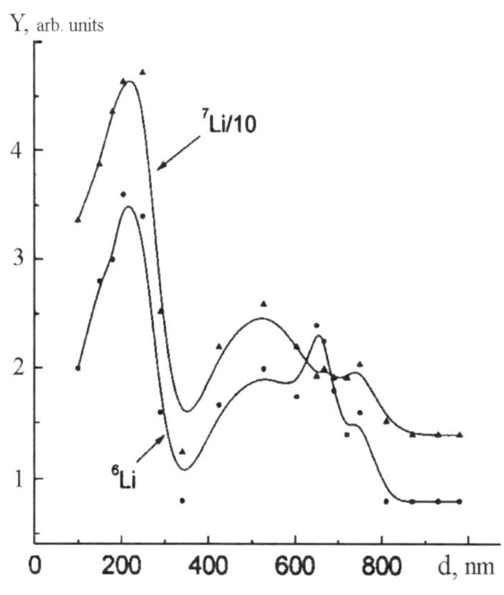

Fig. 13. The layer by layer profiles of the secondary lithium ions yields from the Ti/Al system after its saturation by hydrogen isotopes.

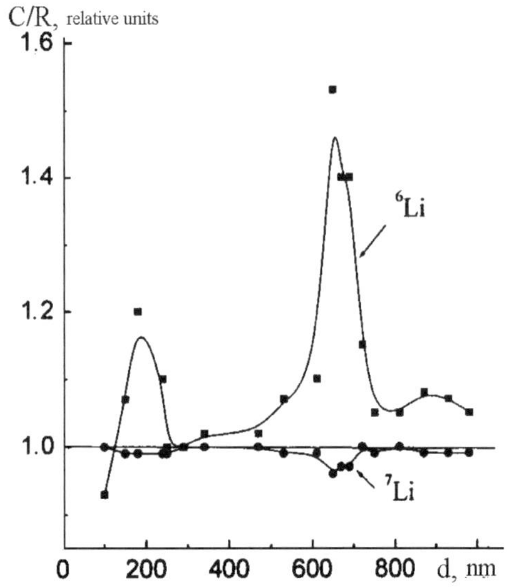

Fig. 14. The layer by layer (normalized to the natural occurrence) lithium isotope composition in the Ti/Al system, saturated by hydrogen isotopes.

In order to find the isotope ratio along the depth, the intensity ratio of $I_{(^6Li^+)}/I_{(^6Li^+ + ^7Li^+)}$ and $I_{(^7Li^+)}/I_{(^6Li^+ + ^7Li^+)}$ for all experimental spectra were measured. The quantities measured were divided by the corresponding tabulated isotope natural abundances. The results are shown in Fig. 14. It follows that, for ^{6}Li, a maximum excess above the natural abundance is observed.

Fig. 15 shows an example of the change in isotopic composition of a titanium film on ceramic, i.e., of the primary element in the film after saturation with a total dose of 12000 mA min cm^{-2}. In the case of a titanium film on aluminum, these differences (Fig. 15) are more obvious.

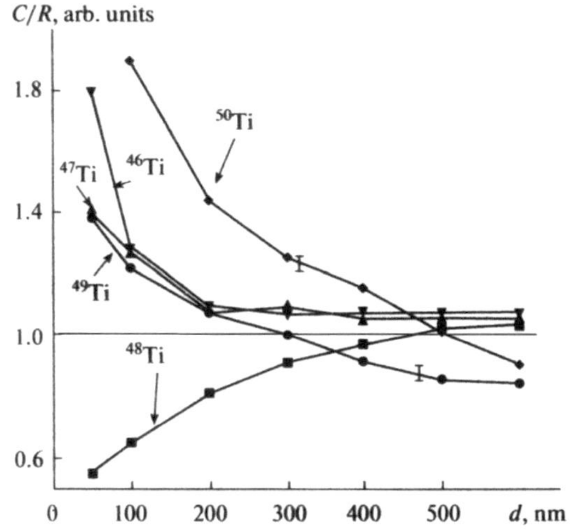

Fig. 15. Layer-by-layer difference in the relative concentrations of titanium isotopes after electrolytic saturation of the Ti/ceramic system.

This is apparently due to quite effective interdiffusion in the metal–metal system compared to that in the metal–ceramic system.

We will consider the observed effects in the lithium isotope separation (^6Li and ^7Li). On the basis of the theory and practice of isotope separation (Larionov, 2010), we will distinguish three possible processes leading to lithium isotope separation in the circumstances underconsideration: distribution with respect to mobilities in ion transport to the cathode; thermodynamic exchange in the solution–metal cathode system, specifically isotope diffusion in the cathode matrix; and lithium ion diffusion stimulated by hydrogen ions in the cathode metal. This last effect is corroborated by a rather high value of the diffusion coefficient in the cathode materials (D=3x10^{-9} cm^2/s). The separation factor for lithium isotopes with respect to nobilities does not exceed 1.003–1.006, and thus this process has little influence on the magnitude of the observed effect. We will consider the steps in the separation process, which most likely lead to isotope effects (from here on, isotope effect is understood to mean the difference in the lithium isotope concentration from that in the original composition). As the means for analyzing the effect of isotope separation based on thermodynamic exchange, we will use an estimate of β-factors of isotopes of the same metal (the β-factor is the ratio of the statistical sums over the states of the isotopic forms of molecular complexes). In this case the isotope effect or separation factor is equal to β-factor of the electrolyte/β-factor of the cathode metal.

Lithium ions in solution contain hydration layers. According to structural investigations, the coordination number for lithium ions in the primary layer, as well as in the secondary layer, is equal to 4; i.e., the total number of water molecules can be more than 8, depending on the lithium ion concentration in aqueous solution. As a lithium ion is implanted in the metal, its hydration shell is completely destroyed, and without this shell, the lithium atom forms a characteristic defect in the matrix of the primary metal (cathode). The free energy difference (^6Li and ^7Li) in the metal matrix is equal to

$$\Delta F = -kT\beta(Me), \tag{7}$$

where k – Boltzmann constant, T – absolute temperature, and β(Me) – β-factor of lithium isotopes in the metals.

$$\beta(Ti) = \exp\left[\frac{1}{2}kT\ln\left(\frac{K'}{K}\right)\frac{m}{m^*}\right] + kT\left\{\begin{matrix}\frac{1}{24}\left[\frac{K}{K^*}\left(1+\frac{m}{m^*}\right)-1\right]\mathrm{x} \\ \mathrm{x}\left\{\left(\frac{\hbar\omega_p^2}{kT}\right)^2 - \frac{1}{2880}\right\}\left[\left(\frac{K'}{2K}\right)^2\left(\frac{m}{m^*}+\frac{m^2}{m^{*2}}\right)\right]+\mathrm{N}\end{matrix}\right\}, \tag{8}$$

were $N = \frac{1}{8}\left(1+\frac{K'}{K}\right)^2 - \frac{1}{4}\left(\frac{\hbar\omega_p^2}{kT}\right)^4 + \cdots$

In these expressions: K, K' – force constants of the isotopic bond in the metal matrix, m, m* masses of the light and heavy isotopes, respectively, implanted in the metal matrix (electrode), and ω_p – vibrational frequency of the matrix–electrode lattice. Equation (8) was obtained from the formal apparatus developed for defects in crystals (Maradudin et al,

1965). Defect interaction in the "implanted metal-cathode matrix" structure can be estimated from the isotopic change in the vibration frequency of "defects"-atoms (^{6}Li and ^{7}Li). The calculated value for the isotopic shift in frequency is about 0.7–0.9 cm^{-1}. From Eq. (8), the β-factor of lithium isotopes in titanium is 1.088. The β-factor of hydration complexes of lithium isotopes is calculated from the expression:

$$\beta = \prod_{i=1}^{n} \left(\frac{U_i^*}{U_i} e^{-\frac{U_i^*}{U_i}} \cdot \frac{1-e^{-U_i}}{1-e^{-U_i^*}} \right)^{1/2}, \tag{9}$$

where n - the number of equivalent atoms of a given isotope available in the molecule or complex under consideration, and U_i, U^*_i - the vibrational frequencies of molecules with the light and heavy lithium isotopes, respectively $U = \frac{\omega ch}{kT}$, h - Planck's constant, ω - the vibrational frequency of an isotope in a molecule (complex), c is the speed of light in a vacuum, and T is the absolute temperature. The vibrational frequencies of isotopic lithium ions in the molecule of an aquo complex were calculated by solving the secular equations. The β-factor of a lithium aquo complex with four molecules of water is 1.136. The effect of the secondary hydration layer on the value of the β-factor is shown in the Table 3. We note that a sharp change in the β-factor is typical for 1-4 molecules of water, while it changes smoothly for n = 5–8. Further increase in the number of water molecules in the hydration sphere does not lead to a change in the β-factor. Thus, the significant change in the isotopic composition (isotope effect) of lithium implanted in the cathode can be explained by the destruction of the hydration layer as the ion enters the metal. Furthermore, this effect increases with the induced diffusion of lithium atoms through the cathhode in the flow of hydrogen atoms (Fig. 3, depth of 700 nm). This is supported by the fact that there is almost no hydrogen in the aluminum substrate, while its concentration is significant in the titanium. Isotope effects occurring during SIMS analysis (see part 2.1 of this chapter) can somewhat decrease the observed isotope effect during lithium implantation in the cathode. In a series of articles devoted to the electrolytic implantation of heavy and light hydrogen in different metals from solutions, the authors suggest that the change in isotopic composition in the metal-cathode (titanium) in which implantation took place, in impurities in the cathode, and in the ions responsible for electrolyte conductivity was the result of cold nuclear transmutation (Mizuno et al, 1996), (Beaudette, 2002). However, our experimental data point primarily to the effects of isotope disproportionation, which accompany electrolytic ion implantation in the cathode.

nH_2O	5	6	7	8
β - factor	1.146	1.155	1.162	1.170

Table 3. Calculated values of the β-factor for aquo complexes of lithium isotopes (n represents the number of water molecules in the aquo complex).

So, we have established that the electrochemical implantation of lithium in metals and composite materials is accompanied by isotope separation. This is caused by both a thermodynamic effect and the diffusion processes involving hydrogen, which accompany the process of implantation.

2.3 The isotope effect by thermo diffusion of copper atoms in nickel

The isotope effect during thermo-diffusion of copper atoms in nickel under the conditions of material creep has been studied (Puchkareva et al, 2002). When this condition holds, grain boundary diffusion of copper atoms in nickel occurs in conditions an external electric field and temperature-time non-stationary regime under an appreciate inner strain (osmotic pressure). According to these conditions (especially, the inner strain), the situation when the grain boundary diffusion occurs, is analogues to that for hydrogen saturation, because the hydrogen charge in solid matrix is accompanied by considerable strains.

Fig. 16 shows the isotope composition depth distribution of the copper atoms in nickel after diffusion and creep occurring over four hours at a temperature 600 K and strain of 10 kg/mm^2. An analogous isotope composition change was registered for other samples. Under these conditions, appreciable isotope redistribution is observed. It has been shown that there is a qualitative resemblance of this effect with the titanium isotope depth composition change while hydrogen saturates a titanium-ceramic system and bulk titanium and palladium samples. In all cases, the effect happens only in the near-surface layers (diffusion zone). In an isotope mixture, a concentration decrease of the most widely distributed isotope in the diffusion surface zone has been observed. A specific depth (about the influence of the surface on the diffusion redistribution of isotopes. The Cu/Ni samples, which were not influenced by an external force, do not show an isotope composition change.

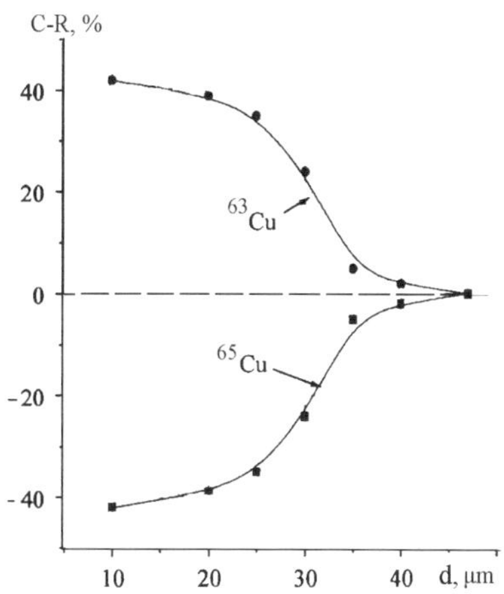

Fig. 16. Lateral view of the isotope composition diversions from natural of copper in nickel after diffusion in conditions of a creep.

2.4 Isotope effects during action of chemical active solutions on the solid surface

The influence of electrochemically activated sulphuric acid solutions on silicon wafers which are used for integrated circuits has been studied (Nikitenkov et al, 1993). The sulphuric acid solution was activated in an electrolyzer with a platinum anode and graphite cathode. The mirror silicon wafer surfaces were treated by an activated solution in a fluorine plastic cup at room temperature and various time intervals. After this procedure, the wafers were analyzed by secondary ion energy mass spectrometers. The experimental data on the silicon composition on the silicon wafers after various treatments are shown in Table 4. The samples are polished silicon wafer (up to mirror finish). Treatment 1 - was in a mixture H_2SO_4:H_2O_2.

Isotope A.m.u. Si	Natural occurrence %	Silicon isotope concentration, %		
		Initial sample	Treatment 1	Treatment 2
28	92.27	84.4	72.8	76.9
29	4.68	12.0	23.7	19.8
30	3.05	3.6	3.5	3.3

Table 4. Silicon isotope concentration in a Si^+ flux while sputtering the silicon target surface after various treatments.

Treatment 2 - was in electro-chemically activated H_2SO_4. The Table shows that the mechanical buffing and subsequent chemical treatment caused an appreciable increase in the secondary heavy ion yields.

This increasing is not due to the isotopic effect in the secondary ion emission (part 2.1 of this chapter), because this effect decreases the heavy ion yield. The peak overlapping (^{28}SiH upon ^{29}Si) and ($^{28}SiH+^{29}Si$ upon ^{30}Si) can partially be responsible for this increase. However, such an explanation contradicts the Table 1 data: the ^{29}Si concentration after mechanical and chemical treatment is several times greater than the natural one, while the ^{30}Si concentration is only 15–20 % greater. If this effect were due to the peaks overlapping, the opposite behaviour should be observed. We assume that the surface treatment is responsible for the natural isotope concentration violation. This violation mechanism (after chemical treatment) is probably caused by the isotopic displacement of the silicon valence electron energy levels analogous to the ionization level shift in the secondary ion emission.

Another isotopic effect while treating the sample surface by chemical solutions consists in changing the $LiNbO_3$ crystal surface after proton-ion doping. The studies (Sergeev, et al, 1991) were made using energy-mass-spectrometry method. In the proton-ion doping process, the sample surface is treated in a benzoic acid (C_6H_5COOH) melt with doping ingredients (salts or oxides) at the temperature 200–440 °C during the time interval up to 10 hours. Table 2 shows the lithium isotope concentrations in the secondary lithium ion flux while adding various doping ingredients.

Lithium isotope a.m.u	Natural occurrence, %	Doping agent					
		Be	Ca	Mn	Ni	Sr	Ba
		Lithium isotope concentration , %					
6	7.4	4.8	5.7	5.5	6.5	6.8	6.2
7	92.6	95.2	94.3	94.5	93.5	93.2	93.8

Table 5. The lithium isotope concentration in the secondary Li^+ flux while sputtering the $LiNbO_3$ crystal surfaces doped by various elements.

The Table 5 data are similar to those ones in the Table 4: the enrichment of the lithium secondary ion flux by the heavier lithium isotope is observed. The effect tends to grow with the doping ingredient mass. In the Table, they are arranged in order of their mass. As for silicon samples (Tab. 4), the natural abundance violation cannot be explained by the experimental errors. While proton-ion doping, the doping atom replaces the lithium or niobium atom in $LiNbO_3$ lattice, thus the effect is probably caused by the isotopic displacement of valence electron energy levels.

3. Conclusion

The experimental data presented may be divided into three groups in accordance with the physical process.

The first one is associated with surface sputtering, surface scattering, and ion implantation. The main features observed in these processes is an enrichment of surface isotope composition by heavy isotopes reflected or sputtered from the surface. The features of atomic collision and electron interaction in the atom-surface system are responsible for these regularities.

The second one is associated with hydrogen saturation metals and thermo diffusion from external sources. In this case the main features are a decreasing concentration of the most prevalent isotope in a diffusion surface zone. This is caused by an entropy concentration levelling. For the thin-film systems, the accumulation of lighter isotopes in defect regions is due to preferential pinning of the smaller radius atom on defects.

The third one is less studied, and is associated with the chemical reactions on the surface and in the surface layers. The main features, as in the first group, are the enrichment by heavier but not always the heaviest isotopes: ^{29}Si (Tab. 4) and ^{7}Li (Tab. 5). That is in this case the effect is observed for nuclear which possess by the magnet moment (nuclear do not possess even values of a charge and mass number). Therefore, the mechanism of this effect is probably fractionation of nuclear-isotopes on a nuclear magnetic moment by chemical reactions (Buhanchenko A.L., 2007).

4. References

Aksenov, A.I.; Bugaev, S.P.; Emelyanov, V.A.; et al. (1987). Deriving of the wide-aperture ions beams of metals. *Devices and experimental technique;* No 3, pp. 139–142, ISSN 0032-8162 [Russian].

Buhanchenko A.L. *New isotopy in chemistry and biochemistry.* (2007). Moscow. Nauka. ISBN 5-02-035617-4 [Russian].

Galimov E.M. (1981). *The nature of a biological fractionation of carbon isotopes.* Moscow. Nauka. ISBN 5-02-035617-4 [Russian].

Chernov, I.P.; Nikitenkov, N.N.; Puchkareva, L.N.; et al. (1999) Change in isotopic composition of metals enriched in hydrogen. *Russian Phys. Journ.*, No 4, pp. 61–65, ISSN 0367-6755 [Russian].

Chernov, I.P.; Nikitenkov, N.N.; Puchkareva, L.N.; Kolobov, Yu.R. (1998). Isotopic Composition of Metals at Deuterium Charge *Proceedings of the seventh international Conference. Cold fusion (ICCF-7).* Vancouver. pp. 441–446 [Copyright 1998 ENECO. - Inc. Salt Lake City, Utah, USA]

Chernov, I.P.; Nikitenkov, N.N.; Krening, M.; Baumbach, H. (1999). Changes in lithium isotopic composition in hydrogen-enriched Ti/Al systems. *Russian Phys. Journ.*, Vol. 42. - № 11. - pp. 947–951. ISSN 0021-3411

Chernov, I.P.; Nikitenkov, N.N.; Puchkareva, L.N. et al. (1999). Change in isotopic composition of metals enriched in hydrogen *Russian Phys. Journ.*, Vol.42, No 4, pp.427–430. ISSN 0021-3411

Chernov, I.P.; Nikitenkov, N.N.; Kroning, M.; Baumbah, H. (2000). Change in lithium isotope composition in thin-film structures at hydrogen charge. *Izv Russ Acad Sci. Phys.*, Vol.64, No11, pp.2181–2185, ISSN 0367-6755 [Russian].

Chernov, I.P.; Nikitenkov, N.N.; Kröning, M.; Baumbah, H. (2000). Investigation of the modification mechanism of metals isotope composition at hydrogen charge. *Proc Tomsk Polytech Uni.,* Vol. 303, Is.3, pp.62–71, ISSN1684-8519 [Russian].

Erikson, R.I.; Smith, D.P. (1975). Electronic processes in low-energy ion-surface scattering. *Phys Rev Lett,* 34, pp. 297–300 ISSN 0031-9007.

Gangrskiy, Yu. P.; Markov, B.N. (1984). *Nuclei in laser beams*. Ed. Kutuzova, K.A.; Moscow: Knowledge, 64p. ISSN 0373-2444. [Russian].

Gnaser, H. & Hutcheon, J.D. (1987). Velocity-dependent isotope fractionation in secondary ion emission. *Phys Rev B*; Vol.35, No1, pp.877–879. ISSN 1098-0121.

Gnaser, H. & Hutcheon, J.D. (1988). Preferential emission of light isotopes in the initial stage of sputtering. *Surf Sci*; Vol.195, pp. 499–511. ISSN 0039-6028 .

Gnaser, H. & Oechsner, H. (1990). Isotopic mass effect in sputtering: dependence on fluency and emission angle. *Nucl Instrum Methods Phys Res;* B48, pp. 544–8. ISSN 0168-9002

Helbig, H.F. & Adelman, P.J. (1977). Isotope Effects in Ion-Scattering Double Spectroscopy. *J Vac Sci Technol*, Vol. 14, pp. 488–93. ISSN

Helbig, H.F. & Orvek, K. (1980). Isotope Effects in Elastic Ion-Surface Scattering; He^+ and Ne^+ on Solid and Liquid Ga. *J Nucl Instrum Methods*, Vol. 170, pp. 505–507.

Kaplan, J. (1955). Nuclear Physics. Addison-Wesley. Cambridge.Mass.

Larionov, V.V.; Nikitenkov, N.N. & Tyurin, Yu.I. (2010). Isotope Distribution in Electrochemical Ion Implantation in Metals and Composite Materials. *Russian Journal of Electrochemistry*, Vol. 46, No. 1, pp. 112–115. ISSN 1023-1935

Lipson, A.G.; Maili, J.H.; Kuznetsov, V.A.; Asami, N. (2003). Isotope separation of a palladium under the influence of the strong mechanical stresses originating at loading Pd-folg by a deuterium. *Soli-state physics*, V. 45, Is. 8, pp. 1345–1350.ISSN: 0367-3294(Russian)

Lorin, J.C.; Havette, A. & Slodzian, G. (1982). *Abstract of conference on secondary ion mass spectrometry*. Berlin: Springer, p. 140.

Maradudin, A; Montrell, E & Weiss J. 1965. The dynamic theory of a crystal lattice in the harmonious approach. Moscow. Mir, 362 p.

Mizuno, T; Ohmori, T & Enyo, M. (1996). Isotopic changes on the reaction products induced by cathode electrolysis in Pd. *Journal of New Energy*, V.1, No 3, p.31.

Beaudette, C.G. (2002). *Excess Heat. Why Cold Fusion Research Prevailed.* OAK GROVE PRESS, LLC. South Bristol, Maine, USA. 410p.

Nikitenkov, N.N.; Kositsin, L.G.; Markova, N.M.; Shulepov, I.A. (1987). About an isotope effect in the secondary ion emission. *Abstract of XVII international conference Interaction the Charged Particles with Crystals*, MSU, p. 110 [Russian].

Nikitenkov, N.N. (1987). Theoretical and experimental studies of the secondary ion distribution while sputtering the complex composition targets by the Kev ions. *The author's abstract of doc. phys-math. sci. thesis.* M. MSU; 17p [Russian].

Nikitenkov, N.N.; Kositsin, L.G.; Markova, N.M.; Shulepov, I.A. (1988). About an isotope effect in secondary ion emission. *The secondary ion and photon emission. Reports of All-Union seminar.* Charkov: ChSU; Part 1. pp. 69–71 [Russian].

Nikitenkov, N.N.; Chernova, E.E.; Markova, N.M.; et al. (1993). Estimate of inhomogeneity of impurities distribution on a silicon plates surfaces. *Surface. Phys Chem Mech,* No. 10, pp. 74–78 [Russian].

Nikitenkov, N.N.; Chernova, E.E.; Karbainov, Yu.A.; et al. (1996). Some regularity of a real surface silicon modification by electrochemically activation in solutions of sulphuric acid. *Surface. X-ray Synchrotron Neutron Stud,* No. 11, pp. 45–50 ISSN: 1027-4510 [Russian].

Nikitenkov, N.N.; Chernov, I.P. &Tyurin, Yu.I. (2004). Isotope effects at technogenic affectings on a solid surface. Part 1: *Proc Tomsk Polytech Uni,* V. 307, No2. pp. 9–14; Part 2: *Proc Tomsk Polytech Uni,*V. 307, No. 3, pp. 26–29. ISSN: 1684-8519 [Russian].

Nikitenkov, N.N.; Kolokolov, D.Yu.; Chernov, I.P.; Tyurin, Yu.I. (2006). SIMS investigations of isotope effects at a processed solid surface. *Vacuum,* V.81, Is.2, pp. 202–210. ISSN: 0042-207X.

Poluhin, P.I.; Gorelik, S.S.; Vorontsov, V.K. (1982). Physics basis of plastic strain. Moscow. Metallurgija, 584 p.

Puchkareva, L.N.; Chernov, I.P. & Nikitenkov, N.N. (2002). Modifications of isotope composition of copper at the ionic implantation in nickel. *Izv Russ Acad SciPhys,* V.66, No. 8, pp. 1219–22, ISSN: 0367-6755 [Russian].

Plekhanov V.G. (2003). Isotopic effects in lattice dynamics, *Successes of physical sciences* (Russian), V. 173, No 7, pp. 711–738. (For the English speaking reader *Physics–Uspekhi* 46 (7) (2003) ISSN: 1063-7869 (Print), 1468-4780 (Online))

Slodzian, G.; Lorin, J.C. & Havette, (1980). A. Isotopic effect the ionization probabilities in secondary ion emission. *J Phys Lett,* V. 41, pp. L555–558. ISSN: 0370-2693

Shapiro, M.H.; Haff, P.K.; Tombrello, T.A.; Harrison, D.E. (1985). Simulation of isotopic mass effects in sputtering *Nucl Instrum Methods Phys Res;* V. B12, pp. 137–145. ISSN: 0168-9002

Shimizu, N. & Hart S.R. (1982). Isotope fractionation in secondary ion mass spectrometry. *J Appl Phys,* V. 53(3), pp. 1303–1311. ISSN: 0021-8979

Shwarz, S.A. (1987). Measurement of the secondary ion mass spectrometry isotope effect. *J Vac Sci Technol,* V. A5(3), pp. 308–312. ISSN: 0734-211X

Smith, D.P. (1971). Analysis of surface composition with low-energy backscattered ions. *Surf Sci,* V. 25, pp. 171–191. 0039-6028, ISSN 0039-6028

Sergeev, A.N.; Bamburov, V.G.; Nikitenkov, N.N.; Shveikin, G.P. (1991). The Oxide Surface Proton–Ion Doping. *Soviet Union Science Academy Reports.* Sverdlovsk: Chemistry Institute Preprint; 90p [Russian].

Shukolyukov Yu. A. (1996). Star dust in the hand. *Soros Educat. Journ.* №7, pp.74-80 ISSN 1684-9876 [Russian].

Tolk, N.H.; Tully, J.C.; Kraus J. et al. (1976). Angular dependence of oscillatory structure in low-energy ion-surface scattering. *Phys Rev Lett,* V. 36, pp. 747–750. ISSN. 0031-9007.

Wilson, R.G.; Stevie, F.A.; MageeL.W. (1989). Secondary Ion Mass Spectrometry. A Practical Handbook for Depth Profiling and Bulk Impurity Analysis. J. Wiley, N.Y.

24

Cavity Enhanced Laser Cooling of Solids

Biao Zhong, Youhua Jia and Jianping Yin
State Key Laboratory of Precise Spectroscopy, Department of Physics,
East China Normal University, Shanghai 200062
P. R. China

1. Introduction

Since the successful realization of laser cooling of solids in 1995[1], the optical cooling techniques[2-5] have made fast development, and laser cooling experiments have obtained some important progress. However, how to improve the absorption of pump laser and enhance the temperature drops so as to realize the practical optical refrigeration, it has become a hot research point in applied physics. As early as 2002, the Heeg et. al proposed an extra- or intra-cavity enhancement scheme to improve the laser cooling effect of solids[2], and experimentally demonstrated their intra-cavity enhancement scheme in 2004, and found that there is no large temperature drops due to the experimental parameters were not optimized[3]. In 2003, Hoyt proposed a multiple-pass scheme to enhance the absorption of pump laser and did a series of experimental researches about the multiple-pass scheme[4]. As expected, a greater temperature drop was obtained. However, due to some limitation in the early work, it is difficulty to cool the sample to the cryogenics temperature, such as low absorption of cooling materials[5], the coating of the sample surfaces[3], the purity of the sample[6], the load from the ambient[1,7], etc. Therefore, people have done much work to improve the record of temperature drop. For example, to avoid the coating absorption deposited on the sample surface, they cut the sample with a Brewster angle along with the cavity axis[8], or improve the purity of cooling material[9-11], and so on. Mansoor. Seletskiy used the multi-pass scheme to do some meaningful experiments and obtain a milestone progress, that is, they cool a high pure Yb^{3+}-doped YLF to cryogenic temperature[12] in this year. To obtain an efficient absorption, another scheme was proposed to insert a Brewster-cut sample into a resonant cavity and make the cavity to be resonated with the pump laser, and then the maximum absorption of the pump laser can be achieved [2, 13-18]. Seletskiy demonstrated this scheme in the experiment[13]. However, this scheme depends on the power of the pump laser. Fortunately, the intra-cavity scheme proposed by Heeg can solve this problem in the case of low absorption[2]. He also suggested that inserting a cooling sample into the resonant cavity of a laser with a gain (so-called intra-cavity enhancement scheme) to realize the maximum absorption of the pump laser[2, 14, 19] and carried out it in the experiment[3].

Besides of enhancement laser cooling of bulk solids using resonant cavity, we propose another new scheme to optically cool some film materials (i.e., filmier solids) by using a high fineness planar waveguide, which is similar to the cavity enhanced cooling scheme. In our laser cooling scheme, the solid film as a high-index gap waveguide is separated from a prism by a low-index gap one, and an incident laser beam is resonantly coupled into a waveguide mode through the gap, by frustrating total internal reflection. This will result in a large enhancement of the pump laser in the solid film and strengthen the effect of laser cooling.

In this chapter, some multi-pass, extra- and intra-cavity enhanced schemes for laser cooling of solids are reviewed and some experimental details of these schemes are discussed. In section 2, multi-pass laser pump scheme for laser cooling of solids are presented. In section 3, the basic principles of extra- and intra-cavity enhanced resonant absorption for laser cooling of solids are described, and the laser cooling effects between the multi-pass laser pump and cavity-enhancement schemes are compared. In section 4, some experimental results on laser cooling of solids using cavity-enhanced technologies and their recent progress are introduced in some detail and reviewed. Some main conclusions and outlook are summarized in the final section.

2. Multi-pass laser pump scheme for laser cooling of solids

In this section, the basic principles of the multi-pass laser pump scheme to increase absorption for the laser cooling of solids are briefly described, and the corresponding experimental results are introduced in some detail.

For the single pass pump geometry, the net power transferred to the sample for the laser radiation can be expressed as[8]

$$P_{net} = P_{abs}\eta_{cool} = [P_{in}(1 - e^{-\alpha_{total}(\nu)L})][\frac{\alpha_b + (1-\widetilde{\eta_q})\alpha_r(\nu) - \alpha_r(\nu)\widetilde{\eta_q}\frac{h\nu_f - h\nu}{h\nu}}{\alpha_{total}(\nu)}]. \quad (2.1)$$

The negative P_{net} corresponds to laser cooling for solids. In Eq.(2.1), P_{in} is the incident laser power, P_{abs} is the absorption power of the sample, and η_{cool} is the cooling efficiency, α_b is the background absorption coefficient, which includes both the material's inherent absorption and the frequency-independent heat-generated absorption. α_r is the resonant absorption coefficient, $\alpha_{total}(\nu)$ is the total absorption, which includes a generalized background absorption (α_b) contributing to heating the sample, resonant absorption (α_r)and an effective absorption due to scattering losses (α_s), which describe the Fresnel reflections losses from the sample facets[20]. L is the physical path length of the laser beam in the sample. And ν and ν_f are the laser frequency and the mean fluorescent (which represents the mean fluorescent photon energy), respectively. $\widetilde{\eta_q}$ is the external quantum efficiency which represents the fraction of fluorescent emitting from the sample[21] and given by

$$\widetilde{\eta_q} \equiv \frac{\eta_e W_{rad}}{\eta_e W_{rad} + W_{nr}}. \quad (2.2)$$

The principle scheme of the multi-pass pump laser cooling for the cuboid solid sample[4] is shown in Fig 2.1. An incident pump laser beam goes into a solid sample by a hole on the first mirror, and is reflected back at a slightly sloped angle (or along the slightly deviating the incident direction) by the second concave mirror. Taking account of the first bracket term in Eq. 2.1, we can see that if we adopt the multiple pump passes geometry, it will be increased with the number of passes. If the P_{in} is the laser pump power before the first mirror, then the absorbed power after multiply passing the sample with $\alpha_{total}(\nu)$ and the length L is expressed as[16]

$$
\begin{aligned}
P_{abs} &= P_{in}[1-\exp(-\alpha L)][1+R_2\exp(-\alpha L)+R_2R_1\exp(-2\alpha L)+ \\
&\quad +R_2{}^2R_1\exp(-3\alpha L)+R_2{}^2R_1{}^2\exp(-4\alpha L)+\ldots] \\
&\approx P_{in}[1-\exp(-\alpha L)][\frac{2-2(R_2R_1\exp(-2\alpha L))^n}{1-R_2R_1\exp(-2\alpha L)}] \\
&= P_{in}(1-e^{-\alpha_{total}(\nu)L})\frac{2-2S^N}{1-S}
\end{aligned}
\quad , \tag{2.3}
$$

where N is number of passes, $S = R_2R_1\exp(-2\alpha L)$, R_1 and R_2 are the reflectance of the first and second mirrors, respectively.

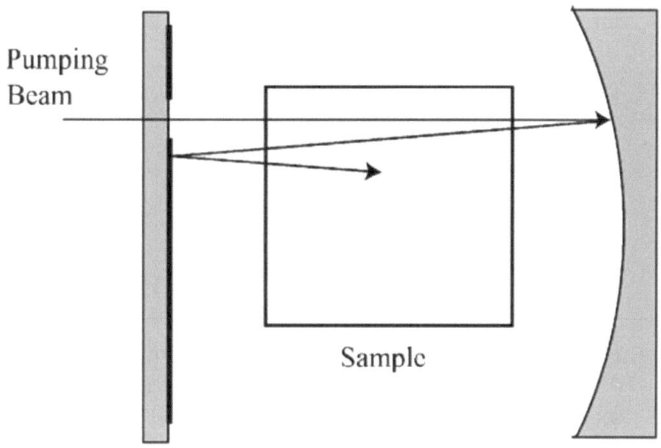

Fig. 2.1. Multiple-pass illustration scheme of cuboid sample laser cooling (figure from ref. [4])

To simplify equation 2.3, we can express the net power transferred to the sample as[3]

$$P_{net} = MP_{in}(1-e^{-\alpha_{total}(\nu)L})\eta_{cool}\,, \tag{2.4}$$

where $M=(2-2S^N)/(1-S)$, when η_{cool} is negative, P_{net} is also negative and represents the cooling power.

In recent years, Hoyt et al proposed and demonstrated three multiple-pass cooling schemes for solid sample and obtained some good results, which will be introduced as follows: The first is the Brewster sample multi-pass illustration scheme (see Fig.22). In this scheme, the solid sample is placed between the two mirrors, and the sample is cut as a cuboid geometric shape. It is clear that there are some disadvantages in this scheme, such as the Fresnel reflection losses from the sample facets, or the absorption pump laser contributing to heating due to the anti-reflection coating on the sample, etc. To avoid these disadvantages above mentioned, Hoyt et al used the Brewster-cutting geometry to reduce the Fresnel reflection losses from the sample facets by adjusting the polarization[4] of the incident laser beam.

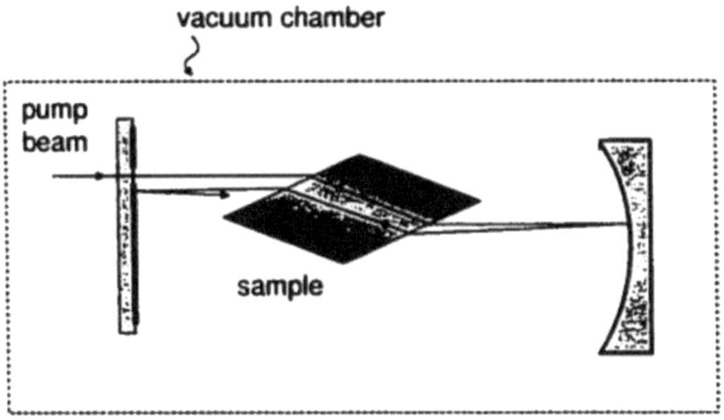

Fig. 2.2. Multi-pass illustration scheme of Brewster cut sample (figure from ref. [4])

The second scheme is the mirrored sample multiple-pass illumination (see Fig.2.3). In this scheme, the two mirrors are deposited directly on the samples, and the coating is exaggerated. There is a pore on one facet of the sample and the incident laser beam irradiates the sample by this hole, and the orientation of the pump beam deviates slightly the normal of the sample facet so as to multiply reflect the pump light in the sample back and forth. In this case, the absorption of the pump laser can be maximized.

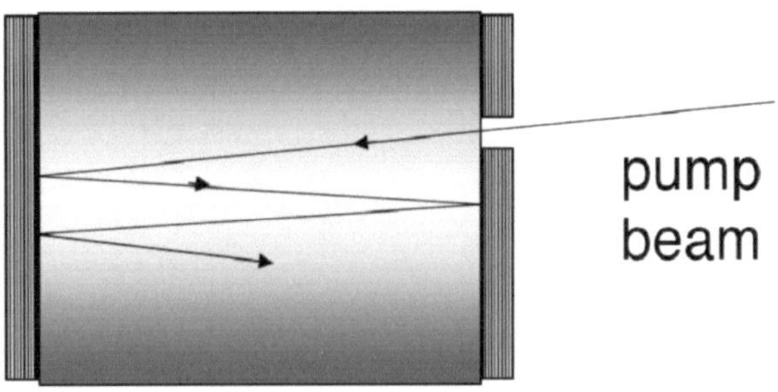

Fig. 2.3. Mirrored sample multiple-pass scheme (figure from ref. [4])

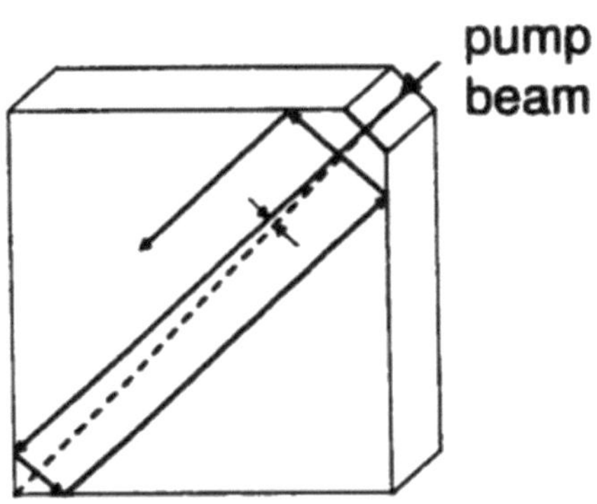

Fig. 2.4. Total internal reflection scheme of pump beam (figure from ref. [3])

The last scheme is the total internal reflection one of the pump laser beam, as shown in the Fig. 2.4. In this scheme, the total internal reflection is used to multiply reflect the pump beam in the sample and absorb the pump light as much as possible. For this, one vertex angle of the cubic sample at a diagonal is cut as a small square facet, and the pump beam goes into the sample from this square facet with a small departure, and then the pump laser beam can realize the total internal reflection in the sample, and then the pump beam can be almost absorbed.

On the other hand, the above three schemes have individual advantages. From the right term of Eq.2.5, we can see that if we want to maximize the net cooling power P_{net} deposited on the sample, we must maximize the factor *M*. Therefore in the second scheme when the mirrors are deposited directly on both sides of the sample, the factor *M* can be effectively maximized[3] because of a potential large number of passes (*N*). However, this will be detrimental for obtaining a potential large α_b due to the mirror absorption of the pump laser and resulting in a heating. This can be avoided in the first scheme because the mirrors are placed at the outsides of the cooling sample, and the effective background absorption will leave the material alone. Unfortunately, *M* is no longer maxima due to the finite scattering losses (e.g. Fresnel reflections), and there is also the difficulty of the maximizing *N*, but if adopting the Brewster-cut sample, the facet reflections are negligible and the factor *M* can be increased.

Hoyt et al made a cylindrical 1 wt% Tm^{3+}: ZBLANP sample with a radius of ~3mm and did some experimental researches for the above scheme[4, 8]. In this experiment, the highly-reflecting dielectric mirrors (Cascade optical) are deposited directly on both sides of the sample, and a small entrance aperture for the pump beam is made in one of the mirrors. After coupling the pump laser beam into the sample, unfortunately, the heating effect was observed at all pump wavelengths instead of cooling in the expected wavelength. The experimental results in Fig. 2.5 tell us that the heating magnitude varies with the pump wavelength[22]. From Fig. 2.6, it is obvious that there is the water absorption in the mirrors, and this water absorption in the mirrors are contained in the effective total absorption[22] (i.e. added to α_b =0.0002 cm^{-1}) distributed across the sample length in the mathematics.

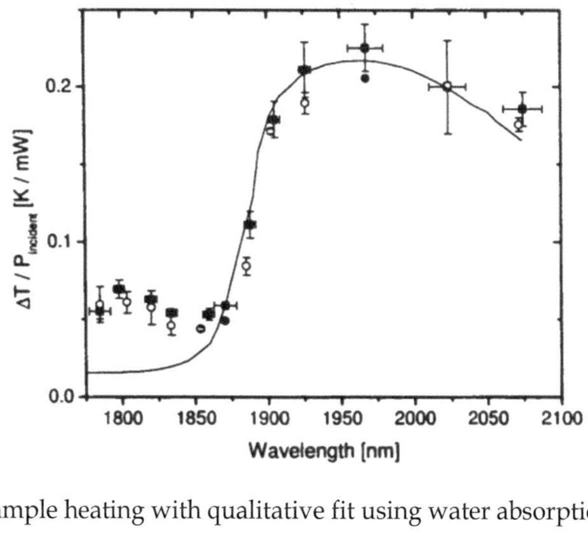

Fig. 2.5. Mirrored sample heating with qualitative fit using water absorption (figure from ref. [4])

If using Eq. 2.6[8] with an additional absorption due to water in the mirrors to fit the experimental data in Fig 2.5, we can obtain the expression of the temperature drop of the sample as follows

$$\frac{\Delta T}{P_{in}} \approx \kappa[\alpha_b + \alpha_r(\nu)(1-\tilde{\eta}_q) - \alpha_r(\nu)\tilde{\eta}_q \frac{\lambda - \lambda_f}{\lambda_f}] \tag{2.6}$$

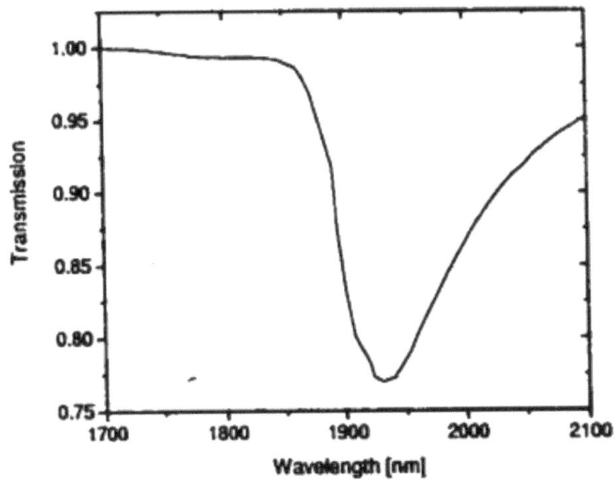

Fig. 2.6. Transmission spectrum of the cooling sample with a length of 20 μ m (figure from ref. [4])

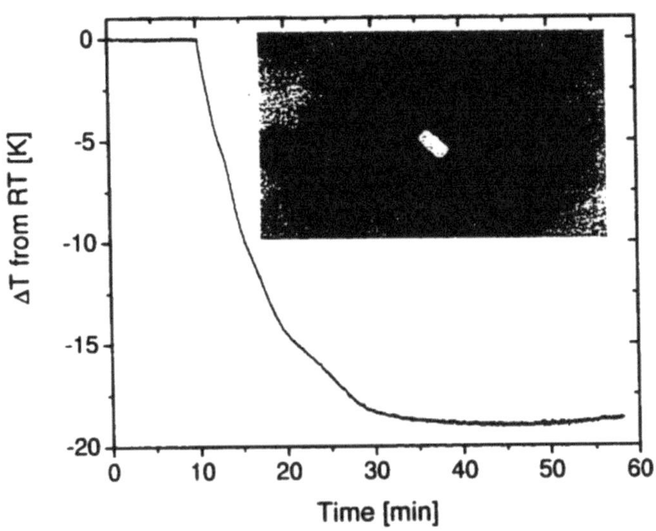

Fig. 2.7. The temperature drops from the room temperature. The bright areas correspond to cooling and the dark to heating (figure from ref. [4])

The solid line in Fig. 2.5 represents the quantitative fitting. From Fig.2.5 we can see that the change of the sample temperature is dependent on the wavelength of the pump laser. Fig. 2.7 shows the cooling results of the multiple-pass pumped cuboid sample in a non-resonant cavity configuration. The pump power is about 3.5W at a wavelength of 1.9μm. We can see from Fig. 2.7 that after the sample is illuminated by the pump laser, the sample temperature starts to be dropped at about 10min, and the sample was cooled to 19 K below the room temperature at the 30min and the sample temperature is no longer changed after ~ 30min. This shows that the net cooling is obvious because the heating from the mirrors that directly deposited on the sample is eliminated, however, the parameters M is not optimized because the pump beam undergoes 4% Fresnel losses at one facet of the sample at each pass[20].
To avoid the Fresnel losses, Hoyt et al. used the Brewster-cut sample in the non-resonant cavity to improve experimental results, and the corresponding experimental setup and results are shown in Fig. 2.8 and table 2.1. Here the OPO beam is the pump source with a power of 4.35W and 4.5W, which are coupled into the first facet of the two samples, respectively. In the two experiments, the cavity length is about 11cm and the second concave mirrors has a radius of 20 cm, and the change of the sample temperature was measured by recording the phase change in a calibrated Mach-Zehnder interferometer.

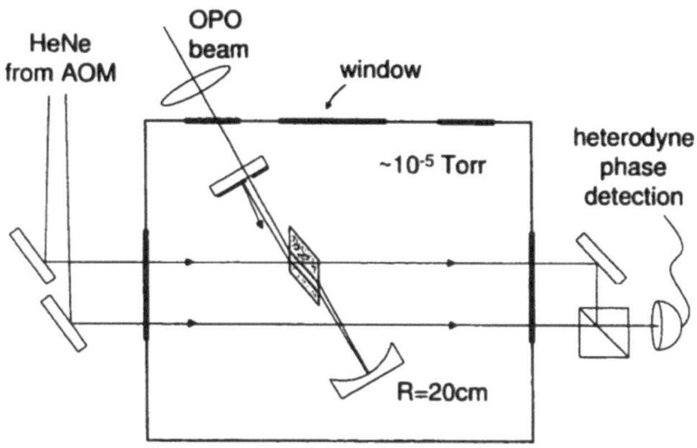

Fig. 2.8. Laser cooling setup using Brewster-cut sample in non-resonant cavity (figure from ref. [4])

Sample	ΔT from RT [K]	P_{abs} [W]	P_{cool} [mW]	τ_{fit} [s]	τ_{calc}
Tm C	-19	2.23	73	13.8	12
Tm D	-24	1.22	40	9	7.8

Table 2.1. The cooling results of Brewster-cut sample (table from ref. [4])

In the table 2.1, τ_{fit} is the fitted time constant, and τ_{calc} is the theoretical time constant from the equation[4]

$$\tau = \frac{c_m \rho V_s (1+\chi)}{4\varepsilon_s \sigma A_s T_C^3}, \tag{2.7}$$

where c_m is the heat capacity per volume, ρ is the sample density, V_s is the sample volume, T_c is the chamber temperature, A_s is the sample surface area, and σ is the Stefan-Boltzmann constant. In addition, the χ in Eq. 2.7 is defined as

$$\chi \equiv (\varepsilon_s A_s / \varepsilon_c A_c)(1-\varepsilon_c), \tag{2.8}$$

where $\varepsilon_{s,c}$ is the emissivity of the sample and chamber. Assuming the unity emissivity for sample and chamber, we can find from table 2.1 that τ_{fit} are 13.8min and 12min for the two samples, while τ_{calc} are 12min and 7.8min, respectively. So the difference between τ_{fit} and τ_{calc} are approximately 15%, but if presuming the sample emissivity is 0.9, the difference between τ_{fit} and τ_{calc} is reduced to about 5% in the both cases.
Hoyt et al studied the dependences of laser cooling power on the number of passes *N* for three different multi-pass schemes: cuboid sample with Fresnel losses in the non-resonant cavity, Brewster-cut sample with the same condition as cuboid sample, and a similar sample with the mirrors deposited directly on its facets, and the experimental results are shown in Fig. 2.9.

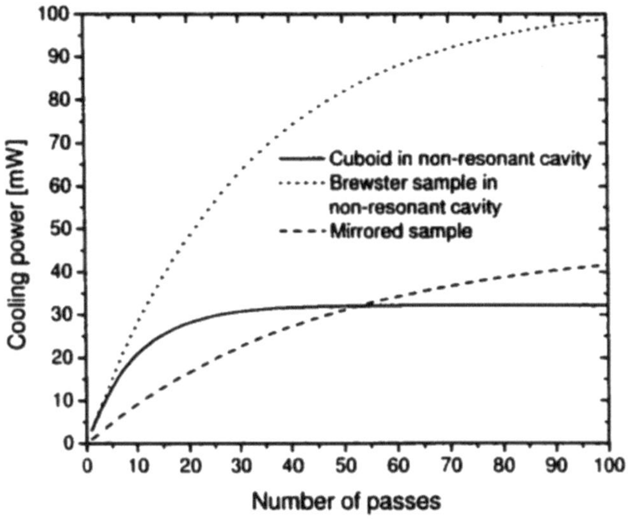

Fig. 2.9. Cooling power modeling for samples with cuboid and Brewster respectively, placed in a non-resonant cavity and a sample with mirrors deposited directly on its facets. (Figure from ref. [4])

In Fig. 2.9, the solid line represents the cuboid sample in non-resonant cavity according to Eq.2.5, and setting the parameters as follows: α_b =0.0002 cm^{-1}, α_r =0.025 cm^{-1}, $\widetilde{\eta_q}$ =0.99 and the sample length is 0.84 cm, the scattering losses α_s and Fresnel loss about 4% at each facet per pass are included in the total effective absorptivity α_b. The power of pump beam is 1.5W at a wavelength of 1.9μm, and the mean fluorescent wavelength is 1.803μm. The doted line shows the cooling power with the same condition as cuboid sample except without the Fresnel losses due to Brewster-cut surface. Taking into account the little loss on the Brewster cut facets, so setting that a loss values is 0.5% per pass in each facets. The dashed line is also the same calculation other than the mirrors deposited directly on the surface of the sample. Then in the case of α_s =0, the absorption in the mirrors can be added into the background absorption distributed across the sample length[4], so we can set α_b=0.0002 cm^{-1}. The parameters for calculating the cooling power for the mirrors attached directly on the two sides of the samples are arbitrary because it lies on the background absorption in the mirrors[4]. So it is the most important for cooling power that reducing the mirror's absorption. If the reflectance of the mirror deposited directly on the sample is 99.95%, then the remaining 0.05% power was absorbed by the mirrors. As the above discussed, we add this absorption into the background absorption distributed across the sample length. So each mirror has an optical density (OD) of $(\alpha L)_{mirror}$ =0.0005.

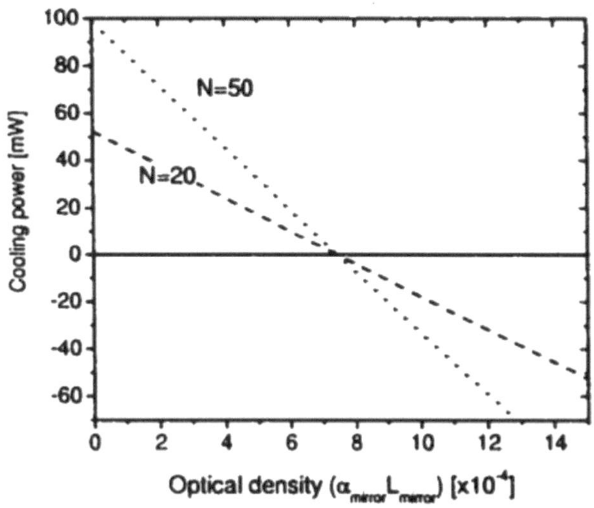

Fig. 2.10. The relationship between the cooling power and optical density (OD) in the scheme of mirror deposited directly on the samples. (Figure from ref. [4])

Fig. 2.10 describes the relationship between the optical density (OD) and the cooling power for N=50 and 20, respectively. From Fig. 2.10, we can find that the OD must less than the certain value otherwise the cooling power will be become a negative value, which can also be seen from Fig. 2.10. If the value of the OD is larger than 0.00078, there will be no cooling was observed.

In 2010, D. Seletskiy's group prepared the home-prepared high-purity 5% ytterbium-doped YLF crystal and cooled it down to the cryogenic temperatures, and the absorption, emission and cooling efficiency spectra of the 5% doped Yb^{3+} YLF crystal is shown in Fig.2.11.

In Fig. 2.11, figure (a) represents the absorption and emission spectra, and (b) describes the cooling efficiency. Fitting the experimental cooling efficiency using Eq. 2.9 [23], we can obtain $\eta_{ext} = 0.995$ and $\alpha_b = 4.2\times10^{-4}cm^{-1}$.

$$\eta_c(\nu,T) = \eta_{ext}[\frac{1}{1+\alpha_b/\alpha(\nu,T)}]\frac{\tilde{\nu}_f(T)}{\nu} - 1, \tag{2.9}$$

where η_{ext} is the external quantum efficiency, α_b is the background absorption, $\alpha(\nu,T)$ is the resonant absorption coefficient of the active ion, and $\tilde{\nu}_f$ is the frequency of the escaped mean luminescent photon, including the reabsorption effect of the emitted fluorescent photon.

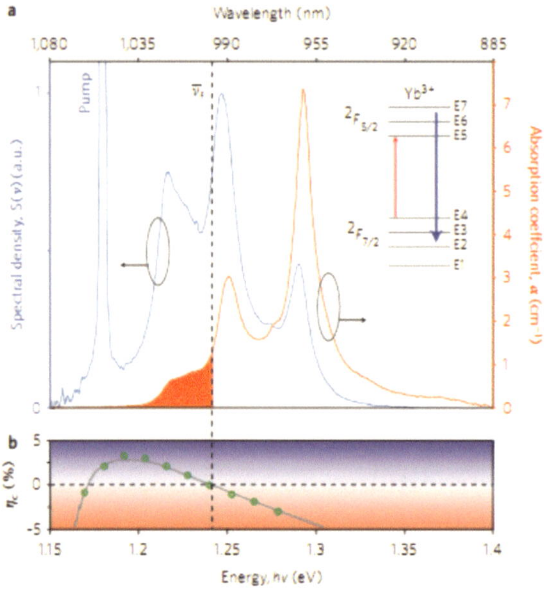

Fig. 2.11. Absorption, emission and cooling efficiency spectra of the 5% doped Yb^{3+} YLF crystal. The red and blue lines are the absorption and emission spectra respectively for a pump laser with λ=1055nm and polarization along the c-axis. The shade of the red line is the cooling tail. Inset: $^2F_{5/2}$ and $^2F_{7/2}$ levels in the crystal position containing seven stark manifolds. Figure b, cooling efficiency curve, the open circles is the experimental data and the line is the mode fitting one. (Figure from ref. [12])

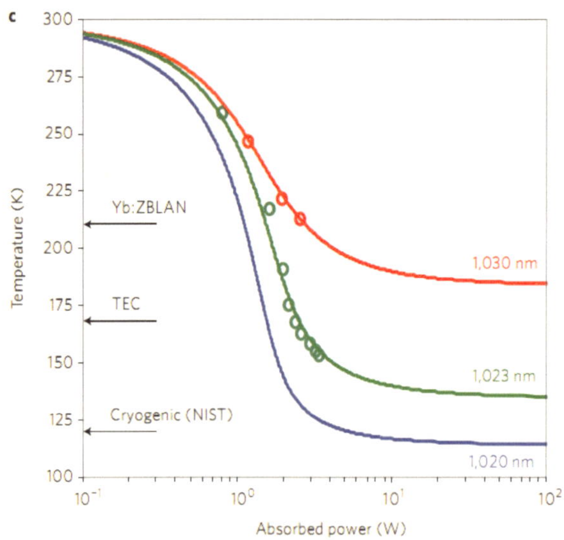

Fig. 2.12. Dependences of the sample cooling temperature on the absorbed power for different pump laser wavelength. The open circles are the experimental data and the solid lines are the mode fitting curves from Eq. 2.10. The blue line indicates the sample used in the experiment can achieve the minimum temperature. (Figure from ref. [23])

Fig. 2.12 shows the dependences of the cooling temperature of Yb: YLF sample for the pump wavelength at $\lambda = 1020$nm, 1023nm and 1030nm, respectively. According to the experimental data in Fig. 2.12 and the Stefan-Boltzmann law associated with the measured $\eta_{abs}(\nu,T)$ and η_{ext}, the dependence of the sample cooling temperature on the absorbed power can be modeled[23] as

$$\eta_c(\nu,T)p_{abs}(\nu,T) = \kappa(T_C^4 - T^4), \tag{2.10}$$

where T_c is the chamber temperature, and κ is the radiation load constant depending on the geometry and joint thermal emissivities of chamber and sample.

In Fig. 2.12, the red, green and blue lines represent the cases of the pump laser wavelength of 1030nm, 1023nm and 1020nm, respectively. The above arrow indicates the previous cooling record in ytterbium-doped ZBLAN glass, the middle arrow shows the standard thermoelectric coolers (TEC) and the below arrow indicates NIST-defined cryogenic temperature. In this experiment at the maximum available absorbed power of 3.5W at 1023nm, the sample is cooled to $155K \pm 1K$ absolute temperature, and the heating is lifted up to 90mW.

3. Resonant cavity-enhancement schemes for laser cooling of solids

In this section, we review the basic principles and theoretical researches of laser cooling of solid in resonant extra- or intra-cavity, and compare the laser cooling results in the above schemes. Furthermore, we compare the cavity-enhancement schemes with the multi-pass enhancement absorption one.

3.1 Extra-cavity enhancements for laser cooling of solids

Fig. 3.1 shows the scheme of the extra-cavity enhancement for laser cooling of solids. As shown in Fig. 3.1, the two mirrors with a reflectivity R_1 and R_2 are composed of a resonant extra-cavity, and the cooling medium with an absorption coefficient α is placed between the mirrors. According to the Beer Law, the loss of the pump beam is proportional to $e^{-\alpha d}$ in each pass, and in the same time the other losses including the reflection on the surface of sample and its scattering is defined as x, the pump laser with an intensity of I_0 is coupled into the cavity, and pump/illuminate the sample with a length of L and a lower absorption. The pump beam is reflected on the two cavity mirrors back and forth. If the cavity is resonant with the pump laser, consequently, the beam intensity in the cavity will be resonantly enhanced, and the absorption of cooling sample will be also enhanced.

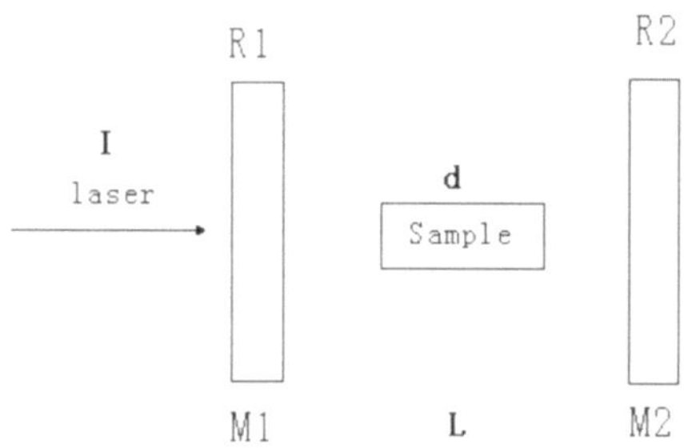

Fig. 3.1. The principle scheme of extra-cavity enhancement for laser cooling of solids (Figure from ref. [16])

Here we assume that the amplitude of the pump laser is E in the cavity and the corresponding intensity is I, while the amplitude of the incident laser beam before the first mirror R_1 is E_0 and the corresponding intensity is I_0, and then the amplitude of the laser after each time reflection from R_1 and R_2 are given by

$$E^2 \times (1 - R_1) = E_0^{\ 2}, \tag{3.1}$$

$$E_1 = E_0 r_1 r_2 \exp(ik \times 2L), \tag{3.2}$$

$$E_2 = E_1 r_1 r_2 \exp(ik \times 2L), \tag{3.3}$$

$$E_n = E_{n-1} r_1 r_2 \exp(ik \times 2L), \tag{3.4}$$

where $r_{1,2}$ are the amplitudes of the reflectivity of the first and second mirrors respectively, and then the total amplitude of the pump beam forward to the right can be expressed as

$$E_a = E_0 \sum_{n=1}^{\infty} (r_1 r_2)^{2(n-1)} \exp[ik \times 2(n-1)L] = \frac{E_0}{1 - r_1 r_2 \exp(ik \times 2L)}, \tag{3.5}$$

and the corresponding intensity is

$$I_a = E_a \times E_a^* = \frac{I_0}{(1 - r_1 r_2 e^{i\delta})(1 - r_1 r_2 e^{-i\delta})} = \frac{I_0}{1 + (r_1 r_2)^2 - 2 r_1 r_2 \cos\delta}, \tag{3.6}$$

where $\delta = 2kL$. When the laser is resonant with the cavity, we have

$$\cos\delta = 1 \text{，} I_a = \frac{I_0}{1 + (r_1 r_2)^2 - 2 r_1 r_2}. \tag{3.7}$$

Assuming the axis of optical cavity is the x axis, the total amplitude of pump laser in the cavity is given by

$$E = (E_a e^{jkx} + E_{a'} e^{-jkx}) e^{-jwt} = 2E_a j \sin kx e^{-jwt}, \tag{3.8}$$

and the corresponding intensity of the pump beam is

$$I = \tilde{E} \times \tilde{E}^* = I_a (1 - \cos 2kx). \tag{3.9}$$

Then the mean intensity of the pump beam can be written as[16]

$$\bar{I} = \frac{I_0}{1 + [r_1 r_2 (1 - x) e^{-\alpha l}]^2 - 2 r_1 r_2 (1 - x) e^{-\alpha l}}. \tag{3.10}$$

In final, the mean enhancement factor is defined as a ratio of the mean intensity of the pump beam in the cavity to the initial intensity of the incident laser beam and given by [16]

$$E_{cav} = \frac{p_{circl}}{P_{in}} = \frac{\bar{I}}{I} = \frac{(1 - r_1^{\ 2})}{1 + [r_1 r_2 (1 - x) e^{-\alpha l}]^2 - 2 r_1 r_2 (1 - x) e^{-\alpha l}}. \tag{3.11}$$

We take the ZBLAN for an example and set x= 0.05, l=3mm, and α =0.09cm^{-1}, and study the dependence of the enhancement factor E_{cav} of the pump laser in the cavity on both the amplitude r_1 of the reflectivity of the first mirror R_1, and the sample length, and the results are shown in Fig.3.2.

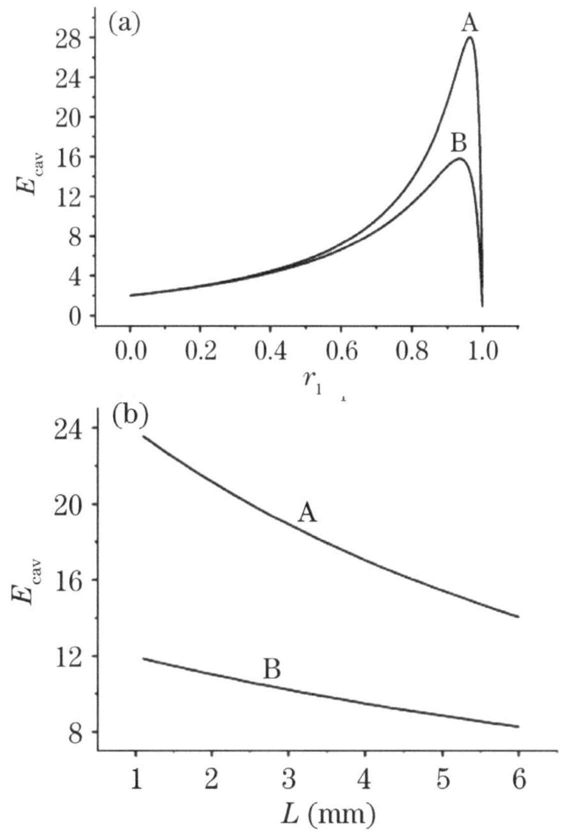

Fig. 3.2. (a) The relationship between the enhancement factor E_{cav} of the pump laser in the cavity and the amplitude of the reflectivity of the first mirror. (b) The relationship between the enhancement factor E_{cav} and the sample length L. (Figure from ref. [16])

In Fig. 3.2 (a), the curve A is the case of the sample surface with an AR coating, which can be used to reduce the losses of reflection on the sample surfaces so as to enhance the intensity of the pump beam in the cavity. The curve B is the case of the sample's surface without an AR coating. It can be seen from the curves A and B that when the reflection loss from the sample surfaces is reduced from the 0.04 to the 0.01, the enhancement factor of the pump laser in the cavity is enhanced from the 15 to 28. From Fig. 3.2(b), we can see that with the increasing of the sample length, the enhancement factor will be reduced. This is because the absorption of sample will follow the Beer Law. This shows that by reducing the sample length, one can obtain higher enhancement factor of the pump beam in the cavity, but it also reduce the cooling power. So it is must be taken into account an optimal sample length and a cooling power as high as possible.

Then we use the two energy level system to analyze the Yb^{3+} level transition between $^2F_{7/2}\rightarrow{}^2F_{5/2}$, and obtain an expression of the cooling power in the Extra-cavity enhancement scheme [24] as follows

$$P_{cool} = \frac{N\sigma_{abs}L\alpha_{eff}I_s(\lambda/\lambda_{F*} - 1)}{1 + \sigma_{se}/\sigma_{abs} + \alpha_{eff}I_s/P}, \tag{3.12}$$

where α_{eff} is the effective irradiation area of the sample, P is the power of pump laser, and I_s is the characteristic wavelength-dependent saturation intensity[19], which is defined as

$$I_s = hc\gamma_{rad}/\lambda\sigma_{abs}, \tag{3.13}$$

where σ_{abs} and σ_{se} are the absorption and stimulated emission cross-sections respectively, λ is the wavelength of the pump laser, and λ_{F*} is the effective mean fluorescent wavelength[7], which can be defined as

$$\lambda_{F*} = (\frac{1}{\lambda_F} - \frac{\kappa}{hc\gamma_{rad}})^{-1}. \tag{3.12}$$

According to Eq 3.12, we study the dynamic process of the cooling temperature for different pump power, and the results are shown in Fig.3.3. In our calculation, the used parameters come from Ref. [7].

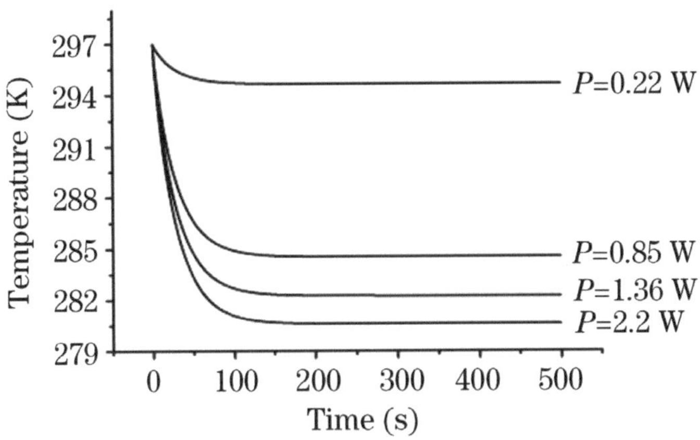

Fig. 3.3. Temperature drop with different pump laser power (Figure from ref. [16])

From Fig.3.3, we can find that with the increase of the pump power from 0.22W to 2.2W, the temperature of the cooling sample will be reduced from 294.6K to 280.5K. This shows that the extra-cavity enhanced laser cooling for solid is also very effective.

In 2006, D. Seletskiy et al proposed another method to describe the extra-cavity resonant enhancement for optical refrigerator. They assumed that the reflectivity R_2 of the second mirror of the cavity is approximately equal to 1, the absorption coefficient A of the cavity is defined as[13]

$$A = 1 - (T + R), \tag{3.13}$$

or

$$A = \frac{1-\Delta}{1+F\sin^2\delta}, \tag{3.14}$$

where

$$\Delta = \frac{(\sqrt{R_{ic}} - e^{-\alpha l})^2}{(1-e^{-\alpha l}\sqrt{R_{ic}})^2}, \tag{3.15}$$

$$F = \frac{4e^{-\alpha l}\sqrt{R_{ic}}}{(1-e^{-\alpha l}\sqrt{R_{ic}})^2}, \tag{3.16}$$

and R_{ic} is the input coupling reflectivity and $\delta = kl$, k is the wave vector inside the absorbing material here assuming that the material fills up the whole cavity, and l is the length of cooling material, and α is the absorption coefficient. From Eq. 3.14, it is clear that A can be achieve to be unity when Δ and $\sin\delta$ are set to be zero, i.e., when $R_{ic} = e^{-2\alpha l}$ and $\delta = m\pi$.
From the above two methods, we find that when the reflectivity R_1 of the in-couple mirror is about 95% in the case of setting the same sample absorption, the intensity of the pump beam in the cavity and the absorption of the pump laser can achieve the maximum value.
Similar to the extra-cavity resonant enhancement laser cooling scheme for solids, we proposed a resonant waveguide structure (see Fig.3.4) to effectively cool solid sample, which will enhance the intensity of an evanescent wave by more than two orders of magnitude in the doped Yb^{3+} YAG film with a refractive index n_3. The waveguide structure for this scheme is consisted of a prism and two waveguides, which form three layer films with different refraction index n_1, n_2 and n_3 from the bottom to up, here $n_2 < n_1 < n_3$. The enhancement principle of this resonant waveguide scheme is similar to a Fabry-Perot cavity with a high finenss[25]. On the doped Yb^{3+}YAG film-vacuum interface, the light is totally reflected, corresponds to the second mirror of the cavity with a reflection coefficient of 1.0. We thus have a Fabry-Perot cavity with a high fineness: for given values of incident light parameters (e.g. the wavelength), the light field amplitude inside the waveguide will experience a strong resonant enhancement.

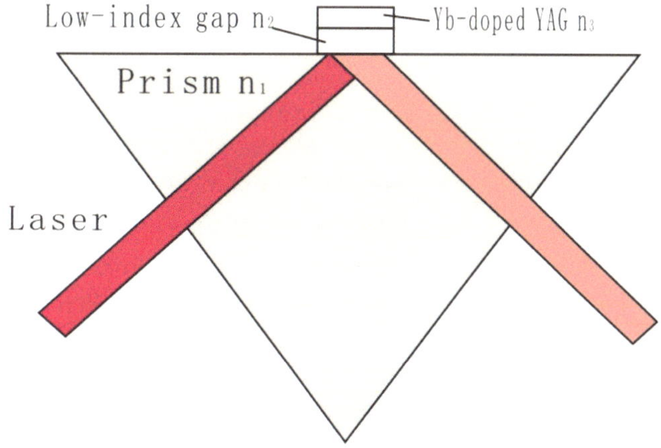

Fig. 3.4. Scheme of resonant waveguide enhancement for laser cooling of solids

We calculate the enhancement factor of the pump laser in the resonant waveguide structure for the different thickness of doped Yb^{3+}YAG film, and the results are shown in Fig.3.5.

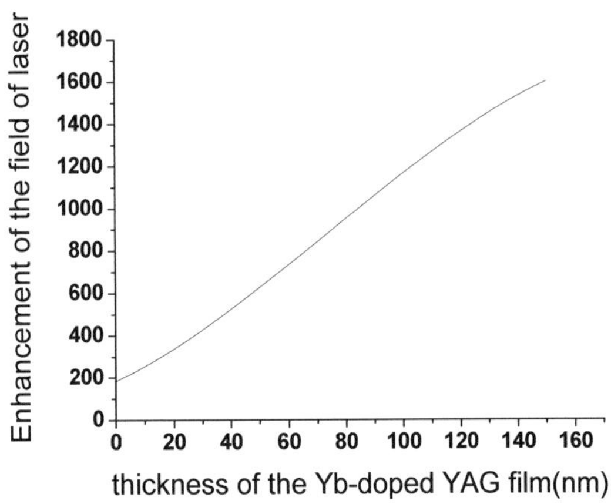

Fig. 3.5. Dependence of enhancement factor on the thickness of Yb-doped YAG film

It is clear from Fig.3.5 that with the increase of the thickness of Yb-doped YAG film, the enhancement factor of the intensity inside Yb-doped YAG film will be greatly increased. In particular, when the thickness of Yb-doped YAG film is increased from 10nm to 160nm, the enhancement factor will be increased from about 200 to 1600 times. This shows that our proposed resonant waveguide enhancement scheme can be used to efficiently cool the solid film material.

3.2 Intra -cavity enhancements for laser cooling of solids

The extra-cavity resonant enhancement technology is useful to laser cooling of solids due to the greatly-increased pump laser intensity. However, because the partial transparence of in-couple mirror M_1 ($R_1 < 1$) confines the pump power coupled into the cavity (that is, there is a coupling loss from M_1), the enhancement of pump-beam intensity and the cooling effect of the sample is limited. However, if placing a cooling sample into a resonant cavity of with a gain, we can obtain a greater pump-laser power for laser cooling of solids. This is so-called the intra-cavity cooling scheme, which is shown in Fig. 3.6. In this scheme, the cooled sample is inserted in the resonant cavity of laser with a gain, and the two mirrors (M_1 and M_2) with a reflectivity R_1 and R_2 are the two resonant-cavity mirrors of laser with a gain coefficient g_0. In this case, the two main losses, the absorption loss of sample and the reflection loss of sample surfaces, are added into the original loss δ_0 of the used laser, which forms a total loss δ for the laser. As long as $g_0 \geq \delta$, the used laser can be still satisfied its resonant condition and form a stable laser oscillation. It is clear that the laser intensity in such a laser cavity is far higher than one in a conventional laser because the reflectivity R (= $R_1 = R_2$) of the mirrors in our proposed intra-cavity cooling scheme can be higher than one R_2 of the output mirror M_2 in a conventional laser. So the intensity of the pump laser in intra - cavity resonant enhancement scheme will be greatly amplified, and an efficient laser cooling for solids will be obtained.

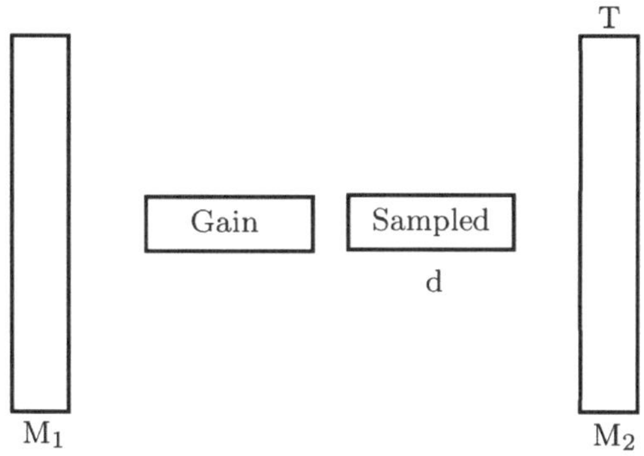

Fig. 3.6. Intra-cavity cooling scheme for solids (figure from ref. [14])

It is well known that the output power of a conventional laser system is given by [26]

$$P_{out} = \frac{1}{2} A I_s T \left(\frac{2g_0}{T+\delta_0} - 1\right), \tag{3.17}$$

where I_s is the saturation parameter of laser system, A is the average effective cross section, g_0 is the gain coefficient of laser, δ_s is the loss on the surface of sample, δ_0 is the original loss in laser cavity. When the cooling sample is placed in a resonant cavity of laser, the circular power in the cavity is given by [26]

$$P_{circ} = \frac{1}{2} A I_s \left(\frac{2g_0}{T+\delta_0+\delta_S+2\alpha d} - 1\right), \tag{3.18}$$

and then the enhancement factor can be expressed as[14]

$$M = \frac{P_{circ}}{P_{out}} = \frac{\dfrac{2g_0}{T+\delta_0+\delta_S+2\alpha d} - 1}{T\left(\dfrac{2g_0}{T+\delta_0} - 1\right)}. \tag{3.19}$$

In the low gain case, assuming $g_0 = 0.084$, $\delta_0 = 0.076$, $\delta_S = 0.004$, and the sample length is d=3cm and $\alpha = 0.09cm^{-1}$, for Yb^{3+}-doped material. When d=2mm, 3mm, 4mm, according to Eq. 3.19, we study the dependence of the enhancement factor M on the transmittance T of the mirror M_2, and the results are shown in Fig. 3.7.

From Fig.3.7, we can see that with increasing of the sample length, the enhancement factor will be reduced because the longer the sample length is, the more the sample absorption of the pump laser is according to Beer Law. However, for a given sample length, the enhancement of the pump beam intensity will be reduced with the increase of the transmissivity T of the output mirror M_2 due to the increase of the output loss from the mirror M_2. The cooling power of the perfect sample placed in the intra-cavity can be described as[1]

$$P_{cool} = P_{abs}\eta = P_{abs}\frac{\lambda - \lambda_f}{\lambda}, \tag{3.20}$$

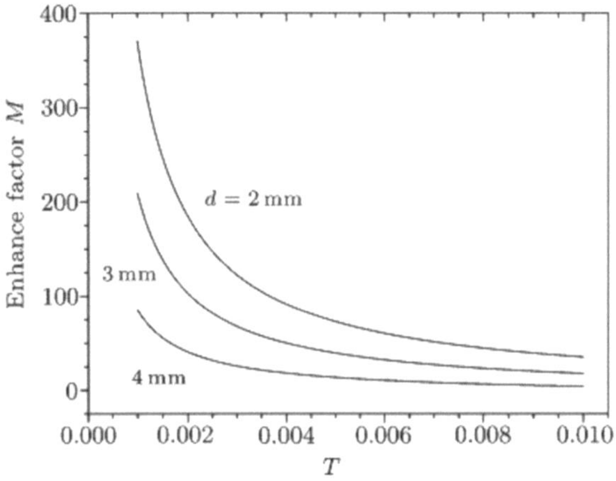

Fig. 3.7. Enhancement factor as a function of transmittance T of the mirror M_2 for different sample length (Figure from ref. [14])

where P_{abs} is the absorption power of the sample, η is the quantum efficiency that will be slightly reduced during the temperature drop of the sample, simultaneously the mean fluorescent wavelength will be red shift[26], and we have

$$P_{abs} = P_{circ} \times (e^{\alpha d} - 1)\,. \tag{3.21}$$

Substituting Eq.3.21 into Eq.3.20, and making the assuming that αd is small, Eq. 3.21 can be approximately as[14]

$$P_{abs} = \frac{1}{2}\alpha d A I_s \left(\frac{2g_0}{T + \delta_0 + \delta_S + 2\alpha d} - 1\right). \tag{3.22}$$

From Eq.3.22, it is obvious that for the some sample the absorption power of the sample depends on both the sample length and the transmissivity of the output mirror M_2. It is clear in physics that the longer the sample length is, the more the absorption power of the sample is, and then the increasing cavity loss will result in the reduction of the laser oscillating intensity in the cavity. So there must be the optimal length, which can lead to the maximum P_{abs}.

By solving the steady-state solution of Eq.(3.22), that is, by solving $\frac{d(P_{abs})}{d(\alpha d)} = 0$, we can obtain the relationship between the optimal sample length d_{opt} and the transmissivity T of the output mirror as follows[14]

$$\alpha d_{opt} = \frac{1}{2}\left[\sqrt{2g_0(T + \delta_0 + \delta_S)} - (T + \delta_0 + \delta_S)\right], \tag{3.23}$$

and the corresponding maximum absorption power is[14]

$$P_{abs,\max} = \frac{1}{4} A I_s \left(\sqrt{2g_0} - \sqrt{T + \delta_0 + \delta_S}\right)^2. \tag{3.24}$$

Fig. 3.8 shows the dependence of the optimal sample length on the transmittance T of the output cavity mirror M_2.

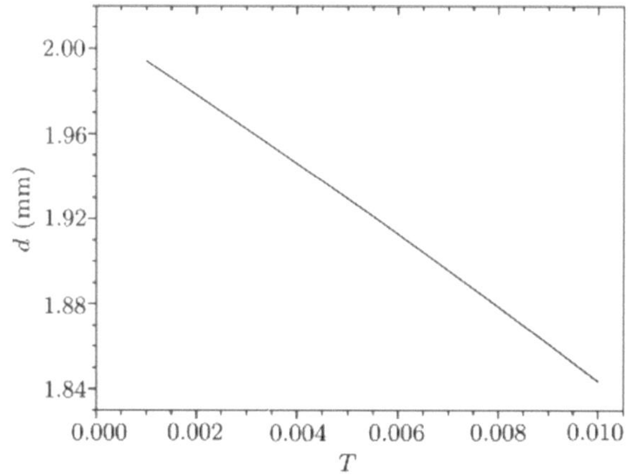

Fig. 3.8. Optimal sample length as a function of transmittance T of the output mirror M_2 (Figure from ref. [14])

It is clear from Fig.3.8 that the optional sample length will be nearly linearly reduced with increasing the transmittance T of the output mirror M_2. In particular, the sample length does not exceed 2mm in order to obtain a higher enhancement factor and better cooling effect.

3.3 Comparison between multi-passes pump and cavity-enhancement schemes

The enhancement of three schemes, multi-passes, extra-cavity and intra-cavity, are analyzed in the section 2, sub-section 3.1 and 3.2 respectively, and these schemes have their different advantages. In this section, we will compare make the comparison between the multi-pass pump scheme and two cavity-enhancement ones, and hope that one can learn some guiding sense.

Firstly, we compare the multi-passes pump scheme with the extra-cavity enhancement one. According to Eq. 2.3, when the reflection times N of the pump beam approach an infinite value, we can obtain the maximum absorption power as follows [14]

$$P_{abs} = P_{in}[1-\exp(-\alpha L)][\frac{2}{1-R_2R_1\exp(-2\alpha L)}] = 9.7\times P_{in}[1-\exp(-\alpha L)], \qquad (3.25)$$

where assuming $R_2 = R_1$ =99.9%. However, in the extra-cavity enhancement scheme, the absorption power can be described as

$$\mathrm{P}_{abs} = P_{circ}[1-\exp(-\alpha L)], \qquad (3.26)$$

where $P_{circ} = E_{cav}P_{in}$, according to Eq.3.26 and setting the same parameters, we can obtain $P_{circ} = 13.8P_{in}$. It is clear that the enhancement factor in the extra-cavity resonant (absorption) scheme is larger than that in the multi-passes one. By comparing the temperature drop in the above two schemes, one can easily observe the advantages of the cavity-enhancement. The temperature drop of the cooling sample can be given by [6]

$$\frac{C_m \rho_m \pi D^2}{4} \frac{d(T_S - T_r)}{dt} = P_{load}(T_S) - P_{cool}(T_S), \tag{3.27}$$

and

$$P_{load} = A \int \varepsilon_v \pi \frac{\partial B_v(T)}{\partial T} \Delta T dv = 4\pi D \varepsilon \sigma_B T_r^3 \Delta T = \pi D \varepsilon \sigma_B (T_r^4 - T_S^4), \tag{3.28}$$

wgere S is the cross section of the sample, P_{load} is the radiative and conductive load, C_m is the specific heat, ρ_m is the mass density, T_S and T_r are the sample and room temperatures, respectively. Fig. 3.9 is the results of the temperature drop of the multi-passes pump and extra-cavity enhancement schemes.

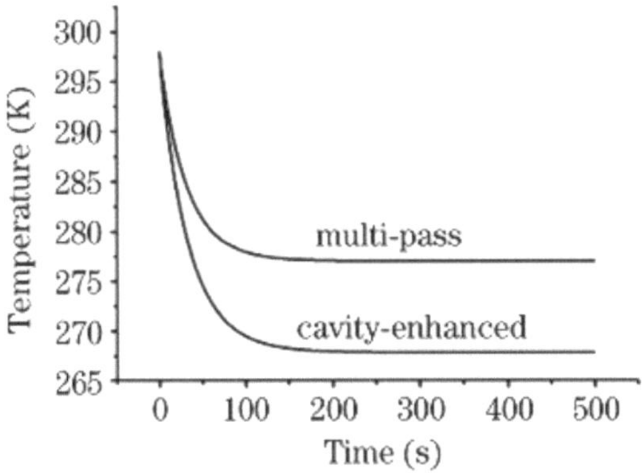

Fig. 3.9. the temperature drop comparison between the multi-pass pump and the extra-cavity enhancement schemes (Figure from ref. [14])

From Fig. 3.9, we can find that under the same parameters, the limit of temperature drop in the multi-pass pump scheme is smaller than that in the extra-cavity enhancement one. That is, the cooling effect of the extra-cavity enhancement scheme is better than one of the multi-pass pump one.

Usually, in the resonant extra-cavity enhancement scheme, the absorption power of the sample for the pump can be easily calculated by the multi-absorption. However, in the intra-cavity enhancement scheme, before calculating the absorption power of of the sample for the pump laser, we first have to calculate the laser power in the cavity.

In order to compare the intra-cavity enhancement scheme with the extra-cavity one, we define the input power of the extra-cavity by the gain-saturation formula[26]

$$P_{in} = \frac{1}{2} A I_s T \left(\frac{g_0}{T + \delta_0} - 1\right), \tag{3.29}$$

where T is the transmittance of the couple mirror M_1, I_s is the system saturation parameter, and g_0 is the gain of laser gain medium.

Once the pump beam passes the sample, the absorption power of the sample for the pump laser can be expressed as[2, 19]

$$P_{abs,s}^{ext} = (1 - e^{-\alpha L})(1 - R_s) P_{in} = \frac{1}{2}(1 - T_s)(1 - R_s) A I_s T \left(\frac{g_0}{T + \delta_0} - 1\right). \tag{3.30}$$

After the pump beam Nth passes through the sample, the total effective absorption power of the sample is given by [2, 19]

$$P_{abs}^{ext} = \frac{1}{2}(1-e^{-\alpha L})(1-R_s)AI_sT(\frac{g_0}{T+\delta_0}-1)\sum_{n=1}^{N}(\theta T_S)^{n-1} \quad (3.31)$$

$$= \frac{1}{2}(1-e^{-\alpha L})(1-R_s)AI_sT(\frac{g_0}{T+\delta_0}-1)\frac{1-(\theta T_S)^N}{1-\theta T_S}. \quad (3.32)$$

By optimizing the output coupler ($d P_{in} / dT=0$) for maximum P_{in}, we have

$$T_{opt} = \sqrt{g_0\delta_0} - \delta_0 . \quad (3.33)$$

So the maximum of P_{abs}^{ext} can be obtained by [2, 19]

$$P_{abs,opt}^{ext} = \frac{1}{2}(1-e^{-\alpha L})(1-R_s)AI_sT(\sqrt{g_0}-\delta_0)^2\frac{1-(\theta T_S)^N}{1-\theta T_S}. \quad (3.34)$$

For the sake of convenience, Eq. 3.34 and 3.21 are written in dimensionless, respectively [2, 19]

$$P_{abs,opt}^{ext} = (1-e^{-\alpha L})(1-R_s)(\sqrt{g_0}-\delta_0)^2\frac{1-(\theta T_S)^N}{1-\theta T_S}, \quad (3.35)$$

and

$$P_{abs} = \alpha d(\frac{2g_0}{T+\delta_0+\delta_S+2\alpha d}-1). \quad (3.36)$$

Under the low gain conditions, and according to Ref [3], we have g_0=0.089 and δ_0=0.078, which were measured for a practical cw pumped dye laser system by placing an acousto-optic modulator (EOM) inside the cavity. We used δ_S=0.004, R_s=0.001 to calculate and analyze the dependence of the absorption power of the sample on the optical density for the different numbers N of passes, and the results are shown in Fig. 3.10.

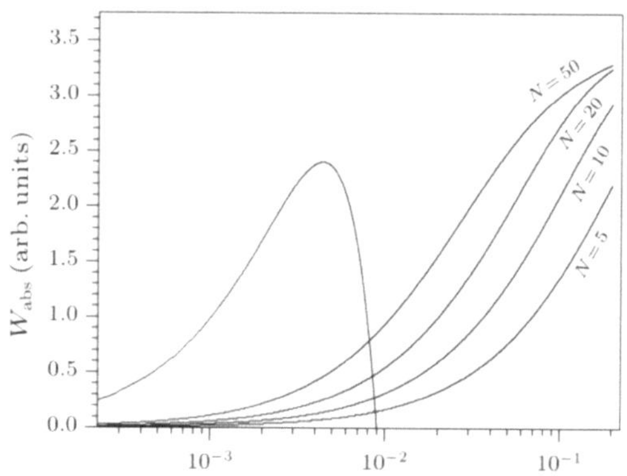

Fig. 3.10. Absorption power of the sample in the external cavity scheme with different numbers N of passes and the intra-cavity scheme as a function of the optical density. (Figure from ref. [19])

From Fig.3.10, we can find that for the low optical density, specially as $\alpha d < 0.008$, the absorption power in the intra-cavity scheme is far high than that in the extra-cavity one. However, when $\alpha d > 0.01$, because the losses from the sample approach the laser threshold, the absorption power will be sharply decreased. So in the region of the large optical density, the absorption power in the extra-cavity scheme is far higher than one in the intra-cavity one. As shown in Fig. 3.10, if we increase the pass number N of the pump beam, the absorption power in the extra-cavity scheme will increase to even exceed one in the intra-cavity scheme.

4. Experiment studies on laser cooling of solids using cavity-enhanced technologies

In the above section, we discuss and compare the enhancement effect of laser cooling of solids in the multi-pass pump, intra-cavity and resonant extra-cavity schemes in some detail. In this section, we will review experimental researches on laser cooling of solids using the above cavity-enhanced schemes and its recent progress.
As early as 2003 years, Hoyt et al. at University of New Mexico[4] and LANL group did detailed experimental research on laser cooling of solids [8] by using the multi-pass scheme after proposing how to improve the absorption power of the pump laser. At the same year, Heeg proposed the two methods using the intra-cavity and extra-cavity scheme to enhance the absorption power for laser cooling of solids and comparing laser cooling efficiencies between the intra-cavity and extra-cavity schemes. Afterwards, Heeg' group demonstrates the intra-cavity solid-state laser cooling of solids in experiment[3]. Seletskiy's group and LANL did a series of experimental researches on the cavity-enhanced resonant absorption for laser cooling of solids and obtained some important progress. More recently, they have successfully completed a milestone work and realized laser cooling of doped-Yb^{+3} YLF crystal materials with a cryogenic temperature.

4.1 Cooling of Yb: ZBLAN using intra-cavity enhanced absorption

When a diode laser is used to pump a solid laser, the laser radiation will be oscillated in an optical resonant cavity, at the same time the intensity of laser in the cavity will be amplified due to the simulation emission of laser gain medium. After the intensity of laser in the optical resonant cavity exceeds its threshold, the laser will be output from the cavity[26]. If inserting a solid sample with a low loss into the optical cavity, but the excited wavelength can be sustained to be oscillated in the cavity, and then the intra-cavity enhanced laser will increase the resonant absorption of the sample, which will result in an efficient laser cooling for solids in the laser cavity. As discussed in section 3.1, the almost materials used to form optical refrigeration has a low optical density (i.e., a low absorption), so one can use this intra-cavity enhanced scheme to realize laser cooling of solids. By choosing some optimal parameters, the intra-cavity enhancement scheme will become an efficient laser cooling method. In 2004, Heeg' group demonstrated the intra-cavity solid-state laser cooling scheme for solids experimentally [3] and the corresponding experiment setup is shown Fig. 4.1. In their experiment, the Yb^{3+}KY $(WO_4)_2$ (KYW) gain medium (10% doped, 1mm length, antireflection coated) was pumped along its b axis in the longitudinal direction by either one or two counter-propagating diode laser ($1\times 100\mu m$) at 981.2nm. The diode laser was focused into the gain medium, which was placed at the confocal center of a Z cavity, by using two concave mirrors with a high reflectivity at 1020-1060nm and a high transmission at 981.2nm. About 3W laser power per diode was illuminated on the gain medium due to the reflection losses from the pump optics. A Brewster prism was used to tune the excited wavelength from about 1020nm to 1040nm. A spatial filter was utilized here to minimize the residual radiation and super fluorescence on the cooling medium. An output coupling mirror with a 2% transmission was used to output a laser beam with the maximum power of up to 800mw at 1032nm when the power each diode laser is 1.2W. The intracavity circulating powers was be increased to 180W by using a high reflectivity mirror.

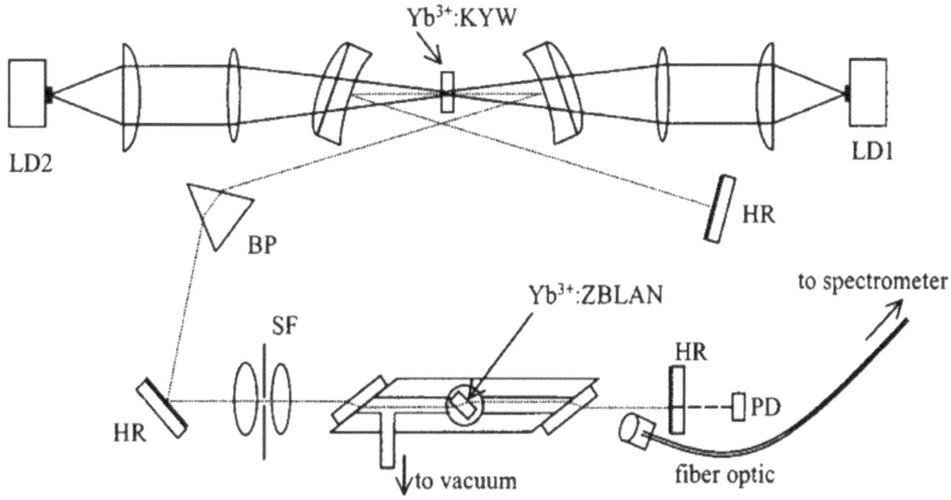

Fig. 4.1. Schematic of intra-cavity enhancement laser cooling of solids. LD is the laser diode, PD is the photodiode, HR is the high reflector, BP is the Brewster tuning system, and SF is the spatial filter. (Figure from ref. [3])

The cooling medium, a 2% doped Yb^{3+}: ZBLAN rectangular glass with dimensions $2\times2\times3$ mm, is used in this experiment, and the end-faces of the cooling medium is cut as a Brewster's angle in order to reduce the reflection losses of the sample surfaces. In the experiment, the temperature change of the sample is measured by a noncontact fashion according to the fluorescent spectral narrowing with decreased homogeneous broadening at lower temperatures. They measured the temperature drop spectrum of the cooling sample with a 2mm length, and the results are shown in Fig. 4.2.

In this experiment, the gain medium was pumped by two diode laser and the pump power in the cavity was up to 125W. After 10 minutes the sample reached the new thermal equilibrium and its temperature dropped to 6 K from the ambient temperature. At the same time, the heat sink's temperature rises by 3K due to absorbing the scatter laser and fluorescent emitted by the sample.

This experiment demonstrated the intra-cavity enhancement for laser cooling of solids. Because the absorption of cooling medium is low and the laser power for the optical refrigerator is not very high, so this approach is a prospect method for future optical refrigerator. The temperature drop of the cooling medium is not large in this experiment due to not enough pure of the sample and resulting in parasitic heating[27], the thermal radiation from the chamber, the conductive thermal from the supporter, the reapportion fluorescent[28, 29, 30], however, these problems will be improved in the future along with the advance of material fabrication[9], low emissivity coating and so on. Furthermore, introducing the Brewster prism to tune the wavelength is very important for cooling the sample to the cryogenic temperature. This is because with the temperature drop the population in the ground upper multi-levels will be decreased, so it must adjust the wavelength to the short wavelength so as to weaken the decreasing absorption of the pump beam, and is able to obtain the cryogenic temperature[11,31]. Finally, the intra-cavity cooling approach has prosperously perspective for the practical optical refrigerator.

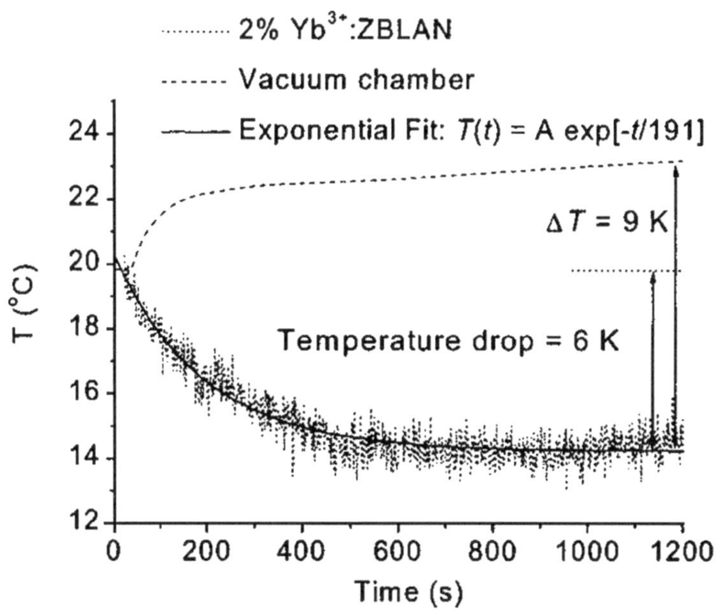

Fig. 4.2. The temperature change of the sample with a 2mm length and the heat sink as function of time (Figure from ref. [3])

4.2 Cooling of Yb: ZBLAN using extra-cavity enhanced resonant absorption

The multi-pass scheme make great successful in laser cooling of solids in Ref [12] and obtain 143K temperature drop from room temperature, but it has own disadvantage. One of the fundamental defects is a hole on the end-face of the sample because it will introduce light scattering (parasitic loss) and leak the pump light. Another shortcoming is the mirror deposited on another end-face of the sample, which will lead to the parasitic heating. The intra-cavity scheme has some advantages, such as the higher pump laser intensity in the cavity and an adjustable pump wavelength by the Brewster tuning prism[3], but it has not been adopted usually due to more optical components in the cavity and the complexity of operation. So it goes against to develop a practical optical refrigeration due to these disadvantage of the above two schemes. Fortunately, the extra-cavity resonant enhancement laser cooling scheme for solids offers a relatively simple and feasible technology for optical refrigeration.

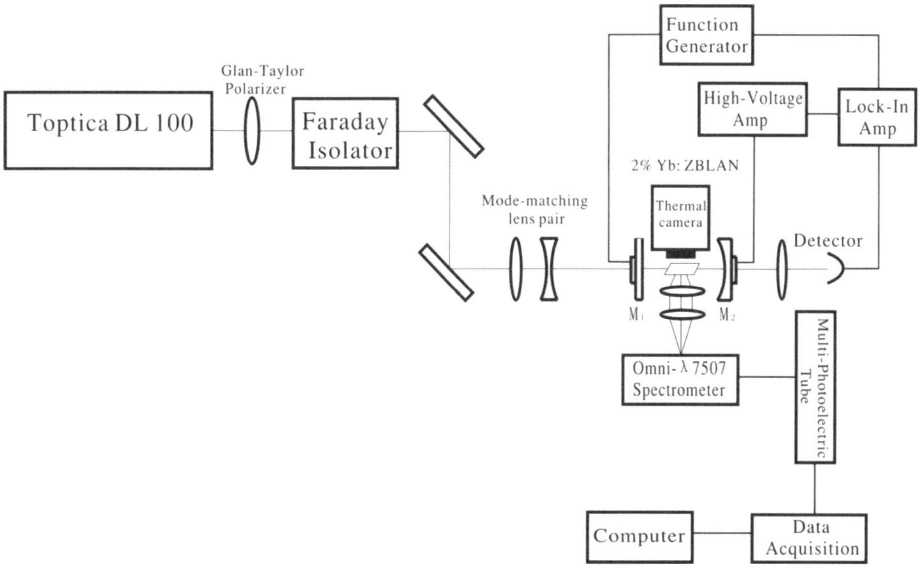

Fig. 4.3. Schematic experimental setup of the extra-cavity resonant enhancement laser cooling

The schematic experimental setup for extra-cavity enhancement resonant absorption for laser cooling of solids is shown in Fig. 4.3. The cavity is made up of two mirrors, and the first mirror M_1 is the input coupling mirror with a reflectivity of 95%, which can be adjusted to the proper value so as to obtain the maximum resonant absorption of cooling medium according to the absorption of the cooling material and the loss in the cavity. The reflectivity of the second mirror M_2 is almost equal to unity. In this experiment, the laser wavelength should be resonant with the optical cavity with a length of l so that one can obtain the maximum power in the cavity and the maximum absorption of cooling medium. A small amplitude signal generated by the function generator is used to dither the cavity, and then the intensity of laser is modulated, simultaneously, this signal is also input into Lock-In Amplifier as a reference signal. The dithered transmission is monitored by the photo-detector and mixed it with the reference signal from the function generator in Lock-In Amplifier. Then the error signal demodulated by Lock-In Amplifier was feedback to the PZT on the mirror after amplified by the High-Voltage Amplifier. Thus one can keep the laser frequency to be resonant with the frequency of cavity. The temperature of solid sample can be monitored by the thermal camera or Differential luminescence thermometry[32-35]. In 2006, D. Seletskiy et al studied the cavity enhanced resonant absorption for laser cooling of Yb^{3+} ZBLAN glass in experiment[13]. A commercial laser (ELS, VeraDisk Yb: YAG) with the maximum output power of about 45W at 1030nm was used. To avoid the laser unstable due to the optical feedback, a Faraday isolator (EOT, Model# 4I1030) with > 30dB isolation was used. The MML_2 and MML_1 (MML_1 with f1=50 mm, MML_2 with f2=100mm) coating with anti-reflection coat for 1030nm on their sides was used to enable the laser beam has a proper divergence and a proper transverse mode inside the cavity[26]. The input–coupler was mounted on the 3-axis piezo-controllers (Thorlabs, Part#MDT693) to control the transverse mode of the cavity to match with that of the laser. The reflectivity of the coating on the sample end-face serve as high reflection mirror is higher than 99.9%.

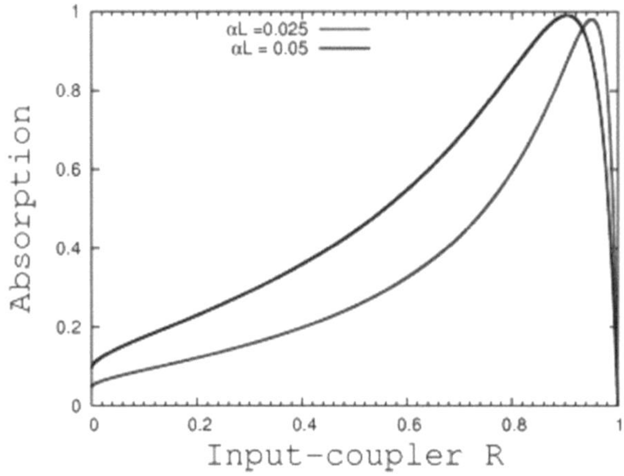

Fig. 4.4. The relationship between the cavity-enhanced absorption and the input-coupler reflectivity for two different values of αl (Figure from ref. [13])

From the extra-cavity enhancement theory in section 3.1, D. Seletskiy calculated the input-coupler reflectivity as a function of the sample absorption for different optical density αl, and the results are shown in Fig. 4.4. In the experiment, D. Seletskiy et al demonstrated the absorption enhancement for two different input-coupler reflectivities, and the results are shown in Fig. 4.5. We can see from the theoretical calculation (see Fig.4.4) and the experimental results (see Fig.4.5) that the experimental data are in good agreement with the theoretical results, which can be further found in Fig. 4.6. In this work, D. Seletskiy et.al did not give the temperature drop of the sample, but just to prove the enhancement of resonant absorption of Yb: ZBLAN glass sample by a factor of 11 compared to the double-pass configuration. This is the great significance meaningful work because it has relatively simple optical arrangement and great enhancement of the absorption for the cooling sample.

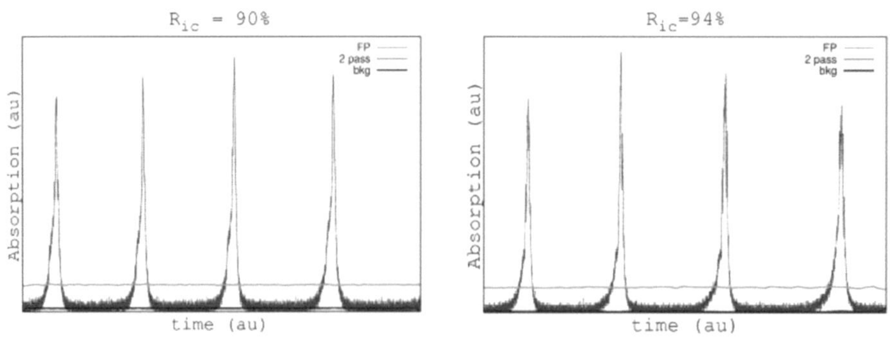

(a) $R_{ic} = 90\%$, γ=10.7 (b) $R_{ic} = 94\%$, γ=11.1

Fig. 4.5. Experimental demonstration of the absorption enhancement for two different input-coupler reflectivities $R_{ic} = 90\%$ and 94%, respectively (Figure from ref. [13])

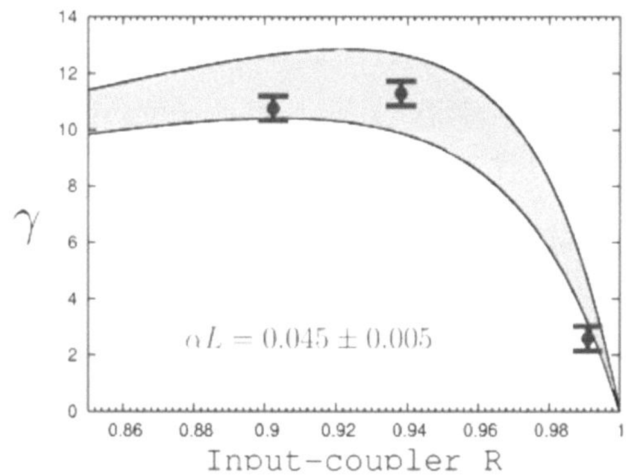

Fig. 4.6. The dependence of the enhancement absorption factor on the reflectivity of Input-coupler, and the shaded region represents uncertainty in the measured value of αl=0.0045 ± 0.005. The enhancement factors γ are plotted in the same scale, which is obtained from the experiments with different reflectivities (90%, 94%, 99%) and show very good agreement with the expected values, calculated from Eq 3.14. (Figure from ref. [13])

After this work, D. Seletskiy et al did a series of work to improve the extra-cavity enhancement for laser cooling of solids and made the great achievements, and nearly 20 fold of enhancement of the resonant absorption are obtained.

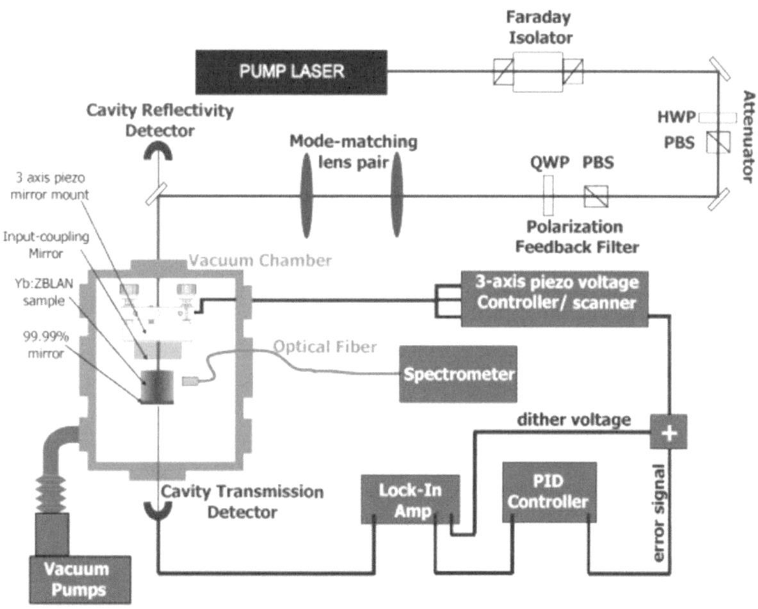

Fig. 4.7. Experimental setup of extra-cavity resonant enhancement scheme for laser cooling of solids. The thin lines are the optical beam path, and the thicker lines are electronic connections. (Figure from ref. [10])

Fig. 4.7 is the improved experimental setup. In this experiment, the reflectivity of the dielectric mirror is greater than 99.99%, and the sample was supported by 4 quartz 600μm diameter fiber. The sample was placed in a clam-shell, which was placed in the aluminum vacuum chamber with 8 vacuum ports, and the windows were coated by AR coating to increase the transmission of the pump laser. The inside surfaces of clam-shell chamber are painted with a preferential low-emissivity Maxorb coating, which is a gold-plated copper. The cavity is kept on working resonant with the pump wavelength and is stabilized via an electronic feedback loop.

In this experiment, they obtained the absorption efficiency up to $91 \pm 2\%$, and the temperature drop was about 3K, which was measured by a noncontact method, so-called differential luminescence. This measurement method is so sensitive that can resolve a 0.1K temperature change by resolving the temperature change from the reference spectra.

In 2008, D. Seletskiy et al studied the cavity resonant-enhanced laser cooling of Yb^{3+}: YLF crystal and obtained the temperature drop of about 70 degrees from the ambient temperature. In the experiment, the sample was cut with a Brewster angle to reduce the Fresnel loss in the end-faces of the sample, and the cavity is also formed in Brewster geometry for the p-polarized pump light. Fig. 4.8 is the photograph of Brewster geometry cavity. From Fig.4.8, we can see that in order to reduce the loss of cavity, the mirrors should be almost contacted with the sample.

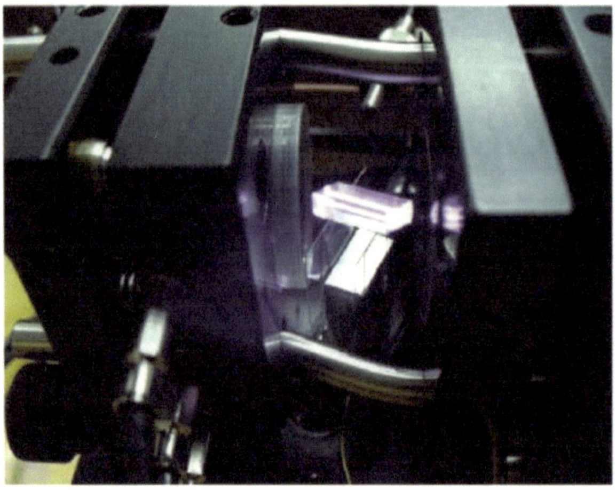

Fig. 4.8. Photograph of the Brewster geometry cavity. The left mirror of the image is the input coupler mounted on the 3-axis piezo mirror mount. The fiber above the sample is used for collecting luminescence. False color luminescence is visible in the sample, induced by pump light trapped on resonance. (Figure from ref. [17])

Fig. 4.9 shows the experiment results on the temperature drop of the sample for the different input power. Due to the sample so close to the mirror, the fluorescent emitted from the sample will be absorbed by the mirror. Simultaneously, the mirror with a nearly unity emissivity at the thermal wavelengths radiates on the sample and heats it. So the temperature drops of the Yb: YLF crystal is up to 69K from the room temperature. This is the cooling record for crystalline solids, but it not achieves the minimum achievable temperature (MAT). However, under the ideal conditions, the MAT could be achieved to $170 \pm 10K$ [12].

In this year, D. V. Seletskiy investigate the resonant cavity-enhanced absorption for optical refrigeration with a low-power laser and find the 20-fold increase over the single path absorption with the low loss medium (Yb^{3+}-doped ZBLAN glass[23]) placed in the resonant cavity mentioned. Their experiment setup is shown in Fig. 4.10. The 2% Yb^{3+}-doped ZBLAN glass was placed in the cavity, and one end-face of the sample was coated as a high reflectivity of mirror (R>0.999), and another face of the sample was deposited a anti-reflection coating to decrease Fresnel loss on the surface.

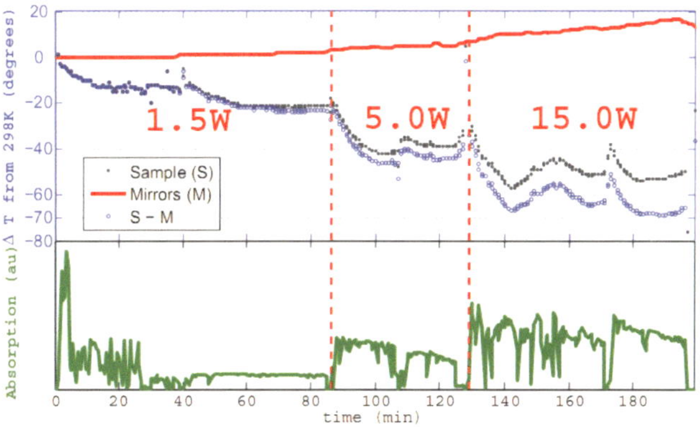

Fig. 4.9. The top graph shows the temperature dynamics Δ T(offset to zero at 298K). The grey closed circles, the red solid line and the blue open circles represent the temperature change of the sample, cavity mirrors and the difference temperature of the chamber and sample, respectively. The bottom graph is the absorption on linear scale in arbitrary units. (Figure from ref. [17])

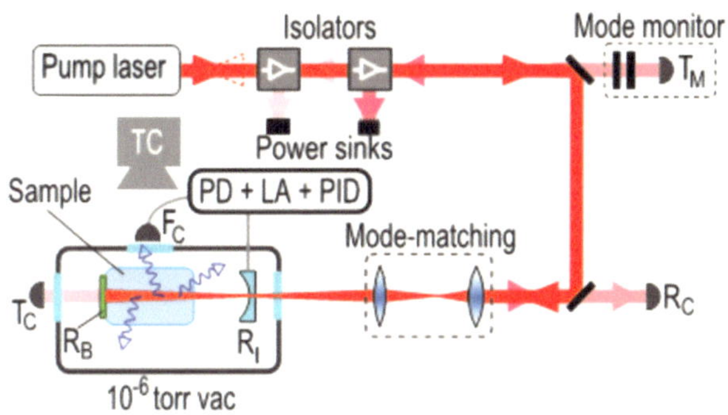

Fig. 4.10. The arrangement of experimental setup. PD+LA+PID is the proportional -integral-derivative circuit, PD is the piezo driving circuit, LA is the lock-in amplifier, which are used to control the cavity length to keep stabilization. TC is the thermal camera for monitoring the cooling process. (Figure from ref. [23])

To optimize the maximum pump laser coupled into the cavity, the reflectivity of the input mirror is usually equal to the value of $R_{ic} = e^{-\alpha l}$. The input mirror with a reflectivity of R=94% is housed in a piezo-actuated mount, allowing for cavity length scanning and stabilization. Moreover, to minimum the heat load adding to the sample, thin quartz fibers used to support the sample, which is placed in a tightly fitted shell with a low-emissivity coating. The cooling process is recorded by the thermal camera. Due to the transmission and losses are minimized by design, so the absorption A can be deduced from A=1-R_c. In the Gires-Tournois limit, the pump absorption at resonance can be expressed as[23]

$$A = 1 - \left(\frac{\sqrt{R_I} - e^{-\alpha l}}{1 - e^{-\alpha l}\sqrt{R_I}}\right). \quad (4.1)$$

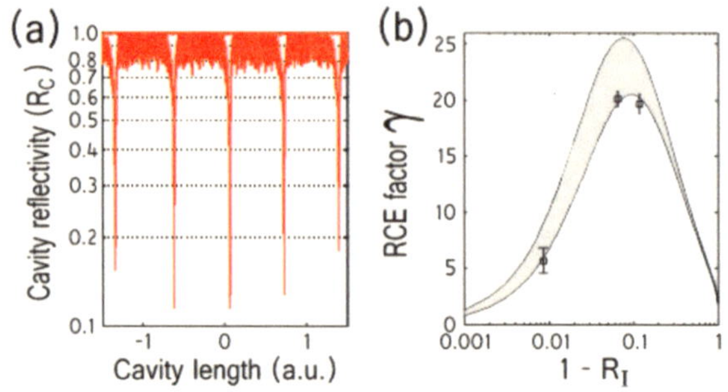

Fig. 4.11. (a) Resonant 89% absorption of the cavity plotted on a semi-log scale. (b) The experiment data agree with model fitting results, according to Eq. 4.1. (Figure from ref. [23])

Fig. 4.11 (a) shows the relationship between the cavity reflectivity and the cavity length. From Fig.4.11(a), we can find that the maximum absorption reaches to $89 \pm 3\%$ when the cavity length are scanned. From Fig. 4.11 (b), it is clear that the resonant cavity enhancement factor is up to 20 fold compare with single-time passing the sample. Fig. 4.10 is the experimental results of cooling process and the temperature drop of the sample. To avoid the saturation of the thermal camera image, the sample was radiated by the pump laser with a 1W power for 40 minute.

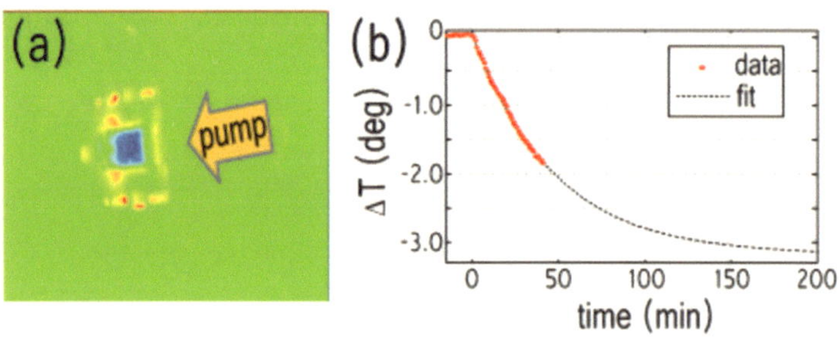

Fig. 4.12. (a) Top view thermal image of the cavity at resonance. Blue area represents the sample, the green area represents the ambient and light gray structure around the sample is the shell, which is heat by the fluorescence absorption. (b) The red line is the experiment measurement data, and the gray dot line is the fitting one. (Figure from ref. [23])

By fitting the experimental data to $\Delta T(t) = \Delta T_f[1 - \exp(-t/\tau)]$, they obtain the final temperature drop of $T_f = -3.2 \pm 0.1K$ from the ambient. Form the fitting formula, the thermal time constant is $\tau = 45$ min. In the case of small input power, the temperature drop linearly scales to be imperfect efficiencies. The ratio of absorption, stabilization, radiation shielding are 90%, 60%, and 30% respectively. It can be estimated that the final temperature drop is $3.2 \times [0.9 \times 0.6 \times 0.3]^{-1} = 20K/W$. The difference between the above mentioned final temperature drop per unity power of incident laser (20K/W) and the best cooling performance of $\sim 22K/W$ in Yb: ZBLAN glass can be ascribed to the variation of the impurity of the different sample.

5. Conclusions and outlook

In this chapter, we have reviewed various enhancement means of the pump beam to facilitate laser cooling of solids, and especially discussed the multi-pass pump, extra-cavity and intra-cavity resonant enhancement absorption in some detail from their principles to the experiment technologies. Although the multi-pass pump scheme has a lower absorption enhancement relative to two other schemes, it has simple operation and less optical elements. Therefore one hope to find some more efficient material for laser cooling or develop some high purity of solids, and first using the multi-pass pump scheme to do experimental researches on laser cooling of solids, such as using Yb^{3+} -doped ZBLAN and Yb^{3+} -doped YLF materials in the experiments. So the multi-pass scheme can compete with other two cavity-enhanced ones in the factual optical refrigerator. The intra-cavity enhancement scheme can greatly enhance the intensity of the pump laser in the cavity when the low-loss cavity and low absorption of cooling medium is used, however, due to the complication of technology and more optical elements, it will lead to some obstacles for forming a feasible optical refrigerator. In particular, it will be no suitable for some cooling materials with a large absorption. Since the extra-cavity enhancement scheme is a relatively simple technology and has a large enhancement of resonant absorption, it would be chosen as one of the practical or useful optical refrigerators.
The scientists have used the three different schemes mentioned above to investigate laser cooling of solids in experiment and hoped to find a more feasible technology for the future optical refrigeration. By optimizing the experiment parameters, one hope to obtain the maximum temperature drops of laser cooling samples. Furthermore, by probing some new coating materials to shield the thermal radiation more effectively, adjusting the geometry of the sample, and reducing the impurity of the sample, one hope to find some new paths to a practical optical refrigerator. Recently, D. V. Seletskiy utilizing the extra-cavity resonant enhancement scheme to improve the pump laser absorption by up to 20 folds compared with the single pass pump one. In this year, especially, laser cooling of solids to cryogenic temperatures using the multi-pass enhancement absorption shows a new dawn to develop a promising optical refrigerator. Since such an all-optical refrigerator has some novel and unique advantages, such as a mini volume, lightly weight, electromagnetic radiation free, non-vibration, non-noise, and long lifetime, etc, it will become an irreplaceable cryogenics in the fields of military, space technology, integrated optics, micro-electronics, remote sensing and telemetering, and so on.

6. Acknowledgments

We acknowledge the support from the National Natural Science Foundation of China under Grant No. 10974055.

7. References

[1] R. I. Epstein, M. I. Buchwald, B. C. Edwards, T. R. Gosnell. C. E. Mungan, *Nature*, 1995, Vol. 377(12): 500-502.
[2] B. Heeg, G. Rumbles, A. Khizhnyak, P. A. DeBarber, *J. Appl. Phys*, 2002, Vol. 91(5): 3356-3362
[3] B. Heeg, M. D. Stone, A. Khizhnyak, G. Rumbles, G. Mills, P. A. DeBarber, *Phys. Rev.* A, 2004, Vol. 70: 021401(1-4)
[4] Chad. William. Hoyt, The University of New Mexico, PHD May 2003

[5] C. W. Hoyt, M. Sheik-Bahae, R. I. Epstein, B. C. Edwards, and J. E. Anderson, *Phys. Rev. Lett*, 2000, Vol. 85 (17): 3600-3603
[6] T. R. Gosnell, *Opt. Lett.* 1999, Vol. 24: 1041-1043
[7] X. Luo, M. D. Eisaman, and T. R. Gosnell, *Opt. Lett.* 1998, Vol. 23: 639-641
[8] C. W. Hoyt, M. P. Hasselbeck, M. Sheik-Bahae, *J. Opt. Soc. Am.* B, 2003, Vol. 20, (5) 1066-1074
[9] M. P. Hehlen, R. I. Epstein, *SPIE, 2007,* Vol. 6461, 646103(1-6)
[10] W. M. Pattersona, M. P. Hehlenb, M. Sheik-Bahaea, R. I. Epstein, *SPIE,* Vol. 7228: 72280C(1-10)
[11] R. I. Epstein, M. Sheik-Bahae, Optical Refrigeration science and applications of laser cooling of solids, 2010
[12] D. V. Seletskiy, S. D. Melgaard, S. Bigotta, A. D. Lieto, M. Tonelli, M. Sheik-Bahae, *Nature Photonics*, 2010, 17: 161-164
[13] D. Seletskiy, M. P. Hasselbeck, M. Sheik-Bahae, J. Thiede, R. I. Epstein, *SPIE 2006,* Vol 6130: 61300p(1-8)
[14] Y. H. Jia, B. Zhong, J. P. Yin, Chin. *Phys. Lett*, 2008, Vol. 25(1): 85-88
[15] D. Seletskiy, M. P. Hasselbeck, M. Sheik-Bahae, R. I. Epstein, *SPIE, 2007,* Vol. 6461: 646104(1-8)
[16] Y. H. Jia, B. Zhong, J. P. Yin, *Chin. Opt. Lett*, 2008, Vol. 6(11): 848-851
[17] D. Seletskiy, M. P. Hasselbeck, M. Sheik-Bahae, R. I. Epstein, S. Bigottac, M. Tonelli. *SPIE 2008*, Vol. 6907: 69070B(1-9)
[18] D. V. Seletskiy, S. D. Melgaard, M. P. Hasselbeck, M. Sheik-Bahae, R. I. Epstein, S. Bigottac, M. Tonelli, *SPIE, 2009,* Vol. 7228: 72280F(1-4)
[19] Y. H. Jia, B. Zhong, J. P. Yin, *Chin. Phys. Lett*, 2010, Vol 7: 073201(1-4)
[20] M. Born, E. Wolf, *Principled of Optics* (Seventh Edition)
[21] A. Rayner, N. R. Heckenberg, H. Rubisztein-dunlop, *J. Opt. Soc. Am.* B, 2003, Vol. 20(5): 1037-1052
[22] L. H. Kou, D. Labrie, P. Chylek, *Applied. Optics* 1993, Vol. 32(19): 3531-3540
[23] D. V. Seletskiy, M. P. Hasselbeck, M. Sheik-Bahae, *App. Phy. Lett,* 2010, Vol. 96: 181106(1-4)
[24] C. E. Mungan, M. I. Buchwald, B. C. Edwards, R. I. Epstein and T. R. Gosnell, *Phys. Rev. Lett,* 1997, Vol. 78(6), 1030-1033
[25] G. Labeyrie, A. Landragin, J. V. Zanthier, R. Kaiser, N. Vansteenkiste, C. Westbrook. A. Aspect, *Quantum. Semiclass. Opt*, 1996, Vol. 8: 603–627.
[26] A. E. Siegman, Lasers, University Science Books, Sausalito, 1986
[27] J. C. Fajardo,G. H. Sigel Jr, B. C. Edwards, R. I. Epstein, T. R. Gosnell, C. E. Mungan, *J. Non-Cryst. Solids*, 1997, Vol. 213&214: 95-100
[28] B. Heeg, G. Rumbles, *J. Appl. Phys.* 2003, Vol. 93(4): 1966-1973
[29] B. Heeg, P. A. DeBarber, and Garry Rumbles, *Applied. Optics*, 2005, Vol. 44(15): 3117-3124
[30] X. F. Wang, S. L. Chang, J. K. Yang, M. Zhou, D. X. Cao, J. C. Tan, *Applied. Optics*, 2007, Vol. 46(35): 8446-84452
[31] M. Sheik-Bahae, R. I. Epstein, Nature. *Photonics*, 2007, Vol. 1: 693-699
[32] B. Imangholi, M. P. Hasselbeck, D. A. Bender, C. G. Wang, M. Sheik-Bahae, R. I. Epstein, S. Kurtz, *SPIE, 2006*, Vol. 6115: 61151C(1-6)
[33] W. Patterson, E. Soto, M. Fleharty, M. Sheik-Bahae, *SPIE, 2007,* Vol. 6461: 64610B(1-6)
[34] D. V. Seletskiy, M. P. Hasselbeck, M. Sheik-Bahae, R. I. Epstein, *SPIE, 2009,* Vol. 7228: 72280K(1-5)
[35] W. M. Patterson, D. V. Seletskiy, M. Sheik-Bahae, R. I. Epstein, M. P. Hehlen, 2010, *J. Opt. Soc. Am.* B, Vol. 27 (3): 611-618